*To my friend, my love, my companion, my wife, Joanne,
without whom this book and the first edition never would have been written.*

PREFACE

APPROACH

When dealing with a design or redesign problem the design engineer knows that the design must (1) function according to some prescribed requirements and (2) have an acceptable level of structural integrity. *Stress analysis* is that part of the design process which strives to ensure that each element of a given system will not fail to meet the structural requirements of the design throughout the specified life of the system. Stress analysis tools range from the countless theoretical techniques employed in elementary and advanced mechanics of materials and the mathematical theory of elasticity, to computer-based numerical procedures such as the finite element method and to the wide variety of available experimental techniques. Whole industries exist which are exclusively devoted to supplying experimental or numerical materials to the stress analysis community. This book was originally devoted to expand the theoretical, experimental, and numerical background of the reader beyond that covered in a first course in elementary mechanics of materials. This edition continues to adhere to this purpose where the author has attempted to improve the presentation of the practical topics of the first edition while infusing important additional subject matter and procedures.

Practical stress analysis problems range from problems which closely resemble simple models which have known closed-form solutions to complex problems which are not easily adaptable to the various classical techniques. For the complex problem, the analyst must either obtain an approximation from a theoretical analysis, seek a numerical solution, perform an experimental analysis, or carry out a combination of these approaches. In order to be versatile, the analyst should have a good working knowledge of the theoretical, numerical, and experimental procedures employed in the field of stress analysis. No single book could ever completely cover this wide range of concepts. However, this book covers the fundamental aspects of each area with a sufficient amount of depth at the mathematical level of a junior undergraduate to first-year graduate engineering student. References are cited throughout the book where, if desired, the reader can obtain additional information in a specific area.

CONTENTS

Although it is assumed that the reader has completed a first course in elementary mechanics of materials, it cannot be assumed that the reader is comfortable exploring a more advanced book straightaway. The object of an advanced book is to build upon elementary concepts and processes that have been *mastered.* Many of the subjects in Chapters One and Three should be a review for the reader. However, many of the examples and problems in these chapters are designed to test the mastery of the subject material.

Chapter One provides the basic definitions and relationships between the fundamental state properties of force, stress, strain and displacement. Chapter Two extends the basic concepts by presenting the various three-dimensional stress and strain relationships which include transformations, equilibrium, and compatibility. Chapters Four and Five extend the elementary theoretical processes of Chapter Three to techniques from the mathematical theory of elasticity and advanced mechanics of materials, respectively. Topics in Chapter Five include single- and multiple-celled thin-walled tubes in torsion, bending of unsymmetrical beams, shear center, composite beams, curved beams, plates, thick-walled cylinders, contact stress, and stress concentrations. Chapter Six provides a very extensive presentation of energy methods pertaining to deflection analysis. Chapter Seven is devoted to strength theories and design methods which include the concept of strength; the design factor and uncertainty; strength theories for ductile and brittle materials; introductions to fracture mechanics, fatigue analysis, structural stability, and inelastic behavior; and engineering approximations used in statically indeterminate problems. Chapter Eight is devoted to experimental stress analysis which includes dimensional analysis, analysis techniques, strain gages and instrumentation, and transmission and reflection photoelastic techiniques. Finally, Chapters Nine and Ten provide an introduction to the numerical finite method and finite element modeling techniques used in practice.

The author has attempted to present a modern and concise development of the fundamental areas of stress analysis using consistent notation with well-defined coordinate systems. In addition, an ample and comprehensive collection of examples and exercise problems, equally in SI and U.S. customary units, are included.

NEW TO THIS EDITION

Two changes in the sequence of material from that of the first edition have been incorporated in this edition. Based on reviewer input, (1) three-dimensional transformations, equilibrium equations, and compatibility were moved to the beginning of the book (Chapter Two) as a part of the fundamental subject material, and (2) the concepts from the theory of elasticity were also moved forward (Chapter Four) to provide an earlier insight to the more mathematical approach of the concepts and to provide some results which can be used with some of the topics of advanced mechanics of materials in Chapter Five. If using this book at the junior undergraduate level, the instructor may wish to postpone or selectively organize the coverage of the more mathematical rigor of the material provided in Chapters Two and Four.

The significant changes incorporated in this edition include

- A rearrangement of all the basic concepts including three-dimensional considerations to the beginning of the text in Chapters One and Two.

- A significant elevation in the degree of difficulty of some of the examples and problems in the review of elementary mechanics of materials covered in Chapter Three. The intent is to test the mastery of the prerequisite material.

- Addition of several classical problems in the theory of elasticity using the Airy stress function in polar coordinates (Chapter Four).

- Included in the significant changes in Chapter Five, *Topics from Advanced Strength of Materials,* are the additional topics: torsion of multiple cell thin-walled tubes, shear center for open unsymmetric and closed thin-walled beams, approximate and numerical calculation of the offset of the centroidal and neutral axes of curved beams, and contact stresses between curved surfaces. The section on the bending of unsymmetrical beams has been significantly clarified and expanded and the section on the bending of thin flat plates completely rewritten presenting many cases of rectangular and annular plates.

- Chapter Seven, *Strength Theories and Design Methods,* includes new sections on fracture mechanics and structural stability.

- Chapter Nine, *Introduction to the Finite Element Method,* has been completely rewritten with improved notation and *additional* topics such as the Rayleigh-Ritz formulation, the assembly partitioning process, skew supports, two- and three-dimensional transformations, distributed and thermal loading, hinges, the frame element in two- and three-dimensional space, load-stiffening and buckling of beams, and the two-dimensional isoparametric quadrilateral element.

- Chapter Ten, *Finite Element Modeling Techniques,* is a new chapter which provides insight into the application of commercial finite element software to static structural problems. Commercial finite element software has become quite commonplace in the engineering environment, yet there is very little literature, notwithstanding manufacturer's literature, which discloses actual modeling practices. Topics include: the planning and creation of the finite element model (preprocessing), element selection and mesh strategy, load application, constraints, preprocessing checks, processing, and postprocessing.

- A significant increase in the number of problems are included at the end of each chapter. In addition to the standard closed-form solutions, computer-oriented problems are also provided at the end of the problem sections of most chapters.

SUPPLEMENT

An Instructor's Solutions Manual is available to adopters. It contains detailed solutions to all the end-of-chapter problems except most of the open-ended computer problems.

ACKNOWLEDGMENTS

A great many of the changes included in this edition came about from the many suggestions made by readers over the years for which the author is most appreciative. A special thanks to Dr. David Lineback of Measurements Group, Inc., Raleigh, North Carolina, for his review and input on the treatment of strain gages and instrumentation given in Chapter Eight. The author is also thankful for the many helpful suggestions made by the following reviewers:

Daniel Suchora
Youngstown State University

Scott Burns
University of Illinois, Urbana

Jennifer Cordes
Steven Institute of Technology

Herman Migliore
Portland State University

Isaac Elishakoff
Naval PostGraduate School

Arturs Kalmins
Lehigh University

Robert L. Bedore
San Diego State University

Sam Y. Zamrik
Pennsylvania State University

Raghu B. Agarwal
San Jose State University

John R. Berger
Colorado School of Mines

Ronald E. Smelser
University of Idaho

Stephen F. Felszeghy
California State University, Los Angeles

Last, but not least, I lovingly dedicate this book to my wife, Joanne, for her continued understanding and support during the writing of this and the first edition.

Richard G. Budynas

CONTENTS

CHAPTER NINE

INTRODUCTION TO THE FINITE ELEMENT METHOD   673

CHAPTER TEN

FINITE ELEMENT MODELING TECHNIQUES   789

LIST OF SYMBOLS

A	area, area of cross section, light-wave vector amplitude, analyzer filter axis
$\overline{A}$	area bounded by perimeter centerline of a thin-walled tube
a	dimension, crack width, varying amplitude of a light-wave vector
b	dimension, beam width
b_e	equivalent width of a composite beam section
C	material calibration constant for transmission photoelasticity
C_1	correction factor for photoelastic coatings
c	distance from the neutral axis to an outer beam fiber, speed of light, strain pulse speed
D	flexural rigidity of a thin plate, diameter
d	diameter
d_x, d_y, d_z	directional numbers of a plane
E	modulus of elasticity, voltage
$\mathbf{E}$	modulus of elasticity scale factor
e	eccentricity, distance from the centroidal axis to the neutral axis of a curved beam
F	concentrated force
$\mathbf{F}$	force scale factor
F_e	equivalent concentrated force
$\overline{F}_x, \overline{F}_y, \overline{F}_z$	body forces per unit volume
f	material calibration constant for photoelastic coatings, finite element nodal force, coefficient of friction
f_s	strength uncertainty factor
f_σ	stress uncertainty factor
G	shear modulus, Griffith energy release rate
g	gravitational constant
h	dimension, depth of beam
h_p	depth of plastic region
I, I_y, I_z, I_{yz}	second-area moments (area moments of inertia) of a cross section
I_m, I_n	principal second-area moments (area moments of inertia) of a cross section
I_1, I_2, I_3	stress invariants
$\mathbf{i}, \mathbf{j}, \mathbf{k}$	unit vectors in the x, y, z directions respectively
J	polar second-area moment (polar moment of inertia) of a cross section
J_e	equivalent polar second-area moment (polar moment of inertia) of a cross section
K	column support factor
K_f	fatigue stress concentration factor
K_I	stress intensity factor
K_{Ic}	critical stress intensity factor
K_t	static stress concentration factor, strain gage transverse sensitivity factor
k	spring constant, form correction factor for shear, stress optic coefficient

L	length
$\mathbf{L}$	length scale factor
l, m, n	directional cosines
l_g	strain gage length
l_p	strain pulse length
M	applied or reaction concentrated bending moment (couple)
$\mathbf{M}$	moment scale factor
M_P	limit (plastic) moment
M_r, M_θ	plate bending moments per unit length in polar coordinates
M_x, M_y, M_{xy}	plate bending and twisting moments per unit length in rectangular coordinates
M_Y	yield moment
M_y, M_z	net internal bending moments about axes parallel to the y and z axes respectively
m	mass, margin of safety
m, n	axes of the principal second-area moments
N	normal force, number of cycles, photoelastic isochromatic fringe order, shape functions
N_θ	photoelastic isochromatic fringe order at an angle of incidence θ
n	design factor, index of refraction, angular speed (rpm)
n_x, n_y, n_z	directional cosines
P	concentrated force, polarizer filter axis
$\mathbf{P}$	pressure scale factor
P_{cr}	critical buckling load
P_L	limit force
p	pressure, press-fit interference pressure
Q	first-area moment of a partial area of a beam section
q	shear force per unit length, distributed load intensity, notch sensitivity factor
R	reaction force, radius, crack resistance force, electrical resistance
R_g	strain gage nominal resistance
r	radius
r, θ, z	cylindrical coordinates
r_c	the distance from the center of curvature to the centroidal axis of a curved beam
r_g	radius of gyration
r_n	the distance from the center of curvature to the neutral axis of a curved beam
r_p^*	plastic zone radius
S	elastic section modulus
$\overline{S}$	perimeter length of the centerline of a closed thin-walled tube
S_a	strain gage axial sensitivity
S_E	endurance strength
S_F	fatigue strength
S_g	strain gage factor
S_t	strain gage transverse sensitivity
S_Y	yield strength
S_U	ultimate strength
s	position, curvilinear coordinate

SF	beam shape factor
T	torsional moment (couple), temperature
t	thickness, time
t_p	pulse time
t_x, t_y, t_z	directional cosines of the net shear stress on an isolated surface
U	strain energy
u	strain energy per unit volume
u, v, w	displacements in the x, y, z directions respectively
u_r, u_θ, u_z	displacements in the r, θ, z directions respectively
V	net internal shear force, input voltage
V_r	plate shear force per unit length in polar coordinates
V_x, V_y	plate shear forces per unit length in rectangular coordinates
V_y, V_z	beam shear forces in rectangular coordinates
v_c	displacement of beam centroidal axis
W	weight, work
W_c	complementary work
W_p	work potential
w	force per unit length, work per unit volume, width, weighting factor
x, y, z	rectangular coordinates
Z	inelastic section modulus

Greek

α	angle, coefficient of thermal expansion, angular location of the neutral plane for unsymmetrical bending relative to the m axis
β	angle, angular location of the neutral plane for unsymmetrical bending relative to the y axis
Δ	change, shift in phase of light waves
δ	deflection, press-fit radial interference
$\boldsymbol{\delta}$	deflection scale factor
δ_{ij}	Kronecker delta
ε	normal strain
$\boldsymbol{\varepsilon}$	strain scale factor
ε_a	axial strain
ε_t	transverse strain
γ	shear strain, weight density, temperature coefficient of resistivity
Θ	angular deflection
θ	angle, angular deflection
θ'	angular deflection per unit length
λ	wavelength, Lamè constant
μ	Lamè constant
v	Poisson's ratio
ξ, η	natural coordinates
Π	potential energy
ρ	radius of curvature, mass density, resistivity
ρ'	Neuber constant
σ	normal stress
σ_a, σ_m	alternating and mean stresses in fatigue applications
$\boldsymbol{\sigma}$	stress scale factor
$\sigma_1, \sigma_2, \sigma_3$	principal stresses

σ_{cr}	critical buckling stress
σ_{vM}	equivalent von Mises stress
τ	shear stress
τ_{Tresca}	equivalent Tresca shear stress
Φ	Airy stress function, complementary strain energy
Ψ	Prandtl's stress function
ϕ	angle, helix angle of twist
ω	angular speed (rad/s)

Mathematical

∇^2	Laplacian operator
$[\mathbf{J}]$	Jacobian matrix
$[\mathbf{T}]$	Transformation matrix

ADVANCED STRENGTH AND APPLIED STRESS ANALYSIS

chapter

ONE

Basic Concepts of Force, Stress, Strain, and Displacement

1.0   Introduction

The concepts of force and force distributions are introduced in a fundamental course in statics, whereas the concepts of stress, strain, and elastic displacements are presented in a first course in mechanics of materials. The importance of the basic ideas developed in these courses cannot be overemphasized. To understand more advanced formulations, experimental methods, and the implementation of design improvements, a comprehensive knowledge of the fundamentals of statics and strength of materials is essential.

This chapter reviews the fundamental properties of state in the traditional ordering; i.e., beginning with external forces, the discussion then leads into the definition of internal force distributions (stress). Once this definition is made, strain is related to stress, and finally displacement is related to strain. When approaching simple structural problems, however, stress or strain formulations are first based on an analysis of deformation or deflections. For example, the linear bending-stress distribution for straight beams is based on the first assumption that plane surfaces initially perpendicular to the bending axis remain plane as the beam undergoes bending. Once this assumption is made, the equations for the equilibrium of forces and moments are utilized, and the bending equation is then derived. Thus in a physical sense applied forces generate elastic deflections which can be analytically related to strains and subsequently to internal stresses. Likewise, in the numerical technique called the *finite-element method* (see Chaps. 9 and 10), structural models are formulated through the unknown deflections. The stresses are then evaluated after the deflection solution is obtained. The material in this chapter emphasizes the physical understanding of the important state properties and should be a review for

1

readers, who are urged to reexamine their own course notes and textbook on strength of materials upon completion of these sections.

The theoretical relations discussed in this chapter will be restricted to simple linear, homogeneous, and isotropic materials undergoing small deflections.

1.1   FORCE DIAGRAMS

Free-body diagrams are a necessary tool in the solution of any stress-analysis problem, and any reader who is weak on this point should attempt to correct this deficiency early, as drawing free-body diagrams is generally the first and most important step of an analysis.

Generally, an element under analysis is supported or attached to another element. In order to analyze the element, the element is completely isolated from the supports; then all the applied loads *on* the element are shown on the element. Finally, every type of internal force and moment which possibly can be transmitted through the supports is shown on the isolated element. The values of the support forces and moments are then obtained using the appropriate equations of motion based on the dynamic state of the element within the overall structure. For example, any element within a structure is always in a specific state of dynamic equilibrium. Hence, when the element is isolated from the structure, the applied and support loads acting on the element must be such that the isolated element is in the same state of dynamic equilibrium as that specified within the overall structure. Several examples of element isolation and support analysis are shown in Fig. 1.1-1, where the structures given are in a state of static equilibrium. Naturally, these examples are not meant to be comprehensive and are only given to illustrate the method of element isolation.

In the method of element isolation, it is necessary to examine each support point and determine exactly the types of forces and moments the support can transmit. Note that if the 500-lb force shown in Fig. 1.1-1(a) were not present, the support force N would be zero and it would obviously be unnecessary to show N on the diagram. However, in many cases it is a safe procedure to show first that the support is capable of providing a possible reaction. Then, by using the equilibrium equations, the value of the reaction is determined. For example, an error might arise in problems like those shown in Fig. 1.1-1(c) and (d) if all the possible reactions were not shown.

1.2   FORCE DISTRIBUTIONS

One is generally quite confident when dealing with concentrated forces and moments like those shown in Sec. 1.1. However, when forces are distributed over lengths or areas, they seem to become more abstract to the student. This is unfortu-

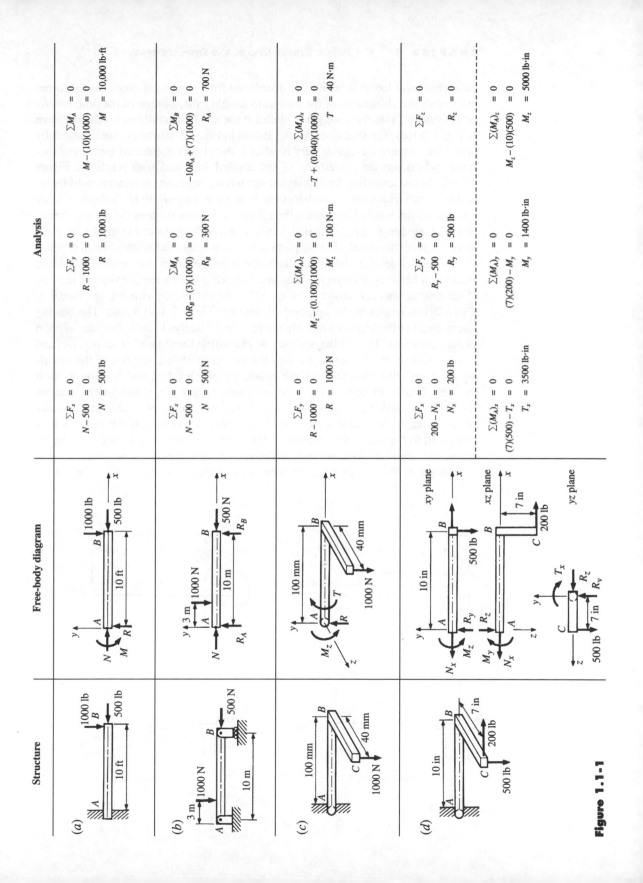

Figure 1.1-1

nate, since *real* forces in nature are distributed forces, whereas concentrated forces are merely the abstraction of the analyst to simplify the solution of the problem. As an illustration, a uniform weight is placed at the end of a cantilever beam, as shown in Fig. 1.2-1(*a*). The load distributions shown in Fig. 1.2-1(*b*) to (*d*) are all statically equivalent. Figure 1.2-1(*b*) is very idealized, showing concentrated forces and moments which are the *net* effects of the applied load and wall reactions. Figure 1.2-1(*c*) is less idealized, force distributions having replaced the concentrated forces and moments. However, the distributions have been *assumed* to be uniform. A more realistic model would incorporate the effects of *deformations* of the weight, beam, and wall, as shown in Fig. 1.2-1(*d*). One could take further steps to model the exact structure more accurately, but the point to be made is that the model chosen depends on the results desired. To survey the stresses in the beam, the analyst should be aware that bending, transverse shear, and compressive bearing stresses are present. If the analyst was concerned about the effect of the bearing stresses, the model of Fig. 1.2-1(*b*) would not be appropriate; whereas Fig. 1.2-1(*d*) would. The bearing stress distribution decreases quickly the further the analysis site is from the support or load locations. Thus, if the analyst was evaluating the stresses at section *c-c*, and if the section is at a sufficient distance from these locations, any one of the models would essentially provide the same results for the bending and transverse shear stresses. This is an application of *Saint-Venant's principle*, which asserts that the contact stresses due to support or load application become negligible at a distance of the order of the lateral dimensions of the beam or greater. If the beam is long compared to the lateral dimensions, and the length of the contact zone is of the order of the lateral dimensions, it is common engineering practice to ignore the precise shape of the contact stress distribution completely. If this is the case, then it

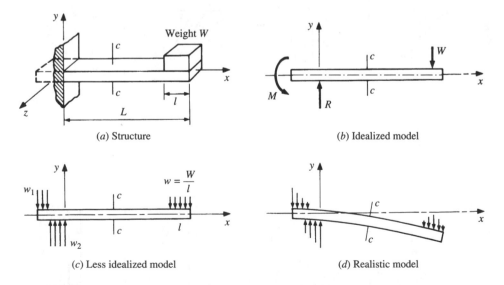

(*a*) Structure

(*b*) Idealized model

(*c*) Less idealized model

(*d*) Realistic model

Figure 1.2-1

makes sense to use the simplest model [Fig. 1.2-1(b)]. However, it is important that the analyst always be *aware* of the nature of any *approximation*.

Note also that as a structure deforms or deflects, the net reactions will change. This means that in order to perform a precise calculation of the reactions, the final deformed geometry must be known before the equilibrium equations are used. When deformations are small, however, the deformed structure is almost identical to the undeformed geometry, and the errors in evaluating forces and moments are negligible. The majority of structural problems fall in this category, but in a small class of problems the coupling between deflections and the corresponding reaction forces is quite large and the deformations affect the force analysis considerably.

These problems include large deflections of flexible structures, structures constructed of nonlinear materials, elastic stability problems (buckling), contact stresses, combined bending and axial loading of beams, and structures with deflection-dependent boundary conditions. Some but not all of these problem types are considered in this text.

If the load distribution is known or assumed, a method which simplifies the analysis is the technique of determining an "equivalent" or equipollent concentrated force. Consider the beam shown in Fig. 1.2-2(a), which is carrying a distributed load $w(x)$, in units of force per unit length, which varies along the length. The actual wall reaction at point A will look as shown in Fig. 1.2-2(b), but what is normally wanted first is the net force and moment exerted on the beam by the wall. Thus, the force R and moment M shown in Fig. 1.2-2(c) are the net result of the force distributions w_1 and w_2. For the beam to be in equilibrium R must be equal and opposite to the total force exerted by $w(x)$, and M must be equal and opposite to the total moment exerted by $w(x)$. To simplify the determination of the support reactions, an equivalent force F_e is determined, where F_e is the total force exerted by $w(x)$. The equivalent force must be applied at a particular point x_e such that the force exerts the

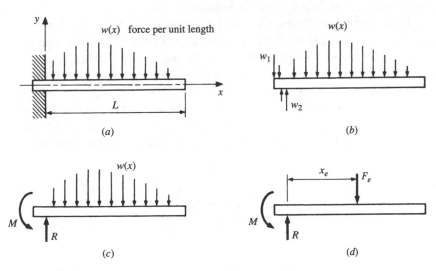

(a)

(b)

(c)

(d)

Figure 1.2-2

same net moment as $w(x)$ applies [see Fig. 1.2-2(d)]. The method of determining F_e and x_e is quite simple and is normally developed in a first course on statics, but because of its importance, it will be repeated here.

Examine a small portion of the load acting at x over an infinitesimal span of dx. The load distribution can be considered constant over this very small distance equal to $w(x)$, as shown in Fig. 1.2-3. The force exerted on dx is then $w(x)\,dx$. The total force due to the distribution $w(x)$ for every value of x from $x = 0$ to L is the sum, or integral, of each $w(x)\,dx$. Thus

$$F_e = \int_0^L w(x)\,dx \qquad\qquad \textbf{[1.2-1]}$$

and the total moment exerted at $x = 0$ is

$$M_e = \int_0^L xw(x)\,dx \qquad\qquad \textbf{[1.2-2]}$$

However, $M_e = F_e x_e$; thus

$$x_e = \frac{M_e}{F_e} = \frac{\displaystyle\int_0^L xw(x)\,dx}{\displaystyle\int_0^L w(x)\,dx} \qquad\qquad \textbf{[1.2-3]}$$

It can be seen from Eq. (1.2-1) that the equivalent force is simply the "area" of the force distribution over its length of application, and the position of application of the equivalent force is at the centroid of this area [Eq. (1.2-3)]. For simple load distributions like those shown in Fig. 1.2-4, the integrations in Eqs. (1.2-1) and (1.2-3) are unnecessary, as the areas and centroids are known. When the distribution is discontinuous, like that in Fig. 1.2-5, equivalent forces for each zone where the loading is continuous can be found, as shown in Fig. 1.2-5(b).

It should be obvious that the equivalent force of the total load distribution is used only to find the support reactions and cannot be used for analyzing internal forces or deflections throughout the basic element. To analyze the internal forces within the element, it is first necessary to isolate that portion of the element which exposes the surface to be examined. The equivalent-force approach can then be reapplied to whatever load distribution remains on that portion of the element. For example, consider the uniformly loaded cantilever beam shown in Fig. 1.2-6(a). The reader should be able to prove that the wall supplies a total force on the beam of

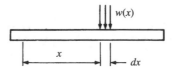

Figure 1.2-3

Load distribution   Equivalent concentrated force

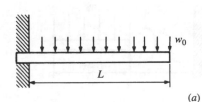
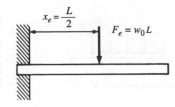

(a)

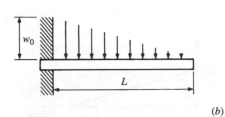
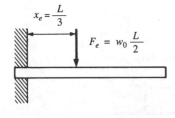

(b)

Figure 1.2-4

Load distribution   Equivalent concentrated force

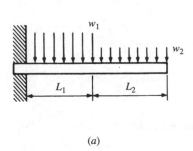
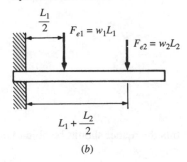

(a)   (b)

Figure 1.2-5

$w_0 L$ upward and a total moment reaction of $0.5w_0 L^2$ counterclockwise. If one wants to find the net internal force and moment occurring at a surface located at x, the equivalent force employed in solving the wall reactions *cannot* be used again. First, make a "break" at x and isolate the right-hand portion of the beam [see Fig. 1.2-6(b), showing the actual load distribution that exists on that portion]. In order to determine the internal reactions R and M it is necessary to determine the equivalent force of the load distribution acting on the portion of the element under consideration. Thus, from Fig. 1.2-6(b),

$$F_e = w_0(L - x) \qquad \text{and} \qquad x_e = \frac{L + x}{2}$$

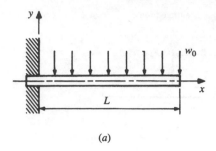

(a)

Distribution load

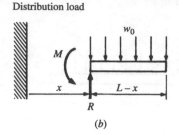

(b)

Equivalent concentrated force

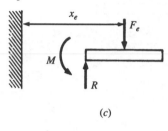

(c)

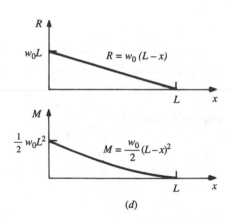

(d)

Figure 1.2-6

From this the reader should be able to obtain R and M: [†]

$$R = w_0(L - x) \qquad \text{and} \qquad M = \frac{w_0}{2}(L - x)^2$$

It can be seen that the internal force and moment in the beam depend on the value of x, which is illustrated in Fig. 1.2-6(d).

In reality an applied force distribution is a force per unit area. When dealing with beams in bending, it is common practice to express the distribution in units of force per unit length, where the force per unit area is multiplied by the thickness in the z direction. It is assumed that the force per unit area is not varying in the z direction. It is often necessary to integrate force distributions over an area in order to obtain the net force and moment acting on that surface.

[†] The notation used here for internal shear force and bending moments does not necessarily agree with the standard conventions used in strength of materials. This is discussed in Chap. 3.

Figure 1.2-7(*a*) illustrates the cross section of a rectangular beam in bending. From elementary strength of materials, the internal force distribution corresponding to bending about the *z* axis is given by $\sigma = -M_z y/I_z$, in units of force per unit area in the *x* direction, where M_z is the bending moment about the *z* axis (force times length) and I_z is the area moment of inertia about the *z* axis (given by $I_z = \frac{1}{12} bh^3$). Using area integration, prove that the net force in the *x* direction is zero and that the net moment about the *z* axis from this distribution is M_z.

Example 1.2-1

Solution:

Since the stress does not vary with respect to the *z* direction, isolate a *dA* area where $dA = b\,dy$, as shown in Fig. 1.2-7(*b*). If σ were positive in the *x* direction, as shown, the force on the *dA* area in the *x* direction would be $\sigma\,dA$. The total force is found by integration across the entire area:

$$F_x = \int \sigma\,dA = \int_{-h/2}^{h/2} -\frac{M_z y}{I_z}(b\,dy) = -\frac{M_z b}{2I_z}\left[\left(\frac{h}{2}\right)^2 - \left(-\frac{h}{2}\right)^2\right] = 0$$

The net moment about the *z* axis due to the force on the *dA* area is $-y\sigma\,dA$. Integration across the entire area yields the net moment about the *z* axis.

$$M_z = \int -y\sigma\,dA = \int_{-h/2}^{h/2} -y\,\frac{-M_z y}{I_z}(b\,dy) = \frac{M_z}{I_z}\frac{bh^3}{12}$$

However, $I_z = bh^3/12$, and the moment about the *z* axis reduces to $M_z = M_z$.

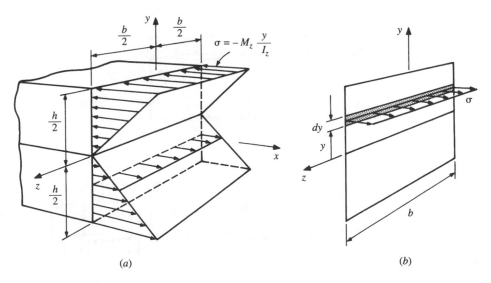

(*a*) (*b*)

Figure 1.2-7

1.3   STRESS

Stress is simply an internally distributed force within a body. To obtain a physical understanding of this idea, consider being submerged in water at a particular depth. The force one feels at this depth is a pressure (or compressive *stress*) and is not a finite number of "concentrated" forces. Other types of force distributions (stress) can occur in a liquid or solid. Tensile (pulling) and shear (rubbing or sliding) force distributions can also exist. As an example of a compressive force distribution within a solid, a uniform weight is placed on the short block B shown in Fig. 1.3-1(a). To determine the force distribution acting at section c-c, the first step is to make a break at section c-c and isolate the top portion of the element as shown in Fig. 1.3-1(b). The weight can be replaced by its equivalent force F_e, where $F_e = W$. If the portion left is to be in equilibrium, there must be an internal force F at section c-c. If the weight of the block is considered negligible, $F = F_e = W$. However, the material cannot exert a concentrated force such as F. Like all real forces in nature, the internal force F will actually be distributed across the surface in the form of a pressure p, as shown in Fig. 1.3-1(c). This pressure is pushing against the surface; hence it is a compressive distributed force, or simply a compressive stress.

The total force F is the equivalent concentrated force of the force distribution p. Since the pressure is a distributed force, it can be examined at a particular point over an infinitesimal span of area, say dA. Consider the pressure at this point to have a value of p. The force due to the pressure p acting over an area of dA is simply $p\,dA$. The total force acting on the cross section then is the sum of $p\,dA$ across the entire surface, i.e., simply the integral of $p\,dA$. Thus

$$F = \int_A p\,dA \qquad\qquad \textbf{[1.3-1]}$$

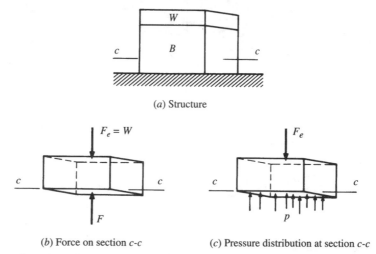

(a) Structure

(b) Force on section c-c (c) Pressure distribution at section c-c

Figure 1.3-1

If the pressure, p, or stress is uniform or constant across the area, the integral in Eq. (1.3-1) becomes simply pA, where A is the total cross-sectional area. Thus

$$p = \frac{F}{A} \qquad\qquad \text{[1.3-2]}$$

This should be very familiar to the reader, but note that the main *assumption* is that the stress is constant across the area. Compressive stresses are normally designated as negative quantities to differentiate them from tensile stresses which are considered positive. Thus the stress in the above example is

$$\sigma = -\frac{F}{A} \qquad\qquad \text{[1.3-3]}$$

where instead of p to designate pressure σ is used to designate *normal stress*.

An example of a pure tensile stress is shown in Fig. 1.3-2, where a weight W is welded to bar B. If the stress is desired at section c-c, a break is made there and the equivalent applied force and the internal reaction are shown in Fig. 1.3-2(b). If the weight of bar B is negligible, then for equilibrium, $F = F_e = W$. If the force distribution of F across section c-c (the stress) is uniform, as in the previous example, integration of σ over the area yields

$$\sigma = \frac{F}{A} \qquad\qquad \text{[1.3-4]}$$

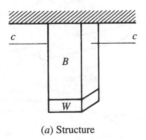

(a) Structure

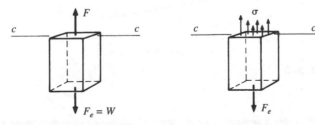

(b) Force on section c-c (c) Stress distribution at section c-c

Figure 1.3-2

where σ in this example is positive, or a tensile stress where the internal force is pulling at the surface.

An example of pure shear stress is illustrated in Fig. 1.3-3, where a tensile force P is transmitted through a yoke-and-tongue assembly. Isolating the tongue and pin within requires breaks of the pin above and below the tongue. If no friction exists between the yoke and tongue, then for equilibrium, internal tangential forces at the exposed pin surfaces are present, as shown in Fig. 1.3-3(b). From symmetry, these forces are equal and for equilibrium are $F = P/2$. These forces are tangential to the surfaces, and, as before, the force manifests itself as a force distribution, or stress. This stress, since it is tangential to the surface, is called a *shear stress* τ .[†] As before,

$$F = \int \tau \, dA$$

and if the shear stress is assumed constant across the surface, [‡]

$$\tau = \frac{F}{A} \qquad\qquad \textbf{[1.3-5]}$$

The sign convention for shear stress will be defined later.

The three examples of uniform stress help clarify the concept of stress. However, stress should not be thought of as necessarily being constant across a fi-

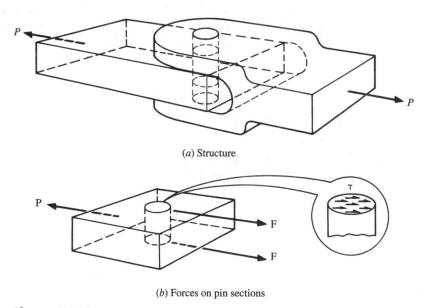

(a) Structure

(b) Forces on pin sections

Figure 1.3-3

[†] Note, the symbols being used here agree with standard engineering convention, where σ is used for normal stresses and τ for shear stresses.

[‡] If the clearances between the elements are not close, the pin tends to bend. The bending stresses cause the transverse shear stress to become nonuniform. Transverse shear stresses in beams in bending are discussed in Sec. 3.4-3.

nite surface. Stress is a *point function.* That is, stress is actually a force distribution, and in general it can be considered constant only over an infinitesimal area at a specific point located on a specific surface.

Consider a general solid body loaded as shown in Fig. 1.3-4(a). P_i are applied forces and R_i are possible support forces. To determine the state of stress at point Q in the body, it is necessary to expose a surface containing point Q. This is done by making a planar slice, or break, through the body intersecting Q. The orientation of this slice is arbitrary, but it is generally made in a convenient plane where the state of stress can be determined easily or where certain geometric relations can be utilized. The first slice, illustrated in Fig. 1.3-4(b), is arbitrarily oriented by the surface normal x. This establishes the yz plane. The external forces on the portion of the remaining body are shown, as well as the internal stress distribution across the exposed surface. In the general case, the stress distribution will not be uniform across

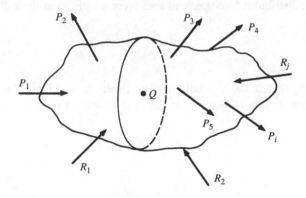

(a) Structural member

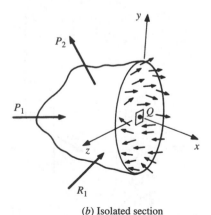

(b) Isolated section

Figure 1.3-4

the surface, and the stresses will be neither normal nor tangential to the surface at a given point. However, the stress distribution at a point will have components in the normal and tangential directions giving rise to a normal stress (tensile or compressive) and a tangential stress (shear).

The slice establishes the normal direction to the slice, the x direction. It is next left to establish the y direction, which again is arbitrary. Once this is established, the z direction follows directly for a right-handed rectangular coordinate system.

Examine an infinitesimal area $\Delta y \, \Delta z$ surrounding point Q, as shown in Fig. 1.3-5(a). The equivalent concentrated force due to the force distribution across this area is ΔF_x, which in general is neither normal nor tangential to the surface (the subscript x is used to designate the normal direction of the first slice). The force ΔF_x has components in the x, y, and z directions, which are ΔF_{xx}, ΔF_{xy}, ΔF_{xz}, respectively, as shown in Fig. 1.3-5(b). Note that the first subscript gives the direction normal to the surface and the second gives the direction of the force component. These forces are actually due to the distributed forces in the respective directions. The average distributed force per unit area (average stress) in the x direction is

$$\overline{\sigma}_{xx} = \frac{\Delta F_{xx}}{\Delta A}$$

where $\Delta A = \Delta y \, \Delta z$.

Recalling that stress is actually a point function, we obtain the exact stress in the x direction at point Q by allowing ΔA to approach zero. Thus

$$\sigma_{xx} = \lim_{\Delta A \to 0} \frac{\Delta F_{xx}}{\Delta A}$$

or

$$\sigma_{xx} = \frac{dF_{xx}}{dA} \qquad\qquad \textbf{[1.3-6]}$$

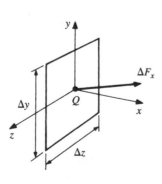

(a) Force on the ΔA surface

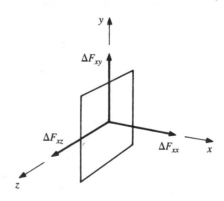

(b) Force components

Figure 1.3-5

Equation (1.3-6) can be written in integral form as

$$F_{xx} = \int_A \sigma_{xx}\, dA \qquad\qquad \textbf{[1.3-7]}$$

where F_{xx} is the total force on the entire exposed surface of Fig. 1.3-4(b) in the x direction. F_{xx} can normally be found from the equations of motion for the isolated section. The stress σ_{xx} on ΔA as ΔA approaches zero is illustrated in Fig. 1.3-6.

Distributed forces or stresses arise from the tangential forces ΔF_{xy} and ΔF_{xz} as well, and since these stresses are tangential, they are shear stresses. The procedure for establishing the shear stresses is the same as before, and the exact values at point Q are

$$\tau_{xy} = \frac{dF_{xy}}{dA} \qquad\qquad \textbf{[1.3-8]}$$

$$\tau_{xz} = \frac{dF_{xz}}{dA} \qquad\qquad \textbf{[1.3-9]}$$

The corresponding integral relations are

$$F_{xy} = \int_A \tau_{xy}\, dA \qquad\qquad \textbf{[1.3-10]}$$

$$F_{xz} = \int_A \tau_{xz}\, dA \qquad\qquad \textbf{[1.3-11]}$$

As before, F_{xy} and F_{xz} are the total forces on the entire exposed surface of Fig. 1.3-4(b) (but in the y and z directions, respectively) and can normally be found from the equations of motion. The shear stresses for the isolated section can be illustrated by Fig. 1.3-7(a) and (b).

The three stresses existing on the exposed surface at point Q can be illustrated together, but to avoid a confusing picture, each stress will be depicted by only one arrow, as shown in Fig. 1.3-8(a). It must be kept in mind that each arrow represents a

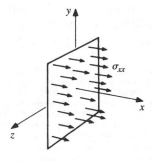

Figure 1.3-6

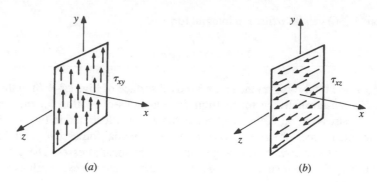

Figure 1.3-7

stress (force per unit area) and not a concentrated force. In addition, since by definition σ is a normal stress pointing in the same direction as the corresponding surface
normal, double subscripts are redundant.[†] Thus it is common practice to replace σ_{xx}
by the simpler notation σ_x, and this convention will be adopted from now on.

On a given surface only one normal stress and *one* shear stress exist. The net
tangential force on the surface is $\sqrt{(\Delta F_{xy})^2 + (\Delta F_{xz})^2}$; the net shear stress can
therefore be shown to be[‡]

$$(\tau_x)_{\text{net}} = \sqrt{\tau_{xy}^2 + \tau_{xz}^2} \qquad \textbf{[1.3-12]}$$

[see Fig. 1.3-8(*b*)].

To describe the state of stress at point Q completely it would be necessary to
examine other surfaces by making different planar slices. Since different planar
slices would necessitate different coordinates and different free-body diagrams, the
stresses on each slice would be, in general, quite different. Then to understand the
complete state of stress at point Q, every possible surface intersecting point Q
should be examined. However, this would require an infinite number of slices surrounding point Q. That is, if point Q were completely isolated from the body, it
would be described by an infinitesimal sphere. Naturally, this would be impossible
to do, and it is also unnecessary since there is a simple method for accomplishing
the same result, called *coordinate transformation,* which is described in Chap. 2.

Returning to the general body of Fig. 1.3-4(*a*), a planar slice can be made
through point Q perpendicular to the y direction. The procedure for describing the
state of stress on the infinitesimal area $\Delta z\, \Delta x$ is identical to the technique used on

---

[†] In the mathematical theory of elasticity, it is common practice to use only one symbol for stress. Here the
complete subscript notation indicates the stress type and is used for the more advanced tensor mathematical
operations (see Appendix I). For example, if the single symbol for stress was σ, then the normal and shear
stresses σ_x, τ_{xy}, and τ_{xy}, defined in Fig. 1.3-8(a), would be represented by σ_{xx}, σ_{xy}, and σ_{xz}, respectively. For
tensor mathematical operations it is more convenient to use numerical subscripts rather than letters where the
numbers 1, 2, and 3 replace x, y, and z coordinates, respectively. Thus, for example, the stress σ_x, τ_{xy}, and
τ_{xy}, would be written as σ_{11}, σ_{12}, and σ_{13}, respectively.

[‡] It is important to note that *vector operations* on stresses are *restricted* to the common area the stresses act
over. Since τ_{xy} and τ_{xz} act over the *same* surface, they can be added like vectors.

the first slice. However, the stresses on the surface whose normal is in the y direction will be designated as σ_y, τ_{yz}, and τ_{yx}. The third orthogonal slice is made perpendicular to the z direction, and the resulting stresses are σ_z, τ_{zx}, and τ_{zy}. Thus the state of stress at point Q described by only three mutually perpendicular surfaces is shown in Fig. 1.3-9(a). It will be shown later through coordinate transformation that this is sufficient to determine the state of stress on any surface intersecting point Q.

To isolate the point Q from the body completely using the orthogonal surfaces, consider the hidden faces of a rectangular parallelepiped of dimensions Δx, Δy, Δz, as shown in Fig. 1.3-9(b). If $\Delta x \, \Delta y \, \Delta z$ approaches zero, the stresses on the hidden faces must be equal in magnitude and opposite in direction to the stresses on the visible faces. Thus the state of stress at a point using three mutually orthogonal

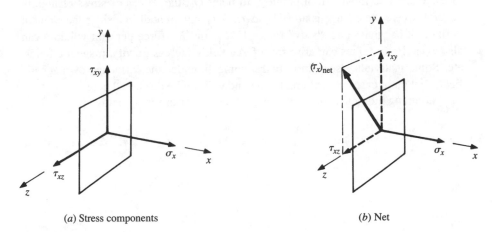

(*a*) Stress components   (*b*) Net

Figure 1.3-8

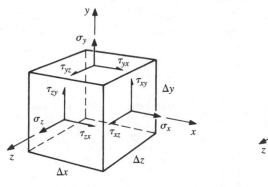

(*a*) Stresses on visible faces   (*b*) Equilibrant stresses on back faces

Figure 1.3-9

planes can be described by nine distinct values of stress. This state of stress can be written in matrix form, where the stress matrix $[\boldsymbol{\sigma}]$ is given by

$$[\boldsymbol{\sigma}] = \begin{bmatrix} \sigma_x & \tau_{xy} & \tau_{xz} \\ \tau_{yx} & \sigma_y & \tau_{yz} \\ \tau_{zx} & \tau_{zy} & \sigma_z \end{bmatrix} \qquad \text{[1.3-13]}$$

If the element has *finite* dimensions, however small, the stresses on opposing faces are not necessarily the same since we are no longer dealing with a point. In Fig. 1.3-10 consider two points in a body, Q_1 and Q_2, where point Q_2 is located Δx, Δy, and Δz from point Q_1. The state of stress for point Q_1 is shown on the hidden faces. When one moves from point Q_1 to point Q_2, the values of stress change in general, where the change is noted by $\Delta\sigma_x$, $\Delta\sigma_y$, etc. In addition, since the element is finite, it has mass and "body" forces $\overline{F}_x$, $\overline{F}_y$, and $\overline{F}_x$ (force per unit volume) can also exist. Body forces can arise from force fields such as gravity, magnetic fields, etc. Some materials also exhibit body moments under the influence of magnetic fields, but these cases are extremely rare and will not be discussed here.

Summing moments about the z axis, at the center of mass, yields

$$\tau_{xy}\,\Delta y\,\Delta z\,\frac{\Delta x}{2} + (\tau_{xy} + \Delta\tau_{xy})\,\Delta y\,\Delta z\,\frac{\Delta x}{2} - \tau_{yx}\,\Delta z\,\Delta x\,\frac{\Delta y}{2} - (\tau_{yx} + \Delta\tau_{yx})\Delta z\;\Delta x\,\frac{\Delta y}{2} = 0$$

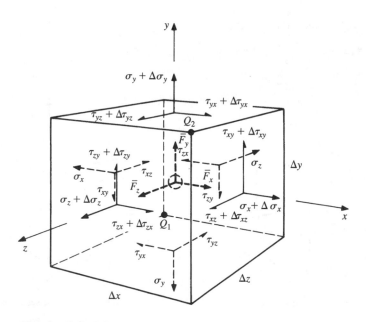

Figure 1.3-10

Dividing all terms by $\Delta x\, \Delta y\, \Delta z$ results in

$$\tau_{xy} + \frac{\Delta\tau_{xy}}{2} = \tau_{yx} + \frac{\Delta\tau_{yx}}{2}$$

If $\Delta x\, \Delta y\, \Delta z$ is allowed to approach zero, $\Delta\tau_{xy}$ and $\Delta\tau_{yz}$ will approach zero; thus

$$\tau_{xy} = \tau_{yx} \qquad\qquad\qquad \textbf{[1.3-14a]}$$

Summing moments about the other axes will yield similar relationships for the other shear stresses:

$$\tau_{yz} = \tau_{zy} \qquad\qquad\qquad \textbf{[1.3-14b]}$$

$$\tau_{zx} = \tau_{xz} \qquad\qquad\qquad \textbf{[1.3-14c]}$$

Thus since the cross shears are equal, it is only necessary to specify six quantities to establish the state of stress for a point, and the stress matrix is

$$[\boldsymbol{\sigma}] = \begin{bmatrix} \sigma_x & \tau_{xy} & \tau_{zx} \\ \tau_{xy} & \sigma_y & \tau_{yz} \\ \tau_{zx} & \tau_{yz} & \sigma_z \end{bmatrix} \qquad\qquad \textbf{[1.3-15]}$$

There are many practical problems where the stresses in the z direction are zero, that is, $\sigma_z = \tau_{yz} = \tau_{zx} = 0$. This is referred to as the state of *plane stress*. The stress matrix can then be written

$$[\boldsymbol{\sigma}] = \begin{bmatrix} \sigma_x & \tau_{xy} \\ \tau_{xy} & \sigma_y \end{bmatrix} \qquad\qquad \textbf{[1.3-16]}$$

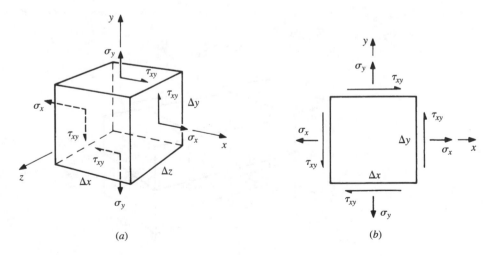

(a)   (b)

Figure 1.3-11   Plane stress.

and the element surrounding point Q is shown in Fig. 1.3-11(a). The element under-going plane stress is often shown with a two-dimensional diagram, as in Fig. 1.3-11(b), but it is important to realize that the state of stress is two-dimensional whereas the element is still three-dimensional; i.e., the depth of the element shown in Fig. 1.3-11(b) is Δz, and the stresses shown are acting over the $\Delta x \, \Delta z$ and $\Delta y \, \Delta z$ surfaces.

1.4   STRAIN, STRESS-STRAIN RELATIONS

1.4.1   NORMAL STRAINS

Stresses on a rectangular parallelepiped $\Delta x \, \Delta y \, \Delta z$ element cause a change of the element both dimensionally and in shape. The normal stresses cause the element to grow and/or shrink in the x, y, and z directions so that the element remains a rectangular parallelepiped. The shear stresses basically do not cause dimensional changes, but shear causes the element to change shape from a rectangular parallelepiped to a rhombohedron.

Initially, consider only one normal stress σ_x applied to the element as shown in Fig. 1.4-1. We see that the element increases in length in the x direction and de-creases in length in the y and z directions. The dimensionless rate of increase in length is defined as the *normal strain* where $\varepsilon_x, \varepsilon_y,$ and ε_z represent the normal strains in the x, y, and z directions, respectively. Thus the new length in any direction is equal to its original length plus the rate of increase (normal strain) times its original length. That is,

$$\Delta x' = \Delta x + \varepsilon_x \, \Delta x \qquad \Delta y' = \Delta y + \varepsilon_y \, \Delta y \qquad \Delta z' = \Delta z + \varepsilon_z \, \Delta z \quad \textbf{[1.4-1]}$$

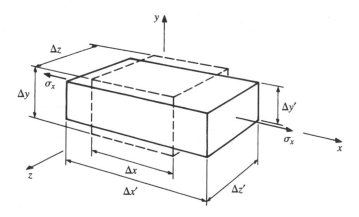

Figure 1.4-1   Deformations attributed to σ_x.

There is a direct relationship between strain and stress. *Hooke's law* for a linear material is simply that the normal strain is directly proportional to the normal stress; i.e.,

$$\varepsilon_x = \frac{\sigma_x}{E}$$

where E is called the *modulus of elasticity* (also referred to as *Young's modulus*). As the element elongates in the direction of the normal stress, contractions in the y and z directions occur. If the material is linear, these contractions are also directly proportional to the normal stress. It is common practice to express the contractions in terms of the primary normal strain, which in this case is ε_x. The proportionality constant relating the contraction to the primary strain is called *Poisson's ratio, v*. If the material is isotropic, i.e., if the material properties are independent of direction, the contractions in the y and z directions are equal and are[†]

$$\varepsilon_y = \varepsilon_z = -v\varepsilon_x = -v\frac{\sigma_x}{E}$$

The normal strains caused by σ_y and σ_z are similar to the strains caused by σ_x. The normal strains caused by σ_y are

$$\varepsilon_y = \frac{\sigma_y}{E} \qquad \varepsilon_x = \varepsilon_z = -v\varepsilon_y = -v\frac{\sigma_y}{E}$$

and the normal strains caused by σ_z are

$$\varepsilon_z = \frac{\sigma_z}{E} \qquad \varepsilon_x = \varepsilon_y = -v\varepsilon_z = -v\frac{\sigma_z}{E}$$

For an element undergoing σ_x, σ_y, and σ_z simultaneously, the effect of each stress can be added using the concept of linear superposition. Thus the general strain-stress relationship for a linear, homogeneous, isotropic material is

$$\varepsilon_x = \frac{1}{E}[\sigma_x - v(\sigma_y + \sigma_z)] \qquad \textbf{[1.4-2a]}$$

$$\varepsilon_y = \frac{1}{E}[\sigma_y - v(\sigma_z + \sigma_x)] \qquad \textbf{[1.4-2b]}$$

$$\varepsilon_z = \frac{1}{E}[\sigma_z - v(\sigma_x + \sigma_y)] \qquad \textbf{[1.4-2c]}$$

Example 1.4-1

The stress at a point in a body is

$$[\sigma] = \begin{bmatrix} 5 & 3 & 2 \\ 3 & -1 & 0 \\ 2 & 0 & 4 \end{bmatrix} \text{kpsi}$$

Determine the normal strains in the x, y, and z directions if $E = 10$ Mpsi and $v = 0.30$.

[†] Nonisotropic behavior is discussed in Sec. 2.3.

Solution:

From Eqs. (1.4)

$$\varepsilon_x = \frac{1}{10 \times 10^6} [5 - 0.3(-1 + 4)] \times 10^3 = 410 \times 10^{-6} = 410\,\mu$$

$$\varepsilon_y = \frac{1}{10 \times 10^6} [-1 - 0.3(5 + 4)] \times 10^3 = -370 \times 10^{-6} = -370\,\mu$$

$$\varepsilon_z = \frac{1}{10 \times 10^6} [4 - 0.3(5 - 1)] \times 10^3 = 280 \times 10^{-6} = 280\,\mu$$

If the strains are known, Eqs. (1.4-2) represent three simultaneous equations in σ_x, σ_y, and σ_z. Solving for the stresses yields

$$\sigma_x = \frac{E}{(1 + v)(1 - 2v)} [(1 - v)\varepsilon_x + v(\varepsilon_y + \varepsilon_z)] \qquad \textbf{[1.4-3a]}$$

$$\sigma_y = \frac{E}{(1 + v)(1 - 2v)} [(1 - v)\varepsilon_y + v(\varepsilon_z + \varepsilon_x)] \qquad \textbf{[1.4-3b]}$$

$$\sigma_z = \frac{E}{(1 + v)(1 - 2v)} [(1 - v)\varepsilon_z + v(\varepsilon_x + \varepsilon_y)] \qquad \textbf{[1.4-3c]}$$

For *plane stress*, $\sigma_z = 0$ and Eqs. (1.4-2) reduce to

$$\varepsilon_x = \frac{1}{E}(\sigma_x - v\sigma_y) \qquad \textbf{[1.4-4a]}$$

$$\varepsilon_y = \frac{1}{E}(\sigma_y - v\sigma_x) \qquad \textbf{[1.4-4b]}$$

$$\varepsilon_z = -\frac{v}{E}(\sigma_x + \sigma_y) \qquad \textbf{[1.4-4c]}$$

It can be seen from Eqs. (1.4-4) that for plane stress it is only necessary to know two of the strains in order to evaluate the stresses σ_x and σ_y. Normally the two known strains are ε_x and ε_y. Thus solving for the stresses from Eqs. (1.4-4a) and (1.4-4b) yields

$$\sigma_x = \frac{E}{1 - v^2}(\varepsilon_x + v\varepsilon_y) \qquad \textbf{[1.4-5a]}$$

$$\sigma_y = \frac{E}{1 - v^2}(\varepsilon_y + v\varepsilon_x) \qquad \textbf{[1.4-5b]}$$

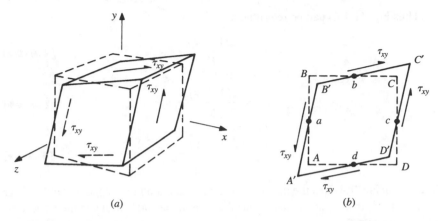

(a)   (b)

Figure 1.4-2   Pure shear.

1.4.2   SHEAR STRAINS

The change in shape of the element caused by shear stresses can be illustrated first by examining the effect of τ_{xy} alone, as shown in Fig. 1.4-2. The shear strain γ_{xy} is a measure of the deviation of the stressed element from a rectangular parallelepiped. The shear strain γ_{xy} is defined by

$$\gamma_{xy} = -\Delta\angle BAD = \angle BAD - \angle B'A'D'$$

where γ_{xy} is in dimensionless radians. For a linear, homogeneous, isotropic material, the shear strain is directly related to the shear stress by

$$\gamma_{xy} = \frac{\tau_{xy}}{G} \qquad\qquad \textbf{[1.4-6a]}$$

where G is the *shear modulus*. Similarly, the remaining shear strains are related to the corresponding shear stresses. Thus

$$\gamma_{yz} = \frac{\tau_{yz}}{G} \qquad\qquad \textbf{[1.4-6b]}$$

$$\gamma_{zx} = \frac{\tau_{zx}}{G} \qquad\qquad \textbf{[1.4-6c]}$$

It can be shown that for a linear, homogeneous, and isotropic material the shear modulus is related to the modulus of elasticity and Poisson's ratio by[†]

$$G = \frac{E}{2(1+v)} \qquad\qquad \textbf{[1.4-7]}$$

[†] This relationship is proved later in Example 2.2-1 (Sec. 2.2).

Thus Eqs. (1.4-6) can be rewritten as

$$\gamma_{xy} = \frac{2(1 + v)}{E} \tau_{xy}$$ [1.4-8a]

$$\gamma_{yz} = \frac{2(1 + v)}{E} \tau_{yz}$$ [1.4-8b]

$$\gamma_{zx} = \frac{2(1 + v)}{E} \tau_{zx}$$ [1.4-8c]

Note that in Fig. 1.4-2(b) the distances ac and bd remain unchanged when the shear stress τ_{xy} is applied to the element since there are no normal strains in the x and y directions due to τ_{xy}.

Example 1.4-2 | Determine the shear strains in Example 1.4-1.

Solution:

$$\gamma_{xy} = \frac{2(1 + 0.3)}{10 \times 10^6} \, 3000 = 780 \times 10^{-6} = 780 \, \mu$$

$$\gamma_{yz} = 0$$

$$\gamma_{zx} = \frac{2(1 + 0.3)}{10 \times 10^6} \, 2000 = 520 \times 10^{-6} = 520 \, \mu$$

1.4.3 THERMAL STRAINS

When an unconstrained solid experiences a temperature change, normal strains develop. Thermal strains for a homogeneous and isotropic material are given by

$$\varepsilon_x = \varepsilon_y = \varepsilon_z = \alpha \, \Delta T$$ [1.4-9]

where

α = coefficient of linear expansion, in/(in)(°F) or m/(m)(°C)

ΔT = change in temperature, °F or °C

1.5 DISPLACEMENTS, STRAIN-DISPLACEMENT RELATIONS

1.5.1 RECTANGULAR COORDINATES

The cumulative effect of the strains caused by the varying stresses throughout a structural member causes gross deflections of the points within the member. Thus

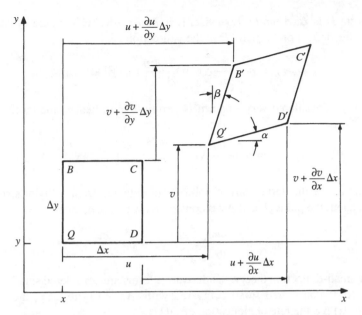

Figure 1.5-1   Rigid body and elastic deflections of an infinitesimal rectangular element.

the deflections are directly related to the strains. Since in most practical situations, deflection analyses are restricted to small deflection theory, where two-dimensional analysis in each of the three planes is valid, in this section the discussion is limited to small-deflection theory. Consider point Q in a member where the position of point Q before loading of the member is located by the coordinates x, y, and z with respect to an arbitrary origin (see Fig. 1.5-1). An element of infinitesimal dimensions Δx, Δy, and Δz originating from point Q can be constructed where the corners of the initially *undeformed* element are indicated by $QBCD$. Stresses, which in turn cause strains, cause point Q to deflect and the element to change geometrically. The deflections of point Q in the x and y directions are denoted by u and v, respectively. The corresponding deflections of points B, C, and D would be identical if the element were rigid and did not rotate. However, since we are dealing with an elastic member, the element will rotate and change shape geometrically.

The deflection of point Q can be described by continuous functions of x and y. Considering deflections in the xy plane

$$u = u(x, y) \qquad v = v(x, y)$$

The functions can be expanded about point Q in terms of a Taylor's series expansion. If u, $\partial u/\partial x$, $\partial^2 u/\partial x^2$, etc., are evaluated at point Q, the deflection for point D in the x direction will be

$$u_D = u + \frac{\partial u}{\partial x} \Delta x + \frac{1}{2!} \frac{\partial^2 u}{\partial x^2} (\Delta x)^2 + \cdots$$

since point D is Δx from Q. Likewise, if v, $\partial v/\partial x$, $\partial^2 v/\partial x^2$, etc., are evaluated for point Q, the deflection for point D in the y direction is[†]

$$v_D = v + \frac{\partial v}{\partial x}\Delta x + \frac{1}{2!}\frac{\partial^2 v}{\partial x^2}(\Delta x)^2 + \cdots$$

If Δx is considered very small, it is permissible to neglect the terms $(\Delta x)^2$ or higher. Thus

$$u_D = u + \frac{\partial u}{\partial x}\Delta x \qquad v_D = v + \frac{\partial v}{\partial x}\Delta x$$

Similarly, if the deflections of point B are obtained from a Taylor's series expansion about the point Q, and Δy is considered very small, then

$$u_B = u + \frac{\partial u}{\partial y}\Delta y \qquad v_B = v + \frac{\partial v}{\partial y}\Delta y$$

For small-deflection theory, the derivative terms are considered small.[‡] Thus, if $(\partial v/\partial x)\,\Delta x$ is considered small compared with $\Delta x + (\partial u/\partial x)\,\Delta x$, then $Q'D' \approx \Delta x + (\partial u/\partial x)\,\Delta x$, the rate of elongation of QD is

$$\varepsilon_x = \frac{Q'D' - QD}{QD} = \frac{[\Delta x + (\partial u/\partial x)\Delta x] - \Delta x}{\Delta x} = \frac{\partial u}{\partial x} \qquad \textbf{[1.5-1a]}$$

and the strain in the y direction of point Q is the rate of elongation of QB, or

$$\varepsilon_y = \frac{\partial v}{\partial y} \qquad \textbf{[1.5-1b]}$$

The reduction in angle BQD is defined as the shear strain at the point Q and is $\gamma_{xy} = \alpha + \beta$. From Fig. 1.5-1 it can be seen that

$$\tan \alpha = \frac{(\partial v/\partial x)\Delta x}{\Delta x} = \frac{\partial v}{\partial x} \qquad \text{and} \qquad \tan \beta = \frac{(\partial u/\partial y)\Delta y}{\Delta y} = \frac{\partial u}{\partial y}$$

However, if the strains are small, $\tan \alpha \approx \alpha$, and $\tan \beta \approx \beta$. Thus the shear strain can be represented by[§]

$$\gamma_{xy} = \frac{\partial v}{\partial x} + \frac{\partial u}{\partial y} \qquad \textbf{[1.5-1c]}$$

The rigid-body rotation of a line segment Θ_{xy} at point Q can be found from the average rotations of the line segments QD and QB. This can be accomplished by de-

---

[†] The derivatives are still in terms of x since point D is Δx from Q.

[‡] This brings up one of the many misnomers in stress analysis. Small-deflection theory actually means small-*strain* theory, since the derivative terms are measures of strain.

[§] This definition of shear strain is referred to as the *engineering shear strain* which is defined differently than the *elasticity shear strain*. This is discussed further in Sec. 2.0.

termining the rotation of the bisector of QD and QB. The initial angle of the bisector of angle BQD relative to the x axis is $\pi/4$. The final angle that the bisector of angle $B'Q'D'$ makes with the x axis is

$$\alpha + \frac{1}{2}\left[\frac{\pi}{2} - (\alpha + \beta)\right] = \frac{\pi}{4} + \frac{1}{2}(\alpha - \beta)$$

Subtracting the initial angle of the bisector $\pi/4$ from this yields the rigid-body rotation

$$\Theta_{xy} = \frac{1}{2}(\alpha - \beta)$$

But since $\alpha \approx \partial v/\partial x$ and $\beta \approx \partial u/\partial y$ then

$$\Theta_{xy} = \frac{1}{2}\left(\frac{\partial v}{\partial x} - \frac{\partial u}{\partial y}\right) \qquad \textbf{[1.5-1d]}$$

Considering w to be the deflection of point Q in the z direction and performing a similar analysis in the yz and zx planes results in

$$\varepsilon_z = \frac{\partial w}{\partial z} \qquad \textbf{[1.5-1e]}$$

$$\gamma_{yz} = \frac{\partial w}{\partial y} + \frac{\partial v}{\partial z} \qquad \textbf{[1.5-1f]}$$

$$\gamma_{zx} = \frac{\partial u}{\partial z} + \frac{\partial w}{\partial x} \qquad \textbf{[1.5-1g]}$$

$$\Theta_{yz} = \frac{1}{2}\left(\frac{\partial w}{\partial y} - \frac{\partial v}{\partial z}\right) \qquad \textbf{[1.5-1h]}$$

$$\Theta_{zx} = \frac{1}{2}\left(\frac{\partial u}{\partial z} - \frac{\partial w}{\partial x}\right) \qquad \textbf{[1.5-1i]}$$

If a structural member is very thin in the z direction, the stress field will be plane stress, where $\sigma_z = \tau_{zx} = \tau_{yz} = 0$. Furthermore, the deflections u and v in the x and y directions, respectively, can be considered to be functions of x and y only. If the displacement field $u(x, y)$ and $v(x, y)$ is known for such a member, determination of the strain field in the xy plane is quite straightforward using Eqs. (1.5-1a) to (1.5-1c). Once the strains are known, determination of the stress field is also quite simple using Eqs. (1.4-5) and (1.4-8a) for in this case the problem is one of plane stress.

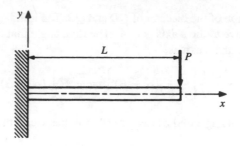

Figure 1.5-2

Example 1.5-1 | The displacement field for the thin beam shown in Fig. 1.5-2 considering bending only is[†]

$$u(x, y) = \frac{Py}{6EI}(6Lx - 3x^2 - vy^2)$$

$$v(x, y) = -\frac{P}{6EI}[3Lx^2 - x^3 + 3vy^2(L - x)]$$

where

P = applied force
I = second-area moment about the bending axis
L = length of beam
E = Young's modulus

Determine (a) the vertical deflection and slope at the centroidal axis $y = 0$ and (b) the entire stress field.

Solution:

(a) The vertical deflection at $y = 0$ is $v(x,0)$

$$v(x, 0) = -\frac{P}{6EI}(3Lx^2 - x^3) = -\frac{Px^2}{6EI}(3L - x)$$

Note that this equation agrees with the deflection of the centroidal axis v_c given in Appendix C (beam C.1).

The rotation at any location of the beam is given by $\Theta_{xy}(x, y)$, which is

$$\Theta_{xy} = \frac{1}{2}\left(\frac{\partial v}{\partial x} - \frac{\partial u}{\partial y}\right)$$

$$= \frac{1}{2}\left[-\frac{P}{6EI}(6Lx - 3x^2 - 3vy^2) - \frac{P}{6EI}(6Lx - 3x^2 - 3vy^2)\right]$$

$$= -\frac{P}{6EI}(6Lx - 3x^2 - 3vy^2)$$

[†] This displacement field is obtained from simple beam theory, which, based on Saint-Venant's principle (see Sec. 3.1), is incorrect in the vicinities of $x = 0$ and L. So, for example, do not be disturbed that deflections exist for $x = 0$ and $y \neq 0$. However, $u = v = \Theta = 0$ at $(x, y) = (0, 0)$.

The slope at $y = 0$ is the rotation $\Theta_{xy}(x, 0)$. Thus,

$$\Theta_{xy}(x, 0) = -\frac{P}{6EI}(6Lx - 3x^2) = -\frac{Px}{2EI}(2L - x)$$

Again, this agrees with the slope equation for the centroidal axis of beam C.1, dv_c/dx, given in Appendix C.

(b) $\varepsilon_x = \dfrac{\partial u}{\partial x} = \dfrac{P}{EI}(L - x)y$ $\varepsilon_y = \dfrac{\partial v}{\partial y} = -\dfrac{vP}{EI}(L - x)y$

$$\gamma_{xy} = \frac{\partial v}{\partial x} + \frac{\partial u}{\partial y} = -\frac{P}{6EI}(6Lx - 3x^2 - 3vy^2) + \frac{P}{6EI}(6Lx - 3x^2 - 3vy^2) = 0$$

Substituting into Eqs. (1.4-5) and (1.4-8a) yields the stress field

$$\sigma_x = \frac{E}{1 - v^2}\left[\frac{P}{EI}(L - x)y - \frac{v^2P}{EI}(L - x)y\right] = \frac{P}{I}(L - x)y$$

$$\sigma_y = \frac{E}{1 - v^2}\left[-\frac{vP}{EI}(L - x)y + \frac{vP}{EI}(L - x)y\right] = 0$$

$$\tau_{xy} = 0$$

NOTE: Using standard convention, the bending moment as a function of x is $M = -P(L - x)$. Hence the result obtained is $\sigma_x = -My/I$. Thus, considering bending only, the three stress fields obtained agree with basic strength of materials (see Chap. 3).

If the stresses or strains are known as functions of x and y, it is a little more difficult to determine the displacement field using Eqs. (1.5-1a) to (1.5-1d). This type of problem is discussed in Chap. 2.

1.5.2 CYLINDRICAL COORDINATES

In many problems the geometry of the component does not lend itself readily to the use of a rectangular coordinate system, and it is more practical to use a different coordinate system. Problems like thin- or thick-walled pressure vessels, circular rings, curved beams, and half-plane problems are more suitable to a cylindrical coordinate system. In Fig. 1.5-3(a), an infinitesimal element is constructed using $r\theta$ coordinates, and the corresponding normal and shear stresses are shown. The depth of the element in the z direction is Δz. As before, it is assumed that both Δr and $\Delta\theta$ are approaching zero.

The strain-stress relations are the same as in rectangular coordinates and are given by

$$\varepsilon_r = \frac{1}{E}[\sigma_r - v(\sigma_\theta + \sigma_z)] \qquad\qquad \textbf{[1.5-2a]}$$

$$\varepsilon_\theta = \frac{1}{E}[\sigma_\theta - v(\sigma_z + \sigma_r)] \qquad\qquad \textbf{[1.5-2b]}$$

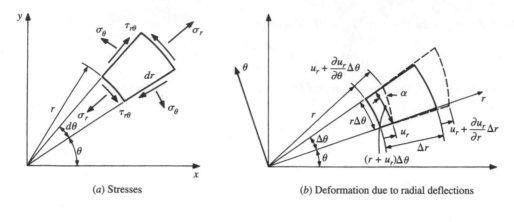

(a) Stresses   (b) Deformation due to radial deflections

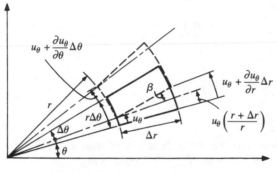

(c) Deformation due to tangential deflections

Figure 1.5-3   Stresses and elastic displacements in cylindrical coordinates (r, θ plane)

$$\varepsilon_z = \frac{1}{E}\left[\sigma_z - v(\sigma_r + \sigma_\theta)\right] \qquad \textbf{[1.5-2c]}$$

and

$$\gamma_{r\theta} = \frac{2(1 + v)}{E}\,\tau_{r\theta} \qquad \textbf{[1.5-3a]}$$

$$\gamma_{\theta z} = \frac{2(1 + v)}{E}\,\tau_{\theta z} \qquad \textbf{[1.5-3b]}$$

$$\gamma_{zr} = \frac{2(1 + v)}{E}\,\tau_{zr} \qquad \textbf{[1.5-3c]}$$

The strain-displacement relations in cylindrical coordinates are determined in a manner similar to the development presented for rectangular coordinates, but due to the complexity of the deformed element, the deformation due to radial and tangential displacements is viewed separately as shown in Fig. 1.5-3(*b*) and (*c*). The initial and deformed elements are indicated by dotted and solid lines, respectively. The net radial strain is due to the radial displacement only and is given by

$$\varepsilon_r = \frac{u_r + (\partial u_r/\partial r)\,\Delta r - u_r}{\Delta r} = \frac{\partial u_r}{\partial r} \qquad \textbf{[1.5-4a]}$$

The net tangential strain is

$$\varepsilon_\theta = \frac{(r + u_r)\,\Delta\theta - r\,\Delta\theta}{r\,\Delta\theta} + \frac{u_\theta + (\partial u_\theta/\partial\theta)\,\Delta\theta - u_\theta}{r\,\Delta\theta}$$

where the first term is due to radial displacement and the second to tangential displacement. Simplifying gives

$$\varepsilon_\theta = \frac{u_r}{r} + \frac{1}{r}\frac{\partial u_\theta}{\partial\theta} \qquad \textbf{[1.5-4b]}$$

The shear strain $\gamma_{r\theta}$ is equal to $\alpha + \beta$. Thus

$$\gamma_{r\theta} = \frac{u_r + (\partial u_r/\partial\theta)\,\Delta\theta - u_r}{r\,\Delta\theta} + \frac{u_\theta + (\partial u_\theta/\partial r)\,\Delta r - u_\theta(r + \Delta r)/r}{\Delta r}$$

where again, the first term is due to radial displacement and the second to tangential displacement. Simplifying gives

$$\gamma_{r\theta} = \frac{1}{r}\frac{\partial u_r}{\partial\theta} + \frac{\partial u_\theta}{\partial r} - \frac{u_\theta}{r} \qquad \textbf{[1.5-4c]}$$

For rotation of the element, similar to the case of rectangular coordinates, we obtain a counterclockwise rotation from $\frac{1}{2}(\beta - \alpha)$. Here, however, we obtain an additional term because the element rotates owing to the translational displacement u_θ. In radians, the angle from this, for small deflections, is u_θ/r. Thus

$$\Theta_{r\theta} = \frac{1}{2}(\beta - \alpha) + \frac{u_\theta}{r}$$

$$= \frac{1}{2}\left\{ \left[\frac{u_\theta + (\partial u_\theta/\partial r)\Delta r - u_\theta\left(\dfrac{r + \Delta r}{r}\right)}{\Delta r} \right] - \frac{u_r + (\partial u_r/\partial\theta)\Delta\theta - u_r}{r\Delta\theta} \right\} + \frac{u_\theta}{r}$$

$$= \frac{1}{2}\left(\frac{\partial u_\theta}{\partial r} + \frac{u_\theta}{r} - \frac{1}{r}\frac{\partial u_r}{\partial\theta} \right) \qquad \textbf{[1.5-4d]}$$

The strains and rotations in the θz and zr planes are developed in a manner similar to that for rectangular coordinates; they are

$$\varepsilon_z = \frac{\partial u_z}{\partial z} \qquad \textbf{[1.5-4e]}$$

$$\gamma_{\theta z} = \frac{1}{r}\frac{\partial u_z}{\partial \theta} + \frac{\partial u_\theta}{\partial z} \qquad\qquad \text{[1.5-4f]}$$

$$\gamma_{zr} = \frac{\partial u_r}{\partial z} + \frac{\partial u_z}{\partial r} \qquad\qquad \text{[1.5-4g]}$$

$$\Theta_{\theta z} = \frac{1}{2}\left(\frac{1}{r}\frac{\partial u_z}{\partial \theta} - \frac{\partial u_\theta}{\partial z} \right) \qquad\qquad \text{[1.5-4h]}$$

$$\Theta_{zr} = \frac{1}{2}\left(\frac{\partial u_r}{\partial z} - \frac{\partial u_z}{\partial r} \right) \qquad\qquad \text{[1.5-4i]}$$

A special case of Eqs. (1.5-4) is encountered for problems symmetric with respect to the longitudinal z axis. These problems are referred to as *axisymmetric* problems (such as pressurized circular disks), where variations with respect to θ are zero and thanks to symmetry $u_\theta = 0$ everywhere. Thus, for axisymmetric problems, Eqs. (1.5-4) reduce to

$$\varepsilon_r = \frac{\partial u_r}{\partial r} \qquad\qquad \text{[1.5-5a]}$$

$$\varepsilon_\theta = \frac{u_r}{r} \qquad\qquad \text{[1.5-5b]}$$

$$\gamma_{r\theta} = 0 \qquad\qquad \text{[1.5-5c]}$$

$$\Theta_{r\theta} = 0 \qquad\qquad \text{[1.5-5d]}$$

$$\varepsilon_z = \frac{\partial u_z}{\partial z} \qquad\qquad \text{[1.5-5e]}$$

$$\gamma_{\theta z} = 0 \qquad\qquad \text{[1.5-5f]}$$

$$\gamma_{zr} = \frac{\partial u_r}{\partial z} + \frac{\partial u_z}{\partial r} \qquad\qquad \text{[1.5-5g]}$$

$$\Theta_{\theta z} = 0 \qquad\qquad \text{[1.5-5h]}$$

$$\Theta_{zr} = \frac{1}{2}\left(\frac{\partial u_r}{\partial z} - \frac{\partial u_z}{\partial r} \right) \qquad\qquad \text{[1.5-5i]}$$

1.6 SUMMARY OF IMPORTANT RELATIONSHIPS

General Three-Dimensional Stress

Strain-stress relationships

$$\varepsilon_x = \frac{1}{E}\left[\sigma_x - v(\sigma_y + \sigma_z)\right] \qquad \textbf{[1.4-2a]}$$

$$\varepsilon_y = \frac{1}{E}\left[\sigma_y - v(\sigma_z + \sigma_x)\right] \qquad \textbf{[1.4-2b]}$$

$$\varepsilon_z = \frac{1}{E}\left[\sigma_z - v(\sigma_x + \sigma_y)\right] \qquad \textbf{[1.4-2c]}$$

$$\gamma_{xy} = \frac{2(1+v)}{E}\,\tau_{xy} \qquad \textbf{[1.4-8a]}$$

$$\gamma_{yz} = \frac{2(1+v)}{E}\,\tau_{yz} \qquad \textbf{[1.4-8b]}$$

$$\gamma_{zx} = \frac{2(1+v)}{E}\,\tau_{zx} \qquad \textbf{[1.4-8c]}$$

Stress-strain relationships

$$\sigma_x = \frac{E}{(1+v)(1-2v)}\left[(1-v)\varepsilon_x + v(\varepsilon_y + \varepsilon_z)\right] \qquad \textbf{[1.4-3a]}$$

$$\sigma_y = \frac{E}{(1+v)(1-2v)}\left[(1-v)\varepsilon_y + v(\varepsilon_z + \varepsilon_x)\right] \qquad \textbf{[1.4-3b]}$$

$$\sigma_z = \frac{E}{(1+v)(1-2v)}\left[(1-v)\varepsilon_z + v(\varepsilon_x + \varepsilon_y)\right] \qquad \textbf{[1.4-3c]}$$

Shear relations are straightforward from Eqs. (1.4-8)

Plane Stress $(\sigma_z = \tau_{yz} = \tau_{zx} = 0)$

Strain-stress

$$\varepsilon_x = \frac{1}{E}\left(\sigma_x - v\sigma_y\right) \qquad \textbf{[1.4-4a]}$$

$$\varepsilon_y = \frac{1}{E}\left(\sigma_y - v\sigma_x\right)\cdot \qquad \textbf{[1.4-4b]}$$

$$\varepsilon_z = -\frac{v}{E}(\sigma_x + \sigma_y)$$ [1.4-4c]

$$\gamma_{xy} = \frac{2(1+v)}{E}\tau_{xy}$$ [1.4-8a]

Stress-strain

$$\sigma_x = \frac{E}{1-v^2}(\varepsilon_x + v\varepsilon_y)$$ [1.4-5a]

$$\sigma_y = \frac{E}{1-v^2}(\varepsilon_y + v\varepsilon_x)$$ [1.4-5b]

$$\tau_{xy} = \frac{E}{2(1+v)}\gamma_{xy}$$ [1.4-8a]

Thermal Strains

$$\varepsilon_x = \varepsilon_y = \varepsilon_z = \alpha\,\Delta T$$ [1.4-9]

Strain-Displacements

$$\varepsilon_x = \frac{\partial u}{\partial x}$$ [1.5-1a]

$$\varepsilon_y = \frac{\partial v}{\partial y}$$ [1.5-1b]

$$\varepsilon_z = \frac{\partial w}{\partial z}$$ [1.5-1e]

$$\gamma_{xy} = \frac{\partial v}{\partial x} + \frac{\partial u}{\partial y}$$ [1.5-1c]

$$\gamma_{yz} = \frac{\partial w}{\partial y} + \frac{\partial v}{\partial z}$$ [1.5-1f]

$$\gamma_{zx} = \frac{\partial u}{\partial z} + \frac{\partial w}{\partial x}$$ [1.5-1g]

$$\Theta_{xy} = \frac{1}{2}\left(\frac{\partial v}{\partial x} - \frac{\partial u}{\partial y}\right)$$ [1.5-1d]

$$\Theta_{yz} = \frac{1}{2}\left(\frac{\partial w}{\partial y} - \frac{\partial v}{\partial z}\right) \qquad \text{[1.5-1}h\text{]}$$

$$\Theta_{zx} = \frac{1}{2}\left(\frac{\partial u}{\partial z} - \frac{\partial w}{\partial x}\right) \qquad \text{[1.5-1}i\text{]}$$

Cylindrical Coordinates

Strain-stress

$$\varepsilon_r = \frac{1}{E}\left[\sigma_r - v(\sigma_\theta + \sigma_z)\right] \qquad \text{[1.5-2}a\text{]}$$

$$\varepsilon_\theta = \frac{1}{E}\left[\sigma_\theta - v(\sigma_z + \sigma_r)\right] \qquad \text{[1.5-2}b\text{]}$$

$$\varepsilon_z = \frac{1}{E}\left[\sigma_z - v(\sigma_r + \sigma_\theta)\right] \qquad \text{[1.5-2}c\text{]}$$

$$\gamma_{r\theta} = \frac{2(1 + v)}{E}\tau_{r\theta} \qquad \text{[1.5-3}a\text{]}$$

$$\gamma_{\theta z} = \frac{2(1 + v)}{E}\tau_{\theta z} \qquad \text{[1.5-3}b\text{]}$$

$$\gamma_{zr} = \frac{2(1 + v)}{E}\tau_{zr} \qquad \text{[1.5-3}c\text{]}$$

Strain-displacement

$$\varepsilon_r = \frac{\partial u_r}{\partial r} \qquad \text{[1.5-4}a\text{]}$$

$$\varepsilon_\theta = \frac{u_r}{r} + \frac{1}{r}\frac{\partial u_\theta}{\partial \theta} \qquad \text{[1.5-4}b\text{]}$$

$$\gamma_{r\theta} = \frac{1}{r}\frac{\partial u_r}{\partial \theta} + \frac{\partial u_\theta}{\partial r} - \frac{u_\theta}{r} \qquad \text{[1.5-4}c\text{]}$$

$$\Theta_{r\theta} = \frac{1}{2}\left(\frac{\partial u_\theta}{\partial r} + \frac{u_\theta}{r} - \frac{1}{r}\frac{\partial u_r}{\partial \theta}\right) \qquad \text{[1.5-4}d\text{]}$$

$$\varepsilon_z = \frac{\partial u_z}{\partial z} \qquad \text{[1.5-4}e\text{]}$$

$$\gamma_{\theta z} = \frac{1}{r}\frac{\partial u_z}{\partial \theta} + \frac{\partial u_\theta}{\partial z} \qquad\qquad \textbf{[1.5-4f]}$$

$$\gamma_{zr} = \frac{\partial u_r}{\partial z} + \frac{\partial u_z}{\partial r} \qquad\qquad \textbf{[1.5-4g]}$$

$$\Theta_{\theta z} = \frac{1}{2}\left(\frac{1}{r}\frac{\partial u_z}{\partial \theta} - \frac{\partial u_\theta}{\partial z}\right) \qquad\qquad \textbf{[1.5-4h]}$$

$$\Theta_{zr} = \frac{1}{2}\left(\frac{\partial u_r}{\partial z} - \frac{\partial u_z}{\partial r}\right) \qquad\qquad \textbf{[1.5-4i]}$$

Axisymmetric Problems

$$\tau_{r\theta} = \tau_{\theta z} = 0 \qquad u_\theta = 0$$

$$\varepsilon_r = \frac{\partial u_r}{\partial r} \qquad\qquad \textbf{[1.5-5a]}$$

$$\varepsilon_\theta = \frac{u_r}{r} \qquad\qquad \textbf{[1.5-5b]}$$

$$\gamma_{r\theta} = 0 \qquad\qquad \textbf{[1.5-5c]}$$

$$\Theta_{r\theta} = 0 \qquad\qquad \textbf{[1.5-5d]}$$

$$\varepsilon_z = \frac{\partial u_z}{\partial z} \qquad\qquad \textbf{[1.5-5e]}$$

$$\gamma_{\theta z} = 0 \qquad\qquad \textbf{[1.5-5f]}$$

$$\gamma_{zr} = \frac{\partial u_r}{\partial z} + \frac{\partial u_z}{\partial r} \qquad\qquad \textbf{[1.5-5g]}$$

$$\Theta_{\theta z} = 0 \qquad\qquad \textbf{[1.5-5h]}$$

$$\Theta_{zr} = \frac{1}{2}\left(\frac{\partial u_r}{\partial z} - \frac{\partial u_z}{\partial r}\right) \qquad\qquad \textbf{[1.5-5i]}$$

1.7   PROBLEMS

1.1   For the beam shown determine the support reactions and draw the internal shear force and bending moment diagrams.

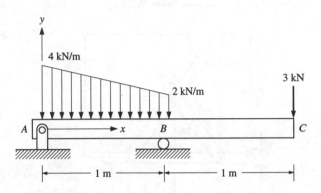

Problem 1.1

1.2   For the bent wire form shown, determine the support reactions at point A.

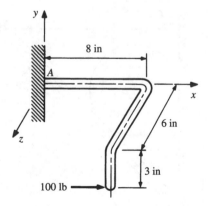

Problem 1.2

1.3 For the frame shown, determine all support and connection forces on each member. The 400-lb·ft couple M is in the plane of the page and applied to member BF. Neglect friction on all mating parts.

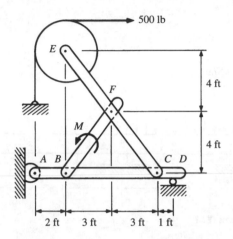

500 lb

Problem 1.3

1.4 A 250-kg drum with a 750-mm diameter is supported by bar AB and cable BC as shown. Determine the forces on each element.

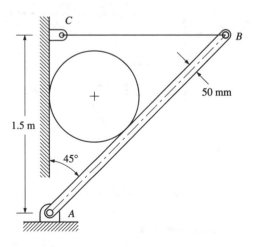

Problem 1.4

1.5   For the figure shown beam AB is 1 m long, supported by a frictionless ball-and-socket joint at point A, and supported by cables attached at point B which is located at (800, 0, 600) mm. The beam supports a load of 3 kN applied at the midpoint. The cables, in turn, are supported at points C and D with coordinates, given in millimeters, of (0, 500, 500) and (0, 500, –500), respectively. Determine the support reactions at points A, C, and D.

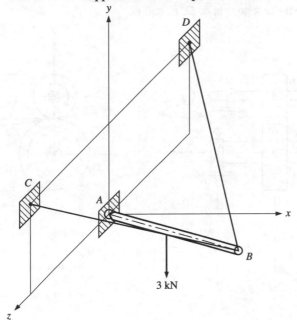

Problem 1.5

1.6   The figure shows a sectional view of a clutch. The pressure on the clutch plate $p = 180$ kPa is distributed over an annular area of inside radius of 75 mm and outer radius of 150 mm. Using appropriate free-body diagrams and equilibrium equations, determine the force F required to maintain the position shown.

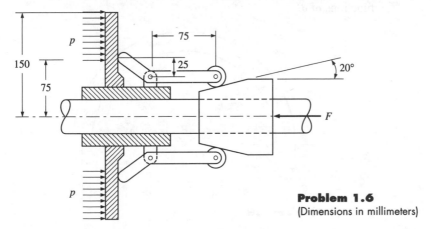

Problem 1.6
(Dimensions in millimeters)

1.7   Assuming no frictional losses, a torque of $T = 750$ lb·in is applied through the shaft of gear F in order to drive the roller chain at point B at a constant speed. The chain sprocket has a pitch diameter of 6 in and transmits a force of 500 lb as shown. The pitch diameters of gears C and F are 10 and 5 in, respectively. The contact force between the gears is transmitted through the pressure angle $\phi = 20°$. Considering the bearing supports to be simple supports, determine the reactions at A, D, E, and G.

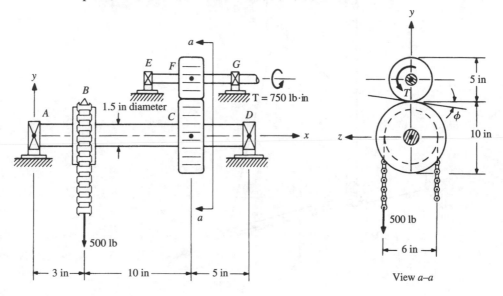

Problem 1.7

1.8   For the semicircular bar shown, determine the internal reactions (normal and shear forces, and bending moment) as functions of P, R, and θ. Plot the results as functions of θ.

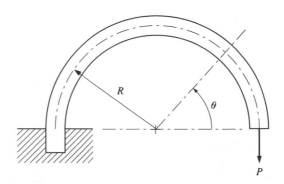

Problem 1.8

1.9 The figure depicts a stress distribution across an internal surface of a rectangular rod of height 3 in and depth 1 in. Assuming that the stress distribution does not vary with respect to z, determine the net force in the x direction and the net moment about the z axis. The stress distribution is $\sigma_x = 2000\, y + 500$ psi.

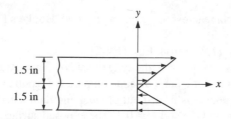

Problem 1.9

1.10 Consider the bar of Problem 1.9. Determine the net force in the x direction and the net moment about the z axis, assuming the stress does not vary in the z direction and is given by

$$\sigma_x = 500 + 2000y - 30y^3$$

1.11 When a rectangular beam of cross-sectional area $2bc$ made of a perfectly elastic-plastic material undergoes plastic deformation, the residual stress distribution after the loading is released appears as shown. Show that the net force and moment of the distribution about the z axis is zero. The stress distribution for *positive* y values is

$$\sigma_{res} = \begin{cases} S_y \dfrac{y}{c}\left[1 + \dfrac{h_p}{c} - \dfrac{1}{2}\left(\dfrac{h_p}{c}\right)^2\right] - S_y\dfrac{y}{c - h_p} & 0 < y < c - h_p \\[3mm] S_y \dfrac{y}{c}\left[1 + \dfrac{h_p}{c} - \dfrac{1}{2}\left(\dfrac{h_p}{c}\right)^2\right] - S_y & c - h_p < y < c \end{cases}$$

where S_y is the yield strength of the material, a constant.

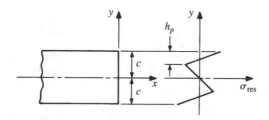

Problem 1.11

1.12 Strains for a state of plane stress are given by $\varepsilon_x = -90\ \mu$, $\varepsilon_y = -30\ \mu$, and $\gamma_{xy} = 120\ \mu$. If the elastic constants for the structure are $E = 209$ GPa and $v = 0.29$, determine the complete strain and stress matrices.

1.13 A state of plane stress exists at a point where $\varepsilon_x = 300\ \mu$, $\varepsilon_y = -100\ \mu$, and $\gamma_{xy} = 200\ \mu$. If $E = 15$ Mpsi and $v = 0.3$, determine the complete strain and stress matrices.

1.14 For a state of plane stress, $\sigma_z = \tau_{yz} = \tau_{zx} = 0$, Hooke's law for stress is written by Eqs. (1.4-5).

(a) Prove Eqs. (1.4-5) from Eqs. (1.4-2).

(b) For $\varepsilon_z = \gamma_{yz} = \gamma_{zx} = 0$, the state is called "plane strain." For this derive the equations for σ_x and σ_y and compare the results with part (a).

1.15 A rectangular pocket is milled out of steel and then filled with a plastic with properties $E = 35$ GPa and $v = 0.4$. The exposed surface is then subjected to a hydrostatic pressure of 6 MPa. Assuming the steel to be rigid, determine σ_x, σ_y, σ_z, and ε_x in the plastic.

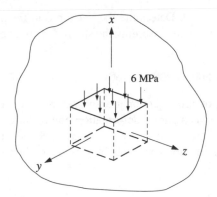

Problem 1.15

1.16 Determine the stress matrix if the modulus of elasticity is 29 Mpsi, Poisson's ratio is 0.3, and the strain matrix is

$$[\varepsilon] = \begin{bmatrix} \varepsilon_x & \gamma_{xy} & \gamma_{zx} \\ \gamma_{xy} & \varepsilon_y & \gamma_{yz} \\ \gamma_{zx} & \gamma_{yz} & \varepsilon_z \end{bmatrix} = \begin{bmatrix} 5 & -2 & 3 \\ -2 & -3 & 1 \\ 3 & 1 & 2 \end{bmatrix} \times 10^{-4}$$

1.17 The strain matrix at a particular point in a structure is

$$[\varepsilon] = \begin{bmatrix} \varepsilon_x & \gamma_{xy} & \gamma_{zx} \\ \gamma_{xy} & \varepsilon_y & \gamma_{yz} \\ \gamma_{zx} & \gamma_{yz} & \varepsilon_z \end{bmatrix} = \begin{bmatrix} 3 & -1 & 2 \\ -1 & 0 & -4 \\ 2 & -4 & 5 \end{bmatrix} \times 10^{-4}$$

Determine the stress matrix if $E = 70$ GPa and $v = 0.33$.

1.18 Given the stress matrix

$$[\sigma] = \begin{bmatrix} -12 & 3 & -4 \\ 3 & 5 & 1 \\ -4 & 1 & -8 \end{bmatrix} \quad \text{kpsi}$$

(a) Determine the magnitude of the net stress on each of the x, y, and z faces.

(b) Determine the net shear stress on each of the x, y, and z faces.

1.19 The thin pressurized ring shown is an *axisymmetric* problem. The normal stresses as functions of r are

$$\sigma_r = 625 - \frac{5625}{r^2} \qquad \sigma_\theta = 625 + \frac{5625}{r^2} \qquad \sigma_z = 0$$

If the material constants are $E = 29 \times 10^6 \ \text{lb/in}^2$ and $\nu = 0.29$, for $1 \ \text{in} \le r \le 3 \ \text{in}$, determine (a) the normal strains as functions of r and (b) the circumference at $r = 3$ in before and after the application of the pressure p.

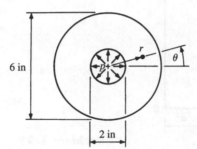

6 in

2 in   **Problem 1.19**

1.20 The deflection equations given in Example 1.5-1 neglect the effect of transverse shear loading. Consider the beam cross section to be rectangular with a height in the y direction to be $2c$. Including the shear loading, the deflection can be written as

$$u(x, y) = \frac{P}{6EI} [3xy(2L - x) - 3(1 + \nu)c^2 y + (2 + \nu)y^3]$$

$$v(x, y) = -\frac{P}{6EI} [3\nu y^2(L - x) + x^2(3L - x) + 3(1 + \nu)c^2 x]$$

(a) Using the given displacement field, determine σ_x, σ_y, and τ_{xy} throughout the beam and verify that these results agree with basic strength of material theory for beams.

(b) Analyze the vertical displacement and slope of the beam along its centerline [i.e., $v(x, 0)$ and $\Theta(x, 0)$]. Compare the results with Example 1.5-1 and discuss any differences.

1.21 A thin, uniformly thick plate of thickness t is hanging under its weight as shown. The corresponding displacement field in the xy plane can be approximated by

$$u(x, y) = \frac{\rho g}{2E} (2bx - x^2 - \nu y^2) \qquad v(x, y) = -\nu \frac{\rho g}{E} y(b - x)$$

where ρ and E are the mass density and Young's modulus of the plate material, respectively, and g is the gravity constant.

(a) Determine the corresponding plane stress field $\sigma_x(x, y)$, $\sigma_y(x, y)$ and $\tau_{xy}(x, y)$ and comment on the validity of the results.

(b) Qualitatively, draw the deformed and undeformed shape of the edges on the same drawing. For the sake of clarity, exaggerate the deflections of the edges.

(c) Determine the rotation of the plate at points A and B. Do the rotations agree with the sketch of part (b)?

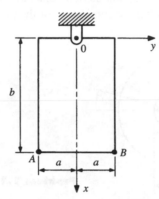

Problem 1.21

Computer Problems

1.22 Link *ABC* weighs 12 lb with the center of mass at point *B*. The link is constrained to move in the horizontal and vertical guides as shown. End *C* is attached to a spring of stiffness $k = 10$ lb/in which is unstretched when $y = 0$. Using appropriate free-body diagrams and equilibrium equations, determine

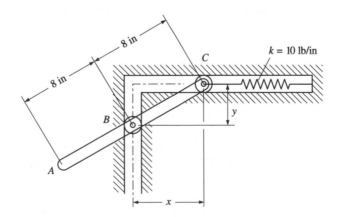

Problem 1.22

the value of y for equilibrium. (HINT: Expressing the equilibrium of the link in terms of x will result in a fourth-order equation in x which can be solved by iterative methods on a spreadsheet, or using a math software package.)

1.23 In Problems 1.16 and 1.17 a stress matrix is sought where material properties and a strain matrix are given. Using a spreadsheet program or a mathematics software package, create a program which performs this function.

1.24 In Problem 1.21, parts (*b*) and (*c*), a qualitative plot of the deformations and the rigid-body rotations at two locations is sought. Using a spreadsheet program or a mathematics software package, make separate plots of the edges:

$$x = 0, \ 0 \le y \le a$$

$$y = a, \ 0 \le x \le b$$

$$x = b, \ 0 \le y \le a$$

in the deformed state. For the problem, let $a = 1, b = 10, v = 0.3$, and $pg/E = 2 \times 10^{-8}$. It may be necessary to exaggerate (scale) the actual deflections to visualize the difference between the deformed and undeformed shapes. If you solved Problem 1.21 earlier, do the results agree?

chapter
TWO

STRESS AND STRAIN. TRANSFORMATIONS, EQUILIBRIUM, AND COMPATIBILITY

2.0   INTRODUCTION

Chapter 1 introduced the fundamental properties of state and defined stress and strain and their basic relationships. This chapter focuses on the behavior of stress and strain in greater detail. Topics include stress and strain transformations, generalized stress-strain relations, the stress equilibrium equations, and the strain compatibility equations.

As stated in Chap. 1, a state of stress not only depends on position within a structure, it also depends on the orientation of the surface containing the stresses. The variance in stress with respect to surface orientation is developed in Sec. 2.1 using the concept of *coordinate transformation*. This relates an arbitrary stress surface to a set of known stresses on three mutually orthogonal surfaces defined by the stress matrix

$$[\boldsymbol{\sigma}] = \begin{bmatrix} \sigma_x & \tau_{xy} & \tau_{xz} \\ \tau_{yx} & \sigma_y & \tau_{yz} \\ \tau_{zx} & \tau_{zy} & \sigma_z \end{bmatrix}$$

[2.0-1]

Strain transformations are presented in Sec. 2.2. Although the strain transformation equations are arrived at in a manner quite different from that for stress, the equations are almost identical in form to the stress transformation equations. The strain defined in Chap. 1 is based on the *engineering* definition of strain and is given by the matrix

$$[\boldsymbol{\varepsilon}]_{\text{eng}} = \begin{bmatrix} \varepsilon_x & \gamma_{xy} & \gamma_{xz} \\ \gamma_{yx} & \varepsilon_y & \gamma_{yz} \\ \gamma_{zx} & \gamma_{zy} & \varepsilon_z \end{bmatrix}$$

[2.0-2]

46

As will be seen later, coordinate transformations cannot be applied to this definition of strain. However, if the shear strain is specified a little differently, the strain matrix will transform. In the mathematical theory of elasticity, the strains are defined such that $\varepsilon_{xx} = \varepsilon_x$, $\varepsilon_{yy} = \varepsilon_y$, $\varepsilon_{zz} = \varepsilon_z$, $\varepsilon_{xy} = \gamma_{xy}/2$, $\varepsilon_{yz} = \gamma_{yz}/2$, $\varepsilon_{zx} = \gamma_{zx}/2$, etc. The elasticity strain matrix given by

$$[\boldsymbol{\varepsilon}]_{\text{elas}} = \begin{bmatrix} \varepsilon_{xx} & \varepsilon_{xy} & \varepsilon_{xz} \\ \varepsilon_{yx} & \varepsilon_{yy} & \varepsilon_{yz} \\ \varepsilon_{zx} & \varepsilon_{zy} & \varepsilon_{zz} \end{bmatrix} = \frac{1}{2}\begin{bmatrix} 2\varepsilon_x & \gamma_{xy} & \gamma_{xz} \\ \gamma_{yx} & 2\varepsilon_y & \gamma_{yz} \\ \gamma_{zx} & \gamma_{zy} & 2\varepsilon_z \end{bmatrix} \qquad \textbf{[2.0-3]}$$

will transform. In this book we will continue with the engineering definition of strain; however, when transforming, we must use $\gamma/2$ rather than γ.

The stress-strain relations (Hooke's law) given in Chap. 1 apply only to isotropic materials. Section 2.3 presents the generalized stress-strain relations.

Stress and strain fields must also satisfy other conditions. One condition is that of *equilibrium,* described in Sec. 2.4; and another, called *compatibility,* which ensures a continuous and consistent displacement field is presented in Sec. 2.5.

2.1   STRESS TRANSFORMATIONS

2.1.1   GENERAL THREE-DIMENSIONAL STRESS TRANSFORMATIONS[†]

Consider the state of stress at a point, as shown in Fig. 2.1-1, where the state of stress is described by the stresses acting on the three mutually orthogonal planar surfaces with normals in the x, y, and z directions. If an arbitrary oblique plane is passed through the solid so that the plane intersects the three mutually perpendicular reference planes, a tetrahedral element about the point will be isolated as shown in Fig. 2.1-2(a). Consider the x' axis to be perpendicular to the oblique plane and the y' and z' axes to be tangent to the plane. The orientation of the perpendicular to the oblique surface can be established by the angles $\theta_{x'x}$, $\theta_{x'y}$, and $\theta_{x'z}$, which relate the x' axis to the x, y, and z axes, respectively, as shown in Fig. 2.1-2(b) .

Given the coordinates of the vertices $A(a, 0, 0)$, $B(0, b, 0)$, and $C(0, 0, c)$ as shown, the equation for the oblique plane is

$$\frac{x}{a} + \frac{y}{b} + \frac{z}{c} - 1 = 0$$

[†] The development of the transformation equations given here is not completely general, as it will be assumed that no internal body moments exist. In this case, as shown in Chap. 1, the stress matrices are symmetric where *cross shears* are equal. That is, $\tau_{yx} = \tau_{xy}$, $\tau_{zy} = \tau_{yz}$, and $\tau_{xz} = \tau_{zx}$.

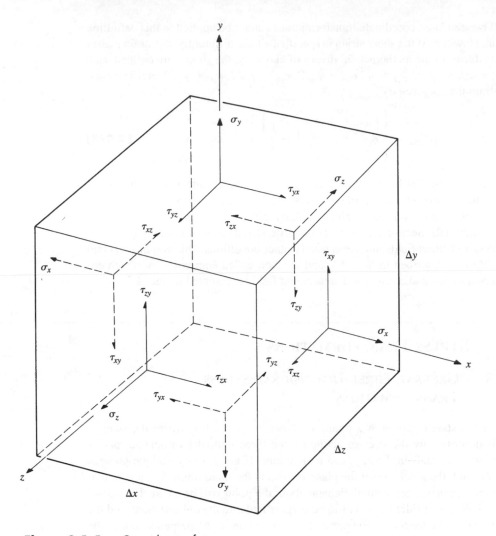

Figure 2.1-1   General state of stress.

The directional cosines of the surface normal are proportional to the gradients (derivatives) of the surface equation with respect to the x, y, and z directions, respectively. For the equation of a plane the derivatives are simply the coefficients of the x, y, and z terms and are called the *directional numbers*. The directional cosines are thus

$$\cos \theta_{x'x} = K\left(\frac{1}{a}\right) \qquad \cos \theta_{x'y} = K\left(\frac{1}{b}\right) \qquad \cos \theta_{x'z} = K\left(\frac{1}{c}\right)$$

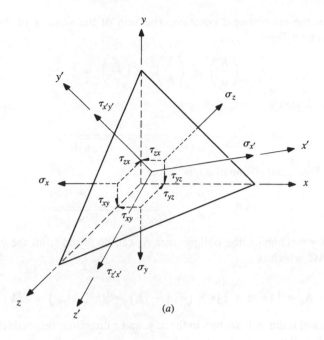

(a)

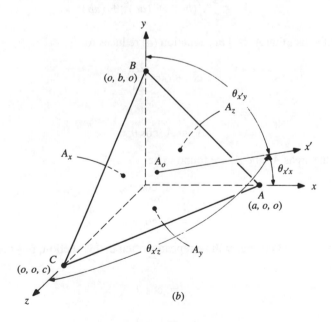

(b)

Figure 2.1-2 Establishment of a general oblique surface.

where K is the proportional constant. The sum of the squares of the directional cosines is unity. Thus

$$\left(\frac{K}{a}\right)^2 + \left(\frac{K}{b}\right)^2 + \left(\frac{K}{c}\right)^2 = 1$$

Solving for K yields

$$K = \frac{abc}{\sqrt{(bc)^2 + (ac)^2 + (ab)^2}}$$

Thus, the directional cosine of $\theta_{x'x}$ is

$$\cos\theta_{x'x} = \frac{K}{a} = \frac{bc}{\sqrt{(bc)^2 + (ac)^2 + (ab)^2}} \qquad \text{[a]}$$

From vector mechanics, the oblique area A_o can be found from the cross product $\frac{1}{2}\,\mathbf{AB} \times \mathbf{AC}$, which is

$$\mathbf{A}_o = \tfrac{1}{2}\left(-a\mathbf{i} + b\mathbf{j}\right) \times \left(-a\mathbf{i} + c\mathbf{k}\right) = \tfrac{1}{2}(bc\mathbf{i} + ac\mathbf{j} + ab\mathbf{k})$$

where $\mathbf{i}$, $\mathbf{j}$, and $\mathbf{k}$ are unit vectors in the x, y, and z direction, respectively. The magnitude of $\mathbf{A}_o$ is thus

$$A_o = \frac{1}{2}\sqrt{(bc)^2 + (ac)^2 + (ab)^2}$$

This, with the fact that A_x is $\frac{1}{2}bc$, equation (a) reduces to

$$\cos\theta_{x'x} = \frac{2A_x}{2A_o}$$

or

$$A_x = A_o \cos\theta_{x'x} \qquad \text{[2.1-1a]}$$

Similarly, the areas A_y and A_z are found to be

$$A_y = A_o \cos\theta_{x'y} \qquad \text{[2.1-1b]}$$

$$A_z = A_o \cos\theta_{x'z} \qquad \text{[2.1-1c]}$$

Since the directional cosines will be repeated often in this section, it is convenient to define them as

$$n_{x'x} = \cos\theta_{x'x} \qquad \text{[2.1-2a]}$$

$$n_{x'y} = \cos\theta_{x'y} \qquad \text{[2.1-2b]}$$

$$n_{x'z} = \cos\theta_{x'z} \qquad \text{[2.1-2c]}$$

Thus, Eqs. (2.1-1) can be rewritten in the form

$$A_x = A_o n_{x'x} \qquad \textbf{[2.1-3a]}$$

$$A_y = A_o n_{x'y} \qquad \textbf{[2.1-3b]}$$

$$A_z = A_o n_{x'z} \qquad \textbf{[2.1-3c]}$$

To determine the normal stress $\sigma_{x'}$ acting on the oblique surface, a summation of forces in the x' direction is performed. The force due to each stress is obtained by multiplying the stress by the area over which the stress acts. Next, the component of the force in the x' direction is found. For example, referring back to Fig. 2.1-2, the force due to σ_x is $\sigma_x A_x = \sigma_x A_o n_{x'x}$. The component of this force in the x' direction is $-(\sigma_x A_o n_{x'x})n_{x'x}$. Summing all forces in the x' direction for equilibrium yields

$$\sigma_{x'} A_o - (\sigma_x A_o n_{x'x})n_{x'x} - (\sigma_y A_o n_{x'y})n_{x'y} - (\sigma_z A_o n_{x'z})n_{x'z}$$

$$- (\tau_{xy} A_o n_{x'x})n_{x'y} - (\tau_{zx} A_o n_{x'x})n_{x'z} - (\tau_{xy} A_o n_{x'y})n_{x'x}$$

$$- (\tau_{yz} A_o n_{x'y})n_{x'z} - (\tau_{zx} A_o n_{x'z})n_{x'x} - (\tau_{yz} A_o n_{x'z})n_{x'y} = 0$$

Factoring A_o and solving for $\sigma_{x'}$ results in

$$\sigma_{x'} = \sigma_x n_{x'x}^2 + \sigma_y n_{x'y}^2 + \sigma_z n_{x'z}^2 + 2\tau_{xy}n_{x'x}n_{x'y} + 2\tau_{yz}n_{x'y}n_{x'z} + 2\tau_{zx}n_{x'z}n_{x'x} \ \textbf{[2.1-4a]}$$

For a complete transformation of stresses with respect to the arbitrary oblique surface, it is necessary to determine the shear stresses $\tau_{x'y'}$ and $\tau_{x'z'}$. If we define the directional cosines of the y' axis relative to the xyz coordinate system as $n_{y'x}, n_{y'y}$, and $n_{y'z}$, a summation of forces in the y' direction for equilibrium yields

$$\tau_{x'y'} A_o - (\sigma_x A_o n_{x'x})n_{y'x} - (\sigma_y A_o n_{x'y})n_{y'y} - (\sigma_z A_o n_{x'z})n_{y'z} - (\tau_{xy} A_o n_{x'x})n_{y'y}$$

$$- (\tau_{zx} A_o n_{x'x})n_{y'z} - (\tau_{xy} A_o n_{x'y})n_{y'x} - (\tau_{yz} A_o n_{x'y})n_{y'z} - (\tau_{zx} A_o n_{x'z})n_{y'x}$$

$$- (\tau_{yz} A_o n_{x'z})n_{y'y} = 0$$

Factoring A_o and solving for $\tau_{x'y'}$ results in

$$\tau_{x'y'} = \sigma_x n_{x'x}n_{y'x} + \sigma_y n_{x'y}n_{y'y} + \sigma_z n_{x'z}n_{y'z} + \tau_{xy}(n_{x'x}n_{y'y} + n_{x'y}n_{y'x})$$

$$+ \tau_{yz}(n_{x'y}n_{y'z} + n_{x'z}n_{y'y}) + \tau_{zx}(n_{x'x}n_{y'z} + n_{x'z}n_{y'x}) \qquad \textbf{[2.1-4b]}$$

The final shear stress on the oblique surface $\tau_{z'x'}$ is found in a similar fashion. Defining the directional cosines of the z' axis relative to the xyz coordinate system as $n_{z'x}, n_{z'y}$, and $n_{z'z}$, equilibrium of forces in the z' direction results in

$$\tau_{z'x'} = \sigma_x n_{x'x}n_{z'x} + \sigma_y n_{x'y}n_{z'y} + \sigma_z n_{x'z}n_{z'z} + \tau_{xy}(n_{x'x}n_{z'y} + n_{x'y}n_{z'x})$$

$$+ \tau_{yz}(n_{x'y}n_{z'z} + n_{x'z}n_{z'y}) + \tau_{zx}(n_{x'x}n_{z'z} + n_{x'z}n_{z'x}) \qquad \textbf{[2.1-4c]}$$

Equations (2.1-4a) to (2.1-4c) are completely sufficient for the determination of the state of stress on any internal surface in which an arbitrarily selected tangential set of coordinates is used (in this case the $y'z'$ coordinates).

For a complete transformation of the stress element of Fig. 2.1-1 to that of a rectangular element oriented by the $x'y'z'$ coordinate system, the six stresses on the two surfaces with normals in the y' and z' directions must also be determined ($\sigma_{y'}$, $\tau_{y'z'}$, $\tau_{y'x'}$, $\sigma_{z'}$, $\tau_{z'y'}$, and $\tau_{z'x'}$). However, throughout this book we will assume that no body moments exist. This, in turn, causes the cross shears to be equal (i.e., $\tau_{y'x'} = \tau_{x'y'}$, $\tau_{z'y'} = \tau_{y'z'}$, $\tau_{x'z'} = \tau_{z'x'}$). Thus, with cross shears being equal, it is only necessary to evaluate $\sigma_{y'}$, $\sigma_{z'}$, and $\tau_{y'z'}$. The analyses for the surfaces containing these stresses are identical to the previous development except that the surface normals are different. For the surface perpendicular to the y' axis, the two additional stresses are

$$\sigma_{y'} = \sigma_x n_{y'x}^2 + \sigma_y n_{y'y}^2 + \sigma_z n_{y'z}^2 + 2\tau_{xy}n_{y'x}n_{y'y} + 2\tau_{yz}n_{y'y}n_{y'z} + 2\tau_{zx}n_{y'z}n_{y'x} \quad \textbf{[2.1-4d]}$$

$$\tau_{y'z'} = \sigma_x n_{y'x}n_{z'x} + \sigma_y n_{y'y}n_{z'y} + \sigma_z n_{y'z}n_{z'z} + \tau_{xy}(n_{y'x}n_{z'y} + n_{y'y}n_{z'x})$$

$$+ \tau_{yz}(n_{y'y}n_{z'z} + n_{y'z}n_{z'y}) + \tau_{zx}(n_{y'x}n_{z'z} + n_{y'z}n_{z'x}) \quad \textbf{[2.1-4e]}$$

Finally, the normal stress $\sigma_{z'}$ is

$$\sigma_{z'} = \sigma_x n_{z'x}^2 + \sigma_y n_{z'y}^2 + \sigma_z n_{z'z}^2 + 2\tau_{xy}n_{z'x}n_{z'y} + 2\tau_{yz}n_{z'y}n_{z'z} + 2\tau_{zx}n_{z'z}n_{z'x} \quad \textbf{[2.1-4f]}$$

Example 2.1-1 | The state of stress at a point relative to an xyz coordinate system is given by the stress matrix

$$[\sigma] = \begin{bmatrix} -8 & 6 & -2 \\ 6 & 4 & 2 \\ -2 & 2 & -5 \end{bmatrix} \text{MPa}$$

Determine the state of stress on an element that is oriented by first rotating the xyz axes 45° about the z axis and then rotating the resulting axes 30° about the new x axis.

Solution:

The surface normals can be found by a series of coordinate transformations for each rotation. From Fig. 2.1-3(a) the vector components for the first rotation can be represented by

$$\begin{Bmatrix} x_1 \\ y_1 \\ z_1 \end{Bmatrix} = \begin{bmatrix} \cos\theta & \sin\theta & 0 \\ -\sin\theta & \cos\theta & 0 \\ 0 & 0 & 1 \end{bmatrix} \begin{Bmatrix} x \\ y \\ z \end{Bmatrix} \quad \textbf{[a]}$$

The last rotation establishes the $x'y'z'$ coordinates as shown in Fig. 2.1-3(b), and they are related to the $x_1y_1z_1$ coordinates by

$$\begin{Bmatrix} x' \\ y' \\ z' \end{Bmatrix} = \begin{bmatrix} 1 & 0 & 0 \\ 0 & \cos\phi & \sin\phi \\ 0 & -\sin\phi & \cos\phi \end{bmatrix} \begin{Bmatrix} x_1 \\ y_1 \\ z_1 \end{Bmatrix} \quad \textbf{[b]}$$

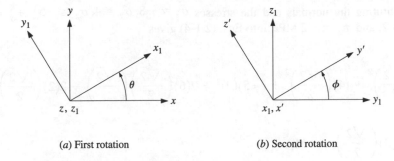

(*a*) First rotation

(*b*) Second rotation

Figure 2.1-3

Substituting Eq. (*a*) in (*b*) gives

$$\begin{Bmatrix} x' \\ y' \\ z' \end{Bmatrix} = \begin{bmatrix} 1 & 0 & 0 \\ 0 & \cos\phi & \sin\phi \\ 0 & -\sin\phi & \cos\phi \end{bmatrix} \begin{bmatrix} \cos\theta & \sin\theta & 0 \\ -\sin\theta & \cos\theta & 0 \\ 0 & 0 & 1 \end{bmatrix} \begin{Bmatrix} x \\ y \\ z \end{Bmatrix}$$

$$= \begin{bmatrix} \cos\theta & \sin\theta & 0 \\ -\sin\theta\cos\phi & \cos\theta\cos\phi & \sin\phi \\ \sin\theta\sin\phi & -\cos\theta\sin\phi & \cos\phi \end{bmatrix} \begin{Bmatrix} x \\ y \\ z \end{Bmatrix} \qquad [\boldsymbol{c}]$$

The final matrix establishes the directional cosines

$$n_{x'x} = \cos\theta \qquad\qquad n_{x'y} = \sin\theta \qquad\qquad n_{x'z} = 0$$

$$n_{y'x} = -\sin\theta\cos\phi \qquad\qquad n_{y'y} = \cos\theta\cos\phi \qquad\qquad n_{y'z} = \sin\phi$$

$$n_{z'x} = \sin\theta\sin\phi \qquad\qquad n_{z'y} = -\cos\theta\sin\phi \qquad\qquad n_{z'z} = \cos\phi$$

Substituting $\theta = 45°$ and $\phi = 30°$ gives

$$n_{x'x} = \frac{\sqrt{2}}{2} \qquad\qquad n_{x'y} = \frac{\sqrt{2}}{2} \qquad\qquad n_{x'z} = 0$$

$$n_{y'x} = -\frac{\sqrt{2}}{2}\frac{\sqrt{3}}{2} = -\frac{\sqrt{6}}{4} \qquad n_{y'y} = \frac{\sqrt{2}}{2}\frac{\sqrt{3}}{2} = \frac{\sqrt{6}}{4} \qquad n_{y'z} = \frac{1}{2}$$

$$n_{z'x} = \frac{\sqrt{2}}{2}\frac{1}{2} = \frac{\sqrt{2}}{4} \qquad\qquad n_{z'y} = -\frac{\sqrt{2}}{2}\frac{1}{2} = -\frac{\sqrt{2}}{4} \qquad\qquad n_{z'z} = \frac{\sqrt{3}}{2}$$

Substituting the normals and the stresses $\sigma_x = -8, \sigma_y = 4, \sigma_z = -5, \tau_{xy} = 6, \tau_{yz} = 2$, and $\tau_{zx} = -2$ MPa into Eqs. (2.1-4) gives

$$\sigma_{x'} = (-8)\left(\frac{\sqrt{2}}{2}\right)^2 + (4)\left(\frac{\sqrt{2}}{2}\right)^2 + (-5)(0)^2 + 2(6)\left(\frac{\sqrt{2}}{2}\right)\left(\frac{\sqrt{2}}{2}\right) + 2(2)\left(\frac{\sqrt{2}}{2}\right)(0)$$

$$+ 2(-2)(0)\left(\frac{\sqrt{2}}{2}\right) = 4 \text{ MPa}$$

$$\tau_{x'y'} = (-8)\left(\frac{\sqrt{2}}{2}\right)\left(-\frac{\sqrt{6}}{4}\right) + (4)\left(\frac{\sqrt{2}}{2}\right)\left(\frac{\sqrt{6}}{4}\right) + (-5)(0)\left(\frac{1}{2}\right)$$

$$+ (6)\left[\left(\frac{\sqrt{2}}{2}\right)\left(\frac{\sqrt{6}}{4}\right) + \left(\frac{\sqrt{2}}{2}\right)\left(-\frac{\sqrt{6}}{4}\right)\right]$$

$$+ (2)\left[\left(\frac{\sqrt{2}}{2}\right)\left(\frac{1}{2}\right) + (0)\left(\frac{\sqrt{6}}{4}\right)\right]$$

$$+ (-2)\left[\left(\frac{\sqrt{2}}{2}\right)\left(\frac{1}{2}\right) + (0)\left(-\frac{\sqrt{6}}{4}\right)\right] = 5.20 \text{ MPa}$$

$$\tau_{z'x'} = (-8)\left(\frac{\sqrt{2}}{2}\right)\left(\frac{\sqrt{2}}{4}\right) + (4)\left(\frac{\sqrt{2}}{2}\right)\left(-\frac{\sqrt{2}}{4}\right) + (-5)(0)\left(\frac{\sqrt{3}}{2}\right)$$

$$+ (6)\left[\left(\frac{\sqrt{2}}{2}\right)\left(-\frac{\sqrt{2}}{4}\right) + \left(\frac{\sqrt{2}}{2}\right)\left(\frac{\sqrt{2}}{4}\right)\right]$$

$$+ (2)\left[\left(\frac{\sqrt{2}}{2}\right)\left(\frac{\sqrt{3}}{2}\right) + (0)\left(-\frac{\sqrt{2}}{4}\right)\right]$$

$$+ (-2)\left[\left(\frac{\sqrt{2}}{2}\right)\left(\frac{\sqrt{3}}{2}\right) + (0)\left(\frac{\sqrt{2}}{4}\right)\right] = -3 \text{ MPa}$$

$$\sigma_{y'} = (-8)\left(-\frac{\sqrt{6}}{4}\right)^2 + (4)\left(\frac{\sqrt{6}}{4}\right)^2 + (-5)\left(\frac{1}{2}\right)^2 + 2(6)\left(-\frac{\sqrt{6}}{4}\right)\left(\frac{\sqrt{6}}{4}\right)$$

$$+ 2(2)\left(\frac{\sqrt{6}}{4}\right)\left(\frac{1}{2}\right) + 2(-2)\left(\frac{1}{2}\right)\left(-\frac{\sqrt{6}}{4}\right) = -4.80 \text{ MPa}$$

$$\tau_{y'z'} = (-8)\left(-\frac{\sqrt{6}}{4}\right)\left(\frac{\sqrt{2}}{4}\right) + (4)\left(\frac{\sqrt{6}}{4}\right)\left(-\frac{\sqrt{2}}{4}\right) + (-5)\left(\frac{1}{2}\right)\left(\frac{\sqrt{3}}{2}\right)$$

$$+ (6)\left[\left(-\frac{\sqrt{6}}{4}\right)\left(-\frac{\sqrt{2}}{4}\right) + \left(\frac{\sqrt{6}}{4}\right)\left(\frac{\sqrt{2}}{4}\right)\right]$$

$$+ (2)\left[\left(\frac{\sqrt{6}}{4}\right)\left(\frac{\sqrt{3}}{2}\right) + \left(\frac{1}{2}\right)\left(-\frac{\sqrt{2}}{4}\right)\right]$$

$$+ (-2)\left[\left(-\frac{\sqrt{6}}{4}\right)\left(\frac{\sqrt{3}}{2}\right) + \left(\frac{1}{2}\right)\left(\frac{\sqrt{2}}{4}\right)\right] = 2.71 \text{ MPa}$$

$$\sigma_{z'} = (-8)\left(\frac{\sqrt{2}}{4}\right)^2 + (4)\left(-\frac{\sqrt{2}}{4}\right)^2 + (-5)\left(\frac{\sqrt{3}}{2}\right)^2 + 2(6)\left(\frac{\sqrt{2}}{4}\right)\left(-\frac{\sqrt{2}}{4}\right)$$

$$+ 2(2)\left(-\frac{\sqrt{2}}{4}\right)\left(\frac{\sqrt{3}}{2}\right) + 2(-2)\left(\frac{\sqrt{3}}{2}\right)\left(\frac{\sqrt{2}}{4}\right) = -8.20 \text{ MPa}$$

The resulting transformed stress matrix is thus

$$[\boldsymbol{\sigma}'] = \begin{bmatrix} 4.00 & 5.20 & -3.00 \\ 5.20 & -4.80 & 2.71 \\ -3.00 & 2.71 & -8.20 \end{bmatrix} \text{MPa}$$

The matrix of the directional cosines in Eq. (c) of Example 2.1-1 establishes a transformation matrix which transforms a vector in the xyz coordinate system to a vector in the $x'y'z'$ coordinate system. That is, a vector with coordinates V_x, V_z, V_z, will transform to $V_{x'}, V_{y'}, V_{z'}$, according to

$$\{\mathbf{V}\}_{x'y'z'} = [\mathbf{T}]\{\mathbf{V}\}_{xyz} \qquad\qquad \textbf{[2.1-5]}$$

where

$$\{V\}_{xyz} = \begin{Bmatrix} V_x \\ V_y \\ V_z \end{Bmatrix} \qquad \{V\}_{x'y'z'} = \begin{Bmatrix} V_{x'} \\ V_{y'} \\ V_{z'} \end{Bmatrix}$$

and the transformation matrix $[T]$ is

$$[T] = \begin{bmatrix} n_{x'x} & n_{x'y} & n_{x'z} \\ n_{y'x} & n_{y'y} & n_{y'z} \\ n_{z'x} & n_{z'y} & n_{z'z} \end{bmatrix} \qquad\qquad \textbf{[2.1-6]}$$

In Appendix I, it is shown that stresses (and strains) transform according to

$$[\sigma]_{x'y'z'} = [T][\sigma]_{xyz}[T]^T \qquad\qquad \textbf{[2.1-7]}$$

where

$$[\sigma]_{x'y'z'} = \begin{bmatrix} \sigma_{x'} & \tau_{x'y'} & \tau_{x'z'} \\ \tau_{y'x'} & \sigma_{y'} & \tau_{y'z'} \\ \tau_{z'x'} & \tau_{z'y'} & \sigma_{z'} \end{bmatrix} \qquad [\sigma]_{xyz} = \begin{bmatrix} \sigma_x & \tau_{xy} & \tau_{xz} \\ \tau_{yx} & \sigma_y & \tau_{yz} \\ \tau_{zx} & \tau_{zy} & \sigma_z \end{bmatrix}$$

and $[T]^T$ is the transpose of the $[T]$ matrix. It is suggested that the reader apply Eq. (2.1-7) to Example 2.1-1 and verify the solution using a spreadsheet or mathematics program.

2.1.2   PLANE STRESS TRANSFORMATIONS

For the state of plane stress shown in Fig. 2.1-4(*a*), $\sigma_z = \tau_{yz} = \tau_{zx} = 0$. Plane stress transformations are normally performed in the *xy* plane as shown in Fig. 2.1-4(*b*). The angles relating the $x'y'z'$ axes to the *xyz* axes are

$$\theta_{x'x} = \theta \qquad\qquad \theta_{x'y} = 90° - \theta \qquad\qquad \theta_{x'z} = 90°$$

$$\theta_{y'x} = -(90° - \theta) \qquad\qquad \theta_{y'y} = \theta \qquad\qquad \theta_{y'z} = 90°$$

$$\theta_{z'x} = 90° \qquad\qquad \theta_{z'y} = 90° \qquad\qquad \theta_{z'z} = 0$$

Thus

$$n_{x'x} = \cos\theta \qquad\qquad n_{x'y} = \sin\theta \qquad\qquad n_{x'z} = 0$$

$$n_{y'x} = -\sin\theta \qquad\qquad n_{y'y} = \cos\theta \qquad\qquad n_{y'z} = 0$$

$$n_{z'x} = 0 \qquad\qquad n_{z'y} = 0 \qquad\qquad n_{z'z} = 1$$

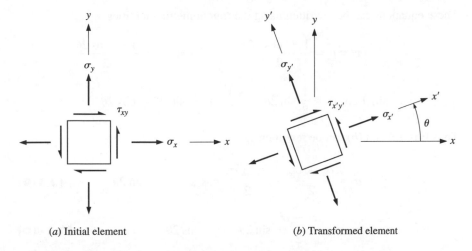

     (a) Initial element   (b) Transformed element

Figure 2.1-4   Plane stress transformations.

Substituting these conditions into Eqs. (2.1-4a), (2.1-4b), and (2.1-4d) results in

$$\sigma_{x'} = \sigma_x \cos^2\theta + \sigma_y \sin^2\theta + 2\tau_{xy} \cos\theta \sin\theta \qquad \textbf{[2.1-8a]}$$

$$\sigma_{y'} = \sigma_x \sin^2\theta + \sigma_y \cos^2\theta - 2\tau_{xy} \sin\theta \cos\theta \qquad \textbf{[2.1-8b]}$$

$$\tau_{x'y'} = -(\sigma_x - \sigma_y) \sin\theta \cos\theta + \tau_{xy}(\cos^2\theta - \sin^2\theta) \qquad \textbf{[2.1-8c]}$$

The transformed element is shown in Fig. 2.1-4(b).

2.1.3   MOHR'S CIRCLE FOR PLANE STRESS[†]

If we were only interested in obtaining the normal stress and shear stress (σ, τ) on a single surface, say the x' surface, then Eqs. (2.1-8a) and (2.1-8c) would provide the necessary equations. That is, σ and τ would be found from

$$\sigma = \sigma_x \cos^2\theta + \sigma_y \sin^2\theta + 2\tau_{xy} \cos\theta \sin\theta \qquad \textbf{[a]}$$

$$\tau = -(\sigma_x - \sigma_y) \sin\theta \cos\theta + \tau_{xy}(\cos^2\theta - \sin^2\theta) \qquad \textbf{[b]}$$

[†] Historically, Mohr's circle was used extensively. Today, however, it is used more as a pedagogical tool which enables one to easily *envision* the transformation process. For actual calculations, the transformation equations are recommended.

These equations can be rewritten using the trigonometric identities

$$\cos^2 \theta = \frac{1 + \cos 2\theta}{2} \qquad\qquad \sin^2 \theta = \frac{1 - \cos 2\theta}{2}$$

$$\sin \theta \cos \theta = \frac{1}{2} \sin 2\theta \qquad \cos^2 \theta - \sin^2 \theta = \cos 2\theta$$

Thus Eqs. (*a*) and (*b*) can be rewritten as

$$\sigma = \frac{\sigma_x + \sigma_y}{2} + \frac{\sigma_x - \sigma_y}{2} \cos 2\theta + \tau_{xy} \sin 2\theta \qquad\qquad \textbf{[2.1-9]}$$

$$\tau = -\frac{\sigma_x - \sigma_y}{2} \sin 2\theta + \tau_{xy} \cos 2\theta \qquad\qquad \textbf{[2.1-10]}$$

Equations (2.1-9) and (2.1-10) are parametric equations which represent a circle in σ, τ space where θ is the parameter. Moving the first term on the right-hand side of Eq. (2.1-9) to the left, squaring, and adding the result to the square of Eq. (2.1-10) gives

$$\left[\sigma - \left(\frac{\sigma_x + \sigma_y}{2} \right) \right]^2 + \tau^2 = \left(\frac{\sigma_x - \sigma_y}{2} \right)^2 + \tau_{xy}^2 \qquad\qquad \textbf{[2.1-11]}$$

Equation (2.1-11) represents the equation of a circle in σ, τ space where the radius is

$$R = \sqrt{\left(\frac{\sigma_x - \sigma_y}{2} \right)^2 + \tau_{xy}^2} \qquad\qquad \textbf{[2.1-12]}$$

and the center of the circle is displaced in the positive σ direction the average of the normal stresses

$$\sigma_{ave} = \frac{\sigma_x + \sigma_y}{2} \qquad\qquad \textbf{[2.1-13]}$$

Equation (2.1-11) can then be rewritten as

$$(\sigma - \sigma_{ave})^2 + \tau^2 = R^2 \qquad\qquad \textbf{[2.1-14]}$$

If one were to plot the circle graphically using Eqs. (2.1-9) and (2.1-10) with the coordinate axes σ and τ positive to the right and up, respectively, the circle would graph 2θ in the clockwise direction as θ increased in the counterclockwise direction. This opposing rotation is undesirable, as it can lead to errors. To circumvent this, one of two methods is practiced: (1) redefine the shear stress as clockwise and counterclockwise (relative to the center of the element) and plot clockwise above the σ axis and counterclockwise below; or (2) plot positive τ below the σ axis. Historically, the first method was used, and it is still practiced today in many textbooks. Here, however, we will adopt the second method, as this does not rely on

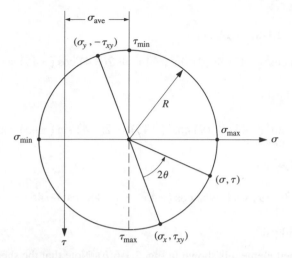

Figure 2.1-5   Mohr's circle for plane stress.

redefining τ. A plot of the circle is shown in Fig. 2.1-5. From the circle it can be observed that the maximum and minimum values of σ and τ are

$$\sigma_{\substack{max \\ min}} = \sigma_{ave} \pm R = \frac{\sigma_x + \sigma_y}{2} \pm \sqrt{\left(\frac{\sigma_x - \sigma_y}{2}\right)^2 + \tau_{xy}^2} \quad \textbf{[2.1-15]}$$

and

$$\tau_{\substack{max \\ min}} = \pm \sqrt{\left(\frac{\sigma_x - \sigma_y}{2}\right)^2 + \tau_{xy}^2} \qquad \textbf{[2.1-16]}$$

It is very important to note here that these are the maximum and minimum values of σ and τ in the *plane of analysis* and may *not* be the absolute maximums or minimums. More is said about this in the discussion of Mohr's circles in three dimensions.

The state of plane stress at a point and relative to an xy coordinate system is given by | **Example 2.1-2**

$$\sigma = \begin{bmatrix} 3 & -4 \\ -4 & 9 \end{bmatrix} \quad \text{kpsi}$$

and is shown in Fig. 2.1-6(*a*). Determine the state of stress on an element rotated 45° clockwise from the initial element. Solve the problem two ways: (*a*) the transformation equations and (*b*) Mohr's circle.

Solution:

(*a*) From Eqs. (2.1-8) we obtain

$$\sigma_{x'} = (3) \cos^2(-45) + (9) \sin^2(-45) + 2(-4) \cos(-45) \sin(-45)$$

$$= 10 \text{ kpsi}$$

$$\sigma_{y'} = (3) \sin^2(-45) + (9) \cos^2(-45) - 2(-4) \sin(-45) \cos(-45)$$

$$= 2 \text{ kpsi}$$

$$\tau_{x'y'} = -(3 - 9) \sin(-45) \cos(-45) + (-4)[\cos^2(-45) - \sin^2(-45)]$$

$$= -3 \text{ kpsi}$$

The transformed element is shown in Fig. 2.1-6(*b*). Note that the shear stress on the surface with the normal in the x' direction is shown in the negative y' direction since $\tau_{x'y'}$ is negative.

(*b*) Mohr's circle can be constructed entirely by graphical means (see Fig. 2.1-7). For accuracy, however, numerical determinations will be obtained from trigonometric calculations. The first point on the circle, P, is established from the stresses acting on the surface perpendicular to the x direction. Using the xy coordinate system to establish the sign of each stress, the (σ, τ) values for point P are $(\sigma_x, \tau_{xy}) = (3, -4)$ kpsi. Next we rotate the xy coordinate system 90° counterclockwise to $x'y'$ such that x' is in the direction of the original y direction and y' is in the negative direction of the original x direction. The state of stress on the surface

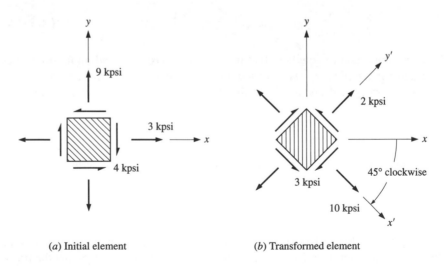

(*a*) Initial element   (*b*) Transformed element

Figure 2.1-6

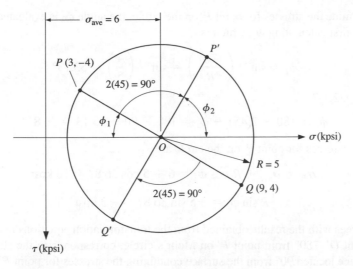

Figure 2.1-7 Mohr's circle for Example 2.1-2.

perpendicular to the x' direction establishes point Q on Mohr's circle. Using the $x'y'$ coordinate system to establish the sign of the stresses for point Q, the (σ, τ) values for point Q are $(\sigma_{x'}, \tau_{x'y'}) = (\sigma_y, -\tau_{xy}) = (9, 4)$ kpsi. Rotations relative to the element (θ) are doubled (2θ) on Mohr's circle. Since points P and Q represent a $90°$ rotation relative to the element, we have rotated $180°$ on Mohr's circle. Thus points P and Q lie diametrically opposite on the circle and the bisector of line PQ must be the center of the circle O. With two points P and Q on the circle and the center O, the circle can now be graphically constructed.

To determine the state of stress for an element rotated $45°$ *clockwise* from the initial element we rotate $90°$ *clockwise* along the circle from P to P' and from Q to Q'. The values of (σ, τ) for these points will establish the states of stress for the corresponding surfaces. As stated earlier, these values can be arrived at in a purely graphical manner using mechanical drawing tools. However, to obtain the coordinates of P' and Q' numerically from the graph we make the following calculations. The center of the circle is located at the average of the normal stresses associated with points P and Q and is

$$\sigma_{\text{ave}} = \frac{\sigma_P + \sigma_Q}{2} = \frac{3 + 9}{2} = 6 \text{ kpsi}$$

The radius of the circle is OP and is given by

$$R = \sqrt{\left(\frac{\sigma_P - \sigma_Q}{2}\right)^2 + |\tau_P|^2} = \sqrt{\left(\frac{3 - 9}{2}\right)^2 + (4)^2} = 5 \text{ kpsi}$$

To determine the stresses for point P' on the circle, the angle ϕ_2 is evaluated. This is done by first calculating ϕ_1, which is

$$\phi_1 = \sin^{-1}\left(\frac{|\tau_{xy}|}{R}\right) = \sin^{-1}\left(\frac{4}{5}\right) = 53.13°$$

From Fig. 2.1-7

$$\phi_2 = 180 - 2(45) - \phi_1 = 180 - 90 - 53.13 = 36.87°$$

Thus the stresses for point P' on the circle are

$$\sigma_{P'} = \sigma_{\text{ave}} + R\cos\phi_2 = 6 + 5\cos 36.87 = 10 \text{ kpsi}$$

$$\tau_{P'} = -R\sin\phi_2 = -5\sin 36.87 = -3 \text{ kpsi}$$

This agrees with the results obtained from the transformation equations in part (a).

Point Q', 180° from point P' on Mohr's circle, corresponds to the stresses on the surface located 90° from the surface containing the stresses for point P'. Thus

$$\sigma_{Q'} = \sigma_{\text{ave}} - R\cos\phi_2 = 6 - 5\cos 36.87 = 2 \text{ kpsi}$$

$$\tau_{Q'} = R\sin\phi_2 = 5\sin 36.87 = 3 \text{ kpsi}$$

This also agrees with the results of part (a), and the final transformed element is the same as shown in Fig. 2.1-6(b).

To clearly establish the true direction of a shear stress obtained from Mohr's circle for a transformed surface, rotate the initial xy coordinate system of the element to $x'y'$ such that x' is normal to the surface containing the stresses corresponding to the point on Mohr's circle. If the shear stress obtained from Mohr's circle for the surface is positive (negative), the direction of the shear stress is in the positive (negative) y' direction.

2.1.4 THREE-DIMENSIONAL STRESS TRANSFORMATION SIMPLIFIED

It is important to keep in mind that only Eqs. (2.1-4a) to (2.1-4c) are necessary to obtain the state of stress on any one given surface. Furthermore, Eqs. (2.1-4b) and (2.1-4c) provide the components of the *total* shear stress on the surface with respect to a particular set of orthogonal axes tangential to the surface. However, the *total* shear stress on the surface depends only on the particular surface as defined by the surface normal. If we redefine the surface normal by the n axis, as shown in Fig. 2.1-8, where the directional cosines of the normal are

$$n_x = \cos\theta_x \qquad\qquad \textbf{[2.1-17a]}$$

$$n_y = \cos\theta_y \qquad\qquad \textbf{[2.1-17b]}$$

$$n_z = \cos\theta_z \qquad\qquad \textbf{[2.1-17c]}$$

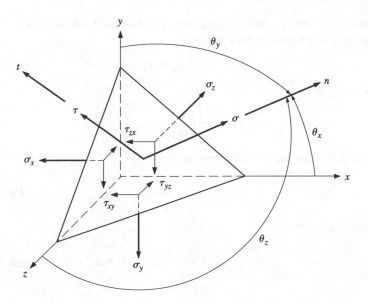

Figure 2.1-8

the normal stress σ is given by Eq. (2.1-4a) and is

$$\sigma = \sigma_x n_x^2 + \sigma_y n_y^2 + \sigma_z n_z^2 + 2\tau_{xy} n_x n_y + 2\tau_{yz} n_y n_z + 2\tau_{zx} n_z n_x \qquad \textbf{[2.1-18]}$$

The total shear stress τ is now the only unknown in Fig. 2.1-8 and can be found from a summation of forces. The total force in the x direction due to the stresses on the orthogonal surfaces is

$$F_x = -\sigma_x A_x - \tau_{xy} A_y - \tau_{zx} A_z$$

However, from Eqs. (2.1-3), $A_x = A_o n_x$, $A_y = A_o n_y$, and $A_z = A_o n_z$. Thus, F_x is

$$F_x = -A_o(\sigma_x n_x + \tau_{xy} n_y + \tau_{zx} n_z) \qquad \textbf{[2.1-19a]}$$

Likewise, the forces in the y and z directions due to the stresses on the orthogonal surfaces are

$$F_y = -A_o(\tau_{xy} n_x + \sigma_y n_y + \tau_{yz} n_z) \qquad \textbf{[2.1-19b]}$$

$$F_z = -A_o(\tau_{zx} n_x + \tau_{yz} n_y + \sigma_z n_z) \qquad \textbf{[2.1-19c]}$$

The magnitude of τ can easily be found since for equilibrium the magnitude of the forces on the oblique surface $A_o(\sigma^2 + \tau^2)^{1/2}$ must balance the magnitude of forces on the orthogonal surfaces $(F_x^2 + F_y^2 + F_z^2)^{1/2}$. Equating the two sets of force magnitudes using Eqs. (2.1-19) and solving for τ results in

$$\tau = [(\sigma_x n_x + \tau_{xy} n_y + \tau_{zx} n_z)^2 + (\tau_{xy} n_x + \sigma_y n_y + \tau_{yz} n_z)^2 + (\tau_{zx} n_x + \tau_{yz} n_y + \sigma_z n_z)^2 - \sigma^2]^{1/2}$$

$$\textbf{[2.1-20]}$$

The directional cosines establishing the direction of τ can be defined as t_x, t_y, and t_z. For equilibrium of forces in the x direction,

$$\tau A_o t_x + \sigma A_o n_x + F_x = 0$$

Solving for t_x yields

$$t_x = -\frac{1}{\tau}\left(\frac{F_x}{A_o} + \sigma n_x\right)$$

Substituting F_x from Eq. (2.1-19a) results in

$$t_x = \frac{1}{\tau}[(\sigma_x - \sigma)n_x + \tau_{xy}n_y + \tau_{zx}n_z] \qquad \textbf{[2.1-21a]}$$

Likewise, t_y and t_z are

$$t_y = \frac{1}{\tau}[\tau_{xy}n_x + (\sigma_y - \sigma)n_y + \tau_{yz}n_z] \qquad \textbf{[2.1-21b]}$$

$$t_z = \frac{1}{\tau}[\tau_{zx}n_x + \tau_{yz}n_y + (\sigma_z - \sigma)n_z] \qquad \textbf{[2.1-21c]}$$

Example 2.1-3 | The state of stress at a particular point relative to the xyz coordinate system is given by the stress matrix

$$[\boldsymbol{\sigma}] = \begin{bmatrix} 14 & 7 & -7 \\ 7 & 10 & 0 \\ -7 & 0 & 35 \end{bmatrix} \text{MPa}$$

Determine the normal stress and the magnitude and direction of the shear stress on a surface intersecting the point and parallel to the plane given by the equation

$$2x - y + 3z = 9$$

Solution:

The perpendicular of the surface is established by the direction numbers of the plane d_x, d_y, and d_z. These are simply the respective coefficients of the x, y, and z terms of the equation of the plane. Thus

$$d_x = 2 \qquad d_y = -1 \qquad d_z = 3$$

The directional cosines are found by simply normalizing the direction numbers so that the sum of their squares is unity. Therefore,

$$n_x = \frac{2}{\sqrt{2^2 + (-1^2) + 3^2}} = \frac{2}{\sqrt{14}}$$

Likewise

$$n_y = \frac{-1}{\sqrt{14}} \qquad n_z = \frac{3}{\sqrt{14}}$$

From the stress matrix, $\sigma_x = 14$, $\tau_{xy} = 7$, $\tau_{zx} = -7$, $\sigma_y = 10$, $\tau_{yz} = 0$, and $\sigma_z = 35$ MPa. Substituting the stresses and direction cosines in Eq. (2.1-18) yields

$$\sigma = 14\left(\frac{2}{\sqrt{14}}\right)^2 + 10\left(\frac{-1}{\sqrt{14}}\right)^2 + 35\left(\frac{3}{\sqrt{14}}\right)^2 + (2)(7)\frac{2}{\sqrt{14}}\frac{-1}{\sqrt{14}}$$

$$+ (2)(0)\frac{-1}{\sqrt{14}}\frac{3}{\sqrt{14}} + (2)(-7)\frac{3}{\sqrt{14}}\frac{2}{\sqrt{14}} = 19.21 \text{ MPa}$$

The shear stress is found using Eq. (2.1-20) and is

$$\tau = \left[\left(14\frac{2}{\sqrt{14}} + 7\frac{-1}{\sqrt{14}} + (-7)\frac{3}{\sqrt{14}}\right)^2\right.$$

$$+ \left(7\frac{2}{\sqrt{14}} + 10\frac{-1}{\sqrt{14}} + 0\frac{3}{\sqrt{14}}\right)^2$$

$$\left. + \left(-7\frac{2}{\sqrt{14}} + 0\frac{-1}{\sqrt{14}} + 35\frac{3}{\sqrt{14}}\right)^2 - (19.2)^2\right]^{1/2} = 14.95 \text{ MPa}$$

From Eqs. (2.1-21), the directional cosines for the direction of τ are

$$t_x = \frac{1}{14.95}\left[(14 - 19.21)\frac{2}{\sqrt{14}} + 7\frac{-1}{\sqrt{14}} + (-7)\frac{3}{\sqrt{14}}\right] = -0.687$$

$$t_y = \frac{1}{14.95}\left[7\frac{2}{\sqrt{14}} + (10 - 19.21)\frac{-1}{\sqrt{14}} + 0\frac{3}{\sqrt{14}}\right] = 0.415$$

$$t_z = \frac{1}{14.95}\left[(-7)\frac{2}{\sqrt{14}} + 0\frac{-1}{\sqrt{14}} + (35 - 19.21)\frac{3}{\sqrt{14}}\right] = 0.596$$

There are two simple checks that can be made on the directional cosines of τ:

1. The sum of the squares of the directional cosines is unity. Thus

$$t_x^2 + t_y^2 + t_z^2 = 1$$

2. The directions of σ and τ are perpendicular. Thus

$$n_x t_x + n_y t_y + n_z t_z = 0$$

2.1.5 PRINCIPAL STRESSES

As indicated in Appendix I, a symmetric 3 × 3 matrix has three eigenvalues with three corresponding orthogonal eigenvectors. For the stress matrix these three eigenvalues correspond to stationary normal stresses, referred to as *principal stresses,* which exist on mutually orthogonal surfaces that contain *no* shear stress. The normals to these surfaces are the aforementioned eigenvectors. The three principal stresses are denoted as σ_1, σ_2, and σ_3, where the ordering is such that $\sigma_1 > \sigma_2 > \sigma_3$. If two of the principal stresses are equal, there will exist an infinite set of surfaces containing these principal stresses, where the normals of these surfaces are perpendicular to the direction of the third principal stress. If all three principal stresses are equal, a *hydrostatic* state of stress exists, and regardless of orientation all surfaces contain the same principal stress with no shear stress. Returning to Fig. 2.1-8, assume that on the oblique surface the shear stress τ equals zero. Then σ will be a principal stress denoted by σ_p. The components of the force on the oblique surface in the x, y, and z directions are

$$F'_x = \sigma_p A_o n_x \qquad \text{[2.1-22a]}$$

$$F'_y = \sigma_p A_o n_y \qquad \text{[2.1-22b]}$$

$$F'_z = \sigma_p A_o n_z \qquad \text{[2.1-22c]}$$

Recall Eqs. (2.1-19), which give the forces in the x, y, and z directions due to the stresses on the orthogonal surfaces. For equilibrium, $F_x + F'_x = 0$, etc. This results in

$$(\sigma_x - \sigma_p)n_x + \tau_{xy}n_y + \tau_{zx}n_z = 0 \qquad \text{[2.1-23a]}$$

$$\tau_{xy}n_x + (\sigma_y - \sigma_p)n_y + \tau_{yz}n_z = 0 \qquad \text{[2.1-23b]}$$

$$\tau_{zx}n_x + \tau_{yz}n_y + (\sigma_z - \sigma_p)n_z = 0 \qquad \text{[2.1-23c]}$$

One possible solution to Eqs. (2.1-23) is $n_x = n_y = n_z = 0$. This cannot occur since

$$n_x^2 + n_y^2 + n_z^2 = 1 \qquad \text{[2.1-24]}$$

In order to avoid the zero solution of the directional cosines the determinant of the coefficients of n_x, n_y, and n_z of Eqs. (2.1-23) is equated to zero. This makes the solution of the directional cosines indeterminate from Eqs. (2.1-23). Thus

$$\begin{vmatrix} \sigma_x - \sigma_p & \tau_{xy} & \tau_{zx} \\ \tau_{xy} & \sigma_y - \sigma_p & \tau_{yz} \\ \tau_{zx} & \tau_{yz} & \sigma_z - \sigma_p \end{vmatrix} = 0$$

Evaluating the determinant results in

$$\sigma_p^3 - (\sigma_x + \sigma_y + \sigma_z)\sigma_p^2 + (\sigma_x\sigma_y + \sigma_y\sigma_z + \sigma_z\sigma_x - \tau_{yz}^2 - \tau_{zx}^2 - \tau_{xy}^2)\sigma_p$$
$$- (\sigma_x\sigma_y\sigma_z + 2\tau_{yz}\tau_{zx}\tau_{xy} - \sigma_x\tau_{yz}^2 - \sigma_y\tau_{zx}^2 - \sigma_z\tau_{xy}^2) = 0 \qquad \text{[2.1-25]}$$

Equation (2.1-25) is a cubic equation in the unknown, σ_p, where three solutions result, the principal stresses σ_1, σ_2, and σ_3. It can be shown that the principal stresses exist on surfaces for which the normal stress is stationary (see Problem 2.24). This means that at least one of the principal stresses is a maximum and at least one is minimum.

To determine the directional cosines for a specific principal stress, the stress is substituted into Eqs. (2.1-23). The three resulting equations in the unknowns n_x, n_y, and n_z will not be independent, as they were used to determine the principal stress. Thus only two of Eqs. (2.1-23) can be used. However, the second-order Eq. (2.1-24) can be used as the third equation for the three directional cosines. Instead of solving one second-order and two linear equations simultaneously, a simplified method is demonstrated in the following example.

For the stress matrix given below, determine the principal stresses and the directional cosines associated with the normals to the surfaces of each principal stress.   **|Example 2.1-4**

$$[\sigma] = \begin{bmatrix} 3 & 1 & 1 \\ 1 & 0 & 2 \\ 1 & 2 & 0 \end{bmatrix} \text{MPa}$$

Solution:

The stresses are $\sigma_x = 3$, $\tau_{xy} = 1$, $\tau_{zx} = 1$, $\sigma_y = 0$, $\tau_{yz} = 2$, and $\sigma_z = 0$ MPa. Substituting this into Eq. (2.1-25) yields

$$\sigma_p^3 - (3 + 0 + 0)\sigma_p^2 + [(3)(0) + (0)(0) + (0)(3) - 2^2 - 1^2 - 1^2]\sigma_p$$

$$- [(3)(0)(0) + (2)(2)(1)(1) - (3)(2^2) - (0)(1^2) - (0)(1^2)] = 0$$

which simplifies to

$$\sigma_p^3 - 3\sigma_p^2 - 6\sigma_p + 8 = 0 \qquad\qquad\text{[a]}$$

The solutions to the cubic equation are $\sigma_p = 4, 1$, and -2 MPa. Thus

$$\sigma_1 = 4 \text{ MPa} \qquad \sigma_2 = 1 \text{ MPa} \qquad \sigma_3 = -2 \text{ MPa}$$

The directional cosines associated with each principal stress are determined independently. First, consider σ_1 and substitute $\sigma_p = 4$ MPa into Eqs. (2.1-23). This results in

$$-n_x + n_y + n_z = 0 \qquad\qquad\text{[b]}$$

$$n_x - 4n_y + 2n_z = 0 \qquad\qquad\text{[c]}$$

$$n_x + 2n_y - 4n_z = 0 \qquad\qquad\text{[d]}$$

Solving Eqs. (b), (c), and (d) simultaneously will not yield a solution, as the three equations are not independent. For example, operating on the equations, one finds

$2(b) + (c) = -(d)$. Thus only two independent equations can be used. In this example any two can be used, and a third equation comes from Eq. (2.1-24). However, instead of solving the three equations simultaneously, let

$$n_x = ac_x \qquad n_y = ac_y \qquad n_z = ac_z$$

where a is an unknown constant. Next let $c_x = 1$ arbitrarily. Substituting this into Eqs. (b) and (c) results in

$$-a + ac_y + ac_z = 0 \qquad \text{and} \qquad a - 4ac_y + 2ac_z = 0$$

or

$$c_y + c_z = 1 \tag{e}$$

and

$$-4c_y + 2c_z = -1 \tag{f}$$

Solving Eqs. (e) and (f) simultaneously yields $c_y = \frac{1}{2}$ and $c_z = \frac{1}{2}$.[†] Now using Eq. (2.1-24),

$$a^2 + \left(\frac{1}{2}a\right)^2 + \left(\frac{1}{2}a\right)^2 = 1$$

results in $a = 2/\sqrt{6}$. Therefore,

$$n_x = ac_x = \frac{2}{\sqrt{6}}(1) = \frac{2}{\sqrt{6}}$$

$$n_y = ac_y = \frac{2}{\sqrt{6}}\left(\frac{1}{2}\right) = \frac{1}{\sqrt{6}}$$

$$n_z = ac_z = \frac{2}{\sqrt{6}}\left(\frac{1}{2}\right) = \frac{1}{\sqrt{6}}$$

Since these directional cosines are associated with the principal stress σ_1, the subscript 1 is used. Thus the directional cosines for the normal to the surface containing the principal stress $\sigma_1 = 4$ MPa are

$$(n_x)_1 = \frac{2}{\sqrt{6}} \qquad (n_y)_1 = \frac{1}{\sqrt{6}} \qquad (n_z)_1 = \frac{1}{\sqrt{6}}$$

Repeating the same procedure for $\sigma_2 = 1$ MPa, we find the directional cosines to be

$$(n_x)_2 = \frac{1}{\sqrt{3}} \qquad (n_y)_2 = \frac{-1}{\sqrt{3}} \qquad (n_z)_2 = \frac{-1}{\sqrt{3}}$$

[†] This technique has one potential problem. If n_x had been zero in this example, then a solution would not have resulted since this would have required a to be zero, thus causing $n_x = n_y = n_z = 0$ and Eq. (2.1-24) could not have been satisfied. If no solution results when this technique is used, simply repeat the solution letting c_y or c_z equal unity.

and for $\sigma_3 = -2$ MPa

$$(n_x)_3 = 0 \qquad (n_y)_3 = \frac{1}{\sqrt{2}} \qquad (n_z)_3 = \frac{-1}{\sqrt{2}}$$

Stress Invariants   Let us return to Eq. (2.1-25). The solutions of σ_p are independent of the coordinate system used to define the coefficients of the cubic equation for σ_p. Therefore, the coefficients of σ_p in Eq. (2.1-25) are constant and are normally referred to as the *stress invariants*. Thus

$$\sigma_x + \sigma_y + \sigma_z = I_1 \qquad\qquad \textbf{[2.1-26a]}$$

$$\sigma_x\sigma_y + \sigma_y\sigma_z + \sigma_z\sigma_x - \tau_{yz}^2 - \tau_{zx}^2 - \tau_{xy}^2 = I_2 \qquad \textbf{[2.1-26b]}$$

$$\sigma_x\sigma_y\sigma_z + 2\tau_{yz}\tau_{zx}\tau_{xy} - \sigma_x\tau_{yz}^2 - \sigma_y\tau_{zx}^2 - \sigma_z\tau_{xy}^2 = I_3 \qquad \textbf{[2.1-26c]}$$

The constants I_1, I_2, and I_3 are referred to as the first, second, and third stress invariants, respectively. Equations (2.1-26) are particularly helpful in checking the results of a stress transformation, as illustrated in the following example.

Returning to Example 2.1-1, verify that the stresses obtained for the $x'y'z'$ coordinate system satisfy Eqs. (2.1-26).

Example 2.1-5

Solution:

From Eq. (2.1-26a),

$$I_1 = \sigma_x + \sigma_y + \sigma_z = -8 + 4 - 5 = -9 \text{ MPa}$$

Relative to the $x'y'z'$ system, the first invariant is

$$I'_1 = \sigma_{x'} + \sigma_{y'} + \sigma_{z'} = 4.00 - 4.80 - 8.20 = -9 \text{ MPa} = I_1$$

The second invariant for the xyz coordinate system is

$$I_2 = \sigma_x\sigma_y + \sigma_y\sigma_z + \sigma_z\sigma_x - \tau_{yz}^2 - \tau_{zx}^2 - \tau_{xy}^2$$

$$= (-8)(4) + (4)(-5) + (-5)(-8) - (2)^2 - (-2)^2 - (6)^2$$

$$= -56 \, (\text{MPa})^2$$

and relative to the $x'y'z'$ system

$$I'_2 = \sigma_{x'}\sigma_{y'} + \sigma_{y'}\sigma_{z'} + \sigma_{z'}\sigma_{x'} - \tau_{y'z'}^2 - \tau_{z'x'}^2 - \tau_{x'y'}^2$$

$$= (4.00)(-4.80) + (-4.80)(-8.20) + (-8.20)(4.00) - (2.71)^2 - (-3.00)^2 - (5.20)^2$$

$$= -56 \, (\text{MPa})^2 = I_2$$

The third invariant is

$$I_3 = \sigma_x \sigma_y \sigma_z + 2\tau_{yz}\tau_{zx}\tau_{xy} - \sigma_x \tau_{yz}^2 - \sigma_y \tau_{zx}^2 - \sigma_z \tau_{xy}^2$$

$$= (-8)(4)(-5) + 2(2)(-2)(6) - (-8)(2)^2 - (4)(-2)^2 - (-5)(6)^2$$

$$= 308 \ (MPa)^3$$

and for the $x'y'z'$ system

$$I_3' = \sigma_{x'} \sigma_{y'} \sigma_{z'} + 2\tau_{y'z'}\tau_{z'x'}\tau_{x'y'} - \sigma_{x'} \tau_{y'z'}^2 - \sigma_{y'} \tau_{z'x'}^2 - \sigma_{z'} \tau_{x'y'}^2$$

$$= (4.00)(-4.80)(-8.20) + 2(2.71)(-3.00)(5.20) - (4.00)(2.71)^2$$

$$- (-4.80)(-3.00)^2 - (-8.20)(5.20)^2 = 308 \ (MPa)^3 = I_3$$

Thus we see that the three stress invariants agree for both coordinate systems.

Plane Stress   For this case, $\sigma_z = \tau_{yz} = \tau_{zx} = 0$, and Eqs. (2.1-23) reduce to

$$(\sigma_x - \sigma_p)n_x + \tau_{xy}n_y = 0 \qquad\qquad \textbf{[2.1-27a]}$$

$$\tau_{xy}n_x + (\sigma_y - \sigma_p)n_y = 0 \qquad\qquad \textbf{[2.1-27b]}$$

$$\sigma_p n_z = 0 \qquad\qquad \textbf{[2.1-27c]}$$

Since there is no shear stress on the surface whose normal is in the z direction, the normal stress on this surface, $\sigma_z = 0$, is a principal stress. Thus, for this case, $\sigma_p = 0$, $n_x = n_y = 0$, and $n_z = 1$, which satisfies Eq. (2.1-27). The directions of the remaining two principal stresses are in the xy plane, making $n_z = 0$ and Eq. (2.1-27c) satisfied. For the remaining two equations, let the counterclockwise angle the direction a principal stress makes with the x axis be θ. Thus $n_x = \cos\theta$, and $n_y = \sin\theta$, and Eqs. (2.1-27a) and (2.1-27b) can be written as

$$(\sigma_x - \sigma_p)\cos\theta + \tau_{xy}\sin\theta = 0 \qquad\qquad \textbf{[2.1-28a]}$$

$$\tau_{xy}\cos\theta + (\sigma_y - \sigma_p)\sin\theta = 0 \qquad\qquad \textbf{[2.1-28b]}$$

These equations amount to analyzing stresses with directions in the xy plane only. As before, we eliminate the trivial solution of Eqs. (2.1-28) by setting the determinant of the coefficients of the directional cosines to zero. That is,

$$\begin{vmatrix} (\sigma_x - \sigma_p) & \tau_{xy} \\ \tau_{xy} & (\sigma_y - \sigma_p) \end{vmatrix} = (\sigma_x - \sigma_p)(\sigma_y - \sigma_p) - \tau_{xy}^2$$

$$= \sigma_p^2 - (\sigma_x + \sigma_y)\sigma_p + (\sigma_x\sigma_y - \tau_{xy}^2) = 0 \quad \textbf{[2.1-29]}$$

Equation (2.1-29) is a quadratic equation in σ_p for which the two solutions are given by

$$\sigma_p = \frac{\sigma_x + \sigma_y}{2} \pm \sqrt{\left(\frac{\sigma_x - \sigma_y}{2}\right)^2 + \tau_{xy}^2} \qquad \text{[2.1-30]}$$

Since for plane stress one of the principal stresses (σ_z) is always zero, numbering of the stresses ($\sigma_1 > \sigma_2 > \sigma_3$) cannot be performed until Eq. (2.1-30) is solved. Each solution of Eq. (2.1-30) can then be substituted into one of Eqs. (2.1-28) to determine the direction of the principal stress. Note that if $\sigma_x = \sigma_y$ and $\tau_{xy} = 0$, then σ_x and σ_y are principal stresses and Eqs. (2.1-28) are satisfied *regardless* of θ. This means that all stresses in the plane of analysis are equal and the state of stress at the point is *isotropic* in the plane. Note further that Eq. (2.1-30) agrees with the maximum and minimum (*in-plane only*) values obtained for Mohr's circle in two dimensions shown in Fig. 2.1-5 and given by Eq. (2.1-15), $\sigma_{\substack{max \\ min}} = \sigma_{ave} \pm R$.

Determine the principal stresses for Example 2.1-2 and show the element containing these stresses properly oriented with respect to the initial *xyz* coordinate system.

Example 2.1-6

Solution:

Recall from Example 2.1-2 the initial stress matrix given by

$$\sigma = \begin{bmatrix} 3 & -4 \\ -4 & 9 \end{bmatrix} \text{kpsi}$$

From Eq. (2.1-30)

$$\sigma_p = \frac{3+9}{2} \pm \sqrt{\left(\frac{3-9}{2}\right)^2 + (-4)^2} = 6 \pm 5 = 11, 1 \text{ kpsi}$$

Thus, the three principal stresses ($\sigma_1, \sigma_2, \sigma_3$) are (11, 1, 0) kpsi, respectively. For directions, first substitute 11 kpsi into either one of Eqs. (2.1-28). Using Eq. (2.1-28a) gives

$$(\sigma_x - \sigma_p) \cos \theta + \tau_{xy} \sin \theta = (3 - 11) \cos \theta + (-4) \sin \theta = 0$$

or

$$\theta|_{\sigma_p = 11 \text{ kpsi}} = \tan^{-1}\left(-\frac{8}{4}\right) = -63.4°, 116.6°$$

Now for the other principal stress, 1 kpsi, Eq. (2.1-28a) gives

$$(\sigma_x - \sigma_p) \cos \theta + \tau_{xy} \sin \theta = (3 - 1) \cos \theta + (-4) \sin \theta = 0$$

or

$$\theta|_{\sigma_p = 1 \text{ kpsi}} = \tan^{-1}\left(\frac{2}{4}\right) = 26.6°, 206.6°$$

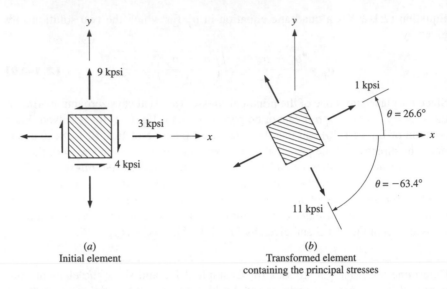

(a)
Initial element

(b)
Transformed element
containing the principal stresses

Figure 2.1-9

The orientation of the element containing the principal stresses is shown in Fig. 2.1-9(b). Note that the results obtained here agree with Mohr's circle shown in Fig. 2.1-7 where

$$\sigma_{\substack{max \\ min}} = \sigma_{ave} \pm R = 6 \pm 5 = 11, 1 \text{ kpsi}$$

σ_{min} is oriented at $\phi_1/2 = 53.13/2 = 26.6°$ counterclockwise from the x axis, whereas σ_{max} is oriented at $(180 - \phi_1)/2 = (180 - 53.13)/2 = 63.4°$ clockwise from the x axis.

2.1.6   MOHR'S CIRCLES IN THREE DIMENSIONS

An element undergoing a general state of three-dimensional stress can be transformed to an element containing only principal stresses σ_1, σ_2, and σ_3 acting along axes 1, 2, and 3, respectively (see Fig. 2.1-10). A transformation of stresses in the 12 plane depends only on σ_1 and σ_2; in the 23 plane depends only on σ_2 and σ_3; and in the 31 plane depends only on σ_1 and σ_3. This means that for each case, a plane stress analysis describes the state of stress in each of the three planes, and three Mohr's circles can be constructed to portray each case as shown in Fig. 2.1-11(a). Furthermore, it will be shown next that *all* possible states of stress (σ, τ) given by Eqs. (2.1-18) and (2.1-20) exist either on the circles or within the shaded area shown in Fig. 2.1-11(b) .

For the element containing the principal stresses shown in Fig. 2.1-10 let n_1, n_2, and n_3 be the directional cosines of an arbitrary surface relative to the 1, 2, and 3

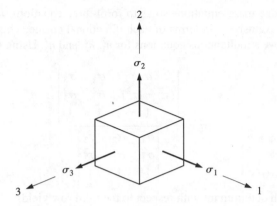

Figure 2.1-10   Principal stress state.

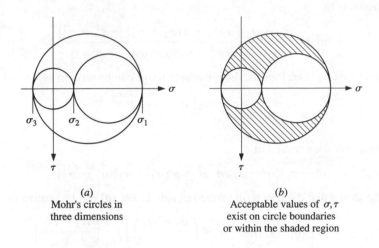

(a)	(b)
Mohr's circles in three dimensions	Acceptable values of σ, τ exist on circle boundaries or within the shaded region

Figure 2.1-11

axes, respectively. Equations (2.1-18) and (2.1-20) together with the relationship of the normals can be written as

$$\sigma_1 n_1^2 + \sigma_2 n_2^2 + \sigma_3 n_3^2 = \sigma \qquad \textbf{[2.1-31]}$$

$$\sigma_1^2 n_1^2 + \sigma_2^2 n_2^2 + \sigma_3^2 n_3^2 = \tau^2 + \sigma^2 \qquad \textbf{[2.1-32]}$$

$$n_1^2 + n_2^2 + n_3^2 = 1 \qquad \textbf{[2.1-33]}$$

We will rearrange these equations so as to form three equations where each equation is written exclusively in terms of each directional cosine. This can be done by solving the above simultaneous equations for n_1^2, n_2^2, and n_3^2. Using Cramer's rule,

$$n_1^2 = \frac{\begin{vmatrix} \sigma & \sigma_2 & \sigma_3 \\ (\tau^2 + \sigma^2) & \sigma_2^2 & \sigma_3^2 \\ 1 & 1 & 1 \end{vmatrix}}{\begin{vmatrix} \sigma_1 & \sigma_2 & \sigma_3 \\ \sigma_1^2 & \sigma_2^2 & \sigma_3^2 \\ 1 & 1 & 1 \end{vmatrix}}$$

Expanding each determinant with respect to the third row yields

$$n_1^2 = \frac{(\sigma_2\sigma_3^2 - \sigma_3\sigma_2^2) - [\sigma\sigma_3^2 - \sigma_3(\tau^2 + \sigma^2)] + [\sigma\sigma_2^2 - \sigma_2(\tau^2 + \sigma^2)]}{(\sigma_2\sigma_3^2 - \sigma_3\sigma_2^2) - (\sigma_1\sigma_3^2 - \sigma_3\sigma_1^2) + (\sigma_1\sigma_2^2 - \sigma_2\sigma_1^2)}$$

which reduces to

$$n_1^2 = \frac{(\sigma_3 - \sigma_2)[\sigma_2\sigma_3 - \sigma(\sigma_2 + \sigma_3) + \tau^2 + \sigma^2]}{(\sigma_3 - \sigma_2)(\sigma_1^2 - \sigma_1\sigma_2 - \sigma_1\sigma_3 + \sigma_2\sigma_3)}$$

Canceling the $(\sigma_3 - \sigma_2)$ terms, the remaining terms can be written as

$$n_1^2 = \frac{(\sigma - \sigma_2)(\sigma - \sigma_3) + \tau^2}{(\sigma_1 - \sigma_2)(\sigma_1 - \sigma_3)}$$

Cross multiplication results in

$$(\sigma - \sigma_2)(\sigma - \sigma_3) + \tau^2 = n_1^2(\sigma_1 - \sigma_2)(\sigma_1 - \sigma_3) \qquad \textbf{[a]}$$

It can be shown that the first term on the left side of Eq. (a) can be written as

$$(\sigma - \sigma_2)(\sigma - \sigma_3) = \left[\sigma - \left(\frac{\sigma_2 + \sigma_3}{2}\right)\right]^2 - \left(\frac{\sigma_2 - \sigma_3}{2}\right)^2$$

Substituting this into Eq. (a) results in

$$\left[\sigma - \left(\frac{\sigma_2 + \sigma_3}{2}\right)\right]^2 + \tau^2 = n_1^2(\sigma_1 - \sigma_2)(\sigma_1 - \sigma_3) + \left(\frac{\sigma_2 - \sigma_3}{2}\right)^2$$

$$\textbf{[2.1-34a]}$$

In a similar fashion, solving for n_2 and n_3 and rearranging, the following equations are obtained:

$$\left[\sigma - \left(\frac{\sigma_3 + \sigma_1}{2}\right)\right]^2 + \tau^2 = n_2^2(\sigma_3 - \sigma_2)(\sigma_1 - \sigma_2) + \left(\frac{\sigma_3 - \sigma_1}{2}\right)^2$$

$$\textbf{[2.1-34b]}$$

$$\left[\sigma - \left(\frac{\sigma_1 + \sigma_2}{2}\right)\right]^2 + \tau^2 = n_3^2(\sigma_1 - \sigma_3)(\sigma_2 - \sigma_3) + \left(\frac{\sigma_1 - \sigma_2}{2}\right)^2$$

[2.1-34c]

From Fig. 2.1-11, the equation for the Mohr's circle connecting σ_2 and σ_3 is given by

$$\left[\sigma - \left(\frac{\sigma_2 + \sigma_3}{2}\right)\right]^2 + \tau^2 = \left(\frac{\sigma_2 - \sigma_3}{2}\right)^2 = R_{23}^2 \qquad [2.1-35]$$

where R_{23} is the radius of the σ_2, σ_3 circle. Observing Equation (2.1-34a), it can be seen that if $n_1^2(\sigma_1 - \sigma_2)(\sigma_1 - \sigma_3) = 0$, then (σ, τ) is *on* the σ_2, σ_3 circle. If $n_1^2(\sigma_1 - \sigma_2)(\sigma_1 - \sigma_3) \neq 0$, then $n_1^2(\sigma_1 - \sigma_2)(\sigma_1 - \sigma_3) > 0$ since $n_1^2 > 0$, $\sigma_1 > \sigma_2$, and $\sigma_1 > \sigma_3$. Thus, for this case, (σ, τ) is *beyond* the σ_2, σ_3 circle. The same argument holds for the σ_1, σ_2 circle equation when compared to Eq. (2.1-34c), meaning that (σ, τ) can be *on* or *beyond* that circle. Finally, when comparing the equation of the σ_1, σ_3 circle with Eq. (2.1-34b), the term $n_2^2(\sigma_3 - \sigma_2)(\sigma_1 - \sigma_2) \leq 0$ meaning that (σ, τ) can be *on* or *within* that circle. Thus it has been proved that all possible values of (σ, τ) exist either on the boundaries of the three Mohr's circles or within the shaded region shown in Fig. 2.1-11(b).

There is a graphical technique to use Mohr's circles in three dimensions to arrive at the values of (σ, τ) for an arbitrary surface. However, the technique does not offer any pedagogical or real application advantages over the use of the transformation equations given by Eqs. (2.1-18), (2.1-20), and (2.1-21).

2.1.7 MAXIMUM SHEAR STRESS

Based on the observations on Mohr's circles in three dimensions the maximum and minimum shear stresses are found from the radius of the largest circle (σ_1, σ_3) and are given by

$$\tau_{\substack{max \\ min}} = \pm \frac{\sigma_1 - \sigma_3}{2} \qquad [2.1-36]$$

The simplest way to determine the orientation of the surfaces containing these stresses is to first locate the principal stress axes and then rotate $\pm 45°$ about principal axis 2. For the *plane stress* state, as stated earlier, two principal stresses are found from Eq. (2.1-30) with the third principal stress being zero. Next, the *three* principal stresses are ordered such that $\sigma_1 > \sigma_2 > \sigma_3$, where σ_1 and σ_3 are subsequently substituted into Eq. (2.1-36).

For Example 2.1-2 determine the maximum and minimum shear stress and show the orientation and complete state of stress of the element containing these stresses.

Example 2.1-7

Solution:

The principal stresses, determined in Example 2.1-6, are

$$\sigma_1 = 11 \text{ kpsi} \qquad \sigma_2 = 1 \text{ kpsi} \qquad \sigma_3 = 0 \text{ kpsi}$$

From Eq. (2.1-36)

$$\tau_{\substack{max \\ min}} = \pm \frac{11 - 0}{2} = \pm 5.5 \text{ kpsi}$$

The three Mohr's circles are shown in Fig. 2.1-12. The orientation of the element containing the principal stresses was determined in Example 2.1-6 and is as shown in Fig. 2.1-13(*b*). Viewing the element along the axis containing $\sigma_2 = 1$ kpsi and rotating the surfaces $\pm 45°$ yields the element containing the maximum shear stresses together with the normal stresses.

The directional cosines associated with the surfaces are found through successive rotations. Rotating the *xyz* axes to the 123 axes yields

$$\begin{Bmatrix} 1 \\ 2 \\ 3 \end{Bmatrix} = \begin{bmatrix} \cos 26.6 & \sin 26.6 & 0 \\ -\sin 26.6 & \cos 26.6 & 0 \\ 0 & 0 & 1 \end{bmatrix} \begin{Bmatrix} x \\ y \\ z \end{Bmatrix}$$

$$= \begin{bmatrix} 0.8944 & 0.4472 & 0 \\ -0.4472 & 0.8944 & 0 \\ 0 & 0 & 1 \end{bmatrix} \begin{Bmatrix} x \\ y \\ z \end{Bmatrix} \qquad \textbf{[a]}$$

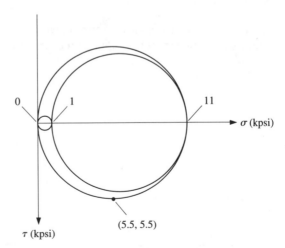

Figure 2.1-12   Mohr's circles for Example 2.1-7.

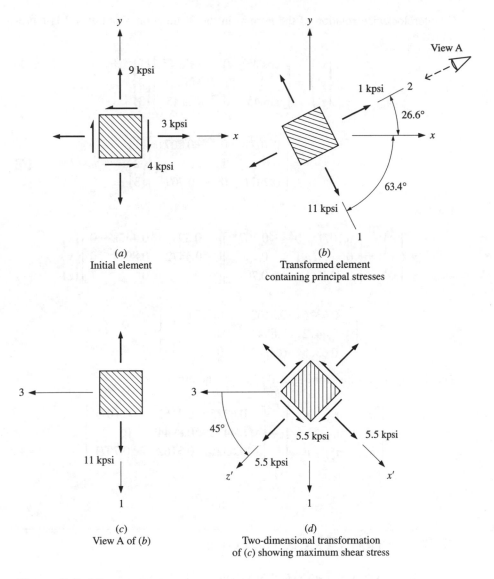

(a)
Initial element

(b)
Transformed element
containing principal stresses

(c)
View A of (b)

(d)
Two-dimensional transformation
of (c) showing maximum shear stress

Figure 2.1-13

Counterclockwise rotation of the normal in the 3 direction about axis 2 is represented by

$$
\left\{ \begin{matrix} x' \\ y' \\ z' \end{matrix} \right\} = \begin{bmatrix} \cos 45 & 0 & -\sin 45 \\ 0 & 1 & 0 \\ \sin 45 & 0 & \cos 45 \end{bmatrix} \left\{ \begin{matrix} 1 \\ 2 \\ 3 \end{matrix} \right\}
$$

$$
= \begin{bmatrix} 0.7071 & 0 & -0.7071 \\ 0 & 1 & 0 \\ 0.7071 & 0 & 0.7071 \end{bmatrix} \left\{ \begin{matrix} 1 \\ 2 \\ 3 \end{matrix} \right\}
$$ [b]

Thus

$$
\left\{ \begin{matrix} x' \\ y' \\ z' \end{matrix} \right\} = \begin{bmatrix} 0.7071 & 0 & -0.7071 \\ 0 & 1 & 0 \\ 0.7071 & 0 & 0.7071 \end{bmatrix} \begin{bmatrix} 0.8944 & 0.4472 & 0 \\ -0.4472 & 0.8944 & 0 \\ 0 & 0 & 1 \end{bmatrix} \left\{ \begin{matrix} x \\ y \\ z \end{matrix} \right\}
$$

$$
= \begin{bmatrix} 0.6325 & 0.3162 & -0.7071 \\ -0.4472 & 0.8944 & 0 \\ 0.6325 & 0.3162 & 0.7071 \end{bmatrix} \left\{ \begin{matrix} x \\ y \\ z \end{matrix} \right\}
$$

The directional cosines for Fig. 2.1-13(c) are therefore

$$
\begin{bmatrix} n_{x'x} & n_{x'y} & n_{x'z} \\ n_{y'x} & n_{y'y} & n_{y'z} \\ n_{z'x} & n_{z'y} & n_{z'z} \end{bmatrix} = \begin{bmatrix} 0.6325 & 0.3162 & -0.7071 \\ -0.4472 & 0.8944 & 0 \\ 0.6325 & 0.3162 & 0.7071 \end{bmatrix}
$$

2.2   STRAIN TRANSFORMATIONS

2.2.1   STRAIN TRANSFORMATIONS, GENERAL

The strain transformation equations are similar to those for stresses. However, arriving at the equations is not. Recall from Sec. 2.1.1, Eq. 2.1-5, that the vector orthogonal transformation is given by

$$
\left\{ \begin{matrix} x' \\ y' \\ z' \end{matrix} \right\} = \begin{bmatrix} n_{x'x} & n_{x'y} & n_{x'z} \\ n_{y'x} & n_{y'y} & n_{y'z} \\ n_{z'x} & n_{z'y} & n_{z'z} \end{bmatrix} \left\{ \begin{matrix} x \\ y \\ z \end{matrix} \right\}
$$ [2.2-1]

It can be shown that for orthogonal transformations, the inverse of the transformation matrix is simply the transpose (see Appendix I). Thus

$$
\begin{Bmatrix} x \\ y \\ z \end{Bmatrix} = \begin{bmatrix} n_{x'x} & n_{y'x} & n_{z'x} \\ n_{x'y} & n_{y'y} & n_{z'y} \\ n_{x'z} & n_{y'z} & n_{z'z} \end{bmatrix} \begin{Bmatrix} x' \\ y' \\ z' \end{Bmatrix}
\qquad \textbf{[2.2-2]}
$$

From Eq. (2.2-1), the displacements relative to the $x'y'z'$ system can be related to that of the xyz system as

$$
\begin{Bmatrix} u' \\ v' \\ w' \end{Bmatrix} = \begin{bmatrix} n_{x'x} & n_{x'y} & n_{x'z} \\ n_{y'x} & n_{y'y} & n_{y'z} \\ n_{z'x} & n_{z'y} & n_{z'z} \end{bmatrix} \begin{Bmatrix} u \\ v \\ w \end{Bmatrix}
\qquad \textbf{[2.2-3]}
$$

Consider the normal strain $\varepsilon_{x'}$ in the $x'y'z'$ system. It can be expanded by the chain rule as

$$
\varepsilon_{x'} = \frac{\partial u'}{\partial x'} = \frac{\partial u'}{\partial x}\frac{\partial x}{\partial x'} + \frac{\partial u'}{\partial y}\frac{\partial y}{\partial x'} + \frac{\partial u'}{\partial z}\frac{\partial z}{\partial x'}
$$

The partial derivatives of u' can be taken from Eq. (2.2-3) and the derivatives of x, y, and z from Eq. (2.2-2). This gives

$$
\varepsilon_{x'} = \frac{\partial u'}{\partial x'} = \left(n_{x'x}\frac{\partial u}{\partial x} + n_{x'y}\frac{\partial v}{\partial x} + n_{x'z}\frac{\partial w}{\partial x} \right)(n_{xx})
$$

$$
+ \left(n_{x'x}\frac{\partial u}{\partial y} + n_{x'y}\frac{\partial v}{\partial y} + n_{x'z}\frac{\partial w}{\partial y} \right)(n_{x'y})
$$

$$
+ \left(n_{x'x}\frac{\partial u}{\partial z} + n_{x'y}\frac{\partial v}{\partial z} + n_{x'z}\frac{\partial w}{\partial z} \right)(n_{x'z})
$$

Rearranging terms gives

$$
\varepsilon_{x'} = \frac{\partial u'}{\partial x'} = \frac{\partial u}{\partial x}n_{x'x}^2 + \frac{\partial v}{\partial y}n_{x'y}^2 + \frac{\partial w}{\partial z}n_{x'z}^2
$$

$$
+ \left(\frac{\partial v}{\partial x} + \frac{\partial u}{\partial y} \right)n_{x'x}n_{x'y} + \left(\frac{\partial w}{\partial y} + \frac{\partial v}{\partial z} \right)n_{x'y}n_{x'z} + \left(\frac{\partial u}{\partial z} + \frac{\partial w}{\partial x} \right)n_{x'z}n_{x'x}
$$

Employing Eqs. (1.5-1), which relate strain to displacement, we obtain the final form of the transformation equation written as

$$
\varepsilon_{x'} = \varepsilon_x n_{x'x}^2 + \varepsilon_y n_{x'y}^2 + \varepsilon_z n_{x'z}^2 + \gamma_{xy}n_{x'x}n_{x'y} + \gamma_{yz}n_{x'y}n_{x'z} + \gamma_{zx}n_{x'z}n_{x'x}
\qquad \textbf{[2.2-4a]}
$$

In a similar manner it can be shown that

$$
\varepsilon_{y'} = \varepsilon_x n_{y'x}^2 + \varepsilon_y n_{y'y}^2 + \varepsilon_z n_{y'z}^2 + \gamma_{xy}n_{y'x}n_{y'y} + \gamma_{yz}n_{y'y}n_{y'z} + \gamma_{zx}n_{y'z}n_{y'x}
\qquad \textbf{[2.2-4b]}
$$

$$\varepsilon_{z'} = \varepsilon_x n_{z'x}^2 + \varepsilon_y n_{z'y}^2 + \varepsilon_z n_{z'z}^2 + \gamma_{xy} n_{z'x} n_{z'y} + \gamma_{yz} n_{z'y} n_{z'z} + \gamma_{zx} n_{z'z} n_{z'x} \qquad \textbf{[2.2-4c]}$$

The shear strain transformations are found in a similar manner, where, for example, $\gamma_{x'y'}$ is found from

$$\gamma_{x'y'} = \frac{\partial v'}{\partial x'} + \frac{\partial u'}{\partial y'} = \left(\frac{\partial v'}{\partial x}\frac{\partial x}{\partial x'} + \frac{\partial v'}{\partial y}\frac{\partial y}{\partial x'} + \frac{\partial v'}{\partial z}\frac{\partial z}{\partial x'} \right) + \left(\frac{\partial u'}{\partial x}\frac{\partial x}{\partial y'} + \frac{\partial u'}{\partial y}\frac{\partial y}{\partial y'} + \frac{\partial u'}{\partial z}\frac{\partial z}{\partial y'} \right)$$

Using Eqs. (2.2-2), (2.2-3), and (1.5-1) as before results in

$$\gamma_{x'y'} = \varepsilon_x n_{x'x} n_{y'x} + \varepsilon_y n_{x'y} n_{y'y} + \varepsilon_z n_{x'z} n_{y'z} + \frac{\gamma_{xy}}{2}\left(n_{x'x} n_{y'y} + n_{x'y} n_{y'x}\right)$$

$$+ \frac{\gamma_{yz}}{2}\left(n_{x'y} n_{y'z} + n_{x'z} n_{y'y}\right) + \frac{\gamma_{zx}}{2}\left(n_{x'x} n_{y'z} + n_{x'z} n_{y'x}\right) \qquad \textbf{[2.2-4d]}$$

In a similar manner

$$\gamma_{y'z'} = \varepsilon_x n_{y'x} n_{z'x} + \varepsilon_y n_{y'y} n_{z'y} + \varepsilon_z n_{y'z} n_{z'z} + \frac{\gamma_{xy}}{2}\left(n_{y'x} n_{z'y} + n_{y'y} n_{z'x}\right)$$

$$+ \frac{\gamma_{yz}}{2}\left(n_{y'y} n_{z'z} + n_{y'z} n_{z'y}\right) + \frac{\gamma_{zx}}{2}\left(n_{y'x} n_{z'z} + n_{y'z} n_{z'x}\right) \qquad \textbf{[2.2-4e]}$$

$$\gamma_{z'x'} = \varepsilon_x n_{x'x} n_{z'x} + \varepsilon_y n_{x'y} n_{z'y} + \varepsilon_z n_{x'z} n_{z'z} + \frac{\gamma_{xy}}{2}\left(n_{x'x} n_{z'y} + n_{x'y} n_{z'x}\right)$$

$$+ \frac{\gamma_{yz}}{2}\left(n_{x'y} n_{z'z} + n_{x'z} n_{z'y}\right) + \frac{\gamma_{zx}}{2}\left(n_{x'x} n_{z'z} + n_{x'z} n_{z'x}\right) \qquad \textbf{[2.2-4f]}$$

Comparing Eqs. (2.2-4) with Eqs. (2.1-4), we see that they are very similar. If in Eqs. (2.1-4) one were to substitute ε for σ and $\gamma/2$ for τ (using the same subscripts) one would arrive at Eqs. (2.2-4).

Example 2.2-1 | Consider an element undergoing a single shear stress τ_{xy}. Using the stress and strain transformations for the state of normal strain at an angle of 45° rotated about the z axis prove Eq. (1.4-7).

Solution:

Given a single shear stress τ_{xy}, the only strain with respect to the xyz system is

$$\gamma_{xy} = \frac{\tau_{xy}}{G} \qquad \textbf{[a]}$$

For a transformation of 45° rotated about the z axis the surface normals are given by

$$\begin{bmatrix} n_{x'x} & n_{x'y} & n_{x'z} \\ n_{y'x} & n_{y'y} & n_{y'z} \\ n_{z'x} & n_{z'y} & n_{z'z} \end{bmatrix} = \begin{bmatrix} \cos 45 & \sin 45 & 0 \\ -\sin 45 & \cos 45 & 0 \\ 0 & 0 & 1 \end{bmatrix}$$

From Eq. (2.2-4a) with all strains being zero excepting γ_{xy}

$$\varepsilon_{x'} = \gamma_{xy} n_{x'x} n_{x'y} = \gamma_{xy} \cos 45 \sin 45 = \frac{\gamma_{xy}}{2}$$

Substituting this into Eq. (a) yields

$$\varepsilon_{x'} = \frac{\tau_{xy}}{2G} \qquad \textbf{[b]}$$

Transforming τ_{xy} to the $x'y'z'$ system, the only nonzero stresses are found to be

$$\sigma_{x'} = 2\tau_{xy} n_{x'x} n_{x'y} = 2\tau_{xy}(\cos 45)(\sin 45) = \tau_{xy} \qquad \textbf{[c]}$$

$$\sigma_{y'} = 2\tau_{xy} n_{y'x} n_{y'y} = 2\tau_{xy}(-\sin 45)(\cos 45) = -\tau_{xy} \qquad \textbf{[d]}$$

Using Eq. (1.4-2a),

$$\varepsilon_{x'} = \frac{1}{E}[\sigma_{x'} - v(\sigma_{y'} + \sigma_{z'})] = \frac{1}{E}[\tau_{xy} - v(-\tau_{xy} + 0)] = \frac{1+v}{E}\tau_{xy} \qquad \textbf{[e]}$$

Equating Eqs. (b) and (e) yields Eq. (1.4-7), which is

$$G = \frac{E}{2(1+v)}$$

Analogous to the matrix stress transformation equation [Eq. (2.1-7)], the strain transformation equations can be written in matrix form. Again, defining the transformation matrix,

$$[\mathbf{T}] = \begin{bmatrix} n_{x'x} & n_{x'y} & n_{x'z} \\ n_{y'x} & n_{y'y} & n_{y'z} \\ n_{z'x} & n_{z'y} & n_{z'z} \end{bmatrix} \qquad \textbf{[2.2-5]}$$

the matrix transformation equation is

$$[\boldsymbol{\varepsilon}]_{x'y'z'} = [\mathbf{T}][\boldsymbol{\varepsilon}]_{xyz}[\mathbf{T}]^T \qquad \textbf{[2.2-6]}$$

where the strain matrices are defined by the elasticity strain matrix, given by Eq. (2.0-3), which can be written in terms of the engineering strains as

$$[\boldsymbol{\varepsilon}]_{xyz} = \frac{1}{2}\begin{bmatrix} 2\varepsilon_x & \gamma_{xy} & \gamma_{xz} \\ \gamma_{yx} & 2\varepsilon_y & \gamma_{yz} \\ \gamma_{zx} & \gamma_{zy} & 2\varepsilon_z \end{bmatrix} \qquad [\boldsymbol{\varepsilon}]_{x'y'z'} = \frac{1}{2}\begin{bmatrix} 2\varepsilon_{x'} & \gamma_{x'y'} & \gamma_{x'z'} \\ \gamma_{y'x'} & 2\varepsilon_{y'} & \gamma_{y'z'} \\ \gamma_{z'x'} & \gamma_{z'y'} & 2\varepsilon_{z'} \end{bmatrix}$$

2.2.2 PRINCIPAL STRAINS

Based on the observation of the similarity of the stress and strain transformation equations, the equations for principal strains will be analogous to the principal stress equations given by Eqs. (2.1-23) and (2.1-25). Again, all that is necessary is to replace σ and τ with ε and $\gamma/2$ (using the same subscripts), respectively. Based on this and physical reasoning, the principal strains will occur on the same axes on which the principal stresses exist provided the material is isotropic.

2.3 GENERALIZED STRESS-STRAIN RELATIONS

The stress-strain equations given in Chap. 1 only apply to homogeneous and isotropic materials. Homogeneous refers to the fact that the elastic properties do not change from point to point in the body, and isotropic means that the properties are not varying with respect to direction. Many engineering materials, however, are neither homogeneous nor isotropic, e.g., rolled or drawn metals, wood, laminates, and fiber-filled epoxy materials. These materials are not isotropic since their elastic properties depend on direction.

In general, each strain is dependent on each stress. For example, the strain ε_x written as a linear function of each stress is

$$\varepsilon_x = C_{11}\sigma_x + C_{12}\sigma_y + C_{13}\sigma_z + C_{14}\tau_{xy} + C_{15}\tau_{yz}$$

$$+ C_{16}\tau_{zx} + C_{17}\tau_{xz} + C_{18}\tau_{zy} + C_{19}\tau_{yx}$$

Each of the remaining eight strains can be written in a similar fashion. If no symmetry is assumed, $9^2 = 81$ independent constants would result. That is, in matrix notation the stress-strain relations would be

$$\begin{Bmatrix} \varepsilon_x \\ \varepsilon_y \\ \varepsilon_z \\ \gamma_{xy} \\ \gamma_{yz} \\ \gamma_{zx} \\ \gamma_{xz} \\ \gamma_{zy} \\ \gamma_{yx} \end{Bmatrix} = \begin{bmatrix} C_{11} & C_{12} & C_{13} & C_{14} & C_{15} & C_{16} & C_{17} & C_{18} & C_{19} \\ C_{21} & C_{22} & C_{23} & C_{24} & C_{25} & C_{26} & C_{27} & C_{28} & C_{29} \\ & & & & & & & & \\ \cdot & \cdot & \cdot & \cdot & \cdot & \cdot & \cdot & \cdot & \cdot \\ & & & & & & & & \\ C_{91} & C_{92} & C_{93} & C_{94} & C_{95} & C_{96} & C_{97} & C_{98} & C_{99} \end{bmatrix} \begin{Bmatrix} \sigma_x \\ \sigma_y \\ \sigma_z \\ \tau_{xy} \\ \tau_{yz} \\ \tau_{zx} \\ \tau_{xz} \\ \tau_{zy} \\ \tau_{yx} \end{Bmatrix} \qquad \textbf{[2.3-1]}$$

However, shear is normally symmetric, where $\tau_{yx} = \tau_{xy}$, etc., and $\gamma_{yx} = \gamma_{xy}$, etc. This reduces the number of stress and strain terms to six each. Thus, there are only $6^2 = 36$ independent elastic constants, i.e.,

$$\begin{Bmatrix} \varepsilon_x \\ \varepsilon_y \\ \varepsilon_z \\ \gamma_{xy} \\ \gamma_{yz} \\ \gamma_{zx} \end{Bmatrix} = \begin{bmatrix} C_{11} & C_{12} & C_{13} & C_{14} & C_{15} & C_{16} \\ C_{21} & C_{22} & C_{23} & C_{24} & C_{25} & C_{26} \\ \cdot & \cdot & \cdot & \cdot & \cdot & \cdot \\ & & & & & \\ & & & & & \\ C_{61} & \cdot & \cdot & \cdot & \cdot & C_{66} \end{bmatrix} \begin{Bmatrix} \sigma_x \\ \sigma_y \\ \sigma_z \\ \tau_{xy} \\ \tau_{yz} \\ \tau_{zx} \end{Bmatrix} \qquad \textbf{[2.3-2]}$$

The matrix containing the elastic constants can be shown to be symmetric where $C_{12} = C_{21}$, $C_{13} = C_{31}$, etc. Consider C_{12}, for example; let σ_x be gradually applied performing work equal to $\frac{1}{2}\sigma_x\varepsilon_x = \frac{1}{2}C_{11}\sigma_x^2$. Next, while σ_x is still present, apply σ_y. This causes additional work of $\frac{1}{2}\sigma_y\varepsilon_y + \sigma_x\varepsilon_x' = \frac{1}{2}C_{11}\sigma_y^2 + C_{12}\sigma_x\sigma_y$ (where $\varepsilon_x' = C_{12}\sigma_y$). Thus the total work is

$$W_1 = \frac{1}{2}C_{11}\sigma_x^2 + \frac{1}{2}C_{22}\sigma_y^2 + C_{12}\sigma_x\sigma_y \qquad [a]$$

Now begin again except change the order of the stress application where we apply σ_y first and then σ_x. Similar to the determination of W_1, the resulting work is

$$W_2 = \frac{1}{2}C_{22}\sigma_y^2 + \frac{1}{2}C_{11}\sigma_x^2 + C_{21}\sigma_y\sigma_x \qquad [b]$$

Now the work performed should be equal regardless of the order of stress application. Equating Eqs. (a) and (b), we find $C_{12} = C_{21}$. This is called *Maxwell's reciprocity theorem*. Thus, in general (except when cross-shear stresses are unequal), there are at most 21 possible independent elastic constants.

If a material has one plane of symmetry, there can be no interaction between the out-of-plane shear stresses and the remaining strains. If the xy plane is assumed to be a plane of symmetry, Eq. (2.3-2) reduces to

$$\begin{Bmatrix} \varepsilon_x \\ \varepsilon_y \\ \varepsilon_z \\ \gamma_{xy} \\ \gamma_{yz} \\ \gamma_{zx} \end{Bmatrix} = \begin{bmatrix} C_{11} & C_{12} & C_{13} & C_{14} & 0 & 0 \\ C_{21} & C_{22} & C_{23} & C_{24} & 0 & 0 \\ C_{31} & C_{32} & C_{33} & C_{34} & 0 & 0 \\ C_{41} & C_{42} & C_{43} & C_{44} & 0 & 0 \\ 0 & 0 & 0 & 0 & C_{55} & C_{56} \\ 0 & 0 & 0 & 0 & C_{65} & C_{66} \end{bmatrix} \begin{Bmatrix} \sigma_x \\ \sigma_y \\ \sigma_z \\ \tau_{xy} \\ \tau_{yz} \\ \tau_{zx} \end{Bmatrix} \qquad [2.3\text{-}3]$$

where $C_{ij} = C_{ji}$.

A material which exhibits symmetry with respect to three mutually orthogonal planes is called an *orthotropic material*. If the xy, yz, and zx planes are considered planes of symmetry, Eq. (2.3-2) reduces to

$$\begin{Bmatrix} \varepsilon_x \\ \varepsilon_y \\ \varepsilon_z \\ \gamma_{xy} \\ \gamma_{yz} \\ \gamma_{zx} \end{Bmatrix} = \begin{bmatrix} C_{11} & C_{12} & C_{13} & 0 & 0 & 0 \\ C_{21} & C_{22} & C_{23} & 0 & 0 & 0 \\ C_{31} & C_{32} & C_{33} & 0 & 0 & 0 \\ 0 & 0 & 0 & C_{44} & 0 & 0 \\ 0 & 0 & 0 & 0 & C_{55} & 0 \\ 0 & 0 & 0 & 0 & 0 & C_{66} \end{bmatrix} \begin{Bmatrix} \sigma_x \\ \sigma_y \\ \sigma_z \\ \tau_{xy} \\ \tau_{yz} \\ \tau_{zx} \end{Bmatrix} \qquad [2.3\text{-}4]$$

where $C_{ij} = C_{ji}$.

If the properties of an orthotropic material are identical in all three directions, the material is said to have a *cubic structure*. Thus, if we let

$$C_{11} = C_{22} = C_{33} = \frac{1}{E}$$

$$C_{12} = C_{13} = C_{23} = C_{21} = C_{31} = C_{32} = -\frac{v}{E}$$

$$C_{44} = C_{55} = C_{66} = \frac{1}{G}$$

where E, v, and G are the material constants normally used in engineering applications, for an orthotropic material with a cubic structure, Eq. (2.3-4) reduces to

$$\begin{Bmatrix} \varepsilon_x \\ \varepsilon_y \\ \varepsilon_z \\ \gamma_{xy} \\ \gamma_{yz} \\ \gamma_{zx} \end{Bmatrix} = \frac{1}{E} \begin{bmatrix} 1 & -v & -v & 0 & 0 & 0 \\ -v & 1 & -v & 0 & 0 & 0 \\ -v & -v & 1 & 0 & 0 & 0 \\ 0 & 0 & 0 & \dfrac{E}{G} & 0 & 0 \\ 0 & 0 & 0 & 0 & \dfrac{E}{G} & 0 \\ 0 & 0 & 0 & 0 & 0 & \dfrac{E}{G} \end{bmatrix} \begin{Bmatrix} \sigma_x \\ \sigma_y \\ \sigma_z \\ \tau_{xy} \\ \tau_{yz} \\ \tau_{zx} \end{Bmatrix}$$ [2.3-5]

If the material properties of a cubic structure are the same for any arbitrary coordinate system, the material is truly isotropic. The derivation given in Example 2.2-1 utilized this fact, and the relationship between E, v, and G resulted, where

$$G = \frac{E}{2(1 + v)}$$

Substituting this into Eq. (2.3-5) gives

$$\begin{Bmatrix} \varepsilon_x \\ \varepsilon_y \\ \varepsilon_z \\ \gamma_{xy} \\ \gamma_{yz} \\ \gamma_{zx} \end{Bmatrix} = \frac{1}{E} \begin{bmatrix} 1 & -v & -v & 0 & 0 & 0 \\ -v & 1 & -v & 0 & 0 & 0 \\ -v & -v & 1 & 0 & 0 & 0 \\ 0 & 0 & 0 & 2(1+v) & 0 & 0 \\ 0 & 0 & 0 & 0 & 2(1+v) & 0 \\ 0 & 0 & 0 & 0 & 0 & 2(1+v) \end{bmatrix} \begin{Bmatrix} \sigma_x \\ \sigma_y \\ \sigma_z \\ \tau_{xy} \\ \tau_{yz} \\ \tau_{zx} \end{Bmatrix}$$

[2.3-6]

Lamé Constants   In the mathematical theory of elasticity, the material constants may be expressed in terms of Lamé's constants μ and λ rather than the engineering constants E and v. In indicial notation, the stress-strain relations can be more conveniently written in terms of Lamé's constants. When indicial notation is used, the stress-strain relation is

$$\sigma_{ij} = 2\mu\varepsilon_{ij} + \lambda\delta_{ij}\varepsilon_{kk} \qquad i, j = 1, 2, 3$$ [2.3-7]

where δ_{ij} is called the *Kronecker delta* and

$$\delta_{ij} = \begin{cases} 1 & \text{when } i = j \\ 0 & \text{when } i \neq j \end{cases}$$

The term ε_{kk} represents a summation term given by

$$\varepsilon_{kk} = \varepsilon_{11} + \varepsilon_{22} + \varepsilon_{33}$$

The σ_{ij} and ε_{ij} terms are[†]

$\sigma_{11} = \sigma_x$	$\sigma_{12} = \tau_{xy}$	$\sigma_{13} = \tau_{xz} = \tau_{zx}$
$\sigma_{21} = \tau_{yx} = \tau_{xy}$	$\sigma_{22} = \sigma_y$	$\sigma_{23} = \tau_{yz}$
$\sigma_{31} = \tau_{zx}$	$\sigma_{32} = \tau_{zy} = \tau_{yz}$	$\sigma_{33} = \sigma_z$
$\varepsilon_{11} = \varepsilon_x$	$\varepsilon_{12} = \frac{1}{2}\gamma_{xy}$	$\varepsilon_{13} = \frac{1}{2}\gamma_{xz} = \frac{1}{2}\gamma_{zx}$
$\varepsilon_{21} = \frac{1}{2}\gamma_{yx} = \frac{1}{2}\gamma_{xy}$	$\varepsilon_{22} = \varepsilon_y$	$\varepsilon_{23} = \frac{1}{2}\gamma_{yz}$
$\varepsilon_{31} = \frac{1}{2}\gamma_{zx}$	$\varepsilon_{32} = \frac{1}{2}\gamma_{zy} = \frac{1}{2}\gamma_{yz}$	$\varepsilon_{33} = \varepsilon_z$

If, for example, we want the equation for τ_{xy}, let $i = 1$ and $j = 2$ in Eq. (2.3-7). Thus $\delta_{12} = 0$, and from Eq. (2.3-7)

$$\sigma_{12} = 2\mu\varepsilon_{12}$$

or

$$\tau_{xy} = 2\mu\frac{\gamma_{xy}}{2} = \mu\gamma_{xy}$$

It can be seen that the Lamé constant μ is equal to the shear modulus G, or, in terms of E and v, it is

$$\mu = \frac{E}{2(1 + v)} \qquad \textbf{[2.3-8]}$$

If σ_y is wanted from Eq. (2.3-7), let $i = j = 2$. Thus $\delta_{22} = 1$, and

$$\sigma_{22} = 2\mu\varepsilon_{22} + \lambda(\varepsilon_{11} + \varepsilon_{22} + \varepsilon_{33}) = (2\mu + \lambda)\varepsilon_{22} + \lambda(\varepsilon_{11} + \varepsilon_{33})$$

or

$$\sigma_y = (2\mu + \lambda)\varepsilon_y + \lambda(\varepsilon_x + \varepsilon_z)$$

If this is compared with Eq. (1.4-3b), the second Lamé constant λ is found to be

$$\lambda = \frac{Ev}{(1 + v)(1 - 2v)} \qquad \textbf{[2.3-9]}$$

[†] Note that the index numbers 1, 2, 3 represent the xyz coordinate system.

2.4   THE EQUILIBRIUM EQUATIONS

In Sec. 1.3, the stresses and body forces on an infinitesimal element were shown in Fig. 1.3-10, repeated here as Fig. 2.4-1 except the changes in stresses (i.e., $\Delta\sigma_x$, $\Delta\sigma_y$, etc.) are replaced by Taylor's series expansion terms (i.e., $\partial\sigma_x/\partial x\ \Delta x$, $\partial\sigma_y/\partial y\ \Delta y$, etc.).

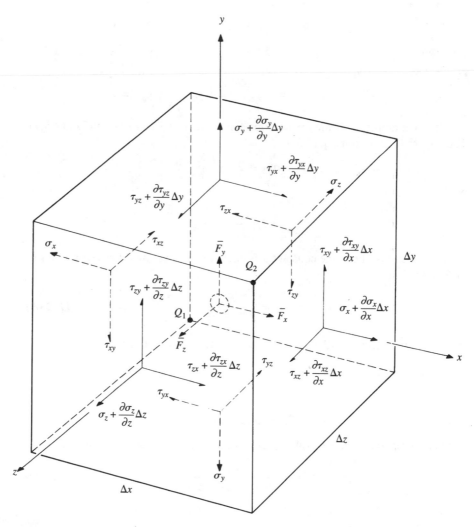

Figure 2.4-1

If we recall that the body forces $\overline{F}_x$, $\overline{F}_y$, and $\overline{F}_z$ are forces per unit volume, we can now write the equilibrium equations. For equilibrium, the sum of the forces in the x direction equals zero. Thus

$$\left(\sigma_x + \frac{\partial \sigma_x}{\partial x}\, \Delta x\right) \Delta y\, \Delta z + \left(\tau_{xy} + \frac{\partial \tau_{xy}}{\partial y}\, \Delta y\right) \Delta z\, \Delta x + \left(\tau_{zx} + \frac{\partial \tau_{zx}}{\partial z}\, \Delta z\right) \Delta x\, \Delta y$$

$$-\sigma_x\, \Delta y\, \Delta z - \tau_{xy}\, \Delta z\, \Delta x - \tau_{zx}\, \Delta x\, \Delta y + \overline{F}_x\, \Delta x\, \Delta y\, \Delta z = 0$$

Simplification of this expression results in

$$\frac{\partial \sigma_x}{\partial x} + \frac{\partial \tau_{xy}}{\partial y} + \frac{\partial \tau_{zx}}{\partial z} + \overline{F}_x = 0 \qquad\qquad \textbf{[2.4-1a]}$$

Similarly, summing forces in the y and z directions, respectively, yields

$$\frac{\partial \tau_{xy}}{\partial x} + \frac{\partial \sigma_y}{\partial y} + \frac{\partial \tau_{yz}}{\partial z} + \overline{F}_y = 0 \qquad\qquad \textbf{[2.4-1b]}$$

$$\frac{\partial \tau_{zx}}{\partial x} + \frac{\partial \tau_{yz}}{\partial y} + \frac{\partial \sigma_z}{\partial z} + \overline{F}_z = 0 \qquad\qquad \textbf{[2.4-1c]}$$

For a state of plane stress, where $\sigma_z = \tau_{zx} = \tau_{yz} = 0$, the equilibrium equations reduce to

$$\frac{\partial \sigma_x}{\partial x} + \frac{\partial \tau_{xy}}{\partial y} + \overline{F}_x = 0 \qquad\qquad \textbf{[2.4-2a]}$$

$$\frac{\partial \tau_{xy}}{\partial x} + \frac{\partial \sigma_y}{\partial y} + \overline{F}_y = 0 \qquad\qquad \textbf{[2.4-2b]}$$

For the beam shown in Fig. 2.4-2, if the weight is neglected, the bending moment as a function of x is $M_z = -\frac{1}{2} wx^2$. From basic mechanics of materials (Sec. 3.4.2) the bending stress is given by $\sigma_x = -M_z y/I_z$, and thus, for this example | **Example 2.4-1**

$$\sigma_x = \frac{w}{2I_z}\, x^2 y \qquad\qquad \textbf{[a]}$$

Starting with Eq. (a), use Eqs. (2.4-2)[†] to determine how τ_{xy} and σ_y vary as functions of x and y.

[†] Using Eqs. (2.4-2) assumes a plane stress field. For a thin rectangular beam, this is a relatively valid assumption since the side faces are free surfaces, i.e., are stress-free, and the stresses in the z direction are zero. If the beam is thin, the stresses in the z direction will not have a chance to become appreciable.

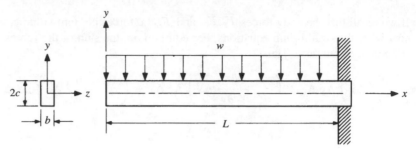

Figure 2.4-2

Solution:

If the weight of the beam is neglected, the body forces are zero. Thus $\overline{F}_x = \overline{F}_y = 0$. Substituting Eq. (a) into Eq. (2.4-2a) results in

$$\frac{\partial \tau_{xy}}{\partial y} = -\frac{w}{I_z} xy$$

Keeping in mind that the derivative in the above equation is a partial derivative, we see that integration of both sides of the equation with respect to y results in

$$\tau_{xy} = -\frac{w}{2I_z} xy^2 + f(x) \qquad\qquad \text{[b]}$$

Instead of a constant of integration, a function of integration is added, where $f(x)$ is a function of x alone [note that $f(x)$ can still contain a constant]. The function is found from the boundary or surface conditions. From Fig. 2.4-2 it can be seen that there is no imposed shear force on the top and bottom surfaces of the beam. For the top surface, the condition is that $\tau_{xy} = 0$ at $y = c$. Solving for $f(x)$ gives

$$f(x) = \frac{wc^2}{2I_z} x$$

Substituting $f(x)$ into Eq. (b) yields

$$\tau_{xy} = \frac{w}{2I_z} x \left(c^2 - y^2 \right) \qquad\qquad \text{[c]}$$

Note that the condition that $\tau_{xy} = 0$ at $y = -c$ is automatically satisfied in Eq. (c). Equation (c) can be arrived at using the equation $\tau_{xy} = V_y Q / (I_z b)$ from basic strength of materials (see Sec. 3.4.3). Next, Eq. (c) is substituted into Eq. (2.4-2b), resulting in

$$\frac{\partial \sigma_y}{\partial y} = -\frac{w}{2I_z} \left(c^2 - y^2 \right)$$

Integrating with respect to y again yields

$$\sigma_y = -\frac{w}{6I_z} y \left(3c^2 - y^2 \right) + g(x) \qquad\qquad \text{[d]}$$

where $g(x)$ is the function of integration. A boundary condition from Fig. 2.4-2 is $\sigma_y = -w/b$ at $y = c$. Thus Eq. (d) is written

$$-\frac{w}{b} = -\frac{w}{6I_z}c(3c^2 - c^2) + g(x) = -\frac{wc^3}{3I_z} + g(x) \qquad [e]$$

The second moment area about the z axis, I_z, is given by $b(2c)^3/12$. Solving for b, we find that $b = 3I_z/(2c^3)$. Substituting this into Eq. (e) results in

$$g(x) = -\frac{wc^3}{3I_z}$$

and Eq. (d) becomes

$$\sigma_y = -\frac{w}{6I_z}(2c^3 + 3c^2y - y^3) \qquad [f]$$

Note that Eq. (f) also satisfies the condition that $\sigma_y = 0$ at $y = -c$.

2.5 COMPATIBILITY

When one is seeking a solution to the stress distribution in a body, the dynamic state of the body must be satisfied. For example, if the body is in equilibrium, any segment of the body together with its corresponding internal-force distribution must maintain the segment in static equilibrium. At any given section it is possible to find many stress distributions which will ensure equilibrium. An *acceptable* stress distribution is one which ensures a piecewise-continuous-deformation distribution of the body. This is the essential characteristic of *compatibility*; i.e., the stress distribution and the resulting deflection distribution must be compatible with boundary conditions and a continuous distribution of deformations so that no "holes" or overlapping of specific points in the body occur.

Recall the simple stress distributions given in basic mechanics of materials (see Chap. 3). For axial loading, torsion, and bending, the stress distributions are initially based on intuitive reasoning of the strain or deflection fields. For example, consider the axial loading shown in Fig. 2.5-1(a). For an isolation made at section c-c, Fig. 2.5-1(b) shows a uniform stress distribution, whereas Fig. 2.5-1(c) shows a parabolic distribution. Both distributions will satisfy equilibrium provided $\int \sigma_x \, dA = P$. As a matter of fact, any stress distribution symmetric with respect to the centroidal axis can ensure equilibrium. Then why is the uniform distribution the correct one? This is where the concept of compatibility becomes important. The uniform stress distribution of Fig. 2.5-1(a) not only ensures equilibrium (provided $\sigma_x = P/A$), it is also compatible with a continuous strain and displacement field consistent with the boundary conditions of the axially loaded member.

If a stress field is assumed, and if a continuous displacement field consistent with the boundary conditions can be found, the stress field is acceptable. The case of axial loading is left as a problem at the end of the chapter. Consider instead

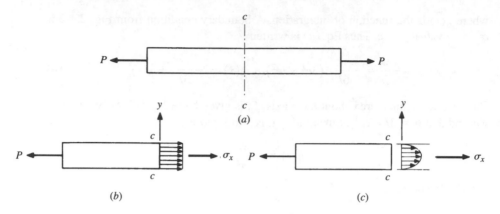

(a)

(b)

(c)

Figure 2.5-1

Example 1.5-1. Recall that the displacement field was given in this example, from which the strain and stress fields were determined. However, in a problem like this, one normally begins with the standard stress formulations which satisfy equilibrium. If a continuous displacement field can then be found which satisfies the geometric boundary conditions of the problem, the stress field is compatible.

Example 2.5-1 | For the beam shown in Fig. 2.5-2, determine the displacement field due to bending only. Consider the cross section of the beam to be rectangular and thin so that deflections are not functions of z. Check the solution against that given in Example 1.5-1. The stiffness of the beam is EI_z, and Poisson's ratio is v.

Solution:

Since only the bending stresses are being considered, only the bending moment as a function of x is necessary. This is

$$M_z = -P(L - x) \qquad 0 < x < L \qquad\qquad \textbf{[a]}$$

Thus the stress field is

$$\sigma_x = -\frac{M_z y}{I_z} = \frac{P}{I_z} y(L - x) \qquad\qquad \textbf{[b]}$$

$$\sigma_y = 0 \qquad\qquad \textbf{[c]}$$

$$\tau_{xy} = 0 \qquad\qquad \textbf{[d]}$$

From Hooke's law, the strain field is

$$\varepsilon_x = \frac{1}{E}(\sigma_x - v\sigma_y) = \frac{P}{EI_z} y(L - x) \qquad\qquad \textbf{[e]}$$

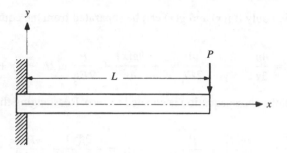

Figure 2.5-2

$$\varepsilon_y = \frac{1}{E}(\sigma_y - v\sigma_x) = -\frac{vP}{EI_z}y(L - x) \qquad [f]$$

and

$$\gamma_{xy} = \frac{2(1 + v)}{E}\tau_{xy} = 0 \qquad [g]$$

Next the normal strains ε_x and ε_y are substituted into the strain-displacement equations, resulting in

$$\frac{\partial u}{\partial x} = \varepsilon_x = \frac{P}{EI_z}y(L - x) \qquad [h]$$

and

$$\frac{\partial v}{\partial y} = \varepsilon_y = -\frac{vP}{EI_z}y(L - x) \qquad [i]$$

Equations (h) and (i) can be integrated, but it must be kept in mind that since these equations contain partial derivatives, appropriate *functions* of integration must be added. Thus

$$u = \frac{P}{2EI_z}xy(2L - x) + f(y) \qquad [j]$$

$$v = -\frac{vP}{2EI_z}y^2(L - x) + g(x) \qquad [k]$$

The functions of integration $f(y)$ and $g(x)$ are pure functions of y and x, respectively, where both functions possibly contain constants. The next step is where the concept of compatibility is used. If Eqs. (j) and (k) are acceptable, they can be substituted into the shear strain-displacement equation, $\gamma_{xy} = \partial v/\partial x + \partial u/\partial y$, and if Eq. ($g$) can be satisfied, a compatible solution will emerge. As will be seen, Eq. (g)

can be satisfied only if $f(y)$ and $g(x)$ can be separated from the equation. From Eqs. (j) and (k)

$$\gamma_{xy} = \frac{\partial v}{\partial x} + \frac{\partial u}{\partial y} = \frac{vP}{2EI_z} y^2 + \frac{\partial g(x)}{\partial x} + \frac{P}{2EI_z} x(2L - x) + \frac{\partial f(y)}{\partial y}$$

From Eq. (g), the above equation is equated to zero. Rearranging the resulting equation yields

$$\frac{\partial g(x)}{\partial x} + \frac{P}{2EI_z} x(2L - x) = -\frac{\partial f(y)}{\partial y} - \frac{vP}{2EI_z} y^2 \qquad [I]$$

Since the beam is not a line in space, a pure function of x [the left-hand side of Eq. (l)] can equal a pure function of y [the right-hand side of Eq. (l)] for a continuous two-dimensional body only if both sides of Eq. (l) equal a constant, say C_1. Setting both sides of Eq. (l) equal to C_1 separates the functions $f(y)$ and $g(x)$, and a compatible solution will result. Thus[†]

$$\frac{dg(x)}{dx} + \frac{P}{2EI_z} x(2L - x) = C_1 \qquad [m]$$

$$\frac{df(y)}{dy} + \frac{vPy^2}{2EI_z} = -C_1 \qquad [n]$$

Solving Eqs. (m) and (n) results in

$$g(x) = -\frac{P}{6EI_z} x^2(3L - x) + C_1 x + C_2 \qquad [o]$$

$$f(y) = -\frac{vPy^3}{6EI_z} - C_1 y + C_3 \qquad [p]$$

where C_2 and C_3 are constants of integration. Substituting Eqs. (o) and (p) into Eqs. (j) and (k) gives

$$u = \frac{P}{2EI_z} xy(2L - x) - \frac{vP}{6EI_z} y^3 - C_1 y + C_3 \qquad [q]$$

$$v = -\frac{vP}{2EI_z} y^2(L - x) - \frac{P}{6EI_z} x^2(3L - x) + C_1 x + C_2 \qquad [r]$$

The constants C_1, C_2, and C_3 are evaluated using the geometric boundary conditions of the particular problem. In this example, there is an infinite set of boundary conditions since the beam cross section is completely fixed at the wall at $x = 0$. Thus all conditions except three must be relaxed, and some error at the wall must be tolerated. This is acceptable since, due to Saint-Venant's principle (see Sec. 3.1), even

---

[†] Once the equations are separated, each equation is written in terms of one independent variable only. This means that the partial derivative can be replaced by an ordinary derivative.

the stress distribution [Eq. (*b*)] is in error there. Probably the most accepted conditions are that of $u = v = \theta_{xy} = 0$ at $x = y = 0$. Recall that $\theta_{xy} = \frac{1}{2}(\partial v/\partial x - \partial u/\partial y)$. Substituting these conditions into Eqs. (*q*) and (*r*) yields $C_1 = C_2 = C_3 = 0$. Thus the final results for the complete deflection field are

$$u = \frac{P}{2EI_z} xy(2L - x) - \frac{vP}{6EI_z} y^3 \qquad [s]$$

and

$$v = -\frac{vP}{2EI_z} y^2(L - x) - \frac{P}{6EI_z} x^2(3L - x) \qquad [t]$$

Referring back to Example 1.5-1, we can see that Eqs. (*s*) and (*t*) agree with what was given there.

The following observations can be made on the results of Example 2.5-1.

1. It can be seen from Eqs. (*s*) and (*t*) that the left surface of the beam is not rigidly held for $x = 0$ and $y \neq 0$, as there are nonzero values for u and v. This is due to the fact that the actual stress distribution at the wall does not behave like that originally given in the problem if the entire left surface is clamped. However, the actual correction would be quite small.

2. The lateral deflection of the neutral axis $v(x, 0)$ is normally discussed in a basic strength of materials course using a purely geometric approach, where for an end-loaded cantilever beam the deflection is given as (see Appendix C)

$$v_c = v(x, 0) = -\frac{Px^2}{6EI}(3L - x)$$

The same result is obtained from Eq. (*t*).

3. The slope or angle that the deflected neutral axis has with respect to the horizontal is given by $\theta_{xy}(x, 0)$. From Eqs. (*s*) and (*t*) the rotation of the neutral axis is

$$\frac{dv_c}{dx} = \theta_{xy}(x, 0) = \frac{1}{2}\left(\frac{\partial v}{\partial x} - \frac{\partial u}{\partial y}\right)\bigg|_{y=0} = -\frac{Px}{2EI}(2L - x)$$

which again agrees with the results obtained from simple deflection theory (see Appendix C).

4. A surface of the beam cross section for any particular value of x does not remain planar. The surfaces warp, as can be seen in Eqs. (*s*) and (*t*). If the surfaces remained planar, u would be a linear function of y and v would not be a function of y.

Sometimes stress distributions are used in the solution of a given problem where the stress distributions are not compatible. In these cases, the incomplete deflection field will not be solvable, and depending on the severity of the incompatibility, the accuracy of the stress distribution may be questionable.

Example 2.5-2 | In Example 2.4-1 the equilibrium equations are used on a uniformly loaded cantilever beam to find the τ_{xy} and σ_y stress distributions using the conventional strength of material equation for σ_x. Each one of these stress distributions ensures equilibrium, but do they provide a compatible displacement field? Using the method in Example 2.5-1, determine whether or not a compatible displacement field can be found for the beam shown in Fig. 2.4-2. Use the following stress fields from Example 2.4-1:

$$\sigma_x = \frac{w}{2I_z} x^2 y \qquad\qquad [a]$$

$$\sigma_y = -\frac{w}{6I_z}(2c^3 + 3c^2 y - y^3) \qquad\qquad [b]$$

$$\tau_{xy} = \frac{w}{2I_z} x(c^2 - y^2) \qquad\qquad [c]$$

Solution:

As in Example 2.5-1, the stresses are substituted into the strain-stress equations,

$$\varepsilon_x = \frac{1}{E}(\sigma_x - v\sigma_y) = \frac{w}{6EI_z}[3x^2 y + v(2c^3 + 3c^2 y - y^3)] \qquad [d]$$

$$\varepsilon_y = \frac{1}{E}(\sigma_y - v\sigma_x) = -\frac{w}{6EI_z}(2c^3 + 3c^2 y - y^3 + 3vx^2 y) \qquad [e]$$

$$\gamma_{xy} = \frac{2(1+v)}{E}\tau_{xy} = \frac{(1+v)w}{EI_z}x(c^2 - y^2) \qquad [f]$$

Since $\partial u/\partial x = \varepsilon_x$ and $\partial v/\partial y = \varepsilon_y$, u and v are found by integrating Eqs. (d) and (e), resulting in

$$u = \frac{w}{6EI_z}x[x^2 y + v(2c^3 + 3c^2 y - y^3)] + f(y) \qquad [g]$$

$$v = -\frac{w}{6EI_z}y\left(2c^3 + \frac{3}{2}c^2 y - \frac{1}{4}y^3 + \frac{3}{2}vx^2 y\right) + g(x) \qquad [h]$$

Next, Eqs. (g) and (h) are substituted into $\gamma_{xy} = \partial v/\partial x + \partial u/\partial y$ and equated to Eq. (f). After simplification this yields

$$\frac{\partial f(y)}{\partial y} + \frac{\partial g(x)}{\partial x} + \frac{w}{6EI_z}[x^3 - 3c^2(2+v)x] + \frac{w}{EI_z}xy^2 = 0 \qquad [i]$$

The last term in Eq. (i) contains an xy^2 term which is a pure function of neither x nor y. Thus Eq. (i) is an invalid equation, and the stress field given by Eqs. (a) to (c) is *not* compatible.

The stress distribution given by Eqs. (*a*) to (*c*) of Example 2.5-2 are fairly accurate, but they are not capable of providing an acceptable displacement field. A correction can be made to the stress distribution so that a compatible displacement field can be determined. One technique is simple to modify the stress equations so as to cancel any undesirable terms.

Determine the simplest correction which will cancel the unwanted xy^2 term obtained in Example 2.5-2.

Example 2.5-3

Solution:

If in Example 2.5-2, σ_x in Eq. (*a*) and consequently ε_x contained a y^3 term, u would contain an xy^3 term. This would then provide an xy^2 term in $\partial u/\partial y$. The addition of y^3 to σ_x would yield no additional term to $\partial v/\partial x$. Thus as a trial, let

$$\sigma_x = F(x)y + K_1 y^3 \qquad [j]$$

where K_1 is a constant and $F(x)$ is an unknown function of x.[†] The moment induced by this stress is M_z. Thus

$$\int_{-c}^{+c} y\sigma_x(b\ dy) = -M_z = \frac{w}{2}x^2$$

Substitution of Eq. (*j*) and integration yield

$$2b\left[F(x)\frac{c^3}{3} + K_1\frac{c^5}{5} \right] = \frac{w}{2}x^2$$

or

$$\frac{2}{3}bc^3\left[F(x) + \frac{3}{5}K_1c^2 \right] = \frac{w}{2}x^2$$

The moment of inertia is $I_z = 2/3\ bc^3$. Thus

$$F(x) + \frac{3}{5}K_1c^2 = \frac{w}{2I_z}x^2 \qquad [k]$$

The additional $K_1 y^3$ term in σ_x creates an additional $K_1 y^3/E$ in ε_x and $K_1 xy^3/E$ in u. Thus the additional term in $\partial u/\partial y$ is $3K_1 xy^2/E$. Equating this to $-(w/EI)xy^2$ will eliminate the unwanted xy^2 term in Eq. (*i*) of Example 2.5-2. Thus K_1 is obtained, where $K_1 = -(w/3I_z)$. Substituting this into Eq. (*k*) yields

$$F(x) = \frac{wx^2}{2I_z} + \frac{wc^2}{5I_z} = \frac{w}{10I_z}(5x^2 + 2c^2)$$

[†] This is fortunate. The additional term contains no function of *x*. Thus Eqs.(*b*) and (*c*) for σ_y and τ_{xy} will be unaffected by the additional term, and equilibrium is still ensured. Also, the additional y^3 term in σ_x gives no net axial force across the section.

The modified stress distribution of Eq. (j) becomes

$$\sigma_x = \frac{w}{10I_z}(5x^2 + 2c^2)y - \frac{w}{3I_z}y^3 \qquad [l]$$

Thus the stress distribution given by the above equation and Eqs. (b) and (c) of Example 2.5-2 will provide a solution to the displacement field of the beam. One point to note, however, is that throughout the beam the stress distribution gives equilibrium but at $x = 0$ there is a distribution of normal stress σ_x. This means that the distribution is still basically incorrect. However, the error is small. Further improvement can be made by assuming series solutions for the stresses. However, these improvements would prove negligible.

Considering the xy plane only, we can define compatibility rigorously as follows. Recall that the strain-displacement equations are given by

$$\varepsilon_x = \frac{\partial u}{\partial x} \qquad \varepsilon_y = \frac{\partial v}{\partial y} \qquad \gamma_{xy} = \frac{\partial v}{\partial x} + \frac{\partial u}{\partial y}$$

Differentiating γ_{xy} with respect to x and then with respect to y yields

$$\frac{\partial^2 \gamma_{xy}}{\partial x\, \partial y} = \frac{\partial^3 u}{\partial x\, \partial y^2} + \frac{\partial^3 v}{\partial x^2\, \partial y}$$

Noting that

$$\frac{\partial^3 u}{\partial x\, \partial y^2} = \frac{\partial^2 \varepsilon_x}{\partial y^2} \qquad \text{and} \qquad \frac{\partial^3 v}{\partial x^2\, \partial y} = \frac{\partial^2 \varepsilon_y}{\partial x^2}$$

we have

$$\frac{\partial^2 \gamma_{xy}}{\partial x\, \partial y} = \frac{\partial^2 \varepsilon_x}{\partial y^2} + \frac{\partial^2 \varepsilon_y}{\partial x^2} \qquad [2.5\text{-}1]$$

Equation (2.5-1), called the *compatibility equation* neglecting z dependence, provides a check on whether a given strain field is compatible in the xy plane.

Example 2.5-4 Show that the strains given by Eqs.(d) to (f) of Example 2.5-2 do not satisfy the compatibility equation.

Solution:

Differentiation of the strains given in Example 2.5-2 results in

$$\frac{\partial^2 \gamma_{xy}}{\partial x\, \partial y} = -\frac{2w(1 + v)}{EI}\, y$$

$$\frac{\partial^2 \varepsilon_x}{\partial y^2} = -\frac{wv}{EI}\, y$$

$$\frac{\partial^2 \varepsilon_y}{\partial x^2} = -\frac{wv}{EI}\, y$$

Substituting these into Eq. 2.5-1 yields

$$\frac{-2w(1+v)}{EI}y = -\frac{wv}{EI}y - \frac{wv}{EI}y$$

Simplifying results in

$$(1+v)y = vy$$

Thus compatibility is valid only at $y = 0$, that is, along the centroidal axis.

There are six compatibility equations, given by

$$\frac{\partial^2 \gamma_{xy}}{\partial x\,\partial y} = \frac{\partial^2 \varepsilon_x}{\partial y^2} + \frac{\partial^2 \varepsilon_y}{\partial x^2} \qquad\qquad \textbf{[2.5-2a]}$$

$$\frac{\partial^2 \gamma_{yz}}{\partial y\,\partial z} = \frac{\partial^2 \varepsilon_y}{\partial z^2} + \frac{\partial^2 \varepsilon_z}{\partial y^2} \qquad\qquad \textbf{[2.5-2b]}$$

$$\frac{\partial^2 \gamma_{zx}}{\partial z\,\partial x} = \frac{\partial^2 \varepsilon_z}{\partial x^2} + \frac{\partial^2 \varepsilon_x}{\partial z^2} \qquad\qquad \textbf{[2.5-2c]}$$

$$2\frac{\partial^2 \varepsilon_x}{\partial y\,\partial z} = \frac{\partial}{\partial x}\left(-\frac{\partial \gamma_{yz}}{\partial x} + \frac{\partial \gamma_{zx}}{\partial y} + \frac{\partial \gamma_{xy}}{\partial z}\right) \qquad\qquad \textbf{[2.5-2d]}$$

$$2\frac{\partial^2 \varepsilon_y}{\partial z\,\partial x} = \frac{\partial}{\partial y}\left(\frac{\partial \gamma_{yz}}{\partial x} - \frac{\partial \gamma_{zx}}{\partial y} + \frac{\partial \gamma_{xy}}{\partial z}\right) \qquad\qquad \textbf{[2.5-2e]}$$

$$2\frac{\partial^2 \varepsilon_z}{\partial x\,\partial y} = \frac{\partial}{\partial z}\left(\frac{\partial \gamma_{yz}}{\partial x} + \frac{\partial \gamma_{zx}}{\partial y} - \frac{\partial \gamma_{xy}}{\partial z}\right) \qquad\qquad \textbf{[2.5-2f]}$$

The first three equations are the in-plane dependence, like Eq. (2.5-1). The last three equations are the out-of-plane dependence and are left for the reader to verify as a problem at the end of the chapter.

2.6 SUMMARY OF IMPORTANT EQUATIONS

Stress Transformations

Three dimensional

$$\sigma_{x'} = \sigma_x n_{x'x}^2 + \sigma_y n_{x'y}^2 + \sigma_z n_{x'z}^2 + 2\tau_{xy}n_{x'x}n_{x'y} + 2\tau_{yz}n_{x'y}n_{x'z} + 2\tau_{zx}n_{x'z}n_{x'x}$$
$$\textbf{[2.1-4a]}$$

$$\tau_{x'y'} = \sigma_x n_{x'x}n_{y'x} + \sigma_y n_{x'y}n_{y'y} + \sigma_z n_{x'z}n_{y'z} + \tau_{xy}(n_{x'x}n_{y'y} + n_{x'y}n_{y'x})$$
$$+ \tau_{yz}(n_{x'y}n_{y'z} + n_{x'z}n_{y'y}) + \tau_{zx}(n_{x'x}n_{y'z} + n_{x'z}n_{y'x}) \qquad \textbf{[2.1-4b]}$$

$$\tau_{z'x'} = \sigma_x n_{x'x} n_{z'x} + \sigma_y n_{x'y} n_{z'y} + \sigma_z n_{x'z} n_{z'z} + \tau_{xy}(n_{x'x} n_{z'y} + n_{x'y} n_{z'x})$$

$$+ \tau_{yz}(n_{x'y} n_{z'z} + n_{x'z} n_{z'y}) + \tau_{zx}(n_{x'x} n_{z'z} + n_{x'z} n_{z'x}) \qquad \textbf{[2.1-4c]}$$

$$\sigma_{y'} = \sigma_x n_{y'x}^2 + \sigma_y n_{y'y}^2 + \sigma_z n_{y'z}^2 + 2\tau_{xy} n_{y'x} n_{y'y} + 2\tau_{yz} n_{y'y} n_{y'z} + 2\tau_{zx} n_{y'z} n_{y'x}$$

$$\textbf{[2.1-4d]}$$

$$\tau_{y'z'} = \sigma_x n_{y'x} n_{z'x} + \sigma_y n_{y'y} n_{z'y} + \sigma_z n_{y'z} n_{z'z} + \tau_{xy}(n_{y'x} n_{z'y} + n_{y'y} n_{z'x})$$

$$+ \tau_{yz}(n_{y'y} n_{z'z} + n_{y'z} n_{z'y}) + \tau_{zx}(n_{y'x} n_{z'z} + n_{y'z} n_{z'x}) \qquad \textbf{[2.1-4e]}$$

$$\sigma_{z'} = \sigma_x n_{z'x}^2 + \sigma_y n_{z'y}^2 + \sigma_z n_{z'z}^2 + 2\tau_{xy} n_{z'x} n_{z'y} + 2\tau_{yz} n_{z'y} n_{z'z} + 2\tau_{zx} n_{z'z} n_{z'x} \quad \textbf{[2.1-4f]}$$

Matrix equation

$$[\sigma]_{x'y'z'} = [\mathbf{T}][\sigma]_{xyz}[\mathbf{T}]^T \qquad \textbf{[2.1-7]}$$

where

$$[\mathbf{T}] = \begin{bmatrix} n_{x'x} & n_{x'y} & n_{x'z} \\ n_{y'x} & n_{y'y} & n_{y'z} \\ n_{z'x} & n_{z'y} & n_{z'z} \end{bmatrix}$$

$$[\sigma]_{x'y'z'} = \begin{bmatrix} \sigma_{x'} & \tau_{x'y'} & \tau_{x'z'} \\ \tau_{y'x'} & \sigma_{y'} & \tau_{y'z'} \\ \tau_{z'x'} & \tau_{z'y'} & \sigma_{z'} \end{bmatrix}$$

$$[\sigma]_{xyz} = \begin{bmatrix} \sigma_x & \tau_{xy} & \tau_{xz} \\ \tau_{yx} & \sigma_y & \tau_{yz} \\ \tau_{zx} & \tau_{zy} & \sigma_z \end{bmatrix}$$

Plane stress

$$\sigma_{x'} = \sigma_x \cos^2 \theta + \sigma_y \sin^2 \theta + 2\tau_{xy} \cos \theta \sin \theta \qquad \textbf{[2.1-8a]}$$

$$\sigma_{y'} = \sigma_x \sin^2 \theta + \sigma_y \cos^2 \theta - 2\tau_{xy} \sin \theta \cos \theta \qquad \textbf{[2.1-8b]}$$

$$\tau_{x'y'} = -(\sigma_x - \sigma_y)\sin\theta\cos\theta + \tau_{xy}(\cos^2\theta - \sin^2\theta) \quad \textbf{[2.1-8c]}$$

Three-dimensional simplified

$$\sigma = \sigma_x n_x^2 + \sigma_y n_y^2 + \sigma_z n_z^2 + 2\tau_{xy} n_x n_y + 2\tau_{yz} n_y n_z + 2\tau_{zx} n_z n_x \qquad \textbf{[2.1-18]}$$

$$\tau = [(\sigma_x n_x + \tau_{xy} n_y + \tau_{zx} n_z)^2 + (\tau_{xy} n_x + \sigma_y n_y + \tau_{yz} n_z)^2$$

$$+ (\tau_{zx} n_x + \tau_{yz} n_y + \sigma_z n_z)^2 - \sigma^2]^{1/2} \qquad \textbf{[2.1-20]}$$

Directional cosines for τ

$$t_x = \frac{1}{\tau}[(\sigma_x - \sigma)n_x + \tau_{xy}n_y + \tau_{zx}n_z] \qquad \text{[2.1-21a]}$$

$$t_y = \frac{1}{\tau}[\tau_{xy}n_x + (\sigma_y - \sigma)n_y + \tau_{yz}n_z] \qquad \text{[2.1-21b]}$$

$$t_z = \frac{1}{\tau}[\tau_{zx}n_x + \tau_{yz}n_y + (\sigma_z - \sigma)n_z] \qquad \text{[2.1-21c]}$$

Principal Stresses
Three dimensional

$$\sigma_p^3 - (\sigma_x + \sigma_y + \sigma_z)\sigma_p^2 + (\sigma_x\sigma_y + \sigma_y\sigma_z + \sigma_z\sigma_x - \tau_{yz}^2 - \tau_{zx}^2 - \tau_{xy}^2)\sigma_p$$
$$- (\sigma_x\sigma_y\sigma_z + 2\tau_{yz}\tau_{zx}\tau_{xy} - \sigma_x\tau_{yz}^2 - \sigma_y\tau_{zx}^2 - \sigma_z\tau_{xy}^2) = 0 \qquad \text{[2.1-25]}$$

Directional cosines[†]

$$(\sigma_x - \sigma_p)n_x + \tau_{xy}n_y + \tau_{zx}n_z = 0 \qquad \text{[2.1-23a]}$$

$$\tau_{xy}n_x + (\sigma_y - \sigma_p)n_y + \tau_{yz}n_z = 0 \qquad \text{[2.1-23b]}$$

$$\tau_{zx}n_x + \tau_{yz}n_y + (\sigma_z - \sigma_p)n_z = 0 \qquad \text{[2.1-23c]}$$

$$n_x^2 + n_y^2 + n_z^2 = 1 \qquad \text{[2.1-24]}$$

Plane stress (one principal stress is zero with the remaining two given by)

$$\sigma_p = \frac{\sigma_x + \sigma_y}{2} \pm \sqrt{\left(\frac{\sigma_x - \sigma_y}{2}\right)^2 + \tau_{xy}^2} \qquad \text{[2.1-30]}$$

Principal angle (use one equation)

$$(\sigma_x - \sigma_p)\cos\theta + \tau_{xy}\sin\theta = 0 \qquad \text{[2.1-28a]}$$

$$\tau_{xy}\cos\theta + (\sigma_y - \sigma_p)\sin\theta = 0 \qquad \text{[2.1-28b]}$$

Maximum shear stress ($\sigma_1 > \sigma_2 > \sigma_3$)

$$\tau_{\substack{max \\ min}} = \pm\frac{\sigma_1 - \sigma_3}{2} \qquad \text{[2.1-36]}$$

[†] Combine any two independent equations from Eqs. (2.1-23) with Eq. (2.1-24).

Strain Transformations and/or Principal Strains

Substitute ε for σ and $\gamma/2$ for τ into the stress transformation and/or principal stress equations.

Equilibrium Equations

Three dimensional

$$\frac{\partial \sigma_x}{\partial x} + \frac{\partial \tau_{xy}}{\partial y} + \frac{\partial \tau_{zx}}{\partial z} + \overline{F}_x = 0 \qquad \text{[2.4-1a]}$$

$$\frac{\partial \tau_{xy}}{\partial x} + \frac{\partial \sigma_y}{\partial y} + \frac{\partial \tau_{yz}}{\partial z} + \overline{F}_y = 0 \qquad \text{[2.4-1b]}$$

$$\frac{\partial \tau_{zx}}{\partial x} + \frac{\partial \tau_{yz}}{\partial y} + \frac{\partial \sigma_z}{\partial z} + \overline{F}_z = 0 \qquad \text{[2.4-1c]}$$

Plane stress

$$\frac{\partial \sigma_x}{\partial x} + \frac{\partial \tau_{xy}}{\partial y} + \overline{F}_x = 0 \qquad \text{[2.4-2a]}$$

$$\frac{\partial \tau_{xy}}{\partial x} + \frac{\partial \sigma_y}{\partial y} + \overline{F}_y = 0 \qquad \text{[2.4-2b]}$$

Compatibility Equations

$$\frac{\partial^2 \gamma_{xy}}{\partial x\, \partial y} = \frac{\partial^2 \varepsilon_x}{\partial y^2} + \frac{\partial^2 \varepsilon_y}{\partial x^2} \qquad \text{[2.5-2a]}$$

$$\frac{\partial^2 \gamma_{yz}}{\partial y\, \partial z} = \frac{\partial^2 \varepsilon_y}{\partial z^2} + \frac{\partial^2 \varepsilon_z}{\partial y^2} \qquad \text{[2.5-2b]}$$

$$\frac{\partial^2 \gamma_{zx}}{\partial z\, \partial x} = \frac{\partial^2 \varepsilon_z}{\partial x^2} + \frac{\partial^2 \varepsilon_x}{\partial z^2} \qquad \text{[2.5-2c]}$$

$$2\frac{\partial^2 \varepsilon_x}{\partial y\, \partial z} = \frac{\partial}{\partial x}\left(-\frac{\partial \gamma_{yz}}{\partial x} + \frac{\partial \gamma_{zx}}{\partial y} + \frac{\partial \gamma_{xy}}{\partial z} \right) \qquad \text{[2.5-2d]}$$

$$2\frac{\partial^2 \varepsilon_y}{\partial z\, \partial x} = \frac{\partial}{\partial y}\left(\frac{\partial \gamma_{yz}}{\partial x} - \frac{\partial \gamma_{zx}}{\partial y} + \frac{\partial \gamma_{xy}}{\partial z} \right) \qquad \text{[2.5-2e]}$$

$$2\frac{\partial^2 \varepsilon_z}{\partial x\, \partial y} = \frac{\partial}{\partial z}\left(\frac{\partial \gamma_{yz}}{\partial x} + \frac{\partial \gamma_{zx}}{\partial y} - \frac{\partial \gamma_{xy}}{\partial z} \right) \qquad \text{[2.5-2f]}$$

2.7   PROBLEMS

2.1 The state of stress at a point in a body relative to the xyz coordinate system is given by

$$[\sigma] = \begin{bmatrix} 0 & -30 & 25 \\ -30 & -40 & -15 \\ 25 & -15 & 10 \end{bmatrix} \text{MPa}$$

Determine the stress matrix relative to a coordinate system defined by first rotating the xyz coordinate system 45° about the x axis, then rotating −45° about the new z axis.

2.2 A point is undergoing plane stress and relative to the xyz and $x'y'z'$ systems $\sigma_x = 40$, $\sigma_y = -30$, and $\tau_{x'y'} = -20$ MPa. Determine $\sigma_{x'}$, $\sigma_{y'}$, and τ_{xy}.

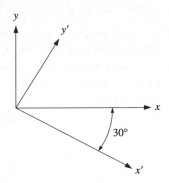

Problem 2.2

2.3 A point in a state of plane stress is isolated by three surfaces as shown. Using the transformation equations determine the values of σ and τ.

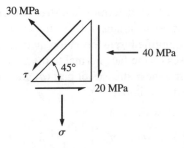

Problem 2.3

2.4   For Mohr's circle for plane stress, described in Sec. 2.1.3, show that when using σ, τ axes as shown in Fig. 2.1-5 counterclockwise transformation rotations of θ relative to the element correspond to counterclockwise rotations of 2θ on Mohr's circle. Use Eqs. (2.1-9) and (2.1-10) for your explanation.

2.5   The state of stress at a point in a structure, relative to an xyz coordinate system, is given by

$$[\sigma] = \begin{bmatrix} 2 & -1 & 1 \\ -1 & -2 & 0 \\ 1 & 0 & 3 \end{bmatrix} \text{kpsi}$$

Given an $x'y'z'$ system defined by

$$\begin{bmatrix} n_{x'x} & n_{x'y} & n_{x'z} \\ n_{y'x} & n_{y'y} & n_{y'z} \\ n_{z'x} & n_{z'y} & n_{z'z} \end{bmatrix} = \frac{1}{6}\begin{bmatrix} 0 & 3\sqrt{2} & -3\sqrt{2} \\ -2\sqrt{3} & 2\sqrt{3} & 2\sqrt{3} \\ 2\sqrt{6} & \sqrt{6} & \sqrt{6} \end{bmatrix}$$

(a) Using Eqs. (2.1-4), determine the stress matrix in the $x'y'z'$ system.

(b) Using the simplified transformation method, determine the normal and shear stresses on the face normal to the x' direction.

(c) Show that the shear stresses of part (a) are consistent with the shear stress determined in part (b).

2.6   Relative to an xyz coordinate system a state of stress at a point in a structure is given by

$$[\sigma] = \begin{bmatrix} 25 & 10 & 15 \\ 10 & 0 & 0 \\ 15 & 0 & -20 \end{bmatrix} \text{MPa}$$

Given an $x'y'z'$ system defined by

$$\begin{bmatrix} n_{x'x} & n_{x'y} & n_{x'z} \\ n_{y'x} & n_{y'y} & n_{y'z} \\ n_{z'x} & n_{z'y} & n_{z'z} \end{bmatrix} = \frac{1}{9}\begin{bmatrix} 1 & -8 & 4 \\ 4 & 4 & 7 \\ -8 & 1 & 4 \end{bmatrix}$$

(a) Using Eqs. (2.1-4), determine the stress matrix in the $x'y'z'$ system.

(b) Check the stress invariants relative to both coordinate systems.

(c) Using the simplified transformation method, determine the normal and shear stresses on the face normal to the x' direction.

(d) Show that the shear stresses of part (a) are consistent with the shear stress determined in part (c).

2.7   Solve Problem 2.6 with

$$\begin{bmatrix} n_{x'x} & n_{x'y} & n_{x'z} \\ n_{y'x} & n_{y'y} & n_{y'z} \\ n_{z'x} & n_{z'y} & n_{z'z} \end{bmatrix} = \frac{1}{11}\begin{bmatrix} 2 & 6 & 9 \\ 6 & 7 & -6 \\ 9 & -6 & 2 \end{bmatrix}$$

2.8 An element can be isolated in which all normal stresses on it are equal (called hydrostatic). The equal stresses are the average of the normal stresses (which is an invariant) given by

$$\sigma_{ave} = \frac{\sigma_1 + \sigma_2 + \sigma_3}{3}$$

The element has eight faces for which the surfaces are symmetric to the principal axes where the normals for the surfaces are eight combinations of $\pm 1/\sqrt{3}$ (e.g., one set is $1/\sqrt{3}, 1/\sqrt{3}, 1/\sqrt{3}$; another is $1/\sqrt{3}, -1/\sqrt{3}, 1/\sqrt{3}$; etc.). The element is therefore an octahedron. The interest in the octahedral surfaces is in a failure theory that states that the failure of this element is strictly dependent on the shear stresses on these surfaces. Show that these shear stresses, called the *octahedral shear stresses,* are given by

$$\tau_{oct} = \frac{1}{3}\sqrt{(\sigma_1 - \sigma_2)^2 + (\sigma_2 - \sigma_3)^2 + (\sigma_3 - \sigma_1)^2}$$

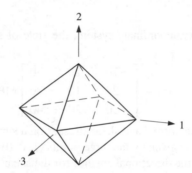

Problem 2.8   Octahedral surfaces.

2.9 The stress at a point relative to an *xyz* coordinate system is

$$[\boldsymbol{\sigma}] = \begin{bmatrix} 6 & 4 & 0 \\ 4 & -3 & 0 \\ 0 & 0 & 3 \end{bmatrix} \text{kpsi}$$

(a) Determine the normal and shear stresses on a surface whose outer normal has the directional cosines $n_x = n_y = 6/11$, and $n_z = 7/11$.
(b) Determine the directional cosines for the shear stress determined in part (a).

2.10 At point Q in a structure the state of stress relative to an *xyz* coordinate system is

$$[\boldsymbol{\sigma}] = \begin{bmatrix} -15 & 0 & -25 \\ 0 & 10 & -20 \\ -25 & -20 & 0 \end{bmatrix} \text{MPa}$$

Using the cube shown, determine the normal and shear stress at point Q for surfaces parallel to the planes (a) *BCGF*, (b) *ABEF*, (c) *BGE*.

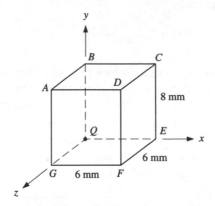

Problem 2.10

2.11 Relative to an *xyz* coordinate system, the state of stress at a point is known to be

$$[\sigma] = \begin{bmatrix} -10 & 20 & 30 \\ 20 & 10 & -20 \\ 30 & -20 & 40 \end{bmatrix} \text{MPa}$$

(a) Evaluate the normal and shear stresses on a surface at the point where the surface is given by the three points $(1, 0, 0)$, $(0, 2, 0)$, $(0, 0, -1)$

(b) Determine the directional cosines for the shear stress found in part (a) and in a rough sketch show the direction of the normal and shear stresses.

2.12 The state of stress at a point relative to an *xyz* coordinate system is given by

$$[\sigma] = \begin{bmatrix} 3 & 0 & -2 \\ 0 & 1 & 2 \\ -2 & 2 & -1 \end{bmatrix} \text{kpsi}$$

Determine the normal and shear stresses and their directional cosines on a surface intersecting the point and parallel to the plane given by the equation

$$x + 2y - 2z = -4$$

2.13 The state of stress at a point within a structure relative to the *xyz* coordinate system is given by

$$[\sigma] = \begin{bmatrix} 20 & 35 & 25 \\ 35 & 0 & -15 \\ 25 & -15 & 10 \end{bmatrix} \text{MPa}$$

If the coordinate location of the point in space is (1, 1, −2) determine the normal and shear stresses at the point and on an internal surface established by a sphere with equation

$$x^2 + (y - 2)^2 + z^2 = 6$$

2.14 to 2.20 For the plane stress state given
  (a) Draw the corresponding stress element properly oriented relative to the *xy* axes.
  (b) Determine the complete stress element associated with an axis system rotated θ (defined positive counterclockwise) using the transformation equations alone.
  (c) Determine the principal stresses and the corresponding stress element containing the stresses properly oriented relative to the *xy* axes using equations only.
  (d) Repeat parts (b) and (c) using Mohr's circle.
  (e) Determine the maximum and minimum shear stresses and show the complete stress element containing these stresses. Show the element properly oriented with respect to the *xyz* coordinate system.

Problem	σ_x	σ_y	τ_{xy}	θ
2.14	10 kpsi	0 kpsi	0 kpsi	30°
2.15	0 MPa	0 MPa	−20 MPa	−20°
2.16	40 MPa	10 MPa	0 MPa	15°
2.17	40 MPa	−10 MPa	0 MPa	−15°
2.18	10 kpsi	10 kpsi	0 kpsi	60°
2.19	80 MPa	0 MPa	−30 MPa	−30°
2.20	−12 kpsi	−20 kpsi	−6 kpsi	25°

2.21 The state of stress at a point in a body relative to the *xyz* coordinate system is given by

$$[\sigma] = \begin{bmatrix} -2 & 3 & -5 \\ 3 & 6 & 2 \\ -5 & 2 & -4 \end{bmatrix} \text{kpsi}$$

  (a) Determine the principal stresses and the directional cosines associated with the directions of each principal stress.
  (b) Determine the maximum shear stress at the point.

2.22 For Problem 2.21 determine the normal and shear stress on a surface intersecting the point where the plane is defined by three points (0, 0, 0), (2, −1, 3), (−2, 0, 1). Give the magnitudes of the stresses and the directional cosines defining their directions.

2.23 Given the state of stress

$$[\sigma] = \begin{bmatrix} 10 & 20 & 0 \\ 20 & 0 & -10\sqrt{2} \\ 0 & -10\sqrt{2} & 10 \end{bmatrix} \text{MPa}$$

(a) Verify that one of the principal stresses is $\sigma_p = 30$ MPa.
(b) Determine the remaining principal stresses.
(c) Determine the directional cosines associated with the principal stress of 30 MPa.
(d) Locate the surface containing the principal stress of 30 MPa by evaluating the coordinates of points B and C of the diagram shown.

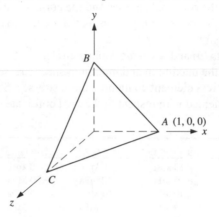

Problem 2.23

2.24 Show that the principal stress has a stationary value. *Recommended procedure:* The normal stress on an oblique surface is given by Eq. (2.1-18). Show that $\partial\sigma/\partial n_x = \partial\sigma/\partial n_y = 0$ ($\partial\sigma/\partial n_z = 0$ is not necessary since n_z is not independent of n_x and n_y). Take the partials of Eq. (2.1-18) relative to n_x and n_y. The results will contain terms related to F_x, F_y, and F_z of Eqs. (2.1-19) and partials of n_z with respect to n_x and n_y. These partials can be found from Eq. (2.1-24), and using Eqs. (2.1-22) with equilibrium ($F_x = -F_{x'}$, etc.) will complete the proof.

2.25 Given the stress matrix

$$[\sigma] = \begin{bmatrix} 1 & 2 & -2 \\ 2 & 4 & -4 \\ -2 & -4 & 4 \end{bmatrix} \text{kpsi}$$

(a) Determine the principal stresses and the directional cosines associated with the directions of each principal stress.
(b) Determine the maximum shear stress at the point.

2.26 Given the stress matrix

$$[\boldsymbol{\sigma}] = \begin{bmatrix} 20 & 10 & 10 \\ 10 & 20 & 10 \\ 10 & 10 & 20 \end{bmatrix} \text{MPa}$$

(a) Determine the principal stresses and the directional cosines associated with the directions of each principal stress.
(b) Determine the maximum shear stress at the point.

2.27 Relative to an *xyz* coordinate system a state of stress at a point in a structure is given by

$$[\boldsymbol{\sigma}] = \begin{bmatrix} 30 & 20 & -20 \\ 20 & 10 & -10 \\ -20 & -10 & 0 \end{bmatrix} \text{MPa}$$

(a) Determine the principal stresses.
(b) Determine the directional cosines for each principal stress.
(c) Determine the maximum shear stress.

2.28 For the plane stress state shown:
(a) Determine the principal stresses and show them on a properly oriented element. For this part, use the transformation equations only and *not* Mohr's circle.
(b) Repeat part (a) using Mohr's circle.
(c) Determine the maximum shear stress and show the complete state of stress on a properly oriented element containing this maximum shear stress.

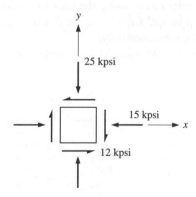

Problem 2.28

2.29 For Problem 2.28 determine the directional cosines for the surfaces containing the maximum and minimum shear stresses.

2.30 The engineering strain matrix relative to an xyz system is defined as

$$[\varepsilon]_{eng} = \begin{bmatrix} \varepsilon_x & \gamma_{xy} & \gamma_{xz} \\ \gamma_{yx} & \varepsilon_y & \gamma_{yz} \\ \gamma_{zx} & \gamma_{zy} & \varepsilon_z \end{bmatrix} = \begin{bmatrix} 2 & -1 & 2 \\ -1 & 0 & 1 \\ 2 & 1 & -4 \end{bmatrix} \times 10^{-4}$$

Determine the engineering strain matrix relative to an $x'y'z'$ system if

$$\begin{bmatrix} n_{x'x} & n_{x'y} & n_{x'z} \\ n_{y'x} & n_{y'y} & n_{y'z} \\ n_{z'x} & n_{z'y} & n_{z'z} \end{bmatrix} = \frac{1}{15}\begin{bmatrix} 14 & -5 & 2 \\ 2 & 10 & 11 \\ -5 & -10 & 10 \end{bmatrix}$$

2.31 The engineering strain matrix at a particular point in a structure is

$$[\varepsilon]_{eng} = \begin{bmatrix} \varepsilon_x & \gamma_{xy} & \gamma_{xz} \\ \gamma_{yx} & \varepsilon_y & \gamma_{yz} \\ \gamma_{zx} & \gamma_{zy} & \varepsilon_z \end{bmatrix} = \begin{bmatrix} -4 & 0 & 2 \\ 0 & 3 & -2 \\ 2 & -2 & -1 \end{bmatrix} \times 10^{-4}$$

Determine the engineering strain matrix relative to a coordinate system defined by first rotating the xyz coordinate system $-30°$ about the x axis, then rotating $40°$ about the new y axis.

2.32 For Problem 2.31 determine the principal strains and their corresponding directional cosines that indicate their directions.

2.33 to 2.39 Each of the following strains result from plane stress. For the strain given:
   (*a*) Determine the strains associated with an axis system rotated θ (defined positive counterclockwise) using the transformation equations alone.
   (*b*) Determine the principal strains and the direction each strain makes with the xy axes using equations only.
   (*c*) Repeat parts (*a*) and (*b*) using Mohr's circle (plot ε versus $\gamma/2$).

Problem	ε_x	ε_y	γ_{xy}	θ
2.33	500 μ	0	0	30°
2.34	0	0	−200 μ	−20°
2.35	400 μ	100 μ	0	15°
2.36	400 μ	−100 μ	0	−15°
2.37	100 μ	100 μ	0	60°
2.38	800 μ	0	−300 μ	−30°
2.39	−120 μ	−200 μ	−60 μ	25°

2.40 In Example 2.5-3, the stress field for σ_x was corrected such that

$$\sigma_x = \frac{w}{10I_z}(5x^2 + 2c^2)y - \frac{w}{3I_z}y^3$$

Using one of Eqs. (2.4-2) together with appropriate boundary conditions, prove that

$$\tau_{xy} = \frac{w}{2I_z}x(c^2 - y^2)$$

2.41 The simply supported beam shown carries a load of w (force per unit length) at an angle θ from the horizontal.
  (a) Using $\sigma_x = -M_z y/I_z + N/A$, where N is the normal force on the section, prove that

$$\sigma_x = -\frac{w}{2I_z}xy(L - x)\sin\theta + \frac{w}{I_z}cxy\cos\theta + \frac{w}{A}x\cos\theta$$

where $I_z = 2bc^3/3$, $A = 2bc$, and the above reduces to

$$\sigma_x = -\frac{w}{2I_z}xy(L - x)\sin\theta + \frac{w}{3I_z}cx(3y + c)\cos\theta \qquad \text{[a]}$$

  (b) Using Eq. (a), determine τ_{xy} and σ_y using Eqs. (2.4-2) and the boundary conditions on the top surface.
  (c) Verify that the solution of part (b) satisfies the boundary conditions on the bottom surface.

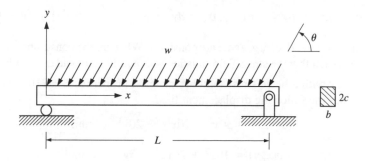

Problem 2.41

2.42 The cantilever beam shown has a rectangular cross section $b \times 2c$.
   (a) Using the equation $\sigma_x = -M_z y / I_z$, prove that $\sigma_x = wx^3 y / (6 I_z L)$.
   (b) Determine $\tau_{xy}(x, y)$ and $\sigma_y(x, y)$ using Eqs. (2.4-2).
   (c) Show whether or not the results of parts (a) and (b) provide a compatible displacement field.

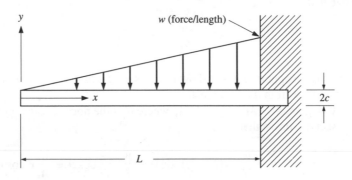

Problem 2.42

2.43 A body with a plane stress field has material properties E, v and has the following displacement field

$$u(x, y) = ax^3 - bxy^2 \qquad v(x, y) = cx^2 y - dy^3$$

where a, b, c, and d are constants. Determine the following:
   (a) σ_x, σ_y, and τ_{xy} as functions of x and y.
   (b) The rotation of the body at $x = 2$, $y = 1$.
   (c) Is the displacement field compatible?

2.44 For a thin plate with no body forces, a plane state of stress exists where

$$\sigma_x = ay^3 + bx^2 y - cx \qquad \sigma_y = dy^3 - e \qquad \tau_{xy} = fxy^2 + gx^2 y - h$$

where a, b, c, d, e, f, g, and h are constants. What are the constraints on the constants such that the stress field satisfies both equilibrium and compatibility?

2.45 A body with a plane stress field has material properties $E = 210$ GPa, $v = 0.3$ and has the following displacement field

$$u(x, y) = 30x^2 - 10x^3 y + 20y^3 \qquad \text{mm}$$

$$v(x, y) = 10x^3 + 20xy^3 + 5y^2 \qquad \text{mm}$$

where x and y are in meters. Determine the stresses and the rotation of the body at $x = 0.050$ m and $y = 0.020$ m. Is the displacement field compatible?

2.46 Repeat the analysis of Example 2.5-1 to determine $u(x, y)$ and $v(x, y)$. Consider the beam cross section to be rectangular of depth $2c$ but include the transverse shear stress given by

$$\tau_{xy} = -\frac{P}{2I_z}(c^2 - y^2)$$

(The derivation of this equation is given in Example 3.4-3, Chap. 3.)

2.47 Determine the deflection field of the uniformly loaded cantilever beam of Fig. 2.4-2 using Eqs. (c) and (f) of Example 2.4-1 and Eq. (1) of Example 2.5-3.

2.48 The cantilever shown has a moment M applied at the free end. Using standard strength of materials stress formulations $(\sigma_x = -M_z y/I_z, \sigma_x = \tau_{xy} = 0)$, determine the displacement fields $u(x, y)$ and $v(x, y)$. Compare $v(x, 0)$ and $\theta_{xy}(x, 0)$ with the beam tables in Appendix C.

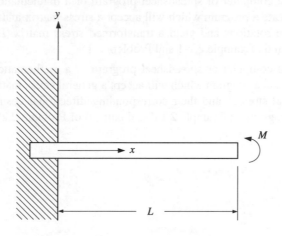

Problem 2.48

2.49 Verify that Eqs. (2.5-2d) to (2.5-2f) are valid.

Computer Problems

2.50 Using a computer or spreadsheet program or a mathematics software package, create a program which will accept a stress and transformation matrix

$$\begin{bmatrix} \sigma_x & \tau_{xy} & \tau_{xz} \\ \tau_{yx} & \sigma_y & \tau_{yz} \\ \tau_{zx} & \tau_{zy} & \sigma_z \end{bmatrix} \qquad \begin{bmatrix} n_{x'x} & n_{x'y} & n_{x'z} \\ n_{y'x} & n_{y'y} & n_{y'z} \\ n_{z'x} & n_{z'y} & n_{z'z} \end{bmatrix}$$

and evaluate the transformed stress matrix according to Eqs. (2.1-4). Demonstrate your program on Example 2.1-1 where

$$[\sigma] = \begin{bmatrix} -8 & 6 & -2 \\ 6 & 4 & 2 \\ -2 & 2 & -5 \end{bmatrix} \text{MPa}$$

$$\begin{bmatrix} n_{x'x} & n_{x'y} & n_{x'z} \\ n_{y'x} & n_{y'y} & n_{y'z} \\ n_{z'x} & n_{z'y} & n_{z'z} \end{bmatrix} = \frac{1}{4} \begin{bmatrix} 2\sqrt{2} & 2\sqrt{2} & 0 \\ -\sqrt{6} & \sqrt{6} & 2 \\ \sqrt{2} & -\sqrt{2} & 2\sqrt{3} \end{bmatrix}$$

2.51 Repeat Problem 2.50 using a spreadsheet program or a mathematics software package with Eq. (2.1-7).

2.52 Using a computer or spreadsheet program or a mathematics software package, create a program which will accept a stress matrix and two arbitrary coordinate rotations and yield a transformed stress matrix. Demonstrate your program on Example 2.1-1 and Problem 2.1.

2.53 Using a computer or spreadsheet program or a mathematics software package, create a program which will accept a general stress matrix and output the principal stresses and their corresponding directional cosines. Demonstrate your program on Example 2.1-4 and part (a) of Problem 2.21.

chapter

THREE

A REVIEW OF THE FUNDAMENTAL FORMULATIONS OF STRESS, STRAIN, AND DEFLECTION

3.0 INTRODUCTION

As stated earlier, it is assumed that the reader has completed a basic course in mechanics of materials and is familiar with the fundamental modes of loading of simple structural elements. The basic topics normally covered in this first course include direct axial and shear loading of prismatic members, torsion of circular shafts, and transverse loading of long, straight, narrow beams. Other topics normally include statically indeterminate problems, superposition, two-dimensional transformations, pressurized thin-walled cylinders, and buckling of long slender columns.

The intent of this chapter is *not* to provide a complete development of the material associated with this first course, as this material is considered to be a *prerequisite* to this textbook. However, the purpose of this chapter is to provide a concise review of the fundamental formulations, a review of the limitations of the formulations, and some challenging applications of the formulations. Furthermore, more advanced treatments of stress and deflections in later chapters rely on the elementary formulations or the methods used to obtain them.

If any of the subjects discussed in this chapter are unfamiliar, the reader should consult a basic text, as complete derivations will not be given here. The basic equations are provided, with comprehensive examples and with an emphasis on the underlying assumptions used in the derivations of the formulations which point out the restrictions to the equations. It is important that analysts know when an equation is appropriate or, if not, to what extent the restrictions are not met. In this way, they can form a relative degree of confidence in their results.

3.1   ASSUMPTIONS AND LIMITATIONS

For the elementary formulations the equations for stress, strain, and deflections are based on the assumptions of an ideal model. Since we do not live in an ideal world, real stress analysis problems will generally not conform to the ideal model. However, as engineers, we have learned to accept this fact and use the formulations if the problem agrees with the model closely. This closeness is always subject to question and dependent on the confidence of the analyst and the critical safety aspect of the application. This is where factors of safety or design factors become necessary. The use of these factors, however, is not intended to mask the ineptness of the analyst but to offset what the analyst knows about the application versus the limitations of the formulations. Further discussion of the design factor is covered in Chap. 7.

Some general statements can be made concerning limitations of the formulations. Other limitations that relate to a specific formulation will be discussed when the formulation is presented. Saint-Venant's principle, discussed in Chap. 1, is associated with the exact modeling of the applied loads on the structure. This principle is included in the list of the general limitations. The idealized stress, strain, and deflection formulations are based on the assumptions that:

1.   The material of the structure is homogeneous and isotropic and obeys the linear strain-stress relations given by Eqs. (1.4-2) and (1.4-8).
2.   The dimensions of the cross section of the structure are exact, and the cross section is constant or gradually varying in the direction normal to the cross section.
3.   The points of load application or support connections are at sufficiently large distances from the point of interest (Saint-Venant's principle).
4.   The applied loads and/or support connections are perfectly positioned geometrically according to the assumptions of the ideal model.
5.   The loads are static and are applied very gradually.
6.   No initial stresses exist from manufacturing assembly operations or residual stresses due to material forming.

3.2   AXIAL LOADING

3.2.1   AXIAL STRESSES

The equations for stress, strain, and deflections of the axially loaded rod of Fig. 3.2-1 (shown in tension) are based on a uniform stress distribution given by

$$\sigma_x = \pm \frac{P_x}{A} \qquad\qquad \textbf{[3.2-1]}$$

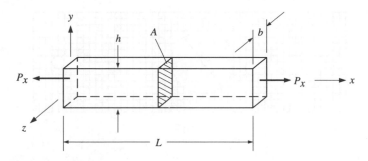

Figure 3.2-1   Axial loading.

where the $+$ and $-$ are used to indicate tension and compression, respectively. All of the limitations discussed in Sec. 3.1 apply to this formulation. Further limitations to Eq. (3.2-1) are:

1. The load P_x must go through the geometric *centroid* of the cross section A. This condition is more serious with compressive than with tensile loading. Loading along an axis offset from the centroid induces additional bending stresses, and for a given magnitude of force, the bending stresses and deflections are greater when in compression versus tension.

2. For compressive loading, a stability problem called *buckling* can occur. If the magnitude of the force reaches the *critical buckling load,* the rod will begin to buckle or bend and Eq. (3.2-1) is no longer valid. The buckling load depends on the length of the rod and the stiffness in bending of the rod relative to a lateral axis, which is discussed in Secs. 3.10 and 7.6.

Some of the limitations discussed in Sec. 3.1 can be illustrated by the plate, of thickness t, shown in Fig. 3.2-2(*a*). The tensile loading of the plate is developed through the pin placed in the hole on the right end of the plate. The plate is supported by rigidly clamping the left end. The plate has a grid marked on it to make the deformations easier to see. An exaggerated view of the deformation after the load is applied is shown in Fig. 3.2-2(*b*). Several observations can be made. In the regions midway between the holes and the larger hole and the left end of the plate, the grids are uniform where $\sigma = P_x/A$ is valid ($A = wt$). Where the load is applied at the right pin, we see the element at D quite elongated compared to its neighboring grids, indicating a nonuniform and higher stress level. At point E we see the element compressing in the axial direction due to the compressive bearing stress exerted by the pin on the plate. At the left of the plate we see relatively uniform behavior indicating P_x/A to be pretty close to the actual stress. However, we see the element at point A is slightly more elongated than its neighbors, indicating a slightly higher stress. This is due to the clamp restricting the Poisson's contraction of the plate being the greatest at point A and the corresponding point at the bottom of the plate. In

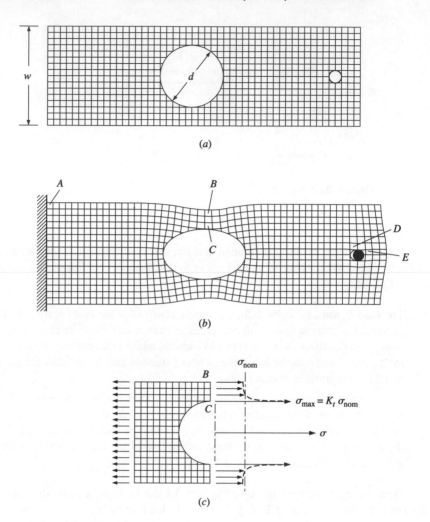

Figure 3.2-2   A plate loaded in tension.

the neighborhood of the large hole we again see a nonuniform behavior, with the element at point C stretching much more than at point B. A plot of the corresponding stress distribution along line BC is shown in Fig. 3.2-2(c). The nominal stress σ_{nom} is determined from P/A_{nom}, where A_{nom} is $(w-d)t$. This nonuniform behavior with a stress much larger than the nominal value is called a *stress concentration,* which occurs because of the rapid change in cross section in the region. Appendix F gives many examples of stress concentrations. Chart F.1, repeated here as Fig. 3.2-3,

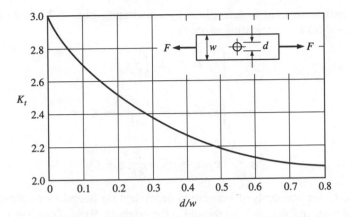

Figure 3.2-3 Bar in tension or simple compression with hole; $\sigma_{nom} = F/A$ where $A = (w - d)t$ and t is the thickness. SOURCE: See R. E. Peterson, "Design Factors for Stress Concentration," *Machine Design*, a Penton publication, vol. 23, no. 7, July 1951, p. 155. Reproduced with the permission of the author and publisher.

applies to the problem being considered here.[†] The chart gives a plot of the stress concentration factor K_t, defined as

$$K_t = \frac{\sigma_{max}}{\sigma_{nom}} \qquad\qquad \textbf{[3.2-2]}$$

3.2.2 AXIAL STRAINS AND DEFLECTIONS

Substituting the uniaxial stress σ_x into Eqs. (1.4-2) yields

$$\varepsilon_x = \frac{\sigma_x}{E} \qquad \varepsilon_y = \varepsilon_z = -v\frac{\sigma_x}{E}$$

or with Eq. (3.2-1)

$$\varepsilon_x = \pm\frac{P_x}{AE} \qquad \varepsilon_y = \varepsilon_z = \mp v\frac{P_x}{AE} \qquad \textbf{[3.2-3]}$$

where the upper sign is used for axial tension and the lower sign for axial compression.

[†] See Ref. 3.1, pt. 5.

The deflection field is determined by substituting Eqs. (3.2-3) into Eqs. 1.5-1(a), (b), and (e) and integrating to obtain

$$u = \int \varepsilon_x \, dx = \pm \frac{P}{AE} x + C_1$$

$$v = \int \varepsilon_y \, dy = \mp v \frac{P}{AE} y + C_2$$

$$w = \int \varepsilon_z \, dz = \mp v \frac{P}{AE} z + C_3$$

where the C_i are constants of integration dependent on boundary conditions. Considering the bar in Fig. 3.2-1, let the end at the origin be fixed. Thus $u = v = w = 0$ at $x = y = z = 0$. This results in $C_1 = C_2 = C_3 = 0$ and

$$u = \pm \frac{P_x}{AE} x \qquad\qquad\qquad \textbf{[3.2-4a]}$$

$$v = \mp v \frac{P_x}{AE} y \qquad\qquad\qquad \textbf{[3.2-4b]}$$

$$w = \mp v \frac{P_x}{AE} z \qquad\qquad\qquad \textbf{[3.2-4c]}$$

The total elongation of the rod is found by substituting in $x = L$, giving

$$u|_{x=L} = \pm \frac{P_x L}{AE} \qquad\qquad\qquad \textbf{[3.2-5]}$$

For this case, the total contraction or expansion in the y and z direction can be shown to be $vP_x h/(AE)$ and $vP_x b/(AE)$, respectively.

Example 3.2-1

One of the limitations of the basic stress formulation is that of homogeneity. Figure 3.2-4(a) shows a bar loaded in tension and consisting of two differing materials bonded together. Let one half be aluminum and the other half be steel with the ratio of their moduli of elasticity being $E_S/E_A = 3$. Determine the stress distribution and the conditions for loading if the bar is to elongate uniformly.

Solution:

For uniform elongation

$$(\varepsilon_x)_{steel} = (\varepsilon_x)_{aluminum}$$

Since $\varepsilon_x = \sigma_x/E$, then

$$\left(\frac{\sigma_x}{E} \right)_{steel} = \left(\frac{\sigma_x}{E} \right)_{aluminum}$$

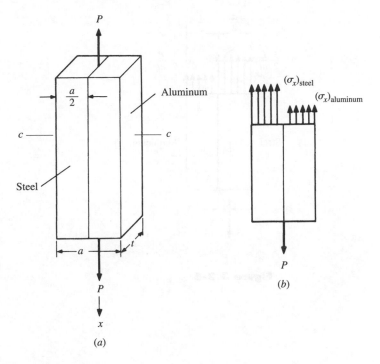

Figure 3.2-4

which can be rewritten as

$$\left(\sigma_x\right)_{\text{steel}} = \frac{E_S}{E_A}\left(\sigma_x\right)_{\text{aluminum}}$$

Since $E_S/E_A = 3$, this gives

$$\left(\sigma_x\right)_{\text{steel}} = 3\left(\sigma_x\right)_{\text{aluminum}} \qquad \textbf{[a]}$$

Thus the stress distribution across section c-c is as shown in Fig. 3.2-4(b). So for uniform elongation, the stress *is not uniform* across the entire section. To determine the values of the stress across the section, the stress distribution is integrated across the section and equated to the net force P. Thus

$$\left(\int \sigma_x \, dA\right)_{\text{steel}} + \left(\int \sigma_x \, dA\right)_{\text{aluminum}} = P$$

Since σ_x is constant in each half section

$$\left(\sigma_x A\right)_{\text{steel}} + \left(\sigma_x A\right)_{\text{aluminum}} = P$$

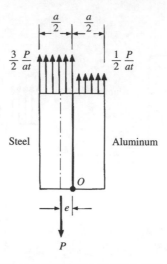

Figure 3.2-5

or

$$(\sigma_x)_{\text{steel}}\,\frac{at}{2} + (\sigma_x)_{\text{aluminum}}\,\frac{at}{2} = P \qquad\qquad \text{[b]}$$

Substituting Eq. (a) into (b) yields

$$(\sigma_x)_{\text{aluminum}} = \frac{1}{2}\frac{P}{at} \qquad (\sigma_x)_{\text{steel}} = \frac{3}{2}\frac{P}{at}$$

If the applied load is centered between the steel and aluminum as shown in Fig. 3.2-4(b), equilibrium in moment will not exist. The application of the force P will need to be offset by e as shown in Fig. 3.2-5. Summing moments about point O gives

$$\sum M_O = (P)(e) + \left(\frac{1}{2}\frac{P}{at}\right)\left(\frac{a}{2}t\right)\left(\frac{a}{4}\right) - \left(\frac{3}{2}\frac{P}{at}\right)\left(\frac{a}{2}t\right)\left(\frac{a}{4}\right) = 0$$

Solving the above yields $e = a/8$.

Example 3.2-2   For the rod shown in Fig. 3.2-6(a), determine the deflection of point B due to the weight of the rod. For the rod, the mass density is ρ, the area is A, and the modulus of elasticity is E.

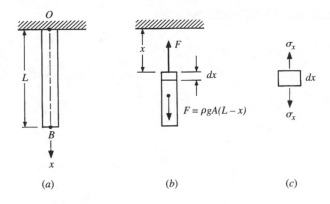

Figure 3.2-6

Solution:

From the force diagram in Fig. 3.2-6(*b*), the force transmitted through a section at *x* is $F = \rho g A(L - x)$, where *g* is the gravitational constant. Thus the stress is

$$\sigma_x = \frac{F}{A} = \rho g(L - x)$$

The stretch of a *dx* element is determined from Eq. (3.2-5), where *dx* is used instead of *L*.

$$du = \frac{F}{AE}\, dx$$

and the total extension of the rod is found by integrating the deflections of each *dx* element, resulting in

$$u_{max} = \int_0^L \frac{F}{AE}\, dx = \int_0^L \frac{\rho g}{E}(L - x)\, dx = \frac{\rho g L^2}{2E}$$

The deflection can be expressed in terms of the total weight *W* of the rod. Since $W = \rho g A L$, we have

$$u_{max} = \frac{1}{2}\frac{WL}{AE}$$

Example 3.2-3

Cables *BC* and *BD* shown in Fig. 3.2-7(*a*) form angles to the vertical of $\alpha = 30°$ and $\beta = 45°$, respectively. The length of cable *BC* is 600 mm. Assuming small deflections, determine the vertical and horizontal displacement of point *B* if a vertical force *P* of 80 kN is applied at *B*. The cables are steel ($E = 207$ GPa), each with an effective cross section of 120 mm².

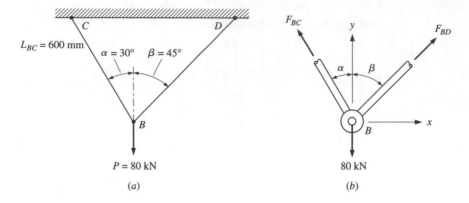

Figure 3.2-7

Solution:

The length of cable BD is found by equating the vertical lengths of each cable. Thus,

$$600 \cos 30 = L_{BD} \cos 45$$

$$L_{BD} = \frac{600 \cos 30}{\cos 45} = 734.85 \text{ mm}$$

For small deflections, the force analysis is performed neglecting the deflections. Isolating point B, Fig. 3.2-7(c), we obtain

$$\sum F_x = F_{BD} \sin \beta - F_{BC} \sin \alpha = 0 \qquad\qquad \textbf{[a]}$$

$$\sum F_y = F_{BC} \cos \alpha + F_{BD} \cos \beta - 80 = 0 \qquad\qquad \textbf{[b]}$$

Solving Eqs. (a) and (b) simultaneously results in

$$F_{BC} = \frac{80 \sin \beta}{\sin (\alpha + \beta)} \qquad F_{BD} = \frac{80 \sin \alpha}{\sin (\alpha + \beta)}$$

Substituting $\alpha = 30°$ and $\beta = 45°$ yields

$$F_{BC} = \frac{80 \sin 45}{\sin (30 + 45)} = 58.56 \text{ kN} \qquad F_{BD} = \frac{80 \sin 30}{\sin (30 + 45)} = 41.41 \text{ kN}$$

The elongation of each cable is given by Eq. (3.2-5). Thus

$$\delta_{BC} = \left(\frac{FL}{AE} \right)_{BC} = \frac{(58.56)(10^3)(600)(10^{-3})}{(120)(10^{-3})^2(207)(10^9)}$$

$$= 1.415(10^{-3}) \text{ m} = 1.415 \text{ mm}$$

$$\delta_{BD} = \left(\frac{FL}{AE}\right)_{BD} = \frac{(41.41)(10^3)(734.85)(10^{-3})}{(120)(10^{-3})^2(207)(10^9)}$$

$$= 1.225(10^{-3}) \text{ m} = 1.225 \text{ mm}$$

Figure 3.2-8 shows the configuration after the load is applied where

$$L_{BC'} = L_{BC} + \delta_{BC} \qquad \qquad \textbf{[c]}$$

$$L_{BD'} = L_{BD} + \delta_{BD} \qquad \qquad \textbf{[d]}$$

The deflection of B in the x and y directions is given by

$$(\delta_B)_x = L'_{BC} \sin \alpha' - 600 \sin 30 \qquad \qquad \textbf{[e]}$$

$$(\delta_B)_y = 600 \cos 30 - L'_{BC} \cos \alpha' \qquad \qquad \textbf{[f]}$$

In order to complete the solution, the angle α' must be determined. The distance CD is constant where from the initial configuration

$$CD = 600 \sin 30 + L_{BD} \sin 45 = 600 \sin 30 + \frac{600 \cos 30}{\cos 45} \sin 45 \qquad \textbf{[g]}$$

$$= 600(\sin 30 + \cos 30)$$

From Fig. 3.2-8,

$$L'_{BD} \cos \beta' = L'_{BC} \cos \alpha' \qquad \qquad \textbf{[h]}$$

$$L'_{BD} \sin \beta' = CD - L'_{BC} \sin \alpha' \qquad \qquad \textbf{[i]}$$

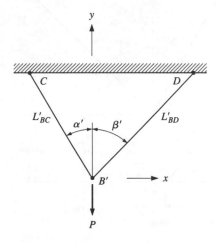

Figure 3.2-8   Final configuration.

Squaring and adding Eqs. (*h*) and (*i*) eliminates β', and solving for α' results in

$$\alpha' = \sin^{-1}\left(\frac{L_{BC}'^2 + (CD)^2 - L_{BD}'^2}{2(CD)L_{BC}'}\right) \qquad \textbf{[}\boldsymbol{j}\textbf{]}$$

To obtain the final solution, great accuracy must be used. L_{BC}' and L_{BD}' are calculated first in Eqs. (*c*) and (*d*) and then substituted into Eq. (*j*) to determine α'. With this, the deflection of *B* is determined from Eqs. (*e*) and (*f*). The result of this is

$$(\delta_B)_x = -0.06281648 = -0.0628 \text{ mm} \qquad \textbf{[}\boldsymbol{k}\textbf{]}$$

$$(\delta_B)_y = -1.66969526 = -1.670 \text{ mm} \qquad \textbf{[}\boldsymbol{l}\textbf{]}$$

The above results were obtained by solving the nonlinear equations (*h*) and (*i*), which is not always as simple as in this case. A linear procedure can be established by returning to where δ_{BC} and δ_{BD} were initially determined. Imagine the pin at point *B* removed and the cables *BC* and *BD* allowed to freely elongate δ_{BC} and δ_{BD}, respectively (see the dotted lines in Fig 3.2-9). Next we rotate the cables to reconnect them. Since the deflections are small, we will *approximate* the rotation by

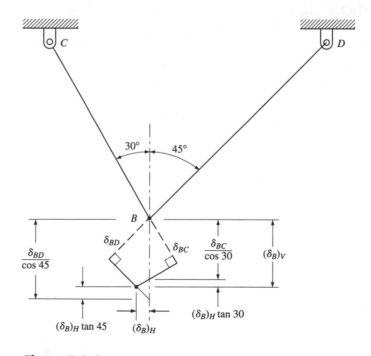

Figure 3.2-9

drawing perpendiculars to each cable. This is where the deflection equations are linearized. From Fig. 3.2-9,

$$(\delta_B)_V - (\delta_B)_H \tan 30 = \frac{\delta_{BC}}{\cos 30} \qquad\qquad [m]$$

$$(\delta_B)_V + (\delta_B)_H \tan 45 = \frac{\delta_{BD}}{\cos 45} \qquad\qquad [n]$$

Substituting in δ_{BC} and δ_{BD} and solving simultaneously yields

$$(\delta_B)_H = 0.06281679 = 0.0628 \text{ mm} \qquad\qquad [o]$$

$$(\delta_B)_V = 1.66969620 = 1.670 \text{ mm} \qquad\qquad [p]$$

Due to the definitions of $(\delta_B)_H$ and $(\delta_B)_V$ the deflection of B is in the negative x and y directions. The differences in the results given in Eqs. (o) and (p) and Eqs. (k) and (l), respectively, are negligible.

This problem will be revisited in Chap. 6, where energy techniques are presented. It will be seen that linear energy techniques give results which are identical to Eqs. (o) and (p). For large deflections, the method in obtaining Eqs. (o) and (p) will not be accurate. The geometric method utilizing Fig. 3.2-8 is acceptable; however, linearization in the previous example also came from the *decoupling* of forces and deflections. That is, it was assumed that the equilibrium state of the cables was in the *undeformed* configuration. Coupling the forces and deflections yields nonlinear equations, and nonlinear analysis is much more complicated to perform. Furthermore, materials that can undergo large deflections normally exhibit a nonlinear stress-strain relationship where E is not constant. This will further complicate the solution procedure. Practical solutions are normally obtained numerically using techniques such as the Newton-Raphson technique[†] or the various types of software packages available that range from solving nonlinear algebraic equations to simulation programs to nonlinear finite element packages. A basic approach in finite elements is to break up the loading into small incremental steps where each step can be solved in a linear fashion, update the geometry, and increment to the next step.

Example 3.2-4

For the previous example let the cables be of a very flexible material with $E = 2.07$ GPa. Solve the problem by iterating the decoupled solution. That is, reevaluate the cable forces after the deflections are determined. Repeat the deflection analysis with the new cable forces. Iterate until reasonable convergence is obtained. Compare the results of the linear analysis with that of the nonlinear solution.

[†] See, for example, Ref. 3.2.

Solution:

From the previous example we have

$$F_{BC} = \frac{80 \sin\beta}{\sin(\alpha + \beta)}$$ [a]

$$F_{BD} = \frac{80 \sin\alpha}{\sin(\alpha + \beta)}$$ [b]

For the first iteration let $\alpha = 30°$ and $\beta = 45°$, resulting in $F_{BC} = 58.56$ kN and $F_{BD} = 41.41$ kN. The corresponding elongation of each cable is

$$\delta_{BC} = \left(\frac{FL}{AE}\right)_{BC} = \frac{58.56(10^3)(0.600)}{120(10^{-3})^2(2.07)(10^9)} = 0.1415 \text{ m} = 141.5 \text{ mm}$$

$$\delta_{BD} = \left(\frac{FL}{AE}\right)_{BD} = \frac{41.41(10^3)(0.73485)}{120(10^{-3})^2(2.07)(10^9)} = 0.1225 \text{ m} = 122.5 \text{ mm}$$

For the final position of B, because of the large deflections, it will be necessary to use the exact Eqs. (h) and (i) from Example 3.2-3 to determine α' and β'. The angle α' was already determined in Example 3.2-3 given by Eq. (j). The angle β' can be determined in a similar manner. The results are

$$\alpha' = \sin^{-1}\left(\frac{L'^2_{BC} + (CD)^2 - L'^2_{BD}}{2(CD)L'_{BC}}\right)$$ [c]

$$\beta' = \sin^{-1}\left(\frac{L'^2_{BD} + (CD)^2 - L'^2_{BC}}{2(CD)L'_{BD}}\right)$$ [d]

For the first iteration, $L'_{BC} = 600 + 141.5 = 741.5$ mm, and $L'_{BD} = 734.9 + 122.5 = 857.4$ mm. With CD given by Eq. (g) of Example 3.2-3, Eqs. (c) and (d) give $\alpha' = 23.59°$ and $\beta' = 37.58°$. These values are then set back into Eqs. (a) and (b) and we find $F_{BC} = 55.69$ kN and $F_{BD} = 36.55$ kN. The process is repeated until the solution converges. This is easily solved using a spreadsheet program. For this problem, the solution converged within engineering accuracy in nine iterations. The final values are $\alpha' = 24.3°$, $\beta' = 37.7°$, $F_{BC} = 55.4$ kN, and $F_{BD} = 37.3$ kN. From Eqs. (e) and (f) of Example 3.2-3, the components of the deflection of point B are $(\delta_B)_x = 2.46$ mm and $(\delta_B)_y = -149.0$ mm. A linear uncoupled analysis would have yielded $\alpha = 30°$, $\beta = 45°$, $F_{BC} = 58.56$ kN, $F_{BD} = 41.41$ kN, $(\delta_B)_x = -6.28$ mm, and $(\delta_B)_y = -167.0$ mm, which are considerably in error.

Solving the coupled equations is much more involved, as is shown in the next example.

Example 3.2-5 | Solve the coupled nonlinear force-deflection equations of the previous example using the Newton-Raphson method.

Solution:

For notational purposes, let α and β be the final angles that cables BC and BD make with the vertical, respectively.

From Example 3.2-3 the force equation Eqs. (a) and (b) together with the geometric equations (h) and (i) are

$$f_1 = F_{BC}\ \sin\ \alpha - F_{BD}\ \sin\ \beta = 0 \qquad\qquad \text{[a]}$$

$$f_2 = F_{BC}\ \cos\ \alpha + F_{BD}\ \cos\ \beta - 80 = 0 \qquad\qquad \text{[b]}$$

$$f_3 = \left(L_{BC} + \frac{F_{BC}L_{BC}}{AE}\right)\cos\ \alpha - \left(L_{BD} + \frac{F_{BD}L_{BD}}{AE}\right)\cos\ \beta = 0 \qquad \text{[c]}$$

$$f_4 = \left(L_{BC} + \frac{F_{BC}L_{BC}}{AE}\right)\sin\ \alpha + \left(L_{BD} + \frac{F_{BD}L_{BD}}{AE}\right)\sin\ \beta - CD = 0 \quad \text{[d]}$$

where L'_{BC} and L'_{BD} were written in terms of the initial lengths plus the elastic deflections. Equations (a) to (d) represent four simultaneous nonlinear equations in the four unknowns α, β, F_{BC}, and F_{BD}. The Newton-Raphson method requires an initial estimate for the unknowns. For this we can use the linear solution with $\alpha = 30°$, $\beta = 45°$, $F_{BC}= 58.56$ kN, and $F_{BD} = 41.41$ kN. The method yields the change in each variable, which provides an improvement to the estimate. The changes are given by the following determinants

$$\Delta\alpha = \frac{1}{|J|}\begin{vmatrix} -f_1 & \partial f_1/\partial\beta & \partial f_1/\partial F_{BC} & \partial f_1/\partial F_{BD} \\ -f_2 & \partial f_2/\partial\beta & \partial f_2/\partial F_{BC} & \partial f_2/\partial F_{BD} \\ -f_3 & \partial f_3/\partial\beta & \partial f_3/\partial F_{BC} & \partial f_3/\partial F_{BD} \\ -f_4 & \partial f_4/\partial\beta & \partial f_4/\partial F_{BC} & \partial f_4/\partial F_{BD} \end{vmatrix} \qquad \text{[e]}$$

$$\Delta\beta = \frac{1}{|J|}\begin{vmatrix} \partial f_1/\partial\alpha & -f_1 & \partial f_1/\partial F_{BC} & \partial f_1/\partial F_{BD} \\ \partial f_2/\partial\alpha & -f_2 & \partial f_2/\partial F_{BC} & \partial f_2/\partial F_{BD} \\ \partial f_3/\partial\alpha & -f_3 & \partial f_3/\partial F_{BC} & \partial f_3/\partial F_{BD} \\ \partial f_4/\partial\alpha & -f_4 & \partial f_4/\partial F_{BC} & \partial f_4/\partial F_{BD} \end{vmatrix} \qquad \text{[f]}$$

$$\Delta F_{BC} = \frac{1}{|J|}\begin{vmatrix} \partial f_1/\partial\alpha & \partial f_1/\partial\beta & -f_1 & \partial f_1/\partial F_{BD} \\ \partial f_2/\partial\alpha & \partial f_2/\partial\beta & -f_2 & \partial f_2/\partial F_{BD} \\ \partial f_3/\partial\alpha & \partial f_3/\partial\beta & -f_3 & \partial f_3/\partial F_{BD} \\ \partial f_4/\partial\alpha & \partial f_4/\partial\beta & -f_4 & \partial f_4/\partial F_{BD} \end{vmatrix} \qquad \text{[g]}$$

$$\Delta F_{BD} = \frac{1}{|J|}\begin{vmatrix} \partial f_1/\partial\alpha & \partial f_1/\partial\beta & \partial f_1/\partial F_{BC} & -f_1 \\ \partial f_2/\partial\alpha & \partial f_2/\partial\beta & \partial f_2/\partial F_{BC} & -f_2 \\ \partial f_3/\partial\alpha & \partial f_3/\partial\beta & \partial f_3/\partial F_{BC} & -f_3 \\ \partial f_4/\partial\alpha & \partial f_4/\partial\beta & \partial f_4/\partial F_{BC} & -f_4 \end{vmatrix} \qquad \text{[h]}$$

where $|J|$, called the determinant of the *Jacobian matrix,* is given by

$$|J| = \begin{vmatrix} \partial f_1/\partial\alpha & \partial f_1/\partial\beta & \partial f_1/\partial F_{BC} & \partial f_1/\partial F_{BD} \\ \partial f_2/\partial\alpha & \partial f_2/\partial\beta & \partial f_2/\partial F_{BC} & \partial f_2/\partial F_{BD} \\ \partial f_3/\partial\alpha & \partial f_3/\partial\beta & \partial f_3/\partial F_{BC} & \partial f_3/\partial F_{BD} \\ \partial f_4/\partial\alpha & \partial f_4/\partial\beta & \partial f_4/\partial F_{BC} & \partial f_4/\partial F_{BD} \end{vmatrix} \qquad \textbf{[i]}$$

The partial derivatives of Eqs. (*a*) to (*d*) are straightforward and are

$$\partial f_1/\partial\alpha = F_{BC}\cos\alpha \qquad \textbf{[j]} \qquad\qquad \partial f_1/\partial F_{BC} = \sin\alpha \qquad \textbf{[r]}$$

$$\partial f_2/\partial\alpha = -F_{BC}\sin\alpha \qquad \textbf{[k]} \qquad\qquad \partial f_2/\partial F_{BC} = \cos\alpha \qquad \textbf{[s]}$$

$$\partial f_3/\partial\alpha = -\left(L_{BC} + \frac{F_{BC}L_{BC}}{AE}\right)\sin\alpha \quad \textbf{[l]} \qquad \partial f_3/\partial F_{BC} = \frac{L_{BC}}{AE}\cos\alpha \qquad \textbf{[t]}$$

$$\partial f_4/\partial\alpha = \left(L_{BC} + \frac{F_{BC}L_{BC}}{AE}\right)\cos\alpha \quad \textbf{[m]} \qquad \partial f_4/\partial F_{BC} = \frac{L_{BC}}{AE}\sin\alpha \qquad \textbf{[u]}$$

$$\partial f_1/\partial\beta = -F_{BD}\cos\beta \qquad \textbf{[n]} \qquad\qquad \partial f_1/\partial F_{BD} = -\sin\beta \qquad \textbf{[v]}$$

$$\partial f_2/\partial\beta = -F_{BD}\sin\beta \qquad \textbf{[o]} \qquad\qquad \partial f_2/\partial F_{BD} = \cos\beta \qquad \textbf{[w]}$$

$$\partial f_3/\partial\beta = \left(L_{BD} + \frac{F_{BD}L_{BD}}{AE}\right)\sin\beta \quad \textbf{[p]} \qquad \partial f_3/\partial F_{BD} = -\frac{L_{BD}}{AE}\cos\beta \qquad \textbf{[x]}$$

$$\partial f_4/\partial\beta = \left(L_{BD} + \frac{F_{BD}L_{BD}}{AE}\right)\cos\beta \quad \textbf{[q]} \qquad \partial f_4/\partial F_{BD} = \frac{L_{BD}}{AE}\sin\beta \qquad \textbf{[y]}$$

As formidable as the problem may look, the solution is easily implemented on a spreadsheet utilizing Eqs. (*a*) to (*y*). As stated earlier, for the initial values let $\alpha = 30°$, $\beta = 45°$, $F_{BC} = 58.56$ kN, and $F_{BD} = 41.41$ kN. Once determined, the Δ values are added to the initial values and a new set of Δ values is calculated. The process is repeated until an acceptable convergence is obtained. In this example, the second set of values is found to be $\alpha = 24.17109°$, $\beta = 37.48898°$, $F_{BC} = 55.34466$ kN, and $F_{BD} = 37.26666$ kN, giving the deflection components of point B of $(\delta_B)_x = 3.5959$ mm and $(\delta_B)_y = -156.829$ mm. The solution oscillates, converging very slowly. The convergence can be accelerated by averaging the new and old values for the trial values. Doing this, the solution converges within four-place accuracy in 12 iterations to $\alpha = 24.34°$, $\beta = 37.72°$, $F_{BC} = 55.40$ kN, $F_{BD} = 37.32$ kN, with $(\delta_B)_x = 2.461$ mm and $(\delta_B)_y = -149.0$ mm. This agrees exactly with the previous solution.

3.3 TORSION OF CIRCULAR SHAFTS

3.3.1 TORSIONAL STRESSES

Figure 3.3-1(*a*) illustrates a circular shaft subjected to a torsional twisting moment T_z. Isolating a surface perpendicular to the z axis at section *c-c*, for equilibrium the resulting internal force distribution on the surface must be tangential to the surface. The shear stress distribution *flows*[†] around the surface such that each shear vector has an equal and opposing shear vector on the opposite end of a diametral line and forming a *couple* as shown in Fig. 3.3-1(*c*). Thus the net force from the shear on the surface is zero, and the net moment is equal to the applied torsional moment T_z. As the shaft twists, points within the shaft strain and deflect directly proportional to the

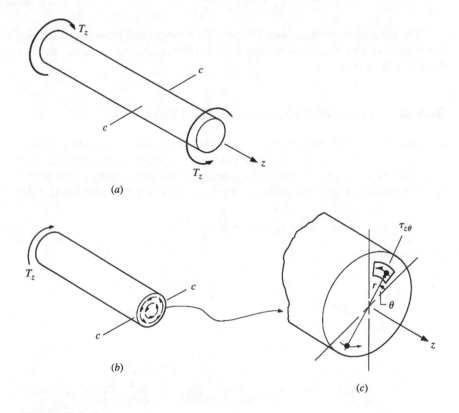

(*a*)

(*b*)

(*c*)

Figure 3.3-1

[†] The concept of stress flow comes up often in stress analysis. The governing equations from the theory of elasticity are quite analogous to those found in fluid mechanics.

distance r from the shaft center. Thus the shear stress is proportional to r and is given by

$$\tau_{z\theta} = \frac{T_z r}{J_z} \qquad \text{[3.3-1]}$$

where J_z is the second-area moment of the cross section relative to the longitudinal z axis, or the *polar second-area moment* (popularly but improperly referred to as the *polar moment of inertia*). For a hollow circular cross section of inner radius r_i and outer radius r_o

$$J_z = \frac{\pi}{2} (r_o^4 - r_i^4) \qquad \text{[3.3-2]}$$

The maximum shear stress occurs at the outer radius and is

$$\tau_{\max} = \frac{T_z r_o}{J_z} \qquad \text{[3.3-3]}$$

The same limitations discussed in Sec. 3.1 apply to this formulation. Buckling instability can also occur in long thin-walled tubes in torsion. However, this is not discussed in this text.

3.3.2   TORSIONAL STRAINS AND DEFLECTIONS

Consider the set of dotted lines scribed on a circular cylinder prior to loading shown in Fig. 3.3-2. After the torque is applied to the free end, the circumferential and radial lines simply rotate, whereas the longitudinal lines become helical. The helix angle ϕ, in radians, is the shear strain $\gamma_{z\theta}$ at $r = r_o$. Thus, together with Eq. (1.5-3b)

$$\phi = \gamma_{z\theta}\big|_{r=r_o} = \frac{2(1 + v)}{E} \tau_{z\theta}\big|_{r=r_o}$$

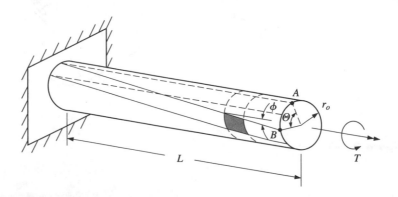

Figure 3.3-2   Twist due to torsion.

Substituting Eq. (3.3-1) gives

$$\phi = \frac{2(1 + v)}{E} \frac{T_z r_o}{J_z} \qquad \textbf{[a]}$$

Arc AB in Fig. 3.3-2 is given by $r_o \Theta$ or $L\phi$. Therefore,

$$\phi = \frac{r_o \Theta}{L}$$

Substituting this into Eq. (a) and solving for Θ gives

$$\Theta = 2(1 + v)\frac{T_z L}{J_z E} \qquad \textbf{[3.3-4]}$$

Example 3.3-1

A step shaft [Fig. 3.3-3(a)] transmits a torque of 5000 lb·in. Determine the maximum shear stress and the relative angle of twist between surfaces located at points A and D ($E = 30 \times 10^6$ psi, $v = 0.3$).

Solution:

For a *solid circular* rod, Eq. (3.3-3) is used together with Eq. (3.3-2), and $r_i = 0$, $r_o = d_o/2$. The resulting equation is

$$\tau_{\max} = \frac{16T}{\pi d_o^3} \qquad \textbf{[3.3-5]}$$

Thus it can be seen that the maximum shear stress will occur in section AB, where d_o is minimum. Therefore,

$$\tau_{\max} = \frac{(16)(5000)}{\pi(1)^3} = 25,465 \text{ psi} \qquad \textbf{[a]}$$

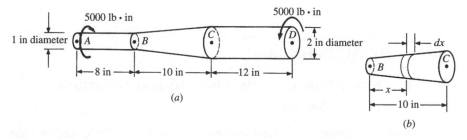

(a)

(b)

Figure 3.3-3

To obtain the total angle of twist, Eq. (3.3-4) can be used for sections AB and CD. The angle of twist of B relative to A is

$$\Theta_{B/A} = \frac{2(1 + 0.3)(5000)(8)}{(30 \times 10^6)(\pi/2)(0.5)^4} = 0.0353 \text{ rad} \qquad [b]$$

The angle of twist of D relative to C is

$$\Theta_{D/C} = \frac{(2)(1 + 0.3)(5000)(12)}{(30 \times 10^6)(\pi/2)(1)^4} = 0.0033 \text{ rad} \qquad [c]$$

Since J varies in section BC, Eq. (3.3-4) must be modified and applied to a differential element as shown in Fig. 3.3-3(b). The radius in section BC varies with respect to x as

$$r_o = 0.05(10 + x) \qquad [d]$$

Thus

$$J_z = \pi r_o^4/2 = 9.83(10^{-6})(10 + x)^4 \qquad [e]$$

Substituting this into Eq. (3.3-4) with $L = dx$, the angle of twist of the dx element is

$$d\Theta = \frac{2(1 + 0.3)(5000)dx}{30(10^6)(9.83)(10^{-6})(10 + x)^4} = \frac{44.1}{(10 + x)^4} dx \qquad [f]$$

The total angle of twist, found by integrating Eq. (f) for section BC, is

$$\Theta_{C/B} = \int_0^{10} \frac{44.1}{(10 + x)^4} dx = -\frac{1}{3} \frac{44.1}{(10 + x)^3} \Big|_0^{10}$$

$$= -14.7\left(\frac{1}{20^3} - \frac{1}{10^3}\right) = 0.0129 \text{ rad} \qquad [g]$$

Adding Eqs. (b), (c), and (g) results in the total angle of twist of the shaft:

$$\Theta_{D/A} = \Theta_{B/A} + \Theta_{C/B} + \Theta_{D/C}$$

$$= 0.0353 + 0.0129 + 0.0033 = 0.0515 \text{ rad}$$

$$= (0.0515)(57.3) = 3.0°$$

3.4   BEAMS IN BENDING

3.4.1   SHEAR FORCE AND BENDING MOMENT EQUATIONS AND DIAGRAMS

Beams loaded as shown in Fig. 3.4-1 transmit transverse shear forces and bending moments along the length of the beam. The corresponding stress, strain, and deflection formulations are based on a sign convention which is defined for the transverse

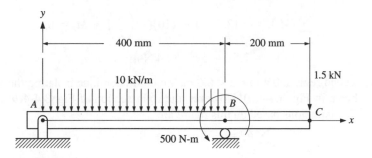

Figure 3.4-1

shear force and bending moment acting on an isolated internal transverse planar surface. Unfortunately, in the literature, there is no consistently defined convention. The most popular convention works well for two-dimensional problems but is not well suited to three-dimensional problems. For this reason it will not be adopted here. The sign convention used in this text is consistent with vector mechanics. If the outward normal of a surface is in positive coordinate direction, then, by convention, all positive forces and moments acting on the surface are in the positive coordinate directions. Conversely, if the outward normal of a surface is in a negative coordinate direction, then, by convention, all positive forces and moments acting on the surface are in the negative coordinate directions.[†] For transverse loading in the xy plane the convention for shear forces and bending moments is as shown in Fig. 3.4-2(a). The two-dimensional representation of this is shown in Fig. 3.4-2(b).

Determine the shear force and bending moment equations for Fig. 3.4-1 and plot the results.   **Example 3.4-1**

Solution:

The reader should verify that the reaction forces at A and B are 2.5 and 3 kN in the y direction, respectively. For Fig. 3.4-1 as x increases from the origin, there is a change in the loading at point B. Thus there are two regions to write the shear force and bending moment equations. The first region, $0 < x < 0.4$ m, is shown in Fig. 3.4-3. Summing forces in the y direction and moments about an axis at x and parallel to the z axis gives

$$\sum F_y = 2.5 - (10)(x) + V_y = 0$$

$$V_y = 10x - 2.5 \text{ kN} \qquad\qquad \textbf{[a]}$$

[†] This convention was actually utilized earlier when defining the stresses on a rectangular parallelepiped element as shown in Fig. 1.3-9.

$$\sum (M_x)_z = -(2.5)(x) + (10)(x)\left(\frac{x}{2}\right) + M_z = 0$$

$$M_z = 2.5x - 5x^2 \qquad \text{kN·m} \tag{b}$$

For the second region, 0.4 m < x < 0.6 m, we could continue isolating the left section as shown in Fig. 3.4-4(a). However, the right section of Fig. 3.4-4(b) is much simpler. The equations based on this figure are

$$\sum F_y = -V_y - 1.5 = 0$$

$$V_y = -1.5 \text{ kN} \tag{c}$$

$$\sum (M_x)_z = -M_z - (1.5)(0.6 - x) = 0$$

$$M_z = 1.5x - 0.9 \qquad \text{kN·m} \tag{d}$$

Equations (a) and (c) are plotted in Fig. 3.4-5(a), whereas Eqs. (b) and (d) are plotted in Fig. 3.4-5(b).

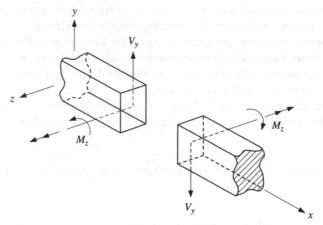

(a) Three-dimensional representation

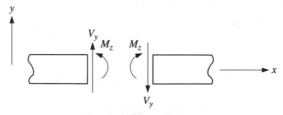

(b) Two-dimensional view

Figure 3.4-2 Convention for positive shear force and bending moment for bending in the xy plane.

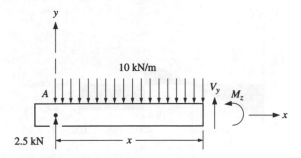

Figure 3.4-3 Isolation at 0 < x < 0.4 m.

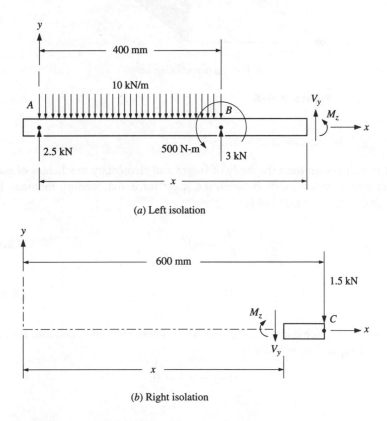

(a) Left isolation

(b) Right isolation

Figure 3.4-4 Isolation at 0.4 m < x < 0.6 m.

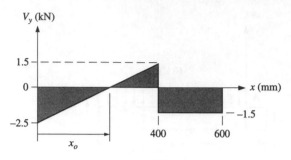

(a) Shear force diagram

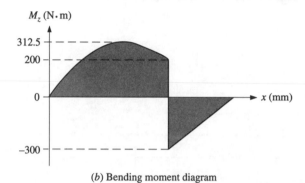

(b) Bending moment diagram

Figure 3.4-5

It is well known from the study of statics and elementary mechanics of materials that a relationship exists between the shear force and bending moment. From Appendix D, Eqs. (D.1-3) and (D.1-4) are

$$\frac{dM_z}{dx} = -V_y \qquad\qquad \textbf{[3.4-1]}$$

$$M_z = -\int V_y \, dx \qquad\qquad \textbf{[3.4-2]}^\dagger$$

In Eq. (3.4-1) we see that when $V_y = 0$, M_z is a maximum or minimum. From Fig. 3.4-5(a) or Eq. (a), $V_y = 0$ when

$$10x_0 - 2.5 = 0$$

---

† NOTE: Based on Eq. (3.4-2), if one were to integrate Fig. 3.4-5(a) directly, the resulting bending moment diagram would be incorrect. The concentrated bending moment would not be accounted for. The 500 N·m counterclockwise moment at B causes the discontinuity step of $-200 - 300 = -500$ N·m as shown.

or $x_o = 0.25$ m $= 250$ mm. From Eq. (3.4-2) the moment at x_o is simply the negative of the area of the V_y curve from $x = 0$ to x_o. This gives

$$M_z(x_o) = -\left(\frac{1}{2}\right)(-2.5)(10^3)(0.25) = 312.5 \text{ N·m}$$

Whenever a change in loading occurs, a new free-body diagram must be drawn and the corresponding shear force and bending moment equations written. This is very tedious if a beam has many loading changes. *Singularity functions*, discussed in Appendix D, provide a much simpler method of modeling highly discontinuous loading. Singularity functions eliminate the need to write separate equations for V_y and M_z. Applying singularity functions to Example 3.4-1, the load intensity is first written as

$$q(x) = 2.5\langle x\rangle^{-1} - 10\langle x\rangle^0 + 10\langle x - 0.4\rangle^0$$

$$+ 3\langle x - 0.4\rangle^{-1} - 0.5\langle x - 0.4\rangle^{-2} - 1.5\langle x - 0.6\rangle^{-1}$$

From Appendix D, $V_y(x) = -\int q(x)\, dx$. With this and Eq. (3.4-2)[†]

$$V_y = -2.5\langle x\rangle^0 + 10\langle x\rangle^1 - 10\langle x - 0.4\rangle^1$$

$$- 3\langle x - 0.4\rangle^0 + 0.5\langle x - 0.4\rangle^{-1} + 1.5\langle x - 0.6\rangle^0$$

$$M_z = 2.5\langle x\rangle^1 - 5\langle x\rangle^2 + 5\langle x - 0.4\rangle^2$$

$$+ 3\langle x - 0.4\rangle^1 - 0.5\langle x - 0.4\rangle^0 - 1.5\langle x - 0.6\rangle^1$$

The $\langle\rangle$ brackets serve a special function. When the argument within the brackets is negative, the function does not exist. When the argument is positive, the function exists. The first two functions in V_y and M_z of this example always exist, so the $\langle\rangle$ brackets are unnecessary. So the equations can be written as

$$V_y = -2.5 + 10x - 10\langle x - 0.4\rangle^1 - 3\langle x - 0.4\rangle^0$$

$$+ 0.5\langle x - 0.4\rangle^{-1} + 1.5\langle x - 0.6\rangle^0 \qquad\qquad \textbf{[e]}$$

$$M_z = 2.5x - 5x^2 + 5\langle x - 0.4\rangle^2 + 3\langle x - 0.4\rangle^1$$

$$- 0.5\langle x - 0.4\rangle^0 - 1.5\langle x - 0.6\rangle^1 \qquad\qquad \textbf{[f]}$$

The equations are interpreted as follows. For $0 < x < 0.4$ m, the singularity functions do not exist and

$$V_y = 10x - 2.5 \text{ kN}$$

$$M_z = 2.5x - 5x^2 \text{ kN·m}$$

[†] Constants of integration are unnecessary for the first two integrations of $q(x)$. See Appendix D.

which agrees with Eqs. (*a*) and (*b*) of Example 3.4-1. When 0.4 m < *x* < 0.6 m,

$$V_y = -2.5 + 10x - 10\langle x - 0.4 \rangle^1 - 3\langle x - 0.4 \rangle^0 + 0.5\langle x - 0.4 \rangle^{-1}$$

The impulse has no physical significance in the V_y equation until it is integrated for the M_z equation. Thus

$$V_y = -2.5 + 10x - 10\langle x - 0.4 \rangle^1 - 3\langle x - 0.4 \rangle^0$$

$$= -2.5 + 10x - 10(x - 0.4) - 3(1) = -1.5 \text{ kN}$$

and

$$M_z = 2.5x - 5x^2 + 5\langle x - 0.4 \rangle^2 + 3\langle x - 0.4 \rangle^1 - 0.5\langle x - 0.4 \rangle^0$$

$$= 2.5x - 5x^2 + 5(x - 0.4)^2 + 3(x - 0.4) - 0.5(1) = 1.5x - 0.9 \text{ kN·m}$$

Thus the shear force and bending moment agree with Eqs. (*c*) and (*d*) of Example 3.4-1.

One final note concerning discontinuities in the shear force and bending moment equations. Concentrated moments give discontinuities in value in the bending moment equation. Concentrated forces cause discontinuities in value and slope in the shear force and bending moment equations, respectively. Segmented uniform loads cause discontinuities in slope in the shear force equations where they begin and end. Since, in general, these ideal loads do not exist, the corresponding discontinuities will not either. Figure 3.4-6(*a*) shows an idealistic load state and the corresponding shear force and bending moment diagrams. Figure 3.4-6(*b*) shows a more realistic model of the loading where the loads are distributed over small but finite span lengths.

3.4.2   B ENDING S TRESSES

The normal stress caused by a pure positive bending moment M_z acting over a transverse surface is given by the *flexure formula*

$$\sigma_x = -\frac{M_z y}{I_z} \qquad \qquad \textbf{[3.4-3]}$$

where y is the lateral position from the x axis, which is the centroidal axis of the cross section and $\int y \, dA = 0$, and $I_z = \int y^2 \, dA$ is the second-area moment of the cross section relative to the bending z axis. The bending stress is represented in Fig. 3.4-7. Again, the derivation of this equation is based on a displacement assumption. Here it is assumed that plane surfaces parallel to the yz plane remain plane after bending. The result is that normal strains and subsequently normal stresses in the x direction are linearly dependent on y. Since, for pure bending, there is no net force

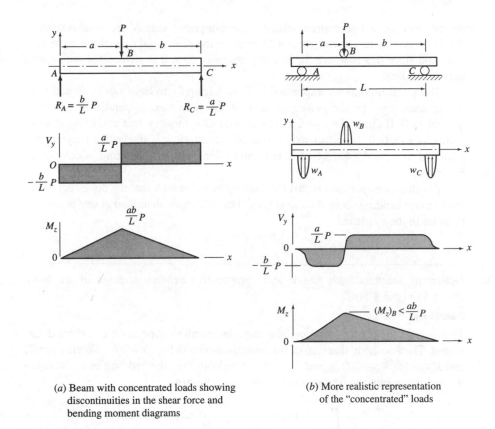

(a) Beam with concentrated loads showing
discontinuities in the shear force and
bending moment diagrams

(b) More realistic representation
of the "concentrated" loads

Figure 3.4-6

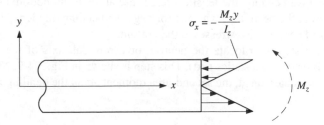

Figure 3.4-7

on the section, a linear stress distribution integrated across the area results in $\int y \, dA = 0$. This means that the x axis must be the *centroidal axis* of the cross section, and since this is where the bending stress is zero, it is also called the *neutral axis* in bending.

The usual limitations, expressed in Sec. 3.1, apply to Eq. (3.4-3). In addition, the yz axes *must* be the *principal axes* of the area (see Appendix E). That is, $\int yz \, dA = 0$. If either the y or z axis is an axis of symmetry, this condition is automatically satisfied. If neither axis is one of symmetry, it is likely that they are not principal axes, and Eq. (3.4-3) is not valid. Nonsymmetric bending is covered in Sec. 5.3.

Another consideration is that the loading must be such that the net effect of the load causes bending about the z axis only. Thus the equivalent load at any position x must lie in the xy plane.[†]

Example 3.4-2 | Determine the maximum tensile and compressive bending stresses in the beam shown in Fig. 3.4-8(*a*).

Solution:

To construct the bending moment diagram, the reactions at points A and B are determined. The free-body diagram of the beam is shown in Fig. 3.4-8(*b*). Solving for R_A and R_B yields $R_A = 600$ lb and $R_B = 3400$ lb. With this, the bending moment equations and diagram can be developed. The equations are

$$M_z = \begin{cases} 600x & 0 < x < 50 \text{ in} \\ -200(7x - 500) & 50 < x < 100 \text{ in} \\ 2000(x - 120) & 100 < x < 120 \text{ in} \end{cases}$$

Using singularity functions, the equation is

$$M_z = 600x - 2000\langle x - 50 \rangle^1 + 3400\langle x - 100 \rangle^1 - 2000\langle x - 120 \rangle^1$$

The bending moment diagram is shown in Fig. 3.4-8(*b*).

For positive bending, the tensile stresses are at the beam bottom and compressive stresses at the beam top; whereas for negative bending, the tensile stresses are at the top and compressive stresses at the bottom.

The next step is to locate the neutral, or centroidal, axis of the section (see Appendix B or review a statics text). This step is shown in Fig. 3.4-9. Now, with the centroidal axis determined, the second-area moment about the bending z axis can be

---

[†] In the case of thin-walled sections where the area is unsymmetric with the xy plane, loading in this plane will cause additional torsional shear stresses. To avoid this, the load is placed at a location called the *shear center*, which is discussed in Chap. 5.

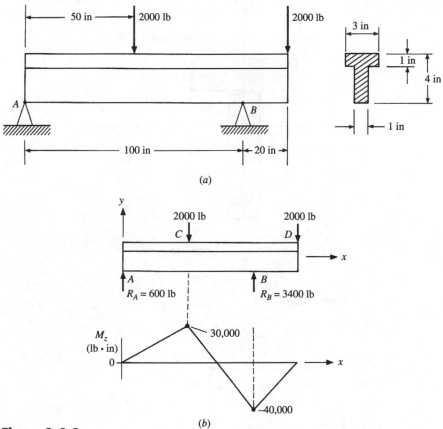

Figure 3.4-8

found as shown in Fig. 3.4-10. For the total section, using the parallel-axis theorem, we obtain[†]

$$I_z = 0.25 + (3)(1.0)^2 + 2.25 + (3)(1.0)^2 = 8.5 \text{ in}^4$$

[†] For rectangular sections, the second-area moment can be determined without resorting to the parallel-axis theorem. A somewhat simpler method for rectangular sections employs the second-area moment about the base of a rectangle where $I_z = 1/3\ bh^3$. This figure illustrates how the section can be arrived at using rectangles with a common base along the z axis. The result is

$$I_z = \frac{1}{3}(1)(2.5)^3 + \frac{1}{3}(3)(1.5)^3 - (2)\left(\frac{1}{3}\right)(1)(0.5)^3 = 8.5 \text{ in}^4$$

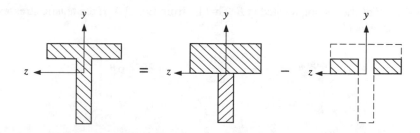

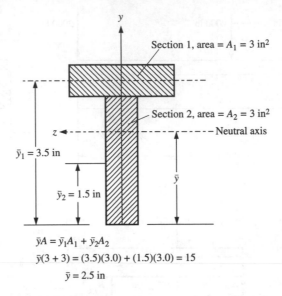

$$\bar{y}A = \bar{y}_1 A_1 + \bar{y}_2 A_2$$
$$\bar{y}(3 + 3) = (3.5)(3.0) + (1.5)(3.0) = 15$$
$$\bar{y} = 2.5 \text{ in}$$

Figure 3.4-9

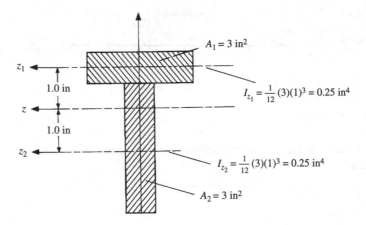

Figure 3.4-10

The maximum tensile stress occurs either at the bottom of the section located at C or the top of the section located at B. That is, from Eq. (3.4-3) the tensile stresses at C and B respectively, are:

$$C: \qquad (\sigma_x)_{y=-2.5 \text{ in}}^{x=50 \text{ in}} = -\frac{(30,000)(-2.5)}{8.5} = 8824 \text{ psi}$$

B: $$(\sigma_x)^{x=100 \text{ in}}_{y=1.5 \text{ in}} = -\frac{(-40,000)(1.5)}{8.5} = 7059 \text{ psi}$$

Obviously, the maximum tensile stress occurs at the bottom of the beam at C. The maximum compression should then be checked at points C and B:

C: $$(\sigma_x)^{x=50 \text{ in}}_{y=1.5 \text{ in}} = -\frac{(30,000)(1.5)}{8.5} = -5294 \text{ psi}$$

B: $$(\sigma_x)^{x=100 \text{ in}}_{y=-2.5 \text{ in}} = -\frac{(-40,000)(-2.5)}{8.5} = -11,765 \text{ psi}$$

Thus the maximum compressive stress will occur at the bottom of the beam at B.

A common expression for the maximum bending stress is

$$\sigma_{max} = \frac{Mc}{I} \qquad \textbf{[3.4-4]}$$

where $M = M_z$, $I = I_z$, and $c = |y_{max}|$. The term I/c is called the *section modulus*, S, with units of length^3. Thus Eq. (3.4-4) can be expressed as

$$\sigma_{max} = \frac{M}{S} \qquad \textbf{[3.4-5]}$$

3.4.3 TRANSVERSE SHEAR STRESSES

The shear force V_y produces transverse shear stresses in beams. Two classes of problems are normally studied in elementary mechanics of materials, solid beams and compound beams constructed with fasteners and/or welds.

Shear Stresses in Solid Beams   The fundamental formulation for the shear stress is based on equilibrium and the relationship between the shear force and the bending moment. This is different from the axial, torsional, and bending stress equations, which are initially based on a physical observation of the deformation. The derivation of the shear stress formulation is included in this section since it provides some additional insight which will aid in its application to the elementary problems here as well as in the advanced problems covered in Chap. 5.

Consider the isolated element of a beam shown in Fig. 3.4-11, which is of infinitesimal length in the x direction (Δx) and of finite length in the y direction. The top of the element is the top of the beam at $y = y_{max}$, and the bottom of the element is at

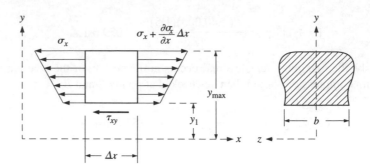

Figure 3.4-11

an isolation break located at $y = y_1$. The right view in the figure shows the beam cross section from y_1 to y_{max}. For clarity, only force distributions in the x direction are shown on the isolated element. The shear stress shown, $\tau_{xy} = \tau_{yx}$, is uniform in the x direction as it is acting over the infinitesimal distance Δx. In the z direction τ_{xy} will be assumed to be uniform; hence τ_{xy} is the average shear stress along width b.

The change in σ_x can be expressed using the bending stress formulation [Eq. (3.4-3)] as

$$\frac{\partial \sigma_x}{\partial x} = \frac{\partial}{\partial x}\left(-\frac{M_z y}{I_z}\right) = -\frac{y}{I_z}\frac{\partial M_z}{\partial x}$$

where it is assumed here that I_z does not vary with respect to x. Since from Eq. (3.4-1), $\partial M_z/\partial x = -V_y$, then

$$\frac{\partial \sigma_x}{\partial x} = \frac{V_y y}{I_z} \qquad \textbf{[a]}$$

Equilibrium of forces in the x direction for the element of Fig. 3.4-11 gives

$$\sum F_x = \int_{y=y_1}^{y=y_{max}} \left(\frac{\partial \sigma_x}{\partial x}\Delta x\right) dA - \tau_{xy} b\, \Delta x = 0$$

Substitution of Eq. (*a*) and recognizing that Δx is independent of the integral, it cancels, resulting in

$$\int_{y=y_1}^{y=y_{max}} \frac{V_y y}{I_z} dA - \tau_x y b = 0$$

The terms V_y and I_z can be factored out of the integral since they can only be functions of x. Solving for τ_{xy} results in

$$\tau_{xy} = \frac{V_y}{I_z b}\int_{y=y_1}^{y=y_{max}} y\, dA$$

The integral is defined as

$$Q = \int_{y=y_1}^{y=y_{max}} y \, dA \qquad \text{[3.4-6]}$$

where Q is the *first-area moment* about the z axis of the cross section isolated between $y_1 < y < y_{max}$.[†] Thus the shear stress can be written as

$$\tau_{xy} = \frac{V_y Q}{I_z b} \qquad \text{[3.4-7]}$$

For long beams undergoing transverse loading, this shear stress is normally quite small when compared to the bending stresses. However, for short beams, beams with narrow widths at the neutral axis, or certain thin-walled beams which are studied in Sec. 5.4, this stress can be significant. Application of Eq. (3.4-7) can be tricky if the analyst is not careful in its interpretation. For this reason, the reader is urged to review the derivation until it is fully understood. Several important aspects of the equation, its derivation, and some observations will be cited here.

1. In Eq. (3.4-7), V_y and I_z are the *total* shear force and second-area moment of the *entire* cross section at position x, respectively.

2. In Eq. (3.4-7), b is the width at the isolation made at $y = y_1$ where τ_{xy} is being evaluated, and Q is the first-area moment of the *completely isolated* portion of the cross section from y_1 to y_{max}.

3. As y_1 increases to y_{max}, the integral in Eq. (3.4-6) vanishes and Q decreases to zero. Thus τ_{xy} goes to zero at the top of the beam where $y_1 = y_{max}$. This agrees with the fact that the top surface is a free surface with no shear forces applied. Likewise, the shear stress is zero at the bottom surface, where integrating from y_{min} to y_{max} Q is zero (since the x axis is the centroidal axis, $\int y \, dA = 0$ for the entire section).

4. Q is *always* a maximum at the centroidal axis where $y_1 = 0$. This does not necessarily mean that the shear stress is maximum at the centroidal axis, but in most cases it is (when Q/b is a maximum). Cross sections efficient in bending generally are thin at the centroidal axes (e.g., wide-flange I-beams). For these cases b is a minimum and Q/b is most certainly a maximum at the centroidal axis.

5. Equation 3.4-6 for Q is based on the area *above* $y = y_1$. If the area *below* $y = y_1$ is simpler, then Q can be evaluated using

$$Q = -\int_{y=y_{min}}^{y=y_1} y \, dA \qquad \text{[3.4-8]}$$

since $\int y \, dA = 0$ for the entire cross section.

[†] For composite sections where the centroid $\bar{y}_i$ and the area A_i of each region of the composite are known, Eq. (3.4-6) can be written as

$$Q = \left(\sum_{i=1}^{n} \bar{y}_i A_i \right)_{y=y_i}^{y=y_{max}}$$

6. The shear stress arises from the bending stress force differential, which is related to the change in the bending moment with x. This idea is helpful when analyzing specific isolations. For example, review the cross section given in Problem 3.10 at the end of this chapter. If one desired to determine the shear stress at some value of y in the vertical webs, one *would not* make a cut at y through all three vertical webs and let the width of the cut b be the sum of the three wall thicknesses. Why not? This would be acceptable if the shear stresses in each wall were equal, but they are *not*. Imagine a complete isolation which only breaks through the center web at y (the isolation cut does not have to go through at a constant value of y). Then imagine a complete isolation which breaks through one outer web wall at y. Consider the *lower* isolations of the two cases. Are the second-moment areas of the isolation the *same*? They are *not*, and therefore the bending stress force differentials are not the same. Consequently, the shear stresses will *not* be equal.

Problem 3.9 is similar where separate isolations are necessary to reveal the shear stresses in the vertical walls. However, for this problem symmetry can be employed where the stresses in each of the outer walls are assumed to be equal. If the outer walls have different thickness, symmetry cannot be utilized, and a direct solution using Eq. (3.4-7) is not possible. The solution of this type of problem is described in Sec. 5.4.4.

Example 3.4-3 | For a beam with a solid rectangular cross section of depth h and width b, derive an expression for the shear-stress distribution if the section is transmitting a transverse shear force V_y.

Solution:

The cross section of the beam is shown in Fig. 3.4-12(a). To determine the shear stress at a particular value of y, a break is made at $y = y_1$. This isolates the shaded area shown in Fig. 3.4-12(a). In order to evaluate the area moment Q, the infinitesimal area $dA = b \, dy$ can be utilized. Thus

$$Q = \int_{y_1}^{h/2} y \, dA = \int_{y_1}^{h/2} y(b \, dy) = \frac{b}{2}\left[\left(\frac{h}{2}\right)^2 - y_1^2\right]^\dagger$$

The width of the section at $y = y_1$ is b. Substituting Q and b into Eq. (3.4-6) results in

$$\tau_{xy} = \frac{V_y Q}{I_z b} = \frac{V_z(b/2)[(h/2)^2 - y_1^2]}{I_z b} = \frac{V_y}{2I_z}\left[\left(\frac{h}{2}\right)^2 - y_1^2\right] \qquad \text{[a]}$$

---

† In this example, it was unnecessary to perform the integration to obtain Q. Using the summation formulation, the centroid and area of the shaded region are $\bar{y}_i = \frac{1}{2}(h/2 + y_1)$ and $A_i = b(h/2 - y_1)$, respectively. This gives

$$Q = \left(\sum_{i=1}^{n} \bar{y}_i A_i\right)_{y=y_1}^{y=\frac{h}{2}} = \bar{y}_1 A_1 = \frac{1}{2}\left(\frac{h}{2} + y_1\right)\left[b\left(\frac{h}{2} - y_1\right)\right] = \frac{b}{2}\left[\left(\frac{h}{2}\right)^2 - y_1^2\right]$$

which agrees with the integral solution.

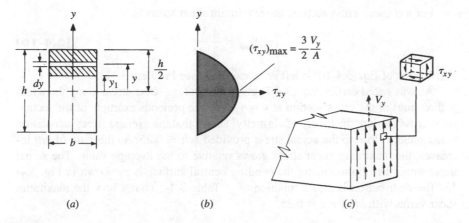

Figure 3.4-12

This equation shows that the transverse shear stress varies in a parabolic fashion with respect to y_1, as shown in Fig. 3.4-12(b). The maximum value of the shear stress occurs when $y_1 = 0$. Substituting $y_1 = 0$ into Eq. (a) yields

$$(\tau_{xy})_{max} = \frac{V_y h^2}{8 I_z}$$

However, since $I_z = bh^3/12$,

$$(\tau_{xy})_{max} = \frac{V_y h^2}{8(bh^3/12)} = \frac{3}{2}\frac{V_y}{bh}$$

Since the total area of the cross section is $A = bh$, the maximum shear stress can be expressed as

$$(\tau_{xy})_{max} = \frac{3}{2}\frac{V_y}{A}$$

Figure 3.4-12(c) illustrates how the shear stress varies across the transverse surface of the beam.

For most cross sections, the ratio of Q/b is maximum at the bending neutral axis. For a *rectangular* cross section, the maximum value of shear stress, as determined in the previous example, is

$$(\tau_{xy})_{max} = \frac{3}{2}\frac{V_y}{A} \qquad\qquad \textbf{[3.4-9]}$$

For a *circular* cross section, the maximum shear stress is

$$(\tau_{xy})_{max} = \frac{4}{3}\frac{V_y}{A}$$
[3.4-10]

Verification of Eq. (3.4-10) is left as an exercise (see Problem 3.11).

As was stated earlier, Eq. (3.4-7) gives the average shear stress along the width (z direction) of the cross section at $y = y_1$. For the previous example of the rectangular cross section, the theory of elasticity[†] shows that the average stress calculation is reasonably close to the actual stress provided $b/h \leq 0.25$. As the ratio of b/h increases, the maximum shear stress grows relative to the average value. The actual shear stress distribution along the bending neutral surface is as shown in Fig. 3.4-13. The distribution is also a function of v. Table 3.4-1 shows how the maximum shear varies with b/h for $v = 0.25$.

Table 3.4-1   Maximum τ_{xy} in a $b \times h$ rectangular cross section

b/h	τ_{max}/τ_{avg}
0	1.000
0.25	1.008
0.5	1.033
1	1.126
2	1.396
5	2.285
10	3.770
25	8.225
50	15.65

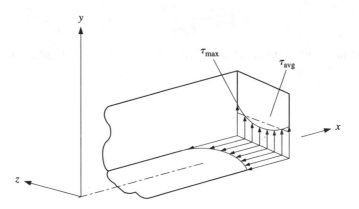

Figure 3.4-13   Shear stress distribution along the neutral surface of a rectangular section.

[†] See Ref. 3.3, pp. 361–66.

When b/h exceeds 15, τ_{zx} is actually larger than τ_{xy} and occurs at the top and bottom of the section. For example, when $b/h = 50$, $(\tau_{zx})_{max} / (\tau_{xy})_{avg} = 19.47$. Thus we see that the state of stress is not as simple as Eq. (3.4-7) depicts. Again, further discussion of the transverse shear stress is presented in Chap. 5.

Determine the maximum average shear stress due to the transverse shear force in the beam given in Example 3.4-2. Plot the average shear stress distribution across the entire section transmitting the maximum transverse shear force. **Example 3.4-4**

Solution:

The shear-force equations are written and the diagram is constructed in order to determine the maximum shear force on an internal surface of the beam. From Fig. 3.4-9, the shear-force equations are

$$V_y = \begin{cases} -600 & 0 < x < 50 \text{ in} \\ 1400 & 50 < x < 100 \text{ in} \\ -2000 & 100 < x < 120 \text{ in} \end{cases}$$

and the resulting diagram is illustrated in Fig. 3.4-14. The maximum shear force occurs throughout the span to the right of B and is $V_{max} = |-2000|$ lb. The maximum shear stress will occur in this span, but at what value of y? The shear stress is greatest with respect to y where Q/b is the greatest value. Since Q is maximum and b is minimum at $y_1 = 0$, Q/b is greatest at $y = 0$. The determination of Q is illustrated in Fig. 3.4-15.

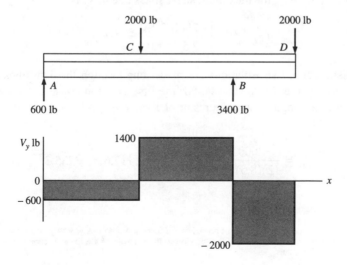

Figure 3.4-14

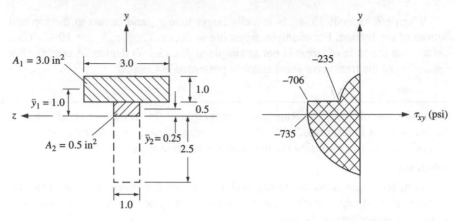

(a) Evaluating Q at $y_1 = 0$ (b) Shear stress distribution across the section

Figure 3.4-15

For the section, at $y_1 = 0$

$$Q_{max} = \int_0^{y_{max}} y\, dA = \sum \bar{y}_i A_i = (1.0)(3.0) + (0.25)(0.5) = 3.13 \text{ in}^{3\dagger}$$

and $b = 1.0$ in and is the minimum value. Thus

$$(Q/b)_{max} = 3.13/1 = 3.13 \text{ in}^3$$

Thus, from Eq. (3.4-7), the maximum shear stress due to V_y is

$$(\tau_{xy})_{max} = \left| \frac{-2000}{8.5} \frac{3.13}{1.0} \right| = 736 \text{ psi}$$

For the distribution, two isolations are made, one through the 1-in thickness and the other through the 3-in thickness. For the first isolation consider the lower section, which is rectangular, and the result of Example 3.4-3 can be used. Thus, for $-2.5 \text{ in} \leq y \leq 0.5$ in,

$$\tau_{xy} = \frac{-2000}{(2)(8.5)}[(-2.5)^2 - y^2] = 117.6y^2 - 735.3 \qquad \text{[a]}$$

[†] Equation (3.4-7) is based on the area above $y = y_1$. Since the area below $y = y_1$ is simpler, Q_{max} could have been evaluated using Eq. (3.4-8). The location to the centroid of the lower section is $\bar{y} = -1.25$ in, and the area is 2.5 in^2; thus

$$Q_{max} = -(-1.25)(2.5) = 3.13 \text{ in}^2$$

For the 3-in thickness consider the upper section, which again is rectangular, and the result of Example 3.4-3 can be used. For 0.5 in $\leq y \leq$ 1.5 in,

$$\tau_{xy} = \frac{-2000}{(2)(8.5)}[(1.5)^2 - y^2] = 117.6y^2 - 264.7 \qquad \text{[b]}$$

Equations (a) and (b) are plotted in Fig. 3.4-15(b).

Again, we have assumed that the shear stress distribution is constant in the z direction, thus giving a distribution of the average shear stress. The actual picture is far more complicated. Fortunately, in most practical situations, the transverse shear stress is small compared to the bending stresses and is not a primary concern. However, there are cases where a better understanding of the shear flow is necessary. One such case involves the concept of the *shear center* and is discussed in Chap. 5.

Shear in Compound Beams   Occasionally, beams are constructed by fastening separate members together. Figure 3.4-16 shows a beam constructed by bolting plates to the top and bottom of a wide-flange I-beam. This has the advantage of drastically increasing the second area moment, thereby reducing bending stresses with a minimal increase in material. However, the shear force at the interface of the plate and the I-beam, due to a differential in the bending moment, must be transmitted by the shear area of the bolts. Since at the interface the bolt shear area is not continuous as in the case of a solid beam, Eq. (3.4-7) must be reformulated. Returning to Fig. 3.4-11, the net shear force on the isolated horizontal surface is $\tau_{xy} b \, \Delta x$. The shear force per unit length, called the *shear flow q*, is $\tau_{xy} b$. Thus, from Eq. (3.4-7),

$$q = \frac{V_y Q}{I_z} \qquad \text{[3.4-11]}$$

This force per unit length must be transmitted through the bolts which are spaced longitudinally at a distance s along the beam where the total force along the spacing is qs. If at an interface there are n fasteners, the force per fastener is qs/n. For the case of Fig. 3.4-16, the number of fasteners at an interface is two. Finally, if the shear area of each fastener is A_s, the average shear stress in the fastener is $qs/(nA_s)$. Thus, from Eq. (3.4-11), the average shear stress in the fastener is

$$\tau_{\text{avg}} = \frac{V_y Q}{I_z} \frac{s}{nA_s} \qquad \text{[3.4-12]}$$

Example 3.4-5

For the I-beam in Fig. 3.4-16 let $h = 210$ mm, $b = 205$ mm, and $I_z = 60.8 \ (10^6)$ mm^4. Also let the thickness of each plate be 15 mm, the diameter of each bolt be 10 mm, and the spacing $s = 75$ mm. Determine the average shear stress in the bolts if the cross section of the beam is transmitting a shear force of $V_y = 4$ kN.

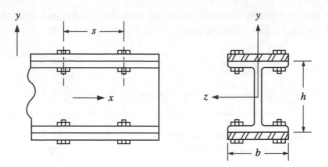

Figure 3.4-16   Compound beam.

Solution:

The area moment Q is evaluated for the section above the interface (the plate) at $y = h/2 = 210/2 = 105$ mm, which is

$$Q = \bar{y}\,A\,\big|_{y=105\text{ mm}}^{y=120\text{ mm}} = \left(105 + \frac{15}{2}\right)[(205)(15)] = 345.9(10^3)\text{ mm}^3$$

Thus the average shear stress in a bolt is

$$\tau_{\text{avg}} = \frac{V_y Q}{I_z}\,\frac{s}{nA_s} = \frac{[4(10^3)][345.9(10^3)(10^{-3})^3]}{60.8(10^6)(10^{-3})^4}\,\frac{0.075}{[2][(\pi/4)(0.010^2)]}$$

$$= 10.87(10^6)\text{ N/m}^2 = 10.87\text{ MPa}$$

3.4.4   BENDING STRAINS AND DEFLECTIONS

The fundamental formulations assume that the beam is narrow and in a state of plane stress. Thus the ideal model is that of a rectangular beam where the width of the beam b is much less than the depth h. For this, the strains are given by

$$\varepsilon_x = -\frac{M_z y}{EI_z} \tag{3.4-13}$$

$$\varepsilon_y = \varepsilon_z = v\frac{M_z y}{EI_z} \tag{3.4-14}$$

$$\gamma_{xy} = \frac{2(1+v)}{E}\,\frac{V_y Q}{I_z b} \tag{3.4-15}$$

In elementary mechanics of materials, the deflection of the centroidal axis due to bending only is considered using analytical geometry.[†] Here the basic formulation for the deflection of the centroidal axis is

$$\frac{1}{\rho} = \frac{d^2v_c/dx^2}{[1 + (dv_c/dx)^2]^{3/2}} = -\frac{\varepsilon_x}{y} = \frac{M_z}{EI_z}$$

[3.4-16]

where ρ and v_c are the radius of curvature and the deflection of the centroidal axis in the y direction, respectively. For small slopes, $dv_c/dx < 0.1$, and Eq. (3.4-16) reduces to

$$\frac{d^2v_c}{dx^2} = \frac{M_z}{EI_z}$$

[3.4-17]

For simple problems, the procedure in establishing the deflection shape is to first integrate Eq. (3.4-17) with respect to x, yielding the equation for the slope of the centroidal axis as a function of x

$$\frac{dv_c}{dx} = \int \frac{M_z}{EI_z} \, dx + C_1$$

[3.4-18]

where the integration is indefinite and C_1 is an integration constant. Integrating once more yields

$$v_c = \int \left(\int \frac{M_z}{EI_z} \, dx \right) dx + C_1 x + C_2$$

[3.4-19]

The integration constants C_1 and C_2 are found using geometric boundary conditions on the beam. Equations (3.4-18) and (3.4-19) are usable for simple beams. However, when a beam is subjected to many discontinuous loads, there are much better approaches such as superposition, energy methods, singularity functions, and the area-moment method, to name a few. Superposition and singularity functions are discussed in this chapter, whereas energy methods are discussed in Chap. 6.

Verify the slope and deflection of the centroidal axis as functions of x for the cantilever beam C.3 given in Appendix C.

Example 3.4-6

Solution:

Referring to beam C.3 of Appendix C, we see that the bending moment equation, which is continuous in x, is given by

$$M_z = -\frac{w}{2} (L - x)^2$$

[a]

[†] Problem 1.20 illustrates deflection due to shear. Also, Chap. 6 presents methods to account for shear deflection.

Thus, from Eq. (3.4-18)

$$\frac{dv_c}{dx} = \int \frac{-(w/2)(L-x)^2}{EI_z}\, dx + C_1 = \frac{-w}{2EI_z}\left(L^2x - Lx^2 + \frac{x^3}{3}\right) + C_1 \qquad [b]$$

and from Eq. 3.4-19

$$v_c = \int \frac{-w}{2EI_z}\left(L^2x - Lx^2 + \frac{x^3}{3}\right) dx + C_1x + C_2$$

$$= -\frac{w}{2EI_z}\left(\frac{L^2x^2}{2} - \frac{Lx^3}{3} + \frac{x^4}{12}\right) + C_1x + C_2 \qquad [c]$$

The end conditions for the beam are $v_c = 0$ at $x = 0$ and $dv_c/dx = 0$ at $x = 0$. Substitution of the end conditions into Eqs. (b) and (c) yields $C_1 = C_2 = 0$. Thus

$$\frac{dv_c}{dx} = -\frac{wx}{6EI_z}(3L^2 - 3Lx + x^2) \qquad [d]$$

and

$$v_c = -\frac{wx^2}{24EI_z}(6L^2 - 4Lx + x^2) \qquad [e]$$

which can be seen to agree with the results given in Appendix C.

Example 3.4-7

Verify the deflection of the centroidal axis as functions of x for the simply supported beam C.5 given in Appendix C (a) using discontinuous equations and (b) using singularity functions.

Solution:

(a) Referring to beam C.5 of Appendix C, the reactions are $R_A = Fb/L$ and $R_B = Fa/L$. The reader should verify that for section AB

$$(M_z)_{AB} = R_Ax = \frac{Fb}{L}x$$

$$\left(\frac{dv_c}{dx}\right)_{AB} = \frac{Fb}{2EIL}x^2 + C_1 \qquad [a]$$

$$(v_c)_{AB} = \frac{Fb}{6EIL}x^3 + C_1x + C_2 \qquad [b]$$

One boundary condition is that $v_c = 0$ at $x = 0$. This gives $C_2 = 0$ and thus

$$(v_c)_{AB} = \frac{Fb}{6EIL}x^3 + C_1x \qquad [c]$$

For section *BC,* from the right isolation

$$(M_z)_{BC} = R_B(L - x) = \frac{Fa}{L}(L - x)$$

$$\left(\frac{dv_c}{dx}\right)_{BC} = \frac{Fa}{EIL}\left(Lx - \frac{x^2}{2}\right) + C_3 = \frac{Fa}{2EIL}(2Lx - x^2) + C_3 \qquad \textbf{[d]}$$

$$(v_c)_{BC} = \frac{Fa}{2EIL}\left(Lx^2 - \frac{x^3}{3}\right) + C_3x + C_4 = \frac{Fa}{6EIL}(3Lx^2 - x^3) + C_3x + C_4 \quad \textbf{[e]}$$

But $v_c = 0$ at $x = L$. This gives

$$C_3L + C_4 = -\frac{FaL^2}{3EI} \qquad \textbf{[f]}$$

We need two additional equations in order to determine C_1, C_3, and C_4. For this we match displacement and slope of the two sets of equations at $x = a$. That is, we first equate Eqs. (*c*) and (*e*) at $x = a$

$$\frac{Fba^3}{6EIL} + C_1a = \frac{Fa}{6EIL}(3La^2 - a^3) + C_3a + C_4$$

Substituting $L - a$ for b and simplifying gives

$$C_1a - C_3a - C_4 = \frac{Fa^3}{3EI} \qquad \textbf{[g]}$$

For slope continuity, equate Eqs. (*a*) and (*d*) at $x = a$

$$\frac{Fba^2}{2EIL} + C_1 = \frac{Fa}{2EIL}(2La - a^2) + C_3$$

Again, substituting $L - a$ for b and simplifying gives

$$C_1 - C_3 = \frac{Fa^2}{2EI} \qquad \textbf{[h]}$$

Solving Eqs. (*f*), (*g*), and (*h*) simultaneously results in

$$C_1 = -\frac{Fa}{6EIL}(2L^2 - 3aL + a^2)$$

$$C_3 = -\frac{Fa}{6EIL}(2L^2 + a^2)$$

$$C_4 = \frac{Fa^3}{6EI}$$

Substituting C_1 into Eq. (c) with $a = L - b$ gives

$$(v_c)_{AB} = \frac{Fb}{6EIL}x^3 + \frac{F(L-b)}{6EIL}[3(L-b)L - 2L^2 - (L-b)^2]x$$

$$= \frac{Fbx}{6EIL}(x^2 + b^2 - L^2)$$

which agrees with that given in Appendix C.

For section BC, substitute C_3 and C_4 into Eq. (e)

$$(v_c)_{BC} = \frac{Fa}{6EIL}(3Lx^2 - x^3) - \frac{Fa}{6EIL}(2L^2 + a^2)x + \frac{Fa^3}{6EI}$$

$$= \frac{Fa}{6EIL}(3Lx^2 - x^3 - 2L^2x - a^2x + a^2L)$$

$$= \frac{Fa(L-x)}{6EIL}(x^2 + a^2 - 2Lx)$$

which again agrees with Appendix C.

(b) Ensuring continuity of displacement and slope as was done in part (a) complicates the analysis. Singularity functions improve the process dramatically. The bending moment equation for the *entire* beam is

$$M_z = R_A\langle x - 0\rangle^1 - F\langle x - a\rangle^1 + R_B\langle x - L\rangle^1$$

$$= \frac{Fb}{L}\langle x - 0\rangle^1 - F\langle x - a\rangle^1 + \frac{Fa}{L}\langle x - L\rangle^1$$

The first function $\langle x - 0\rangle^1$ always exists and can be replaced by x. The last function $\langle x - L\rangle^1$ exists only at $x = L$, where it and its integrals are zero and thus can be dropped. This gives

$$M_z = \frac{Fb}{L}x - F\langle x - a\rangle^1$$

Two integrations yield

$$EI\frac{dv_c}{dx} = \frac{Fb}{2L}x^2 - \frac{F}{2}\langle x - a\rangle^2 + C_1$$

$$EIv_c = \frac{Fb}{6L}x^3 - \frac{F}{6}\langle x - a\rangle^3 + C_1x + C_2 \qquad\qquad \text{[f]}$$

where the constants C_1 and C_2 are different from that of part (a). For the boundary condition $v_c = 0$ at $x = 0$, we find $C_2 = 0$. For $v_c = 0$ at $x = L$,

$$0 = \frac{Fb}{6L}L^3 - \frac{F}{6}\langle L - a\rangle^3 + C_1 L$$

or

$$\frac{Fb}{6}L^2 - \frac{F}{6}(L - a)^3 + C_1 L = 0$$

With $L - a = b$, solving for C_1 gives

$$C_1 = \frac{Fb}{6L}(b^2 - L^2)$$

Substituting this into Eq. (i) results in

$$v_c = \frac{F}{6EIL}\left[bx^3 - L\langle x - a\rangle^3 + bx(b^2 - L^2)\right] \qquad \textbf{[}j\textbf{]}$$

Let us see if this result matches part (a). For section AB, $x < a$, the singularity function $\langle x - a\rangle$ does not exist. Thus

$$(v_c)_{AB} = \frac{F}{6EIL}\left[bx^3 + bx(b^2 - L^2)\right] = \frac{Fbx}{6EIL}(x^2 + b^2 - L^2)$$

which agrees with part (a). For section BC the singularity function does exist and

$$(v_c)_{BC} = \frac{F}{6EIL}\left[bx^3 - L(x - a)^3 + bx(b^2 - L^2)\right]$$

Substituting $b = L - a$ gives

$$(v_c)_{BC} = \frac{F}{6EIL}\left\{(L - a)x^3 - L(x - a)^3 + (L - a)x[(L - a)^2 - L^2]\right\}$$

$$= \frac{Fa}{6EIL}(3Lx^2 - x^3 - 2L^2x - a^2x + a^2L)$$

$$= \frac{Fa(L - x)}{6EIL}(x^2 + a^2 - 2Lx)$$

which again agrees with part (a). Thus Eq. (j) completely describes the displacement of the *entire* beam and is seen to be much simpler to obtain than that for part (a).

3.5   BENDING OF SYMMETRIC BEAMS IN TWO PLANES

In Sec. 3.4 beams bending in one plane are discussed. Design problems often occur where bending is not isolated to one plane and *two-plane analysis* is necessary. The analysis in this section will be restricted to beams with cross sections having at least one axis of symmetry such that $\int yz\, dA = 0$. Nonsymmetric beams are discussed in Sec. 5.3.

Consider a section that is undergoing bending about both the y and z axes, as shown in Fig. 3.5-1, with moments M_y and M_z, respectively. The positive sign convention for bending in two planes is consistent with that given in Sec. 3.4, where the moment vectors are shown positive (according to the right-hand rule) on a surface with a normal in a positive direction; and negative when on a surface with a normal in a negative direction. Similar to Eq. (3.4-3), the bending stress is then given by

$$\sigma_x = \frac{M_y z}{I_y} - \frac{M_z y}{I_z}$$ **[3.5-1]**

Deflections in each plane are found in a manner identical to that of Sec. 3.4-4. The deflection of the centroidal axis in the z direction, w_c, is given by

$$\frac{d^2 w_c}{dx^2} = -\frac{M_y}{EI_y}$$ **[3.5-2]**

The differences in signs associated with M_y versus M_z are due to the differences in the y and z coordinate axes as seen in Fig. 3.5-1(*b*).

Circular Beams   The circular cross section is axisymmetric where all transverse axes are identical. In this case, the bending moments M_y and M_z can be added as vectors giving a total moment of $\sqrt{M_y^2 + M_z^2}$ about a z' axis located at an angle

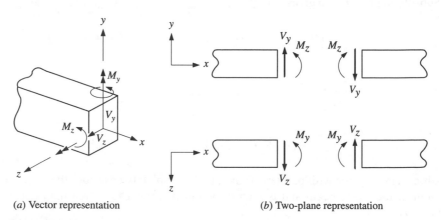

(*a*) Vector representation   (*b*) Two-plane representation

Figure 3.5-1   Two-plane bending convention.

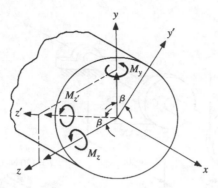

Figure 3.5-2

β of $\tan^{-1}(M_y/M_z)$ from the z axis (see Fig. 3.5-2). Thus the complete bending stress is given by

$$\sigma_x = -\frac{M_{z'}y'}{I_{z'}} \qquad \textbf{[3.5-3]}$$

where $M_{z'} = \sqrt{M_y^2 + M_z^2}$ and $I_{z'} = I_y = I_z$.

Example 3.5-1

Figure 3.5-3 shows a rotating shaft mounted in bearings and a torque being transmitted between pulleys located at points B and C (all dimensions are given in millimeters). The shaft is steel with $E = 200$ GPa and has a diameter of 40 mm. Considering the shaft to be simply supported at the bearings and ignoring dynamic effects, determine for the instant shown

(a) The location and state of stress of the element undergoing the greatest stress conditions due to bending and torsion.

(b) The deflection of the center of the shaft.

Solution:

(a) The net effect of the shaft loading is shown in Fig. 3.5-3(b) where for equilibrium,

$$R_{Ay} = 1.6 \text{ kN} \qquad R_{Az} = 960 \text{ N} \qquad R_{Dy} = 800 \text{ N} \qquad R_{Dz} = 1.92 \text{ kN}$$

Considering bending only, the loading in the xy and xz planes and the corresponding bending moment diagrams are as shown in Fig. 3.5-4. It can be shown that the maximum moment will occur at either point B or C. Since the maximum moment is given by $\sqrt{M_y^2 + M_z^2}$, at points B and C

$$M_B = \sqrt{(-320)^2 + (192)^2} = 373 \text{ N·m}$$

$$M_C = \sqrt{(-160)^2 + (384)^2} = 416 \text{ N·m}$$

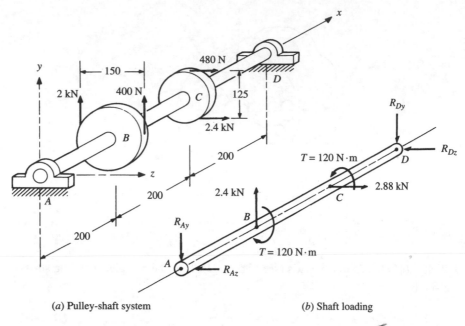

(a) Pulley-shaft system   (b) Shaft loading

Figure 3.5-3

Thus it can be seen that the maximum bending moment occurs at point C. The cross section at point C whose surface normal is in the positive x direction is shown in Fig. 3.5-5(a). This surface is at a value of x slightly less than 400 mm so as to include the transmitted torque T. The z' axis is located by the angle β, where $\beta = \tan^{-1}(M_y/M_z)$. Thus

$$\beta = \tan^{-1}\left(\frac{384}{-160}\right) = 112.6°$$

The $y'z'$ axes are as shown in Fig. 3.5-5(b). From bending, the maximum tensile stress occurs at point Q_1, where $y' = -d/2$, and is

$$(\sigma_x)_{Q_1} = -\frac{(416)(-0.040/2)}{(\pi/64)(0.040)^4} = 66.2 \times 10^6 \text{ N/m}^2 = 66.2 \text{ MPa}$$

The maximum compressive stress due to bending occurs at point Q_2, where $y' = d/2$, and is -66.2 MPa.

The shear stress at points Q_1 and Q_2 due to torque, determined from Eq. (3.3-3), is

$$\tau = \frac{Tr_{\max}}{J_x} = \frac{(120)(0.040/2)}{(\pi/32)(0.040)^4} = 9.55 \times 10^6 \text{ N/m}^2 = 9.55 \text{ MPa}$$

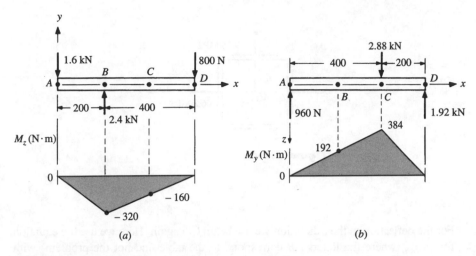

Figure 3.5-4

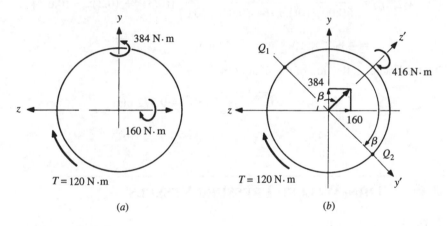

Figure 3.5-5

The state of stress at point Q_1 is shown in Fig. 3.5-6. The state of stress for point Q_2 is similar with the normal stress being compression.

(b) For deflection in the y direction, from Fig. 3.5-4(a), we use beam C.5 with $F = -2400$ N, $a = 0.200$ m, $b = 0.400$ m, $L = 0.600$ m, and $x = L/2 = 0.300$ m. In the table, we use the equation for $(v_c)_{BC}$ (by coincidence the letters match).

$$v_c|_{x=L/2} = \frac{-(2400)(0.2)(0.6 - 0.3)}{6(200)(10^9)(\pi/64)(0.040)^4(0.6)}[(0.3)^2 + (0.2)^2 - 2(0.6)(0.3)]$$

$$= 366.1 \times 10^{-6} \text{ m} = 0.366 \text{ mm}$$

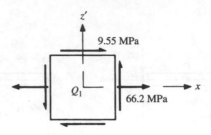

Figure 3.5-6

For the deflection in the z direction we use beam C.5 again. Here we use the equation for $(v_c)_{AB}$ (where the letters AB correspond to the table and not the problem) with $F = -2880$ N, $a = 0.400$ m, $b = 0.200$ m, $L = 0.600$ m, and $x = L/2 = 0.300$ m.

$$w_c|_{x=L/2} = \frac{-(2880)(0.2)(0.3)}{6(200)(10^9)(\pi/64)(0.040)^4(0.6)} [(0.3)^2 + (0.2)^2 - (0.6)^2]$$

$$= 439.3 \times 10^{-6} \text{ m} = 0.439 \text{ mm}$$

The total radial deflection of the shaft at midspan is

$$\delta_{x=L/2} = \sqrt{0.366^2 + 0.439^2} = 0.572 \text{ mm}$$

3.6   THIN-WALLED PRESSURE VESSELS

3.6.1   STRESSES

Consider the doubly curved closed pressurized vessel shown in Fig. 3.6-1. If the wall thickness t of the vessel is small as compared with the radii of curvature of the shell surface r_θ and r_ϕ (i.e., $t < 0.1\, r_\theta$, and $t < 0.1\, r_\phi$), the state of stress at a point can be described by the *membrane equation*.

Isolate an element of the wall using coordinates tangent to the principal arcs of curvature and perpendicular to the surface. Thus the dimensions of the element will be $r_\phi\ \Delta\phi$ by $r_\theta\ \Delta\theta$ by t, and the outward normal of the infinitesimal surface is established by n. The state of stress on the element is shown in Fig. 3.6-1(a), where for equilibrium of forces in the n direction

$$-2\sigma_\theta t r_\phi\ \Delta\phi \sin\frac{\Delta\theta}{2} - 2\sigma_\phi\, t r_\theta\ \Delta\theta \sin\frac{\Delta\phi}{2} + p r_\theta\ \Delta\theta r_\phi\ \Delta\phi = 0$$

Since $\Delta\theta$ and $\Delta\phi$ are infinitesimal, $\sin(\Delta\theta/2) = \Delta\theta/2$ and $\sin(\Delta\phi/2) = \Delta\phi/2$. Simplifying gives

$$\frac{\sigma_\theta}{r_\theta} + \frac{\sigma_\phi}{r_\phi} = \frac{p}{t} \qquad\qquad \textbf{[3.6-1]}$$

The stress in the radial direction σ_r is usually assumed to be zero toward the outer surface of the vessel and $-p$ at the inner surface. To obtain a second relationship for the "membrane" stresses σ_θ and σ_ϕ a symmetric break is usually made in the vessel so that only one of the stresses exists along the break, and the stress is determined using the equilibrium conditions of the isolated element.

Consider the *circular cylinder* shown in Fig. 3.6-2(*a*). One curvature is flat and in the z direction. Consequently we replace the curvilinear coordinate ϕ with the longitudinal coordinate z with $r_\phi = r_z \to \infty$. Letting $r_\theta = r$, the radius of the cylinder, Eq. (3.6-1) yields

$$\frac{\sigma_\theta}{r} + \frac{\sigma_z}{r_z} = \frac{p}{t}$$

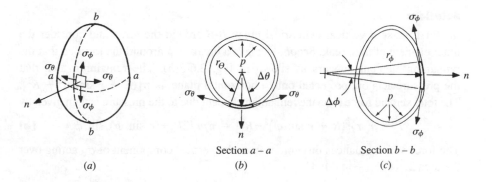

Section $a - a$
(*b*)

Section $b - b$
(*c*)

(*a*)

Figure 3.6-1   Thin-walled pressure vessel.

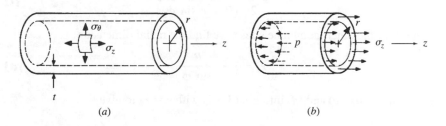

(*a*)

(*b*)

Figure 3.6-2   Circular cylinder.

Solving for σ_θ with $r_z \rightarrow \infty$ yields

$$\sigma_\theta = \frac{pr}{t} \qquad \text{[3.6-2]}$$

where σ_θ is referred to as the *hoop stress*. The longitudinal stress σ_z is found by iso-lating the surface as shown in Fig. 3.6-2(b). Equilibrium of force in the z direction yields $\sigma_z(2\pi r t) = p(\pi r^2)$. Thus

$$\sigma_z = \frac{pr}{2t} \qquad \text{[3.6-3]}$$

For the case of a *spherical* vessel of radius r, $r_\theta = r_\phi = r$, total symmetry exists. Consequently $\sigma_\theta = \sigma_\phi$, and Eq. (3.6-1) yields

$$\sigma_\theta = \sigma_\phi = \frac{pr}{2t} \qquad \text{[3.6-4]}$$

Example 3.6-1   A thin closed toroidal shell of thickness t is shown in Fig. 3.6-3(a). Determine (a) the membrane stresses as a function of ϕ and (b) the membrane stresses at point Q under the following conditions: $p = 105$ kPa, $r = 300$ mm, $R = 1.0$ m, and $t = 10$ mm.

Solution:

(a) First, make a vertical cylindrical break B-B around the torus and consider the outer element of the break. Secondly, make a break at ϕ around the torus so that the remaining element appears as shown in Fig. 3.6-3(b). The remaining area that the pressure acts on, projected onto a horizontal plane, is $\pi[(R + r \sin\phi)^2 - R^2]$. The total vertical force F on the remaining element due to the pressure is then given by

$$F = p(\pi)[(R + r \sin\phi)^2 - R^2] = \pi pr(2R + r \sin\phi) \sin\phi \qquad \text{[a]}$$

The force which balances this comes from the vertical component of σ_ϕ acting over the area $2\pi(R + r \sin\phi)t$. Or

$$F = \sigma_\phi[2\pi(R + r \sin\phi)t] \sin\phi \qquad \text{[b]}$$

Equating Eqs. (a) and (b) and simplifying yields

$$\sigma_\phi = \frac{pr}{2t} \frac{(2R + r \sin\phi)}{(R + r \sin\phi)} \qquad \text{[c]}$$

To determine the other membrane stress, we find the radial dimension

$$r_\theta = r + \frac{R}{\sin\phi} \qquad \text{[d]}$$

and substitute Eqs. (c) and (d) into Eq. (3.6-1) with $r = r_\phi$ to give

$$\sigma_\theta = \frac{pr}{2t} \qquad \text{[e]}$$

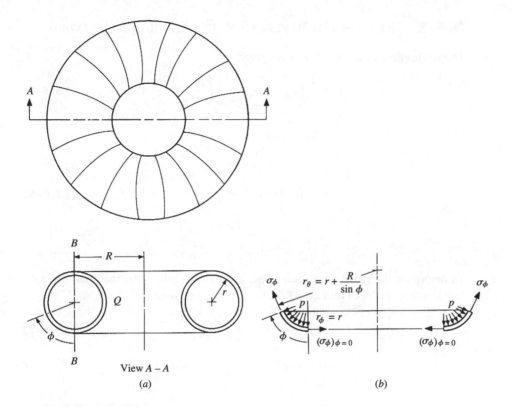

Figure 3.6-3   Toroidal shell.

(b) It can be shown that σ_ϕ is maximum at point Q, where $\phi = 3\pi/2$. For this case, Eq. (c) reduces to

$$\sigma_\phi = \frac{pr}{2t}\frac{2R-r}{R-r} \qquad\qquad \textbf{[f]}$$

Substituting the numerical values of p, R, r, and t into Eqs. (f) and (e) yields

$$\sigma_\phi = \frac{(105)(0.3)[2(1.0) - 0.3]}{(2)(0.01)(1.0 - 0.3)} = 3825 \text{ kPa} = 3.83 \text{ MPa}$$

$$\sigma_\theta = \frac{(105)(0.3)}{(2)(0.1)} = 1575 \text{ kPa} = 1.58 \text{ MPa}$$

3.6.2   STRAINS AND DEFLECTIONS IN A CIRCULAR CYLINDER

The strains for a circular cylinder are given by

$$\varepsilon_\theta = \frac{1}{E}(\sigma_\theta - v\sigma_z) = \frac{1}{E}\left(\frac{pr}{t} - v\frac{pr}{2t}\right) \qquad \text{[3.6-5]}$$

$$= (2 - v)\frac{pr}{2Et}$$

$$\varepsilon_z = \frac{1}{E}(\sigma_z - v\sigma_\theta) = \frac{1}{E}\left(\frac{pr}{2t} - v\frac{pr}{t}\right) \qquad \text{[3.6-6]}$$

$$= (1 - 2v)\frac{pr}{2Et}$$

To determine the radial deflection of the cylinder we utilize the fact that the problem is axisymmetric. From Eq. (1.5-5b), $u_r = r\varepsilon_\theta$, and thus

$$u_r = (2 - v)\frac{pr^2}{2Et} \qquad \text{[3.6-7]}$$

3.7   SUPERPOSITION

Linear elasticity is predicated on the assumption that strains are small (this is sometimes mistakenly referred to as small-deflection theory). For linear systems, the action of each force with respect to a particular effect can be analyzed independently, and the results can be added algebraically or as vectors depending on the situation. The particular effect considered can be either internal or support forces, moments, slopes, deflections, stresses, or strains. The advantage of this is that the results of simple loading configurations are known, and the results of complex loading can be found by adding the results of each individual load or load distribution.

Example 3.7-1 | For the steel beam shown in Fig. 3.7-1 use superposition to determine the support reactions, shear and bending moment diagrams, and deflection of the centroidal axis as functions of x and the maximum tensile stress.

Solution:

The effects of the uniform force distribution and the concentrated force can be viewed independently. The independent results are given in Appendix C and repeated below.

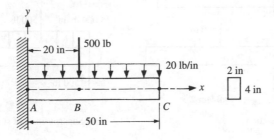

Figure 3.7-1

Support Reactions

See Fig. 3.7-2(*a*) and (*b*).

$$M_1 = 25,000 \text{ lb·in} \qquad M_2 = 10,000 \text{ lb·in}$$

$$R_1 = 1000 \text{ lb} \qquad R_2 = 500 \text{ lb}$$

Superposition

$$M = M_1 + M_2 = 35,000 \text{ lb·in} \qquad R = R_1 + R_2 = 1500 \text{ lb}$$

Shear and Bending Moment Diagrams

See Fig. 3.7-2(*c*) and (*d*).

Deflections

For beam 1:

$$v_{c1} = \frac{wx^2}{24EI}(4Lx - x^2 - 6L^2)$$

For beam 2:

$$(v_{c2})_{AB} = \frac{Px^2}{6EI}(x - 3a) \qquad (v_{c2})_{BC} = \frac{Pa^2}{6EI}(a - 3x)$$

Superposition

$$v_{AB} = \frac{wx^2}{24EI}(4Lx - x^2 - 6L^2) + \frac{Px^2}{6EI}(x - 3a) \qquad \textbf{[a]}$$

$$v_{BC} = \frac{wx^2}{24EI}(4Lx - x^2 - 6L^2) + \frac{Pa^2}{6EI}(a - 3x) \qquad \textbf{[b]}$$

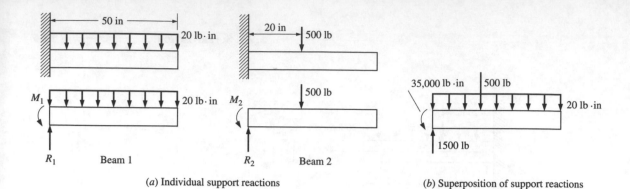

(a) Individual support reactions

(b) Superposition of support reactions

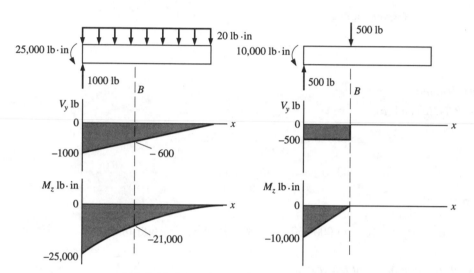

(c) Individual shear force and bending moment diagrams

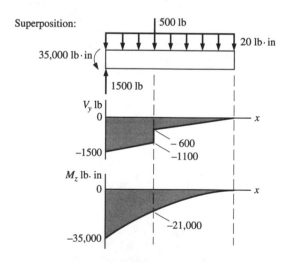

Superposition:

(d) Superposition of shear force and bending moment diagrams

Figure 3.7-2

where $w = 20$ lb/in, $P = 500$ lb, $E = 30 \times 10^6$ psi, $I = 10.7$ in^4, and $a = 20$ in. The maximum deflection occurs at $x = L = 50$ in, where

$$(v_c)_{max} = \frac{(20)(50^2)}{(24)(30 \times 10^6)(10.7)} [(4)(50)(50) - 50^2 - (6)(50^2)]$$

$$+ \frac{(500)(20^2)}{(6)(30 \times 10^6)(10.7)} [20 - (3)(50)] = -0.062 \text{ in}$$

Stress

Maximum tensile stress occurs where the maximum bending moment appears. Since in both beams it occurs at $(x, y) = (0, 2)$, using Eq. (3.4-3) yields

$$(\sigma_x)_{max,1} = -\frac{(-25,000)(2)}{10.7} = 4670 \text{ psi}$$

$$(\sigma_x)_{max,2} = -\frac{(-10,000)(2)}{10.7} = 1870 \text{ psi}$$

Superposition

$$(\sigma_x)_{max} = (\sigma_x)_{max,1} + (\sigma_x)_{max,2} = 6540 \text{ psi}$$

Example 3.7-2

For the beam shown in Fig. 3.7-3 determine the deflection equations using (a) superposition and the tables in Appendix C, and (b) singularity functions.

Solution:

(a) The beam can be visualized as the superposition of the two beams shown in Fig. 3.7-4. The concentrated force in Fig. 3.7-4(a) does not give any deflection. Thus the deflections in region AB arise solely from beam 1, which is a simply sup-

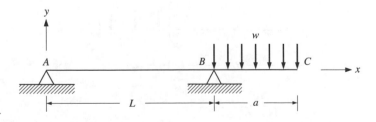

Figure 3.7-3

ported beam with an end moment. This is given in Appendix C as beam C.7 with $M = -wa^2/2$ and $a = L$. Thus

$$(v_c)_{AB} = -\frac{wa^2x}{12EIL}(x^2 + 3L^2 - 6L^2 + 2L^2) = \frac{wa^2x}{12EIL}(L^2 - x^2) \qquad \text{[a]}$$

For region BC, we use superposition of the two beams. For beam 1, deflection is simply due to rotation. The slope of beam 1 at point B, given by beam C.7, is

$$(\theta_B)_1 = -\frac{wa^2}{12EIL}(3L^2 + 3L^2 - 6L^2 + 2L^2) = -\frac{wa^2L}{6EI}$$

The deflection along BC for beam 1 is the slope times the position from point B, $(x - L)$. Thus

$$[(v_c)_{BC}]_1 = -\frac{wa^2L}{6EI}(x - L)$$

The deflection in region BC for beam 2 is found using beam C.3 of Appendix C substituting $x - L$ and a for x and L, respectively. This gives

$$[(v_c)_{BC}]_2 = \frac{w(x - L)^2}{24EI}[4a(x - L) - (x - L)^2 - 6a^2]$$

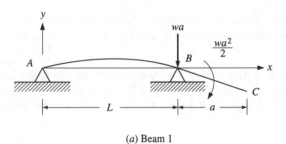

(a) Beam 1

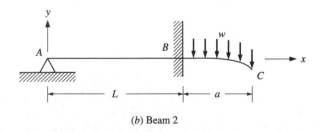

(b) Beam 2

Figure 3.7-4   Beams to use from Appendix C.

For the complete deflection in region BC we simply add $[(v_c)_{BC}]_1$ and $[(v_c)_{BC}]_2$. Thus

$$(v_c)_{BC} = \frac{w(x-L)^2}{24EI}[4a(x-L) - (x-L)^2 - 6a^2] - \frac{wa^2L}{6EI}(x-L)$$

$$= -\frac{w(x-L)}{24EI}[(x-L)^2(x-L-4a) + 2a^2(3x-L)] \qquad [b]$$

(b) The reactions are $R_A = -wa^2/(2L)$ and $R_B = wa[1 + a/(2L)]$. Using singularity functions,

$$M_z = -\frac{wa^2}{2L}x + wa\left(1 + \frac{a}{2L}\right)\langle x - L\rangle^1 - \frac{w}{2}\langle x - L\rangle^2 + \frac{w}{2}\langle x - (L+a)\rangle^2$$

The last term on the right can be omitted as it contributes nothing in the region $0 < x < L + a$. Thus

$$M_z = -\frac{wa^2}{2L}x + wa\left(1 + \frac{a}{2L}\right)\langle x - L\rangle^1 - \frac{w}{2}\langle x - L\rangle^2$$

Integration yields

$$EI\frac{dv_c}{dx} = -\frac{wa^2}{4L}x^2 + \frac{wa}{2}\left(1 + \frac{a}{2L}\right)\langle x - L\rangle^2 - \frac{w}{6}\langle x - L\rangle^3 + C_1$$

$$EIv_c = -\frac{wa^2}{12L}x^3 + \frac{wa}{6}\left(1 + \frac{a}{2L}\right)\langle x - L\rangle^3 - \frac{w}{24}\langle x - L\rangle^4 + C_1x + C_2$$

But $v_c = 0$ at $x = 0$, giving $C_2 = 0$. Also, $v_c = 0$ at $x = L$. Thus

$$0 = -\frac{wa^2}{12L}L^3 + \frac{wa}{6}\left(1 + \frac{a}{2L}\right)(L - L)^3 - \frac{w}{24}(L - L)^4 + C_1L$$

Solving for C_1 gives

$$C_1 = \frac{wa^2L}{12}$$

Substituting this into the deflection equation and simplifying gives the deflection equation for the entire beam

$$v_c = \frac{w}{24EIL}[2a^2x(L^2 - x^2) + 2a(2L + a)\langle x - L\rangle^3 - L\langle x - L\rangle^4] \qquad [c]$$

Let us compare this result with that determined in part (*a*). In region *AB* where $x < L$, the singularity functions in Eq. (*c*) do not exist. Thus in region *AB* Eq. (*c*) is written

$$(v_c)_{AB} = \frac{wa^2x}{12EIL}(L^2 - x^2)$$

which agrees with Eq. (*a*) of part (*a*).

For section *BC*, $x > L$, and the singularity functions become ordinary functions where

$$(v_c)_{BC} = \frac{w}{24EIL}[2a^2x(L^2 - x^2) + 2a(2L + a)(x - L)^3 - L(x - L)^4]$$

After some algebraic manipulation, this can be shown to reduce to

$$(v_c)_{BC} = -\frac{w(x - L)}{24EI}[(x - L)^2(x - L - 4a) + 2a^2L(3x - L)]$$

which again agrees with Eq. (*b*) of part (*a*).

Using superposition, only like stresses can be added algebraically; i.e., normal stresses can be added to normal stresses or shear stresses to shear stresses algebraically. If the two stresses to be superposed are unlike, e.g., a normal stress and a shear stress, they must be shown separately on a diagram.

Example 3.7-3 | The beam shown in Fig. 3.7-5 is fixed at one end and loaded at the other end with a transverse load *P* and a torque *T*. Neglecting the inaccuracies in the stress formulations at the support end, determine the state of stress at points *A*, *B*, *C*, and *D*.

Solution:

The two loading states can be analyzed independently initially. The torque *T* causes a maximum shear stress τ_1 at $r = d/2$, where

$$\tau_1 = \frac{Td/2}{(\pi/32)d^4} = \frac{16T}{\pi d^3} \qquad \text{[a]}$$

The torsional shear stresses at points *A*, *B*, *C*, and *D* are illustrated in Fig. 3.7-6(*a*).

The transverse force *P* causes bending stresses as well as transverse shear stresses. The bending moment at the wall is $M_z = -PL$, and the shear force is $V_y = -P$. The bending stresses are given by

$$\sigma_x = -\frac{M_z y}{I_z}$$

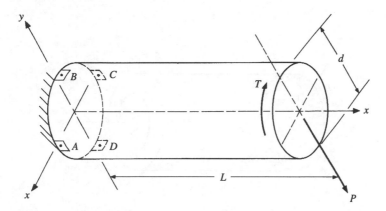

Figure 3.7-5

where $I_z = (\pi/64)d^4$. Thus, at points A, B, C, and D, the bending stresses are

$$(\sigma_x)_A = -\frac{(-PL)(0)}{(\pi/64)d^4} = 0 \qquad (\sigma_x)_B = -\frac{(-PL)(d/2)}{(\pi/64)d^4} = \frac{32PL}{\pi d^3}$$

$$(\sigma_x)_C = -\frac{(-PL)(0)}{(\pi/64)d^4} = 0 \qquad (\sigma_x)_D = -\frac{(-PL)(-d/2)}{(\pi/64)d^4} = -\frac{32PL}{\pi d^3}$$

The shear stresses, given by Eq. (3.4-6), reduce to [see also Eq. (3.4-9)]

$$(\tau_2)_B = (\tau_2)_D = 0 \qquad (\tau_2)_A = (\tau_2)_C = \frac{4}{3}\frac{V_y}{A} = \frac{-16P}{3\pi d^2}$$

The shear and bending stresses induced by the transverse load are illustrated in Fig. 3.7-6(b). The stresses due to the transverse and torsional loading, determined separately, can now be combined by superposition to reveal the complete state of stress. Since the stresses at points A and C are of the same type, namely, shear, the stresses can be added algebraically. Note, however, that at point C the directions of the shear stresses are the same, whereas at point A, the shear stresses are in opposite directions. Since the stresses at points B and D are not of the same type, i.e., both normal and shear stresses exist, the stresses cannot be added algebraically. Thus the combined stresses [illustrated in Fig. 3.7-6(d)] are as follows:

Point A: $\qquad\qquad \sigma_x = 0 \qquad\qquad \tau = \frac{16T}{\pi d^3} - \frac{16P}{3\pi d^2}$

Point B: $\qquad\qquad \sigma_x = \frac{32PL}{\pi d^3} \qquad \tau = \frac{16T}{\pi d^3}$

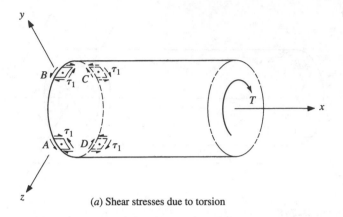

(*a*) Shear stresses due to torsion

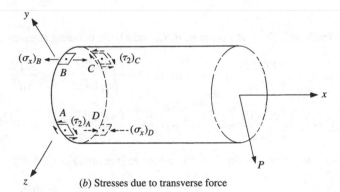

(*b*) Stresses due to transverse force

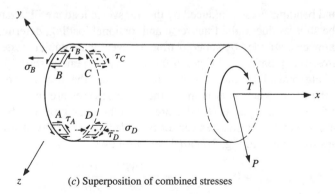

(*c*) Superposition of combined stresses

Figure 3.7-6

Point C: $\sigma_x = 0$ $\tau = \dfrac{16T}{\pi d^3} + \dfrac{16P}{3\pi d^2}$

Point D: $\sigma_x = -\dfrac{32PL}{\pi d^3}$ $\tau = \dfrac{16T}{\pi d^3}$

When the analysis of an element at a point is complete, the state of stress with respect to other internal surfaces at the point can be scrutinized using stress transformations, discussed in detail in Secs. 2.1 and 3.9.

3.8   STATICALLY INDETERMINATE PROBLEMS

The method of superposition is very effective in dealing with simple, statically indeterminate problems where support reactions cannot be completely determined from the equations of equilibrium. When the geometry of the system begins to become complex, superposition becomes difficult and energy techniques become more advantageous (see Chap. 6).

   When a statically indeterminate problem arises, there will be more unknown support or internal reactions than there are equations of equilibrium. The difference will be the *order* for which the structure is redundant or indeterminate. Thus additional equations *equal* to this order of indeterminateness are necessary to complete the solution of the support reactions. The additional equations *must* be arrived at through a *deflection* analysis of the structure. The technique for obtaining these additional deflection equations in this book may differ from the reader's past experience, but it works well for all problems, simple as well as complex, and should be studied carefully. The steps are as follows:

Step 1   Solve for all possible unknown reactions of each member of the structure using the equilibrium equations (even in statically indeterminate problems, some but not all of the reactions may be solvable).

Step 2   Subtract the number of remaining equilibrium equations from the number of remaining unknowns. This is the *order n* for which the structure is indeterminate *and* the number of additional deflection equations necessary.

Step 3   Eliminate *n* of the unknown reactions. At this point, the structure should be statically determinate. Using the applied loads, solve for the deflections and/or rotations of the structure corresponding to the degrees of freedom of the eliminated *n* reactions.

Step 4   Considering the *n* unknown reactions of step 3 as "applied" forces, determine the deflections and/or rotations at the same *n* points of step 3 but this time in terms of the unknown reactions.

Step 5   Using the method of superposition, add the results of steps 3 and 4 at each of the n points such that the n deflections and/or rotations are compatible with the geometric boundary conditions of the problem. This results in a set of n simultaneous equations in terms of the n unknown support reactions.

Step 6   Solve the simultaneous equations to obtain the values of the support reactions.

Step 7   Substitute the results of step 6 into the equilibrium equations of step 1 to solve for the remaining unknown reactions.

It is recommended that these steps be read over again after studying each of the following examples.

Example 3.8-1   A step shaft is rigidly supported at each end through thrust bearings, as shown in Fig. 3.8-1(a). The temperature of the shaft is then increased by 5°C and a force $F = 15$ kN is applied at the step as shown. The stiffness of each bearing is $k = 1.5$ GN/m. The shaft is steel with $E = 207$ GPa and a temperature coefficient of expansion of $\alpha = 11.7\ (10^{-6})\ /°C$.

Solution:

The free-body diagram is shown in Fig. 3.8-1(b).

Step 1   The only equilibrium equation pertinent to the problem is the equilibrium of force in the x direction

$$F_A + F_C = F \qquad\qquad\qquad [a]$$

Step 2   There are two unknowns and one equilibrium equation. Thus the structure is indeterminate of order $n = 2 - 1 = 1$.

Step 3   Since $n = 1$, eliminate *one* of the unknown reactions. Arbitrary selecting of F_A to be eliminated results in the diagram shown in Fig. 3.8-2(a).

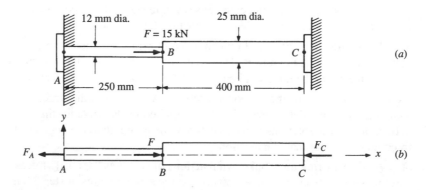

Figure 3.8-1

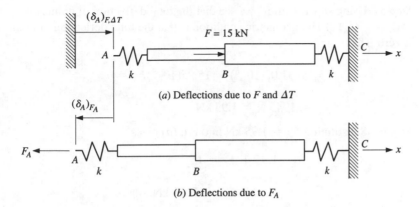

(a) Deflections due to F and ΔT

(b) Deflections due to F_A

Figure 3.8-2

The deflection of point A in the x direction due to the applied force and temperature change is

$$(\delta_A)_{F,\Delta T} = \frac{FL_{BC}}{A_{BC}E} + \frac{F}{k} - \alpha(L_{AB} + L_{BC})\Delta T$$

$$= \frac{(15)(10^3)(0.400)}{(\pi/4)(0.025)^2(207)(10^9)} + \frac{15(10^3)}{1.5(10^9)}$$

$$- 11.7(10^{-6})(0.250 + 0.400)(5)$$

$$= 31.02(10^{-6})\,\text{m} \qquad\qquad\qquad [b]$$

Step 4   The force F_A was eliminated in step 3. The deflection of point A in the positive x direction due to F_A, as shown in Fig. 3.8-2(b), is

$$(\delta_A)_{F_A} = -\frac{F_A L_{AB}}{A_{AB}E} - \frac{F_A L_{BC}}{A_{BC}E} - 2\frac{F_A}{k}$$

$$= -\frac{(F_A)(0.250)}{(\pi/4)(0.012)^2(207)(10^9)} - \frac{(F_A)(0.400)}{(\pi/4)(0.025)^2(207)(10^9)} - 2\frac{F_A}{1.5(10^9)}$$

$$= -15.95(10^{-9})F_A \qquad\qquad\qquad [c]$$

Step 5   Using superposition, we see that the total deflection of point A is $(\delta_A)_{F,\Delta T} + (\delta_A)_{F_A}$. The geometric condition is that the total deflection of point A is zero. Thus

$$\delta_A = 31.02(10^{-6}) - 15.95(10^{-9})F_A = 0$$

$$F_A = 1945 \text{ N} = 1.95 \text{ kN}$$

Step 6   Substituting $F_A = 1.95$ kN into Eq. (*a*) gives

$$1.95 + F_C = 15$$

$$F_C = 13.05 \text{ kN}$$

Example 3.8-2   Determine the support reactions for the beam shown in Fig. 3.8-3(*a*) using
(*a*) beams C.1 and C.3 of Appendix C.
(*b*) beams C.6 and C.7 of Appendix C.

Solution:

(*a*) *Step 1*   Equilibrium in vertical force and moment about point A gives

$$R_A + R_B = wL \qquad R_B L + M_A = \frac{wL^2}{2} \qquad \text{[a]}$$

Step 2   There are three unknowns (R_A, R_B, M_A), and two equilibrium equations. Thus $n = 3 - 2 = 1$.

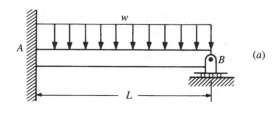

(*a*)

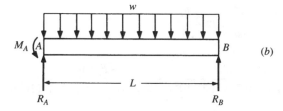

(*b*)

Figure 3.8-3

Step 3 Eliminate one ($n = 1$) unknown. In order to use beams C.1 and C.3, R_B must be eliminated. With R_B eliminated, the deflection of point B due to w is (see beam C.3)

$$(\delta_B)_w = -\frac{wL^4}{8EI} \qquad \text{[b]}$$

Step 4 Considering R_B alone, using beam C.1, we find the deflection of point B to be

$$(\delta_B)_{R_B} = -\frac{(-R_B)L^3}{3EI} = \frac{R_B L^3}{3EI} \qquad \text{[c]}$$

Step 5 Superimposing beams C.1 and C.3 with the restriction that the total deflection of point B must be zero results in

$$\frac{-wL^4}{8EI} + \frac{R_B L^3}{3EI} = 0$$

$$R_B = \frac{3}{8}wL$$

Step 6 Substituting R_B into Eq. (a) yields

$$R_A + \frac{3}{8}wL = wL$$

$$R_A = \frac{5}{8}wL$$

$$\left(\frac{3}{8}wL\right)(L) + M_A = \frac{wL^2}{2}$$

$$M_A = \frac{1}{8}wL^2$$

The results agree with the equations given in the table for beam C.12.

(b) *Step 1* Equations (a) still apply.
   Step 2 Same as part (a), $n = 1$.
   Step 3 In order to use beams C.6 and C.7, M_A must be eliminated. With M_A eliminated, the rotation at point A of beam C.6 due to w is

$$(\theta_A)_w = -\frac{wL^3}{24EI} \qquad \text{[d]}$$

Step 4   Considering M_A alone, using beam C.7 with $x = a = 0$ and $M = M_A$,

$$(\theta_A)_{M_A} = \frac{M_A L}{3EI}$$ [e]

Step 5   Superposing beams C.6 and C.7 with the condition that the net slope at point A is zero gives

$$-\frac{wL^3}{24EI} + \frac{M_A L}{3EI} = 0$$

$$M_A = \frac{1}{8} wL^2$$

Step 6   Substituting M_A into Eqs. (*a*) results in

$$R_B L + \frac{1}{8} wL^2 = \frac{wL^2}{2} \qquad \Rightarrow \qquad R_B = \frac{3}{8} wL$$

$$R_A + \frac{3}{8} wL = wL \qquad \Rightarrow \qquad R_A = \frac{1}{8} wL$$

The results agree with that found in part (*a*).

Example 3.8-3   The steel rods *AD* and *CE* shown in Fig. 3.8-4 each have a diameter of 10 mm. For steel let $E = 200$ GPa. The threads at the ends of the rods are single-threaded with a pitch of 1.5 mm. The nuts are first snugly fit with bar *ABC* horizontal. Next the nut

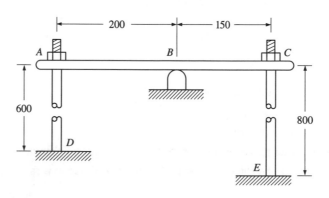

Figure 3.8-4   (Dimensions in millimeters)

at A is tightened one full turn. Determine the resulting tension in each rod and the deflections of points A and C if

(a) bar ABC is considered to be rigid.

(b) bar ABC is steel with a second moment area, $I = 62.5(10^3)$ mm^4.

Solution:

(a) *Step 1*   The free-body diagram of bar ABC is shown in Fig. 3.8-5. Equilibrium of forces and moment about B gives

$$F_A + F_C = F_B \qquad\qquad \textbf{[a]}$$

$$\sum M_B = (200)(F_A) - (150)(F_C) = 0$$

$$F_A = 0.75F_C \qquad\qquad \textbf{[b]}$$

Step 2   With two equations and three unknowns, $n = 3 - 2 = 1$.

Step 3   For this problem, the selection of the one unknown to eliminate is not so arbitrary. Since the turning of the nut at A is where the load is "applied," we do not eliminate F_A. If we eliminate F_C, when the nut is advanced 1.5 mm (the thread pitch times the number of turns), bar ABC simply rotates freely about point B as shown by the dotted line in Fig. 3.8-6, inducing no forces. Point C moves $(150/200)\ 1.5 = 1.125$ mm up.

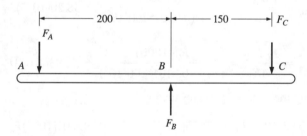

Figure 3.8-5

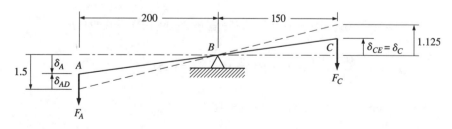

Figure 3.8-6

Step 4   Reapplying the force F_C causes the bar to rotate clockwise to its final position as shown by the solid line in Fig. 3.8-6.

Step 5   The stretch of the rods is the distances δ_{AD} and δ_{CE}. The net deflection (in meters) of points A and C in terms of the stretch of the rods and the turning of the nut is

$$\delta_A = 1.5 \times 10^{-3} - \delta_{AD} \qquad \qquad \text{[c]}$$

$$\delta_C = \delta_{CE} \qquad \qquad \text{[d]}$$

Because the bar is considered rigid

$$\frac{\delta_A}{200} = \frac{\delta_C}{150} \qquad \qquad \text{[e]}$$

Substitution of Eqs. (*c*) and (*d*) gives

$$\frac{1.5(10^{-3}) - \delta_{AD}}{200} = \frac{\delta_{CE}}{150}$$

$$\delta_{CE} = 1.125(10^{-3}) - 0.75\delta_{AD} \qquad \qquad \text{[f]}$$

The stretch in a rod is given by $(FL)/(AE)$. Thus

$$\delta_{AD} = \frac{F_A L_{AD}}{AE} = \frac{F_A(0.600)}{(\pi/4)(0.010)^2 \, 200(10^9)} = 38.20(10^{-9})F_A \qquad \text{[g]}$$

$$\delta_{CE} = \frac{F_C L_{CE}}{AE} = \frac{F_C(0.800)}{(\pi/4)(0.010)^2 \, 200(10^9)} = 50.93(10^{-9})F_C \qquad \text{[h]}$$

Substituting these into Eq. (*f*) results in

$$50.93(10^{-9})F_C = 1.125(10^{-3}) - (0.75)(38.20)(10^{-9})F_A$$

$$F_C = 22.09(10^3) - 0.5625F_A$$

Substituting in Eq. (*b*) gives

$$F_C = 22.09(10^3) - 0.5625(0.75F_C) = 22.09(10^3) - 0.4219F_C$$

$$F_C = \frac{22.09(10^3)}{1.4219} = 15.54(10^3) \text{ N} = 15.54 \text{ kN}$$

Step 6   From Eq. (*b*) the force in rod *AD* is

$$F_A = 0.75F_C = 0.75(15.54) = 11.7 \text{ kN}$$

Deflections. The downward deflection of point A is given by Eq. (*c*).

$$\delta_A = 1.5(10^{-3}) - \frac{F_A L_{AD}}{AE} = 1.5(10^{-3}) - \frac{11.65(10^3)0.600}{(\pi/4)(0.010)^2 200(10^9)}$$

$$= 1.055(10^{-3}) \text{ m} = 1.055 \text{ mm}$$

The upward deflection of point C is given by Eq. (*d*).

$$\delta_C = \frac{F_C L_{CE}}{AE} = \frac{15.54(10^3)0.800}{(\pi/4)(0.010)^2 200(10^9)}$$

$$= 791(10^{-6}) \text{ m} = 0.791 \text{ mm}$$

(*b*) The solid line in Fig. 3.8-6 is the final rotated position of bar ABC if it is rigid. To include bending, first consider how the bar bends without the rotation about point B. This is shown in Fig. 3.8-7. The end deflection of an end-loaded cantilever beam is $(FL^3)/(3EI)$. Thus

$$(\delta_A)_E = \frac{F_A L_{AB}^3}{3EI} = \frac{F_A(0.200)^3}{3(200)(10^9)62.5(10^3)(10^{-3})^4} = 213.3(10^{-9})F_A \qquad \textbf{[i]}$$

$$(\delta_C)_E = \frac{F_C L_{BC}^3}{3EI} = \frac{F_C(0.150)^3}{3(200)(10^9)62.5(10^3)(10^{-3})^4} = 90(10^{-9})F_C \qquad \textbf{[j]}$$

Figure 3.8-8 shows the deflections where $(\delta_A)_R$ and $(\delta_C)_R$ are the rigid-body deflections of bar ABC. Equation (*e*) still applies to them. That is,

$$\frac{(\delta_A)_R}{200} = \frac{(\delta_C)_R}{150}$$

where

$$(\delta_A)_R = 1.5(10^{-3}) - [\delta_{AD} + (\delta_A)_E] \qquad \textbf{[k]}$$

$$(\delta_C)_R = \delta_{CE} + (\delta_C)_E \qquad \textbf{[l]}$$

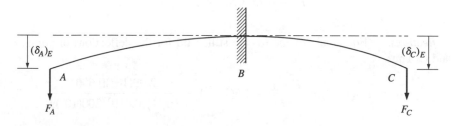

Figure 3.8-7 Bending deflections of ABC without rotation at B.

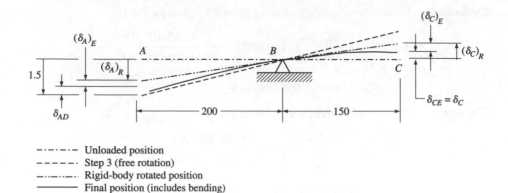

--·--·-- Unloaded position
-- -- -- · Step 3 (free rotation)
--·--·--· Rigid-body rotated position
—————— Final position (includes bending)

Figure 3.8-8

Thus

$$\frac{1.5(10^{-3}) - [\delta_{AD} + (\delta_A)_E]}{200} = \frac{\delta_{CE} + (\delta_C)_E}{150} \qquad [m]$$

Substituting Eqs. (*g*), (*h*), (*i*), and (*j*) gives

$$\frac{1.5(10^{-3}) - [38.2(10^{-9})F_A + 213.3(10^{-9})F_A]}{200} = \frac{50.9(10^{-9})F_C + 90(10^{-9})F_C}{150}$$

Simplification results in

$$F_C = 7983 - 1.339F_A$$

Substituting Eq. (*b*) gives

$$F_C = 7983 - 1.339(0.75F_C)$$

$$F_C = 3984 \text{ N} = 3.98 \text{ kN}$$

Substituting F_C into Eq. (*b*) yields the force in rod *AD*

$$F_A = 0.75(3.98) = 2.99 \text{ kN}$$

Deflections.   Equations (*c*) and (*d*) are still valid for the deflections of points *A* and *C*. Thus

$$\delta_A = 1.5(10^{-3}) - \frac{F_A L_{AD}}{AE} = 1.5(10^{-3}) - \frac{2.99(10^3)0.600}{(\pi/4)(0.010)^2 200(10^9)}$$

$$= 1.386(10^{-3}) \text{ m} = 1.386 \text{ mm}$$

$$\delta_C = \frac{F_C L_{CE}}{AE} = \frac{3.98(10^3)0.800}{(\pi/4)(0.010)^2 200(10^9)} = 202.7(10^{-6}) \text{ m} = 0.203 \text{ m}$$

We see a significant difference between the results of parts (*a*) and (*b*), especially in the forces. As bar *ABC* becomes more flexible, the forces in the rods diminish. As a matter of fact, when the bar becomes very flexible, δ_{AD} and δ_{CE} can be ignored in Eq. (*m*) as the forces in the rods are negligible.

3.9 STRESS AND STRAIN TRANSFORMATIONS

Two-dimensional transformation of the state of plane stress is one of the most important concepts covered in elementary mechanics of materials, as many practical engineering problems involve investigating this state of stress. The general three-dimensional transformation equations are developed in Sec. 2.1.[†]

3.9.1 PLANE STRESS

A given state of plane stress relative to an *xyz* coordinate system is shown in Fig. 3.9-1(*a*). Transforming the stress to an *x'y'z'* coordinate system is shown in Fig. 3.9-1(*b*), where the transformed stresses are given by Eqs. (2.1-8) as

$$\sigma_{x'} = \sigma_x \cos^2 \theta + \sigma_y \sin^2 \theta + 2\tau_{xy} \cos \theta \sin \theta \qquad \textbf{[3.9-1]}$$

$$\sigma_{y'} = \sigma_x \sin^2 \theta + \sigma_y \cos^2 \theta - 2\tau_{xy} \cos \theta \sin \theta \qquad \textbf{[3.9-2]}$$

$$\tau_{x'y'} = -(\sigma_x - \sigma_y) \sin \theta \cos \theta + \tau_{xy}(\cos^2 \theta - \sin^2 \theta) \qquad \textbf{[3.9-3]}$$

The normal stress and shear stress (σ, τ) on a single surface, rotated θ counterclockwise from the *x* surface, are also given by Eqs. (2.1-9) and (2.1-10), which are

$$\sigma = \frac{\sigma_x + \sigma_y}{2} + \frac{\sigma_x - \sigma_y}{2} \cos 2\theta + \tau_{xy} \sin 2\theta \qquad \textbf{[3.9-4]}$$

$$\tau = -\frac{\sigma_x - \sigma_y}{2} \sin 2\theta + \tau_{xy} \cos 2\theta \qquad \textbf{[3.9-5]}$$

[†] For a simpler derivation of the two-dimensional stress transformation equations, see any elementary mechanics of materials textbook.

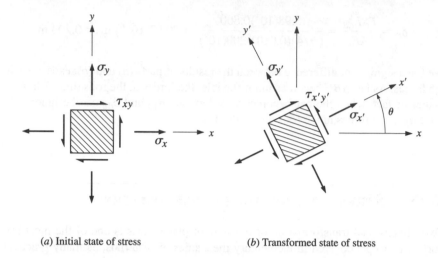

(a) Initial state of stress   (b) Transformed state of stress

Figure 3.9-1   State of plane stress.

Example 2.1-2 illustrates the use of the transformation equations and Mohr's circle for the case of plane stress. Example 2.1-6 illustrates the methods of determining the values and directions of the principal stresses.

3.9.2   PRINCIPAL STRESS

Recall the equation for the two principal stresses in the plane of a two-dimensional analysis given by Eq. (2.1-30) is

$$\sigma_p = \frac{\sigma_x + \sigma_y}{2} \pm \sqrt{\left(\frac{\sigma_x - \sigma_y}{2}\right)^2 + \tau_{xy}^2} \qquad \textbf{[3.9-6]}$$

We will explore the relationship between Eqs. (3.9-4) and (3.9-5) with the principal stress state.

Problem 2.24 deals with the stationary nature of the principal stresses which exist on surfaces containing no shear stress. Here we will demonstrate this for the state of plane stress. For a stationary value, set the derivative of σ in Eq. (3.9-4) with respect to θ equal to zero. This yields

$$\frac{d\sigma}{d\theta} = -(\sigma_x - \sigma_y)\sin 2\theta_p + 2\tau_{xy}\cos 2\theta_p = 0 \qquad \textbf{[a]}$$

where θ_p is the angle to the maximum and minimum normal (principal) stresses. Comparing Eq. (a) with Eq. (3.9-5) we see that $d\sigma/d\theta = 2\tau$. Since the derivative is zero, $\tau = 0$ on surfaces containing the stationary principal stresses. Solving Eq. (a) for θ_p gives

$$\tan 2\theta_p = \frac{2\tau_{xy}}{\sigma_x - \sigma_y}$$ [b]

or

$$\theta_p = \frac{1}{2} \tan^{-1} \frac{2\tau_{xy}}{\sigma_x - \sigma_y}$$ [3.9-7]

The $\tan^{-1}$ can be treated as double-valued with two values being 180° apart. Thus in Eq. (3.9-7) the two surfaces containing the principal stresses are 90° apart. Substituting the two values of θ_p into Eq. (3.9-4) separately yields the two principal stresses and their orientation.[†]

Note that if $\sigma_x = \sigma_y = \sigma$ and $\tau_{xy} = 0$, then σ_x and σ_y are principal stresses and Eq. (3.9-7) is indeterminate. This is because any surface in the plane of analysis, regardless of the value of θ, contains the same principal stress σ. The state of stress at the point is said to be *isotropic* in the plane.

Now, a common mistake is to calculate the two values of θ_p from Eq. (3.9-7) and the two principal stresses σ_p from Eq. (3.9-6) and somehow *magically* match each pair. The substitution of each θ_p into Eq. (3.9-4) will provide the principal stress in the θ_p direction. However, the orientation of the principal stresses obtained from Eq. (3.9-6) can be established if the $\tan^{-1}$ in Eq. (3.9-7) is treated as *single-valued*. This concept will be particularly helpful in Chap. 8 when dealing with strain-gage rosette problems. The single-valued solution of Eq. (3.9-7) is determined by *maintaining* the signs of the numerator and denominator of $\tan^{-1}$ *separately* to determine which of the *four* possible quadrants the $\tan^{-1}$ is located. When using a calculator, one normally loses the separate signs of the numerator and denominator and determines the $\tan^{-1}$ in either the first or fourth quadrant.[‡]

To establish the orientations of the principal stresses obtained from Eq. (3.9-6), we see whether the single-valued result of Eq. (3.9-7) gives a *maximum or minimum*. To do this, we take the second derivative of σ in Eq. (3.9-4) with respect to θ. This yields

$$\frac{d^2\sigma}{d\theta^2} = -2(\sigma_x - \sigma_y) \cos 2\theta_p - 4\tau_{xy} \sin 2\theta_p$$ [c]

[†] Always keep in mind that there will always be three principal stresses. This is discussed in more detail in Sec. 2.1-3. Conventional practice is to label the three principal stresses as σ_1, σ_2, and σ_3 such that $\sigma_1 \geq \sigma_2 \geq \sigma_3$. For the state of plane stress shown in Fig. 3.9-1, the stresses on the surface perpendicular to the z direction are zero. This means no shear stress, and thus the normal stress on this surface, σ_z, is *also* a principal stress with a value of zero. The other two principal stresses are found from Eq. (3.9-6) and the *three* principal stresses can *then* be ordered, $\sigma_1 \geq \sigma_2 \geq \sigma_3$.

[‡] In computer languages such as FORTRAN, Basic, etc., the function ATAN(y/x) determines the $\tan^{-1}$ within the first and fourth quadrant only. The *two-argument inverse tangent function* ATAN2(y,x) determines $\tan^{-1}$ within *all four quadrants*.

From Eq. (b)

$$\sin 2\theta_p = \frac{2\tau_{xy}}{\sqrt{(\sigma_x - \sigma_y)^2 + 4\tau_{xy}^2}} \qquad \cos 2\theta_p = \frac{\sigma_x - \sigma_y}{\sqrt{(\sigma_x - \sigma_y)^2 + 4\tau_{xy}^2}}$$

Substituting this into Eq. (c) gives

$$\frac{d^2\sigma}{d\theta^2} = -\left[\frac{2(\sigma_x - \sigma_y)^2 + 8\tau_{xy}^2}{\sqrt{(\sigma_x - \sigma_y)^2 + 4\tau_{xy}^2}} \right] \qquad \text{[d]}$$

The term within the brackets is always positive. Therefore, the second derivative is always negative and we have a *maximum*. Thus the *single value* θ_p obtained from Eq. (3.9-7) corresponds to the maximum of the two principal stresses determined by Eq. (3.9-6).

Example 3.9-1   Determine the principal stresses for Example 2.1-2 and show the element containing these stresses properly oriented with respect to the initial xyz coordinate system using
(a) Eqs. (3.9-7) and (3.9-4).
(b) Eq. (3.9-6) and the single-valued solution of Eq. (3.9-7).

Solution:

Recall from Example 2.1-2 the initial stress matrix given by

$$\sigma = \begin{bmatrix} 3 & -4 \\ -4 & 9 \end{bmatrix} \text{ kpsi}$$

(a) From Eq. (3.9-7)

$$\theta_p = \frac{1}{2} \tan^{-1} \frac{(2)(-4)}{3 - 9} = \frac{1}{2} \tan^{-1} \frac{-8}{-6} = 26.6°, -63.4° \qquad \text{[a]}$$

For $\theta_{p1} = 26.6°$ Eq. (3.9-4) gives

$$\sigma_{p1} = \frac{3 + 9}{2} + \left(\frac{3 - 9}{2} \right) \cos\left[(2)(26.6) \right] + (-4) \sin\left[(2)(26.6) \right] = 1 \text{ kpsi}$$

For $\theta_{p2} = -63.4°$ Eq. (3.9-4) gives

$$\sigma_{p2} = \frac{3 + 9}{2} + \left(\frac{3 - 9}{2} \right) \cos\left[(2)(-63.4) \right] + (-4) \sin\left[(2)(-63.4) \right] = 11 \text{ kpsi}$$

The initial and principal states of stress are shown in Fig. 3.9-2. The three principal stresses ($\sigma_1, \sigma_2, \sigma_3$) are (11, 1, 0).

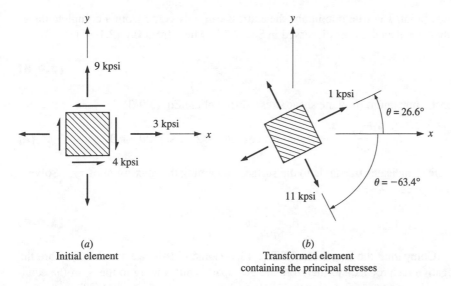

(a)
Initial element

(b)
Transformed element
containing the principal stresses

Figure 3.9-2

(b) From Eq. (3.9-6)

$$\sigma_p = \frac{3+9}{2} \pm \sqrt{\left(\frac{3-9}{2}\right)^2 + (-4)^2} = 6 \pm 5 = 11, 1 \text{ kpsi}$$

In part (a), Eq. (a), the numerator and denominator of the argument of the $\tan^{-1}$ are both negative. Thus the single-valued solution of $\tan^{-1}$ is in the *third quadrant*. That is,

$$\theta_p = \frac{1}{2} \tan^{-1} \frac{(2)(-4)}{3-9} = \frac{1}{2} \tan^{-1} \frac{-8}{-6} = \frac{1}{2}(233.13°) = 116.6°$$

This surface, which is the same as $\theta_p = -63.4°$, contains the maximum principal stress calculated by Eq. (3.9-6), 11 kpsi. The minimum principal stress, 1 kpsi, is on the surface 90° from the maximum. This solution is also represented by Fig. 3.9-2(b).

3.9.3   MAXIMUM IN-PLANE SHEAR STRESS

The transformation equations given in this section, Eqs. (3.9-1) to (3.9-3), restrict the transformation process to the determination of stresses on surfaces perpendicular to the *xy* plane. The largest shear stress found under these conditions is called the *maximum in-plane shear stress* and may not be the absolute maximum shear stress at the point. The true maximum shear stress can only come from a complete three-dimensional analysis as discussed in Sec. 2.1.7, where from Eq. (2.1-36)

$$\tau_{max} = \frac{\sigma_1 - \sigma_3}{2}$$ [3.9-8]

For the maximum in-plane shear stress, differentiate Eq. (3.9-3)

$$\frac{d\tau_{x'y'}}{d\theta} = -(\sigma_x - \sigma_y) \cos 2\theta_s - 2\tau_{xy} \sin 2\theta_s = 0$$ [a]

where θ_s indicates the angle to the surface containing the maximum of $\tau_{x'y'}$. Solving for θ_s gives

$$\theta_s = \frac{1}{2} \tan^{-1} \frac{\sigma_y - \sigma_x}{2\tau_{xy}}$$ [3.9-9]

Comparing the arguments of Eqs. (3.9-7) and (3.9-9) we see that they are the negative reciprocals and therefore are 90° apart. Thus, owing to the $\frac{1}{2}$ in the equations, the surfaces containing the in-plane shear stress maxima are 45° from the surfaces containing the principal stresses.

Considering the $\tan^{-1}$ in Eq. (3.9-9) to be doubled-valued, two values of θ_s can be determined which are 90° apart. Substituting the two values of θ_s into Eq. (3.9-4) separately yields the same normal stress, which is the average of σ_x and σ_y. Thus on the two orthogonal surfaces containing the maximum in-plane shear stress, the normal stresses are equal and are given by

$$\sigma_{av} = \frac{\sigma_x + \sigma_y}{2}$$ [3.9-10]

Substituting either value of θ_s into Eq. (3.9-5) will yield the state of stress for the element containing the maximum in-plane shear stress.

Example 3.9-2 | For the state of stress of Example 3.9-1 determine (a) the state of stress on the element containing the maximum in-plane shear stress and (b) the absolute maximum shear stress.

Solution:

(a) Again, the initial state of stress is $\sigma_x = 3$, $\sigma_y = 9$, and $\tau_{xy} = -4$ kpsi. From Eq. (3.9-9)

$$\theta_s = \frac{1}{2} \tan^{-1} \frac{9 - 3}{2(-4)} = 71.6°, -18.4°$$

The normal stresses on the element containing the maximum in-plane shear stress are

$$\sigma_{av} = \frac{3 + 9}{2} = 6 \text{ kpsi}$$

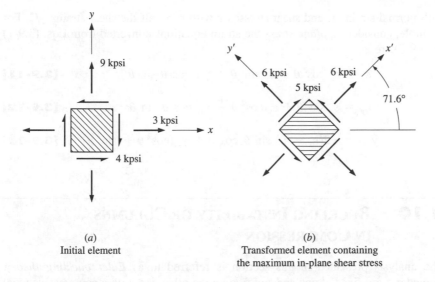

(a)
Initial element

(b)
Transformed element containing
the maximum in-plane shear stress

Figure 3.9-3

Arbitrarily selecting $\theta_s = 71.6°$ and substituting this into Eq. (3.9-5) gives

$$\tau = -\frac{3-9}{2}\sin\left[(2)(71.6)\right] + (-4)\cos\left[(2)(71.6)\right] = 5 \text{ kpsi}$$

Thus the shear stress on this surface is positive based on a coordinate rotation of 71.6° counterclockwise. With cross shears being equal, the shear stress on the adjacent orthogonal surface can be established. Thus the state of stress of the element containing the maximum in-plane shear stress is as shown in Fig. 3.9-3(b).

(b) Since from Example 3.9-1, σ_1, σ_2, and σ_3 are 11, 1, and 0, respectively, from Eq. (3.9-8)

$$\tau_{max} = \frac{11-0}{2} = 5.5 \text{ kpsi}$$

The surface containing this shear stress is not perpendicular to the xy plane and can be found by a two-dimensional transformation viewing the element containing the principal stresses σ_1 and σ_3 (see Example 2.1-7).

3.9.4   STRAIN TRANSFORMATIONS

As discussed in Sec. 2.2, the stress transformation equations can be directly converted to strain transformation equations by simply replacing normal stresses σ

with normal strains ε, and shear stresses τ with one-half the shear strains $\gamma/2$. For example, considering *plane stress* the strain equations converted from Eqs. (3.9-1) to (3.9-3) are

$$\varepsilon_{x'} = \varepsilon_x \cos^2 \theta + \varepsilon_y \sin^2 \theta + \gamma_{xy} \cos \theta \sin \theta \qquad \textbf{[3.9-11]}$$

$$\varepsilon_{y'} = \varepsilon_x \sin^2 \theta + \varepsilon_y \cos^2 \theta - \gamma_{xy} \cos \theta \sin \theta \qquad \textbf{[3.9-12]}$$

$$\gamma_{x'y'} = -2(\varepsilon_x - \varepsilon_y) \sin \theta \cos \theta + \gamma_{xy}(\cos^2 \theta - \sin^2 \theta) \qquad \textbf{[3.9-13]}$$

3.10   BUCKLING INSTABILITY OF COLUMNS IN COMPRESSION

The analysis presented in this section is referred to as *Euler buckling theory*. Consider a rod fixed at one end and free on the other end with a compressive axial force, as shown in Fig. 3.10-1(*a*). Linear elasticity theory will predict no bending deflection in this case. However, since buckling is an instability phenomenon, it is necessary to assume the rod to be perturbed slightly in the lateral direction. Thus the beam at the instant of perturbation would appear as in Fig. 3.10-1(*b*). If the load *P* is small, the disturbance will diminish and the rod will return to its original state as shown in Fig. 3.10-1(*a*). However, if the load is large enough, the disturbance will grow as the force causes increased bending and the rod begins to buckle. Linear elasticity theory will predict the value of the load which causes buckling, as well as the initial deflection shape. However, the actual magnitude of the deflection as a function of load must be arrived at through a nonlinear analysis. In the discussion which follows, it is assumed that the *yz* axes are principal axes of the second-area

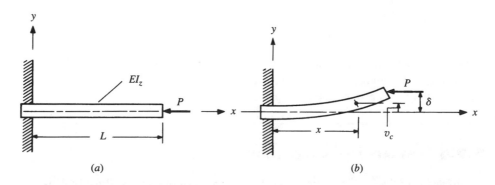

(*a*)   (*b*)

Figure 3.10-1

moment (see Sec. 5.3 or Appendix E) and that the moment of inertia I_y about the y axis is greater than I_z. Thus, if buckling occurs, the rod will first bend about the z axis.

Returning to Fig. 3.10-1(b) and neglecting the weight of the rod, we see that the bending moment at any point in the rod is[†]

$$M_z = P(\delta - v_c)$$

Using the deflection equation $d^2v_c/dx^2 = M_z/(EI_z)$ results in

$$\frac{d^2v_c}{dx^2} + \frac{P}{EI_z}v_c = \frac{P}{EI_z}\delta \qquad \textbf{[3.10-1]}$$

This equation represents a second-order linear nonhomogeneous differential equation for which the solution is

$$v_c = C_1 \sin kx + C_2 \cos kx + \delta$$

where C_1 and C_2 are constants of integration and $k = \sqrt{P/(EI_z)}$.

To determine C_1 and C_2, geometric conditions that $v_c = 0$ and $dv_c/dx = 0$ at $x = 0$ yield $C_1 = 0$ and $C_2 = -\delta$. Thus

$$v_c = \delta(1 - \cos kx) \qquad \textbf{[3.10-2]}$$

This equation gives the buckling shape of the rod. However, we still have two unknowns δ and k (since P is unknown). When $x = L$, $v_c = \delta$, and there are two cases in which Eq. (3.10-2) is valid, namely, $\delta = 0$ and $\cos kL = 0$. If $\delta = 0$, the rod will return from the perturbation to its original nonbending shape. If $\cos kL = 0$, a bending deflection value of δ is possible for equilibrium. There are distinct forces which will cause this. For $\cos kL = 0$,

$$kL = \frac{n\pi}{2} \qquad n = 1, 3, 5, \ldots$$

or, since $k = n\pi/(2L) = \sqrt{P/(EI_z)}$, then

$$P = \left(\frac{n\pi}{2L}\right)^2 EI_z \qquad n = 1, 3, 5, \ldots$$

The first opportunity for the rod to buckle is when $n = 1$. Thus the critical load is

$$P_{cr} = \frac{\pi^2 EI_z}{4L^2} \qquad \textbf{[3.10-3]}$$

and is called the *Euler load*.

[†] Note that the bending moment is no longer a function of the initial geometry but is strongly coupled to deflections.

With $k = \pi/(2L)$, the deflection curve of the rod is

$$v_c = \delta\left(1 - \cos\frac{\pi x}{2L}\right) \qquad \textbf{[3.10-4]}$$

However, no additional information can be found from linear theory, and the value of δ cannot be determined.

The critical load depends on the geometric end conditions of the rod. Equations (3.10-3) and (3.10-4) can be rewritten for various end conditions. The Euler load is

$$P_{cr} = \frac{\pi^2 EI_z}{(KL)^2} \qquad \textbf{[3.10-5]}$$

where K and the buckling shape are given in Table 3.10-1.

Table 3.10-1

Constraint		K	v_c
Fixed-Free		2.0	$\delta_{max}\left(1 - \cos\dfrac{\pi x}{2L}\right)$
Pinned-Pinned		1.0	$\delta_{max}\sin\dfrac{\pi x}{L}$
Fixed-Pinned		0.699	$0.715\,\delta_{max}\left(1 - \cos\dfrac{ax}{L} + \dfrac{1}{a}\sin\dfrac{ax}{L} - \dfrac{x}{L}\right)$ $a = 4.493$
Fixed-Fixed		0.5	$0.5\,\delta_{max}\left(1 - \cos\dfrac{2\pi x}{L}\right)$

The value of the stress corresponding to the critical load is called the critical stress or critical unit load and is given by

$$\sigma_{cr} = \frac{P_{cr}}{A} = \frac{\pi^2 E I_z}{(KL)^2 A} \qquad \textbf{[3.10-6]}$$

The term $\sqrt{I_z/A}$ is called the *radius of gyration* r_g. Thus, in Eq. (3.10-6), $I_z/A = r_g^2$, which gives

$$\sigma_{cr} = \frac{\pi^2 E}{(KL/r_g)^2} \qquad \textbf{[3.10-7]}$$

The dimensionless term L/r_g is referred to as the *slenderness ratio*. A plot of σ_{cr} as a function of the slenderness ratio, called *Euler's curve,* is given in Fig. 3.10-2(*a*). There is an upper bound on σ_{cr} based on the failure criterion of yield strength or ultimate strength under pure compression. Figure 3.10-2(*b*) corresponds to the yield strength S_y being the upper limit.

Section 7.6 provides further discussion of column buckling including buckling modes, inelastic buckling, column-beam behavior, and design equations. A short section on plate buckling is also included.

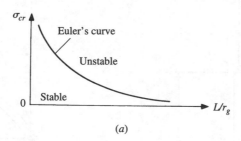

(*a*)

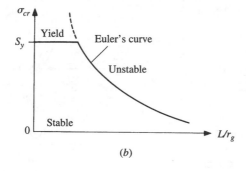

(*b*)

Figure 3.10-2

Example 3.10-1

The column shown in Fig. 3.10-3(a) is pinned at points A and B with a compressive force P at the free end C. Assuming that the column will buckle in the xy plane, determine the critical buckling force P_{cr} as a function of the ratio of a/L. Determine P_{cr} if $EI_z = 1.2$ kN·m^2, $L = 750$ mm, and $a = 500$ mm.

Solution:

Figure 3.10-3(b) shows the free-body diagram of the column in its perturbed state. The bending moment throughout the column is given by

$$M_z = -Pv_c - \frac{P\delta}{a}x \qquad 0 \le x \le a \qquad \text{[a]}$$

$$M_z = -P(\delta + v_c) \qquad a \le x \le L \qquad \text{[b]}$$

Setting $M_z = EId^2v_c/dx^2$ for Eq. (a) yields

$$\frac{d^2v_c}{dx^2} + k^2v_c = -k^2\frac{\delta}{a}x$$

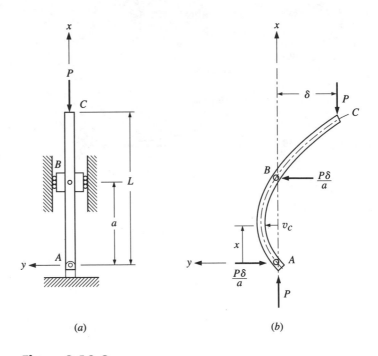

(a)   (b)

Figure 3.10-3

where $k^2 = P/(EI)$. Thus the solution of the differential equation for region AB is

$$(v_c)_{AB} = A_1 \sin kx + A_2 \cos kx - \frac{\delta}{a}x$$

where A_1 and A_2 are constants. Substituting the boundary conditions $v_c = 0$ at $x = 0$ and a results in

$$(v_c)_{AB} = \delta\left(\frac{\sin kx}{\sin ka} - \frac{x}{a}\right) \qquad \text{[c]}$$

From Eq. (b) the differential equation for region BC is

$$\frac{d^2 v_c}{dx^2} + k^2 v_c = -k^2 \delta$$

with the solution

$$(v_c)_{BC} = B_1 \sin kx + B_2 \cos kx - \delta \qquad \text{[d]}$$

The first boundary condition is $v_c = 0$ at $x = a$. This yields

$$B_1 \sin ka + B_2 \cos ka = \delta \qquad \text{[e]}$$

Next we need to match slopes for the two sections. That is, at $x = a$, $(dv_c/dx)_{AB} = (dv_c/dx)_{BC}$. This results in

$$B_1 \cos ka - B_2 \sin ka = \frac{\delta}{ka}\left(ka\,\frac{\cos ka}{\sin ka} - 1\right) \qquad \text{[f]}$$

Solving Eqs. (e) and (f) simultaneously gives

$$B_1 = \frac{\delta}{ka \sin ka}(ka - \sin ka \cos ka) \qquad \text{[g]}$$

$$B_2 = \frac{\delta}{ka} \sin ka \qquad \text{[h]}$$

Finally, we substitute the condition that $v_c = -\delta$ at $x = L$ in Eq. (d). This gives

$$-\delta = B_1 \sin kL + B_2 \cos kL - \delta$$

Substituting B_1 and B_2 from Eqs. (g) and (h) and simplifying gives the equation

$$(ka - \sin ka \cos ka)\sin kL + \sin^2 ka \cos kL = 0 \qquad \text{[i]}$$

Equation (*i*) can be solved more easily by substituting *cL* for *a* in Eq. (*i*) where *c* is a known fraction of the total length of the beam. Thus

$$F(kL) = (ckL - \sin ckL \cos ckL) \sin kL + \sin^2 ckL \cos kL = 0 \qquad [j]$$

For a given value of *c*, the lowest value of *kL* that satisfies Eq. (*j*) provides the value for the critical buckling force, P_{cr}. Equation (*j*) can be solved easily using a root searching method. Starting at *kL* = 0, increment *kL* until *F(kL)* changes sign. Then using smaller and smaller increments, the first root of *kL* can be found to whatever accuracy desired. This can be done relatively quickly using a programmable calculator or a spreadsheet program. Table 3.10-2 gives values of *kL* for a range of *a*/*L*.

Table 3.10-2   Fundamental roots of Eq. (*j*)

c = *a*/*L*	*kL*	Comments based on Table 3.10-1
0.001	1.5718	Approaching the fixed-free case where $kL = \pi/2$
0.01	1.5813	
0.1	1.6829	
0.2	1.8119	
0.3	1.9609	
0.4	2.1333	
0.5	2.3311	
0.6	2.5516	
0.7	2.7796	
0.8	2.9789	
0.9	3.1043	
1.0	3.1416	Pinned-pinned case where $kL = \pi$

NOTE: The ratio *a*/*L* = 0 corresponds to the fixed-free case where $kL = \pi/2$. Equation (*i*) will not yield a value of *kL*, since for this case, the force $P\delta/b$ approaches infinity as *a*/*L* approaches zero. However, Table 3.10-2 shows that as *a*/*L* approaches zero, *kL* approaches $\pi/2$. Also, it can be seen that for *a*/*L* = 1, $kL = \pi$ which corresponds to the pinned-pinned case.

For the case given in this example, *a*/*L* = 500/750 = 0.667, and Eq. (*j*) gives a fundamental root of 2.7045 (linear interpolation of the chart would have given 2.7036). Thus

$$kL = L\sqrt{\frac{P_{cr}}{EI_z}} = 2.7045$$

$$P_{cr} = \frac{(2.7045)^2 EI_z}{L^2} = \frac{(2.7045)^2 (1.2)}{(0.750)^2} = 15.60 \text{ kN}$$

3.11 PROBLEMS

3.1 A composite bar of rectangular cross section, 20 × 24 mm, is loaded in tension by a force $P = 1$ kN. The shaded part of the bar is made of a material where Young's modulus is $E_2 = 210$ GPa. The remaining part is made of a material with $E_1 = 105$ GPa. If the bar is to uniformly deflect in the x direction, determine the stresses in each material and the location of the loading axis relative to the center of the bar.

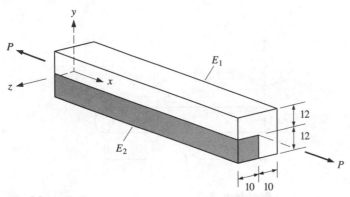

Problem 3.1

3.2 A pin is slightly tapered and placed in a hole with the same taper angle. A force F is then applied to the top of the pin as shown. The force is resisted entirely by friction f (force per unit length) along the pin. If the friction force is assumed to vary with x as shown, establish the constant k as a function of F and L; then determine and plot:
(a) the displacement of the pin as a function of x, $u(x)$.
(b) the normal stress in the pin as a function of x, $\sigma_x(x)$.

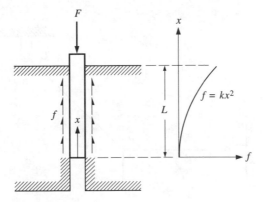

Problem 3.2

3.3 A steel rod of length L rotates with a constant angular velocity ω in a horizontal plane about the y axis.

(*a*) Prove that the internal axial stress in the rod due to dynamic effects is given by $\rho\omega^2 (L^2 - x^2)/2$, where ρ is the mass density of the rod.

(*b*) From part (*a*) determine the deflection of the rod as a function of x and the variables ρ, ω, E, and L.

(*c*) Plot the results of parts (*a*) and (*b*) if the *weight* density of steel $\gamma = 0.284$ lb/in^3, $L = 30$ in, $E = 29$ Mpsi, and $\omega = 100$ rad/s.

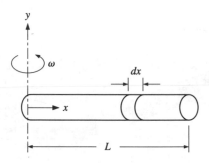

Problem 3.3

3.4 Cables AB and BC each have an effective area of 125 mm^2. The lengths of cables AB and BC are 500 and 750 mm, respectively. A vertical force of $P = 8$ kN is applied to joint B. If the cables are steel with $E = 200$ GPa, determine the stresses in the cables and the deflection of point B assuming small deflections.

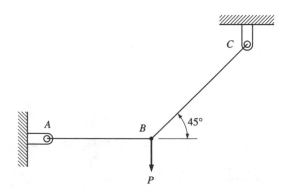

Problem 3.4

3.5   The steel shaft shown is loaded in torsion. A 0.5-in-diameter, 4.0-in-deep flat-bottomed hole is in the 1.0-in-diameter section. If E = 29 Mpsi and v = 0.29, determine (*a*) the maximum shear stresses in the four differing cross sections neglecting any stress concentrations, and (*b*) the total angle of twist of end E relative to end A.

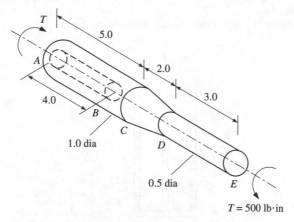

Problem 3.5   (All dimensions are in inches)

3.6   For Problem 1.4 determine the maximum bending stress in member AB if it has a rectangular cross section of 25×50 mm.

3.7   A cantilever beam 1.5 m long has a square box cross section with the outer width and height being 100 mm and a wall thickness of 8 mm. The beam carries a uniform load of 6.5 kN/m along the entire length and, in the same direction, a concentrated force of 4 kN at the free end.
(*a*) Determine the maximum bending stress.
(*b*) Determine the maximum transverse shear stress.
(*c*) Determine the maximum shear stress in the beam.

3.8   The cross section shown is that of a beam in bending about the z axis and transmitting a transverse shear force of $V_y = 2$ kN. The section is symmetric about the y axis, and all dimensions are in millimeters. Determine the shear stresses at $y = 0$ in each wall. All wall thicknesses are 4 mm.

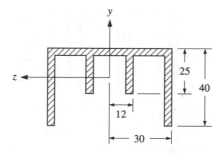

Problem 3.8

3.9 The location of the centroidal z axis for the section of a beam is shown. The beam is transmitting a maximum bending moment of 2000 lb·in about the z axis and a maximum transverse shear force of 500 lb in the y direction. All wall thicknesses are 0.125 in.

(a) Determine the maximum transverse shear stress in each of the vertical walls.

(b) Determine the maximum shear stress in the beam.

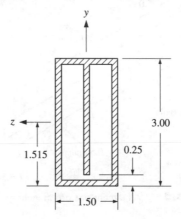

Problem 3.9   (Dimensions in inches)

3.10 An aluminum cantilever beam is 500 mm long and has a cross section as shown. All wall thicknesses are 5 mm. A lateral force of 2.0 kN in the negative y direction is applied to the free end of the beam. For aluminum let $E = 69$ GPa. Determine

(a) the maximum tensile and compressive bending stresses.

(b) the maximum shear stresses in each web wall due to the transverse shear force in the section.

(c) the maximum bending deflection of the beam in the negative y direction.

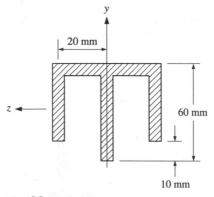

Problem 3.10

3.11 Consider a beam with a solid circular cross section of diameter d transmitting a transverse shear force V_y.
   (*a*) Derive an expression for the average shear-stress distribution across the transverse surface.
   (*b*) Verify that the maximum shear stress occurs at $y_1 = 0$ and is given by Eq. (3.4-10).
   (*c*) Sketch the shear stresses on the transverse surface and discuss any possible inconsistencies in the distribution as related to geometry.

3.12 The maximum average transverse shear stress in a beam is given by

$$\tau_{max} = k\frac{V_y}{A}$$

where $k = 1.5$ for a rectangular section and 1.333 for a solid circular section. Show that $k = 2.0$ for a *thin-walled* circular tube of radius r and wall thickness t (where $r \gg t$).

3.13 The cross section of a beam is constructed by bolting two W 8 $\times$ 40 I-beams together using 0.5-in-diameter bolts spaced longitudinally in the x direction every 5 in. Determine the average shear stress in the bolts if the section is transmitting a transverse shear force of $V_y = 12$ kips. For an 8 $\times$ 40 I-beam the area is 11.7 in^2 and the second-area moment about its own *centroidal* axis parallel to the z axis is $I = 146$ in^4.

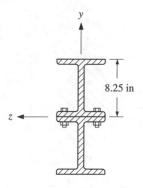

Problem 3.13

3.14 Two 1.5 $\times$ 7.5 in wood planks are nailed together. The composite beam is then cantilevered and an end force of 100 lb is applied as shown. Determine the resulting maximum shear stress in the nails if the diameter of each nail is $\frac{1}{8}$ in and the nails are double rows with a spacing of 6 in.

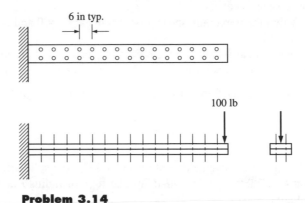

Problem 3.14

3.15 For the diamond-shaped cross section shown, bending occurs about the z axis, and the surface transmits a net shear force V_y in the positive y direction. Since the maximum τ_{xy} shear stress due to V_y occurs where Q/b is a maximum, prove that τ_{xy} is maximum at $y = \pm h/8$.

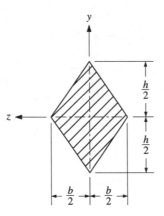

Problem 3.15

3.16 Using the double-integration method, verify the deflection and slope equations as functions of x for beam C.6 given in Appendix C.

3.17 For the beam shown, in addition to the bending and transverse shear stresses, the stress σ_y also exists. This stress was derived in Example 2.4-2 using the equilibrium equations. Here, show that

$$\sigma_y = -\frac{w(x)}{6I_z}(2c^3 + 3c^2y - y^3)$$

by isolating the same element used in Fig. 3.4-11 except showing all forces in the y direction which include $w(x)$ and σ_y.

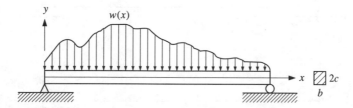

Problem 3.17

3.18 Using superposition, determine the deflection and slope equations for the beam shown.

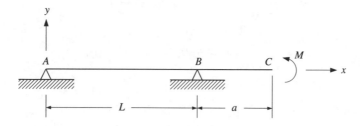

Problem 3.18

3.19 Solve Problem 3.18 using singularity functions. If you have solved Problem 3.18, show that the results are the same.

3.20 Using the double-integration method, verify the deflection and slope equations as functions of x for beam C.7 given in Appendix C.

3.21 Using singularity functions, verify the deflection and slope equations as functions of x for beam C.7 given in Appendix C.

3.22 Using superposition and the concepts given for statically indeterminate problems, verify the support reactions and the deflection and slope equations as functions of x for beam C.12 given in Appendix C.

3.23 For beam C.12 given in Appendix C, using singularity functions only and the three displacement and slope boundary conditions verify the support reactions and the deflection and slope equations as functions of x.

3.24 Determine the deflection equations for the beam shown using (a) superposition, and (b) singularity functions. Verify that the results are the same.

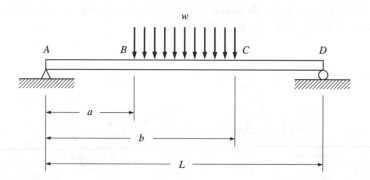

Problem 3.24

3.25 For the beam shown prove that the slope and deflection are given by

$$\theta = \frac{w}{24EIL} (x^4 - 4x^3L + 6x^2L^2 - 4xL^3)$$

$$v_c = \frac{w}{120EIL} (x^5 - 5x^4L + 10x^3L^2 - 10x^2L^3)$$

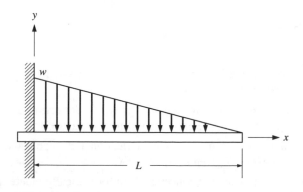

Problem 3.25

3.26 The beam shown is uniformly loaded from $0 \le x \le a$. From $a \le x \le L$ the load reduces linearly to zero. Determine the deflection equation(s) for $0 \le x \le L$: (*a*) using the results of Prob. 3.25 and, if necessary, the tables in Appendix C, (*b*) using singularity functions. Verify that the results of parts (*a*) and (*b*) are the same.

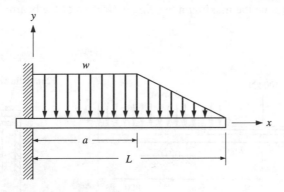

Problem 3.26

3.27 For the beam shown determine the slope and deflection equations for $0 < x < L$ using the double-integration method.

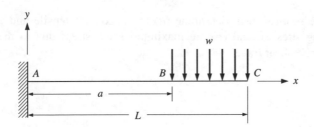

Problem 3.27

3.28 Solve Problem 3.27 using the tables in Appendix C with the method of super-position. If you have worked out Problem 3.27, show that the results are the same.

3.29 Solve Problem 3.27 using singularity functions. If you have worked out Problem 3.27 and/or 3.28, show that the results are the same.

3.30 The steel beam shown is constructed of a channel with a center wall and a flat plate bolted to the top. The wall thickness of the channel is 15 mm. The longitudinal spacing of the bolts is 200 mm. All dimensions are given in millimeters. Let $E = 209$ GPa.

(a) Determine the maximum tensile, compressive, and transverse shear stresses in the beam.

(b) Determine the maximum shear stress in the beam.

(c) Determine the maximum average shear stress in the bolts.

(d) Determine the maximum bending deflection of the beam.

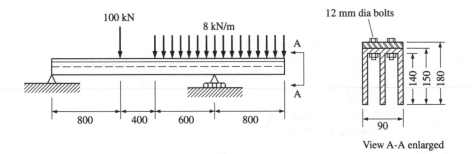

Problem 3.30

3.31 For the beam shown, determine (a) the maximum tensile and compressive bending stresses, and (b) the maximum shear stress due to the maximum transverse shear force.

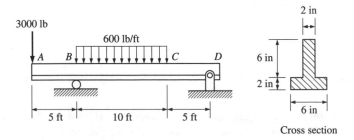

Problem 3.31

3.32 The tube shown has an outer diameter of 1.0 in and a wall thickness of 0.0625 in. The force applied at the free end is

$$F = 32\,\mathbf{i} - 22\,\mathbf{j} + 16\,\mathbf{k}\ \text{lb}$$

where **i**, **j**, and **k** are unit vectors in the x, y, and z directions, respectively. Using the thin-walled equations for geometric properties and the result of Problem 3.12, determine the maximum shear stresses at points A and B.

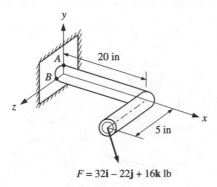

$F = 32\mathbf{i} - 22\mathbf{j} + 16\mathbf{k}$ lb

Problem 3.32

3.33 The 2.0-in-diameter solid steel shaft shown is simply supported at the ends. Two pulleys are keyed to the shaft where pulley B is of diameter 4.0 in and pulley C is of diameter 8.0 in.
   (a) Neglecting stress concentrations, determine the locations and magnitudes of the greatest tensile, compressive, and shear stresses in the shaft.
   (b) Determine the bending deflection of point B on the shaft.

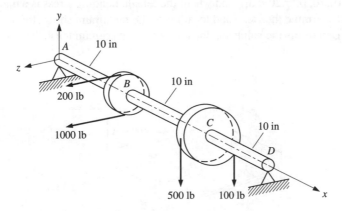

Problem 3.33

3.34 The 30-mm-diameter solid steel shaft shown is simply supported at its ends. Three pulleys are keyed to the shaft where pulley C is of diameter 100 mm and pulleys B and D are of diameter 200 mm. The shaft lengths are given in millimeters. The forces on pulleys C and D are in the z and negative y directions, respectively. The forces on pulley B are 45° from the negative z direction.

(a) Neglecting stress concentrations, determine the locations and magnitudes of the greatest tensile, compressive, and sheer stresses in the shaft.

(b) Determine the bending deflection of point C on the shaft. Let $E = 210$ GPa.

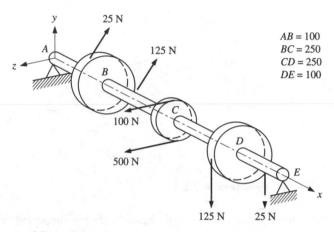

AB = 100
BC = 250
CD = 250
DE = 100

Problem 3.34

3.35 The *closed* tube shown has an outside diameter of 40 mm and a wall thickness of 2 mm. The tube is subjected to the two loads shown and an internal pressure of 3 MPa.

(a) Assuming an ideal model, determine the states of stress at (0, ±20 mm, 0), (0, 0, ±20 mm), and where the tensile bending stress is a maximum.

(b) Determine the value and location of the maximum shear stress.

(c) Determine the value and location of the maximum tensile stress.

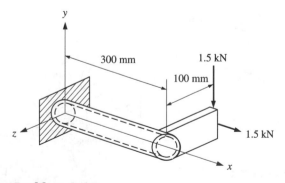

Problem 3.35

3.36 The force of 1 kN shown is applied through the centroid of the beam cross section. Determine the locations and values of the maximum tensile and compressive bending stresses.

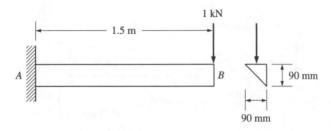

Problem 3.36

3.37 A pressure vessel consists of a circular cylinder capped at both ends by welding on two hemispherical caps. The vessel is then subjected to an internal pressure of 7.0 MPa. If the inner diameter and thickness of each member are 500 mm and 12 mm, respectively, determine the membrane stresses of each member.

3.38 A thin stainless steel cylinder with an outer diameter of 500 mm and an inner diameter of 480 mm just fits over an aluminum cylinder with an outer diameter of 480 mm and an inner diameter of 460 mm. Determine the tangential stresses in the cylinders if the temperature is increased by 100°C. Assume that the longitudinal expansion induces no stresses [NOTE: Equation (3.6-7) will need to be rederived for zero longitudinal stress.] For stainless steel, let $E_s = $ 190 GPa and $\alpha_s = 17.3 \ (10^{-6})/°C$, and for aluminum $E_a = 72$ GPa and $\alpha_a = 23.9(10^{-6})/°C$.

3.39 Using the method of linear superposition and the appropriate beams in Appendix C, determine the deflection and slope at point A of the beam in Problem 3.31. The material of the beam is steel, where $E_s = 30 \times 10^6$ psi. HINT: The slope in section AB due to the uniform load of 600 lb/ft only is constant.

3.40 For Problem 1.7, locate the point on the simply supported main shaft $ABCD$ which contains the maximum tensile bending stress and the maximum torsional shear stress. Draw the corresponding stress element indicating the stress values and directions. From this, determine and show the stress states corresponding to (*a*) the maximum tensile stress, and (*b*) the maximum shear stress.

3.41 For the step shaft shown, $E = 200$ GPa. The shaft is supported by the rigid wall at A. Forces are applied at B and C.
(*a*) Determine the force F required for the shaft to deflect to the rigid wall at C.

(*b*) Determine the reactions at *A* and *C* if *F* is 20 kN above the value found in part (*a*).

(*c*) Determine the stresses in each section and the deflection of point *B* for parts (*a*) and (*b*).

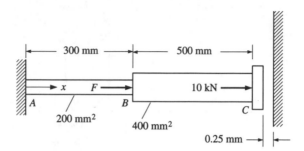

Problem 3.41

3.42 If $F = 20$ kN in Problem 3.41 and the temperature of the shaft increases by 120°C, determine the reactions at *A* and *C*, the deflection of point *B*, and the stresses in each section. The coefficient of expansion for the material is $\alpha = 11.7(10^{-6})/$°C.

3.43 The stiffness of the beam shown is $EI = 200$ kN·m² and the section modulus is $35(10^3)$ mm³.

(*a*) Determine the deflection of *B* and the bending stresses at *A* and *B*.

(*b*) Repeat part (*a*) but let the translational stiffness of the supports at *A* and *C* be 10 and 5 MN/m, respectively, and the rotational stiffness at *A* be 8 MN·m.

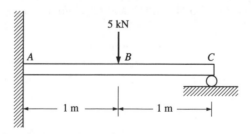

Problem 3.43

3.44 Each of the three solid steel shafts has a diameter of 15 mm. Shafts *AB* and *EF* are 500 mm long, whereas shaft *CD* is 250 mm long. The pitch radii of gears *B*, *C*, and *F* are 25, 100, and 50 mm, respectively. For steel let $E = 200$

GPa and $v = 0.29$. If a torque of 10 N·m is applied to shaft AB as shown, determine the shear stresses in each shaft and the angle through which each gear rotates.

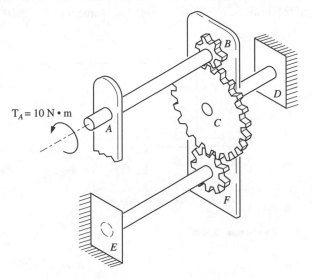

$T_A = 10$ N • m

Problem 3.44

3.45 Two wood beams each with an actual cross section of 2×4 in are loaded as shown. If there is light contact between the beams prior to load application, determine the support reactions after the 500-lb force is applied. Assume that only a vertical force is transmitted between the beams at point B.

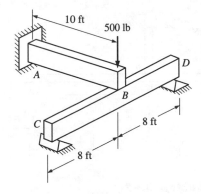

10 ft

500 lb

8 ft

8 ft

Problem 3.45

3.46 The steel beam *ABCD* shown is simply supported at *A* and supported at *B* and *D* by steel cables each having a diameter of 12 mm. A force of 20 kN is applied at point *C*. For steel, let $E = 209$ GPa. Determine the stresses in the cables and the deflections of points *B*, *C*, and *D* if (*a*) beam *ABCD* is considered to be rigid, and (*b*) beam *ABCD* has a second-area moment of $I = 8(10^5)$ mm⁴.

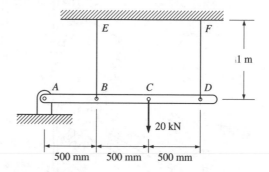

Problem 3.46

3.47 The steel beam *ABCD* shown is supported at *C* as shown and supported at *B* and *D* by steel bolts each having a diameter of 8 mm. The lengths of *BE* and *DF* are 50 and 60 mm, respectively. All dimensions are given in millimeters. Prior to loading, the nuts are just in contact with the beam, which is horizon-

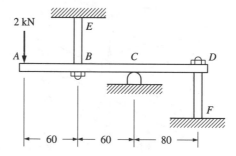

Problem 3.47

tal. A force of 2 kN is applied at point A. For steel, let $E = 209$ GPa. Determine the stresses in the cables and the deflections of points A, B, and D if (a) beam $ABCD$ is considered to be rigid, and (b) beam $ABCD$ is elastic with a second-area moment of $I = 200$ mm⁴.

3.48 Solve parts (a) to (c) using transformation equations *only* (not Mohr's circle), and show the complete state of stress with the appropriate stress element properly oriented relative to the given xy coordinate system. Given the state of plane stress shown,

   (a) determine the state of stress for an element oriented 20° clockwise relative to the given xy coordinate system.

   (b) determine the state of stress for the element containing the principal stresses.

   (c) determine the state of stress for the element containing the maximum in-plane shear stresses.

   (d) determine the state of stress for the element containing the maximum shear stresses.

   (e) draw Mohr's circle indicating the states of stress and angular orientation found in parts (a) to (c).

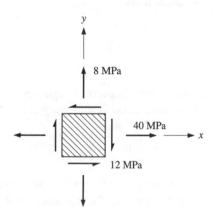

Problem 3.48

3.49 A state of plane stress at a point is shown.

   (a) Considering only surfaces with normals in the xy plane use the stress transformation equations to determine the maximum and minimum normal stresses acting at the point.

(b) Show the element containing these stresses properly oriented to the given xy coordinate system.

(c) What are the stresses obtained in part (a) called?

(d) Determine the *magnitude* of the maximum shear stress at the point.

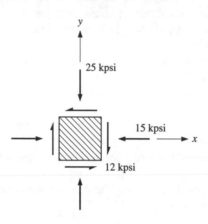

Problem 3.49

3.50 For a state of plane stress use the transformation equations [Eqs. (3.9-1) to (3.9-3)] to show that the shear stress is stationary at

$$\theta_s = \frac{1}{2} \tan^{-1} \frac{\sigma_y - \sigma_x}{2\tau_{xy}}$$

3.51 For Problem 3.50, prove that Eq. (3.9-3) gives the maximum shear stress in the plane of analysis if the $\tan^{-1}$ in the equation for θ_s is treated as *single-valued*.

3.52 A plate of uniform thickness t has a tapered section as shown. For small θ, the normal stress in the x direction of the tapered section can still be represented by Eq. (3.2-1). That is, $\sigma_x = P/ht$, where h is the width of the plate and a function of x in the tapered section. At point A show that (a) the maximum principal stress is

$$\sigma_1 = \sigma_x / \cos^2 \theta$$

and (b) the stresses σ_y and τ_{xy} are given by

$$\sigma_y = \sigma_x \tan^2 \theta \qquad \tau_{xy} = \sigma_x \tan \theta$$

HINT: In addition to an element with surfaces perpendicular to the x and y directions, consider an element at point A with surfaces parallel and perpendicular to the *free* surface.

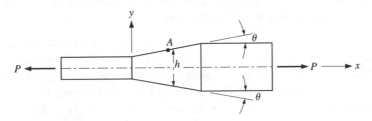

Problem 3.52

3.53 A transversely loaded beam commonly referred to as a *constant-strength beam* is actually a beam designed such that the maximum stress is constant along its length. Consider a cantilever beam of length L with a concentrated transverse force F applied at the free end. We desire to design the beam such that the maximum shear stress[†] is constant at a value of τ_{allow} along its length. Let the beam longitudinal axis x begin at the free end and proceed toward the wall. Design the beam with a rectangular cross section such that (*a*) the beam is of constant thickness b in the bending axis direction, and (*b*) the beam is of constant depth h in the direction of the applied force.

3.54 A state of plane stress exists at a point where

$$\varepsilon_x = 300\mu \qquad \varepsilon_y = -100\mu \qquad \gamma_{xy} = -200\mu$$

For the material, $E = 15$ Mpsi and $v = 0.3$. Directly from the given strains determine the principal strains and their orientation relative to the xy coordinate system.

3.55 For the pinned-pinned column in Table 3.10-1 verify the values of K and v_c.

3.56 For the fixed-fixed column in Table 3.10-1 verify the values of K and v_c.

Computer Problems

3.57 Consider the cables in Problem 3.4 are made of a material with $E = 2$ GPa.
  (*a*) Determine the stresses in the cables and the deflection of point B assuming large deflections and using the technique employed in Example 3.2-4.
  (*b*) Solve the problem using linear techniques and discuss the accuracy as compared with the results of part (*a*).

[†] The maximum-shear-stress theory of failure is covered in Sec. 7.3.

3.58 Consider the cables in Problem 3.4 are made of a material with $E = 2$ GPa.
 (a) Determine the stresses in the cables and the deflection of point B assuming large deflections and using the Newton-Raphson method. If solved, compare your results with part (a) of Problem 3.57.
 (b) Solve the problem using linear techniques and discuss the accuracy as compared with the results of part (a).

3.59 Consider the cables in Problem 3.4 to be made of a nonlinear material with the stress-strain relationship $\sigma = E\varepsilon - K\varepsilon^2$ where $E = 2$ GPa and $K = 700$ GPa. Determine the stresses in the cables and the deflection of point B assuming large deflections. HINT: Use strain instead of writing the equations in terms of the unknown cable forces, where $F = \sigma A = (E\varepsilon - K\varepsilon^2)A$.

3.60 Using spreadsheet or mathematics computer software, program the equations used in the solution of part (b) of Example 3.8-3. Define each input parameter as a variable so they may be changed easily to see their effects on the solution.
 (a) Experiment by changing each parameter one at a time from its original value.
 (b) Set each parameter at its original value save the second-moment area I of the beam. Vary I and plot the deflection and internal force solutions as functions of I. Determine the values of I where the solutions are within 5 percent of the cases in which (i) the beam is rigid and (ii) the rods are rigid.

3.61 Using spreadsheet or mathematics computer software, program the equations used in the solution of part (b) of Problem 3.47. Define each input parameter as a variable so it may be changed easily to see its effects on the solution.
 (a) Experiment by changing each parameter one at a time from its original values.
 (b) Set each parameter at its original value save the second-moment area I of the beam. Vary I and plot the deflection and internal force solutions as functions of I. Determine the values of I where the solutions are within 5 percent of the cases in which (i) the beam is rigid, and (ii) the rods are rigid.

3.62 Using spreadsheet or mathematics computer software, duplicate Table 3.10-2.

3.63–3.65 In viewing the state of stress at a point, an infinitesimal $dx\ dy\ dz$ or $dx'\ dy'\ dz'$ rectangular element is normally defined. However, the stresses at a point change continuously as we rotate the defining surface. Thus to completely define the state of stress at a point, an infinitesimal *sphere* of radius dr about the point is necessary. For two-dimensional analyses of plane stress an infinitesimal cylinder of radius dr about the point can be used to describe the stress in the plane of analysis.

 For the plane stress state given, use a spreadsheet program to determine the normal and shear stresses on the surface of the cylinder shown below at $22.5°$ intervals of θ. Give the stress magnitudes and show the stresses in the correct directions in a diagram as shown below. Using finer increments for θ, plot the normal and shear stress as functions of θ.

Problem	σ_x	σ_y	τ_{xy}
3.63	10 MPa	0	0
3.64	0	0	10 MPa
3.65	20 kpsi	10 kpsi	0

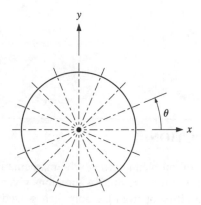

Problems 3.63–3.65   Infinitesimal cylinder of radius *dr*.

3.12   REFERENCES

3.1   Peterson, R. E. "Design Factors for Stress Concentration," *Machine Design*, a Penton publication, vol. 23, pt.1, no. 2, p. 169, February 1951; pt. 2, no. 3, p. 161, March 1951; pt. 3, no. 5, p. 159, May 1951; pt. 4, no. 6, p. 173, June 1951; pt. 5, no. 7, p. 155, July 1951.

3.2   Chapra, S. C., and R. P. Canale. *Numerical Methods for Engineers,* 3rd ed. Burr Ridge, IL: WCB/McGraw-Hill, 1998.

3.3   Timoshenko, S. P., and J. N. Goodier. *Theory of Elasticity*, 3rd ed. New York: McGraw-Hill, 1970.

FOUR

CONCEPTS FROM THE
THEORY OF ELASTICITY[†]

4.0   INTRODUCTION

The solutions of structural problems through the mathematical theory of elasticity are obtained from the use of the constitutive equations given in Chap. 2. The methods employed are quite different from the approaches used in basic mechanics of materials, where the governing equations are normally arrived at by means of physical observations and reasoning, combined with incomplete applications of the procedures of solid-body mechanics. Some of the basic concepts developed in the theory of elasticity will be introduced in this chapter and applied to some classical yet practical problems.

   Plane elastic problems are defined in this chapter, and the solutions of several cases are presented using Airy's stress function in both rectangular and polar coordinates. The cases include beams in bending, curved beams, a plate containing a hole and loaded in tension, a concentrated force on a straight boundary, and a disk with opposing concentrated forces. In addition to plane elastic problems, torsion of a bar with a noncircular solid cross section is presented in Sec. 4.3 using Prandtl's torsional stress function.

   The selection of problems in this chapter is based on their being well-suited for design applications or their being utilized later in the text for related applications or comparisons with the results of other methods. For example, later in Chap. 5, topics such as curved beams, stress concentrations, and contact stresses are again presented.

[†] See Refs. 4.1–4.3.

4.1 PLANE ELASTIC PROBLEMS

4.1.1 DEFINITION

To be classified as a plane elastic problem, the problem must have the following characteristics.

Geometry As shown in Fig. 4.1-1, a plane body consists of a region of uniform thickness t, bounded by two parallel planes, parallel to the xy plane, and by any closed surface. If the thickness t is small compared to the dimensions in the parallel planes, the problem is classified as a *plane stress* problem. If the thickness is large compared to the dimensions in the parallel planes, the problem is classified as a *plane strain* problem.

Loading Applied surface loads and/or body forces must be nonvarying in the z direction and cannot have components in the z direction. Applied loads cannot exist on the top and bottom parallel surfaces.

Plane Stress The assumption for plane stress problems is that since the stresses in the z direction are zero on the parallel free surfaces, and t is small, the stresses in the z direction cannot grow to any appreciable values (the limiting case would be $t \rightarrow 0$). Thus for plane stress

$$\sigma_z = \tau_{yz} = \tau_{zx} = 0 \qquad \textbf{[4.1-1]}$$

Furthermore, the stresses in the x and y directions are not functions of z.

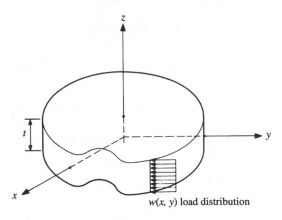

$w(x, y)$ load distribution

Figure 4.1-1   Plane elastic body.

Plane Strain   For plane strain problems, t is large (the limiting case being $t \to \infty$). In this state the strains in the z direction are small, and for the limiting case

$$\varepsilon_z = \gamma_{yz} = \gamma_{zx} = 0 \qquad \textbf{[4.1-2]}$$

and the strains in the x and y directions are not functions of z.

A simple example which illustrates the difference between a plane stress and plane strain problem is that of a narrow-beam problem versus a wide-beam problem. If a beam with a rectangular cross section is narrow, it is considered to be a plane stress problem. If a rectangular beam is wide (platelike), it is considered to be a plane strain problem.

4.1.2   Governing Equations

For plane elastic problems, the strain-stress, equilibrium, and compatibility equations can be combined to produce a useful differential equation in stress.

Plane Stress   From Eqs. (1.4-4) and (1.4-8a) the strain-stress equations for plane stress are

$$\varepsilon_x = \frac{1}{E}(\sigma_x - v\sigma_y) \qquad \textbf{[4.1-3a]}$$

$$\varepsilon_y = \frac{1}{E}(\sigma_y - v\sigma_x) \qquad \textbf{[4.1-3b]}$$

$$\gamma_{xy} = \frac{2(1 + v)}{E}\tau_{xy} \qquad \textbf{[4.1-3c]}$$

The compatibility equation of interest is given by Eq. (2.5-2a) as

$$\frac{\partial^2 \varepsilon_x}{\partial^2 y} + \frac{\partial^2 \varepsilon_y}{\partial^2 x} = \frac{\partial^2 \gamma_{xy}}{\partial x \partial y} \qquad \textbf{[4.1-4]}$$

Substituting Eqs. (4.1-3) into (4.1-4) gives

$$\frac{\partial^2}{\partial y^2}(\sigma_x - v\sigma_y) + \frac{\partial^2}{\partial x^2}(\sigma_y - v\sigma_x) = 2(1 + v)\frac{\partial^2 \tau_{xy}}{\partial x \partial y} \qquad \textbf{[4.1-5]}$$

With no variations in the z direction, the equilibrium equations in the x and y directions, from Eqs. (2.4-2), are

$$\frac{\partial \sigma_x}{\partial x} + \frac{\partial \tau_{xy}}{\partial y} + \overline{F}_x = 0 \qquad \textbf{[4.1-6a]}$$

$$\frac{\partial \tau_{xy}}{\partial x} + \frac{\partial \sigma_y}{\partial y} + \overline{F}_y = 0 \qquad \textbf{[4.1-6b]}$$

Differentiating the first equation by x and the second by y and rearranging yields

$$\frac{\partial^2 \tau_{xy}}{\partial x \partial y} = -\frac{\partial^2 \sigma_x}{\partial x^2} - \frac{\partial \overline{F}_x}{\partial x} \qquad \textbf{[4.1-7a]}$$

$$\frac{\partial^2 \tau_{xy}}{\partial y \partial x} = -\frac{\partial^2 \sigma_y}{\partial y^2} - \frac{\partial \overline{F}_y}{\partial y} \qquad \textbf{[4.1-7b]}$$

Assuming τ_{xy} to be continuous in x and y,

$$\frac{\partial^2 \tau_{xy}}{\partial x \partial y} = \frac{\partial^2 \tau_{xy}}{\partial y \partial x}$$

Thus adding Eqs. (4.1-7a) and (4.1-7b) yields

$$\frac{\partial^2 \tau_{xy}}{\partial x \partial y} = -\frac{1}{2}\left(\frac{\partial^2 \sigma_x}{\partial x^2} + \frac{\partial^2 \sigma_y}{\partial y^2} + \frac{\partial \overline{F}_x}{\partial x} + \frac{\partial \overline{F}_y}{\partial y} \right)$$

Substituting this into Eq. (4.1-5) results in

$$\frac{\partial^2}{\partial y^2}(\sigma_x - v\sigma_y) + \frac{\partial^2}{\partial x^2}(\sigma_y - v\sigma_x) = -(1 + v)\left(\frac{\partial^2 \sigma_x}{\partial x^2} + \frac{\partial^2 \sigma_y}{\partial y^2} + \frac{\partial \overline{F}_x}{\partial x} + \frac{\partial \overline{F}_y}{\partial y} \right)$$

Simplification yields

$$\frac{\partial^2}{\partial x^2}(\sigma_x + \sigma_y) + \frac{\partial^2}{\partial y^2}(\sigma_x + \sigma_y) = -(1 + v)\left(\frac{\partial \overline{F}_x}{\partial x} + \frac{\partial \overline{F}_y}{\partial y} \right) \qquad \textbf{[4.1-8]}$$

In operator notation, $\partial^2/\partial x^2 + \partial^2/\partial y^2$ is called the *Laplacian* operator, ∇^2. Thus Eq. (4.1-8) can be expressed as

$$\nabla^2(\sigma_x + \sigma_y) = -(1 + v)\left(\frac{\partial \overline{F}_x}{\partial x} + \frac{\partial \overline{F}_y}{\partial y} \right) \qquad \textbf{[4.1-9]}$$

For problems in which the body forces are zero, Eq. (4.1-9) reduces to

$$\nabla^2(\sigma_x + \sigma_y) = 0 \qquad \textbf{[4.1-10]}$$

Plane Strain   The strain-stress equations for plane strain differ from that for plane stress. To see this, return to the full three-dimensional strain-stress equations given by Eqs. (1.4-2)

$$\varepsilon_x = \frac{1}{E}[\sigma_x - v(\sigma_y + \sigma_z)] \qquad \textbf{[4.1-11a]}$$

$$\varepsilon_y = \frac{1}{E}[\sigma_y - v(\sigma_x + \sigma_z)] \qquad \textbf{[4.1-11b]}$$

$$\varepsilon_z = \frac{1}{E}[\sigma_z - v(\sigma_x + \sigma_y)] \qquad \textbf{[4.1-11c]}$$

For plane strain $\varepsilon_z = 0$. Thus Eq. (4.1-11c) gives $\sigma_z = v(\sigma_x + \sigma_y)$. Substituting this into Eqs. (4.1-11a) and (4.1-11b) yields

$$\varepsilon_x = \frac{1 - v^2}{E}\left(\sigma_x - \frac{v}{1 - v}\sigma_y\right) \qquad \text{[4.1-12a]}$$

$$\varepsilon_y = \frac{1 - v^2}{E}\left(\sigma_y - \frac{v}{1 - v}\sigma_x\right) \qquad \text{[4.1-12b]}$$

Equations (4.1-12) can be rewritten as

$$\varepsilon_x = \frac{1 + v}{E}\left[(1 - v)\sigma_x - v\sigma_y\right] \qquad \text{[4.1-13a]}$$

$$\varepsilon_y = \frac{1 + v}{E}\left[(1 - v)\sigma_y - v\sigma_x\right] \qquad \text{[4.1-13b]}$$

Thus we see that the normal strain-stress equations for plane strain [Eqs. (4.1-13)] differ from that for plane stress, [Eqs. (4.1-3)]. If we now repeat the derivation used for plane stress except using Eqs. (4.1-13) in place of Eqs. (4.1-3a) and (4.1-3b), we obtain

$$\nabla^2(\sigma_x + \sigma_y) = -\frac{1}{1 - v}\left(\frac{\partial \overline{F}_x}{\partial x} + \frac{\partial \overline{F}_y}{\partial y}\right) \qquad \text{[4.1-14]}$$

4.1.3   CONVERSION BETWEEN PLANE STRESS AND PLANE STRAIN PROBLEMS

In the governing equations the difference between plane stress and plane strain problems arises from the differing strain-stress relations as illustrated in the previous section. If a solution for σ_x and σ_y is obtained for a plane stress problem, then Eqs. (4.1-3a) and (4.1-3b) are appropriate. If the class of problem is that of plane strain, then Eqs. (4.1-12) or (4.1-13) are appropriate. A solution obtained from one class of problem can easily be converted to the other. Equations (4.1-3a) and (4.1-3b) can be converted to Eqs. (4.1-12) by replacing $1/E$ with $(1 - v^2)/E$ and v with $v/(1 - v)$. The reverse is not as obvious and is left for a problem at the end of the chapter. The conversion rules are shown in Table 4.1-1.

Table 4.1-1   Conversions for plane elastic solutions

Solution	Convert to	Substitute	For
Plane stress	Plane strain	$\dfrac{E}{1 - v^2}, \dfrac{v}{1 - v}$	E, v
Plane strain	Plane stress	$\dfrac{1 + 2v}{(1 + v)^2}E, \dfrac{v}{1 + v}$	E, v

4.2   THE AIRY STRESS FUNCTION

4.2.1   RECTANGULAR COORDINATES

In this section we will look at problems which do not have any body forces. Thus, for either plane stress or plane strain Eq. (4.1-10), $\nabla^2 (\sigma_x + \sigma_y) = 0$ applies.

Let us assume that there exists a function $\Phi(x,y)$ such that

$$\sigma_x = \frac{\partial^2 \Phi}{\partial y^2} \qquad\qquad \textbf{[4.2-1a]}$$

$$\sigma_y = \frac{\partial^2 \Phi}{\partial x^2} \qquad\qquad \textbf{[4.2-1b]}$$

$$\tau_{xy} = -\frac{\partial^2 \Phi}{\partial x \partial y} \qquad\qquad \textbf{[4.2-1c]}$$

Substituting these into the equilibrium equations [Eqs. (4.1-6)] with $\overline{F}_x = \overline{F}_y = 0$ shows that equilibrium is satisfied. For compatibility, substitute Eqs. (4.2-1) into Eq. (4.1-10). This gives

$$\frac{\partial^4 \Phi}{\partial x^4} + 2\frac{\partial^4 \Phi}{\partial x^2 \partial y^2} + \frac{\partial^4 \Phi}{\partial y^4} = 0 \qquad\qquad \textbf{[4.2-2]}$$

Using the notation $(\partial^2/\partial x^2 + \partial^2/\partial y^2)^2 = (\nabla^2)^2 = \nabla^4$, we can rewrite Eq. (4.2-2) as[†]

$$\nabla^4 \Phi(x,y) = 0 \qquad\qquad \textbf{[4.2-3]}$$

Equations (4.2-2) and (4.2-3) are called the *biharmonic equation,* and the function $\Phi(x, y)$ is referred to as the *Airy stress function.*

Determine the stress fields that arise from the following stress functions: **Example 4.2-1**

(a) $$\Phi = Cy^2$$

(b) $$\Phi = Ax^2 + Bxy + Cy^2$$

(c) $$\Phi = Ax^3 + Bx^2y + Cxy^2 + Dy^3$$

where A, B, C, and D are constants.

Solution:

(a) $$\sigma_x = \frac{\partial^2 \Phi}{\partial y^2} = 2C$$

[†] The superscript 2 on the expression in parentheses is an operation and does not imply squaring. That is, $\nabla^4\Phi = (\partial^2/\partial x^2 + \partial^2/\partial y^2)^2 \Phi$ implies that $\nabla^4\Phi = (\partial^2/\partial x^2 + \partial^2/\partial y^2)(\partial^2\Phi/\partial x^2 + \partial^2\Phi/\partial y^2)$.

$$\sigma_y = \frac{\partial^2 \Phi}{\partial x^2} = 0$$

$$\tau_{xy} = -\frac{\partial^2 \Phi}{\partial x \, dy} = 0$$

This function is suitable for a bar or plate in a uniform tensile or compressive state of stress.

(b)
$$\sigma_x = \frac{\partial^2 \Phi}{\partial y^2} = 2C$$

$$\sigma_y = \frac{\partial^2 \Phi}{\partial x^2} = 2A$$

$$\tau_{xy} = \frac{-\partial^2 \Phi}{\partial x \, \partial y} = -B$$

This function is suitable for a general uniform stress field over the entire body.

(c)
$$\sigma_x = \frac{\partial^2 \Phi}{\partial y^2} = 2Cx + 6Dy$$

$$\sigma_y = \frac{\partial^2 \Phi}{\partial x^2} = 6Ax + 2By$$

$$\tau_{xy} = \frac{-\partial^2 \Phi}{\partial x \, \partial y} = -(2Bx + 2Cy)$$

All stresses vary linearly with respect to x and y. Note that if $A = B = C = 0$, the state of stress would correspond to a beam in pure bending. Thus the function $\Phi = Dy^3$ can be used for bending. Also, note that compatibility for the above cases is automatically satisfied since $\nabla^4 \Phi = 0$.

Stress functions with constants and linear terms of x and y will give zero stresses, so that only functions of xy, x^2, y^2, and higher are usable. To see how the Airy stress function is generated and used we return to the familiar problem of the uniformly loaded cantilever beam of Example 2.4-1.

Example 4.2-2 | Determine a stress function for the stress field of Fig. 4.2-1 and evaluate the stress field.

Solution:

For Fig. 4.2-1 the conditions of stress at the boundaries are:
At $x = 0$:

$$\sigma_x = 0 \qquad\qquad\qquad\qquad \text{[a]}$$

$$\tau_{xy} = 0 \qquad\qquad\qquad\qquad \text{[b]}$$

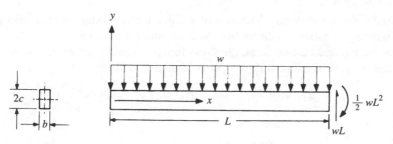

Figure 4.2-1

At $x = L$:

$$\int_{-c}^{c} \tau_{xy} b \ dy = wL \qquad\qquad\qquad \text{[c]}$$

$$\int_{-c}^{c} \sigma_x b \ dy = 0 \qquad\qquad\qquad \text{[d]}$$

$$\int_{-c}^{c} \sigma_x yb \ dy = \frac{1}{2} wL^2 \qquad\qquad \text{[e]}$$

At $y = c$:

$$\sigma_y = -\frac{w}{b} \qquad\qquad\qquad \text{[f]}$$

$$\tau_{xy} = 0 \qquad\qquad\qquad\quad \text{[g]}$$

At $y = -c$:

$$\sigma_y = 0 \qquad\qquad\qquad\quad \text{[h]}$$

$$\tau_{xy} = 0 \qquad\qquad\qquad\quad \text{[i]}$$

It is impossible to satisfy condition (a) with a simple stress function. Recall Example 2.5-3, where it was found that the stress distribution was not zero at the free end. Thus condition (a) will be ignored.[†]

Some observations should be made before attempting to establish the stress function. Since the cross section is symmetric about the y axis and $\sigma_y = -w/b$ at $y = c$ and $\sigma_y = 0$ at $y = -c$, σ_y should be an odd function of y. Thus the stress function should contain odd functions of y. This is substantiated by the fact that

---

[†] However, at $x = 0$, the conditions that $\int_{-c}^{c} \sigma_x b \ dy = 0$ and $\int_{-c}^{c} \sigma_x yb \ dy = 0$ must hold for equilibrium.

σ_x should also contain odd functions of y since the net axial force is zero [this automatically satisfies condition (d)]. Second, since σ_y is constant as a function of x on the top and bottom faces, the stress function should not contain powers of x greater than x^2. Thus the following stress function will be tried:

$$\Phi = Axy + Bx^2 + Cx^2y + Dy^3 + Exy^3 + Fx^2y^3 + Gy^5 \qquad [j]^\dagger$$

Since $\nabla^4\Phi = 0$,

$$\nabla^4\Phi = 24Fy + 120Gy = 0$$

Thus

$$F = -5G$$

Substituting this into Eq. (j) and evaluating the stress field using Eqs. (4.2-1) results in

$$\sigma_x = \frac{\partial^2\Phi}{\partial y^2} = 6Dy + 6Exy - 30Gx^2y + 20Gy^3 \qquad [k]$$

$$\sigma_y = \frac{\partial^2\Phi}{\partial x^2} = 2B + 2Cy - 10Gy^3 \qquad [l]$$

$$\tau_{xy} = -\frac{\partial^2\Phi}{\partial x\partial y} = -(A + 2Cx + 3Ey^2 - 30Gxy^2) \qquad [m]$$

Dealing with τ_{xy} first, we see that condition (b) at $x = 0$ requires that

$$A + 3Ey^2 = 0$$

The only way this can be true for all values of y is

$$A = E = 0$$

When either one of conditions (g) or (i) is used at $y = \pm c$, Eq. (m) becomes

$$0 = -(2Cx - 30Gc^2x)$$

or

$$C = 15Gc^2 \qquad [n]$$

Substituting this into Eq. (l) yields

$$\sigma_y = 2B + 30Gc^2y - 10Gy^3 \qquad [o]$$

---

† Note that with the omission of condition (a), there are eight conditions and one equation remaining, Eq. (4.2-3). Thus it would seem that nine constants are in order. However, by the selection of odd functions of y, condition (d) is automatically satisfied. In addition, τ_{xy} will be found to contain even functions of y, making τ_{xy} symmetric with respect to y. Thus only one of conditions (g) or (i) can be used. This explains the selection of only seven unknown constants.

From conditions (*f*) and (*h*) at $y = c$ and $y = -c$, respectively, Eq. (*o*) becomes .

$$-\frac{w}{b} = 2B + 30Gc^3 - 10Gc^3 = 2B + 20Gc^3$$

and

$$0 = 2B - 30Gc^3 + 10Gc^3 = 2B - 20Gc^3$$

Solving the two simultaneous equations for *B* and *G* yields

$$B = -\frac{w}{4b} \qquad G = -\frac{w}{40bc^3}$$

The area moment of inertia is $I_z = \frac{2}{3}bc^3$. Thus the above terms can be rewritten as

$$B = -\frac{wc^3}{6I_z} \qquad G = -\frac{w}{60I_z}$$

Then, from Eq. (*n*)

$$C = 15Gc^2 = -\frac{wc^2}{4I_z}$$

Substituting *E* and *G* into Eq. (*k*) results in

$$\sigma_x = 6Dy + \frac{w}{2I_z}x^2y - \frac{w}{3I_z}y^3 \qquad \textbf{[p]}$$

From condition (*e*) at $x = L$, Eq. (*p*) becomes

$$\int_{-c}^{c}\left(6Dy + \frac{w}{2I_z}L^2y - \frac{w}{3I_z}y^3\right)yb\,dy = \frac{1}{2}wL^2$$

Solving for *D* yields

$$D = \frac{wc^2}{30I_z}$$

Substituting the constants back into the equations for the stresses gives

$$\sigma_x = \frac{w}{10I_z}(5x^2 + 2c^2)y - \frac{w}{3I_z}y^3 \qquad \textbf{[q]}$$

$$\sigma_y = -\frac{w}{6I_z}(2c^3 + 3c^2y - y^3) \qquad \textbf{[r]}$$

$$\tau_{xy} = \frac{w}{2I_z}x(c^2 - y^2) \qquad \textbf{[s]}$$

Comparing these results with Eq. (*l*) of Example 2.5-3 and Eqs. (*b*) and (*c*) of Example 2.5-2 shows exact agreement.

4.2.2   P OLAR C OORDINATES

There are many practical plane elastic problems where polar coordinates are better suited. If Airy's stress function is to be used, then it must be written in terms of r and θ instead of x and y. To change variables, consider the chain rule in matrix form

$$
\left\{ \begin{array}{c} \dfrac{\partial \Phi}{\partial r} \\[2mm] \dfrac{\partial \Phi}{\partial \theta} \end{array} \right\} = \begin{bmatrix} \dfrac{\partial x}{\partial r} & \dfrac{\partial y}{\partial r} \\[2mm] \dfrac{\partial x}{\partial \theta} & \dfrac{\partial y}{\partial \theta} \end{bmatrix} \left\{ \begin{array}{c} \dfrac{\partial \Phi}{\partial x} \\[2mm] \dfrac{\partial \Phi}{\partial y} \end{array} \right\}
\qquad \textbf{[4.2-4]}
$$

Relating polar to rectangular coordinates, $x = r \cos \theta$ and $y = r \sin \theta$, then

$$
\frac{\partial x}{\partial r} = \cos \theta \qquad \frac{\partial x}{\partial \theta} = -r \sin \theta
$$

$$
\frac{\partial y}{\partial r} = \sin \theta \qquad \frac{\partial y}{\partial \theta} = r \cos \theta
$$

Substitution into Eq. (4.2-4) gives

$$
\left\{ \begin{array}{c} \dfrac{\partial \Phi}{\partial r} \\[2mm] \dfrac{\partial \Phi}{\partial \theta} \end{array} \right\} = \begin{bmatrix} \cos \theta & \sin \theta \\ -r \sin \theta & r \cos \theta \end{bmatrix} \left\{ \begin{array}{c} \dfrac{\partial \Phi}{\partial x} \\[2mm] \dfrac{\partial \Phi}{\partial y} \end{array} \right\}
\qquad \textbf{[4.2-5]}
$$

To determine the inverse equation we need the inverse of the matrix in the [ ] brackets (see Appendix I). This is

$$
\begin{bmatrix} \cos \theta & \sin \theta \\ -r \sin \theta & r \cos \theta \end{bmatrix}^{-1} = \frac{1}{r} \begin{bmatrix} r \cos \theta & -\sin \theta \\ r \sin \theta & \cos \theta \end{bmatrix}
$$

Thus the inverse of Eq. (4.2-5) is

$$
\left\{ \begin{array}{c} \dfrac{\partial \Phi}{\partial x} \\[2mm] \dfrac{\partial \Phi}{\partial y} \end{array} \right\} = \frac{1}{r} \begin{bmatrix} r \cos \theta & -\sin \theta \\ r \sin \theta & \cos \theta \end{bmatrix} \left\{ \begin{array}{c} \dfrac{\partial \Phi}{\partial r} \\[2mm] \dfrac{\partial \Phi}{\partial \theta} \end{array} \right\}
\qquad \textbf{[4.2-6]}
$$

Dropping the matrix form, Eqs. (4.2-6) can be written as

$$
\frac{\partial \Phi}{\partial x} = \frac{\partial \Phi}{\partial r} \cos \theta - \frac{1}{r} \frac{\partial \Phi}{\partial \theta} \sin \theta
\qquad \textbf{[4.2-7a]}
$$

$$
\frac{\partial \Phi}{\partial y} = \frac{\partial \Phi}{\partial r} \sin \theta + \frac{1}{r} \frac{\partial \Phi}{\partial \theta} \cos \theta
\qquad \textbf{[4.2-7b]}
$$

To determine second derivatives, consider $\partial^2\Phi/\partial x^2$ first. Substitute $\partial\Phi/\partial x$ for Φ in Eq. (4.2-7a). This gives

$$\frac{\partial^2\Phi}{\partial x^2} = \frac{\partial}{\partial r}\left[\frac{\partial\Phi}{\partial r}\cos\theta - \frac{1}{r}\frac{\partial\Phi}{\partial\theta}\sin\theta\right]\cos\theta - \frac{1}{r}\frac{\partial}{\partial\theta}\left[\frac{\partial\Phi}{\partial r}\cos\theta - \frac{1}{r}\frac{\partial\Phi}{\partial\theta}\sin\theta\right]\sin\theta$$

Taking derivatives and simplifying results in

$$\frac{\partial^2\Phi}{\partial x^2} = \frac{\partial^2\Phi}{\partial r^2}\cos^2\theta - \frac{2}{r}\frac{\partial^2\Phi}{\partial r\partial\theta}\sin\theta\cos\theta + \frac{1}{r^2}\frac{\partial^2\Phi}{\partial\theta^2}\sin^2\theta$$

$$+ \frac{1}{r}\frac{\partial\Phi}{\partial r}\sin^2\theta + \frac{2}{r^2}\frac{\partial\Phi}{\partial\theta}\sin\theta\cos\theta \qquad \textbf{[4.2-8a]}$$

Similarly, substituting $\partial\Phi/\partial y$ for Φ in Eq. (4.2-7b) gives

$$\frac{\partial^2\Phi}{\partial y^2} = \frac{\partial^2\Phi}{\partial r^2}\sin^2\theta + \frac{2}{r}\frac{\partial^2\Phi}{\partial r\partial\theta}\sin\theta\cos\theta + \frac{1}{r^2}\frac{\partial^2\Phi}{\partial\theta^2}\cos^2\theta$$

$$+ \frac{1}{r}\frac{\partial\Phi}{\partial r}\cos^2\theta - \frac{2}{r^2}\frac{\partial\Phi}{\partial\theta}\sin\theta\cos\theta \qquad \textbf{[4.2-8b]}$$

Finally, substituting $\partial\Phi/\partial y$ for Φ in Eq. (4.2-7a) gives

$$\frac{\partial^2\Phi}{\partial x\partial y} = \frac{\partial^2\Phi}{\partial r^2}\sin\theta\cos\theta - \frac{1}{r}\frac{\partial^2\Phi}{\partial r\partial\theta}(\sin^2\theta - \cos^2\theta)$$

$$- \frac{1}{r^2}\frac{\partial^2\Phi}{\partial\theta^2}\sin\theta\cos\theta - \frac{1}{r}\frac{\partial\Phi}{\partial r}\sin\theta\cos\theta$$

$$+ \frac{1}{r^2}\frac{\partial\Phi}{\partial\theta}(\sin^2\theta - \cos^2\theta) \qquad \textbf{[4.2-8c]}$$

Adding Eqs. (4.2-8a) and (4.2-8b) results in

$$\frac{\partial^2\Phi}{\partial x^2} + \frac{\partial^2\Phi}{\partial y^2} = \frac{\partial^2\Phi}{\partial r^2} + \frac{1}{r}\frac{\partial\Phi}{\partial r} + \frac{1}{r^2}\frac{\partial^2\Phi}{\partial\theta^2}$$

Thus, the biharmonic equation in polar coordinates is

$$\nabla^4\Phi = \left(\frac{\partial^2}{\partial r^2} + \frac{1}{r}\frac{\partial}{\partial r} + \frac{1}{r^2}\frac{\partial^2}{\partial\theta^2}\right)^2\Phi = 0 \qquad \textbf{[4.2-9]}$$

For stresses, from the transformation equations [Eqs. (2.1-8)]

$$\sigma_r = \sigma_x\cos^2\theta + \sigma_y\sin^2\theta + \tau_{xy}\sin 2\theta$$

$$\sigma_\theta = \sigma_x\sin^2\theta + \sigma_y\cos^2\theta - \tau_{xy}\sin 2\theta$$

$$\tau_{r\theta} = -\frac{1}{2}(\sigma_x - \sigma_y)\sin 2\theta + \tau_{xy}\cos 2\theta$$

Note, when $\theta = 0$, $\sigma_r = \sigma_x$, $\sigma_\theta = \sigma_y$, and $\tau_{r\theta} = \tau_{xy}$. Thus, from Eqs. (4.2-1) and (4.2-8) for $\theta = 0$

$$\sigma_r = \left(\frac{\partial^2 \Phi}{\partial y^2}\right)_{\theta=0} = \frac{1}{r^2}\frac{\partial^2 \Phi}{\partial \theta^2} + \frac{1}{r}\frac{\partial \Phi}{\partial r} \qquad \text{[4.2-10a]}$$

$$\sigma_\theta = \left(\frac{\partial^2 \Phi}{\partial x^2}\right)_{\theta=0} = \frac{\partial^2 \Phi}{\partial r^2} \qquad \text{[4.2-10b]}$$

$$\tau_{r\theta} = -\left(\frac{\partial^2 \Phi}{\partial x \partial y}\right)_{\theta=0} = \frac{1}{r^2}\frac{\partial \Phi}{\partial \theta} - \frac{1}{r}\frac{\partial^2 \Phi}{\partial r \partial \theta} \qquad \text{[4.2-10c]}$$

Axisymmetric Problems   For this case, there is no variation with respect to θ and partial derivatives with respect to r can be replaced by ordinary derivatives. Thus Eq. (4.2-9) reduces to

$$\left(\frac{d^2}{dr^2} + \frac{1}{r}\frac{d}{dr}\right)^2 \Phi = \frac{d^4\Phi}{dr^4} + \frac{2}{r}\frac{d^3\Phi}{dr^3} - \frac{1}{r^2}\frac{d^2\Phi}{dr^2} + \frac{1}{r^3}\frac{d\Phi}{dr} = 0 \qquad \text{[4.2-11]}$$

which is in the form of *Euler's equation*. Equations (4.2-10) reduce to

$$\sigma_r = \frac{1}{r}\frac{d\Phi}{dr} \qquad \text{[4.2-12a]}$$

$$\sigma_\theta = \frac{d^2\Phi}{dr^2} \qquad \text{[4.2-12b]}$$

$$\tau_{r\theta} = 0 \qquad \text{[4.2-12c]}$$

If we multiply Eq. (4.2-11) by r^4 we get

$$r^4\frac{d^4\Phi}{dr^4} + 2r^3\frac{d^3\Phi}{dr^3} - r^2\frac{d^2\Phi}{dr^2} + r\frac{d\Phi}{dr} = 0 \qquad \text{[4.2-13]}$$

The solution to this is of the form $\Phi = Cr^m$. Substituting this into Eq. (4.2-13) yields

$$m^4 - 4m^3 + 4m^2 = 0$$

The solutions to this are $m = 0, 0, 2, 2$. For a repeated root of m, the solution has the form $Cr^m \ln r$. Thus, the solution of Eq. (4.2-13) is

$$\Phi = C_1 + C_2\ln r + C_3 r^2 + C_4 r^2 \ln r \qquad \text{[4.2-14]}$$

This function is ideal for thick-walled pressurized cylinders, which is left as a problem at the end of the chapter. The topic of thick-walled cylinders is also covered in the next chapter using a slightly different method of solution. Here, we will demonstrate the function in the next section on a curved beam in bending.

4.2.3   CURVED BEAM IN BENDING

Consider the beam of constant curvature shown in Fig. 4.2-2. With a pure moment applied, there is no variation in θ, so Eqs. (4.2-12) and (4.2-14) are valid. Substituting Eq. (4.2-14) into (4.2-12) gives

$$\sigma_r = \frac{1}{r}\frac{d\Phi}{dr} = \frac{C_2}{r^2} + 2C_3 + 2C_4\ln r + C_4 \qquad \textbf{[4.2-15a]}$$

$$\sigma_\theta = \frac{d^2\Phi}{dr^2} = -\frac{C_2}{r^2} + 2C_3 + 2C_4\ln r + 3C_4 \qquad \textbf{[4.2-15b]}$$

$$\tau_{r\theta} = 0 \qquad \textbf{[4.2-15c]}$$

To evaluate the constants we use the conditions

$$\sigma_r = 0 \qquad \text{at} \qquad r = r_i \qquad \textbf{[a]}$$

$$\sigma_r = 0 \qquad \text{at} \qquad r = r_o \qquad \textbf{[b]}$$

for the radial stresses on the stress-free inner and outer surfaces, and

$$\int_{r_i}^{r_o} \sigma_\theta b \, dr = 0 \qquad \textbf{[c]}$$

$$\int_{r_i}^{r_o} r\sigma_\theta b \, dr = M \qquad \textbf{[d]}$$

for a net force of zero and a net moment M on the isolated surfaces. Substituting Eq. (4.2-15a) into conditions (a) and (b) gives

$$\frac{C_2}{r_i^2} + 2C_4\ln r_i + 2C_3 + C_4 = 0 \qquad \textbf{[e]}$$

$$\frac{C_2}{r_o^2} + 2C_4\ln r_o + 2C_3 + C_4 = 0 \qquad \textbf{[f]}$$

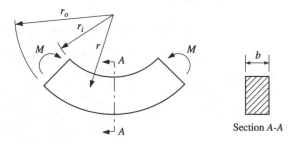

Figure 4.2-2   Curved beam.

Subtracting the equations and solving for C_2 in terms of C_4 gives

$$C_2 = \frac{2r_i^2 r_o^2}{r_o^2 - r_i^2} \ln \frac{r_o}{r_i} C_4 \qquad \text{[g]}$$

Substituting this back into Eq. (e) and rearranging yields

$$C_3 = -\frac{r_o^2(1 + 2\ln r_o) - r_i^2(1 + 2\ln r_i)}{2(r_o^2 - r_i^2)} C_4 \qquad \text{[h]}$$

Condition (c) is automatically satisfied since the stress function satisfies equilibrium. To verify this, perform the integration of condition (c), and using Eqs. (e) and (f) it will be seen that the condition is satisfied.

For condition (d) we have

$$\int_{r_i}^{r_o} \left[-\frac{C_2}{r} + 2C_4 r \ln r + (2C_3 + 3C_4)r \right] b \, dr = M$$

Integration and simplification results in

$$-C_2 \ln \frac{r_o}{r_i} + C_4(r_o^2 \ln r_o - r_i^2 \ln r_i) + (C_3 + C_4)(r_o^2 - r_i^2) = \frac{M}{b} \qquad \text{[i]}$$

Substitution of Eqs. (g) and (h) and solving for C_4 results in

$$C_4 = \frac{2M}{bK}(r_o^2 - r_i^2)$$

where

$$K = (r_o^2 - r_i^2)^2 - 4r_i^2 r_o^2 \left(\ln \frac{r_o}{r_i} \right)^2 \qquad \text{[4.2-16]}$$

Substituting C_4 back into Eqs. (g) and (h) yields

$$C_2 = \frac{4M}{bK} r_i^2 r_o^2 \ln \frac{r_o}{r_i}$$

and

$$C_3 = -\frac{M}{bK}[r_o^2(1 + 2\ln r_o) - r_i^2(1 + 2\ln r_i)]$$

Substituting C_2, C_3, and C_4 into Eqs. (4.2-15) and simplifying yields the final results

$$\sigma_r = \frac{4M}{bK} \left(\frac{r_i^2 r_o^2}{r^2} \ln \frac{r_o}{r_i} - r_o^2 \ln \frac{r_o}{r} - r_i^2 \ln \frac{r}{r_i} \right) \qquad \text{[4.2-17a]}$$

$$\sigma_\theta = -\frac{4M}{bK} \left[\frac{r_i^2 r_o^2}{r^2} \ln \frac{r_o}{r_i} + r_o^2 \ln \frac{r_o}{r} + r_i^2 \ln \frac{r}{r_i} - (r_o^2 - r_i^2) \right] \qquad \text{[4.2-17b]}$$

$$\tau_{r\theta} = 0 \qquad \text{[4.2-17c]}$$

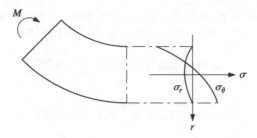

Figure 4.2-3   Stress distribution for a curved beam with a rectangular cross section.

Figure 4.2-3 shows a plot of σ_r and σ_θ. We see that the bending stress σ_θ is not linear as with straight beams. Therefore, the neutral axis in bending does not occur at the centroidal axis, and the maximum bending stress acts at the inner radius r_i.

   Curved beams are also covered in Chap. 5 using a different technique which is applicable to general cross sections. A comparison is made in Example 5.6-2 between the two methods.

4.2.4 CIRCULAR HOLE IN A PLATE LOADED IN TENSION[†]

Consider the plate shown in Fig. 4.2-4 loaded in tension by a force per unit area σ. Although not drawn to scale, it will be assumed that the outer dimensions of the plate are very large compared to the diameter of the hole $2a$. The approach is to use a stress function which will yield stresses consistent with pure tension as $r \to \infty$

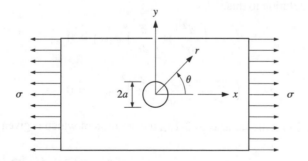

Figure 4.2-4   Circular hole in a plate loaded in tension.

[†] This is referred to as *Kirsch's problem*, first solved by G. Kirsch in 1898, but not by using the stress function approach.

(i.e., $\sigma_x = \sigma$ and $\sigma_y = \tau_{xy} = 0$). For this case the stresses in polar coordinates can be found from the transformation equations [Eqs. (2.1-8)] and are

$$\sigma_r = \sigma_x \cos^2\theta = \frac{1}{2}\sigma(1 + \cos 2\theta) \qquad \textbf{[4.2-18a]}$$

$$\sigma_\theta = \sigma_x \sin^2\theta = \frac{1}{2}\sigma(1 - \cos 2\theta) \qquad \textbf{[4.2-18b]}$$

$$\tau_{r\theta} = -\sigma_x \sin\theta \cos\theta = -\frac{1}{2}\sigma \sin 2\theta \qquad \textbf{[4.2-18c]}$$

The stress function for the case of pure tension was given in Example 4.2-1, Eq. (a), where $C = \sigma_x/2$. That is,

$$\Phi = \frac{1}{2}\sigma_x y^2$$

In polar coordinates, $y = r \sin \theta$; therefore,

$$\Phi = \frac{1}{2}\sigma_x r^2 \sin^2\theta = \frac{1}{4}\sigma_x r^2 (1 - \cos 2\theta) \qquad \textbf{[4.2-19]}$$

For the plate with the hole we will assume a similar stress function. That is,

$$\Phi = \phi_1(r) - \phi_2(r)\cos 2\theta \qquad \textbf{[4.2-20]}$$

where $\phi_1(r)$ and $\phi_2(r)$ are functions of r only. Substitution of Φ into Eq. (4.2-9) results in

$$\nabla^4\Phi = \left(\frac{d^2}{dr^2} + \frac{1}{r}\frac{d}{dr}\right)^2 \phi_1(r) + \left(\frac{d^2}{dr^2} + \frac{1}{r}\frac{d}{dr} - \frac{4}{r^2}\right)^2 \phi_2(r)\cos 2\theta = 0$$

For a general solution to this,

$$\left(\frac{d^2}{dr^2} + \frac{1}{r}\frac{d}{dr}\right)^2 \phi_1(r) = 0 \qquad \textbf{[4.2-21]}$$

$$\left(\frac{d^2}{dr^2} + \frac{1}{r}\frac{d}{dr} - \frac{4}{r^2}\right)^2 \phi_2(r) = 0 \qquad \textbf{[4.2-22]}$$

Equation (4.2-21) is the same as (4.2-13), the solution of which is given by Eq. (4.2-14), which is

$$\phi_1(r) = C_1 + C_2 \ln r + C_3 r^2 + C_4 r^2 \ln r \qquad \textbf{[4.2-23]}$$

Expanding Eq. (4.2-22) gives

$$\frac{d^4\phi_2}{dr^4} + \frac{2}{r}\frac{d^3\phi_2}{dr^3} - \frac{9}{r^2}\frac{d^2\phi_2}{dr^2} + \frac{9}{r^3}\frac{d\phi_2}{dr} = 0 \qquad \textbf{[4.2-24]}$$

This equation is similar to Eq. (4.2-13), and its solution is also of the form Cr^m. Substituting this into Eq. (4.2-24) gives

$$m^4 - 4m^3 - 4m^2 + 16m = 0$$

which has the solution $m = 0, -2, 2, 4$. Thus

$$\phi_2(r) = C_5 + \frac{C_6}{r^2} + C_7 r^2 + C_8 r^4 \qquad \text{[4.2-25]}$$

Substituting $\phi_1(r)$ and $\phi_2(r)$ into Eq. (4.2-20) and differentiating according to Eqs. (4.2-10) gives

$$\sigma_r = \frac{C_2}{r^2} + 2C_3 + C_4(1 + 2\ln r) - \left(\frac{4C_5}{r^2} + \frac{6C_6}{r^4} + 2C_7 \right) \cos 2\theta$$

$$\sigma_\theta = -\frac{C_2}{r^2} + 2C_3 + C_4(3 + 2\ln r) + \left(\frac{6C_6}{r^4} + 2C_7 + 12C_8 \right) \cos 2\theta$$

$$\tau_{r\theta} = \left(-\frac{2C_5}{r^2} - \frac{6C_6}{r^4} + 2C_7 + 6C_8 r^2 \right) \sin 2\theta$$

As $r \to \infty$ the above equations must approach Eqs. (4.2-18). For this to be true C_4 and C_8 must be zero. Thus

$$\sigma_r = \frac{C_2}{r^2} + 2C_3 - \left(\frac{4C_5}{r^2} + \frac{6C_6}{r^4} + 2C_7 \right) \cos 2\theta \qquad \text{[4.2-26a]}$$

$$\sigma_\theta = -\frac{C_2}{r^2} + 2C_3 + \left(\frac{6C_6}{r^4} + 2C_7 \right) \cos 2\theta \qquad \text{[4.2-26b]}$$

$$\tau_{r\theta} = \left(-\frac{2C_5}{r^2} - \frac{6C_6}{r^4} + 2C_7 \right) \sin 2\theta \qquad \text{[4.2-26c]}$$

The stresses must also satisfy the boundary conditions at the free surface of the hole. That is, at $r = a$,

$$\sigma_r = \tau_{r\theta} = 0$$

and, according to Eqs. (4.2-18), as $r \to \infty$,

$$\sigma_r = \frac{1}{2} \sigma(1 + \cos 2\theta)$$

$$\sigma_\theta = \frac{1}{2} \sigma(1 - \cos 2\theta)$$

$$\tau_{r\theta} = -\frac{1}{2} \sigma \sin 2\theta$$

These conditions yield five equations:

$$\frac{C_2}{a^2} + 2C_3 = 0 \qquad \frac{4C_5}{a^2} + \frac{6C_6}{a^4} + 2C_7 = 0 \qquad \frac{2C_5}{a^2} - \frac{6C_6}{a^4} + 2C_7 = 0$$

$$2C_3 = \frac{\sigma}{2} \qquad \text{and} \qquad -2C_7 = \frac{\sigma}{2}$$

Solving these yields $C_2 = -\sigma a^2/2$, $C_3 = \sigma/4$, $C_5 = \sigma a^2/2$, $C_6 = -\sigma a^4/4$, and $C_7 = -\sigma/4$. Substituting these into Eqs. (4.2-26) results in

$$\sigma_r = \frac{\sigma}{2}\left[1 - \frac{a^2}{r^2} + \left(1 - 4\frac{a^2}{r^2} + 3\frac{a^4}{r^4} \right)\cos 2\theta \right] \qquad \textbf{[4.2-27a]}$$

$$\sigma_\theta = \frac{\sigma}{2}\left[1 + \frac{a^2}{r^2} - \left(1 + 3\frac{a^4}{r^4} \right)\cos 2\theta \right] \qquad \textbf{[4.2-27b]}$$

$$\tau_{r\theta} = -\frac{\sigma}{2}\left(1 + 2\frac{a^2}{r^2} - 3\frac{a^4}{r^4} \right)\sin 2\theta \qquad \textbf{[4.2-27c]}$$

Note, as $r \to \infty$, Eqs. (4.2-27) reduce to Eqs. (4.2-18).

The maximum stress is $\sigma_\theta = 3\sigma$ at $r = a$ and $\theta = \pm 90°$. Figure 4.2-5 shows how the tangential stress varies along the x and y axes of the plate. For the top (and bottom) of the hole we see the stress gradient is extremely large and hence the term *stress concentration* applies.[†] Along the surface of the hole the tangential stress is $-\sigma$ at $\theta = 0°$ and $180°$ and increases, as θ increases, to 3σ at $\theta = 90°$ and $270°$.

4.2.5 CONCENTRATED FORCE ON A FLAT BOUNDARY (FLAMANT SOLUTION)

Consider the force P acting on the flat boundary, called a half-space, as shown in Fig. 4.2-6a. Isolate a semicircular surface of radius r as shown in Fig. 4.2-6b. In 1892, Flamant showed that the only stresses on the surface were radial stresses proportional to $(\cos \theta)/r$. A simple stress function which provides this and satisfies the biharmonic equation is

$$\Phi = Cr\theta \sin\theta \qquad \textbf{[4.2-28]}$$

From Eqs. (4.2-10), the corresponding stresses are

$$\sigma_r = \frac{1}{r^2}\frac{\partial^2 \Phi}{\partial \theta^2} + \frac{1}{r}\frac{\partial \Phi}{\partial r} = 2C\frac{\cos\theta}{r} \qquad \textbf{[4.2-29a]}$$

$$\sigma_\theta = \frac{\partial^2 \Phi}{\partial r^2} = 0 \qquad \textbf{[4.2-29b]}$$

[†] See Table F.1, Appendix F. As $d/w \to 0$, $K_t \to 3$.

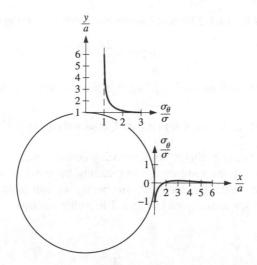

Figure 4.2-5   Tangential stress distribution for $\theta = 0°$ and $90°$.

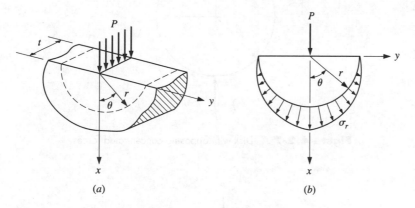

(a)   (b)

Figure 4.2-6   Concentrated force on a flat boundary.

$$\tau_{r\theta} = \frac{1}{r^2}\frac{\partial \Phi}{\partial \theta} - \frac{1}{r}\frac{\partial^2 \Phi}{\partial r\partial \theta} = 0 \qquad \text{[4.2-29c]}$$

The radial force on an $r\,d\theta$ element is given by $\sigma_r tr d\theta$, the component of which in the x direction is $(\sigma_r tr\,d\theta)\cos\theta$. Therefore, for equilibrium of force in the x direction

$$P = -\int_{-\pi/2}^{\pi/2} \sigma_r \cos\theta tr\,d\theta$$

Substitution of Eq. (4.2-29a) and integrating yields $C = P/(\pi t)$. Thus

$$\sigma_r = -\frac{2P}{\pi t}\frac{\cos\theta}{r}$$

[4.2-30]

This formulation will be used in Chap. 5 in the section on contact stresses.

4.2.6   DISK WITH OPPOSING CONCENTRATED FORCES

Consider the disk of radius R with opposing concentrated forces as shown in Fig. 4.2-7. To approach the solution, let us return to the previous section with the concentrated force on the flat boundary. To begin, we will isolate a disk of radius R within the half-space as shown in Fig. 4.2-8. At the position r, θ there exists a radial

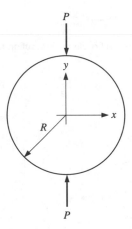

Figure 4.2-7   Disk with opposing concentrated forces.

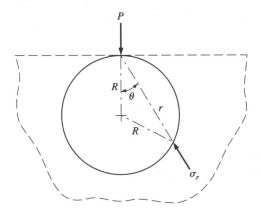

Figure 4.2-8   A disk isolated within a half-space.

stress σ_r given by Eq. (4.2-30). However, from Fig. 4.2-8 it can be seen that $r = 2R \cos \theta$, or $\cos \theta / r = 1/(2R)$. Substituting this into Eq. (4.2-30) results in

$$\sigma_r = -\frac{P}{\pi Rt} \qquad \text{[4.2-31]}$$

Thus it can be seen that the radial stress along the circular surface is constant.

Next we superimpose a disk with an opposing force as shown in Fig. 4.2-9(a). As before, the stresses along the disk are

$$\sigma_{r1} = \sigma_{r2} = -\frac{P}{\pi Rt} \qquad \text{[4.2-32]}$$

From geometry, angle ABC is 90°; hence σ_{r1} and σ_{r2} act on perpendicular surfaces as shown. This is an isotropic state of stress in the plane of analysis. Mohr's circle of stress for this reduces to a point, which means the state of stress on any surface is $\sigma = -P/(\pi Rt)$. Thus the resultant state of stress along the surface of the circle is as shown in Fig. 4.2-9(b). Since the hydrostatic stress, σ, is not present in the isolated disk in Fig. 4.2-7, it must be eliminated. This is a simple task, as we simply add a positive hydrostatic stress state of $\sigma_r = \sigma_\theta = P/(\pi Rt)$. Determining the stresses at an arbitrary point within the disk is another matter, as the stresses from the opposing forces are given relative to separate polar coordinate systems r_1, θ_1 and r_2, θ_2. Thus to combine the stresses we will transform them to the xy coordinate system

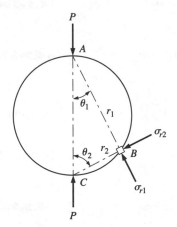

(a) Superposition of stresses due to the opposing forces from Eq. (4.2-32)

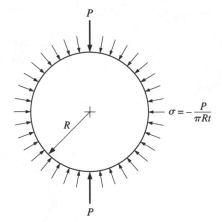

(b) Net result of stresses due to opposing forces from Eq. (4.2-32)

Figure 4.2-9

given in Fig. 4.2-7. For the top concentrated force σ_{r1} is written in terms of Eq. (4.2-30) and transformed through the angle θ_1 to give

$$\sigma_x = -\frac{2P\cos\theta_1}{\pi r_1 t}\sin^2\theta_1 \qquad\qquad \textbf{[4.2-33a]}$$

$$\sigma_y = -\frac{2P\cos\theta_1}{\pi r_1 t}\cos^2\theta_1 \qquad\qquad \textbf{[4.2-33b]}$$

$$\tau_{xy} = \frac{2P\cos\theta_1}{\pi r_1 t}\sin\theta_1\cos\theta_1 \qquad\qquad \textbf{[4.2-33c]}$$

where

$$r_1 = \sqrt{x^2 + (R-y)^2} \qquad\qquad \textbf{[4.2-34a]}$$

$$\sin\theta_1 = \frac{x}{\sqrt{x^2 + (R-y)^2}} \qquad\qquad \textbf{[4.2-34b]}$$

$$\cos\theta_1 = \frac{R-y}{\sqrt{x^2 + (R-y)^2}} \qquad\qquad \textbf{[4.2-34c]}$$

Similarly, for the lower concentrated force[†]

$$\sigma_x = -\frac{2P\cos\theta_2}{\pi r_2 t}\sin^2\theta_2 \qquad\qquad \textbf{[4.2-35a]}$$

$$\sigma_y = -\frac{2P\cos\theta_2}{\pi r_2 t}\cos^2\theta_2 \qquad\qquad \textbf{[4.2-35b]}$$

$$\tau_{xy} = \frac{2P\cos\theta_2}{\pi r_2 t}\sin\theta_2\cos\theta_2 \qquad\qquad \textbf{[4.2-35c]}$$

where

$$r_2 = \sqrt{x^2 + (R+y)^2} \qquad\qquad \textbf{[4.2-36a]}$$

$$\sin\theta_2 = \frac{x}{\sqrt{x^2 + (R+y)^2}} \qquad\qquad \textbf{[4.2-36b]}$$

$$\cos\theta_2 = \frac{R+y}{\sqrt{x^2 + (R+y)^2}} \qquad\qquad \textbf{[4.2-36c]}$$

Finally, we add the isotropic stress to cancel the hydrostatic boundary stress shown in Fig. 4.2-9(b), which transforms directly to $\sigma_x = \sigma_y = P/(\pi R t)$. Adding this to-

---

[†] The angle θ_2 is clockwise, hence negative. However, it can be treated as positive since the net effect of the trigonometric terms in Eqs. (4.2-35) is insensitive to the sign of θ_2.

gether with Eqs. (4.2-33) and (4.2-35), using Eqs.(4.2-34) and (4.2-36), and simplifying gives

$$\sigma_x = -\frac{2P}{\pi t}\left\{\frac{(R-y)x^2}{[x^2+(R-y)^2]^2}+\frac{(R+y)x^2}{[x^2+(R+y)^2]^2}-\frac{1}{2R}\right\}$$ [4.2-37a]

$$\sigma_y = -\frac{2P}{\pi t}\left\{\frac{(R-y)^3}{[x^2+(R-y)^2]^2}+\frac{(R+y)^3}{[x^2+(R+y)^2]^2}-\frac{1}{2R}\right\}$$ [4.2-37b]

$$\tau_{xy} = \frac{2P}{\pi t}\left\{\frac{(R-y)^2x}{[x^2+(R-y)^2]^2}-\frac{(R+y)^2x}{[x^2+(R+y)^2]^2}\right\}$$ [4.2-37c]

The disk is sometimes used as a test specimen in the area of photoelasticity (covered in Chap. 8) where the stress along the x axis is used for calibration purposes. Substituting $y=0$ in Eqs. (4.2-37) and simplifying yields

$$\sigma_x = \frac{P}{\pi Rt}\left(\frac{x^2-R^2}{x^2+R^2}\right)^2$$ [4.2-38a]

$$\sigma_y = -\frac{P}{\pi Rt}\frac{(3R^2+x^2)(R^2-x^2)}{(x^2+R^2)^2}$$ [4.2-38b]

$$\tau_{xy} = 0$$ [4.2-38c]

Figure 4.2-10 gives a plot of Eqs. (4.2-38a) and (4.2-38b) in dimensionless form. We see that along the x axis the greatest compressive value occurs at $x=0$ where σ_y

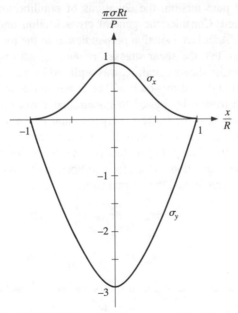

Figure 4.2-10 σ_x and σ_y stresses along the x axis.

is $-3P/(\pi Rt)$. Likewise at this point the maximum tensile value of σ_x is $P/(\pi Rt)$. The stresses along the y axis are found by substituting $x = 0$ in Eqs. (4.2-37), giving

$$\sigma_x = \frac{P}{\pi t R} \qquad\qquad \textbf{[4.2-39a]}$$

$$\sigma_y = -\frac{P}{\pi Rt} \frac{3R^2 + y^2}{R^2 - y^2} \qquad\qquad \textbf{[4.2-39b]}$$

$$\tau_{xy} = 0 \qquad\qquad \textbf{[4.2-39c]}$$

We see that σ_x is constant along the y axis, whereas σ_y is $3P/(\pi Rt)$ at the origin and approaches $-\infty$ as y approaches $\pm R$. This makes sense, as we are modeling the applied load as a contact force. Stresses in the contact area are covered in the next chapter.

4.3   PRANDTL'S STRESS FUNCTION FOR TORSION

4.3.1   GENERAL FORMULATION

The stress function approach is also useful when dealing with torsion on a prismatic element with a noncircular cross section. Airy's stress function cannot be used here since the general torsion problem does not fall into the category of a plane elastic problem. Thus, for pure torsion, the equations of equilibrium and compatibility must be reformulated. Consider the general cross section undergoing torsion as shown in Fig. 4.3-1. A surface isolation perpendicular to the rod axis is also shown, where it can be seen that the shear stresses τ_{xy} and τ_{zx} act over a $dy\,dz$ element. (NOTE: Assuming cross shears equal, τ_{xz} was replaced by τ_{zx}.)

For equilibrium in the x direction it will be assumed that σ_x is negligible. Note, however, that when torsion is applied to noncircular cross sections without constraint in the x direction, plane surfaces perpendicular to the longitudinal axis will not remain plane and the surfaces *warp*. If the rod is constrained from warping in the x direction, σ_x will develop. In this section, the discussion will be restricted to free constraint in the x direction.[†] Thus, for equilibrium of forces in the x direction, neglecting the body force $\overline{F}_x$ and the normal stress σ_x,

$$\frac{\partial \tau_{xy}}{\partial y} + \frac{\partial \tau_{zx}}{\partial z} = 0 \qquad\qquad \textbf{[4.3-1]}$$

[†] For further discussion of constrained warp of open, thin-walled members, see Ref. 4.4 (pp. 365–84) and Ref. 4.5 (pp. 318–25).

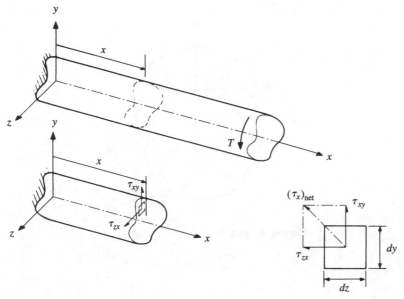

Enlarged view of element

Figure 4.3-1

If a stress function $\Psi(y, z)^\dagger$ exists such that

$$\tau_{xy} = \frac{\partial \Psi}{\partial z}$$ [4.3-2a]

$$\tau_{zx} = -\frac{\partial \Psi}{\partial y}$$ [4.3-2b]

Equation (4.3-1) (equilibrium) is automatically satisfied.

When deflections in terms of the angle of twist θ are considered, the deflection of a point on the isolated surface is as shown in Fig. 4.3-2, from which it can be seen that for small θ

$$w = \theta y \qquad v = -\theta z$$

Assuming that the rod is fixed at the origin of xyz (see Fig. 4.3-1), we can represent the angle of twist by the angle of twist per unit length θ', where $\theta = \theta'x$, and

$$w = \theta'xy$$ [4.3-3a]

$$v = -\theta'xz$$ [4.3-3b]

---

† Ψ is called Prandtl's stress function, differing from Airy's stress function.

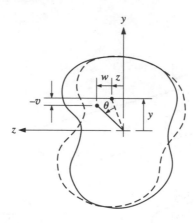

Figure 4.3-2

For noncircular cross sections, the surface warps, and in general the point will also deflect in the x direction u. However, this is independent of x and $u = u(y, z)$. The strains γ_{xy} and γ_{zx} are

$$\gamma_{xy} = \frac{\partial v}{\partial x} + \frac{\partial u}{\partial y} = -\theta'z + \frac{\partial u}{\partial y} \qquad \gamma_{zx} = \frac{\partial w}{\partial x} + \frac{\partial u}{\partial z} = \theta'y + \frac{\partial u}{\partial z}$$

Since $\tau = [E/2(1 + v)]\gamma$, we have

$$\tau_{xy} = \frac{E}{2(1 + v)}\left(-\theta'z + \frac{\partial u}{\partial y}\right) \qquad \text{[a]}$$

$$\tau_{zx} = \frac{E}{2(1 + v)}\left(\theta'y + \frac{\partial u}{\partial z}\right) \qquad \text{[b]}$$

Writing the expression $\partial\tau_{zx}/\partial y - \partial\tau_{xy}/\partial z$ using Eqs. (a) and (b) yields

$$\frac{\partial\tau_{zx}}{\partial y} - \frac{\partial\tau_{xy}}{\partial z} = \frac{E}{1 + v}\theta'$$

Substitution of Eqs. (4.3-2) results in

$$\frac{\partial^2\Psi}{\partial y^2} + \frac{\partial^2\Psi}{\partial z^2} = \frac{-E}{1 + v}\theta' \qquad \text{[4.3-4]}$$

Equation (4.3-4), called *Poisson's equation*, is the governing equation for the torsional stress function Ψ.

At the boundary, the net shear stress must be tangent to the boundary. Thus, from Fig 4.3-1

$$\frac{\tau_{xy}}{\tau_{zx}} = \frac{dy}{dz} \qquad \text{or} \qquad \tau_{zx}\,dy - \tau_{xy}\,dz = 0$$

Substituting in Eqs. (4.3-2) results in

$$\frac{\partial\Psi}{\partial y}\,dy + \frac{\partial\Psi}{\partial z}\,dz = 0$$

However, since $\Psi = \Psi(y, z)$, the above can be written

$$d\Psi = 0 \qquad\qquad\qquad \textbf{[4.3-5]}$$

Since Eq. (4.3-5) applies to the boundary, Ψ is constant along the boundary of the cross section. The value of this constant is arbitrary and is normally chosen to be zero. If the boundary of the cross section is a well-behaved function of y and z such as a circle, ellipse, etc., the equation of the boundary becomes an excellent stress function. This is demonstrated in Example 4.3-1, which follows.

To relate the stress function to the transmitted torque T return to Fig. 4.3-1. The net torque about the x axis due to the stresses on the $dy\,dz$ element is $(y\tau_{zx} - z\tau_{xy})dy\,dz$. Thus the total torque is

$$T = \iint (y\tau_{zx} - z\tau_{xy})dy\,dz$$

Substitution of Eqs. (4.3-2) gives

$$T = \iint \left(-y\frac{\partial\Psi}{\partial y} - z\frac{\partial\Psi}{\partial z} \right)dy\,dz \qquad\qquad \textbf{[4.3-6]}$$

Consider the first term in the integral. Integrate this with respect to y by parts first. This gives

$$\iint -y\frac{\partial\Psi}{\partial y}\,dy\,dz = -\int \left(\int y\frac{\partial\Psi}{\partial y}\,dy \right)dz = -\int \left(\Psi y\Big|_{y_1}^{y_2} - \int \Psi\,dy \right)dz$$

where y_1 and y_2 are boundary points for the dz slice. However, as stated earlier, Ψ is zero at the boundary. Thus Ψ is zero at y_1 and y_2, and the first term within the integral disappears, resulting in

$$\iint -y\frac{\partial\Psi}{\partial y}\,dy\,dz = \iint \Psi\,dy\,dz$$

In a similar fashion, the second integral in Eq. (4.3-6) gives identical results so that Eq. (4.3-6) reduces to

$$T = 2\iint \Psi\,dy\,dz \qquad\qquad \textbf{[4.3-7]}$$

Example 4.3-1 | A solid circular shaft of radius r_o is transmitting a torque T. Determine the corresponding shear-stress distribution.

Solution:

The equation for the boundary of a circle of radius r_o in the xz plane is $y^2 + z^2 = r_o^2$. Try the following stress function:

$$\Psi = k(y^2 + z^2 - r_o^2) \qquad \text{[a]}$$

where k is a constant. Note that $\Psi = 0$ along the entire boundary. To establish the value of k, Eq. (4.3-7) is used. From Eq. (a) it can be seen that polar coordinates are more suitable to the problem. Let $r^2 = y^2 + z^2$, where r is a variable radial position. The infinitesimal area $dy\, dz$ can be replaced by $2\pi r\, dr$ since at any given position r the stress function is constant. Thus, for this problem the double integral of Eq. (4.3-7) reduces to a single integral, written

$$T = 4\pi k \int_0^{r_o} (r^2 - r_o^2) r\, dr$$

Integrating and solving for k results in

$$k = -\frac{T}{\pi r_o^4}$$

The polar moment of inertia of a circular cross section is $J = (\pi/2) r_o^4$. Thus

$$k = -\frac{T}{2J}$$

Upon substituting k into Eq. (a), the stress function is found to be

$$\Psi = -\frac{T}{2J} (y^2 + z^2 - r_o^2) \qquad \text{[b]}$$

The stresses are found from Eqs. (4.3-2):

$$\tau_{xy} = -\frac{Tz}{J} \qquad \text{[c]}$$

and

$$\tau_{zx} = \frac{Ty}{J} \qquad \text{[d]}$$

Note that at any given point, the net shear stress is given by $(\tau_x)_{\text{net}} = \sqrt{\tau_{xy}^2 + \tau_{zx}^2}$ (see Fig. 4.3-1).

$$(\tau_x)_{\text{net}} = \frac{T}{J}\sqrt{z^2 + y^2} = \frac{Tr}{J} \qquad \text{[e]}$$

which is identical to that used in elementary strength of materials.

For the angle of twist, substitution of Eq. (*b*) into Eq. (4.3-4) yields

$$-\frac{T}{J} - \frac{T}{J} = -\frac{E}{1+v}\theta'$$

or

$$\theta' = 2(1+v)\frac{T}{EJ}$$

If the total length of the bar is *L*, the angle of twist across the entire length is $\theta = \theta' L$, and thus

$$\theta = 2(1+v)\frac{TL}{EJ}$$

which again agrees with the elementary strength of materials solution.

Consider the equilateral cross section with sides of length *a* shown in Fig. 4.3-3. Determine the shear stress distribution if the section is transmitting a torque *T*.   **Example 4.3-2**

Solution:

The equations for the boundary are given by

Right line:
$$y = \frac{2h}{3a}(3z + a) \qquad\qquad \textbf{[a]}$$

Left line:
$$y = -\frac{2h}{3a}(3z - a) \qquad\qquad \textbf{[b]}$$

Bottom line:
$$y = -\frac{h}{3} \qquad\qquad \textbf{[c]}$$

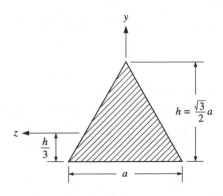

Figure 4.3-3

Based on these equations let the stress function be

$$\Psi = C\left\{\left[y - \frac{2h}{3a}(3z + a)\right]\left[y + \frac{2h}{3a}(3z - a)\right]\left(y + \frac{h}{3}\right)\right\}$$ [d]

where C is a constant. In light of Eqs. (a) to (c) this function is zero along the boundary. Multiplying the terms out yields

$$\Psi = C\left(y^3 - \frac{4h^2}{a^2}z^2y - hy^2 - \frac{4h^3}{3a^2}z^2 + \frac{4}{27}h^3\right)$$ [e]

Substitution of Eq. (e), with $a = 2h/\sqrt{3}$, into Eq. (4.3-4) yields

$$C = \frac{E\theta'}{4h(1 + v)}$$ [f]

To relate the torque T, we use Eq. (4.3-7) integrating with respect to z first. Equations (a) and (b) written for z are

Right line: $$z = \frac{a}{6h}(3y - 2h)$$ [g]

Left line: $$z = -\frac{a}{6h}(3y - 2h)$$ [h]

Substituting Eq. (e) into (4.3-7) and rearranging terms gives

$$I = 2C\int_{-h/3}^{2h/3}\left\{\int_{a(3y-2h)/6h}^{-a(3y-2h)/6h}\left[\left(y^3 - hy^2 + \frac{4}{27}h^3\right) - 4\frac{h^2}{a^2}\left(y + \frac{h}{3}\right)z^2\right]dz\right\}dy$$ [i]

Integration results in $T = Cah^4/15$. Substituting $h = \sqrt{3}a/2$ and solving for C results in

$$C = \frac{80T}{3a^5}$$ [j]

Substituting this into Eq. (f) with $h = \sqrt{3}a/2$ and solving for θ' gives

$$\theta' = \frac{160}{\sqrt{3}}\frac{(1 + v)T}{Ea^4}$$ [k]

for the angle of twist per unit length.

The stresses are arrived at by substituting Eq. (e) into Eq. (4.3-2):

$$\tau_{xy} = \frac{\partial\Psi}{\partial z} = C\left(-8\frac{h^2}{a^2}zy - \frac{8}{3}\frac{h^3}{a^2}z\right)$$

$$\tau_{zx} = -\frac{\partial\Psi}{\partial y} = -C\left(3y^2 - 2hy - 4\frac{h^2}{a^2}z^2\right)$$

Substituting Eq. (*j*) into this with $h = \sqrt{3}a/2$ and simplifying yields

$$\tau_{xy} = -\frac{160T}{3a^5}\left(3y + \frac{\sqrt{3}}{2}a\right)z \qquad [l]$$

$$\tau_{zx} = -\frac{80T}{3a^5}\left(3y^2 - \sqrt{3}ay - 3z^2\right) \qquad [m]$$

Along the *y* axis where $z = 0$, $\tau_{xy} = 0$, and τ_{zx} varies as shown in Fig. 4.3-4. The maximum shear stress at point *A* has a magnitude of

$$\tau_{max} = 20\frac{T}{a^3} \qquad [n]$$

Owing to cyclic symmetry, the stresses at points *B* and *C* are the same as at point *A*. Thus we see that the maximum stress occurs where the largest circumscribed circle (shown dotted) is tangent to the boundary of the triangle. From the plot we also see that the stress is zero at the vertices where two free surfaces intersect at an external corner.

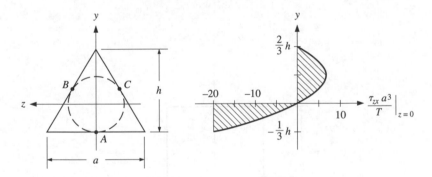

Figure 4.3-4

4.3.2   TORSION ON A RECTANGULAR CROSS SECTION

The solution for the equilateral cross section in the previous section was rather fortuitous. Normally the determination of the stress function for cross sections whose boundaries are not well defined by continuous functions is a more complicated matter. When this occurs, the stress functions employed are generally in the form of an infinite series. Such a case is that of a rectangular cross section where the technique employed for the equilateral cross section will not yield an acceptable solution to Eq. (4.3-4). Using a series solution, it can be shown that the stress function is[†]

$$\Psi = \frac{4}{\pi^3} \frac{E\theta'}{1+v} \sum_{n=1,3,5,\ldots}^{\infty} \frac{1}{n^3}(-1)^{(n-1)/2}\left[1 - \frac{\cosh{(n\pi y/b)}}{\cosh{(n\pi h/2b)}}\right]\cos\left(\frac{n\pi z}{b}\right) \qquad \textbf{[4.3-8]}$$

Substituting this into Eq. (4.3-7) the angle of twist per unit length is found to be

$$\theta' = \frac{2(1+v)T}{k_1 Ehb^3} \qquad \textbf{[4.3-9]}$$

where

$$k_1 = \frac{1}{3}\left[1 - \frac{192}{\pi^5}\frac{b}{h}\sum_{n=1,3,5,\ldots}^{\infty}\frac{1}{n^5}\tanh\left(\frac{n\pi h}{2b}\right)\right] \qquad \textbf{[4.3-10]}$$

The shear stress is maximum at the midpoint of the longest sides (points A and B of Fig. 4.3-5, where again the largest circumscribed circle is tangent to the boundary). Here the absolute value of τ_{xy} is

$$\tau_{max} = \frac{T}{k_2 hb^2} \qquad \textbf{[4.3-11]}$$

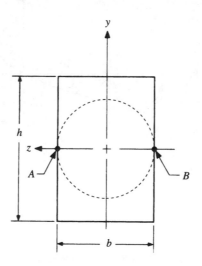

Figure 4.3-5

| [†] See Ref. 4.3, p. 309.

where

$$k_2 = \frac{k_1}{k} \qquad \textbf{[4.3-12]}$$

The factor k_1 is given by Eq. (4.3-10) and k is given by

$$k = 1 - \frac{8}{\pi^2} \sum_{n=1,3,5,\dots}^{\infty} \frac{1}{n^2 \cosh\left(n\pi h/2b\right)} \qquad \textbf{[4.3-13]}$$

The factors k_1 and k_2 are tabulated in Table 4.3-1 in terms of the ratio of h/b.

Table 4.3-1   Stress and deflection factors for torsion of rectangular bars

h/b	k_1	k_2	h/b	k_1	k_2
1.0	0.1406	0.208	3.0	0.263	0.267
1.2	0.166	0.219	4.0	0.281	0.282
1.5	0.196	0.231	5.0	0.291	0.291
2.0	0.229	0.246	10.0	0.312	0.312
2.5	0.249	0.258	∞	0.333	0.333

Thin Rectangular Shapes   It can be seen that when $h \gg b$, $1/k_1 = 1/k_2 \to 3$. Thus for a thin rectangular cross section, Eqs. (4.3-11) and (4.3-9) reduce to

$$\tau_{max} \approx \frac{3T}{hb^2} \qquad \textbf{[4.3-14]}$$

$$\theta \approx \frac{6(1+v)TL}{Ehb^3} \qquad \textbf{[4.3-15]}$$

Equations (4.3-14) and (4.3-15) can also be used to approximate the shear stress in "open sections," like that shown in Fig. 4.3-6, where the wall thickness is b and the total extended length of the section is h.

Equation (4.3-14) can be modified to handle sections in which the thickness varies. Rewrite Eq. (4.3-14) in the form

$$\tau_{max} = \frac{3T}{hb^2} = \frac{Tb}{\frac{1}{3}hb^2}$$

The term in the denominator is called the equivalent polar second-area moment of the cross section J_e. Thus

$$\tau_{max} = \frac{Tb}{J_e} \qquad \textbf{[4.3-16]}$$

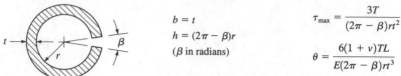

$b = t$

$h = (2\pi - \beta)r$

(β in radians)

$$\tau_{max} = \frac{3T}{(2\pi - \beta)rt^2}$$

$$\theta = \frac{6(1 + v)TL}{E(2\pi - \beta)rt^3}$$

(a) Open cylindrical section

$b = t$

$h = h_1 + h_2$

$$\tau_{max} = \frac{3T}{(h_1 + h_2)t^2}$$

$$\theta = \frac{6(1 + v)TL}{E(h_1 + h_2)t^3}$$

(b) Angle section

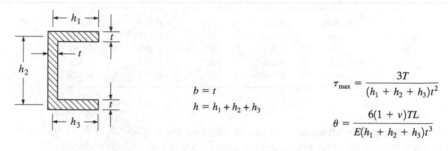

$b = t$

$h = h_1 + h_2 + h_3$

$$\tau_{max} = \frac{3T}{(h_1 + h_2 + h_3)t^2}$$

$$\theta = \frac{6(1 + v)TL}{E(h_1 + h_2 + h_3)t^3}$$

(c) Channel section

Figure 4.3-6   Torsion equations for open thin-walled sections.

If a section has varying thickness,

$$J_e = \sum \tfrac{1}{3} hb^3 \qquad\qquad \textbf{[4.3-17]}$$

and b in Eq. (4.3-16) corresponds to the thickness of the section where the shear stress is being evaluated.

Example 4.3-3

Estimate the maximum shear stress on the section shown in Fig. 4.3-7 if a torque of 1000 lb·in is applied to the section.

Solution:

The polar second-area moment of the horizontal leg is

$$J_{e1} = (\tfrac{1}{3})(3)(0.5)^3 = 0.125 \text{ in}^4$$

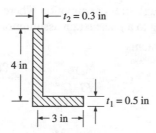

Figure 4.3-7

and in the vertical leg

$$J_{e2} = (\tfrac{1}{3})(4)(0.3)^3 = 0.036 \text{ in}^4$$

Thus

$$J_e = 0.125 + 0.036 = 0.161 \text{ in}^4$$

Since $t_1 > t_2$, the maximum shear occurs on the lower leg where $b = t_1$ and is

$$\tau_{max} \approx \frac{Tb}{J_e} = \frac{(1000)(0.5)}{0.161} = 3106 \text{ psi}$$

4.4 DISCUSSION

The treatment of the topics of elasticity in this chapter could certainly continue, but beyond this point, the mathematical capabilities of the reader must be considerably greater than the level intended for this book. Keep in mind, however, that the subject of elasticity is invaluable in continuing study and understanding in the field of stress analysis. As the reader's mathematical background increases, he or she will discover a vast warehouse of advanced problems which have been solved by the various mathematical techniques of the theory of elasticity.

4.5 PROBLEMS

4.1 Prove that to convert a plane strain solution to a plane stress solution you substitute $\dfrac{1 + 2v}{(1 + v)^2} E$ and $\dfrac{v}{1 + v}$ for E and v, respectively.

4.2 Show that Eqs. (4.2-1) satisfy the equilibrium equations with no body forces.

4.3 The radial deflection of a pressurized circular cylinder is given by Eq. (3.6-7)

$$u_r = (2 - v)\frac{pr^2}{2Et}$$

This is based on a plane stress assumption where t is small. What is the deflection corresponding to a plane strain assumption for t being large?

4.4 For narrow beam problems

$$\varepsilon_x = -\frac{M_z y}{EI_z} \qquad \varepsilon_y = v\frac{M_z y}{EI_z} \qquad \gamma_{xy} = \frac{2(1+v)}{E}\frac{V_y Q}{I_z b}$$

and

$$\frac{d^2 v_c}{dx^2} = \frac{M_z}{EI_z}$$

(a) Convert these equations to wide beam (plate) problems.
(b) How can the formulas in Appendix C be used for wide beam problems?

4.5 Determine the resulting stress equations for σ_x, σ_y, and τ_{xy} given the Airy stress function

$$\Phi = a_1 x^4 + a_2 x^3 y + a_3 x^2 y^2 + a_4 xy^3 + a_5 y^4$$

where the a_i are constants. Also determine, if any, the restrictions between the constants.

4.6 Using the stress function of Example 4.2-2, Eq. (j), obtain the stress field for the simply supported beam shown. Compare the results with the elementary mechanics of materials formulations.

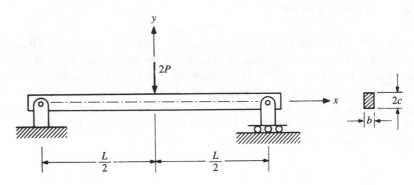

Problem 4.6

4.7 Consider a thick-walled cylinder with a pressure p_i on the inner surface where $r = r_i$, and a pressure p_o on the outer surface where $r = r_o$. Using the stress function given by Eq. (4.2-14) and the conditions that the radial stress σ_r is $-p_i$ and $-p_o$ at r_i and r_o, respectively, prove that

$$\sigma_r = \frac{1}{r_o^2 - r_i^2}\left[p_i r_i^2 - p_o r_o^2 + \left(\frac{r_i r_o}{r}\right)^2 (p_o - p_i)\right]$$

$$\sigma_\theta = \frac{1}{r_o^2 - r_i^2}\left[p_i r_i^2 - p_o r_o^2 - \left(\frac{r_i r_o}{r}\right)^2 (p_o - p_i)\right]$$

HINT: You will find that another condition is necessary. Recall Eqs. (1.5-5) where for axisymmetric problems, $\varepsilon_r = \partial u_r/\partial r$ and $\varepsilon_\theta = u_r/r$. This, together with Hooke's law for plane stress (or strain) provides the necessary additional condition.

4.8   For Fig. 4.2-2 of Sec. 4.2.3 let $M = 500$ lb·in, $r_i = 4$ in, $r_o = 6$ in, and $b = 0.5$ in.
(a) Determine the maximum and minimum tangential and radial stresses.
(b) Plot σ_r and σ_θ as functions of r.
(c) On the plot of σ_θ of part (b) plot the bending stress that would be obtained from the straight beam formulation. Discuss the differences in the results of parts (b) and (c).

4.9   Consider Fig. 4.2-4 in Sec. 4.2.4. Apply a compressive load of $-\sigma$ to the top and bottom of the plate. Together with the tensile load this will be equivalent to an application of a pure shear stress field on the plate.
(a) Determine the resulting stress field for the entire plate.
(b) Plot σ_θ/σ as a function of θ around the hole.

4.10 Show that Eq. (4.2-28), $\Phi = Cr\theta \sin \theta$, satisfies the biharmonic equation, Eq. (4.2-9).

4.11 Determine the stresses in a semi-infinite plate due to a shear load as shown. The net force P is applied at $x = y = 0$. HINT: Try the stress function $\Phi = Cr\theta \cos \theta$.

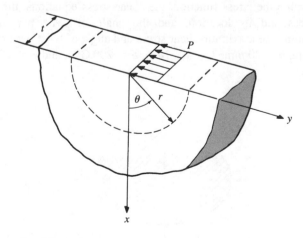

Problem 4.11

4.12 Determine the stresses in the semi-infinite plate of thickness t subjected to the oblique force P at an angle ϕ from the horizontal. HINT: Problem 4.11 must first be solved.

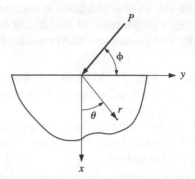

Problem 4.12

4.13 For Fig. 4.2-7 of Sec. 4.2.6 place a polar coordinate system at the center of the disk and determine the equations for σ_r and σ_θ as functions of θ at $r = R$.

4.14 The disk in Fig. 4.2-7 of Sec. 4.2.6 is sometimes used as a calibration specimen when using the experimental technique involving photoelastic fringes (see Chap. 8). The fringes depend on the difference in the principal stresses $\sigma_1 - \sigma_2$. Owing to symmetry, $\tau_{xy} = 0$ along the x axis. Thus σ_x and σ_y are the principal stresses along the x axis.
(a) Using Eqs. (4.2-37) with $y = 0$, reduce $\sigma_x - \sigma_y$ to the simplest form.
(b) With $P = 1.0$ kN, $R = 50$ mm, and $t = 6$ mm plot $\sigma_x - \sigma_y$, for $y = 0$, as functions of x.

4.15 The elliptical cross section shown is transmitting a torsional moment T.
(a) Determine the stress function, the shear stress equations, the maximum shear stress and its location, and the angle of twist per unit length.
(b) Determine the maximum shear stress and angle of twist per unit length if $a = 10$ mm, $b = 20$ mm, $T = 400$ N·m, $E = 200$ GPa, and $v = 0.3$.

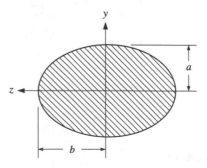

Problem 4.15

4.16 For a section undergoing torsion, the value of τ_{max} for a unit area and torque will give a measure of the efficient use of material for a section. The lower this value, the more efficient the section is. Determine τ_{max} with $A = 1.0$ unit and $T = 1$ unit for each section shown. Normalize your results with respect to case (c), tabulate them, and discuss their relative values.

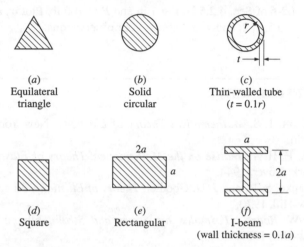

(a)	(b)	(c)
Equilateral triangle	Solid circular	Thin-walled tube $(t = 0.1r)$

(d)	(e)	(f)
Square	Rectangular	I-beam (wall thickness $= 0.1a$)

Problem 4.16

4.17 A rectangular bar with a cross section 25 mm × 50 mm is 500 mm long. Determine the maximum shear stress and the total angle of twist if a torque of 600 N·m is transmitted through the shaft. Let $E = 200$ GPa and $v = 0.3$.

4.18 Using Eqs. (4.3-10), (4.3-12), and (4.3-13), verify the factors k_1 and k_2 in Table 4.3-1 for an h/b of 1.5.

4.19 Each of the sections shown is transmitting a torque of 200 lb·in. Estimate the maximum shear stress in each section and the angle of twist per unit length.

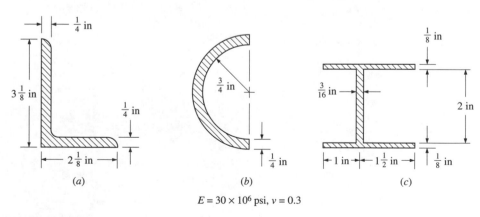

(a)   (b)   (c)

$E = 30 \times 10^6$ psi, $v = 0.3$

Problem 4.19

Computer Problems

4.20 For Fig. 4.2-4 of Sec. 4.2.4 let $a = 25$ mm and $\sigma = 10$ MPa plot σ_r and σ_θ as functions of θ for $r = 25$, 50, 75, and 100 mm. Then plot σ_r and σ_θ as functions of θ for the plate without the hole and discuss your observations with that obtained in the previous plots.

4.21 For Fig. 4.2-6 of Sec. 4.2.5 let $t = 1$ in and $P = 100$ lb. Plot σ_r as a function of θ at $r = 0.5$, 1, 2, and 4 in. Discuss your observations.

4.6   REFERENCES

4.1 Sokolnikoff, I. S. *Mathematical Theory of Elasticity*. New York: McGraw-Hill, 1956.

4.2 Love, A. E. H. *A Treatise on the Mathematical Theory of Elasticity*, 4th ed. New York: Dover, 1944.

4.3 Timoshenko, S. P., and J. N. Goodier. *Theory of Elasticity*, 3d ed. New York: McGraw-Hill, 1970.

4.4 Young, W. *Roark's Formulas for Stress and Strain*, 6th ed. New York: McGraw-Hill, 1989.

4.5 Cook, R. D., and W. C. Young. *Advanced Mechanics of Materials*. New York: Macmillan, 1985.

FIVE

TOPICS FROM ADVANCED
MECHANICS OF MATERIALS

5.0   INTRODUCTION

This chapter is devoted to extending the analyst's knowledge and capability further
into the area of classical problems and their solutions. The techniques used in elementary mechanics of materials of Chap. 3 and some of the results of the theory of
elasticity of Chap. 4 are expanded to cover additional practical problems of a more
advanced nature. The topics of torsion, bending, and pressurized cylinders are extended to include torsion of thin-walled tubes; unsymmetric bending; further discussion of transverse shear stresses including shear flow in thin sections, and the
shear center; composite beams; curved beams; plates in bending; and thick-walled
cylinders and rotating disks. Various specialized topics such as contact stresses and
stress concentrations are also presented.

5.1   SHEAR FLOW

The shear stresses that develop in torsion and direct shear problems behave quite
similarly to that of fluid flow. The differential equations for the normal and shear
stresses discussed in Chap. 4 are quite similar to those found in fluid flow, heat
transfer, etc. The equations developed for the stress functions are referred to as
scalar field equations. For torsion and direct shear problems, the shear stresses flow
tangent to free surfaces. Some examples of the shear stress distributions for various
sections in torsion are shown in Fig. 5.1-1. For the solid circular section, the shear
flow is uniform in a circular path at a given radius where the shear stress linearly

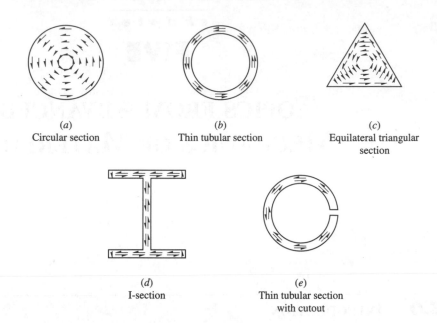

(a)
Circular section

(b)
Thin tubular section

(c)
Equilateral triangular
section

(d)
I-section

(e)
Thin tubular section
with cutout

Figure 5.1-1   Shear flow due to torsion.

increases from zero at the center to a maximum at the outer edge. With the stresses being very small near the center, the solid circular section is not a very efficient use of material for torsion. The closed, thin-walled, circular section is, since the stress is nearly constant throughout the tube. Since the net force due to the torsional shear stresses is zero, *shear pairs* will exist which are equal and in opposite directions. These shear pairs create moment couples which are proportional to the applied torsion. For this reason *open* thin-walled sections are not very good for transmitting torsion. As can be seen in Fig. 5.1-1(*d*) and (*e*), the distance between the shear pairs is very small, of the order of the wall thickness, and thus the corresponding shear stresses must be very large. Compare this to the closed thin-walled circular tube where the shear pair distance is of the order of the diameter of the tube. Solid and open thin-walled sections were covered in Chap. 4. Since closed thin-walled sections are structurally very effective, they are covered in the next section.

5.2   TORSION OF CLOSED THIN-WALLED TUBES

Closed thin-walled tubes are very effective when a torsion application is being considered and weight is at a premium. There are two classes of problems, that of torsion transmitted through single cell, and multiple cell cross sections.

5.2.1 SINGLE CELL SECTIONS IN TORSION

Consider the tube shown in Fig. 5.2-1. An element is isolated as shown in Fig. 5.2-1(b), where the possibility of a varying thickness and shear stress along the perimeter of the tube has been assumed. Assuming that the cross shear stresses on the $t_i \, dx$ areas are uniform and $\sigma_x \approx 0$ (or constant), equilibrium of forces in the x direction gives $\tau_2 t_2 \, dx - \tau_1 t_1 \, dx = 0$, or

$$\tau_2 t_2 = \tau_1 t_1 = \tau t = q \qquad \textbf{[5.2-1]}$$

This indicates that at any point along the tube wall q, called the *shear flow* (force per unit length), is constant.

   To relate the stresses to the torque, sum moments about an arbitrary point O as shown in Fig. 5.2-2(a). The net force on the ds element is $\tau t \, ds = q \, ds$. The torque about O is $(r \cos \phi)q \, ds$. Integrating this from $\phi = 0$ to 2π gives

$$T = \int r \cos\phi \, q \, ds = q \int r \cos\phi \, ds$$

where, as noted earlier, q is constant. The term in the integral, $r \cos \phi \, ds$, can be seen to be twice the shaded area $d\bar{A}$ shown in Fig. 5.2-2(b). Summing the $d\bar{A}$ areas

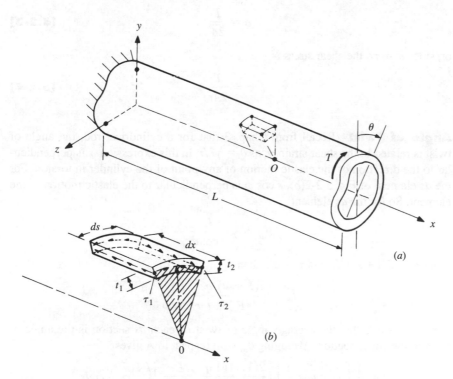

Figure 5.2-1   Closed, single cell tube in torsion.

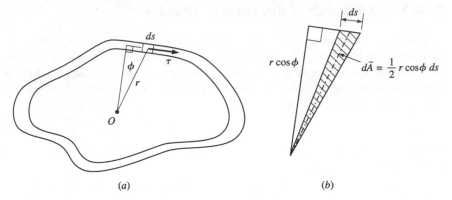

(a) (b)

Figure 5.2-2

yields the total area enclosed by the median line of the tube wall, $\bar{A}$. Thus $\int r \cos\phi\, ds = 2\bar{A}$, and the total torque is given by

$$T = 2q\bar{A} \tag{5.2-2}$$

The shear flow is then

$$q = \frac{T}{2\bar{A}} \tag{5.2-3}$$

or, since $q = \tau t$, the shear stress is

$$\tau = \frac{T}{2\bar{A}t} \tag{5.2-4}$$

Angle of Twist   Recall from Chap. 3 that for a cylindrical rod, the angle of twist is related to the shear strain γ, as $\theta = \gamma L/r$. In this expression, r is perpendicular to the direction of the elastic motion of any point of the cylinder in torsion. For the ds element of Fig. 5.2-2(b), $r \cos\phi$ is perpendicular to the elastic motion of the element. So for the ds element

$$\theta_{ds} = \gamma \frac{L}{r \cos\phi}$$

Since $\gamma = 2(1+v)\,\tau/E$, and $\tau = q/t$, then

$$\theta_{ds} = \frac{2(1 + v)}{E}\frac{q}{t}\frac{L}{r \cos\phi}$$

We will assume that the average angle of twist of each ds section is the angle of twist of the entire section. Averaging θ_{ds} with respect to $\bar{A}$ gives

$$\theta = \frac{1}{\bar{A}}\int \theta_{ds}\, d\bar{A} = \frac{1}{\bar{A}}\int \left[\frac{2(1 + v)}{E}\frac{q}{t}\frac{L}{r \cos\phi}\right]\left(\frac{1}{2}r\cos\phi\, ds\right)$$

Assuming v, q, L, E to be constants yields

$$\theta = \frac{(1 + v)qL}{E\bar{A}} \int_0^{\bar{S}} \frac{ds}{t} \qquad \text{[5.2-5]}$$

where $\bar{S}$ is the total perimeter length of the median line of the wall. Equation (5.2-5) can be written in terms of the applied torque by substituting Eq. (5.2-3) into (5.2-5) to give

$$\theta = \frac{(1 + v)TL}{2E\bar{A}^2} \int_0^{\bar{S}} \frac{ds}{t} \qquad \text{[5.2-6]}$$

For simple torsion, the angle of twist is given by $2(1+v)TL/EJ$. Equating this to Eq. (5.2-6), one can solve for J and obtain an equivalent second-area polar moment J_e. This results in

$$J_e = \frac{4\bar{A}^2}{\displaystyle\int_0^{\bar{S}} \frac{ds}{t}} \qquad \text{[5.2-7]}$$

Estimate the shear stress and total angle of twist of the thin-walled circular cylinder shown in Fig. 5.2-3. The steel tube is 0.5 m long and is transmitting a torque of 1 kN·m. The material constants are $E = 200$ GPa and $v = 0.29$.

Example 5.2-1

Solution:

The mean radius r is constant at $r = (125 - 5)/2 = 60$ mm. Thus

$$\bar{A} = \pi r^2 = \pi(0.060)^2 = 1.131 \times 10^{-2}\,\text{m}^2$$

Substituting this into Eq. (5.2-3) yields

$$\tau = \frac{T}{2\bar{A}t} = \frac{1000}{(2)(0.01131)(0.005)} = 8.84 \times 10^6\,\text{N/m}^2 = 8.84\ \text{MPa}$$

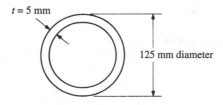

$t = 5$ mm

125 mm diameter

Figure 5.2-3

Since the thickness t is constant, the angle of twist is

$$\theta = \frac{(1 + v)TL}{2E\bar{A}^2t} \int_0^{\bar{S}} ds$$

The integral is simply $\int_0^{\bar{S}} ds = 2\pi r$. Thus

$$\theta = \frac{\pi(1 + v)TLr}{E\bar{A}^2t} = \frac{\pi(1 + 0.29)(1000)(0.5)(0.06)}{(200)(10^9)[(1.131)^2 \times 10^{-4}](0.005)} = 9.50 \times 10^{-4} \text{ rad}$$

The reader is urged to check the solution against that given in elementary strength of materials using the conventional torsion equations.[†] The exact equation for the shear stress in a circular cylinder undergoing torsion gives a maximum value on the outer radius of 9.19 MPa and an average value at the mean radius of 8.83 MPa.

Example 5.2-2 | The cross section shown in Fig. 5.2-4 is subjected to a torque of 50 kip·in. Estimate the shear stress in each wall and determine the integral $\int ds/t$ for the cross section.

Solution:

The area enclosed by the wall median line is

$$\bar{A} = \left(8 - \frac{1}{8} - \frac{1}{4}\right)\left(6 - \frac{1}{8} - \frac{1}{8}\right) = 43.8 \text{ in}^2$$

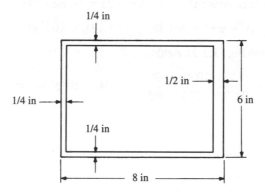

Figure 5.2-4

[†] For a thin-walled cylinder, using $J = 2\pi tr^3$, the stress at the *center* of the wall agrees exactly with Eq. (5.2-4), and the angle of twist agrees exactly with Eq. (5.2-6).

The shear stress on the $\frac{1}{2}$-in wall is

$$(\tau)_{1/2\,\text{in}} = \frac{T}{2\bar{A}t} = \frac{50(10^3)}{(2)(43.8)(1/2)} = 1140 \text{ psi}$$

and on the $\frac{1}{4}$-in walls

$$\tau_{1/4\,\text{in}} = \frac{50(10^3)}{(2)(43.8)(1/4)} = 2280 \text{ psi}$$

In evaluating the integral $\int ds/t$ note that t is constant for each wall; thus

$$\int \frac{ds}{t} = \frac{2(8 - \frac{1}{8} - \frac{1}{4})}{\frac{1}{4}} + \frac{6 - \frac{1}{8} - \frac{1}{8}}{\frac{1}{4}} + \frac{6 - \frac{1}{8} - \frac{1}{8}}{\frac{1}{2}} = 95.5$$

5.2.2   MULTIPLE CELL SECTIONS IN TORSION

Figure 5.2-5 shows a closed, thin-walled, multiple cellular cross section transmitting a torque T. Conventional practice is to treat the torsion contribution of each cell separately and combine the results using superposition. This is done subject to the condition that the angle of twist of each cell is the same.

The torsion contribution of the ith cell is given by Eq. (5.2-3) as $T_i = 2q_i\bar{A}_i$. Summing the contribution of the n cells yields

$$T = \sum_{i=1}^{n} T_i = 2\sum_{i=1}^{n} q_i\bar{A}_i \qquad\qquad \textbf{[5.2-8]}$$

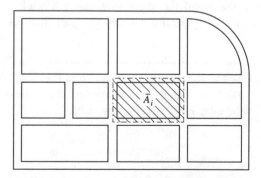

Figure 5.2-5   Multiple cell cross section.

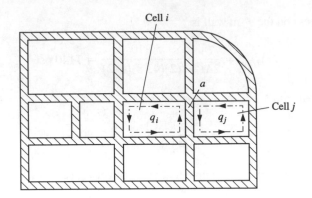

Figure 5.2-6   Positive shear flow in cells i and j.

where we have n unknown values of q_i. The condition that each cell has the same angle of twist θ establishes n equations of twist for each cell. Equation (5.2-5) applies to a single cell where q is the same for all walls of the cell. For a cell within a multiple cell structure, the shear flow q in each wall is the superposition of the shear flow of adjacent cells. Thus in this case, q can vary throughout the walls of a given cell and q must be placed within the integral of Eq. (5.2-5). As an example, the angle of twist of cell i is

$$\theta = \frac{(1 + v)L}{E\bar{A}_i}\left(\int \frac{q}{t}\, ds\right)_i \qquad\qquad \textbf{[5.2-9]}$$

The sign of the shear flow is extremely important. Here we will consider counter-clockwise flow within a cell and its walls to be positive. For example, the twist equation of wall a of cell i shown in Fig. 5.2-6 is $(q_a)_i = q_i - q_j$. For the twist equation of wall a of cell j, $(q_a)_j = q_j - q_i$.

Equation (5.2-9) represents n equations, one for each cell. This, together with Eq. (5.2-8), represents $n + 1$ equations in terms of the $n + 1$ unknowns; θ, q_1, q_2, ..., q_n. Once the q_i are found the shear flow in each wall can be determined. Dividing the shear flow in a specific wall by the wall thickness yields the shear stress in that wall.

Example 5.2-3 | A steel tube 1.5 m long has the cross section shown in Fig. 5.2-7. The tube is transmitting a torque of 200 N·m. Determine the average shear stress in each wall and the angle of twist of the tube. $E = 210$ GPa, $v = 0.29$.

Solution:
From Eq. (5.2-8),

$$T = 2\left[q_1\left(\frac{\pi}{4}\, a^2\right) + q_2(a)(c - a) + q_3(bc)\right]$$

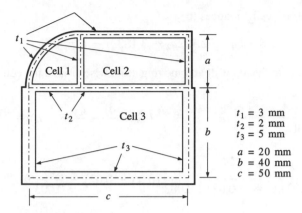

Figure 5.2-7

Substitution of values gives

$$200 = 2\left[q_1\left(\frac{\pi}{4}0.02^2\right) + q_2(0.02)(0.05 - 0.02) + q_3(0.04)(0.05)\right]$$

Simplifying results in

$$3.142q_1 + 6q_2 + 20q_3 = 1(10^6) \qquad\qquad \textbf{[a]}$$

For the angle of twist of each cell, consider cell 1 first. Starting at the radial wall and considering each wall of the cell in a clockwise order, Eq. (5.2-9) for cell 1 is

$$\theta = \frac{(1 + v)L}{E[(\pi/4)a^2]}\left[\frac{q_1}{t_1}\left(\frac{\pi}{2}\right)(a) + \frac{q_1 - q_2}{t_1}(a) + \frac{q_1 - q_3}{t_2}(a)\right]$$

$$= \frac{(1 + v)L}{E}\frac{4}{\pi(0.02^2)}\left[\frac{q_1}{0.003}\left(\frac{\pi}{2}\right)(0.02) + \frac{q_1 - q_2}{0.003}(0.02) + \frac{q_1 - q_3}{0.002}(0.02)\right]$$

Placing the term involving v, L, and E on the left side of this equation together with θ and simplifying gives

$$\frac{\theta E}{(1 + v)L} = (8.638q_1 - 2.122q_2 - 3.183q_3)(10^4)$$

Let

$$C = \frac{\theta E}{(1 + v)L}(10^{-4}) \qquad\qquad \textbf{[b]}$$

Thus Eq. (5.2-9) for cell 1 reduces to

$$8.638q_1 - 2.122q_2 - 3.183q_3 - C = 0 \qquad \textbf{[c]}$$

For cell 2, starting at the top wall and moving clockwise around the cell

$$\frac{\theta E}{(1+v)L} = \frac{1}{(a)(c-a)}\left[\frac{q_2}{t_1}(c-a) + \frac{q_2}{t_1}(a) + \frac{q_2-q_3}{t_2}(c-a) + \frac{q_2-q_1}{t_1}(a)\right]$$

$$= \frac{1}{(0.02)(0.05-0.02)}\left[\frac{q_2}{0.003}(0.05-0.02) + \frac{q_2}{0.003}(0.02)\right.$$

$$\left. + \frac{q_2-q_3}{0.002}(0.05-0.02) + \frac{q_2-q_1}{0.003}(0.02)\right]$$

Using the term C as before, we obtain

$$-1.111q_1 + 6.389q_2 - 2.5q_3 - C = 0 \qquad \textbf{[d]}$$

For cell 3, combining the right, bottom, and left walls, and then separating the top two sections gives

$$\frac{\theta E}{(1+v)L} = \frac{1}{bc}\left[\frac{q_3}{t_3}(2b+c) + \frac{q_3-q_1}{t_2}(a) + \frac{q_3-q_2}{t_2}(c-a)\right]$$

$$= \frac{1}{(0.04)(0.05)}\left[\frac{q_3}{0.005}(0.08+0.05) + \frac{q_3-q_1}{0.002}(0.02) + \frac{q_3-q_2}{0.002}(0.05-0.02)\right]$$

As with the other cells, simplification yields

$$-0.5q_1 - 0.75q_2 + 1.754q_3 - C = 0 \qquad \textbf{[e]}$$

Solving Eqs. (a) and (c) to (e) simultaneously yields

$$q_1 = 24.62 \text{ kN/m}$$

$$q_2 = 25.11 \text{ kN/m}$$

$$q_3 = 38.60 \text{ kN/m}$$

$$C = 36.56 \text{ kN/m}^3$$

Figure 5.2-8 shows the shear flow and shear stress definitions for this example. The shear stresses are:

$$\tau_1 = \frac{q_1}{t_1} = \frac{24.62}{0.003} = 8.21(10^3)\text{kN/m}^2 = 8.21\text{MPa}$$

$$\tau_2 = \frac{q_2}{t_1} = \frac{25.11}{0.003} = 8.37(10^3)\text{kN/m}^2 = 8.37\text{MPa}$$

$$\tau_3 = \frac{q_3}{t_3} = \frac{38.60}{0.005} = 7.72(10^3)\text{kN/m}^2 = 7.72\text{MPa}$$

$$\tau_4 = \frac{q_3 - q_1}{t_2} = \frac{38.60 - 24.62}{0.002} = 6.99(10^3)\text{kN/m}^2 = 6.99\text{MPa}$$

$$\tau_5 = \frac{q_2 - q_1}{t_1} = \frac{25.11 - 24.62}{0.003} = 0.163(10^3)\text{kN/m}^2 = 0.163\text{MPa}$$

$$\tau_6 = \frac{q_3 - q_2}{t_2} = \frac{38.60 - 25.11}{0.002} = 6.75(10^3)\text{kN/m}^2 = 6.75\text{MPa}$$

The angle of twist is found from Eq. (b) as

$$\theta = \frac{(1 + v)L(10^4)}{E} C$$

$$= \frac{(1 + 0.29)(1.5)(10^4)}{210(10^9)} 36.56(10^3) = 3.37(10^{-3}) \text{ rad} = 0.193°$$

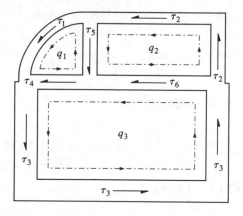

Figure 5.2-8   Shear flow for Example 5.2-3.

5.3   BENDING OF UNSYMMETRICAL BEAMS

5.3.1   STRESSES

Chapter 3 covers two-plane bending where the cross-sectional area is in or parallel to the yz plane and the y and z axes are *principal axes* of the second-area moment. That is, $I_{yz} = \int_A yz\, dA = 0$, where I_{yz} is called the *mixed second-area moment.*[†] For this case, Eq. (3.5-1) is valid, and for pure bending in two planes,

$$\sigma_x = \frac{M_y z}{I_y} - \frac{M_z y}{I_z}$$ [5.3-1]

For unsymmetrical cross sections, I_{yz} is generally not zero and Eq. (5.3-1) is invalid. However, for any given cross section there exists an axis system, the *principal axes,* where the mixed second-area moment is zero.[‡] Denoting the principal axes as the m, n axes, M_m, M_n as the components of the bending moment, and I_m, I_n as the second-moment areas about the m, n axes, respectively, Eq. (5.3-1) can be rewritten as[§]

$$\sigma_x = \frac{M_m n}{I_m} - \frac{M_n m}{I_n}$$ [5.3-2]

Example 5.3-1

For the beam shown in Fig. 5.3-1 neglect Saint-Venant's effect at the rigid wall and determine the bending stresses at $(0, 25, 13)$ and $(0, -25, -13)$ mm.

Solution:

Dividing the area into three rectangles as shown in Fig. 5.3-2, the second-area moments are

$$I_y = I_{1y} + I_{2y} + I_{3y} = I_{1y} + 2I_{2y}$$

$$= \frac{1}{12}(50)(10)^3 + 2\left[\frac{1}{12}(15)(8)^3 + (8)(15)(9)^2\right]$$

$$= 24.89 \times 10^3 \text{ mm}^4$$

and

$$I_z = I_{1z} + I_{2z} + I_{3z} = I_{1z} + 2I_{2z}$$

$$= \frac{1}{12}(10)(50)^3 + 2\left[\frac{1}{12}(8)(15)^3 + (8)(15)(17.5)^2\right]$$

$$= 182.17 \times 10^3 \text{ mm}^4$$

[†] Sometimes referred to as the *product of inertia.*
[‡] The method of locating the principal axes of the second area moment is presented in Appendix E.
[§] When dealing with standard unsymmetrical sections, such as angle sections, the location of the principal axes can be found in handbooks [such as the *AISC Handbook* (American Institute of Steel Construction)].

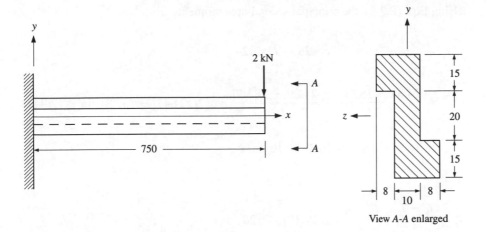

Figure 5.3-1 (Dimensions in millimeters.)

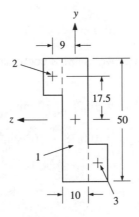

Figure 5.3-2

and

$$I_{yz} = I_{2yz} + I_{3yz} = (9)(17.5)(8)(15) + (-9)(-17.5)(8)(15)$$

$$= 37.8 \times 10^3 \text{ mm}^4$$

From Appendix E, Eq. (E.2-4), the angle to the principal axes is

$$\theta = \frac{1}{2}\tan^{-1}\left(\frac{2I_{yz}}{I_z - I_y}\right) = \frac{1}{2}\tan^{-1}\left(\frac{2(37.8)}{182.17 - 24.89}\right)$$

$$= 12.84°$$

Using Eqs. (E.2-3), the principal second-area moments are

$$I_{y'} = \frac{I_y + I_z}{2} + \frac{I_y - I_z}{2}\cos 2\theta - I_{yz}\sin 2\theta$$

$$= \frac{24.89 + 182.17}{2} + \frac{24.89 - 182.17}{2}\cos[2(12.84)] - (37.8)\sin[2(12.84)]$$

$$= 16.28 \text{ k–mm}^4 = 16.28 \times 10^3 \text{ mm}^4$$

and

$$I_{z'} = \frac{I_y + I_z}{2} - \frac{I_y - I_z}{2}\cos 2\theta + I_{yz}\sin 2\theta$$

$$= \frac{24.89 + 182.17}{2} - \frac{24.89 - 182.17}{2}\cos[2(12.840)] + (37.8)\sin[2(12.840)]$$

$$= 190.78 \text{ k–mm}^4 = 190.78 \times 10^3 \text{ mm}^4$$

Instead of using y', z' coordinates we use m, n coordinates as shown in Fig. 5.3-3, where

$$I_m = I_{y'} = 16.28 \times 10^3 \text{ mm}^4$$

$$I_n = I_{z'} = 190.78 \times 10^3 \text{ mm}^4$$

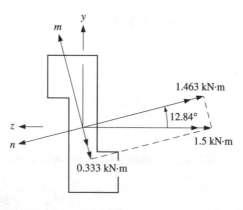

Figure 5.3-3 Applied moment and components.

The bending moment at the wall ($x = 0$) is negative about the z axis and is

$$M_z = -(0.750)(2) = -1.5 \text{ kN·m}$$

The components in the m and n directions, shown in Fig. 5.3-3, are

$$M_m = -1.5 \sin 12.84 = -0.333 \text{ kN·m}$$

$$M_n = -1.5 \cos 12.84 = -1.463 \text{ kN·m}$$

To use Eq. (5.3-2) we transform y, z components of position to m, n components. Referring to Fig. 5.3-4, we see that the m component of the y value for point P is $y \cos \theta$ and the m component of the z value is $z \sin \theta$. Thus

$$m = y \cos \theta + z \sin \theta \qquad \text{[5.3-3a]}$$

Likewise the n components are given by

$$n = -y \sin \theta + z \cos \theta \qquad \text{[5.3-3b]}$$

For point $(0, 25, 13)$

$$m = 25 \cos 12.84 + 13 \sin 12.84 = 27.26 \text{ mm}$$

$$n = -25 \sin 12.84 + 13 \cos 12.84 = 7.12 \text{ mm}$$

From Eq. (5.3-2)

$$\sigma_x = \frac{(-0.333)(10^3)(7.12)(10^{-3})}{16.28(10^3)(10^{-3})^4} - \frac{(-1.463)(10^3)(27.26)(10^{-3})}{190.78(10^3)(10^{-3})^4}$$

$$= 63.4 \times 10^6 \text{ N/m}^2 = 63.4 \text{ MPa}$$

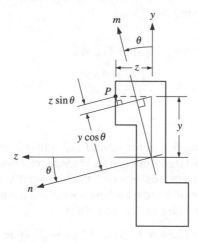

Figure 5.3-4

For point $(0, -25, -13)$ it can be shown that $m = -27.26$ mm and $n = -7.12$ mm, resulting in $\sigma_x = -63.4$ MPa.

If one were to erroneously use Eq. (5.3-1) for this problem with $M_z = -1.5$ kN·m and $M_y = 0$, say, for point $(0, 25, 13)$, the result would be

$$\sigma_x = -\frac{(-1.5)(10^3)(25)(10^{-3})}{(182.17)(10^3)(10^{-3})^4}$$

$$= 205.8 \times 10^6 \, \text{N/m}^2 = 206 \, \text{MPa}$$

which is considerably different from the correct value of 63.4 MPa.

Maximum and Minimum Bending Stresses   In the previous example problem, the stresses at the points investigated are not the maximum and minimum bending stresses. It is not always obvious where they occur. However, since the bending stress is linear and changes from tension to compression through the cross section, there exists an axis for which the stress is zero. This is called the *neutral axis* in bending, and points farthest from it undergo the maximum and minimum bending stresses. To locate the neutral axis we set Eq. (5.3-2) to zero. Thus

$$\frac{M_m n}{I_m} - \frac{M_n m}{I_n} = 0 \qquad \text{or} \qquad \frac{n}{m} = \frac{M_n}{M_m} \frac{I_m}{I_n}$$

This locates the neutral axis with respect to the m axis. Defining an angle α *counterclockwise* from the m axis, we have $\tan \alpha = n/m$, or

$$\tan \alpha = \frac{M_n}{M_m} \frac{I_m}{I_n} \qquad\qquad \text{[5.3-4]}$$

Example 5.3-2 | Locate the neutral axis and the points undergoing the maximum and minimum bending stresses in Example 5.3-1.

Solution:

The angle α is determined from Eq. (5.3-4) and is

$$\tan \alpha = \frac{-1.463}{-0.333} \frac{16.28}{190.78} = 0.3749$$

$$\alpha = 20.6°$$

Figure 5.3-5 depicts the location of the neutral axis where it can be seen that points A and B are the farthest distance from it. The shaded area represents that part of the section undergoing tensile stresses, whereas the unshaded area is in compression. We see that for tension, point A is the farthest point from the neutral axis where the coordinates, from Fig. 5.3-1, are $(0, 25, -5)$. Thus

$$m = 25 \cos 12.84 + (-5) \sin 12.84 = 23.26 \, \text{mm}$$

$$n = -25 \sin 12.84 + (-5) \cos 12.84 = -10.43 \, \text{mm}$$

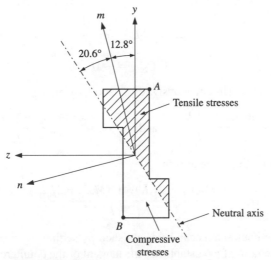

Figure 5.3-5

and the corresponding maximum tensile stress, using Eq. (5.3-2), is

$$\sigma_x = \frac{(-0.333)(10^3)(-10.43)(10^{-3})}{16.28(10^3)(10^{-3})^4} - \frac{(-1.463)(10^3)(23.26)(10^{-3})}{190.78(10^3)(10^{-3})^4}$$

$$= 391.9 \times 10^6 \text{ N/m}^2 = 392 \text{ MPa}$$

Note that this is still considerably different from the incorrect result Eq. (5.3-1) yields.

Likewise, the minimum bending stress (the maximum compressive stress) occurs at point B, where $\sigma_x = -392$ MPa.

Equation (5.3-2) has the disadvantage of requiring a transformation of the coordinates and resolving the bending moments to the m, n coordinate system. Recognizing that the bending stress is linear in m and n coordinates, the stress must also be linear in y and z coordinates. Thus

$$\sigma_x = C_1 y + C_2 z \qquad \textbf{[a]}$$

where C_1 and C_2 are constants. The stress is related to the bending moments as

$$\int_A z\sigma_x \, dA = M_y \qquad \textbf{[b]}$$

$$\int_A y\sigma_x \, dA = -M_z \qquad \textbf{[c]}$$

Substituting Eq. (a) into (b) and (c), with

$$I_y = \int_A z^2 \, dA \qquad I_z = \int_A y^2 \, dA \qquad \text{and} \qquad I_{yz} = \int_A yz \, dA$$

gives

$$C_1 I_{yz} + C_2 I_y = M_y \qquad\qquad \text{[d]}$$

$$C_1 I_z + C_2 I_{yz} = -M_z \qquad\qquad \text{[e]}$$

Solving Eqs. (d) and (e) simultaneously for C_1 and C_2 yields

$$C_1 = \frac{M_y I_{yz} + M_z I_y}{I_{yz}^2 - I_y I_z} \qquad \text{and} \qquad C_2 = \frac{-M_z I_{yz} - M_y I_z}{I_{yz}^2 - I_y I_z}$$

Substituting this into Eq. (a) results in

$$\sigma_x = \frac{(-M_y I_{yz} - M_z I_y)y + (M_y I_z + M_z I_{yz})z}{I_y I_z - I_{yz}^2} \qquad\qquad \text{[5.3-5]}$$

As before, the neutral axis can be established by setting $\sigma_x = 0$ in Eq. (5.3-5), and solving for the ratio of z/y establishes the tangent of the counterclockwise angle the neutral plane makes with the y axis. Thus

$$\tan \beta = \frac{z}{y} = \frac{M_y I_{yz} + M_z I_y}{M_y I_z + M_z I_{yz}} \qquad\qquad \text{[5.3-6]}$$

Example 5.3-3 | For Examples 5.3-1 and 5.3-2 use Eqs. (5.3-6) and (5.3-5) to locate the neutral plane and determine the stress at point A at the wall (0, 25, -5), respectively.

Solution:

For the problem $M_y = 0$, $M_z = -1.5$ kN·m, $I_y = 24.89 \times 10^3$ mm^4, $I_z = 182.17 \times 10^3$ mm^4, and $I_{yz} = 37.8 \times 10^3$ mm^4. Thus

$$\tan \beta = \frac{M_y I_{yz} + M_z I_y}{M_y I_z + M_z I_{yz}} = \frac{I_y}{I_{yz}} = \frac{24.89}{37.8}$$

$$\beta = 33.4°$$

which agrees with Fig. 5.3-5 (20.6° + 12.8° = 33.4°). The stress at point A is

$$\sigma_x = \frac{(-M_y I_{yz} - M_z I_y)y + (M_y I_z + M_z I_{yz})z}{I_y I_z - I_{yz}^2}$$

$$= \frac{-(-1.5)(10^3)(24.89)(10^3)(10^{-3})^4(25)(10^{-3}) + (-1.5)(10^3)(37.8)(10^3)(10^{-3})^4(-5)(10^{-3})}{[(24.89)(182.47) - (37.8)^2][(10^3)(10^{-3})^4]^2}$$

$$= 391.9 \times 10^6 \text{ N/m}^2 = 392 \text{ MPa}$$

which agrees with the result using Eq. (5.3-2).

5.3.2 DEFLECTIONS

The most straightforward approach to deflections is to resolve the loading into components in the directions of each principal axis, determine the deflection in the direction of each principal axis using elementary beam deflection methods, and then add the results vectorially.

Determine the deflection of the end of the beam of Example 5.3-1. Let $E = 200$ GPa. **Example 5.3-4**

Solution:

The 2-kN force is resolved into components in the m and n directions

$$F_m = -2\cos 12.84 = -1.950 \text{ kN}$$

$$F_n = 2\sin 12.84 = 0.444 \text{ kN}$$

The deflection of the end of an end-loaded cantilever beam is given by $FL^3/(3EI)$. Thus the deflections of the free end of the beam in the m and n directions are

$$(\delta_c)_m = \frac{F_m L^3}{3EI_n} = \frac{(-1.950)(10^3)(0.750^3)}{3(200)(10^9)(190.78)(10^3)(10^{-3})^4}$$

$$= -7.187 \times 10^{-3}\text{m} = -7.187 \text{ mm}$$

$$(\delta_c)_n = \frac{F_n L^3}{3EI_m} = \frac{(0.444)(10^3)(0.750^3)}{3(200)(10^9)(16.28)(10^3)(10^{-3})^4}$$

$$= 19.18 \times 10^{-3}\text{m} = 19.18 \text{ mm}$$

respectively. Generally, the y, z components would be of more physical interest. In a fashion similar to Eqs. (5.3-6a) and (5.3-6b) in Example 5.3-1, we can write

$$y = m \cos\theta - n \sin\theta \qquad\qquad \textbf{[5.3-7a]}$$

$$z = m \sin\theta + n \cos\theta \qquad\qquad \textbf{[5.3-7b]}$$

Thus, at $x = 750$ mm, with $\theta = 12.84°$

$$v_c = -7.187 \cos 12.84 - 19.18 \sin 12.84 = -11.27 \text{ mm}$$

$$w_c = -7.187 \sin 12.84 + 19.18 \cos 12.84 = 17.10 \text{ mm}$$

The total end deflection would be

$$\delta_{max} = \sqrt{(-7.187)^2 + (19.18)^2} = \sqrt{(-11.27)^2 + (17.10)^2} = 20.5 \text{ mm}$$

Note that, although the load on the cantilever beam is in the y direction, there is actually more deflection occurring in the z direction.

5.4   FURTHER DISCUSSION OF TRANSVERSE SHEAR STRESSES

5.4.1   SHEAR FLOW IN OPEN THIN-WALLED BEAMS

For transverse loading of beams bending in the xy plane, the shear stress resulting from the transverse force V_y was given in Chap. 3 as Eq. (3.4-7), repeated here as

$$\tau_{xy} = \frac{V_y Q}{I_z b} \qquad \text{[5.4-1]}$$

where Q is the area moment $\int y\, dA$, integrated from $y = y_1$ to y_{max}; and b is the width of the beam at $y = y_1$. For thin-walled beams Eq. (5.4-1) is used in a slightly different manner than in Chap. 3. Consider, for example, the I-beam shown in Fig. 5.4-1(a). The shear stress distribution predicted by Eq. (5.4-1) is shown on the exposed surface in Fig. 5.4-1(a) and is plotted as a function of y in Fig. 5.4-1(b).The maximum stress occurs at $y = 0$, where the area moment Q is a maximum and the beam width $b = t_w$ is a minimum. At $y = \pm(h/2 - t_f)$ a discontinuity in the shear stress seems to occur where the beam width changes abruptly from t_w to b_f. In reality, there is no discontinuity here. Recall for Eq. (5.4-1) it is assumed that the shear stress is uniform in the z direction. Thus within the flange at $|y|$ slightly greater than $(h/2 - t_f)$, Eq. (5.4-1) is predicting a uniform shear stress across the entire flange. However, on the free surfaces of the flange where $y = \pm h/2$ and $\pm(h/2 - t_f)$, τ_{xy} must be zero as shown in Fig. 5.4-1(c). If the flange thickness is small and τ_{xy} is zero along the free surfaces of the flange, τ_{xy} will remain small throughout the flange [certainly much smaller than the already small value predicted by Eq. (5.4-1)]. Thus τ_{xy} is mostly contained within the web, and since $2(h/2 - t_f) \gg t_w$, Eq. (5.4-1) is fairly accurate in the web (see Table 3.4-1 for rectangular sections).

Since the free surfaces of the flange are horizontal (in the z direction), shear stress τ_{zx} can be present. Consider the section of the flange shown in Fig. 5.4-1(d). Recalling the derivation of Eq. (5.4-1), the shear stress arises from a differential in the bending stress σ_x with respect to x occurring on the isolated element segment. Thus, in a similar manner

$$\tau_{zx} = \frac{1}{t_f} \int_{A_1} \frac{d\sigma_x}{dx}\, dA$$

Again, since

$$\frac{d\sigma_x}{dx} = \frac{d}{dx}\left(-\frac{M_z y}{I_z} \right) = -\frac{dM_z}{dx}\frac{y}{I_z} = \frac{V_y y}{I_z}$$

Then

$$\tau_{zx} = \frac{V_y Q}{I_z t_f} \qquad \text{[5.4-2]}$$

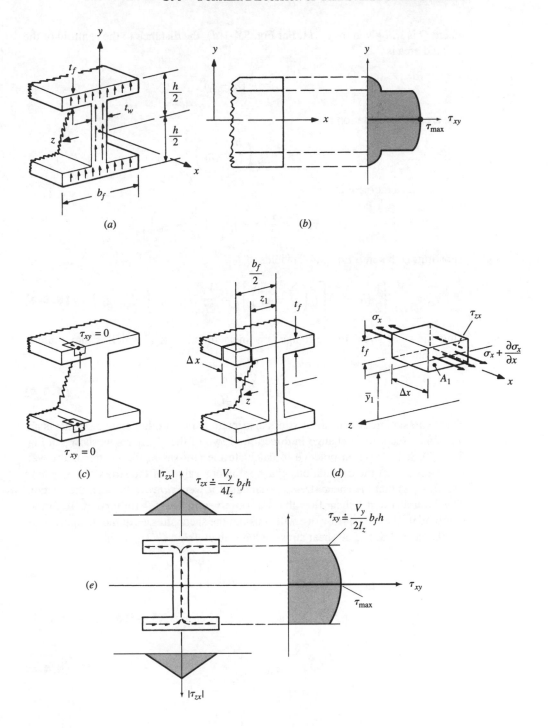

Figure 5.4-1

where Q is *still* given by $\int y\, dA$. For Fig. 5.4-1(d), the distance to the centroid of the isolated area is

$$\bar{y}_1 = \frac{1}{2}(h - t_f)$$

The area of the isolation is

$$A_1 = t_f\left(\frac{b_f}{2} - z_1\right)$$

Thus the area moment is

$$Q = \bar{y}_1 A_1 = \frac{1}{2}(h - t_f)\left[t_f\left(\frac{b_f}{2} - z_1\right)\right]$$

Substituting this into Eq. (5.4-2) yields

$$\tau_{zx} = \frac{V_y}{I_z t_f}\left\{\frac{1}{2}(h - t_f)\left[t_f\left(\frac{b_f}{2} - z_1\right)\right]\right\} = \frac{V_y}{2I_z}(h - t_f)\left(\frac{b_f}{2} - z_1\right) \qquad \text{[5.4-3]}$$

Thus the shear stress in the flange is a linear function of z, which is maximum at $z_1 = 0$ and is

$$\tau_{zx} = \frac{V_y b_f (h - t_f)}{4I_z} \qquad \text{[5.4-4]}$$

Thanks to symmetry, the shear stress distributions in the other flange segments are the same except for a change in direction of two of the segments as shown in Fig. 5.4-1(e). It is common practice to calculate the stresses at the center of the wall thickness so that the calculations give continuous values of the stress. Also, note in Fig. 5.4-1(e) that the shear stress appears as if it were *flowing* through the section. This is similar to the shear flow that was observed in Sec. 5.1 for torsion. To further substantiate the idea of flow we will consider the shear stresses at the intersection of the flange and web. The shear stress in the web at $y = \pm(h/2 - t_f)$ is

$$\tau_{xy} = \frac{V_y Q}{I_z t_w} = \frac{V_y}{I_z t_w}\left\{\frac{b_f}{2}\left[\left(\frac{h}{2}\right)^2 - \left(\frac{h}{2} - t_f\right)^2\right]\right\} = \frac{V_y}{2I_z}\frac{t_f}{t_w}b_f(h - t_f) \qquad \text{[5.4-5]}$$

For a simpler case let $t_f = t_w = t$, and $t_f \ll h$. Then Eqs. (5.4-5) and (5.4-4) can be written as

$$\tau_{xy} = \frac{V_y}{2I_z}b_f h \qquad \text{[5.4-6]}$$

and

$$\tau_{zx} = \frac{V_y}{4I_z}b_f h \qquad \text{[5.4-7]}$$

respectively. We observe that the shear stress in the flange at the intersection with the web is one-half that of the shear stress in the web at that location. This is what would be expected in fluid channel flow, where the flow would divide equally in half if the channels were of equal thickness.

From what we have learned from the I-beam, we can modify Eq. (5.4-1) to apply to thin-walled beams. In general, a wall need not be parallel to either the y or the z axis, but Eq. (5.4-1) can be used to determine the net average shear stress in a wall $(\tau_x)_{ave}$. For simplicity, this will be referred to as τ, and given by the equation

$$\tau = \frac{V_y Q}{I_z t} \qquad \textbf{[5.4-8]}$$

where t is the wall thickness at the location where τ is being evaluated.

The direction of τ is found easily by observing the direction of the shear stress caused by V_y and being careful in the determination of the sign of Q. The shear stress will always flow in the same direction unless Q changes sign. In the I-beam example, the shear stress in the web is upward in the direction of V_y. The direction of the shear stress in each of the flanges is consistent with the direction of the shear flow.

5.4.2   SHEAR CENTER FOR OPEN THIN-WALLED BEAMS WITH ONE AXIS OF SYMMETRY

If the cross section of a thin-walled beam is not symmetric relative to the xy loading plane, Eq. (5.4-8) may not completely provide equilibrium of torsion about the

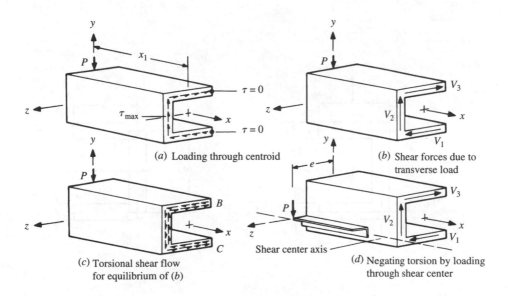

(a) Loading through centroid

(b) Shear forces due to transverse load

(c) Torsional shear flow for equilibrium of (b)

(d) Negating torsion by loading through shear center

Figure 5.4-2   Shear behavior when loading thin-walled beams.

x axis. As an example, consider the cantilever beam with a channel cross section shown in Fig. 5.4-2. The cross section is symmetric about the *z* axis. Hence the centroidal *y, z* axes are principal axes. Beams are normally loaded through the centroidal axis to avoid twisting. However, if the load *P* is applied through the unsymmetric *y* axis of the thin-walled channel as shown in Fig. 5.4-2(*a*), the beam will twist. To see why, consider the section of channel isolated at $x = x_1$ where for clarity the bending stresses are not shown and only the shear stresses predicted by Eq. (5.4-8) are portrayed. Note the shear flow direction is consistent with equilibrium of forces on the beam segment. However, this distribution does not provide equilibrium of torsion about the *x* axis. Integrating the shear stress distributions in the flanges and web provides the net forces in the sections as shown in Fig. 5.4-2(*b*). Once the forces are determined, the net torque about the centroidal axis can be evaluated. To obtain equilibrium in torsion we must then superpose a torsional shear stress distribution as shown in Fig. 5.4-2(*c*), where the value is given by Eq. (4.3-14). The maximum shear stress occurs at the inner edge of the web at $y = 0$ and is the superposition of the stresses determined in Fig. 5.4-2(*a*) and (*c*).

As stated in Sec. 5.1, open thin-walled sections are not recommended for the transmission of torsion. If the load *P* is moved, the torsional stresses can be avoided. This is shown in Fig. 5.4-2(*d*), where *P* is placed so the torsion due to V_1, V_2, and V_3 is canceled. The axis for which the load is moved to cancel twisting is called the *shear center axis*.

Example 5.4-1 | The channel section shown in Fig. 5.4-3(*a*) is bending about the *z* axis and undergoes a net shear force V_y in the *y* direction. Determine (*a*) the shear flow throughout

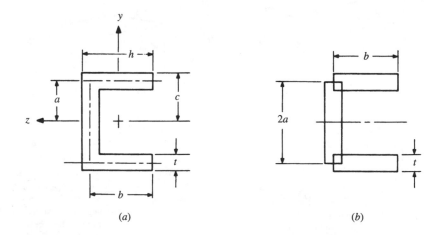

(*a*)   (*b*)

Figure 5.4-3

the section, (b) the total shear force in each section, and (c) the shear center through which the plane of loading should intersect.

Solution:

The second-area moment about the z axis is approximated by considering the section made up of three rectangles, as shown in Fig. 5.4-3(b)[†]; then

$$I_z = \frac{2}{3} a^3 t + 2\left(\frac{1}{12} bt^3 + bta^2\right)$$

However, if t is much less than a and b, then I_z can be approximated by

$$I_z \approx \frac{2}{3} a^3 t + 2a^2 bt$$

For the shear-stress distribution in the top flange, consider the variable position s, as shown in Fig. 5.4-4(a), where $0 \leq s \leq b$. This isolates the shaded area shown in Fig. 5.4-4(b), of which the area moment is

$$Q = \bar{y}_1 A_1 = a(st)$$

Substituting this into Eq. (5.4-8), we find the shear-stress distribution in the top flange to be

$$\tau = \frac{V_y}{I_z} \frac{Q}{t} = \frac{V_y}{I_z} \frac{a(st)}{t} = \frac{V_y a}{I_z} s \qquad\qquad\text{[a]}$$

Thus the stress distribution in the top flange is linear, as shown in Fig. 5.4-4(c).

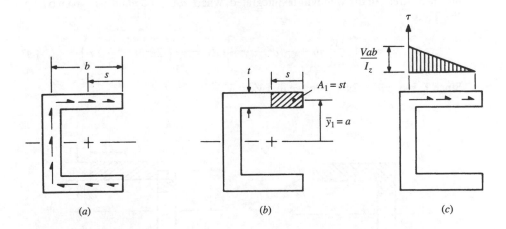

Figure 5.4-4   Determination of the shear stress distribution in the flange.

[†] This models the thin-walled section perfectly as $t \to 0$; it will ensure continuity of the shear stress at the intersections of the web and flanges and provide simple calculations for the net force in a wall for checks on equilibrium.

The stress distribution in the side web can be found by determining Q for the section shown in Fig. 5.4-5(a). Here, for convenience, the variable is changed to y instead of s. Q is approximated by using the two rectangles shown in Fig. 5.4-5(b), where

$$Q = \sum \bar{y} A = a(bt) + \frac{a+y}{2}(a-y)t$$

or

$$Q = abt + \frac{t}{2}(a^2 - y^2)$$

Thus the stress distribution in the side web is

$$\tau = \frac{V_y}{I_z}\left[ab + \frac{1}{2}(a^2 - y^2)\right]\qquad\text{[b]}$$

which can be seen to be a parabolic function, as shown in Fig. 5.4-6(a). Thanks to symmetry about the z axis, the distribution of shear stress is symmetric.

To determine the net shear force across each section, as shown in Fig. 5.4-6(b), it is necessary to integrate $\int \tau \, dA$ across the surface. By symmetry, $V_1 = V_3$. The area dA for the top flange is $t \, ds$, and the shear stress $\tau = V_y as/I_z$. Thus the shear force on the top flange is

$$V_3 = \int_0^b \left(\frac{V_y}{I_z} as\right) t \, ds = \frac{V_y}{2I_z} ab^2 t \qquad\text{[c]}$$

It is not necessary to determine V_2 to locate the shear center. However, as a check, the shear stress in the web will be integrated, where $dA = t \, dy$ and τ is given by Eq. (b). Thus

$$V_2 = 2\int_0^a \frac{V_y}{I_z}\left[ab + \frac{1}{2}(a^2 - y^2)\right] t \, dy = \frac{V_y}{I_z}\left(2a^2 bt + \frac{2}{3} a^3 t\right) \qquad\text{[d]}$$

Since $I_z \approx \frac{2}{3} a^3 t + 2a^2 bt$, we see that $V_2 = V_y$.

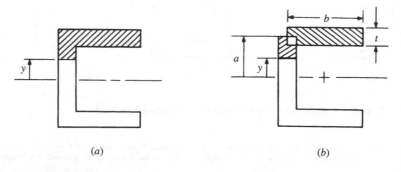

(a)   (b)

Figure 5.4-5   Determination of Q in vertical wall.

The shear center can be located by summing moments about point D of Fig. 5.4-7. For equilibrium,

$$V_y e = 2\left(\frac{V_y}{2I_z} ab^2 t\right)a$$

Thus

$$e = \frac{a^2 b^2 t}{I_z}$$

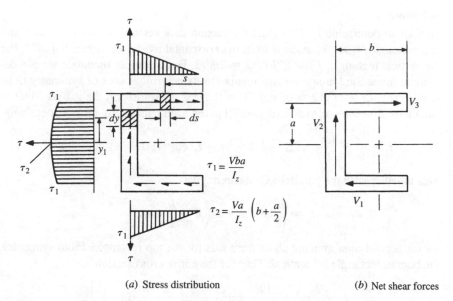

(a) Stress distribution   (b) Net shear forces

Figure 5.4-6

$$\tau_1 = \frac{Vba}{I_z}$$

$$\tau_2 = \frac{Va}{I_z}\left(b + \frac{a}{2}\right)$$

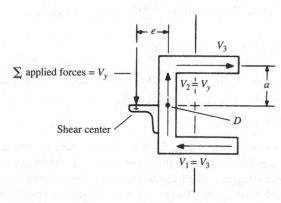

Figure 5.4-7

However, since $I_z \approx \frac{2}{3} a^3 t + 2a^2 bt$,

$$e = \frac{a^2 b^2 t}{\frac{2}{3} a^3 t + 2a^2 bt} = \frac{3}{2} \frac{b^2}{a + 3b}$$

Example 5.4-2 | The section shown in Fig. 5.4-8(a) is undergoing a transverse shear force V_y. Assuming t small, determine the shear stress distribution in the section. Also, determine the location of the shear center. For the section, let $c = 80$ mm, $b = 60$ mm, $h = 50$ mm, $t = 6$ mm, and $\alpha = 30°$.

Solution:

First let us determine I_z. We model the section as a vertical rectangle $2h \times t$, and two rectangles $b \times t$ at angle α from the horizontal as shown in Fig. 5.4-8(b). For the vertical rectangle, $I_{z1} = t(2h)^3/12 = 2th^3/3$. For the upper rectangle we can determine the second-area moments relative to its centroidal axes of symmetry to be $I_{z'} = tb^3/12$, $I_{y'} = bt^3/12 \approx 0$, and $I_{y'z'} = 0$. Next we transform the z' axis counterclockwise $\beta = 90 - \alpha$ to an axis parallel to the z axis, using Eq. (E.2-3), obtaining

$$I_z = \frac{I_{z'}}{2} + \frac{I_{z'}}{2} \cos 2\beta = I_{z'} \cos^2 \beta = I_{z'} \cos^2 (90 - \alpha) = I_{z'} \sin^2 \alpha$$

This together with the parallel-axis theorem gives

$$I_z = \frac{tb^3}{12} \sin^2\alpha + bt\left(c - \frac{b}{2} \sin\alpha \right)^2$$

for the second-area moment about the z axis for the top rectangle. From symmetry, the bottom rectangle is identical. Thus for the entire cross section

$$I_z = \frac{2}{3} th^3 + 2\left[\frac{tb^3}{12} \sin^2\alpha + bt\left(c - \frac{b}{2} \sin\alpha \right)^2 \right]$$

Simplifying gives

$$I_z = \frac{2}{3} t \left(h^3 + 3bc^2 - 3b^2 c \sin\alpha + b^3 \sin^2\alpha \right) \qquad \text{[a]}$$

Substituting values gives

$$I_z = \frac{2}{3} (6)[50^3 + 3(60)(80^2) - 3(60^2)(80) \sin 30 + (60^3) \sin^2 30]$$

$$= 3.596(10^6) \text{ mm}^4$$

The shear flow is shown in Fig. 5.4-8(c). If one were to sum moments about point F to determine the location of the shear center, it would not be necessary to determine the shear stresses in the flanges. However, determining the stress is more difficult in the web than in the flanges, so we will sum moments about point G.

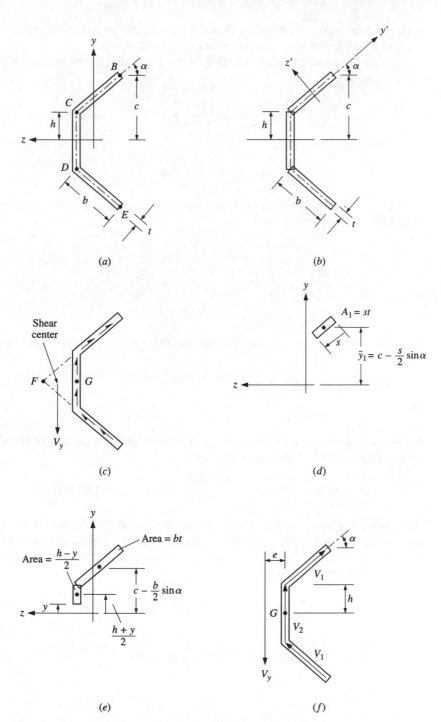

Figure 5.4-8

Although unnecessary for the shear center determination, we will still determine the shear in the web, as it will serve as a check on our calculations.

For the flange we use the variable s from the free end as shown in Fig. 5.4-8(d). The area moment is

$$Q = \bar{y}_1 A_1 = \left(c - \frac{s}{2} \sin \alpha \right) st$$

Thus, from Eq. (5.4-8),

$$\tau = \frac{V_y s}{I_z} \left(c - \frac{s}{2} \sin \alpha \right)$$

To determine the net force on the flange, we integrate this across the flange area, obtaining

$$V_1 = \int \tau t \, ds = \int_0^b \frac{V_y s}{I_z} \left(c - \frac{s}{2} \sin \alpha \right) t \, ds$$

$$= \frac{V_y}{6I_z} t b^2 (3c - b \sin \alpha) \qquad\qquad [b]$$

From symmetry, this also will apply to the lower flange. Substituting values gives

$$V_1 = \frac{V_y}{6(3.596)(10^{-6})} (0.006)(0.06)^2 [3(0.08) - (0.06) \sin 30]$$

$$= 0.2102 \, V_y \qquad\qquad [c]$$

Summing moments about point G in Fig. 5.4-8(f) yields the location of the shear center. That is, $V_y \, e = 2 \, V_1 h \cos \alpha$. Solving for e yields

$$e = 2 \frac{V_1}{V_y} h \, \cos \alpha = 2 \frac{0.2102 V_y}{V_y} (5) \, \cos 30 = 1.82 \text{ mm}$$

Although unnecessary for the shear center determination, the determination of the shear in the web will serve as a check on our result for V_1. For shear in the web we will use the variable y. The area moment for an isolation in the web is found from the composite section shown in Fig. 5.4-8(e):

$$Q = \sum \bar{y} A = \frac{h + y}{2} (h - y)t + \left(c - \frac{b}{2} \sin \alpha \right) bt$$

$$= \frac{1}{2} (h^2 - y^2)t + \left(c - \frac{b}{2} \sin \alpha \right) bt$$

From Eq. (5.4-8)

$$\tau = \frac{V_y}{2I_z} [(h^2 - y^2) + b(2c - b \sin \alpha)]$$

To determine the net force in the web, we integrate $\tau \, dA$ along the web. Utilizing symmetry in the web

$$V_2 = \int \tau(t \, dy) = 2 \int_0^h \frac{V_y}{2I_z} [(h^2 - y^2) + b(2c - b \sin \alpha)]t \, dy$$

$$= \frac{V_y t}{3I_z} (2h^3 + 6bch - 3b^2 h \sin \alpha) \qquad\qquad \textbf{[d]}$$

Substitute $h = c - b \sin \alpha$ for the linear terms in h in Eq. (d):

$$V_2 = \frac{V_y t}{3I_z} (2h^3 + 6bc^2 - 9b^2 c \sin \alpha + 3b^3 \sin^2 \alpha) \qquad\qquad \textbf{[e]}$$

Now, as a check on V_1 and V_2, we sum forces in the y direction on the section. This should yield V_y. From Eqs. (b) and (e) we have

$$\sum F_y = 2V_y \sin \alpha + V_2$$

$$= 2 \frac{V_y}{6I_z} tb^2(3c - b \sin \alpha) \sin \alpha + \frac{V_y t}{3I_z} (2h^3 + 6bc^2 - 9b^2 c \sin \alpha + 3b^3 \sin^2 \alpha)$$

$$= \frac{2V_y t}{3I_z} (h^3 + 3bc^2 - 3b^2 c \sin \alpha + b^3 \sin^2 \alpha) = V_y$$

where Eq. (a) was used for I_z in the last step. Obtaining V_y improves our confidence in the calculations.

5.4.3   SHEAR CENTER FOR OPEN UNSYMMETRIC THIN-WALLED BEAMS

Similar to the bending of unsymmetrical beams, one can deal with the behavior relative to the principal axes or relative to the unsymmetrical axes using the mixed second-area moment. In this section we will use the principal axes. Considering the components of the shear force on the section in the directions of the principal axes, the shear center location for each component is found using the methods of the previous section.

Determine the shear center for the cross section shown in Fig. 5.4-9. Dimensions are given from the wall centers where appropriate, and the thickness of each wall is $t = 5$ mm.

| **Example 5.4-3**

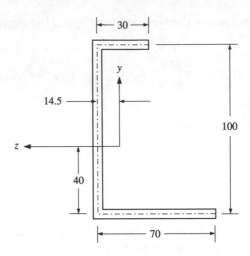

Figure 5.4-9   (All dimensions are in millimeters.)

Solution:

We first determine the location of the principal axes. The reader should verify that the y, z axes shown in Fig. 5.4-9 are the centroidal axes. The second-area moments relative to these axes are (ignoring terms of order t^3)

$$I_y = \frac{1}{12}(5)(70)^3 + (5)(70)(35 - 14.5)^2 + \frac{1}{12}(5)(30)^3 + (5)(30)(15 - 14.5)^2 + (5)(100)(14.5)^2$$

$$= 406.4 \times 10^3 \text{ mm}^4$$

$$I_z = \frac{1}{12}(5)(100)^3 + (5)(100)(50 - 40)^2 + (5)(30)(100 - 40)^2 + (5)(70)(40)^2$$

$$= 1.5667 \times 10^6 \text{ mm}^4$$

$$I_{yz} = (5)(30)(100 - 40)(14.5 - 15) + (5)(100)(50 - 40)(14.5) + (5)(70)(-40)(14.5 - 35)$$

$$= 355.0 \times 10^3 \text{ mm}^4$$

From Appendix E, Eq. (E.2-4), the angle to the principal axes is

$$\theta = \frac{1}{2}\tan^{-1}\left(\frac{2I_{yz}}{I_z - I_y}\right) = \frac{1}{2}\tan^{-1}\left[\frac{2(0.355)}{1.5667 - 0.4064}\right] = 15.73°$$

Using Eqs. (E.2-3) results in

$$I_m = \left[\left(\frac{0.4064 + 1.5667}{2} \right) + \left(\frac{0.4064 - 1.5667}{2} \right) \cos(31.46) - 0.355 \sin(31.46) \right] (10^6)$$

$$= 0.3054 \times 10^6 \text{ mm}^4$$

$$I_n = \left[\left(\frac{0.4064 + 1.5667}{2} \right) - \left(\frac{0.4064 - 1.5667}{2} \right) \cos(31.46) + 0.355 \sin(31.46) \right] (10^6)$$

$$= 1.667 \times 10^6 \text{ mm}^4$$

The principal axes together with the component of the *applied* shear force which gives rise to a positive shear V_m on the surface are shown in Fig. 5.4-10. To determine the location of the shear center in the n direction, we will sum moments about point O. For this it is only necessary to determine the shear in the upper flange.

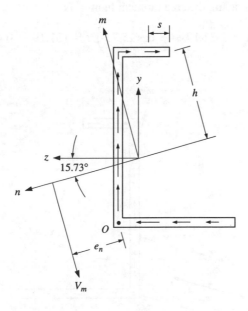

Figure 5.4-10   Shear center in the n direction.

The dimension h is given by

$$h = 60\cos 15.73 - (30 - 14.5)\sin 15.73 = 53.55 \text{ mm}$$

Using the variable s from the free end of the flange, the area moment in mm³ is

$$Q_n = \overline{m}_1 A_1 = \left(53.55 + \frac{s}{2}\sin 15.73 \right) st = (53.55 + 0.1356s)st$$

Integrating $\tau \, dA$ from $0 \le s \le 30$ mm gives the net shear force in the upper flange

$$V_1 = \int \tau t \, ds = \int_0^{30} \frac{V_m}{I_n t} [(53.55 + 0.1356s)st]t \, ds$$

$$= \frac{V_m}{1.667(10^6)} \left[(53.55)\left(\frac{30^2}{2}\right) + (0.1356)\left(\frac{30^3}{3}\right) \right] \tag{5}$$

$$= 0.07594 V_m$$

Summing moments about point O yields $e_n V_m - 100 V_1 = 0$ or

$$e_n = \frac{(100)(0.07594 V_m)}{V_m} = 7.59 \text{ mm}$$

Figure 5.4-11 shows the component of the applied shear force which gives rise to a positive shear V_n on the surface. The distance b is

$$b = 60\sin 15.73 + (30 - 14.5)\cos 15.73 = 31.19 \text{ mm}$$

Using the variable s again, the area moment in mm³ is

$$Q_m = \overline{n}_1 A = \left(31.19 - \frac{s}{2}\cos 15.73 \right) st = (31.19 - 0.4813s)st$$

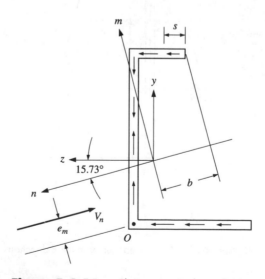

Figure 5.4-11   Shear center in the m direction.

The shear force on the upper flange is thus

$$V_2 = \int \tau t \, ds = \int_0^{30} \frac{V_n}{I_m t} [(31.19 - 0.4813s)st]t \, ds$$

$$= \frac{V_n}{0.3054(10^6)} \left(31.19 \frac{30^2}{2} - 0.4813 \frac{30^3}{3} \right)(5)$$

$$= 0.1589 V_n$$

Again, summing moments about point O yields $e_m V_n - 100 V_2 = 0$ or

$$e_m = \frac{(100)(0.15894 V_n)}{V_n} = 15.89 \text{ mm}$$

The shear center relative to point O in the y and z directions is given by the transformation equations [Eqs. (5.3-7)]. Thus, with $\theta = 15.73°$,

$$e_y = e_m \cos 15.73 - e_n \sin 15.73$$

$$= 15.89 \cos 15.73 - 7.594 \sin 15.73 = 13.24 \text{ mm}$$

$$e_z = e_m \sin 15.73 + e_n \cos 15.73$$

$$= 15.89 \sin 15.73 + 7.594 \cos 15.73 = 11.62 \text{ mm}$$

The shear center, point C, is shown in Fig. 5.4-12.

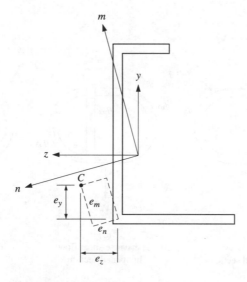

Figure 5.4-12

5.4.4   SHEAR IN CLOSED THIN-WALLED SECTIONS

Determination of the shear flow in open sections is achieved by making a *complete* isolation of a part of the beam section which exposes only *one* surface containing the desired shear stress. For closed sections, this can only be done in the case of a one-cell, *symmetric* closed section. For the general closed single or multiple cell section, a different approach must be taken. Here the shear flow will be represented by the superposition of transverse and torsional shear flow.

Consider the two-cell section shown in Fig. 5.4-13(*a*), where the *applied* load is represented by the vector V_y. This, in turn, develops a net shear force across the section of V_y in the positive y direction. If the cells are opened by the cuts as shown in Fig. 5.4-13(*b*), the corresponding transverse shear flow or stresses in each wall can be determined by the methods of Secs. 5.4.2 or 5.4.3. For Fig. 5.4-13(*b*), let the shear flow in each open wall be q_{o1}, q_{o2}, and q_{o3} and vary according to $q = V_y Q / I_z$.

Next the shear flow that was released by the cuts is considered to be torsional shear flow as discussed in Sec. 5.2 and shown in Fig. 5.4-13(*c*). The shear flow in each cell is assumed to be constant in a counterclockwise direction and is given by q_{c1} and q_{c2}. Superposition of the shear flow of Figs. 5.5-13(*b*) and (*c*) gives the shear flow in each wall, consistent with Fig. 5.4-13(*a*) to be

$$q_1 = q_{c1} - q_{o1} \qquad\qquad \textbf{[5.4-9a]}$$

$$q_2 = q_{c1} - q_{c2} + q_{o2} \qquad\qquad \textbf{[5.4-9b]}$$

$$q_3 = q_{c2} + q_{o3} \qquad\qquad \textbf{[5.4-9c]}$$

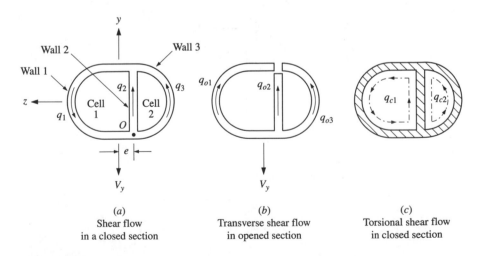

(*a*)	(*b*)	(*c*)
Shear flow	Transverse shear flow	Torsional shear flow
in a closed section	in opened section	in closed section

Figure 5.4-13   Two-cell closed section.

For the moment summation, pick an arbitrary point on the cross section O. Summing moments in the counterclockwise direction yields

$$\sum M_O = V_y e + M_{q1} + M_{q2} + M_{q3} = 0 \qquad \textbf{[5.4-10]}$$

where M_{q1}, M_{q2}, and M_{q3} are the moments caused by the shear flow. The moment from the torsional shear flow can be represented by Eq. (5.2-8). Thus Eq. (5.4-10) becomes

$$V_y e + M_{q_{o1}} + M_{q_{o2}} + M_{q_{o3}} + 2\sum_{i=1}^{3} q_{ci}\bar{A}_{ci} = 0 \qquad \textbf{[5.4-11]}$$

where $\bar{A}_{ci}$ is the area enclosed by the median line of the walls of closed cell i, and M_{qo1}, M_{qo2}, and M_{qo3} are the moments caused by the transverse shear flow in Fig. 5.4-13(b).

The angle of twist for each cell is given by Eq. (5.2-9) and is repeated here as

$$\theta = \frac{(1+v)L}{E\bar{A}_{ci}} \int \frac{q}{t}\, ds$$

The angle of twist for each cell, with counterclockwise being positive, is

Cell 1: $\qquad \theta = \dfrac{(1+v)L}{E\bar{A}_{c1}}\left(\displaystyle\int_{\text{wall }1} \frac{q_1}{t_1}\, ds + \int_{\text{wall }2} \frac{q_2}{t_2}\, ds \right) \qquad \textbf{[5.4-12a]}$

Cell 2: $\qquad \theta = \dfrac{(1+v)L}{E\bar{A}_{c2}}\left(\displaystyle\int_{\text{wall }3} \frac{q_3}{t_3}\, ds - \int_{\text{wall }2} \frac{q_2}{t_2}\, ds \right) \qquad \textbf{[5.4-12b]}$

where t_1, t_2, and t_3 are the thicknesses of walls 1, 2, and 3, respectively. To complete the solution, substitute Eqs. (5.4-9) with the determined values of q_{o1}, q_{o2}, and q_{o3}, into Eqs. (5.4-12). The two resulting equations together with Eq. (5.4-11) will provide three equations for q_{c1}, q_{c2}, θ, and e. If the load is to be applied at the shear center, then $\theta = 0$, and q_{c1}, q_{c2}, and e can be determined. If the load is applied at a specified position, e will be known, and q_{c1}, q_{c2}, and θ can be determined.

The cross section shown in Fig. 5.4-14 is that of a beam in bending and transmitting a transverse shear force of $V_y = 10$ kN. If the transverse shear load is applied at the shear center, determine the shear-stress distributions in each wall and the location of the shear center. All dimensions, between wall centers, are in millimeters. The thickness of each wall, AB, CD, and DA is 3 mm, whereas wall BC is 6 mm.

Example 5.4-4

Solution:

The second-area moment about the z axis (neglecting terms of the order of thickness^3) is

$$I_z = \frac{1}{12}(6)(100)^3 + \frac{1}{12}(3)(100)^3 + 2(3)(40)(50)^2 = 1.350 \times 10^6 \text{ mm}^4$$

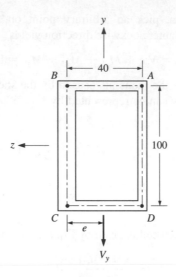

Figure 5.4-14

To begin, we will "open" the single cell at point B as shown in Fig. 5.4-15(a). Wall 1 (side AB) is basically the same as the flange of the channel in Example 5.4-1, where the shear flow is given by

$$q_{o1} = \tau t = \frac{V_y a t}{I_z} s = \frac{10(10^3)(0.05)(0.003)}{1.35(10^{-6})} s = 1.111 \times 10^6 s \qquad [a]$$

with s being in meters.

Since the top of wall 2 (side BC) is now open and the wall is symmetric with the z axis, the wall behaves as a rectangular cross section. The shear flow for this is given in Example 3.4-3, where

$$q_{o2} = \tau t = \frac{V_y}{2I_z}\left[\left(\frac{h}{2}\right)^2 - y^2\right]t = \frac{10(10^3)}{2(1.35)(10^{-6})}\left[\left(\frac{0.100}{2}\right)^2 - y^2\right](0.006)$$

$$= 55.56(10^3) - 22.22(10^6)y^2 \qquad [b]$$

where y is in meters.

For the lower wall 3 (side CD), make a break in the wall and isolate the left side of the break. The entire wall 2 will be a part of the isolation. However, since $Q = 0$ for wall 2 in its entirety, the value of Q for the isolation will be due only to the part of wall 3 that is isolated. Thus Q for wall 3 is the same as that of wall 1, giving

$$q_{o3} = q_{o1} = 1.111 \times 10^6 s \qquad [c]$$

where s is measured the same as for wall 1.

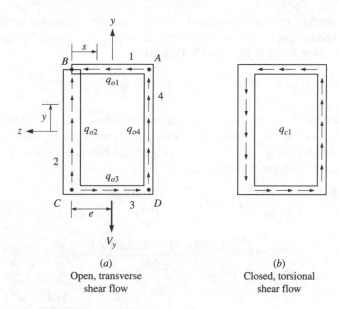

(a)
Open, transverse
shear flow

(b)
Closed, torsional
shear flow

Figure 5.4-15

For wall 4 (side DA), we isolate the upper section by making a cut at y. The resulting area moment for the top wall and the segment of wall 4 is

$$Q = (0.05)(0.003)(0.04) + \frac{1}{2}\left[\left(\frac{0.1}{2}\right)^2 - y^2\right](0.003)$$

$$= 9.75(10^{-6}) - 1.5(10^{-3})y^2$$

given in m^3. Thus the shear flow in wall 4 is

$$q_{o4} = \frac{V_y}{I_z}Q = \frac{10(10^3)}{1.35(10^{-6})}[9.75(10^{-6}) - 1.5(10^{-3})y^2]$$

$$= 72.22(10^3) - 11.111(10^6)y^2 \qquad\qquad \text{[d]}$$

The open shear flow is shown in its proper direction in Fig. 5.4-15(a).

Next we will include the closed cell torsional shear flow q_{c1} as shown in Fig.
4.4-15(b) and assumed in the counterclockwise direction. With the load applied at
the shear center, the angle of twist for the cell is given by

$$\theta = \frac{(1 + v)L}{E\bar{A}_{c1}}\left(\sum_{i=1}^{4}\int_{\text{wall }i}\frac{q_i}{t_i}\,ds\right) = 0$$

The terms outside the summation are not zero. Thus the summation must be zero. Assuming counterclockwise flow to be positive, each q_i is replaced by the superposition of the shear flows in Fig. 5.4-15. This gives

$$\int_{\text{wall} 1} \frac{q_{c1} + q_{o1}}{t_1} ds + \int_{\text{wall} 2} \frac{q_{c1} - q_{o2}}{t_2} ds + \int_{\text{wall} 3} \frac{q_{c1} + q_{o3}}{t_3} ds + \int_{\text{wall} 4} \frac{q_{c1} + q_{o4}}{t_4} ds = 0$$

Since q_{c1} and the thickness in each integral are constants, their part within the integrals can be factored out, leaving $\int ds$, the length of each wall. Equations (a) to (d) are substituted for the q_{oi} terms, and recognizing the first and third walls are equal and the vertical walls are symmetric yields

$$\left(\frac{40}{3} + \frac{100}{6} + \frac{40}{3} + \frac{100}{3} \right) q_{c1} + 2 \int_0^{0.04} \frac{1.111(10^6)s}{0.003} ds$$

$$- 2 \int_0^{0.05} \frac{55.56(10^3) - 22.22(10^6)y^2}{0.006} dy$$

$$+ 2 \int_0^{0.05} \frac{72.22(10^3) - 11.111(10^6)y^2}{0.003} dy = 0$$

Integration yields

$$q_{c1} = -27.05 \times 10^3 \text{ N/m}$$

Thus the closed cell shear flow is actually clockwise. Superposing this with the open shear flow found earlier and dividing by the wall thickness gives the shear stress in each wall. The results are:

Wall 1 (assuming ← positive) and *wall 3* (assuming positive →)

$$q_1 = 1.111(10^6)s - 27.05(10^3)$$

$$\tau_1 = \frac{q_1}{0.003} = 370.4(10^6)s - 9.017(10^6) \text{ N/m}^2$$

Wall 2 (assuming positive ↑)

$$q_2 = 55.56(10^3) - 22.22(10^6)y^2 + 27.05(10^3)$$

$$= 82.61(10^3) - 22.22(10^6)y^2$$

$$\tau_2 = \frac{q_2}{0.006} = 13.77(10^6) - 3.704(10^9)y^2 \text{ N/m}^2$$

Wall 4 (assuming positive ↑)

$$q_4 = 72.22(10^3) - 11.111(10^6)y^2 - 27.05(10^3)$$

$$= 45.17(10^3) - 11.111(10^6)y^2$$

$$\tau_4 = \frac{q_4}{0.003} = 15.06(10^6) - 3.704(10^9)y^2 \text{ N/m}^2$$

A plot of the shear-stress distributions is shown in Fig. 5.4-16, where the actual directions of the stresses are shown within the walls.

To determine the location of the shear center, we sum moments about point C. For this single cell problem it is then only necessary to determine the forces in walls 1 and 4. Equation (5.4-11) is meant for a more complicated multiple cell problem. As a check, we will also determine the force in wall 2. For the force in wall 1 we integrate τ_1 over the wall, obtaining

$$F_1 = \int_0^{0.04} [370.4(10^6)s - 9.017(10^6)](0.003)ds = -193.1 \text{ N}$$

This means that the net force is actually toward the right. For wall 4,

$$F_4 = 2\int_0^{0.05} [15.06(10^6) - 3.704(10^9)y^2](0.003)\,ds = 3590 \text{ N}$$

Summing moments about point C, with $V_y = 10$ kN, yields

$$\sum M_C = -e(10)(10^3) + (40)(3590) - (100)(193.08) = 0$$

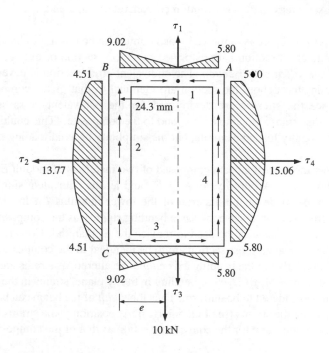

Figure 5.4-16   Shear stresses (given in MPa).

Solving for e gives $e = 12.4$ mm.

As a check on the forces, the force in wall 2 is

$$F_2 = 2 \int_0^{0.05} [13.77(10^6) - 3.704(10^9)y^2](0.006)\, dy = 6410 \text{ N}$$

Note that $F_2 + F_4 = 10,000 \text{ N} = 10 \text{ kN}$, which agrees with the applied force V_y. This serves as a check on the calculations.

5.5   COMPOSITE BEAMS IN BENDING

Recall Example 3.2-1, where axial loading of a composite bar is investigated. In order to avoid bending of the bar, it is necessary to apply the loading through an axis which is not the geometrical centroid of the total cross section. This is due to the discontinuity of the stress distribution that arises from the differing moduli of elasticity of the composite materials. For the same strain, higher stresses develop in the material with the higher modulus. This is advantageous in many cases, as some structural materials are inexpensive but have certain strength shortcomings. With composites, large quantities of the lower-modulus material can be used in the lower-stress areas, and small quantities of the higher-modulus material can be used in the higher-stressed areas, e.g., steel reinforced concrete beams and steel-clad wooden beams.

A simple procedure is employed when composite beams are in bending where the composite cross section is transformed to a cross section of only one material, called the *equivalent cross section*. The equivalent cross section is developed such that the strains throughout the section are identical to that of the composite cross section. Once the stresses are determined for the equivalent cross section, the stresses are then transformed to correspond to the composite. (This could have been done for the axially loaded example, but the simplicity of axial loading made it unnecessary.)

Consider the case of a beam composed of two materials of moduli E_1 and E_2 as shown in Fig. 5.5-1. Arbitrarily, let $E_2 > E_1$, and let the equivalent single-material cross section be made of the material of the lower modulus E_1. In order for the equivalent cross section to have the same bending rigidity as the composite it is necessary that *more* E_1 material be used to replace the E_2 material. However, since the strains in the equivalent cross section are to be identical to the composite, the *depth* of the section relative to the bending axis *cannot* be altered, as strains are a function of y, namely, $\varepsilon_x = -M_z y/(EI_z)$. For bending in the xy plane, strains in the z direction are generally considered to be uniform. Thus the *width* of the beam can be increased without affecting the strains [see Fig. 5.5-1(c)]. At position y, the strains in the x direction must be the same for the equivalent section as that of the composite. Thus

$$(\varepsilon_x)_e = (\varepsilon_x)_c$$

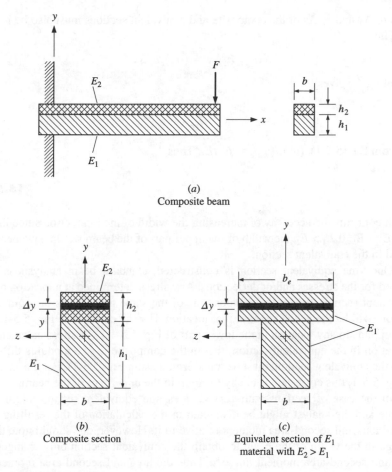

(a)
Composite beam

(b)
Composite section

(c)
Equivalent section of E_1
material with $E_2 > E_1$

Figure 5.5-1

where the subscripts e and c represent the equivalent and composite sections, respectively. From Hooke's law

$$\frac{(\sigma_x)_e}{E_1} = \frac{(\sigma_x)_c}{E_2}$$

Once the stresses are determined for the equivalent section $(\sigma_x)_e$, the actual stresses in the E_2 composite material are given by

$$(\sigma_x)_c = \frac{E_2}{E_1}(\sigma_x)_e \qquad \textbf{[5.5-1]}$$

To determine the stresses, we need to determine the width of the equivalent E_1 material we are replacing the E_2 material with. For equilibrium, the forces on the areas $b_c \Delta y$ and $b_e \Delta y$ of the composite and equivalent sections must also be identical. Thus

$$(\sigma_y)_c \, b_c \Delta y = (\sigma_x)_e \, b_e \Delta y$$

canceling Δy gives

$$b_e = \frac{(\sigma_x)_c}{(\sigma_x)_e} \, b_c$$

but from Eq. (5.5-1), $(\sigma_x)_c/(\sigma_x)_e = E_2/E_1$. Thus

$$b_e = \frac{E_2}{E_1} \, b_c \qquad\qquad\qquad \textbf{[5.5-2]}$$

which confirms the necessity of increasing the width of the top section since in this case $E_2 > E_1$. If $E_1 > E_2$, the width of the upper part of the beam would have been reduced in the equivalent section.

Once the equivalent section is constructed, standard beam analysis is performed for the stresses and/or deflections. Any stress determined at a location of the equivalent section, that in the composite is of the same modulus as the equivalent section, will be the actual stress at that location. For the example in Fig. 5.5-1, this corresponds to any location in the lower part of Fig. 5.5-1(c). A stress calculated at a location in the equivalent section, that in the composite is of a modulus different from the equivalent section, must be transformed using Eq. (5.5-1). For the example in Fig. 5.5-1, this corresponds to any location in the upper part of the beam.

In the case of sections with curved sides, the equivalent section might look strange and the analyst might be distracted in the calculation of the resulting centroidal axis and second-area moment relative to it. However, one should note that a change in the width of the beam to obtain the equivalent section only changes the area and second-area moment *linearly.* Thus the area and second-area moment of the equivalent section are obtained by multiplying the actual area and second-area moment of the corresponding composite section by the ratio E_2/E_1. Thus, in the case of Fig. 5.5-1(c), the section of the area that is being transformed from the composite to the equivalent area is

$$A_e = \frac{E_2}{E_1} \, A_c \qquad\qquad\qquad \textbf{[5.5-3]}$$

and the corresponding second-area moment of the transformed part of the section is

$$(I_z)_e = \frac{E_2}{E_1} \, (I_z)_c \qquad\qquad\qquad \textbf{[5.5-4]}$$

Deflections   The bending deflections of the composite beam are determined directly from the beam using the equivalent cross section and the modulus of elasticity of the equivalent cross section.

Example 5.5-1

A beam section composed of aluminum ($E_a = 10$ Mpsi) and steel ($E_s = 30$ Mpsi) are bonded together as shown in Fig. 5.5-2(a). If the section is undergoing a positive bending moment of $M_z = 10$ kip·in about a horizontal axis, determine the resulting stress distribution.

Solution:

Let the total cross section be completely made of aluminum. Thus, for the bottom section, using Eq. (5.5-2) with $E_2/E_1 = 3$, the equivalent width is

$$b_e = 3(1) = 3 \text{ in}$$

and thus the equivalent aluminum section is as shown in Fig. 5.5-2(b). Next, the centroidal axis of the equivalent cross section is determined. From Fig. 5.5-2(b)

$$d(6 + 2) = (1)(6) + (3)(2)$$

$$d = 1.50 \text{ in}$$

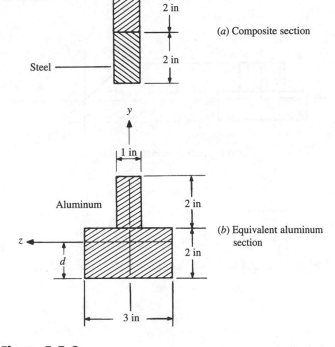

(a) Composite section

(b) Equivalent aluminum section

Figure 5.5-2

Now, the moment of inertia of the totally aluminum cross section about the centroidal axis $(I_z)_e$ is determined. Using the parallel-axis theorem, we have

$$(I_z)_e = \left(\frac{1}{12}\right)(3)(2^3) + (6)(0.50)^2 + \left(\frac{1}{12}\right)(1)(2^3) + (2)(1.50)^2 = 8.667 \text{ in}^4$$

The stress distribution in the equivalent section is determined by using the flexure formula $\sigma_x = -M_z y/(I_z)_e$, where $M_z = 10{,}000$ lb·in. The stress at the top surface is

$$(\sigma_x)_e|_{y=2.5} = -\frac{10(10^3)(2.5)}{8.667} = -2885 \text{ psi}$$

at the interface

$$(\sigma_x)_e|_{y=0.5} = -\frac{10(10^3)(0.5)}{8.667} = -577 \text{ psi}$$

and at the bottom

$$(\sigma_x)_e|_{y=-1.5} = -\frac{10(10^3)(-1.5)}{8.667} = 1731 \text{ psi}$$

The stress distribution for the equivalent section is shown in Fig. 5.5-3(a). However, the stress in the lower half is for the equivalent material and must be transformed

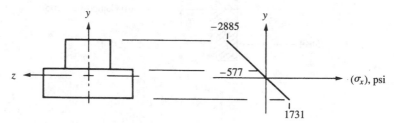

(a) Equivalent aluminum section

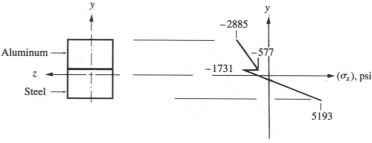

(b) Composite section

Figure 5.5-3

using Eq. (5.5-1) to correspond to the actual material. Thus for the steel at the interface with $E_2/E_1 = 3$

$$\sigma_x|_{y=0.5} = (3)(-577) = -1731 \text{ psi}$$

and at the bottom

$$\sigma_x|_{y=-1.5} = (3)(1731) = 5193 \text{ psi}$$

The actual stress distribution is as shown in Fig. 5.5-3(b).

Rotating the section investigated in Example 5.5-1 90°, as shown in Fig. 5.5-4(a), **Example 5.5-2** determine the stress distribution for bending about a horizontal axis of the section if $M_z = 10 \text{ kip·in}$.

Solution:

The moment of inertia of the equivalent section is

$$(I_z)_e = \left(\frac{1}{12}\right)(8)(1^3) = 0.667 \text{ in}^4$$

Thus

$$(\sigma_x)_e|_{y=\pm0.5} = -\frac{10(10^3)(\pm0.5)}{0.667} = \mp7500 = \mp7.5 \text{ ksi}$$

This corresponds to the half made of aluminum. For the steel

$$(\sigma_x)_s|_{y=\pm0.5} = (3)(\mp7.5) = \mp22.5 \text{ ksi}$$

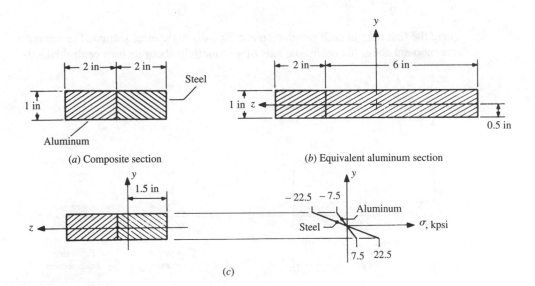

(a) Composite section

(b) Equivalent aluminum section

(c)

Figure 5.5-4

The stress distributions in each half are presented in Fig. 5.5-4(c).

NOTE: In this problem the loading plane xy must be located so that the moment about the y axis is zero. The location of the centroidal y axis of the composite is found by considering bending about the y axis. This was actually done in Example 5.5-1. The y axis should be located 1.5 in from the right-hand side of Fig. 5.5-4(c).

Example 5.5-3

A beam section, shown in Fig. 5.5-5, is made of steel (E_s = 210 GPa) and brass (E_b = 105 GPa) welded together. The beam, 1 m long, is cantilevered with a force at the free end of 100 N. Determine the maximum tensile stress in the steel, the maximum compressive stress in the brass, and the deflection of the free end. Make the equivalent section out of brass.

Solution:

To determine the centroid of the equivalent cross section entirely made of brass, we will make use of Eq. (5.5-3) when dealing with the steel part. The centroid of a semicircle from its base is given by $\bar{y} = 4r/3\pi$, where r is the radius. Thus the location of the centroid for the equivalent section from the geometric z' axis (with $E_2/E_1 = 210/105$) is

$$\bar{y} = \frac{\sum\limits_{i=1}^{2}\bar{y}_i A_i}{\sum\limits_{i=1}^{2} A_i} = \frac{\left[\dfrac{4(20)}{3\pi}\right]\left(\dfrac{210}{105}\right)\left[\dfrac{\pi(20^2)}{2}\right] + \left[-\dfrac{4(20)}{3\pi}\right]\left[\dfrac{\pi(20^2)}{2}\right]}{\left(\dfrac{210}{105}\right)\left[\dfrac{\pi(20^2)}{2}\right] + \left[\dfrac{\pi(20^2)}{2}\right]}$$

$$= \frac{80}{9\pi} = 2.829 \text{ mm}$$

where the first term in each summation corresponds to the steel section. The second-area moment about the centroidal axis of a semicircle about its own centroidal axis

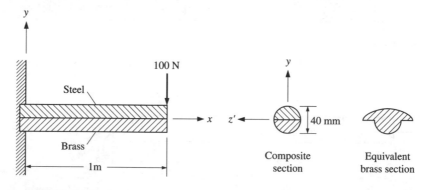

Steel

100 N

Brass

1m

z' 40 mm

Composite section

Equivalent brass section

Figure 5.5-5

is $(\pi/8 - 8/9\pi)r^4$. Thus, using Eq. (5.5-4) and the parallel-axis theorem, the second-area moment about the centroidal axis for the equivalent section is

$$(I_z)_e = \left(\frac{210}{105}\right)\left(\frac{\pi}{8} - \frac{8}{9\pi}\right)(20^4) + \left(\frac{210}{105}\right)\left[\frac{\pi(20^2)}{2}\right]\left[\frac{(4)(20)}{3\pi} - \frac{80}{9\pi}\right]^2$$

$$+ \left(\frac{\pi}{8} - \frac{8}{9\pi}\right)(20^4) + \left[\frac{\pi(20^2)}{2}\right]\left[\frac{(4)(20)}{3\pi} + \frac{80}{9\pi}\right]^2 = 173.41 \times 10^3 \text{ mm}^4$$

The bending moment at the wall is $-(1)(100) = -100$ N·m. Using the bending stress equation, with Eq. (5.5-1), the maximum tensile stress in the steel is

$$(\sigma_x)_{steel} = \left(\frac{E_2}{E_1}\right)\left[-\frac{M_z y_{max}}{(I_z)_e}\right] = \left(\frac{210}{105}\right)\left[-\frac{(-100)(20 - 2.829)(10^{-3})}{173.41(10^{-9})}\right]$$

$$= 19.80 \times 10^6 \text{ N/m}^2 = 19.8 \text{ MPa}$$

The maximum compressive stress in the brass is

$$(\sigma_x)_{brass} = \left[-\frac{M_z y_{min}}{(I_z)_e}\right] = \left[-\frac{(-100)(-20 - 2.829)(10^{-3})}{173.41(10^{-9})}\right]$$

$$= -13.16 \times 10^6 \text{ N/m}^2 = -13.16 \text{ MPa}$$

The deflection of the free end is given by $v_c = -FL^3/(3EI_z)$. If for I_z we use $(I_z)_e$ for the equivalent section in brass, we must use E_b for E. Thus

$$(v_c)\bigg|_{x=1m} = -\frac{(100)(1^3)}{3(105)(10^9)(173.41)(10^{-9})} = -1.831 \times 10^{-3} \text{ m} = -1.831 \text{ mm}$$

5.6 CURVED BEAMS

5.6.1 TANGENTIAL STRESSES

Consider the segment of the curved beam undergoing pure bending shown in Fig. 5.6-1. For simple bending in the $r\theta$ plane, the cross section must have an axis of symmetry in this plane. Assume, as in straight-beam theory, that a plane cross section remains plane after bending. Thus the tangential strain at any position r (or y) is given by

$$\varepsilon_\theta = -\frac{y d\theta}{r\theta} = -\frac{(r_n - r)d\theta}{r\theta}$$

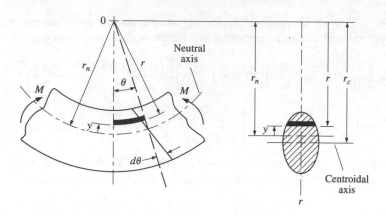

Figure 5.6-1

where r_n is the distance from the center of curvature to the neutral axis where the strain and stress are zero. If we assume that the only stress present is σ_θ,[†] then the corresponding tangential stress is given by

$$\sigma_\theta = E\varepsilon_\theta = -\frac{E(r_n - r)}{r}\frac{d\theta}{\theta} \qquad\qquad \textbf{[5.6-1]}$$

Since the net force across the cross section in the tangential direction must be zero,

$$\int_A \sigma_\theta \, dA = -\int_A \frac{E(r_n - r)}{r}\frac{d\theta}{\theta} \, dA = -\frac{E \, d\theta}{\theta}\int_A \frac{r_n - r}{r} \, dA = 0$$

The terms E, $d\theta$, and θ are independent of the integration. The term $E \, d\theta/\theta$ will not be zero. Therefore, the integral must be zero, giving

$$\int_A \frac{r_n - r}{r} \, dA = r_n\int_A \frac{dA}{r} - A = 0$$

or

$$r_n = \frac{A}{\displaystyle\int_A \frac{dA}{r}} \qquad\qquad \textbf{[5.6-2]}$$

[†] This is not true. As will be seen in the next section, radial stresses σ_r also develop. However, the resulting error in the development of σ_θ is typically small since σ_θ is maximum when σ_r is zero and vice versa.

The location of the centroidal axis is given by $(\int r \, dA)/A$. Thus, unlike a straight beam, the neutral axis and centroidal axes of a curved beam are not the same. Furthermore, for a curved beam, it can be shown that $r_n < r_c$. The *eccentricity* of the neutral axis for pure bending is defined as $e = r_c - r_n$.

Determine the location of the neutral axis and the eccentricity e for the curved bar **Example 5.6-1**
of rectangular cross section shown in Fig. 5.6-2.

Solution:

The location of the centroidal axis is $r_c = 175$ mm. Establishing $dA = 10dr$ and $A = (10)(50) = 500 \text{ mm}^2$ leads to

$$r_n = \frac{500}{10 \displaystyle\int_{150}^{200} \frac{dr}{r}} = \frac{50}{\ln 200 - \ln 150} = \frac{50}{\ln 1.333} = 173.803 \text{ mm}$$

Thus

$$e = 175.000 - 173.803 = 1.197 = 1.20 \text{ mm}^\dagger$$

Table 5.6-1 provides formulas for calculating $\int dA/r$ of cross sections commonly used for curved beams.

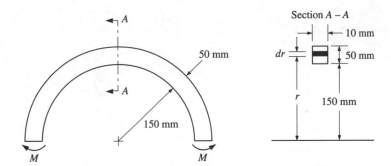

Figure 5.6-2

---

† Note that in this example, r_c and r_n must be calculated to six-place accuracy in order for e to be determined to within three-place accuracy.

Table 5.6-1

Cross Section	$\int_A \dfrac{dA}{r}$
	$b \ln\left(\dfrac{r_o}{r_i}\right)$
	$b_i \ln\left(\dfrac{r_i + h_i}{r_i}\right) + b_o \ln\left(\dfrac{r_o}{r_i + h_i}\right)$
	$2\pi\left(r_c - \sqrt{r_c^2 - R^2}\right)$
	$\left(\dfrac{b_i r_o - b_o r_i}{h}\right)\ln\left(\dfrac{r_o}{r_i}\right) - b_i + b_o$

Table 5.6-1 (concluded)

Cross Section	$\displaystyle\int_A \frac{dA}{r}$

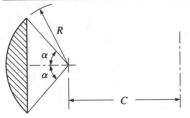

For $C \geq R$

$$2C\alpha - 2R\sin\alpha - \sqrt{C^2 - R^2}\left[\pi - 2\sin^{-1}\left(\frac{R + C\cos\alpha}{C + R\cos\alpha}\right)\right]$$

For $C < R$

$$2C\alpha - 2R\sin\alpha + 2\sqrt{R^2 - C^2}\ln\left(\frac{R + C\cos\alpha + \sqrt{R^2 - C^2}\sin\alpha}{C + R\cos\alpha}\right)$$

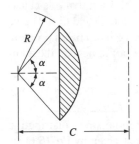

$$2C\alpha + 2R\sin\alpha - \sqrt{C^2 - R^2}\left[\pi + 2\sin^{-1}\left(\frac{R - C\cos\alpha}{C - R\cos\alpha}\right)\right]$$

In order to develop the stress distribution, the net moment from σ_θ is equated to M.[†] Thus

$$M = -\int_A y\sigma_\theta\, dA = \frac{E\, d\theta}{\theta}\int_A \frac{(r_n - r)^2}{r}\, dA$$

$$= \frac{E\, d\theta}{\theta}\left(r_n^2\int_A \frac{dA}{r} - 2r_n\int_A dA + \int_A r\, dA\right)$$

[†] From Fig. 5.6-1, M is defined positive such that the radius of curvature decreases with the application of the moment.

But $\int_A dA = A$, $\int_A r\, dA = r_c A$, and from Eq. (5.6-2), $\int_A dA/r = A/r_n$. Therefore,

$$M = \frac{E\, d\theta}{\theta}(-r_n A + r_c A) = \frac{E\, d\theta}{\theta}A(r_c - r_n)$$

Solving for $E\, d\theta/\theta$, and letting $r_c - r_n = e$, yields

$$\frac{E\, d\theta}{\theta} = \frac{M}{Ae}$$

Substituting this into Eq. (5.6-1) results in

$$\sigma_\theta = -\frac{M(r_n - r)}{Aer} \qquad \textbf{[5.6-3]}$$

Equation (5.6-3) can also be written in terms of y. Substitution of $r = r_n - y$ gives

$$\sigma_\theta = -\frac{My}{Ae(r_n - y)} \qquad \textbf{[5.6-4]}$$

Note, as shown in Example 5.6-1, the term e in Eqs. (5.6-3) and (5.6-4) is usually a very low value obtained from the difference of two much higher valued terms $r_c - r_n$. Thus calculation of r_c and r_n must be accurate enough such that e is calculated to at least three significant figures. Note also that Eqs. (5.6-3) and (5.6-4) give the normal stresses due to bending effects only. Normally in a curved beam problem there is an additional axial force as well. Care must be taken not to neglect the additional normal stress that arises from this force.

Example 5.6-2 | Consider Fig. 5.6-2 with $M = 250$ N·m. Compare the tangential stress distribution obtained from the theory of elasticity, Eq. (4.2-17b), with that given by Eq. (5.6-3).

Solution:

Recall Eq. (4.2-17b), which is given by

$$\sigma_\theta = -\frac{4M}{bK}\left[\left(\frac{r_i r_o}{r}\right)^2 \ln \frac{r_o}{r_i} + r_o^2 \ln \frac{r_o}{r} + r_i^2 \ln \frac{r}{r_i} - (r_o^2 - r_i^2)\right]$$

where

$$K = (r_o^2 - r_i^2)^2 - 4r_i^2 r_o^2\left(\ln \frac{r_o}{r_i}\right)^2$$

Substitution of $M = 250$ N·m and $r_i = 0.150$, $r_o = 0.200$, and $b = 0.010$ m gives

$$\sigma_\theta = 210.58 - \frac{3.1155}{r^2} - 481.32\ln\left(\frac{0.2}{r}\right) - 270.74\left(\frac{r}{0.15}\right) \qquad \textbf{[a]}$$

where σ_θ is in MPa and r is in meters.

For Eq. (5.6-3)

$$\sigma_\theta = -\frac{(-250)(0.1738 - r)(10^{-6})}{(0.050)(0.010)(1.197)(10^{-3})r}$$ [b]

$$= \frac{417.71}{r}(0.1738 - r)$$

where again σ_θ is in MPa and r is in meters.

Equations (a) and (b) can be plotted together to see their differences. However, the differences are so small the two plots would be basically the same. Instead of a plot, Table 5.6-2 tabulates the values and differences.

Table 5.6-2   Stresses in MPa

r	σ_θ (Elasticity)	σ_θ [Eq. (5.6-3)]	% Difference
0.150	66.357	66.284	−0.11
0.155	50.663	50.671	0.02
0.160	35.999	36.035	0.10
0.165	22.255	22.285	0.13
0.170	9.336	9.344	0.09
0.175	−2.840	−2.857	0.60
0.180	−14.345	−14.381	0.25
0.185	−25.242	−25.281	0.15
0.190	−35.586	−35.608	0.06
0.195	−45.425	−45.405	−0.04
0.200	−54.802	−54.713	−0.16

The clamp body shown in Fig. 5.6-3 undergoes a force F of 1000 lb. Determine the normal-stress distribution across section a-a and the maximum tensile and compressive stresses.

Example 5.6-3

Solution:

First, locate the centroidal axis. Considering the web and flange as separate rectangles, we have

$$0.5625r_c = (0.25)(1.25)(4.875) + (0.25)(1.00)(4.125)$$

$$r_c = 4.54167 \text{ in}^\dagger$$

After making a break at section a-a, the direct transmitted force F is placed *on the centroidal axis as shown*. This is done so that a uniform stress distribution can be

† It will be shown later that six significant figures will be necessary for e to be accurate to within three significant places.

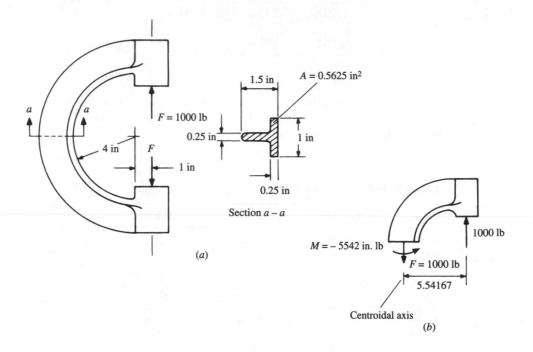

(a)

Section a – a

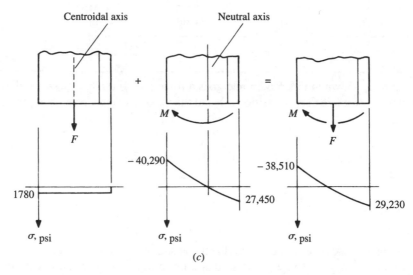

(b)

(c)

Figure 5.6-3

substituted for the force F. The neutral axis in pure bending is found from Table 5.6-1:

$$r_n = \frac{(1.00)(0.25) + (0.25)(1.25)}{1.00 \ln[(4.00 + 0.25)/4.00] + 0.25 \ln[5.5/(4.00 + 0.25)]} = 4.49705 \text{ in}$$

Thus

$$e = 4.54167 - 4.49705 = 0.04461 = 0.0446 \text{ in}$$

Because of bending, the maximum tensile stress occurs at the point closest to the center of curvature, where $y = 4.497 - 4.000 = 0.497$; from Eq. (5.6-4) this maximum is

$$(\sigma_\theta)_{r_i} = -\frac{(-5542)(0.497)}{(0.5625)(0.0446)(4.497 - 0.497)} = 27{,}450 \text{ psi} = 27.45 \text{ kpsi}$$

The maximum compressive stress due to bending occurs at $y = 4.497 - 5.500 = -1.003$ and is

$$(\sigma_\theta)_{r_o} = -\frac{(-5542)(-1.003)}{(0.5625)(0.0446)(4.497 + 1.003)} = -40{,}290 \text{ psi} = -40.29 \text{ kpsi}$$

The direct stress due to F is $\sigma = F/A$, or

$$\sigma = \frac{1000}{0.5625} = 1780 \text{ psi} = 1.78 \text{ kpsi}$$

The *net* stresses at these points are found by superposition of the axial and bending stresses and are

$$(\sigma_\theta)_{r_i} = 27.45 + 1.78 = 29.23 \text{ kpsi}$$

$$(\sigma_\theta)_{r_o} = -40.29 + 1.78 = -38.51 \text{ kpsi}$$

5.6.2 APPROXIMATE AND NUMERICAL CALCULATIONS OF e

Calculating r_n and r_c numerically and subtracting the difference can lead to large errors since r_n and r_c are typically large values as compared to e. Since e is in the denominator of Eqs. (5.6-3) and (5.6-4), a large error in e can lead to an inaccurate stress calculation. Furthermore, if one has a complex section that the tables do not handle, alternative methods for determining e are needed.

Approximate Calculation of e[†] For a quick and simple approximation of e let us return to the definition of r_n. Recall that just prior to obtaining Eq. (5.6-2)

$$\int_{\text{area}} \frac{r_n - r}{r} \, dA = 0 \qquad\qquad \textbf{[5.6-5]}$$

[†] As will be seen later, the approximate technique yields accurate results provided the radius of curvature of the beam is large relative to the depth of the section.

Consider a new coordinate variable s taken from the *centroidal axis*, where $s = r_c - r$. Substituting $r = r_c - s$ and $r_n = r_c - e$ into Eq. (5.6-5) results in

$$\int_{area} \frac{r_c - e - (r_c - s)}{r_c - s}\, dA = \int_{area} \frac{s - e}{r_c - s}\, dA = 0$$

The integral can be separated, and solving for e results in

$$e = \frac{\displaystyle\int_{area} \frac{s}{r_c - s}\, dA}{\displaystyle\int_{area} \frac{dA}{r_c - s}} \qquad\qquad \textbf{[5.6-6]}$$

The term in the denominator of both integrals, $(r_c - s)^{-1}$, can be expanded binomially as

$$(r_c - s)^{-1} = r_c^{-1}\left(1 - \frac{s}{r_c}\right)^{-1}$$

$$= r_c^{-1}\left[1 + \frac{s}{r_c} + \left(\frac{s}{r_c}\right)^2 + \left(\frac{s}{r_c}\right)^3 + \cdots \right] \qquad \text{for } s < r_c$$

If $s \ll r_c$, only the linear term in s/r_c need be retained. Thus

$$e \approx \frac{\displaystyle\int_{area} s\left(1 + \frac{s}{r_c}\right) dA}{\displaystyle\int_{area} \left(1 + \frac{s}{r_c}\right) dA}$$

Since s is taken from the centroidal axis, $\int s\, dA = 0$ and $I = \int s^2\, dA$ is the second-moment area about the *centroidal* axis of the section. Thus the above reduces to

$$e \approx \frac{I}{r_c A} \qquad\qquad \textbf{[5.6-7]}$$

This approximation is good for a large curvature where e is small *with* $r_n \approx r_c$ and $s \approx y$. In this case, Eq. (5.6-3) can be approximated by $\sigma_\theta \approx -Ms/(Aer)$. Substituting e from Eq. (5.6-7) yields

$$\sigma_\theta \approx -\frac{Ms}{I}\frac{r_c}{r} \qquad\qquad \textbf{[5.6-8]}$$

As $r_c \to \infty$, $r \to r_c$ and $s \to y$, and we arrive at the straight beam bending stress equation $-My/I$.

Keep in mind that Eqs. (5.6-7) and (5.6-8) are estimates and degrade significantly for beams as the radii of curvature decrease.

Numerical Calculation of e   The integrals of Eq. (5.6-6) can be evaluated numerically using a spreadsheet, creating a simple program or a mathematics pro-

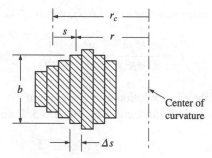

Figure 5.6-4

gram such as Matlab, Mathcad, Maple, or Mathematica. The area can be broken up into infinitesimal rectangles of thickness Δs and width b evaluated at s as shown in Fig. 5.6-4. The resulting equations for r_c and e would then be

$$r_c = \frac{\sum rb\Delta s}{\sum b\Delta s} \qquad \text{[5.6-9]}$$

$$e = \frac{\sum \dfrac{s}{r_c - s} b\Delta s}{\sum \dfrac{b\Delta s}{r_c - s}} \qquad \text{[5.6-10]}$$

Consider the circular section in Table 5.6-1 with $r_c = 3$ and $R = 1$. Determine e using the formula from the table, approximately by Eq. (5.6-7), and numerically using Eq. (5.6-9). Compare the results of the three solutions.

Example 5.6-4

Solution:

Using the formula from Table 5.6-1,

$$\int \frac{da}{r} = 2\pi\left(r_c - \sqrt{r_c^2 - R^2} \right) = 2\pi\left(3 - \sqrt{3^2 - 1^2} \right) = 1.078024169$$

Thus

$$r_n = \frac{A}{\displaystyle\int \frac{dA}{r}} = \frac{\pi(1)^2}{1.078024169} = 2.914213562$$

This gives an eccentricity of

$$e = 3 - 2.914213562 = 0.085786438 \qquad \text{[a]}$$

The approximate method, using Eq. (5.6-7), yields

$$e \approx \frac{\pi R^4 / 4}{r_c(\pi R^2)} = \frac{R^2}{4 r_c} = \frac{1}{4(3)} \approx 0.08333 \qquad \text{[b]}$$

This differs from the exact solution by -2.9 percent.

For the numerical solution, consider Fig. 5.6-5. Using the equation for a circle,

$$b = 2\sqrt{R^2 - s^2}$$

Substituting this into Eq. (5.6-10) gives

$$e = \frac{\displaystyle\sum \frac{s\sqrt{R^2 - s^2}}{r_c - s}\Delta s}{\displaystyle\sum \frac{\sqrt{R^2 - s^2}}{r_c - s}\Delta s} \qquad \text{[c]}$$

Although this may look complicated, it is rather easy to solve numerically. Figure 5.6-6 shows a simple Visual Basic program which can be implemented in the spreadsheet program, EXCEL. In the program, SUM1 and SUM2 are the numerator and denominator of Eq. (c), respectively. Figure 5.6-7 shows the entries made into row 1 of the spreadsheet. In the first row, substituting $r_c = 3$ into column A, $R = 1$ into column B, and the expression "$= ecc(A1, B1)$" into column C will yield

$$e = 0.085777387 \qquad \text{[d]}$$

which is lower than the exact by 0.011 percent.

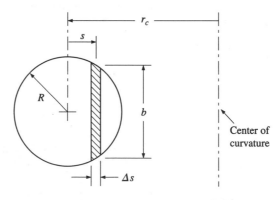

Figure 5.6-5

```
Function ecc(rc, R)
DS = 2 * R / 1000 'Divide the diameter into 1000 increments
S = -R 'Start the summation at s = -R
          'for the calculation of the numerator (SUM1)
          'and the denominator (SUM2)
   SUM1 = 0 'Initialize the summations
   SUM2 = 0
   For I = 1 To 999 Step 1 'Summation loop
     S = S + DS          'increment S
     SUM1 = SUM1 + DS * (S * Sqr(R^2 - S^2))/(rc - S)
     SUM2 = SUM2 + DS * (Sqr(R^2 - S^2)) / (rc - S)
   Next I
   ecc = SUM1/SUM2 'Determine e
   End Function
```

Figure 5.6-6   Visual Basic program to determine e.

	A	B	C
1	3	1	= ecc(A1,B1)

Figure 5.6-7   Spreadsheet entries.

Table 5.6-3 shows the error in the approximate and numerical solutions as $r_c \to R$. One can see that for a large radius of curvature the approximate equation is quite adequate. However, as $r_c \to R$ the error becomes appreciable, whereas the error in the numerical method remains very low.

Table 5.6-3   Values of e for circular section with varying R/r_c

r_c	R	Exact e	Approximate e	% Error	Numerical e	% Error
10	1	0.025063	0.025000	−0.25	0.025060	−0.010
8	1	0.031373	0.031250	−0.39	0.031370	−0.010
6	1	0.041960	0.041667	−0.70	0.041956	−0.010
4	1	0.063508	0.062500	−1.59	0.063502	−0.010
3	1	0.085786	0.083333	−2.86	0.085777	−0.011
2	1	0.133975	0.125000	−6.70	0.133959	−0.011
1.5	1	0.190983	0.166667	−12.73	0.190958	−0.013
1.25	1	0.250000	0.200000	−20.00	0.249959	−0.016

5.6.3   RADIAL STRESSES

Radial stresses σ_r also develop in curved beams and may be significant if the width of the cross section near the neutral axis of the tangential stress is small. Considering pure bending again, isolate a section of Fig. 5.6-1, as shown in Fig. 5.6-8. If σ_θ is taken as positive, the net force in the radial direction that σ_θ exerts on each surface is

$$\Delta F_r = -\sin\left(\frac{\Delta\theta}{2}\right)\int_{A_1}\sigma_\theta\, dA$$

where the integral is evaluated from r_i to r_1. Since $\Delta\theta$ is infinitesimal, $\sin(\Delta\theta/2) = \Delta\theta/2$ and the force reduces to

$$\Delta F_r = -\frac{\Delta\theta}{2}\int_{A_1}\sigma_\theta\, dA$$

The force in the radial direction due to σ_r is $\sigma_r\, br_1\Delta\theta$, where b is the width of the section at $r = r_1$. Summing forces on the three surfaces in the radial direction yields

$$\sigma_r = \frac{1}{br_1}\int_{r_i}^{r_1}\sigma_\theta\, dA$$

Substitution of Eq. (5.6-3) and simplifying results in

$$\sigma_r = -\frac{M}{Abr_1e}\left(r_n\int_{r_i}^{r_1}\frac{dA}{r} - A_1\right) \qquad \textbf{[5.6-11]}$$

where A is the total cross sectional area, and A_1 is the area from r_i to r_1.

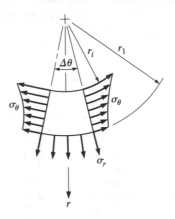

Figure 5.6-8

Determine the radial stress distribution in zone *a-a* of the clamp of Example 5.6-3.   **Example 5.6-5**

Solution:

For the inner flange, evaluate the integral $\int dA/r$ with $r_i = 4$ in, $r_o = r_1$, and $b = 1$ in using the rectangular cross section in Table 5.6-1:

$$\int \frac{dA}{r} = 1 \ln \frac{r_1}{4}$$

The area from r_i to r_1 is

$$A_1 = (1)(r_1 - 4)$$

Substituting these terms into Eq. (5.6-11) with $b = 1$ results in the radial stress distribution in the flange:

$$\sigma_r = -\frac{M}{Ar_1e}\left(r_n \ln \frac{r_1}{4} - r_1 + 4\right) \qquad [a]$$

In the web area, the section isolated at r_1 is a T section (see Table 5.6-1) with $b_i = 1$ in, $h_i = 0.25$ in, $r_i = 4$ in, $r_o = r_1$, and $b = b_o = 0.25$ in. Thus

$$\int \frac{dA}{r} = 1 \ln\left(\frac{4 + 0.25}{4}\right) + (0.25) \ln \frac{r_1}{4 + 0.25} = 0.06062 + \frac{1}{4} \ln \frac{r_1}{4.25}$$

and

$$A_1 = (1)(0.25) + (r_1 - 4.25)(0.25) = \frac{r_1}{4} - 0.8125$$

Substitution of $\int dA/r$ and A_1 into Eq. (5.6-11) with $b = 0.25$ yields the radial stress distribution in the web,

$$\sigma_r = -\frac{4M}{Ar_1e}\left(\frac{r_n}{4} \ln \frac{r_1}{4.25} + 0.06062r_n - \frac{r_1}{4} + 0.8125\right) \qquad [b]$$

From Example 5.6-3 $M = -5542$ lb·in, $A = 0.5625$ in², $e = 0.0446$ in, and $r_n = 4.4971$ in. Thus Eqs. (*a*) and (*b*) reduce to

Flange: $\sigma_r = (2.21 \times 10^5)\left(\dfrac{4.4971}{r_1} \ln \dfrac{r_1}{4} + \dfrac{4}{r_1} - 1\right)$ **[c]**

Web: $\sigma_r = (8.84 \times 10^5)\left(\dfrac{1.1244}{r_1} \ln \dfrac{r_1}{4.25} + \dfrac{1.0850}{r_1} - \dfrac{1}{4}\right)$ **[d]**

The maximum radial stress in the flange occurs at $r_1 = 4.25$ and is †

$$(\sigma_r)_{\text{flange max}} = 1180 \text{ psi}$$

---

† This value is actually low, because in the derivation it is assumed that the radial stress acts over the entire thickness. As in shear flow in an I-beam, the boundary conditions at the free surface of the flange dictate that $\sigma_r = 0$ on the free surfaces of the flange at $r_1 = 4.25$ in.

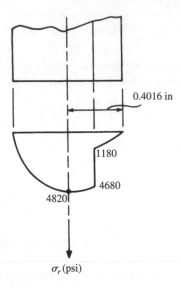

Figure 5.6-9

The maximum radial stress in the web occurs at $r_1 = 4.4016$ in where $d\sigma_r/dr = 0$. This value is

$$(\sigma_r)_{\text{web max}} = 4820\,\text{psi}$$

The stress distribution is shown in Fig. 5.6-9.

5.7   BENDING OF THIN FLAT PLATES

Entire books have been devoted to this subject[†] and an inordinate amount of space would be necessary here to present the many techniques and solutions available. Thus we will only discuss the basic governing equations in rectangular and cylindrical coordinates (for axisymmetric problems), boundary conditions, and some introductory examples, and will tabulate the solutions to some basic configurations. For more detailed tabulations, see Ref. 5.3.

5.7.1   GOVERNING EQUATIONS IN RECTANGULAR COORDINATES

Here we will follow the notation given in most of the literature which deviates from the notation used earlier in this book for beams in bending. For flat plates it is common practice to set the x, y axes in the midplane of the plate and let the bending de-

[†] For example, see Refs. 5.1 and 5.2.

flections be in the z direction [see Fig. 5.7-1(a)]. It is assumed that this deflection is a function of x and y, i.e., $w = w(x,y)$. Furthermore, the derivations in this section are restricted to deflections of the order of the *thickness* of the plate. Unlike linear theory for beams, where small slopes are assumed, in-plane membrane stresses develop in plates, even under small slopes.[†] These membrane stresses, if included, force the governing equations to become nonlinear, which is beyond the level of this textbook. For small deflections, however, as with simple beam theory, it is assumed that plane surfaces remain plane. Line cd in Fig. 5.7-1(b) deflects w in the z direction and rotates $\partial w/\partial x$. This, in turn, causes a point initially at z to deflect in the x direction, $u = -z\, \partial w/\partial x$. Likewise, if one were to view the yz plane, a deflection of a point at z in the y direction is $v = -z\, \partial w/\partial y$. From the strain-displacement relations

$$\varepsilon_x = \frac{\partial u}{\partial x} = -z\frac{\partial^2 w}{\partial x^2} \qquad \textbf{[5.7-1a]}$$

$$\varepsilon_y = \frac{\partial v}{\partial y} = -z\frac{\partial^2 w}{\partial y^2} \qquad \textbf{[5.7-1b]}$$

$$\gamma_{xy} = \frac{\partial v}{\partial x} + \frac{\partial u}{\partial y} = -2z\frac{\partial^2 w}{\partial x\, \partial y} \qquad \textbf{[5.7-1c]}$$

Loads will be applied to the top and/or bottom surfaces and give rise to σ_z contact stresses. However, as with beams, this stress is usually localized or small compared

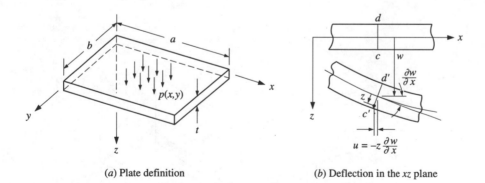

(*a*) Plate definition (*b*) Deflection in the xz plane

Figure 5.7-1

[†] Similar behavior occurs in beams when constrained axially (see Secs. 6.13 and 9.3.7).

to the σ_x and σ_y bending stresses, and therefore will be ignored. Substituting the strains of Eqs. (5.7-1) into the stress-strain relations for plane stress gives

$$\sigma_x = \frac{E}{1-v}(\varepsilon_x + v\varepsilon_y) = -\frac{E}{1-v}z\left(\frac{\partial^2 w}{\partial x^2} + v\frac{\partial^2 w}{\partial y^2}\right) \qquad \text{[5.7-2a]}$$

$$\sigma_y = \frac{E}{1-v}(\varepsilon_y + v\varepsilon_x) = -\frac{E}{1-v}z\left(\frac{\partial^2 w}{\partial y^2} + v\frac{\partial^2 w}{\partial x^2}\right) \qquad \text{[5.7-2b]}$$

$$\tau_{xy} = \frac{E}{2(1+v)}\gamma_{xy} = -\frac{E}{2(1+v)}z\frac{\partial^2 w}{\partial x\,\partial y} \qquad \text{[5.7-2c]}$$

Next, in a manner similar to simple beam theory, we will relate the internal moments to the stresses. Figure 5.7-2 shows the net moments and shear forces *per unit length* on the $t\,dx$ and $t\,dy$ faces, where again the notation is peculiar to that used in the literature such as Refs. 5.1 and 5.2.

The moment per unit length M_x is related to σ_x by

$$M_x\,dy = \left(\int_{-t/2}^{t/2} z\sigma_x\,dz\right)dy$$

$$M_x = \int_{-t/2}^{t/2} z\sigma_x\,dz \qquad \text{[5.7-3a]}$$

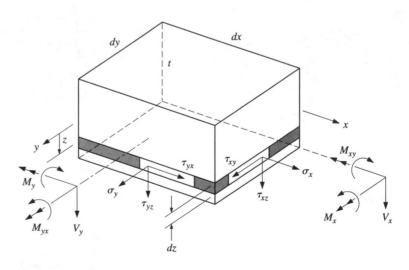

Figure 5.7-2 Rectangular plate element definition.

where it is assumed that σ_x is constant in the y direction over the dy length. Similarly,

$$M_y = \int_{-t/2}^{t/2} z\sigma_y \, dz \qquad \text{[5.7-3b]}$$

$$M_{xy} = M_{yx} = \int_{-t/2}^{t/2} z\tau_{xy} \, dz \qquad \text{[5.7-3c]}$$

where for Eq. (5.7-3c) it was assumed that $\tau_{yx} = \tau_{xy}$.

Substituting Eqs. (5.7-2) and noting that w is not a function of z yields

$$M_x = -D\left(\frac{\partial^2 w}{\partial x^2} + v\frac{\partial^2 w}{\partial y^2}\right) \qquad \text{[5.7-4a]}$$

$$M_y = -D\left(\frac{\partial^2 w}{\partial y^2} + v\frac{\partial^2 w}{\partial x^2}\right) \qquad \text{[5.7-4b]}$$

$$M_{xy} = -D(1 - v)\frac{\partial^2 w}{\partial x \partial y} \qquad \text{[5.7-4c]}$$

where

$$D = \frac{Et^3}{12(1 - v^2)} \qquad \text{[5.7-5]}$$

and is called the *flexural rigidity* of the plate. Combining Eqs. (5.7-2) and (5.7-4) with the use of Eq. (5.7-5) results in

$$\sigma_x = 12\frac{M_x z}{t^3} \qquad \text{[5.7-6a]}$$

$$\sigma_y = 12\frac{M_y z}{t^3} \qquad \text{[5.7-6b]}$$

$$\tau_{xy} = 12\frac{M_{xy} z}{t^3} \qquad \text{[5.7-6c]}$$

Now, the maximum and minimum stresses occur at $z = \pm t/2$, given by

$$(\sigma_x)_{\text{max, min}} = \pm 6\frac{M_x}{t^2} \qquad \text{[5.7-7a]}$$

$$(\sigma_y)_{\text{max, min}} = \pm 6\frac{M_y}{t^2} \qquad \text{[5.7-7b]}$$

$$(\tau_{xy})_{\text{max, min}} = \pm 6\frac{M_{xy}}{t^2} \qquad \text{[5.7-7c]}$$

In simple beam theory, moments are easily determined from equilibrium. However, for plates, this is not an easy matter. Consider the $t\, dx\, dy$ element in Fig. 5.7-3 where now the lateral load $p(x,y)$ and the changes in the forces and moments per unit lengths are shown. Also, we have utilized the fact that $M_{yx} = M_{xy}$. Summing forces in the z direction (keeping in mind that the V's are force per unit length) yields

$$\sum F_z = \left(\frac{\partial V_x}{\partial x}dx\right)dy + \left(\frac{\partial V_y}{\partial y}\, dy\right)dx + p(x,y)dx\, dy = 0$$

Dividing by $dx\, dy$ gives

$$\frac{\partial V_x}{\partial x} + \frac{\partial V_y}{\partial y} + p(x,y) = 0 \qquad \textbf{[5.7-8]}$$

Summing moments about an axis parallel to the x axis through the element center gives

$$\sum M_{\bar{x}} = -\left(\frac{\partial M_{xy}}{\partial x}\, dx\right)dy - \left(\frac{\partial M_y}{\partial y}\, dy\right)dx + (V_y\, dx)dy = 0$$

or

$$V_y = \frac{\partial M_{xy}}{\partial x} + \frac{\partial M_y}{\partial y} \qquad \textbf{[5.7-9]}$$

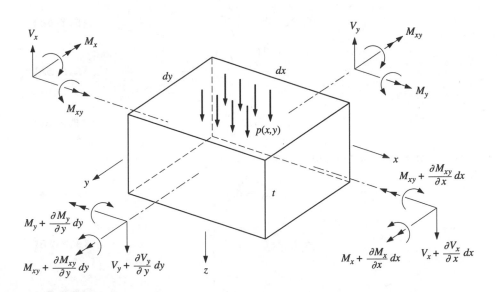

Figure 5.7-3   Variation in moments and force.

where the higher-order term $1/2\,[(\partial V_y/\partial y)dy\,dx)]dy$ term was ignored. In a similar fashion, summing moments about an axis parallel to the y axis through the element center yields

$$V_x = \frac{\partial M_{xy}}{\partial y} + \frac{\partial M_x}{\partial x} \qquad \textbf{[5.7-10]}$$

Substitution of Eqs. (5.7-9) and (5.7-10) into (5.7-8) gives

$$\frac{\partial^2 M_x}{\partial x^2} + 2\frac{\partial^2 M_{xy}}{\partial x\,\partial y} + \frac{\partial^2 M_y}{\partial y^2} = -p(x,y) \qquad \textbf{[5.7-11]}$$

Finally, substituting Eqs. (5.7-4) into Eq. (5.7-11) and simplifying results in

$$\frac{\partial^4 w}{\partial x^4} + 2\frac{\partial^4 w}{\partial x^2 \partial y^2} + \frac{\partial^4 w}{\partial y^4} = \nabla^4 w(x,y) = \frac{p(x,y)}{D} \qquad \textbf{[5.7-12]}$$

The solution of Eq. (5.7-12) is no easy matter. However, once $w(x,y)$ is determined, the moments are found from Eqs. (5.7-4) and the bending and twisting stresses are obtained using Eqs. (5.7-6) or (5.7-7).

5.7.2 TABULATED SOLUTIONS OF UNIFORMLY LOADED RECTANGULAR PLATES

This section provides a table containing only a very small sampling of the basic configurations contained in Ref. 5.3. For Table 5.7-1 the following notation is used:

1. The terms a and b are the dimensions of the sides of the plate.
2. The load p_o is a uniform pressure applied along the entire plate.
3. The maximum stress and deflection are given by the formulas

$$\sigma_{\max} = k_1 p_o \left(\frac{b}{t}\right)^2 \qquad w_{\max} = k_2 \frac{p_o b^4}{E t^3}$$

Table 5.7-1 Uniformly loaded flat plates with straight boundaries and constant thickness t

Case 1. *All Edges Simply Supported ($v = 0.3$). See Ref. 5.1*

a/b	1.0	1.2	1.4	1.6	1.8	2.0	3.0	4.0	5.0	∞
k_1	0.2874	0.3762	0.4530	0.5172	0.5688	0.6102	0.7134	0.7410	0.7476	0.7500
k_2	0.0444	0.0616	0.0770	0.0906	0.1017	0.1110	0.1335	0.1400	0.1417	0.1421

Case 2. *All Edges Fixed ($v = 0.3$). See Ref. 5.1*

a/b	1.0	1.2	1.4	1.6	1.8	2.0	∞
k_1	0.3078	0.3834	0.4356	0.4680	0.4872	0.4974	0.500
k_2	0.0138	0.0188	0.226	0.0251	0.0267	0.0277	0.0284

Table 5.7-1 (concluded)

Case 3. Three Edges Simply Supported, One Edge (b Length) Free ($v = 0.3$). See Ref. 5.3

a/b	0.5	0.667	1.0	1.5	2.0	4.0
k_1	0.36	0.45	0.67	0.77	0.79	0.8
k_2	0.080	0.106	0.140	0.160	0.165	0.167

Case 4. Three Edges Simply Supported, One Edge (b Length) Fixed ($v = 0.3$). See Ref. 5.3

a/b	1	1.5	2.0	2.5	3.0	3.5	4.0
k_1	0.50	0.67	0.73	0.74	0.75	0.75	0.75
k_2	0.030	0.071	0.101	0.122	0.132	0.137	0.139

Case 5. Three Edges Simply Supported, One Edge (a Length) Fixed ($v = 0.3$). See Ref. 5.3

a/b	1	1.5	2.0	2.5	3.0	3.5	4.0
k_1	0.50	0.66	0.73	0.74	0.74	0.75	0.75
k_2	0.030	0.046	0.054	0.056	0.057	0.058	0.058

Case 6. Two Opposing Sides (a Length) Simply Supported, Two Opposing Sides (b Length) Fixed ($v = 0.3$). See Ref. 5.1

a/b	1.0	1.2	1.4	1.6	1.8	2.0	∞
k_1	0.4182	0.5208	0.5988	0.6540	0.6912	0.7146	0.750
k_2	0.0210	0.0349	0.0502	0.0658	0.0800	0.0922	0.1422

Case 7. Two Opposing Sides (b Length) Simply Supported, Two Opposing Sides (a Length) Fixed ($v = 0.3$). See Ref. 5.1

a/b	1.0	1.2	1.4	1.6	1.8	2.0	∞
k_1	0.4182	0.4626	0.4860	0.4968	0.4971	0.4973	0.500
k_2	0.0210	0.0243	0.0262	0.0273	0.0280	0.0283	0.0285

5.7.3 GOVERNING EQUATIONS FOR AXISYMMETRIC CIRCULAR PLATES IN BENDING

The development of the governing equations for axisymmetric circular plates in bending is very similar to that of the previous section. The element moments and shear force per unit length and the corresponding stresses are shown in Fig. 5.7-4. Because of symmetry,[†] (1) $\tau_{r\theta}$ does not exist; thus there is no $M_{r\theta}$; and (2) $\tau_{\theta z}$ does not exist; thus there is no V_θ.

Viewing the displacement in the rz plane as we viewed displacements in the xz plane in Fig. 5.7-1(*b*), the radial displacement is

$$u_r = -z \frac{dw}{dr}$$

[†] For further discussion on conditions of symmetry, see Sec. 8.2.

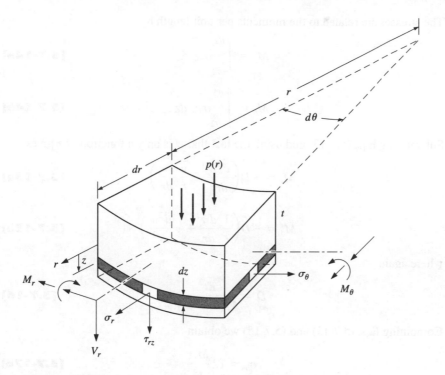

Figure 5.7-4   Axisymmetric plate element definition.

where partial derivatives are unnecessary as r is the only independent variable. The strains, for the axisymmetric case, are given by

$$\varepsilon_r = \frac{du_r}{dr} = -z\frac{d^2w}{dr^2}$$

$$\varepsilon_\theta = \frac{u_r}{r} = -\frac{z}{r}\frac{dw}{dr}$$

Substituting this into the stress-strain relations with $\sigma_z = 0$ results in

$$\sigma_r = -\frac{E}{1-v^2}z\left(\frac{d^2w}{dr^2} + \frac{v}{r}\frac{dw}{dr}\right) \qquad \textbf{[5.7-13a]}$$

$$\sigma_\theta = -\frac{E}{1-v^2}z\left(\frac{1}{r}\frac{dw}{dr} + v\frac{d^2w}{dr^2}\right) \qquad \textbf{[5.7-13b]}$$

The stresses are related to the moments per unit length by

$$M_r = \int_{-t/2}^{t/2} \sigma_r z \, dz \qquad \text{[5.7-14a]}$$

$$M_\theta = \int_{-t/2}^{t/2} \sigma_\theta z \, dz \qquad \text{[5.7-14b]}$$

Substituting Eqs. (5.7-13) and using the fact that w is only a function of r gives

$$M_r = -D\left(\frac{d^2 w}{dr^2} + \frac{v}{r}\frac{dw}{dr}\right) \qquad \text{[5.7-15a]}$$

$$M_\theta = -D\left(\frac{1}{r}\frac{dw}{dr} + v\frac{d^2 w}{dr^2}\right) \qquad \text{[5.7-15b]}$$

where again

$$D = \frac{Et^3}{12(1 - v^2)} \qquad \text{[5.7-16]}$$

Combining Eqs. (5.7-13) and (5.7-15) we obtain

$$\sigma_r = 12\frac{M_r z}{t^3} \qquad \text{[5.7-17a]}$$

$$\sigma_\theta = 12\frac{M_\theta z}{t^3} \qquad \text{[5.7-17b]}$$

where, as with the rectangular plate, the maximum and minimum values are at $\pm\, t/2$, giving

$$(\sigma_r)_{\text{max, min}} = \pm 6\frac{M_r}{t^2} \qquad \text{[5.7-18a]}$$

$$(\sigma_\theta)_{\text{max, min}} = \pm 6\frac{M_\theta}{t^2} \qquad \text{[5.7-18b]}$$

The variations in the shear force and moment per unit length are shown in Fig. 5.7-5. Owing to axisymmetry, M_θ is only a function of r and does not vary with respect to θ. Summing moments about an axis through the element center and perpendicular to the rz plane,

$$\left(M_r + \frac{dM_r}{dr}dr\right)(r + dr)\,d\theta - M_r r\,d\theta - 2(M_\theta dr)\sin\left(\frac{d\theta}{2}\right)$$

$$-\left(V_r + \frac{dV_r}{dr}dr\right)(r + dr)d\theta\frac{dr}{2} - (V_r r\,d\theta)\frac{dr}{2} = 0$$

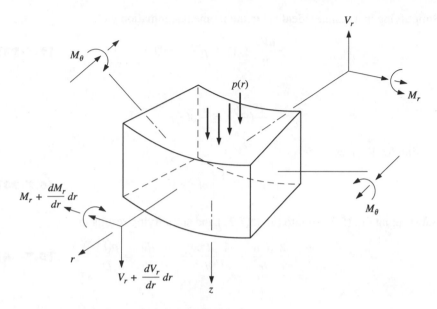

Figure 5.7-5   Variations in moment and force.

Note that each of the two M_θ moment per unit length vectors has a component in the r direction. For small $d\theta$, $\sin d\theta/2 \approx d\theta/2$, and ignoring differential terms of order higher than 2 gives

$$M_r \, dr \, d\theta + r \frac{dM_r}{dr} dr \, d\theta - M_\theta \, dr \, d\theta - V_r r \, dr \, d\theta = 0$$

Canceling $dr \, d\theta$ and solving for V_r gives

$$V_r = \frac{M_r - M_\theta}{r} + \frac{dM_r}{dr} \qquad \textbf{[5.7-19]}$$

Substituting Eqs. (5.7-15) and simplifying results in

$$V_r = -D\left(\frac{d^3w}{dr^3} + \frac{1}{r} \frac{d^2w}{dr^2} - \frac{1}{r^2} \frac{dw}{dr} \right) \qquad \textbf{[5.7-20]}$$

This can also be expressed in the form

$$\frac{d}{dr}\left[\frac{1}{r} \frac{d}{dr}\left(r \frac{dw}{dr} \right) \right] = -\frac{V_r}{D} \qquad \textbf{[5.7-21]}$$

Summing forces in the z direction gives

$$\left(V_r + \frac{dV_r}{dr} dr \right)(r + dr)\, d\theta - V_r r \, d\theta + p(r) r \, d\theta \, dr = 0$$

Simplifying in a manner identical to the moment summation yields

$$r\frac{dV_r}{dr} + V_r + p(r)r = 0 \qquad\qquad \textbf{[5.7-22]}$$

or

$$\frac{d}{dr}(rV_r) = -p(r)r$$

Solving for V_r gives

$$V_r = -\frac{1}{r}\int p(r)r\,dr \qquad\qquad \textbf{[5.7-23]}$$

Substituting Eq. (5.7-20) into Eq. (5.7-22) and simplifying results in

$$\frac{d^4w}{dr^4} + \frac{2}{r}\frac{d^3w}{dr^3} - \frac{1}{r^2}\frac{d^2w}{dr^2} + \frac{1}{r^3}\frac{dw}{dr} = \frac{p(r)}{D} \qquad\qquad \textbf{[5.7-24]}$$

or

$$\nabla^4 w(r) = \left(\frac{d^2}{dr^2} + \frac{1}{r}\frac{d}{dr}\right)\left(\frac{d^2w}{dr^2} + \frac{1}{r}\frac{dw}{dr}\right) = \frac{p(r)}{D} \qquad\qquad \textbf{[5.7-25]}$$

which agrees with the deflection equation obtained for rectangular plates [Eq. (5.7-12)].

Solutions to the governing equations of axisymmetric circular plates are more readily obtainable, and several examples will be presented here.

Example 5.7-1 | Determine the equations for the displacement, moments, reactions, and maximum stresses for the uniformly loaded, simply supported plate of radius a shown in Fig. 5.7-6.

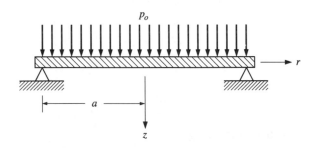

Figure 5.7-6

Solution:

From Eq. (5.7-23)

$$V_r = -\frac{1}{r}\int p_o r\,dr = -\frac{p_o r}{2} + C$$

where C is a constant of integration. As $r \to 0$, $V_r \to 0$. Thus $C = 0$, and $V_r = -p_o r/2$. Substituting this into Eq. (5.7-21) gives

$$\frac{d}{dr}\left[\frac{1}{r}\frac{d}{dr}\left(r\frac{dw}{dr}\right)\right] = \frac{p_o r}{2D}$$

Integrating with respect to r gives

$$\frac{1}{r}\frac{d}{dr}\left(r\frac{dw}{dr}\right) = \frac{p_o r^2}{4D} + C_1$$

Multiplying by r and integrating again gives

$$r\frac{dw}{dr} = \frac{p_o r^4}{16D} + C_1\frac{r^2}{2} + C_2$$

or

$$\frac{dw}{dr} = \frac{p_o r^3}{16D} + C_1\frac{r}{2} + \frac{C_2}{r}$$

At $r = 0$, the last term is infinite, which is not physically acceptable. Thus $C_2 = 0$, and we see that $dw/dr = 0$ at $r = 0$. This makes sense, since we have symmetry. Thus

$$\frac{dw}{dr} = \frac{p_o r^3}{16D} + C_1\frac{r}{2}$$

Integrating a final time gives

$$w = \frac{p_o r^4}{64D} + C_1\frac{r^2}{4} + C_3 \qquad\qquad \text{[a]}$$

Next we apply boundary conditions. At the edge where $r = a$, the deflection w and bending moment M_r must be zero. Thus

$$w(a) = 0 \qquad\qquad \text{[b.1]}$$

$$M_r(a) = \left(\frac{d^2w}{dr^2} + \frac{v}{r}\frac{dw}{dr}\right)_{r=a} = 0 \qquad\qquad \text{[b.2]}$$

Substitution of conditions $(b.1)$ and $(b.2)$ into Eq. (a) yields

$$C_1 = -\frac{p_o a^4}{16D} + \frac{(3 + v)p_o a^4}{8(1 + v)D}$$

$$C_3 = -\frac{(3 + v)p_o a^4}{32(1 + v)D}$$

Substituting these back into Eq. (a) and simplifying gives

$$w = \frac{p_o}{64D}(a^2 - r^2)\left(\frac{5 + v}{1 + v}a^2 - r^2\right)$$ [c]

The maximum displacement occurs at $r = 0$ where with the use of Eq. (5.7-16) is

$$w_{max} = \frac{3}{16}\frac{p_o(1 - v^2)(5 + v)a^4}{(1 + v)Et^3}$$ [d]

The bending moments per unit length are given by substituting Eq. (c) into Eqs. (5.7-15). This yields, after simplification,

$$M_r = \frac{3 + v}{16}p_o(a^2 - r^2)$$ [e]

$$M_\theta = \frac{1}{16}p_o[(3 + v)a^2 - (1 + 3v)r^2]$$ [f]

Both are maximum at $r = 0$ and are

$$(M_r)_{max} = (M_\theta)_{max} = \frac{3 + v}{16}p_o a^2$$ [g]

Thus the maximum bending stresses also occur at $r = 0$, given by Eqs. (5.7-18),

$$(\sigma_r)_{max} = (\sigma_\theta)_{max} = \frac{3}{8}(3 + v)p_o\left(\frac{a}{t}\right)^2$$ [h]

The reaction force per unit length V_r evaluated at $r = a$ is found using Eq. (5.7-23) and is

$$(V_r)_{r=a} = -\frac{1}{a}p_o\int_0^a r\,dr = -\frac{1}{2}p_o a$$ [i]

5.7.4   TABULATED SOLUTIONS OF CIRCULAR PLATES

As stated earlier, Ref. 5.3 contains an extensive tabulation of configurations of circular plates. Here we will provide some very basic cases (Table 5.7-2). Other configurations can be developed by superposition techniques illustrated in Sec. 5.7.5.

Table 5.7-2

Case 1. Circular Plate, Pinned at Center, Loaded by a Moment $M_r = M_o$ Uniformly Applied along the Outer Edge

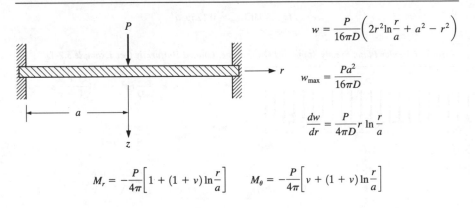

$$w = -\frac{M_o r^2}{2(1 + v)D}$$

$$\frac{dw}{dr} = -\frac{M_o r}{(1 + v)D}$$

$$M_r(r) = M_\theta(r) = M_o$$

Case 2. Circular Plate, Fixed Outer Edge, Loaded by Central Concentrated Force[†]

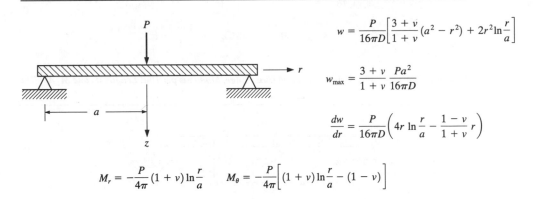

$$w = \frac{P}{16\pi D}\left(2r^2\ln\frac{r}{a} + a^2 - r^2\right)$$

$$w_{\max} = \frac{Pa^2}{16\pi D}$$

$$\frac{dw}{dr} = \frac{P}{4\pi D}r\ln\frac{r}{a}$$

$$M_r = -\frac{P}{4\pi}\left[1 + (1 + v)\ln\frac{r}{a}\right] \qquad M_\theta = -\frac{P}{4\pi}\left[v + (1 + v)\ln\frac{r}{a}\right]$$

Case 3. Circular Plate, Simply Supported Outer Edge, Loaded by Central Concentrated Force[†]

$$w = \frac{P}{16\pi D}\left[\frac{3 + v}{1 + v}(a^2 - r^2) + 2r^2\ln\frac{r}{a}\right]$$

$$w_{\max} = \frac{3 + v}{1 + v}\frac{Pa^2}{16\pi D}$$

$$\frac{dw}{dr} = \frac{P}{16\pi D}\left(4r\ln\frac{r}{a} - \frac{1 - v}{1 + v}r\right)$$

$$M_r = -\frac{P}{4\pi}(1 + v)\ln\frac{r}{a} \qquad M_\theta = -\frac{P}{4\pi}\left[(1 + v)\ln\frac{r}{a} - (1 - v)\right]$$

[†] Note that as $r \to 0$; M_r, $M_\theta \to \infty$, and σ_r, $\sigma_\theta \to \infty$. This is physically unacceptable and is due to the modeling of a concentrated force. All real forces must be distributed over an area, no matter how small. Thus the equations given for M_r and M_θ are not usable for small values of r. Example 5.7-4 illustrates a distributed force over a small radius c which is much larger than the plate thickness. For a smaller radius, see Ref. 5.1.

Table 5.7-2 (continued)

Case 4. Circular Plate, Fixed Outer Edge, Loaded Uniformly

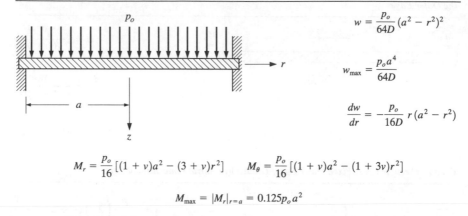

$$w = \frac{p_o}{64D}(a^2 - r^2)^2$$

$$w_{max} = \frac{p_o a^4}{64D}$$

$$\frac{dw}{dr} = -\frac{p_o}{16D}r(a^2 - r^2)$$

$$M_r = \frac{p_o}{16}[(1+v)a^2 - (3+v)r^2] \qquad M_\theta = \frac{p_o}{16}[(1+v)a^2 - (1+3v)r^2]$$

$$M_{max} = |M_r|_{r=a} = 0.125 p_o a^2$$

Case 5. Circular Plate, Simply Supported Outer Edge, Loaded Uniformly (see Example 5.7-1)

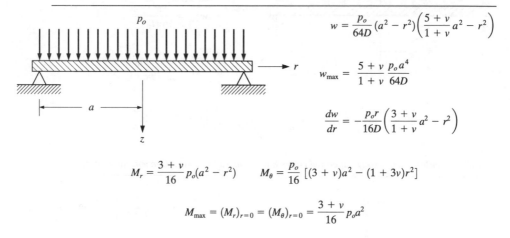

$$w = \frac{p_o}{64D}(a^2 - r^2)\left(\frac{5+v}{1+v}a^2 - r^2\right)$$

$$w_{max} = \frac{5+v}{1+v}\frac{p_o a^4}{64D}$$

$$\frac{dw}{dr} = -\frac{p_o r}{16D}\left(\frac{3+v}{1+v}a^2 - r^2\right)$$

$$M_r = \frac{3+v}{16}p_o(a^2 - r^2) \qquad M_\theta = \frac{p_o}{16}[(3+v)a^2 - (1+3v)r^2]$$

$$M_{max} = (M_r)_{r=0} = (M_\theta)_{r=0} = \frac{3+v}{16}p_o a^2$$

Case 6. Annular Plate, Simply Supported Outer Edge, Moments per Unit Length M_1 and M_2 Applied at Edges

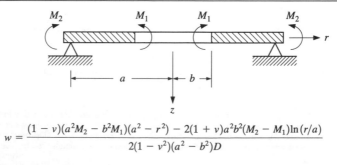

$$w = \frac{(1-v)(a^2 M_2 - b^2 M_1)(a^2 - r^2) - 2(1+v)a^2 b^2(M_2 - M_1)\ln(r/a)}{2(1-v^2)(a^2 - b^2)D}$$

Table 5.7-2 (continued)

$$\frac{dw}{dr} = -\left[\frac{(1-v)(a^2M_2 - b^2M_1)r + (1+v)a^2b^2(M_2 - M_1)(1/r)}{(1-v^2)(a^2-b^2)D}\right]$$

$$M_r = \frac{a^2M_2 - b^2M_1 - (ab/r)^2(M_2 - M_1)}{a^2-b^2} \qquad M_\theta = \frac{a^2M_2 - b^2M_1 + (ab/r)^2(M_2 - M_1)}{a^2-b^2}$$

Case 7. Annular Plate, Simply Supported Outer Edge, Shear Force per Unit Length Applied at the Inner Edge

$$w = \frac{V_1 b}{8D}\left\{(a^2 - r^2)\left(\frac{3+v}{1+v} - \frac{2b^2}{a^2-b^2}\ln\frac{b}{a}\right) + 2\left[r^2 + 2\frac{1+v}{1-v}\frac{a^2b^2}{a^2-b^2}\ln\frac{b}{a}\right]\ln\frac{r}{a}\right\}$$

$$\frac{dw}{dr} = \frac{V_1 b}{2D}\left[\left(\ln\frac{r}{a} + \frac{b^2}{a^2-b^2}\ln\frac{b}{a} - \frac{1}{1+v}\right)r + \left(\frac{1+v}{1-v}\frac{a^2b^2}{a^2-b^2}\ln\frac{b}{a}\right)\frac{1}{r}\right]$$

$$M_r = \frac{(1+v)V_1 b}{2}\left[\frac{a^2-r^2}{a^2-b^2}\left(\frac{b}{r}\right)^2\ln\frac{b}{a} - \ln\frac{r}{a}\right]$$

$$M_\theta = \frac{V_1 b}{2}\left[(1-v) - (1+v)\frac{a^2+r^2}{a^2-b^2}\left(\frac{b}{r}\right)^2\ln\frac{b}{a} - (1+v)\ln\frac{r}{a}\right]$$

$$M_{\max} = (M_\theta)_{r=b} = \frac{V_1 b}{2}\left[(1-v) - (1+v)\frac{2a^2}{a^2-b^2}\ln\frac{b}{a}\right]$$

The maximum deflection, slope, and bending moment are tabulated for various ratios of b/a, where $v = 0.3$.

	b/a	k_1	k_2	k_3
$w_{\max} = k_1 \dfrac{V_1 a^3}{D}$	0.1	0.03639	-0.03707	0.33736
	0.2	0.08097	-0.11118	0.50589
$\left(\dfrac{dw}{dr}\right)_{\max} = k_2 \dfrac{V_1 a^2}{D}$	0.3	0.12656	-0.20472	0.62099
	0.4	0.16601	-0.31087	0.70723
	0.5	0.19335	-0.42622	0.77573
	0.6	0.20362	-0.54895	0.83257
$M_{\max} = k_3 V_1 a$	0.7	0.19273	-0.67802	0.88142
	0.8	0.15714	-0.81287	0.92464
	0.9	0.09381	-0.95321	0.96380

Table 5.7-2 (concluded)

Case 8. Annular Plate, Simply Supported Outer Edge, Uniformly Loaded

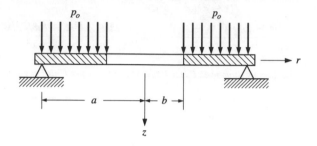

$$w = \frac{P_o}{64D}(a^2 - r^2)\left(\frac{5+v}{1+v}a^2 - 2\frac{3+v}{1+v}b^2 - r^2 + 8\frac{b^4}{a^2-b^2}\ln\frac{b}{a}\right)$$

$$-\frac{P_o}{16D}\left(2b^2r^2 + \frac{3+v}{1-v}a^2b^2 + 4\frac{1+v}{1-v}\frac{a^2b^4}{a^2-b^2}\ln\frac{b}{a}\right)\ln\frac{r}{a}$$

$$\frac{dw}{dr} = -\frac{P_o}{16D}\left(\frac{3+v}{1+v}a^2 - \frac{1-v}{1+v}b^2 + 4\frac{b^4}{a^2-b^2}\ln\frac{b}{a} - r^2 + 4b^2\ln\frac{r}{a}\right)r$$

$$-\frac{P_o}{16D}\left(\frac{3+v}{1-v}a^2b^2 + 4\frac{1+v}{1-v}\frac{a^2b^4}{a^2-b^2}\ln\frac{b}{a}\right)\frac{1}{r}$$

$$M_{\max} = (M_\theta)_{r=b} = \frac{P_o}{8}\left[(3+v)a^2 - (1-v)b^2 + 4(1+v)\frac{a^2b^2}{a^2-b^2}\ln\frac{b}{a}\right]$$

The maximum deflection, slope, and bending moment are tabulated for various ratios of b/a, where $v = 0.3$.

$$w_{\max} = k_1\frac{P_o a^4}{D}$$

$$\left(\frac{dw}{dr}\right)_{\max} = k_2\frac{P_o a^3}{D}$$

$$M_{\max} = k_3 P_o a^2$$

b/a	k_1	k_2	k_3
0.1	0.06866	−0.04357	0.39651
0.2	0.07447	−0.08031	0.36541
0.3	0.07613	−0.10788	0.32723
0.4	0.07206	−0.12530	0.28505
0.5	0.06244	−0.13211	0.24044
0.6	0.04854	−0.12806	0.19423
0.7	0.03245	−0.11298	0.14688
0.8	0.01683	−0.08672	0.09865
0.9	0.00484	−0.04912	0.04967

5.7.5   SUPERPOSITION

Configurations other than those given in Table 5.7-2 can be arrived at using super-position techniques.

Using Table 5.7-2, determine the table for deflection, slope, and moments for the uniformly loaded annular plate with outer edges fixed shown in Fig. 5.7-7(a). Let $v = 0.3$.   **Example 5.7-2**

Solution:

The plate equations can be developed by superposing the plates for cases 6 and 8 (Table 5.7-2) as shown in Fig. 5.7-7(b). For the plate of case 6 we set $M_1 = 0$ and we will determine M_2 by setting $dw/dr = 0$ at $r = a$ for the superposed plates. As an

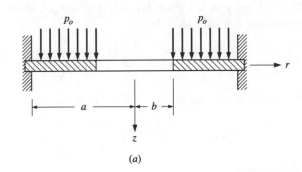

(a)

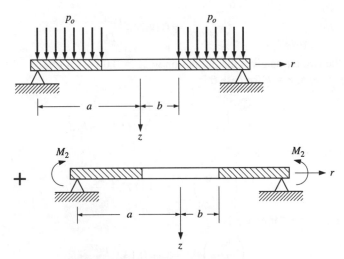

(b) Superposition of cases 6 and 8 from Table 5.7-2

Figure 5.7-7

example, let $b/a = 0.5$. Substituting $r = a$ into the slope equation for case 6 and simplifying gives

$$\left(\frac{dw_{M_2}}{dr}\right)_{r=a} = -\left\{\frac{(1-v) + (1+v)(b/a)^2}{(1-v^2)[1-(b/a)^2]}\right\}\frac{aM_2}{D}$$

For case 8,

$$\left(\frac{dw_{p_o}}{dr}\right)_{r=a} = -\frac{p_o a^3}{8D}\left[\frac{1}{1+v} + \frac{1+3v}{1-v^2}\left(\frac{b}{a}\right)^2 + \frac{4}{1-v}\frac{(b/a)^4}{1-(b/a)^2}\ln\frac{b}{a}\right]$$

Setting the sum of the two terms to zero and solving for M_2 yields

$$M_2 = -\frac{p_o a^2}{8}\left\{\left[1-\left(\frac{b}{a}\right)^2\right]\left[1 - v + (1+3v)\left(\frac{b}{a}\right)^2\right] + 4(1+v)\left(\frac{b}{a}\right)^4\ln\frac{b}{a}\right\}$$

Substituting $b/a = 0.5$ and $v = 0.3$ gives $M_2 = -0.0800\,p_o a^2$. From Table 5.7-2, case 6, the deflection at $r = b$ with $M_1 = 0$ is

$$(w_{M_2})_{r=b} = \frac{M_2 a^2}{2(1-v^2)D}\left[(1-v) - 2\frac{(1+v)(b/a)^2\ln(b/a)}{1-(b/a)^2}\right]$$

Substituting $b/a = 0.5$, $v = 0.3$, and $M_2 = -0.800\,p_o a^2$ gives

$$(w_{M_2})_{r=b} = -0.05717\frac{p_o a^4}{D}$$

For case 8, the deflection is

$$(w_{p_o})_{r=b} = 0.06244\frac{p_o a^4}{D}$$

Adding the two terms gives the deflection $w_{r=b} = 0.00527\,p_o a^4/D$.
    The slope at $r = b$ for case 6 is

$$\left(\frac{dw_{M_2}}{dr}\right)_{r=b} = -\frac{2M_2 a}{(1-v^2)D}\frac{b/a}{1-(b/a)^2}$$

Substituting values gives

$$\left(\frac{dw_{M_2}}{dr}\right)_{r=b} = 0.1172\frac{p_o a^3}{D}$$

The slope at $r = b$ for case 8 is

$$\left(\frac{dw_{p_o}}{dr}\right)_{r=b} = -0.1321\frac{p_o a^3}{D}$$

Adding the two results gives $(dw/dr)_{r=b} = -0.0149\,p_o a^3/D$.

The tangential moment at $r = b$ for case 6 is

$$(M_{\theta, M_2})_{r=b} = \frac{2M_2}{1 - (b/a)^2}$$

Substituting values gives

$$(M_{\theta, M_2})_{r=b} = -0.2133 p_o a^2$$

For case 8, at $r = b$, $M_{\theta, p_o} = 0.2404\, p_o a^2$. Summing the two terms gives $(M_\theta)_{r=b} = 0.0271\, p_o a^2$.

The radial moment at the outer edge is $M_r = M_2 = -0.0800 p_o a^2$.

All of this can be automated using a spreadsheet. The final results are given in Table 5.7-3.

Table 5.7-3

$w_{max} = k_1 \dfrac{p_o a^4}{D}$	b/a	k_1	k_2	k_3	k_4
	0.1	0.01661	−0.01592	0.1448	−0.1246
$\left(\dfrac{dw}{dr}\right)_{max} = k_2 \dfrac{p_o a^3}{D}$	0.2	0.01605	−0.0246	0.1121	−0.1216
	0.3	0.01316	−0.0256	0.0778	−0.1135
	0.4	0.00911	−0.0214	0.0486	−0.0993
	0.5	0.00527	−0.01490	0.0271	−0.0800
	0.6	0.00247	−0.00866	0.01313	−0.0580
$(M_\theta)_{r=b} = k_3 p_o a^2$	0.7	0.000860	−0.00398	0.00517	−0.0361
	0.8	0.0001825	−0.001250	0.001422	−0.01750
	0.9	0.0000120	−0.000162	0.000164	−0.00470
$(M_r)_{r=a} = k_4 p_o a^2$					

Show how case 8 of Table 5.7-2 can be generated from the other cases given.

Example 5.7-3

Solution:

If we remove a disk of radius b from the uniformly loaded circular plate (case 5) shown in Fig. 5.7-8(a), we obtain the annular plate shown in the top of Fig. 5.7-8(b). The shear force and radial moment at $r = b$ for case 5 are

$$V = -\frac{p_o b}{2} \qquad\qquad\qquad\qquad [a]$$

$$M = -\frac{p_o}{16}(3 + v)(a^2 - b^2) \qquad\qquad\qquad [b]$$

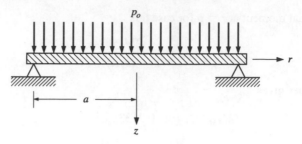

(a) Case 5 of Table 5.7-2

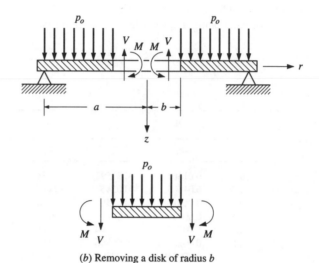

(b) Removing a disk of radius b

Figure 5.7-8

To eliminate the V and M on the annular plate we superpose cases 6 and 7 with $M_1 = M$, $M_2 = 0$, and $V_1 = V$. For example, the deflection for case 8 is

$$w = \frac{p_o}{64D}(a^2 - r^2)\left(\frac{5 + v}{1 + v}a^2 - r^2\right)$$

$$+ \frac{Vb}{8D}\left\{(a^2 - r^2)\left(\frac{3 + v}{1 + v} - \frac{2b^2}{a^2 - b^2}\ln\frac{b}{a}\right)\right.$$

$$\left. + 2\left[r^2 + \frac{2(1 + v)}{1 - v}\frac{a^2b^2}{a^2 - b^2}\ln\frac{b}{a}\right]\ln\frac{r}{a}\right\}$$

$$+ \frac{(1 - v)(-b^2M)(a^2 - r^2) - 2(1 + v)a^2b^2(-M)\ln(r/a)}{2(1 - v^2)(a^2 - b^2)D}$$

Substituting V and M from Eqs. (a) and (b) and simplifying gives

$$w = \frac{p_o}{64D}(a^2 - r^2)\left(\frac{5 + v}{1 + v}a^2 - 2\frac{3 + v}{1 + v}b^2 - r^2 + 8\frac{b^4}{a^2 - b^2}\ln\frac{b}{a}\right)$$

$$- \frac{p_o}{16D}\left(2b^2r^2 + \frac{3 + v}{1 - v}a^2b^2 + 4\frac{1 + v}{1 - v}\frac{a^2b^4}{a^2 - b^2}\ln\frac{b}{a}\right)\ln\frac{r}{a}$$

Thus case 8 is arrived at by the superposition of cases 5, 6, and 7 with $a \geq r \geq b$, as shown in Fig. 5.7-9 with the values of V and M given by Eqs. (a) and (b).

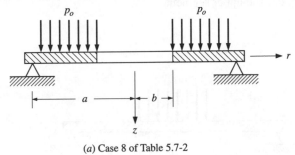

(a) Case 8 of Table 5.7-2

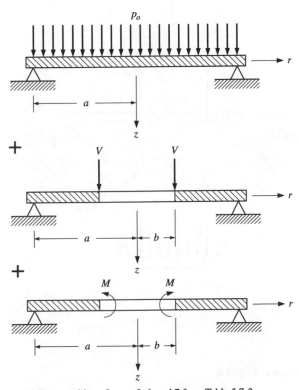

(b) Superposition of cases 5, 6, and 7 from Table 5.7-2
with V and M given by Eqs. (a) and (b), respectively

Figure 5.7-9

Example 5.7-4 Using superposition, determine the deflection equation for the beam with the partial uniform load shown in Fig. 5.7-10(a).

Solution:

The plate can be constructed by superposing the plates shown in Fig. 5.7-10(b). The upper plate can be obtained by superposing case 7 and case 6 with $M_1 = M$ and $M_2 = 0$. The lower plate can be obtained using cases 1 and 5 (substituting b in the place of a).

To determine M, the slope at $r = b$ must be equal for both segments shown in Fig. 5.7-10(b). For the upper segment

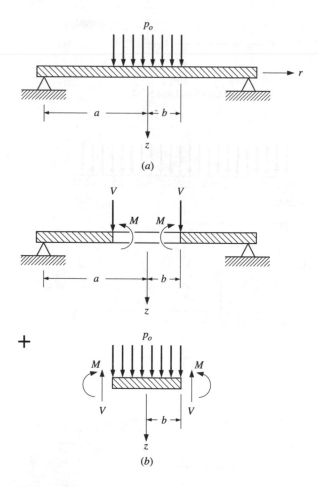

Figure 5.7-10

$$\left(\frac{dw}{dr}\right)_{r=b} = \frac{Vb}{2D}\left[\left(\ln\frac{b}{a} + \frac{b^2}{a^2 - b^2}\ln\frac{b}{a} - \frac{1}{1 + v}\right)b\right.$$

$$\left. + \left(\frac{1 + v}{1 - v}\frac{a^2 b^2}{a^2 - b^2}\ln\frac{b}{a}\right)\frac{1}{b}\right] + \left[\frac{(1 - v)b^2 + (1 + v)a^2}{(1 - v^2)(a^2 - b^2)}\right]\frac{Mb}{D} \quad \textbf{[a]}$$

From a force summation on the lower disk, $V(2\pi b) = p_o \pi b^2$. Thus

$$V = \frac{1}{2}p_o b \qquad \textbf{[b]}$$

Using cases 1 and 5 for the lower segment gives

$$\left(\frac{dw}{dr}\right)_{r=b} = -\frac{Mb}{(1 + v)D} - \frac{p_o b^3}{8(1 + v)D} \qquad \textbf{[c]}$$

Equating the two slopes using Eq. (b), solving for M, and simplifying yields

$$M = \frac{p_o b^2}{16a^2}\left[(1 - v)(a^2 - b^2) - 4(1 + v)a^2\ln\frac{b}{a}\right] \qquad \textbf{[d]}$$

With M known, we can determine w. For $a \geq r \geq b$ we use the equations for cases 6 and 7. Thus

$$w = \frac{Vb}{8D}\left\{(a^2 - r^2)\left(\frac{3 + v}{1 + v} - \frac{2b^2}{a^2 - b^2}\ln\frac{b}{a}\right)\right.$$

$$\left. + 2\left[r^2 + \frac{2(1 + v)}{1 - v}\frac{a^2 b^2}{a^2 - b^2}\ln\frac{b}{a}\right]\ln\frac{r}{a}\right\}$$

$$- \frac{(1 - v)(-b^2 M)(a^2 - r^2) - 2(1 + v)a^2 b^2(-M)\ln(r/a)}{2(1 - v^2)(a^2 - b^2)D}$$

Substituting in Eqs. (b) and (d) and simplifying gives

$$w = \frac{p_o b^2}{32D}\left\{\frac{a^2 - r^2}{1 + v}\left[2(3 + v) - (1 - v)\left(\frac{b}{a}\right)^2\right] + 2(2r^2 + b^2)\ln\frac{r}{a}\right\} \qquad \textbf{[e]}$$

The deflection at $r = b$ is

$$(w)_{r=b} = \frac{p_o b^2}{32D}\left\{\frac{a^2 - b^2}{1 + v}\left[2(3 + v) - (1 - v)\left(\frac{b}{a}\right)^2\right] + 6b^2\ln\frac{b}{a}\right\} \qquad \textbf{[f]}$$

For the inner disk of Fig. 5.7-10(b) we add Eq. (f) to the deflection of the inner disk relative to its outer edge. Case 1 of Table 5.7-2 gives the deflection relative to the center. Thus to obtain the deflection relative to the outer edge we determine the deflection of the outer edge from the table and then subtract the table equation for w. This is

$$w_{rel} = \frac{M}{2(1 + v)D}(b^2 - r^2) \qquad [g]$$

Thus, the deflection equation for $r \le b$ is found by superposing Eqs. (f) and (g) and case 5, resulting in

$$w = \frac{p_o b^2}{32D}\left\{\frac{a^2 - b^2}{1 + v}\left[2(3 + v) - (1 - v)\left(\frac{b}{a}\right)^2\right] + 6b^2 \ln\frac{b}{a}\right\}$$

$$+ \frac{M}{2(1 + v)D}(b^2 - r^2) + \frac{p_o}{64D}(b^2 - r^2)\left(\frac{5 + v}{1 + v}b^2 - r^2\right) \qquad [h]$$

Substituting Eq. (d) and simplifying yields

$$w = \frac{p_o b^2}{64(1 + v)D}\left[4(3 + v)a^2 - (7 + 3v)b^2 - 8r^2 + 2(1 - v)\left(\frac{b}{a}\right)^2 r^2\right]$$

$$+ \frac{p_o}{64D}\left[r^4 + 4b^2(2r^2 + b^2)\ln\frac{b}{a}\right] \qquad [i]$$

The maximum deflection occurs at $r = 0$ and is

$$w_{max} = \frac{p_o b^2}{64(1 + v)D}\left[4(3 + v)a^2 - (7 + 3v)b^2 + 4(1 + v)b^2 \ln\frac{b}{a}\right] \qquad [j]$$

The maximum moment also occurs at $r = 0$ and is determined by the superposition of cases 1 and 5, yielding

$$(M_r)_{max} = (M_r)_{r=0} = M + \frac{3 + v}{16}p_o b^2$$

Substituting Eq. (d) and simplifying results in

$$(M_r)_{max} = \frac{p_o}{16}\left(\frac{b}{a}\right)^2\left[4a^2 - (1 - v)b^2 - 4(1 + v)a^2\ln\frac{b}{a}\right] \qquad [k]$$

There are many additional variations that can be developed, some of which are problems given at the end of this chapter. Again, Ref. 5.3 contains tabulations of a great many configurations.

5.8   THICK-WALLED CYLINDERS AND ROTATING DISKS

Consider a cylinder axially symmetric about the z axis, as shown in Fig. 5.8-1(a). The cylinder could be pressurized at $r = r_i$ and/or $r = r_o$, and/or rotating about the z

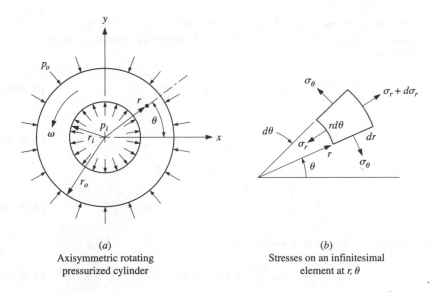

(a)
Axisymmetric rotating
pressurized cylinder

(b)
Stresses on an infinitesimal
element at r, θ

Figure 5.8-1

axis. These conditions cause internal stresses. Thanks to symmetry, however, the shear stresses $\tau_{r\theta}$ will not develop[†] with σ_r and σ_θ only being functions of r. Therefore, an element isolated at position (r, θ) will appear as shown in Fig. 5.8-1(b).

Consider the element with mass density ρ, volume $r\, d\theta\, dr\, dz$, and rotating at a constant angular velocity ω rad/s. The equation of motion in the radial direction is

$$\sum F_r = ma_r = (\rho r\, d\theta\, dr\, dz)(-r\omega^2) = -\rho\omega^2 r^2\, d\theta\, dr\, dz$$

The forces in the r direction are obtained from Fig. 5.8-1(b) by multiplying the stresses by their respective areas. Thus

$$\sum F_r = (\sigma_r + d\sigma_r)(r + dr)d\theta\, dz - \sigma_r r\, d\theta\, dz - 2\sigma_\theta\, dr\, dz\, \sin\frac{d\theta}{2}$$

Equate the two expressions for $\sum F_r$. Since $d\theta$ is infinitesimal, $\sin(d\theta/2) = d\theta/2$, and when the higher-order term $d\sigma_r\, dr\, d\theta\, dz$ is neglected, the result is

$$\sigma_r + r\frac{d\sigma_r}{dr} - \sigma_\theta = -\rho\omega^2 r^2 \qquad\qquad \textbf{[5.8-1]}$$

----

† See Sec. 1.5, $\gamma_{r\theta} = 0$.

Another equation in σ_r and σ_θ is in order. Assuming the disk thin in the z direction (plane stress), $\sigma_z = 0$, and the radial and tangential strains are given by Eqs. (1.5-5). Thus[†]

$$\varepsilon_r = \frac{du_r}{dr} = \frac{1}{E}(\sigma_r - v\sigma_\theta) \qquad \textbf{[5.8-2]}$$

$$\varepsilon_\theta = \frac{u_r}{r} = \frac{1}{E}(\sigma_\theta - v\sigma_r) \qquad \textbf{[5.8-3]}$$

Solving for u_r in Eq. (5.8-3) and differentiating yields

$$\frac{du_r}{dr} = \frac{1}{E}\left[r\frac{d\sigma_\theta}{dr} + \sigma_\theta - v\left(r\frac{d\sigma_r}{dr} + \sigma_r\right)\right] \qquad \textbf{[5.8-4]}$$

Equating Eqs. (5.8-2) and (5.8-4) results in

$$r\frac{d\sigma_\theta}{dr} - vr\frac{d\sigma_r}{dr} + (1 + v)(\sigma_\theta - \sigma_r) = 0 \qquad \textbf{[5.8-5]}$$

From Eq. (5.8-1)

$$\sigma_\theta = r\frac{d\sigma_r}{dr} + \sigma_r + \rho\omega^2 r^2 \qquad \textbf{[5.8-6]}$$

Differentiation with respect to r results in

$$\frac{d\sigma_\theta}{dr} = r\frac{d^2\sigma_r}{dr^2} + 2\frac{d\sigma_r}{dr} + 2\rho\omega^2 r \qquad \textbf{[5.8-7]}$$

Substituting Eqs. (5.8-6) and (5.8-7) into Eq. (5.8-5) yields

$$r\frac{d^2\sigma_r}{dr^2} + 3\frac{d\sigma_r}{dr} = -(3 + v)\rho\omega^2 r \qquad \textbf{[5.8-8]}$$

Case 1 ($\omega = 0$)   If the cylinder is pressurized, with $\omega = 0$, Eq. (5.8-8) reduces to

$$r\frac{d^2\sigma_r}{dr^2} + 3\frac{d\sigma_r}{dr} = 0 \qquad \textbf{[5.8-9]}$$

The form of the solution of Eq. (5.8-9) is $\sigma_r = Cr^n$. Substituting this into Eq. (5.8-9) yields $n = 0$ and $n = -2$. Thus the solution is

$$\sigma_r = C_1 + \frac{C_2}{r^2} \qquad \textbf{[5.8-10]}$$

[†] Since r is the only independent variable, the partial derivatives with respect to r can be replaced by ordinary derivatives.

Substituting this into Eq. (5.8-6) with $\omega = 0$ yields

$$\sigma_\theta = C_1 - \frac{C_2}{r^2} \qquad \textbf{[5.8-11]}$$

From Fig. 5.8-1(a) we have the boundary conditions

$$\sigma_r = \begin{cases} -p_i & \text{at } r = r_i \\ -p_o & \text{at } r = r_o \end{cases}$$

Substituting these into Eq. (5.8-10) results in

$$C_1 = \frac{p_i r_i^2 - p_o r_o^2}{r_o^2 - r_i^2} \qquad C_2 = \frac{(r_i r_o)^2 (p_o - p_i)}{r_o^2 - r_i^2}$$

Substituting these into Eqs. (5.8-10) and (5.8-11) produces the final equations for σ_r and σ_θ

$$\sigma_r = \frac{p_i r_i^2 - p_o r_o^2 + (r_i r_o/r)^2 (p_o - p_i)}{r_o^2 - r_i^2} \qquad \textbf{[5.8-12]}$$

$$\sigma_\theta = \frac{p_i r_i^2 - p_o r_o^2 - (r_i r_o/r)^2 (p_o - p_i)}{r_o^2 - r_i^2} \qquad \textbf{[5.8-13]}$$

Note that these equations agree with that requested in Prob. 4.7 using the stress function approach.

Determine the stress distribution in a cylinder with an inner diameter of 2.0 in and outer diameter of 6.0 in with $p_i = 5000$ psi and $p_o = 0$. | **Example 5.8-1**

Solution:

From Eq. (5.8-12)

$$\sigma_r = \frac{(5000)(1^2) + [(1)(3)/r]^2(-5000)}{3^2 - 1^2} = 625 - \frac{5625}{r^2} \qquad \textbf{[a]}$$

and from Eq. (5.8-13)

$$\sigma_\theta = 625 + \frac{5625}{r^2} \qquad \textbf{[b]}$$

The stress distributions are shown in Fig. 5.8-2.

For the case of a *solid disk*, where $r_i = 0$, it can be shown that Eqs. (5.8-12) and (5.8-13) reduce to

$$\sigma_r = -p_o \qquad \textbf{[5.8-14]}$$

and

$$\sigma_\theta = -p_o \qquad \textbf{[5.8-15]}$$

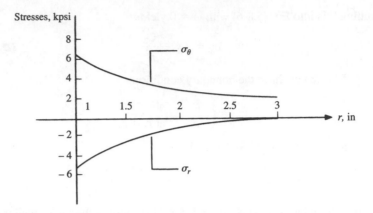

Figure 5.8-2

This means that the state of stress is constant throughout the disk.

If the cylinder is long in the z direction (plane strain), the stress equations do not change as they do not contain ν and E (see Sec. 4.1-3). If the cylinder has closed ends and is under the influence of internal or external pressures p_i and p_o, respectively, a longitudinal stress σ_z will develop.[†] Assuming that this stress is distributed uniformly and that p_o acts on the closed ends, we find from equilibrium in the z direction

$$\sigma_z = \frac{p_i r_i^2 - p_o r_o^2}{r_o^2 - r_i^2}$$ [5.8-16]

Press Fits   The interface pressure between two cylinders press-fitted together cannot be determined from equilibrium equations since the problem is statically indeterminate. Thus deflection analysis must be incorporated in the solution.

Consider two elements to be pressed together, as shown in Fig. 5.8-3, where $b_i > b_o$.

The radial displacement is u_r, which (assuming that $\sigma_z = 0$) is given by

$$u_r = \frac{r}{E}(\sigma_\theta - \nu\sigma_r)$$ [5.8-17]

Let the radial deflection u_r for the inner member at $r = b_i$ be $(u_r)_{b_i}$ and for the outer member at $r = b_o$, be $(u_r)_{b_o}$. The reader should verify that

$$(u_r)_{b_o} - (u_r)_{b_i} = b_i - b_o = \delta$$ [5.8-18]

[†] Up to this point of the analysis, the use of Hooke's law in the determination of σ_r and σ_θ has been on the basis that $\sigma_z = 0$. However, the reader should verify that the inclusion of σ_z does not influence the analysis provided σ_z is constant.

where δ is the *radial* interference. When the two parts are mated, a pressure p develops at their interface. The pressure p corresponds to p_o for the inner element and p_i for the outer element. Thus, for the inner element at $r = b_i$, Eq. (5.8-13) yields

$$(\sigma_\theta)_{b_i} = -p\frac{b_i^2 + a^2}{b_i^2 - a^2}$$ **[5.8-19]**

whereas for the outer element at $r = b_o$

$$(\sigma_\theta)_{b_o} = p\frac{c^2 + b_o^2}{c^2 - b_o^2}$$ **[5.8-20]**

Since b_i and b_o are almost equal, let $b_o = b_i = b$ [except in Eq. (5.8-18)]. Then Eqs. (5.8-19) and (5.8-20) can be rewritten as

$$(\sigma_\theta)_{b_i} = -p\frac{b^2 + a^2}{b^2 - a^2}$$ **[5.8-21]**

and

$$(\sigma_\theta)_{b_o} = p\frac{c^2 + b^2}{c^2 - b^2}$$ **[5.8-22]**

The radial stress at the interface of each cylinder is simply the negative of the interface pressure, namely

$$(\sigma_r)_{b_i} = (\sigma_r)_{b_o} = -p$$ **[5.8-23]**

Substituting Eqs. (5.8-21) and (5.8-23) into Eq. (5.8-17) yields

$$(u_r)_{b_i} = \frac{-b}{E_i}p\left(\frac{b^2 + a^2}{b^2 - a^2} - \nu_i\right)$$ **[5.8-24]**

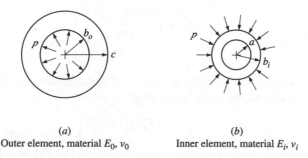

(a)
Outer element, material E_0, ν_0

(b)
Inner element, material E_i, ν_i

Figure 5.8-3   Press fit with a radial interference $\delta = b_i - b_o$.

Likewise, substituting Eqs. (5.8-22) and (5.8-23) into Eq. (5.8-17) yields

$$(u_r)_{b_o} = \frac{b}{E_o} p \left(\frac{c^2 + b^2}{c^2 - b^2} + v_o \right)$$ [5.8-25]

Since $(u_r)_{b_o} - (u_r)_{b_i} = \delta$, subtracting Eq. (5.8-24) from Eq. (5.8-25), setting the result equal to δ, and solving for p yields

$$p = \frac{\delta}{b \left[\dfrac{1}{E_o} \left(\dfrac{c^2 + b^2}{c^2 - b^2} + v_o \right) + \dfrac{1}{E_i} \left(\dfrac{b^2 + a^2}{b^2 - a^2} - v_i \right) \right]}$$ [5.8-26]

Note that diametral terms can be used in Eq. (5.8-26). That is, the terms a, b, and c can be diameters, with δ being the *diametral interference*.

Example 5.8-2 | An aluminum cylinder ($E_a = 70$ MPa, $v_a = 0.33$) with an outer diameter of 150 mm and inner diameter of 100 mm is to be press-fitted over a stainless-steel cylinder ($E_s = 190$ MPa, $v_s = 0.30$) with an outer diameter of 100.20 mm and inner diameter of 50 mm. Determine (*a*) the interface pressure p, and (*b*) the maximum stresses in the cylinders.

Solution:

(*a*) The radial terms in Eq. (5.8-26) can be replaced by diametral terms (*including* δ). Thus

$$p = \frac{100.20 - 100}{100 \left[\dfrac{1}{70(10^9)} \left(\dfrac{150^2 + 100^2}{150^2 - 100^2} + 0.33 \right) + \dfrac{1}{190(10^9)} \left(\dfrac{100^2 + 50^2}{100^2 - 50^2} - 0.30 \right) \right]}$$

$$= 40.78 \times 10^6 \ \text{N/m}^2 = 40.78 \ \text{MPa}$$

(*b*) Calculations for stress can also be made using diametral terms. For the outer cylinder with $p_i = p$ and $p_o = 0$, the maximum tangential stress occurs at the inner radius. From Eq. (5.8-13)

$$(\sigma)_{\text{max,outer}} = (\sigma_\theta)_{r=100/2} = \left\{ \frac{100^2 + \left[\dfrac{(100)(150)}{100} \right]^2}{150^2 - 100^2} \right\} (40.78) = 106.0 \ \text{MPa}$$

For the inner cylinder with $p_i = 0$ and $p_o = p$, the maximum tangential stress is compressive and also occurs at the inner radius

$$(\sigma)_{\text{max,inner}} = (\sigma_\theta)_{r=50/2} = - \left\{ \frac{100^2 + \left[\dfrac{(50)(100)}{50} \right]^2}{100^2 - 50^2} \right\} (40.78) = -108.7 \ \text{MPa}$$

Case 2 ($\omega \neq 0, p_i = p_o = 0$)   The solution to Eq. (5.8-8) is

$$\sigma_r = C_1 + \frac{C_2}{r^2} - \frac{3 + v}{8} \rho \omega^2 r^2 \qquad \text{[5.8-27]}$$

substituting this into Eq. (5.8-6) yields

$$\sigma_\theta = C_1 - \frac{C_2}{r^2} - \frac{1 + 3v}{8} \rho \omega^2 r^2 \qquad \text{[5.8-28]}$$

Applying the conditions that $\sigma_r = 0$ at $r = r_i$ and $r = r_o$ yields

$$\sigma_r = \frac{3 + v}{8} \rho \omega^2 \left[r_i^2 + r_o^2 - \left(\frac{r_i r_o}{r} \right)^2 - r^2 \right] \qquad \text{[5.8-29]}$$

$$\sigma_\theta = \frac{3 + v}{8} \rho \omega^2 \left[r_i^2 + r_o^2 + \left(\frac{r_i r_o}{r} \right)^2 - \frac{1 + 3v}{3 + v} r^2 \right] \qquad \text{[5.8-30]}$$

The maximum radial stress $(\sigma_r)_{\max}$ occurs at $r = \sqrt{r_i r_o}$, where the value is

$$(\sigma_r)_{\max} = \frac{3 + v}{8} \rho \omega^2 (r_o - r_i)^2 \qquad \text{[5.8-31]}$$

The maximum tangential stress $(\sigma_\theta)_{\max}$ occurs at the inner surface, where $r = r_i$, and is

$$(\sigma_\theta)_{\max} = \frac{\rho \omega^2}{4} \left[(3 + v) r_o^2 + (1 - v) r_i^2 \right] \qquad \text{[5.8-32]}$$

If the disk is solid, $r_i = 0$ and Eqs. (5.8-29) and (5.8-30) reduce to

$$(\sigma_r) = \frac{3 + v}{8} \rho \omega^2 (r_o^2 - r^2) \qquad \text{[5.8-33]}$$

$$\sigma_\theta = \frac{\rho \omega^2}{8} \left[(3 + v) r_o^2 - (1 + 3v) r^2 \right] \qquad \text{[5.8-34]}$$

Case 3 ($\omega \neq 0, p_i \neq 0, p_o \neq 0$)   There are examples, e.g., rotating disks press-fitted onto shafts, where the results of cases 1 and 2 can be combined by superposition of Eqs. (5.8-12) and (5.8-13) with Eqs. (5.8-29) and (5.8-30), respectively.

A disk with $r_o = 6.0$ in and $r_i = 1.000$ in is press-fitted onto a shaft of radius 1.003 in. Both members are steel with $E = 30$ Mpsi, $v = 0.29$, and with weight densities of 0.284 lb/in³. Determine the stress distribution of the disk at $n = 5000$ rpm. | **Example 5.8-3**

Solution:

First find the interference at 5000 rpm.

$$\omega = \frac{2\pi}{60} (5000) = 523.6 \text{ rad/s}$$

and the mass density is

$$\rho = \frac{0.284}{386.4} = 7.35 \times 10^{-4} \text{ lb·s}^2/\text{in}^4$$

Because of rotation, the stresses at the outer radius of the shaft are [from Eqs. (5.8-33) and (5.8-34)] $\sigma_r = 0$ and

$$\sigma_\theta = \frac{\rho\omega^2}{4}(1 - v)r_o^2 = \frac{(7.35 \times 10^{-4})(523.6)^2}{4}(1 - 0.29)(1.003)^2 = 36.0 \text{ psi}$$

The radial displacement, given by Eq. (5.8-3), is

$$u_r = \frac{r}{E}(\sigma_\theta - v\sigma_r) = \frac{1.003}{30 \times 10^6}(36.0) = 1.20 \times 10^{-6} \text{ in}$$

The radius of the shaft rotating alone is then approximately[†] $(r_o)_{\text{shaft}} = 1.003$ in.

Next, the same procedure is applied to the disk at its inner radius. Thus at $r = r_i$, $\sigma_r = 0$ and

$$\sigma_\theta = \frac{3 + v}{8}\rho\omega^2\left[r_i^2 + r_o^2 + \left(\frac{r_i r_o}{r_i}\right)^2 - \frac{1 + 3v}{3 + v}r_i^2\right]$$

$$= \frac{3 + 0.29}{8}(7.35 \times 10^{-4})(523.6)^2\left[1^2 + 6^2 + 6^2 - \frac{1 + (3)(0.29)}{3 + 0.29}(1)^2\right]$$

$$= 6000 \text{ psi}$$

The radial displacement at the inner radius is

$$u_r = r\frac{\sigma_\theta}{E} = (1)\frac{6000}{30 \times 10^6} = 2.00 \times 10^{-4} \text{ in}$$

Thus the inner radius of the disk with rotation is $(r_i)_{\text{disk}} = 1 + 2.00 \times 10^{-4} = 1.00020$ in. Hence the radial interference at $n = 500$ rpm is $\delta = 1.003 - 1.00020 = 0.00280$ in. The interference pressure at 5000 rpm is found using Eq. (5.8-26). Thus at 5000 rpm the pressure is

$$p = \frac{0.00280}{1.0\left[\dfrac{1}{30 \times 10^6}\left(\dfrac{6^2 + 1^2}{6^2 - 1^2} + 0.29\right) + \dfrac{1}{30 \times 10^6}\left(\dfrac{1^2 + 0^2}{1^2 - 0^2} - 0.29\right)\right]}$$

$$= 40{,}830 \text{ psi}$$

[†] The assumption of a thin disk is actually being violated, but the error here is negligible since the stresses in the shaft are small and will be localized in a zone comparable to the width of the disk.

With the interference pressure, Eqs. (5.8-12) and (5.8-13) can be used to find the stresses in the disk due to the interference at 5000 rpm. Using $p_i = p$ and $p_o = 0$, we find

$$\sigma_r = \frac{40,830}{6^2 - 1^2}\left[1^2 - \left(\frac{6}{r}\right)^2\right] = 1166 - \frac{42,000}{r^2} \qquad [a]$$

and

$$\sigma_\theta = 1166 + \frac{42,000}{r^2} \qquad [b]$$

The stresses in the disk due to rotation are given by Eqs. (5.8-29) and (5.8-30) and are

$$\sigma_r = \frac{3 + 0.29}{8}(7.35 \times 10^{-4})(523.6)^2\left[1^2 + 6^2 - \left(\frac{6}{r}\right)^2 - r^2\right] \qquad [c]$$

$$= 3066 - \frac{2983}{r^2} - 82.9r^2$$

and

$$\sigma_\theta = 3066 + \frac{2983}{r^2} - 47.1r^2 \qquad [d]$$

Finally, the stresses of Eqs. (a) and (c), and (b) and (d) are combined, resulting in

$$\sigma_r = 4232 - \frac{39,017}{r^2} - 82.9r^2$$

$$\sigma_\theta = 4232 + \frac{39,017}{r^2} - 47.1r^2$$

5.9   CONTACT STRESSES

5.9.1   DISTRIBUTED CONTACT LOADING

When a roller or sphere is in line or point contact with another body, the stresses cannot be determined without considering deflection. In most analyses, deflection does not appreciably alter the results of the analysis of stress. For example, when a beam in bending is considered, normally the force analysis and subsequent stress analysis are made on the model neglecting the fact that the beam deflects. Consider a load P to be transmitted from the roller in Fig. 5.9-1 to the flat surface. If deflections are ignored, the transmitted force area is zero and the stresses would be

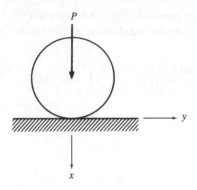

Figure 5.9-1

infinite. Hence the conclusion would be that no force can be transmitted. This is obviously erroneous, which indicates that the deflections of the members must be taken into account.

Contact-stress problems occur often in design where forces are transmitted through contact from one machine element to another, e.g., gearing, roller bearings, cams, and pin joints in linkages [see Fig. 3.2-2(b)]. There are two major types of contact problems, line contact and point contact. Since the analysis of these problems is rather complex, this section gives only a basic introduction to the topic.[†]

Consider a contact pressure p acting over an infinitesimal surface area $t\,dy_1$, as shown in Fig. 5.9-2(a). Isolating a surface at a distance r, as shown in Fig. 5.9-2(b), the radial stress is given by Eq. (4.2-30) with the net force $P = pt\,dy_1$. This is

$$\sigma_r = -\frac{2}{\pi r}\,p\,dy_1\cos\theta \qquad\qquad \textbf{[5.9-1]}$$

The stresses with respect to the xy coordinate system are found using the transformation equations developed in Chap. 2. Thus the radial stress transforms to

$$\sigma_x = \sigma_r\cos^2\theta = -\frac{2}{\pi r}\,p\,dy_1\cos^3\theta \qquad\qquad \textbf{[5.9-2a]}$$

$$\sigma_y = \sigma_r\sin^2\theta = -\frac{2}{\pi r}\,p\,dy_1\sin^2\theta\cos\theta \qquad\qquad \textbf{[5.9-2b]}$$

$$\tau_{xy} = \sigma_r\sin\theta\cos\theta = -\frac{2}{\pi r}\,p\,dy_1\sin\theta\cos^2\theta \qquad\qquad \textbf{[5.9-2c]}$$

[†] See Ref. 5.3, pp. 647–65; Ref. 5.5, pp. 409–20; and Ref. 5.10, pp. 403–37.

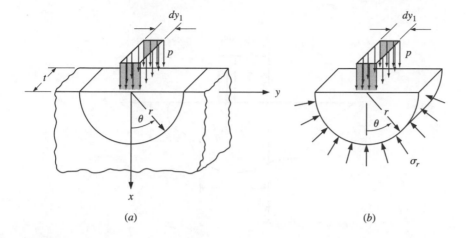

Figure 5.9-2

Using xy coordinates in place of $r\theta$, we can transform Eqs. (5.9-2) using the relations

$$r = \sqrt{x^2 + y^2} \qquad \cos\theta = \frac{x}{\sqrt{x^2 + y^2}} \qquad \sin\theta = \frac{y}{\sqrt{x^2 + y^2}}$$

Thus

$$\sigma_x = -\frac{2}{\pi}\, p\, dy_1\, \frac{x^3}{x^2 + y^2} \qquad\qquad \textbf{[5.9-3a]}$$

$$\sigma_y = -\frac{2}{\pi}\, p\, dy_1\, \frac{xy^2}{x^2 + y^2} \qquad\qquad \textbf{[5.9-3b]}$$

$$\tau_{xy} = -\frac{2}{\pi}\, p\, dy_1\, \frac{x^2 y}{x^2 + y^2} \qquad\qquad \textbf{[5.9-3c]}$$

In order to be able to deal with an arbitrary load distribution across the contact surface, consider the load p to vary as a function of y_1, $p(y_1)$. The force $p(y_1)t\, dy_1$ is shown in Fig. 5.9-3.

The stresses in this case are found simply by substituting $y - y_1$ for y in Eqs. (5.9-3). Thus

$$\sigma_x = -\frac{2}{\pi}\, p\, dy_1\, \frac{x^3}{[x^2 + (y - y_1)^2]^2} \qquad\qquad \textbf{[5.9-4a]}$$

$$\sigma_y = -\frac{2}{\pi}\, p\, dy_1\, \frac{x(y - y_1)^2}{[x^2 + (y - y_1)^2]^2} \qquad\qquad \textbf{[5.9-4b]}$$

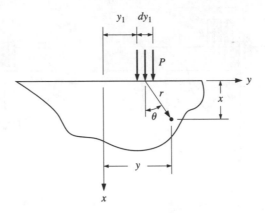

Figure 5.9-3

$$\tau_{xy} = -\frac{2}{\pi} p\, dy_1 \frac{x^2(y - y_1)}{[x^2 + (y - y_1)^2]^2} \qquad \textbf{[5.9-4c]}$$

If the load distribution is known, integration of Eqs. (5.9-4) with respect to y_1 will establish the state of stress at any point (x, y). The integration, however, is not always so simple, and only a uniform load distribution will be given here.

Example 5.9-1 Determine the stress distribution for a constant pressure distribution of $p(y_1) = p_o$ in the region $-a \le y \le a$ as shown in Fig. 5.9-4.

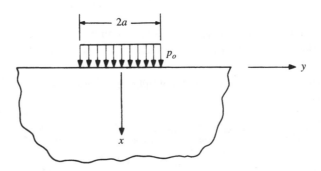

Figure 5.9-4

Solution:

First, from Eq. (5.9-4a)[†]

$$\sigma_x = -\frac{2}{\pi} p_o x^3 \int_{-a}^{a} \frac{dy_1}{[x^2 + (y - y_1)^2]^2}$$

let $u = y - y_1$; thus $du = -dy_1$ and

$$\sigma_x = \frac{2}{\pi} p_o x^3 \int_{y+a}^{y-a} \frac{du}{(x^2 + u^2)^2}$$

Integration yields

$$\sigma_x = \frac{2}{\pi} p_o x \left[\frac{u}{2(x^2 + u^2)} + \frac{1}{2x} \tan^{-1} \frac{u}{x} \right]_{y+a}^{y-a}$$

Substituting the limits of integration and simplifying yields

$$\sigma_x = \frac{p_o}{\pi} \left\{ \frac{2ax(y^2 - x^2 - a^2)}{[x^2 + (y - a)^2][x^2 + (y + a)^2]} + \tan^{-1}\left(\frac{y - a}{x}\right) - \tan^{-1}\left(\frac{y + a}{x}\right) \right\}$$

in a similar fashion,

$$\sigma_y = \frac{p_o}{\pi} \left\{ \frac{2ax(x^2 + a^2 - y^2)}{[x^2 + (y - a)^2][x^2 + (y + a)^2]} + \tan^{-1}\left(\frac{y - a}{x}\right) - \tan^{-1}\left(\frac{y + a}{x}\right) \right\}$$

and

$$\tau_{xy} = -\frac{4p_o}{\pi} \frac{ax^2 y}{[x^2 + (y - a)^2][x^2 + (y + a)^2]}$$

5.9.2 CONTACT BETWEEN CURVED SURFACES[‡]

There are many practical engineering problems that fit into this category, gears, cams, ball and roller bearings, etc. Whenever two bodies having curved surfaces are forced together, the initial point or line contact grows to an elliptical area as shown in Fig. 5.9-5. The stresses that develop, called Hertzian stresses,[§] are three-dimensional. The development of the governing equations for the many possible cases is beyond the scope of this book. However, in this section, the equations for simple roller and spherical contact are given. See Ref. 5.3 for tabulations of other cases.

Figure 5.9-6(a) shows two spherical or cylindrical bodies initially in point or line contact, respectively. With the application of force, the bodies deform as shown in Fig. 5.9-6(b).

[†] Note, relative to the integration x and y are fixed points and integration is with respect to y_1 only.
[‡] See Refs. 5.4–5.7.
[§] Named after H. Hertz, whose work was originally published in 1881.

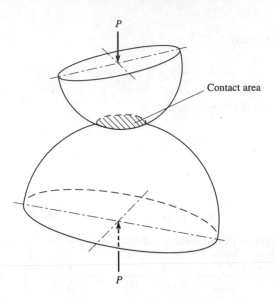

Figure 5.9-5   Contact between curved surfaces.

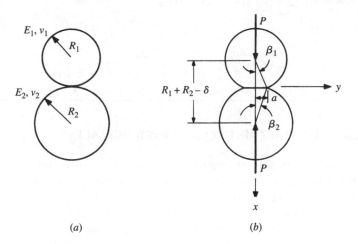

(a)   (b)

Figure 5.9-6

From geometry, $\sin \beta_1 = a/R_1$, and $\sin \beta_2 = a/R_2$. If β_1 and β_2 are considered to be small, $\sin \beta_1 \approx \beta_1$ and $\sin \beta_2 \approx \beta_2$. Thus

$$\beta_1 \approx \frac{a}{R_1} \qquad \beta_2 \approx \frac{a}{R_2} \qquad\qquad \textbf{[5.9-5]}$$

After the load P is applied, the distance between centers is $R_1 + R_2 - \delta$, which can be written as

$$R_1 + R_2 - \delta = R_1 \cos \beta_1 + R_2 \cos \beta_2 \qquad \textbf{[5.9-6]}$$

Again, if β_1 and β_2 are small

$$\cos \beta_1 \approx 1 - \frac{\beta_1^2}{2} \qquad \cos \beta_2 \approx 1 - \frac{\beta_2^2}{2} \qquad \textbf{[5.9-7]}$$

Substituting Eqs. (5.9-5) into (5.9-7), and the result into Eq. (5.9-6) gives

$$\delta \approx \frac{a^2}{2}\left(\frac{1}{R_1} + \frac{1}{R_2}\right) \qquad \textbf{[5.9-8]}$$

Spherical Contact   The equations for the radius of the contact area, maximum contact pressure, and principal stresses along the x axis are

$$a = \sqrt[3]{\frac{3}{4}P\frac{R_1 R_2[(1 - v_2^2)E_1 + (1 - v_1^2)E_2]}{E_1 E_2 (R_1 + R_2)}} \qquad \textbf{[5.9-9]}$$

$$a = 1.109 \sqrt[3]{\frac{P}{E}\frac{R_1 R_2}{(R_1 + R_2)}} \qquad \text{for} \quad v = 0.3 \qquad E_1 = E_2 = E \qquad \textbf{[5.9-10]}$$

$$p_{max} = \frac{3P}{2\pi a^2} \qquad \textbf{[5.9-11]}$$

$$\sigma_x = -\frac{p_{max}}{1 + \zeta^2} \qquad \textbf{[5.9-12]}$$

$$\sigma_y = \sigma_z = -p_{max}\left[(1 + v)\left(1 - \zeta \tan^{-1}\frac{1}{\zeta}\right) - \frac{1}{2(1 + \zeta^2)}\right] \qquad \textbf{[5.9-13]}$$

where $\zeta = x/a$.

If one surface is initially flat, divide the numerator and denominator within the radicals of Eq. (5.9-9) or (5.9-10) by the radius of that element and then allow the radius to approach ∞. If one surface is concave, input the radius of the element as a negative quantity.

The maximum shear stress is given by $\tau_{max} = (\sigma_y - \sigma_x)/2$. Substituting Eqs. (5.9-12) and (5.9-13) and taking the derivative with respect to x reveals that the maximum shear stress is $\tau_{max} = 0.310 p_{max}$ and occurs at $x = 0.481a$ for $v = 0.3$.

For ductile materials, the maximum von Mises stress[†] is more typically used to compare with the tensile yield strength for failure criteria. This is important in

---

[†] The von Mises stress is discussed in more detail in Chap. 7.

fatigue analysis, as most contact problems involve cyclic loading. The von Mises stress is given by Eq. (7.3-12), which is

$$\sigma_{vM} = \sqrt{\frac{1}{2}\left[(\sigma_1 - \sigma_2)^2 + (\sigma_2 - \sigma_3)^2 + (\sigma_3 - \sigma_1)^2\right]} \qquad \textbf{[5.9-14]}$$

For spheres in contact, the maximum von Mises stress also occurs at $x = 0.481a$ for $v = 0.3$ and is $(\sigma_{vM})_{max} = 0.620 p_{max}$.

Cylinders in Contact　　The equations for the radius of the contact area, maximum contact pressure, and principal stresses along the x axis are

$$a = 2\sqrt{\frac{P}{\pi L}\frac{R_1 R_2[(1 - v_2^2)E_1 + (1 - v_1^2)E_2]}{E_1 E_2 (R_1 + R_2)}} \qquad \textbf{[5.9-15]}$$

$$a = 1.522\sqrt{\frac{P}{LE}\frac{R_1 R_2}{R_1 + R_2}} \qquad v = 0.3 \qquad E_1 = E_2 = E \qquad \textbf{[5.9-16]}$$

$$p_{max} = \frac{2P}{\pi a L} \qquad \textbf{[5.9-17]}$$

$$\sigma_x = -\frac{p_{max}}{\sqrt{1 + \zeta^2}} \qquad \textbf{[5.9-18]}$$

$$\sigma_y = -p_{max}\left(\frac{1 + 2\zeta^2}{\sqrt{1 + \zeta^2}} - 2\zeta\right) \qquad \textbf{[5.9-19]}$$

$$\sigma_z = -2vp_{max}(\sqrt{1 + \zeta^2} - \zeta) \qquad \textbf{[5.9-20]}$$

where a and L are the half-width and length of the contact zone, respectively, and $\zeta = x/a$.

The maximum shear stress for $v = 0.3$ is $\tau_{max} = 0.300 p_{max}$ and occurs at $x = 0.786a$. The maximum von Mises stress for $v = 0.3$ is $\sigma_{vM} = 0.558 p_{max}$ and occurs at $x = 0.704\,a$.

5.10　Stress Concentrations

When a large stress gradient occurs in a small, localized area of a structure, the high stress is referred to as a *stress concentration*. This topic is generally discussed in an elementary mechanics of materials course; however, the coverage is typically brief. Several instances of stress concentrations have already been discussed in this book. In Sec. 4.2-4, a wide plate in tension containing a centrally located circular hole is analyzed. Figure 4.2-5 shows the stress distribution where at the edge of the hole the stress gradient is very large and the normal stress reaches a magnitude of *three times* the nominal (average) stress. Contact stresses, as discussed in the previous section of this chapter, also exhibit high stress gradients near the point of contact, which subside quickly as one moves away from the contact area. These are the two most common occurrences of stress concentrations, that is, (1) discontinuities in continuum and (2) contact forces. Discontinuities in continuum include changes in geometry and material properties.

Rapid geometry changes disrupt the smooth flow of stresses through the structure between load application areas. Plates in tension or bending with holes, notches, steps, etc., are simple examples involving direct normal stresses. Shafts in tension, bending and torsion with holes, notches, steps, keyways, etc., are simple examples involving direct and bending normal stresses and torsional shear stresses. Many of these configurations are tabulated or represented by graphs as in Refs. 5.3, 5.8, and 5.9. Other, less obvious geometry changes include rough surface finishes and external and internal cracks.

Changes in material properties were discussed in Secs. 3.1 and 3.2.1 and were demonstrated in Example 3.2-1, where a change in modulus of elasticity drastically changed the stress distribution. Changes in material properties can occur at both macroscopic and microscopic levels, which include alloy formulation, grain size and orientation, foreign materials, etc.

The analysis of the plate in tension with a hole, given in Sec. 4.2.4, is for a very wide plate (infinite in the limit). As the width of the plate decreases, the maximum stress becomes less than three times the nominal stress. Figure 5.10-1(a) shows a plate of thickness $t = 0.125$ in, width $w = 1.50$ in, with a hole of diameter $d = 0.50$ in and an applied uniform stress of $\sigma_o = 320$ psi. A photoelastic[†] model is shown in Fig. 5.10-1(b). From the photoelastic analysis, the stresses at points a, b, and c were found to be

Zone A-A: $\sigma_a = 320$ psi

Zone B-B: $\sigma_b = 280$ psi $\quad \sigma_c = 1130$ psi

If the stress were uniform from b to c, the stress would be 480 psi. However, since the actual stress is much greater at the edge of the hole, and the integral of the stress over the area is 60 lb, the stress at b will be less than 480 psi. The maximum stress is often expressed in terms of the nominal stress using a term called the *static stress concentration factor* K_t, by the equation

$$\sigma_{\max} = K_t \sigma_{\text{nom}} \qquad \text{[5.10-1]}$$

Thus, for this example,

$$K_t = \frac{\sigma_{\max}}{\sigma_{\min}} = \frac{1130}{480} = 2.35$$

[†] Photoelasticity is discussed at some length in Chap. 8.

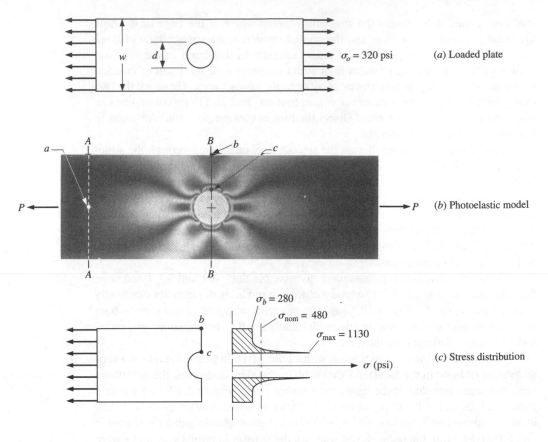

Figure 5.10-1   Stress distribution for a plate in tension containing a centrally located hole.

The static stress concentration factor for a plate containing a centrally located hole in which the plate is loaded in tension depends on the ratio of d/w as shown in Fig. 5.10-2 (see Ref. 5.8 or 5.9, pt. 1). Other charts are available for various geometric and loading configurations. Appendix F gives a few charts that apply to the fundamental forms of geometry and loading conditions.

Consider a part made of a ductile material and loaded by a gradually applied static load such that the stress in an area of a stress concentration goes beyond the yield strength. The yielding will be restricted to a very small region, and the permanent deformation as well as the residual stresses after the load is released will be insignificant and normally can be tolerated. If yielding does occur, the stress distribution changes and tends toward a more uniform distribution (see Sec. 7.7). In the area where yielding occurs, there is little danger of fracture of a ductile material, but if

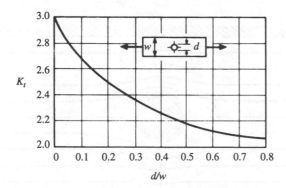

Figure 5.10-2   Plate in tension with a centrally located hole.

SOURCE: See R. E. Peterson, "Design Factors for Stress Concentration," *Machine Design*, a Penton publication, vol. 23, no. 7, July 1951, p. 155; reproduced with the permission of the author and publisher.

the possibility of a brittle fracture exists, the stress concentration must be taken seriously. Brittle fracture is not just limited to brittle materials. Materials often thought of as being ductile can fail in a brittle manner under certain conditions, e.g., any single application or combination of cyclic loading, rapid application of static loads, loading at low temperatures, and parts containing defects in their material structures (see Chap. 7). The effects on a ductile material due to processing such as hardening, hydrogen embrittlement, welding, etc., may also accelerate failure. Thus care should always be exercised when dealing with stress concentrations.

In practice, for general applications, the stress concentration factor is determined numerically or experimentally. A numerical method commonly employed, the *finite element method* (FEM), is discussed in Chaps. 9 and 10. However, in order to obtain accurate FEM results with stress concentration problems, a very fine mesh of elements is necessary, resulting in a model with many elements. This will require more computer resources than a similar problem without a stress concentration. An experimental method commonly used is *photoelasticity* (see Chap. 8).

Intuitive methods such as the *flow analogy* are sometimes helpful to the analyst faced with the task of reducing stress concentrations. When dealing with the situation where it is necessary to reduce the cross section abruptly, the resulting stress concentration can often be minimized by a further reduction of material. This is contrary to the common advice, "If it is not strong enough, make it bigger." This can be explained by examining the flow analogy.

The governing field equations for ideal irrotational fluid flow are quite similar to the stress equations covered in Chap. 4, where there is an analogy with flow lines and stress trajectories, and velocity and pressure gradients with the principal stresses. The flow analogy for the plate in Fig. 5.10-1 is shown in Fig. 5.10-3(*a*)

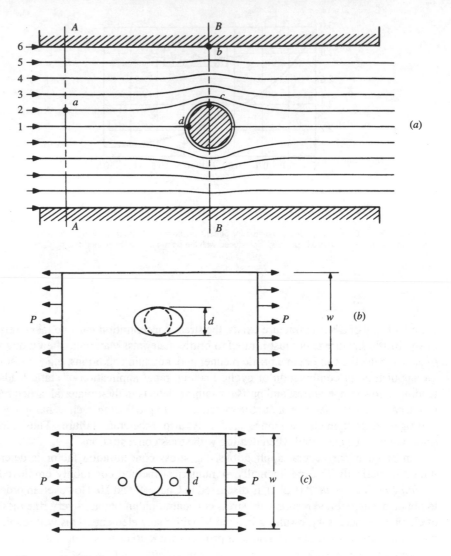

Figure 5.10-3

where stress-free surface boundaries are replaced by solid-channel boundaries (wherever stress cannot exist $\Rightarrow$ fluid flow cannot exist). The uniformly applied loads are replaced by a uniform fluid flow field. Along the entrance section A-A of Fig. 5.10-3 the flow is uniform, and due to symmetry, the flow is uniform at the exit of the channel. However, as the fluid particles approach section B-B, the streamlines need to adjust to move around the circular obstacle. In order to accomplish this, particles close to streamline 1 must make the greatest adjustment and must accelerate until they reach section B-B, where they reach their maximum velocity, and then decelerate to their original uniform velocity some distance from B-B. Thus the velocity at point c is the maximum. The compacting of the streamlines at c will develop a pressure gradient which will actually cause the velocity of point b to be less than that of the incoming velocity of streamline 6 at A-A. Note also that when a particle on streamline 1 reaches point d, the particle theoretically takes on a velocity perpendicular to the net flow. This analogy agrees with that of the plate loaded in tension with a centrally located hole. The stress is a maximum at the edge of the hole, corresponding to point c in Fig. 5.10-3(a). The stress in the plate corresponding to point b is lower than the applied stress, and for point d the stress in the plate is compressive perpendicular to the axial direction.

This analogy can be used to suggest improvements to reduce stress concentrations. For example, for the plate with the hole, the hole can be elongated to an ellipse as shown in Fig. 5.10-3(b), which will improve the flow transition into section B-B (note that this is a reduction of material). An ellipse, however, is not a practical solution, but an ellipse can effectively be approximated by drilling two smaller relief holes in line and in close proximity to the original hole as shown in Fig. 5.10-3(c). The material between the holes, provided the holes are close, will be a stagnation area where the flow (stresses) will be low. Consequently, the configuration acts much like that of an elliptical hole.

At first this might not seem to make sense, since this is a reduction of more material and if one hole weakens the part, obviously more holes will make things worse. One must keep in mind that the first hole increased the stress in two ways: (1) by reducing the cross-sectional area and (2) by changing the shape of the stress distribution. The two additional holes in Fig. 5.10-3(c) do not change the area reduction unless they are larger than the original hole. However, as stated, the additional holes will improve the flow transition and consequently reduce the stress concentration. Another way of improving the plate with the hole is to elongate the hole in the axial direction to a slot.

Some other examples of situations where stress concentrations occur and possible methods of improvement are given in Fig. 5.10-4. Note that in each case improvement is made by reducing material.

This is not a hard and fast rule, however; most reductions in high stress concentrations are made by removing material from adjacent low-stressed areas. This "draws" the high stresses away from the stress concentration area toward the low-stressed area, which decreases the stress in the high-stressed areas.

Problem

Means of Improvement

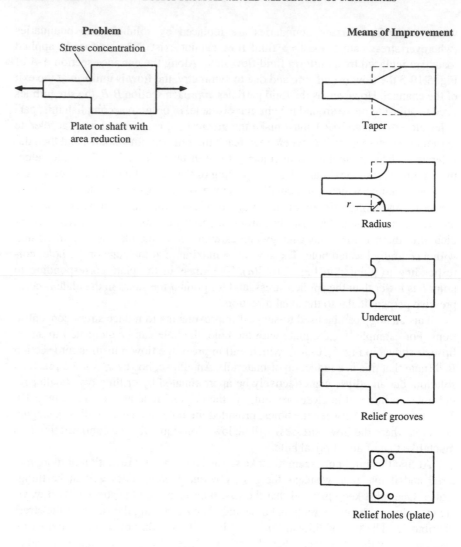

Stress concentration

Plate or shaft with
area reduction

Taper

Radius

Undercut

Relief grooves

Relief holes (plate)

Stress concentrations

Keyway in shaft under
torsion

Drill holes on each side
of keyway

Figure 5.10-4

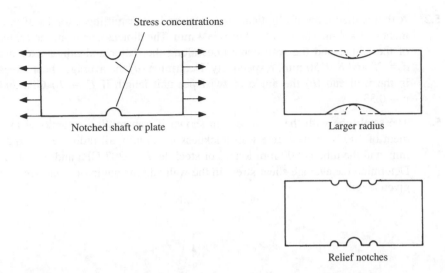

Stress concentrations

Notched shaft or plate

Larger radius

Relief notches

Figure 5.10-4 (concluded)

5.11 PROBLEMS

5.1 A thin-walled steel tube, 1 m long, with the cross section shown, is transmitting a torque of 2 kN·m. The 50-mm dimension is between wall centers. For the material let $E = 210$ GPa and $v = 0.29$. Determine (a) the average shear stress in the wall and (b) the total angle of twist of the tube.

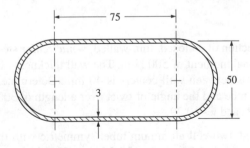

Problem 5.1 (Dimensions in millimeters.)

5.2 A thin-walled tube of elliptical cross section is transmitting a torsional moment of 1 kN·m. The wall thickness is 4 mm. The dimensions from the center of the ellipse and the wall centerlines along the major and minor axes are $a = 75$ and $b = 50$ mm, respectively. Determine (a) the average shear stress in the wall and (b) the angle of twist per unit length if $E = 70$ GPa and $v = 0.33$.

5.3 The steel tube with the cross section shown is transmitting a torsional moment of 100 N·m. The tube wall thickness is 2.5 mm, all radii are $r = 6.25$ mm, and the tube is 500 mm long. For steel, let $E = 207$ GPa and $v = 0.29$. Determine the average shear stress in the wall and the angle of twist over the given length.

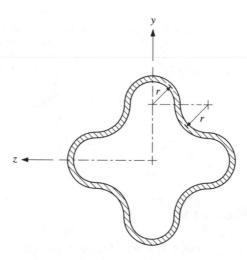

Problem 5.3

5.4 The cross section of a closed, thin-walled, *semicircular* steel tube is transmitting a torsional moment of 500 N·m. The wall thickness of the tube is 3 mm, and the radius between wall centers is 40 mm. Determine the average shear stress in the tube and the angle of twist over a length of 300 mm. For steel let $E = 210$ GPa and $v = 0.29$.

5.5 A thin-walled, two-cell aluminum tube, symmetric with the y and z axes, is transmitting a torsional moment of 2.5 kN·m. The thickness of each wall is 5 mm. For aluminum let $E = 70$ GPa and $v = 0.33$. Determine the average shear stress in each wall and the angle of twist per unit length.

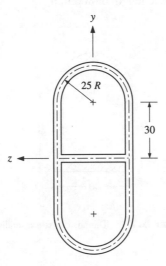

Problem 5.5   (Dimensions are in millimeters.)

5.6 The circular walls of the cross section of the steel tube are not concentric as shown. If the tube is transmitting a torsional moment of 500 N·m, estimate the maximum shear stress in the wall and the angle of twist per unit length.

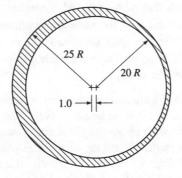

Problem 5.6   (Dimensions are in millimeters.)

5.7   The cross section shown is that of a thin-walled steel tube transmitting a torsional moment of 600 N·m. The wall thickness is 3 mm and dimensions are to the wall centers where shown. Determine the average wall shear stress and the angle of twist for a tube length of 800 mm.

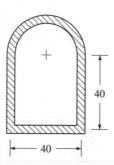

Problem 5.7   (Dimensions are in millimeters.)

5.8   Consider a closed, thin-walled equilateral triangular steel tube transmitting a torsional moment of 5000 lb·in. The length of the centerline of a wall is 1.5 in, and the wall thickness is 0.125 in. Let the material properties be $E = 29$ Mpsi and $v = 0.29$. Determine (a) the average shear stress in the walls and (b) the angle of twist of the tube in degrees if the tube is 20 in long.

5.9   A closed thin-walled tube of wall thickness t is to be used to transmit a torsional moment T. Consider the following cross-sectional shapes: (a) circular, (b) square, and (c) equilateral triangle. Also consider that each section is to be of the same material and have the same wall perimeter length $\bar{S}$ (resulting in the same weight per unit length). Determine the average shear stress and angle of twist per unit of axial length for each section in terms of T, t, and $\bar{S}$. Compare and discuss the results.

5.10 The cross section of the aluminum tube shown transmits a torsional moment of 5000 lb·in. Let $E = 10$ Mpsi and $v = 0.33$. The section, dimensioned in inches, is symmetric with respect to the y and z axes. The 1.0-in dimensions are between wall centerlines, and the centerlines of the 0.25-in vertical walls align with the centerlines of the ends of the circular walls. Determine the average shear stress in each wall and the angle of twist over a length of 24 in.

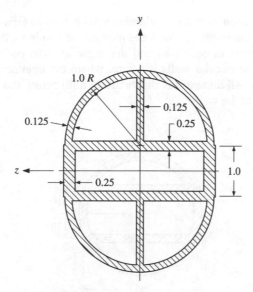

Problem 5.10

5.11 The aluminum tube with the multiple cell cross section shown transmits a torsional moment of 20 kN·m. Let $E = 69$ GPa and $v = 0.33$. The thicknesses of the circular and straight walls are 4 and 3 mm, respectively. Determine the average shear stress in each wall and the angle of twist for a tube length of 500 mm.

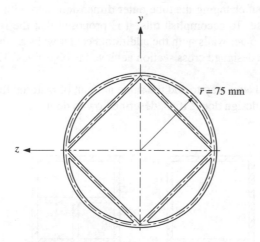

Problem 5.11

5.12 The tubular section shown is of aluminum with $E = 70$ GPa and $v = 0.33$. If the section is transmitting a torsional moment of 150 N·m, determine the average shear stress in each wall and the angle of twist per unit length. The thickness of the circular walls is 4 mm, where the thickness of the straight walls is 3 mm. All dimensions shown are in millimeters and are with respect to the centers of the walls.

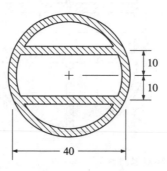

Problem 5.12

5.13 The current design of a square thin-walled tube loaded in torsion has only the outer walls as shown in the figure. The tube needs to be stiffer and the stresses reduced without changing the tube outer dimensions and with a minimum of weight increase. To accomplish this, it is proposed that the redesigned tube have the same outer walls with the additional center walls as shown.

(a) Will the redesigned cross section achieve the objectives? Explain your rationale.

(b) Is there a better way to add the same amount of material than the proposed design does? If so, describe how to do it.

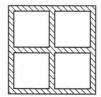

Current design   Proposed redesign

Problem 5.13

5.14 For the beam cross section shown, bending is about the z axis. All dimensions are in inches and, where appropriate, are between the centers of the 0.25-in walls. Determine
   (a) the second-area moments I_y, I_z, and I_{yz}.
   (b) the orientation of the principal axes of the second-area moments.
   (c) the values of the principal second-area moments I_m and I_n.
   (d) the location and magnitudes of the maximum tensile and compressive bending stresses if $M_z = 20$ kip·in.

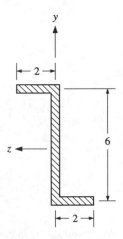

Problem 5.14

5.15 A beam in bending has the cross section of an L3 × 2 × 0.25 in unequal leg steel angle. A bending moment of $M_z = 2$ kip·in is transmitted about the z axis as shown in part (a) of the figure. From a handbook the following is

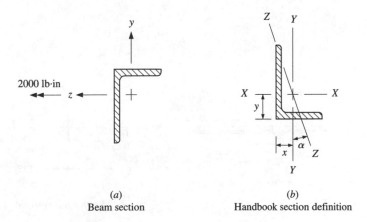

(a)
Beam section

(b)
Handbook section definition

Problem 5.15

given based on part (*b*) of the figure: area = 1.19 in², *x* = 0.493 in, *y* = 0.993 in, I_X = 1.09 in⁴, I_Y = 0.392 in⁴, r_Z = 0.435 in, and tan α = 0.440, where Z-Z is a principal axis and r_Z is the radius of gyration of the cross section relative to the Z-Z axis. Determine the location and magnitudes of the maximum tensile and compressive bending stresses.

5.16 A cantilever beam is loaded as shown. The *yz* axes given are the centroidal axes of the cross section. Determine (*a*) the bending stress at point *B* with co-ordinates (0, 0.833, 1.167) in and (*b*) the deflection of the free end of the beam if *E* = 30 Mpsi.

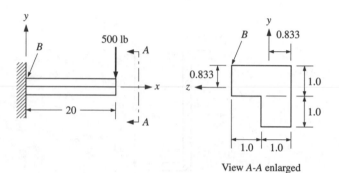

Problem 5.16   (All dimensions are in inches.)

5.17 A uniform load of 100 lb/in is applied as shown. The cross section is antisymmetric. Determine the magnitude and locations of the maximum tensile and compressive bending stresses.

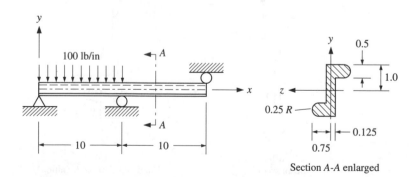

Section A-A enlarged

Problem 5.17   (All dimensions are in inches.)

5.18 The steel cantilever beam shown is 750 mm long. The 2-kN load at the end is in the negative y direction through the centroid of the antisymmetric cross section. For steel, let $E = 210$ GPa.

(*a*) Neglecting Saint-Venant's effect, determine the maximum and minimum bending stresses.

(*b*) Determine the deflection of the free end.

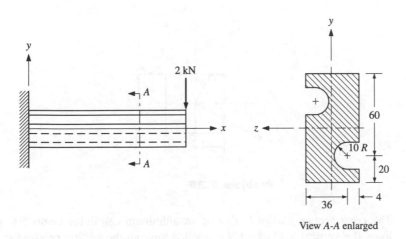

View *A-A* enlarged

Problem 5.18   (All dimensions are in millimeters.)

5.19 For the beam shown, the 1-kN force is applied in the $-y$ direction through the centroid of the cross section. Determine (*a*) the locations and values of the maximum tensile and compressive bending stresses and (*b*) the deflection of the free end of the beam if $E = 70$ GPa.

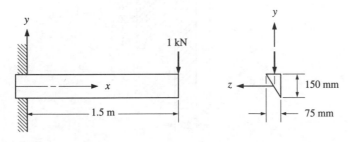

Problem 5.19

5.20 A beam with a square cross section is in bending, where the net bending moment M_z is about the z axis, as shown. Prove that Eq. (3.4-3) can be applied for this case. That is, the stress at any point located a distance y from the z axis is given by $-M_z y/I_z$. Also prove that the maximum bending stress is given by

$$\sigma_{max} = 6\sqrt{2}\,\frac{M_z}{a^3}\cos\left(\frac{\pi}{4} - \beta\right)$$

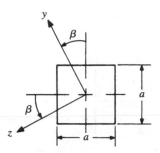

Problem 5.20

5.21 The cross section shown is that of an aluminum cantilever beam 500 mm long. If a vertical load of 4 kN is applied through the section centroid in the negative y direction at the free end of the beam, determine

(a) the locations and values of the maximum tensile and compressive bending stresses in the beam.

(b) the deflection of the free end of the beam. Let $E = 70$ GPa.

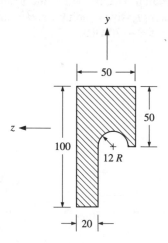

Problem 5.21   (All dimensions are in millimeters.)

5.22 The cross section of a thin-walled beam with a wall thickness of 2 mm is symmetric with respect to the z axis. The section is in bending about the z axis and is supporting a vertical transverse shear force of $V_y = 4$ kN. All dimensions are in millimeters and where appropriate are from wall centers. For wall AB determine how the transverse shear stress varies and the value of the net shear force.

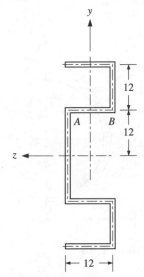

Problem 5.22

5.23 The beam cross section shown is transmitting a bending moment about the z axis and a transverse shear force V_y in the y direction. The thickness of each wall is 0.1 in. Locate the shear center by evaluating e.

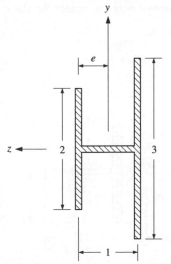

Problem 5.23   (All dimensions are in inches.)

5.24 The cross section of a thin-walled beam supporting a transverse shear force V_y is shown. Determine (a) the shear stress as a function of θ, V_y, r, and t and (b) the location of the shear center e as a function of r.

For I_z and the centroid of a ring element use the thin-walled formulations given in Appendix B.

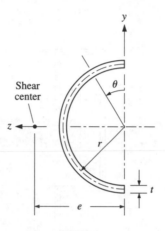

Problem 5.24

5.25 Determine the location of the shear center for Problem 5.22.

5.26 The beam cross section shown transmits a bending moment about the z axis and a transverse shear force in the y direction. The thickness of each wall is 3 mm. Determine the value of e which locates the shear center.

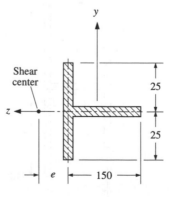

Problem 5.26   (All dimensions are in millimeters.)

5.27 Determine the location of the shear center e for the cross section shown. All dimensions are in millimeters and, where appropriate, are to the center of the 4-mm-thick walls.

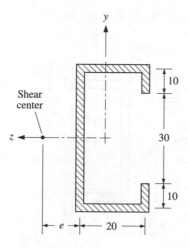

Problem 5.27

5.28 The beam cross section shown is symmetric with respect to the z axis. Determine the location of the shear center e. The wall thickness is 2 mm.

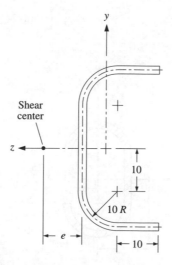

Problem 5.28 (All dimensions are in millimeters.)

5.29 Determine the location of the shear center e of the section shown. All dimensions are in inches and, where appropriate, are from the centers of the 0.1-in-thick walls.

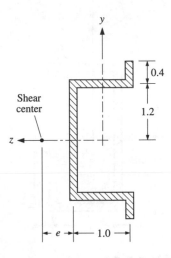

Problem 5.29

5.30 The wall thickness of the section shown is 0.10 in. The *mean* radius of the semicircular walls is 0.5 in. Determine the location of the shear center e.

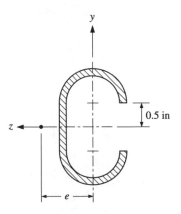

Problem 5.30

5.31 The cross section shown is that for a beam bending about the z axis and transmitting a transverse shear force V_y in the y direction. The thickness of each wall is 3 mm. All dimensions are in millimeters and, where appropriate, are given from wall centers. (*a*) Locate the shear center e, and (*b*) determine the maximum transverse shear stress if $V_y = 500$ N.

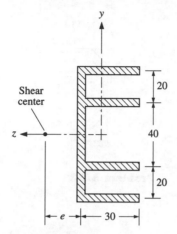

Problem 5.31

5.32 The cross section shown is that for a beam bending about the z axis and transmitting a transverse shear force V_y in the y direction. The thickness of each wall is 4 mm. All dimensions are in millimeters and, where appropriate, are given from wall centers.

(*a*) Show the shear flow in the section due to V_y by drawing the shear stress vectors qualitatively, indicating the magnitude by the length of the vectors.

(*b*) Locate the shear center e.

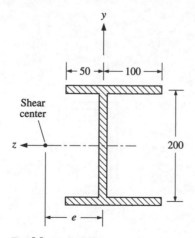

Problem 5.32

5.33 If bending is about the z axis for the cross section shown, determine the location of the shear center from the center of the vertical wall. All dimensions are in inches and, where appropriate, are from the wall centers.

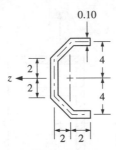

Problem 5.33

5.34 The cross section shown is that for a beam in bending about both the y and z axes. All dimensions are in millimeters and, where appropriate, are from the centerlines of the 2.5-mm-thick walls. The net transverse shear forces on the section are $V_y = 1.8$ kN and $V_z = 2.4$ kN. Determine how the shear stress varies throughout the entire section owing to the transverse shear forces. Assume that the applied forces act through the shear center of the cross section.

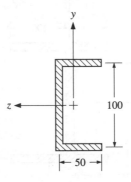

Problem 5.34

5.35 Determine the location of the shear center for the cross section shown. Dimensions, in millimeters, are given from the wall centers where appropriate, and the thickness of each wall is 4 mm.

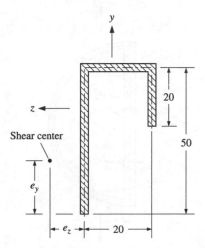

Problem 5.35

5.36 Determine the location of the shear center for the cross section shown. The wall thickness is 0.125 in, and the radial dimension is to the wall center.

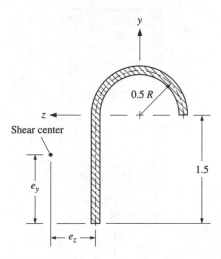

Problem 5.36   (All dimensions in inches.)

5.37 The cross section shown is that of a beam in bending about the z axis and transmitting a transverse shear force V_y. If the transverse shear force is acting through the shear center, determine the shear-stress distributions in each wall and the location of the shear center e. Consider that $t \ll r$.

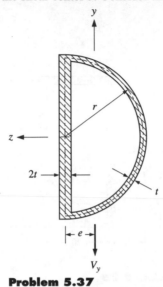

Problem 5.37

5.38 The cross section shown is that of a beam in bending about the z axis and transmitting a transverse shear force $V_y = 2$ kN. If the transverse shear force

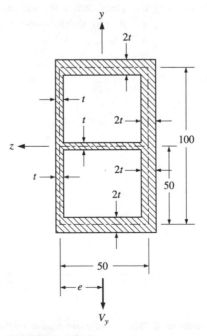

Problem 5.38

is acting through the shear center, determine the shear-stress distributions in each wall and the location of the shear center e. Dimensions in millimeters are given from the wall centers where appropriate, and $t = 3$ mm.

5.39 If for Problem 5.38 all wall thicknesses are t and $V_y = 2$ kN, determine the transverse shear-stress distribution in the entire section and the location of the shear center. If you have solved Problem 5.38, compare the results.

5.40 The cross section shown is that of a beam bending about the z axis and transmitting a transverse shear force $V_y = 5$ kips. If the transverse shear force is acting through the shear center, determine the shear-stress distributions in each wall and the location of the shear center e. Dimensions, in inches, are given from the wall centers where appropriate, and all wall thicknesses are 0.125 in.

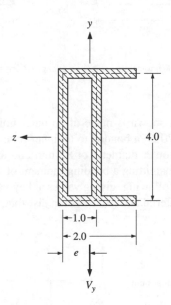

Problem 5.40

5.41 Referring to the composite bar of Fig. 3.2-4 let the force P be applied through the geometric centroid as shown. In terms of P, E_1, E_2, a, and t determine (a) the net bending moment transmitted by the cross section and (b) the maximum normal stresses in the aluminum and steel.

5.42 For the composite beam shown let E_A = 10 Mpsi for aluminum and E_S = 30 Mpsi for steel. Determine

(a) the maximum bending stresses in the aluminum and steel sections.

(b) the location of the loading plane d so that bending is confined to the xy plane.

(c) the maximum vertical deflection of the beam assuming bending is confined to the xy plane.

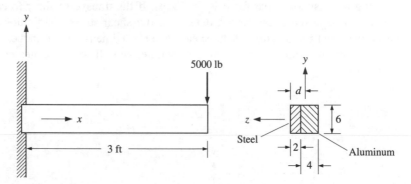

Problem 5.42   (All dimensions in inches except where noted.)

5.43 The composite section shown is made of an outer hollow diamond shape of aluminum with E_a = 70 GPa bonded to an inner tube of steel with E_s = 210 GPa. The tube has an outer diameter of 35 mm and an inner diameter of 20 mm. The section is transmitting a bending moment of M_z = 1 kN·m. In separate graphs, graph the following, giving values at key values of y: (a) the stress distribution in each material and (b) the strain distribution in each material.

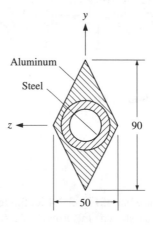

Problem 5.43   (All dimensions in millimeters.)

5.44 A composite beam consists of two cylindrical tubes and one solid cylinder all concentric and bonded together. The inner solid cylinder, with a diameter of 50 mm, is aluminum with $E_a = 70$ GPa. The intermediate tube is brass with an outer diameter of 54 mm, an inner diameter of 50 mm, and $E_b = 105$ GPa. The outer tube is steel with an outer diameter of 58 mm, an inner diameter of 54 mm, and $E_s = 210$ GPa. Determine the maximum bending stresses in each material if the section is transmitting a bending moment of 2 kN·m.

5.45 In Chap. 8, Example 8.8-1, a correction factor is used in calibrating an aluminum beam in bending with a photoelastic coating. The correction factor is obtained from Fig. 8.8-9. The beam cross section with the coating is shown in the figure with this problem. Bending is about the z axis. In calibrating the beam, the bending strain is calculated for the top of the uncoated beam. The actual strain desired is the average strain in the coating, which is the strain at the *midpoint* of the coating thickness. To obtain this, the calculated strain is multiplied by the correction factor. Thus the correction factor is given by

$$C_1 = \frac{\varepsilon_{\text{(coating midpoint)}}}{\varepsilon_{\text{(top of uncoated beam)}}}$$

The strain at the top of the uncoated beam is given by

$$\varepsilon_{\text{(top of uncoated beam)}} = \frac{6M_z}{E_a bt^2}$$

where M_z is the bending moment and E_a is the modulus of elasticity of aluminum. The photoelastic coating is a polymer with a modulus of elasticity of approximately $0.04\,E_a$.

Treating the coated beam as a composite beam, determine C_1 as a function of t_c / t, graph the function for $0 \leq t_c/t \leq 2$, and compare the results with Fig. 8.8-9*b* for aluminum.

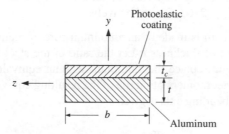

Problem 5.45

5.46 The cross section of the composite beam shown is constructed of a rectangular section of brass bonded to a half-round section of steel. Let $E_s = 30$ Mpsi for the steel and $E_b = 15$ Mpsi for the brass. Determine (*a*) the maximum tensile and compressive bending stresses in the brass and steel, respectively, and (*b*) the deflection of the free end of the beam.

Solve this problem two ways. First, let the equivalent section be of steel. Second, let the equivalent section be of brass.

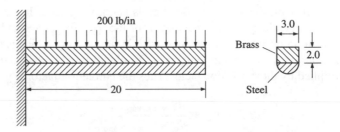

Problem 5.46   (All dimensions are in inches.)

5.47 A composite cantilever beam 2 m long is made of a rectangular aluminum core 50 mm wide and 100 mm deep completely enclosed by steel 10 mm thick all around. The steel and aluminum are totally bonded together. If the maximum allowable stress in the aluminum is 60 MPa, determine the maximum allowable concentrated force which the beam can support at the free end and the corresponding maximum deflection. For the aluminum and steel let $E_a = 70$ and $E_s = 210$ GPa, respectively.

5.48 A composite beam is made of an aluminum core of diameter d bonded to an outer steel shell of thickness t. Let the ratio of the steel to aluminum elastic moduli be 3, and replace the aluminum with an equivalent amount of steel. Show that the net moment of inertia of the resulting equivalent steel cross section about the bending axis is given by

$$I = \frac{\pi}{192} \left[3(d + 2t)^4 - 2d^4 \right]$$

5.49 For the curved beam shown, plot the tangential stresses along line *A-A*. Report the maximum and minimum values.

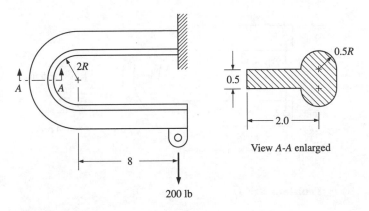

Problem 5.49   (All dimensions are in inches.)

5.50 For the curved beam shown, the load is $P = 30$ kN.
   (*a*) Plot the tangential stress along section *A-A*. Report the maximum and minimum values.
   (*b*) Plot the radial stress along section *A-A*. Report the maximum value.

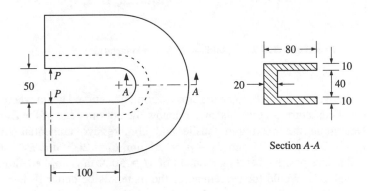

Problem 5.50   (All dimensions in millimeters.)

5.51 For the curved beam shown, the 750-lb force is applied at the pins B and C. Determine (*a*) the maximum and minimum tangential stresses in section A-A and (*b*) the radial stress at the midpoint of section A-A.

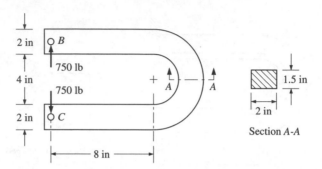

Problem 5.51

5.52 For the curved beam shown, plot the tangential and radial stress distribution along section A-A. Report the maximum and minimum tangential and radial stresses.

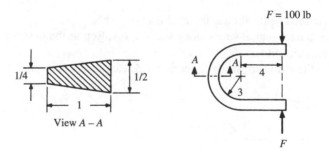

Problem 5.52   (All dimensions are in inches.)

5.53 Consider Fig. 5.5-1(*b*) to be the cross section of a curved beam where the center of curvature is at the distance r_i below the bottom edge of the E_1 material. Determine the maximum tensile and compressive tangential stresses if $r_i = 100$ mm, $b = 25$ mm, $h_1 = h_2 = 10$ mm, $E_1 = 200$ GPa, $E_2 = 100$ GPa, and a pure positive bending moment of $M = 50$ N·m is applied along a horizontal axis. Would the placement of the materials be better if they were exchanged, with the E_2 material in place of the E_1 material and vice versa? Justify your answer.

5.54 Consider the rectangular plate shown in Fig. 5.7-1(a). Show that the displacement field $w = Cxy$ is due to a pure twisting moment M_{xy} and no bending moments M_x and M_y (C is a constant).

Problems 5.55–5.59 demonstrate the *Navier*[†] solution of a thin rectangular plate [see Fig. 5.7-1(a)], simply supported on all edges and subjected to a distributed load $p(x, y)$.

5.55 The Navier solution for a thin rectangular plate [see Fig. 5.7-1(a)], simply supported on all edges and subjected to a distributed load $p(x, y)$, assumes that the loading can be represented by

$$p(x,y) = \sum_{m=1}^{\infty} \sum_{n=1}^{\infty} C_{mn} \sin\left(\frac{m\pi x}{a}\right) \sin\left(\frac{n\pi y}{b}\right) \qquad \textbf{[a]}$$

where C_{mn} are constants.

In a manner like *Fourier* analysis show that

$$C_{mn} = \frac{4}{ab} \int_0^a \int_0^b p(x,y) \sin\left(\frac{m\pi x}{a}\right) \sin\left(\frac{n\pi y}{b}\right) dx\, dy \qquad \textbf{[b]}$$

[HINT: First multiply Eq. (a) by $\sin(n\pi y/b)\, dy$ and integrate from $y = 0$ to $y = b$ using the orthogonality properties of the sin function; then repeat the process relative to the x direction.]

5.56 Consider the first term in Eq. (a) of Problem 5.55. That is, let

$$p(x,y) = p_o \sin\left(\frac{\pi x}{a}\right) \sin\left(\frac{\pi y}{b}\right) \qquad \textbf{[a]}$$

where p_o is a constant. If the displacement $w(x,y)$ is the same form as Eq. (a), that is,

$$w(x,y) = w_o \sin\left(\frac{\pi x}{a}\right) \sin\left(\frac{\pi y}{b}\right) \qquad \textbf{[b]}$$

show that
(a) the conditions for simple supports are satisfied.
(b) w_o is such that

$$w(x,y) = \frac{p_o}{D\pi^4}\left(\frac{a^2 b^2}{a^2 + b^2}\right)^2 \sin\left(\frac{\pi x}{a}\right) \sin\left(\frac{\pi y}{b}\right) \qquad \textbf{[c]}$$

5.57 From Eq. (c) of Problem 5.56 determine
(a) the maximum deflection.
(b) the equations and locations for $(M_x)_{max}$ and $(M_y)_{max}$.
(c) the maximum bending stress if $a > b$.

† See Ref. 5.1.

5.58 If for Prob. 5.55 the plate is subjected to a uniform load $p(x,y) = p_o$, show that

$$C_{mn} = \frac{16p_o}{\pi^2 mn} \qquad m, n = 1, 3, 5 \ldots \qquad [a]$$

Then, similar to that found in Problem 5.56, show that using superposition

$$w(x,y) = \frac{16p_o}{\pi^6 D} \sum_{m=1}^{\infty} \sum_{n=1}^{\infty} \frac{1}{mn} \left[\frac{a^2 b^2}{(bm)^2 + (an)^2} \right]^2 \sin\left(\frac{m\pi x}{a}\right) \sin\left(\frac{n\pi y}{b}\right) \qquad [b]$$

$$m,n = 1,3,5 \ldots$$

where even values of m, n yield zero values.

5.59 Using Eq. (b) of Problem 5.58, verify case 1 of Table 5.7-1 for $a/b = 1.6$.

5.60 Consider a uniformly loaded, 6-mm-thick, steel plate of dimensions 200×400 mm. Let $E = 210$ GPa, $v = 0.3$, and the uniform load be $p_o = 100$ kPa. Determine the maximum deflection and bending stress if
(a) all edges are simply supported.
(b) all edges are fixed.
(c) the two 200-mm sides are simply supported and the two 400-mm sides are fixed.
(d) the two 200-mm sides are simply supported and the two 400-mm sides are free.

(HINT: For an approximation of part (d) consider a beam in plane strain; for a more exact answer see Ref. 5.10.)

5.61 For case 1 of Table 5.7-2 show that all necessary conditions are satisfied by

$$w = -\frac{M_o r^2}{2(1 + v)D}$$

5.62 Completely derive the results of case 4 of Table 5.7-2 using whatever governing Eqs. (5.7-13) to (5.7-25) and equilibrium and boundary conditions necessary. Obtain the equations for w, dw/dr, M_r, M_θ, and the maximum bending moment.

5.63 Consider an axisymmetric annular plate with an outer radius a and inner radius b simply supported at the inner radius and a shear force per unit length V_2 applied in the z direction along the outer radius. Determine the equations for w, dw/dr, M_r, M_θ, and the maximum bending moment. (HINT: Case 7 of Table 5.7-2 can be used to yield the solution.)

5.64 Consider an axisymmetric annular plate with an outer radius a and inner radius b completely fixed at the inner radius and free at the outer radius. A positive bending moment per unit length M_0 is applied at the outer radius. Determine the equations for the maximum displacement and the reactions at the inner radius.

5.65 Determine the equation for the support reaction force per unit length for case 8 of Table 5.7-2.

5.66 Consider an axisymmetric annular plate with an outer radius a and inner radius $0.5a$, completely fixed at the inner radius and free at the outer radius. The plate is to be uniformly loaded by a pressure p_o in the z direction. Using superposition, determine w_{max}, $(dw/dr)_{max}$, and the maximum bending moment. (HINT: Use cases 6 to 8 of Table 5.7-2.) Let $v = 0.3$.

5.67 The axisymmetric annular plate shown is 4 mm thick and has material values $E = 210$ GPa and $v = 0.3$. The shaft is rigidly attached to the plate. Determine the deflection of the applied force and the maximum bending stress.

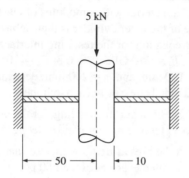

5 kN

50 10

Problem 5.67   (All dimensions are in millimeters.)

5.68 If a simple support is added to the center of case 5 of Table 5.7-2, determine the force reaction at the support.

5.69 Consider case 5 of Table 5.7-2. At the center, place a simple support with a gap e between the plate and the support. Assume that the applied uniform pressure is more than sufficient for contact to occur. Determine the equation for the center support reaction force.

5.70 A steel flywheel 25 mm thick with an outer diameter of 1 m and an inner diameter of 200 mm is rotating at 100 Hz. Determine

(a) the radial and tangential stress distributions as functions of radial position r. Tabulate the values in 50-mm increments of r and plot the distributions.

(b) the radial deflection of the outer radius of the flywheel.

5.71 A cylindrical pressure vessel has an outer diameter of 500 mm and inner diameter of 400 mm. The vessel is subjected to an inner pressure of 500 kPa. Determine the radial and tensile stress distributions as functions of radial position r. Tabulate the values in 5-mm increments of r and plot the distributions.

5.72 Two cylinders are to be press-fitted together. For the outer steel member (E_s = 30 Mpsi, v = 0.29) the outer and inner diameters are 5.0 and 4.000 in, respectively. For the inner brass member (E_b = 15 Mpsi, v = 0.32) the outer and inner diameters are 4.002 and 2.0 in, respectively. Determine
  (a) the interface pressure.
  (b) the radial and tangential stresses as functions of radial position r. Plot the results using a minimum of 10 data points for each plot.
  (c) the radial deflections of the outer and inner surfaces of each cylinder.

5.73 Two thick-walled cylinders are mated with *no* gap *or* interference. Initially the inner cylinder has an inner radius a and outer radius b, whereas the outer cylinder has an inner radius b and outer radius c. The inner cylinder is of a material with a modulus of elasticity E_i and a Poisson's ratio v_i, whereas the outer cylinder is of a material with a modulus of elasticity E_o and a Poisson's ratio v_o. The inside of the inner cylinder is then subjected to a pressure p_i.
  (a) Determine the equation for the resulting interface pressure p at $r = b$.
  (b) If p_i = 1 MPa, E_i = 210 GPa, v_i = 0.3, E_o = 105 GPa, v_o = 0.32, a = 100 mm, b = 150 mm, and c = 200 mm, determine and plot the tangential and radial stresses in each cylinder.

5.74 Repeat Problem 5.73; however, for loading, let the outside of the outer cylinder be subjected to a pressure p_o. For part (b) let p_o = 1 MPa.

5.75 Two cylinders are to be press-fitted together so that the principal stress with the greatest absolute value is not to exceed 60 percent of the yield strength of the material. Both members are made of a steel with a yield strength of 80 ksi, E = 30 Mpsi, and v = 0.29. The nominal diameters of the cylinders are 1.0, 1.5, and 2.0 in.
  (a) Determine the maximum acceptable radial interference and the corresponding interface pressure.
  (b) Under the conditions of part (a) plot the stress distributions for each member.
  (c) If under the conditions of part (a) the assembly is further subjected to an internal pressure of 5000 psi, determine the resulting stress distributions in both members using the method of superposition.

5.76 Due to friction and the interface pressure p between two press-fitted circular members, relative movement in the axial or tangential directions is impeded. Considering dry friction and the coefficient of friction between the mating parts f, show (a) that the force F necessary to cause axial motion between the mating parts is $F = 2\pi ftbp$ and (b) that the maximum allowable torque which can be transmitted from one member to the other is given by $T_{max} = 2\pi ftb^2 p$, where t is the width of the narrowest member and b is the interference radius.

5.77 In Problem 5.76, a steel cylinder is press-fitted over a steel shaft. The assembly can be used to transmit torque as shown. If the torque T at the outer radius of the cylinder can be applied uniformly, prove that the corresponding uniformly distributed tangential load (force per unit area) is $\tau_c = T/2\pi tc^2$. For equilibrium, show that a uniformly distributed tangential load at the inner radius of the cylinder is necessary, where $\tau_b = T/2\pi tb^2$. Finally, prove that the shear-stress distribution in the cylinder is given by $\tau_{r\theta} = T/2\pi tr^2$.

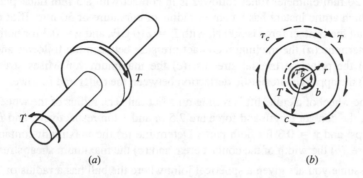

(a)   (b)

Problem 5.77

5.78 Consider a cylinder against a flat plane as in Fig. 5.9-1. As the applied force is increased to P, the cylinder flattens in the contact zone where the interface pressure p becomes an elliptical distribution with $p = 0$ and $p = p_{max}$ at the edges and the center of the contact zone, respectively. Show that the distribution is

$$p = p_{max}\sqrt{1 - \left(\frac{y}{a}\right)^2} \qquad \text{[a]}$$

where a is the half-width of the contact zone. For equilibrium, show that

$$p_{max} = \frac{2}{\pi}\frac{P}{aL} \qquad \text{[b]}$$

which is the same as Eq. (5.9-17) where L is the length of the contact zone.

5.79 As a continuation of Problem 5.78, consider the pressure distribution of Eq. (a) in that problem, and in a manner similar to Example 5.9-1 show that *along the x axis*

$$\sigma_x = -\frac{p_{max}}{\sqrt{1 + (x/a)^2}} \qquad \text{[a]}$$

which is the same as Eq. (5.9-18). HINT: In a manner similar to Example 5.9-1, as part of the solution you should obtain the following integral:

$$\int_0^a \frac{\sqrt{a^2 - y_1^2}}{(x^2 + y_1^2)^2}\,dy_1$$

From a handbook or a mathematics program such as Maple, Mathcad, etc., the integral should reduce to

$$\frac{\pi a^2}{4x^3 \sqrt{a^2 + x^2}}$$

5.80 A 20-mm-diameter roller follower is in contact with a 5 mm thick plate cam which at the instant has a convex radius of curvature of 50 mm. If at this instant the contact force is 500 N, with $E = 210$ GPa and $v = 0.3$ for both parts, determine (a) the maximum contact pressure between the follower and cam, (b) the principal triaxial stresses, (c) the maximum von Mises stress, and (d) the approximate elastic deflection between the roller and follower.

5.81 The wheel of a crane lift is rolling on a flat rail. The radius of the wheel is 4.0 in, the contact length and force are 2.5 in and 1 ton respectively, and $E = 30$ Mpsi and $v = 0.3$ for both parts. Determine (a) the maximum contact pressure, (b) the width of the contact area, and (c) the maximum shear stress.

5.82 Assume you are given a spherical joint where the ball has a radius of 20 mm and the socket has a radius of 22 mm. If the contact load is 50 N, $E = 210$ GPa, and $v = 0.3$, determine (a) the radius of the contact area, (b) the maximum contact pressure, and (c) the maximum shear stress.

5.83 Determine the maximum tensile stress in the 0.25-in-thick slotted plate shown (see Ref. 5.8).

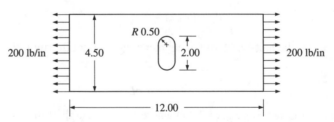

Problem 5.83   (All dimensions are in inches.)

5.84 Using the flow analogy, sketch the stress "flow" lines for the plate shown in Table F.2 of Appendix F. Assume that the end forces are applied uniformly. Qualitatively, indicate how the stress values vary in a relative sense. Where is the maximum stress and why? Where is the stress zero and why? Where are there normal stresses perpendicular to the loading axis and why?

5.85 Figure 3.2-2(b) shows the tensile loading of a plate by means of a pin at the right end of the plate. In this region of the plate show the characteristics of the stress "flow" and some methods to reduce the stress concentration effect at the upper and lower edges of the hole without changing the width of the plate and the diameter of the pin.

5.86 The stress concentration tables for a grooved bar in bending or torsion separately are shown in Tables F.10 or F.11 of Appendix F, respectively. Consider a grooved bar loaded simultaneously in combined bending and torsion. If $M = 50$ N·m, $T = 75$ N·m, $D = 30$ mm, $d = 20$ mm, and $r = 1$ mm, determine the location and value of the maximum shearing stress in the bar.

5.87 For a plate loaded in tension and containing a centrally located hole the stress concentration effect increases as the diameter of the hole decreases relative to the plate. However, the area change has more effect on the change in stress so that the smaller the hole the smaller the stress is. To see this, let $F = 12$ kN, $w = 100$ mm, and $t = 10$ mm. Plot σ_{nom} and σ_{max} as functions of d where $0 \leq d \leq 80$ mm and report on the relative effects of the stress concentration and area change on the stresses.

5.88 For the "T" plate shown with the applied force F, sketch the stress "flow" lines in the plate. Where do the maximum and minimum stresses occur in the plate? Show some methods to reduce the effect of the stress concentration without changing d_1, d_2, or d_3.

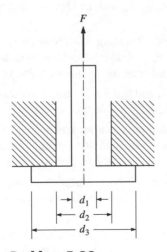

Problem 5.88

Computer Problems

5.89 The section shown is that of a curved beam in bending.
   (*a*) Using tables, determine r_n, r_c, and e.
   (*b*) Using the numerical method, determine r_n, r_c, and e.

(c) If $M = 2000$ lb·in, determine the maximum and minimum bending stresses from parts (a), (b), and the approximate method. Compare the results.

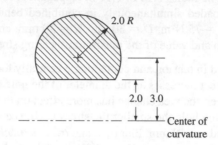

2.0 R

2.0 3.0

Center of curvature

Problem 5.89 (All dimensions are in inches.)

5.90 Develop a spreadsheet or mathematics program for a curved beam in bending with a T-section as shown in Table 5.6-1. Your program should be capable of outputting the tangential and radial stress distributions of a problem such as Examples 5.6-3 and 5.6-5. The input variables should be the applied force F; the offset distance between the force and center of curvature d; and, according to Table 5.6-1, r_i, r_o, h_i, b_i, and b_o. Demonstrate your program on Examples 5.6-3 and 5.6-5.

5.91 If you have completed Problem 5.90, slightly vary one or more of the cross section parameters h_i, b_i, and b_o to achieve equality in the magnitudes of the maximum tensile and compressive tangential stresses of Example 5.6-3.

5.92 Solve Problem 5.54, and using a plotting program, plot the deflections of the edges of the plate and discuss your observations.

5.93 Similar to the tabulations given for cases 7 and 8 of Table 5.7-2, create two tabulations for case 6, one with $M_2 = 0$ and a second with $M_1 = 0$. Tabulate both cases in 0.1 increments of b/a, from 0.1 to 0.9.

5.94 From Eqs. (5.9-12) and (5.9-13) plot the von Mises and τ_{max} stresses as functions of ζ and verify their maximum values and locations for $v = 0.3$ as discussed in Sec. 5.9.

5.95 From Eqs. (5.9-18) to (5.9-20) plot the von Mises and τ_{max} stresses as functions of ζ and verify their maximum values and locations for $v = 0.3$ as discussed in Sec. 5.9.

5.12   REFERENCES

5.1 Timoshenko, S. P., and S. Woinowsky-Krieger. *Theory of Plates and Shells,* 2d ed. New York: McGraw-Hill, 1959.

5.2 Ugural, A. C. *Stresses in Plates and Shells.* New York: McGraw-Hill, 1981.

5.3 Young, W. *Roark's Formulas for Stress and Strain,* 6th ed. New York: McGraw-Hill, 1989.

5.4 Hertz, H. *Miscellaneous Papers.* New York: Macmillan, 1896.

5.5 Timoshenko, S. P., and J. N. Goodier. *Theory of Elasticity*, 3d ed. New York: McGraw-Hill, 1970.

5.6 Belyayev, N. M. *Work in the Theory of Elasticity and Plasticity.* Moscow: Stat Press for Technical Theoretical Literature, 1957.

5.7 Thomas, H. R., and V. A. Hoersch. "Stresses Due to the Pressure of One Elastic Solid on Another." Urbana/Champaign, IL: University of Illinois Engineering Experiment Station, Bulletin 212, 1930.

5.8 Peterson, R. E. *Stress Concentration Factors.* New York: Wiley, 1974.

5.9 Peterson, R. E. "Design Factors for Stress Concentration." *Machine Design*, a Penton publication, vol. 23, pt.1, no. 7, p. 169, February 1951; pt. 2, no. 3, p. 161, March 1951; pt. 3, no. 5, p. 159, May 1951; pt. 4, no. 6, p. 173, June 1951; pt. 5, no. 7, p. 155, July 1951.

5.10 Pilkey, W. D. *Formulas for Stress, Strain, and Structural Matrices.* New York: Wiley, 1994.

chapter

SIX

ENERGY TECHNIQUES IN STRESS ANALYSIS

6.0   INTRODUCTION

Many practical engineering problems involve the combination of a large system of simple elements in a complex and often highly statically indeterminate structure. If a structure is statically indeterminate, it is necessary to use deflection theory in order to first determine the complete support reactions. Once the reactions are determined, a complete analysis of the structure can be performed. Even if the structure is statically determinate, it may be necessary to perform a deformation analysis, as deflections at various points of the structure may pertain to the overall design requirements.

Deflection analysis using geometric approaches and superposition techniques are advantageous for simple systems, but if the system is complex, these approaches become difficult and cumbersome. To illustrate this point a simple example is analyzed.

Example 6.0-1

Cables BC and BD, shown in Fig. 6.0-1, form angles of 30° and 45°, respectively, to the vertical. Determine the vertical and horizontal displacement of point B if a vertical force P of 2000 lb is applied at B. The cables can be considered to be of solid steel ($E = 30$ Mpsi) each with a cross-sectional area of 0.2 in².

Solution:

The first step is to determine the forces in each cable. From Fig. 6.0-1(b), equilibrium of forces in the x and y directions result in

$$F_{BC} = 1464 \text{ lb} \qquad F_{BD} = 1035 \text{ lb} \qquad \textbf{[a]}$$

404

The next step is to imagine the pin at B removed and cables BC and BD allowed to stretch under the internal forces F_{BC} and F_{BD}, respectively. Then, by rotating the cables, they can be "rejoined." This must be where point B actually displaces to. If the rotations are very small, rotation of BC and BD can be approximated by a perpendicular movement, as shown in Fig. 6.0-2. The elongations of each cable are

$$\delta_{BC} = \frac{F_{BC}L_{BC}}{AE} = \frac{(1464)(24)}{(0.2)(30 \times 10^6)} = 5.86 \times 10^{-3} \text{ in}$$

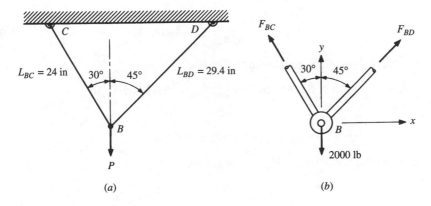

(a)   (b)

Figure 6.0-1

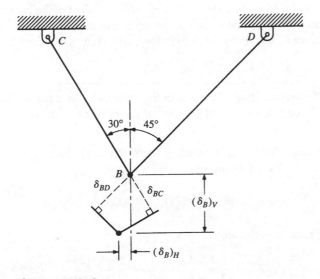

Figure 6.0-2

and

$$\delta_{BD} = \frac{F_{BD}L_{BD}}{AE} = \frac{(1035)(29.4)}{(0.2)(30 \times 10^6)} = 5.07 \times 10^{-3} \text{ in}$$

to obtain the deflection of point B, relationships between δ_{BC}, δ_{BD} and $(\delta_B)_V$, $(\delta_B)_H$ must be developed. Obtaining the necessary relationships is not always simple, but after some head scratching one can see that

$$(\delta_B)_V = \frac{\delta_{BC}}{\cos 30} + (\delta_B)_H \tan 30 \qquad \text{[b]}$$

and

$$(\delta_B)_V = \frac{\delta_{BD}}{\cos 45} - (\delta_B)_H \tan 45 \qquad \text{[c]}$$

Setting the two equations equal with

$$\delta_{BC} = 5.86 \times 10^{-3} \text{ in} \qquad \delta_{BD} = 5.07 \times 10^{-3} \text{ in}$$

yields

$$(\delta_B)_H = 0.260 \times 10^{-3} \text{ in}$$

Substituting this into either of Eqs. (b) or (c) yields

$$(\delta_B)_V = 6.91 \times 10^{-3} \text{ in}$$

Thus it can be seen that even problems which appear rather simple can be aggravating when analyzed using geometry of deformation. Most of the energy techniques do not utilize overall geometry and, in practice, can simplify the analysis tremendously. To illustrate a simple energy technique, let us return to Example 6.0-1.

Example 6.0-2   Determine the vertical drop of point B in Example 6.0-1 by equating the total work done by the force P to the sum of the work P performs on each cable.

Solution:

Assuming a linear force-displacement process, the work done by gradually applying the force P to 2000 lb is simply

$$W = \frac{1}{2} P(\delta_B)_V = 1000(\delta_B)_V$$

The work applied to each cable is

$$W_{BC} = \frac{1}{2} F_{BC} \delta_{BC} = \frac{1}{2} \frac{F_{BC}^2 L_{BC}}{AE} = \frac{1}{2} \frac{(1464)^2(24)}{(0.2)(30 \times 10^6)} = 4.29 \text{ in·lb}$$

$$W_{BD} = \frac{1}{2} F_{BD}\delta_{BD} = \frac{1}{2}\frac{F_{BD}^2 L_{BD}}{AE} = \frac{1}{2}\frac{(1035)^2(29.4)}{(0.2)(30 \times 10^6)} = 2.62 \text{ in·lb}$$

The total work is the sum of the work applied to each cable, and so

$$W = W_{BC} + W_{BD}$$

$$1000(\delta_B)_V = 4.29 + 2.62$$

$$(\delta_B)_V = 6.91 \times 10^{-3} \text{ in}$$

Thus it can be seen that a work or energy approach can simplify things considerably.[†]

In order to determine the horizontal reaction of point B, an artifice must be employed to avoid the fact that no work is done by P moving horizontally.

Determine the horizontal deflection of point B in Example 6.0-1 by the approach used in Example 6.0-2.

Example 6.0-3

Solution:

Since there is no horizontal force at point B, there is no work due to the horizontal deflection $(\delta_B)_H$. However, it is possible first to place an exceedingly small horizontal force H at point B where the force is so small that it causes negligible deflections (see Fig. 6.0-3). Then, when the vertical force P is placed on the structure, the constant horizontal force H will do work. The work due to H is[‡]

$$W_H = H(\delta_B)_H$$

The forces in cables BC and BD due to H alone are found by static analysis and are

$$(F_{BC})_H = 0.732H \qquad (F_{BD})_H = -0.897H^{[§]}$$

The additional work obtained from these forces on the cables arises from the deflections caused by the actual applied force. The additional work terms are[¶]

$$(W_{BC})_H = (F_{BC})_H\,\delta_{BC} = (0.732H)(5.86 \times 10^{-3}) = 4.29H \times 10^{-3}$$

and

$$(W_{BD})_H = (F_{BD})_H\,\delta_{BD} = (-0.897H)(5.07 \times 10^{-3}) = -4.55H \times 10^{-3}$$

[†] The answer obtained in Example 6.0-2 is identical to that of Example 6.0-1. Recall that in Example 6.0-1 an approximation for small rotations was made. This same assumption applies to the approach used in Example 6.0-2. For example, the work in cable BC is not exactly $1/2 F_{BC}\,\delta_{BC}$ but is slightly less due to the rotation. However, as long as rotations are small, this error is negligible.

[‡] The $1/2$ is omitted since the force H remains constant throughout the deflection $(\delta_B)_H$.

[§] Do not be concerned here with a compressive force in a cable. Recall that H is actually an artificial force and is basically infinitesimal in value.

[¶] Again, since the forces $(F_{BC})_H$ and $(F_{BD})_H$ are present before deflections, the work expression will not contain the $1/2$ term.

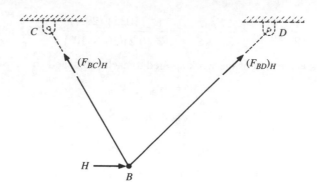

Figure 6.0-3

The work due to H equals the work H performs on the elements of the system

$$W_H = (W_{BC})_H + (W_{BD})_H$$

$$H(\delta_B)_H = 4.29H \times 10^{-3} - 4.55H \times 10^{-3}$$

$$(\delta_B)_H = -0.260 \times 10^{-3} \text{ in}^\dagger$$

Note that the procedure followed in Examples 6.0-2 and 6.0-3 avoided analyzing the overall geometry of deformation of the structure, which was necessary in Example 6.0-1, where Eqs. (b) and (c) were developed.

Since energy techniques are quite powerful, this chapter will attempt to provide a detailed exploration of the major techniques applied to the field of stress analysis. In order to develop and illustrate the techniques, the analyses will be demonstrated on some simple problems in the beginning to avoid overburdening the reader with complex systems at the start. Thus, in some cases, the particular energy technique may not appear that impressive, since, as stated earlier, geometric and superposition approaches are generally more practical on simple, single-element systems.

The major energy techniques discussed in this chapter are:

1. Castigliano's first theorem.
2. The complementary-energy theorem.
3. Special applications of the complementary-energy theorem.
   a. Castigliano's second theorem applied to linear systems.
   b. The virtual load method.
4. Rayleigh's technique.
5. The Rayleigh-Ritz technique.

---

† The negative sign indicates that the deflection is in the direction opposite H.

6.1   WORK

When a force P is gradually applied at a particular point Q on a structure (see Fig. 6.1-1), the deflection of the point δ in the direction of the applied force increases at a rate directly related to P .

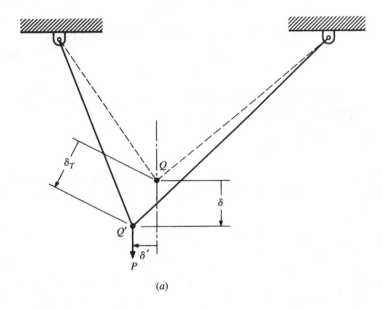

(a)

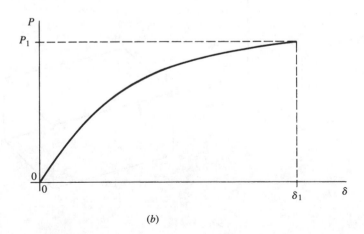

(b)

Figure 6.1-1

The total deflection of point Q is δ_T, the vector sum of δ and δ'. However, the work performed by P is

$$W = \int_0^{\delta_1} P\, d\delta \qquad\qquad \textbf{[6.1-1]}$$

where δ is in the direction of P and the integral of Eq. (6.1-1) is merely the area under the P-vs.-δ curve evaluated up to δ_1. If the relationship between P and δ is linear, that is, $P = k\delta$, where k is a constant, then from Eq. (6.1-1) the work is

$$W = \frac{1}{2}k\delta_1^2 = \frac{1}{2}P_1\delta_1 \qquad\qquad \textbf{[6.1-2]}$$

If P is say, constant, $P = P_1$; while if a deflection process takes place, then

$$W = P_1\delta_1 \qquad\qquad \textbf{[6.1-3]}$$

6.2   STRAIN ENERGY

6.2.1   UNIAXIAL CASE

Consider an infinitesimal element within a member which develops a final uniaxial stress of σ_x (see Fig. 6.2-1). If the behavior of the material is linear, the force due to

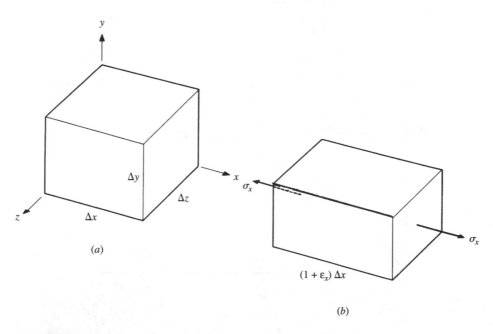

(a)

(b)

Figure 6.2-1

$\sigma_x(F_x = \sigma_x\,\Delta y\,\Delta z)$ displaces in a linear fashion the amount $\delta_x = \varepsilon_x\,\Delta x$. From Eq. (6.1-2) the total work on the element is

$$W = \frac{1}{2}F_x\delta_x = \frac{1}{2}\sigma_x\varepsilon_x(\Delta x\,\Delta y\,\Delta z)$$

The work per unit volume w is determined by dividing the work by the volume $\Delta x\,\Delta y\,\Delta z$. Thus

$$w = \frac{1}{2}\sigma_x\varepsilon_x \qquad \textbf{[6.2-1]}$$

where the units of w are inch-pounds per cubic inch (USCU) and newton-meters per cubic meter or joules per cubic meter (SI). If the material is perfectly elastic where it exhibits no energy losses, the total work w (or work per unit volume w) increases the potential energy of the element U, called the strain energy (or strain energy per unit volume u). Thus

$$u = w = \frac{1}{2}\sigma_x\varepsilon_x \qquad \textbf{[6.2-2]}$$

Since the stress is uniaxial, in the elastic range $\varepsilon_x = \sigma_x/E$, the resulting strain energy per unit volume is

$$u = \frac{1}{2E}\sigma_x^2 \qquad \textbf{[6.2-3]}$$

6.2.2 ADDITIONAL NORMAL STRESSES

If the normal stresses σ_y and σ_z are also present,

$$u = \frac{1}{2}\sigma_x\varepsilon_x + \frac{1}{2}\sigma_y\varepsilon_y + \frac{1}{2}\sigma_z\varepsilon_z \qquad \textbf{[6.2-4]}$$

and using the general strain-stress relationships [Eqs. (1.4-2)], we see that the strain energy per unit volume is

$$u = \frac{1}{2E}[\sigma_x^2 + \sigma_y^2 + \sigma_z^2 - 2\nu(\sigma_x\sigma_y + \sigma_y\sigma_z + \sigma_z\sigma_x)] \qquad \textbf{[6.2-5]}$$

6.2.3 SHEAR STRESS

The work due to τ_{xy} can be seen in Fig. 6.2-2. The force $\tau_{xy}\,\Delta x\,\Delta z$ moves a distance $\gamma_{xy}\,\Delta y$, which results in a work of $\tau_{xy}\gamma_{xy}\,\Delta x\,\Delta y\,\Delta z$, or an increase in strain energy per unit volume of

$$u = \frac{1}{2}\gamma_{xy}\tau_{xy} \qquad \textbf{[6.2-6]}$$

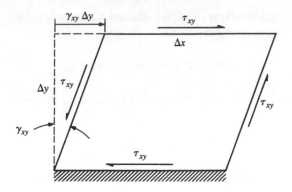

Figure 6.2-2

Substituting Eq. (1.4-8a) for γ_{xy} yields

$$u = \frac{1 + v}{E}\,\tau_{xy}^2 \qquad \text{[6.2-7]}$$

6.2.4   GENERAL STATE OF STRESS

Considering an element with the normal stresses σ_x, σ_y, σ_z, and shear stresses τ_{xy}, τ_{yz}, τ_{zx}, the strain energy per unit volume can be expressed as

$$u = \frac{1}{2}\left(\sigma_x \varepsilon_x + \sigma_y \varepsilon_y + \sigma_z \varepsilon_z + \tau_{xy}\gamma_{xy} + \tau_{yz}\gamma_{yz} + \tau_{zx}\gamma_{zx}\right) \qquad \text{[6.2-8]}$$

Using matrix notation,[†] this can be expressed in shorthand form as

$$u = \frac{1}{2}\{\sigma\}^T\{\varepsilon\} \qquad \text{[6.2-9]}$$

where

$$\{\sigma\}^T = \{\sigma_x \ \sigma_y \ \sigma_z \ \tau_{xy} \ \tau_{yz} \ \tau_{zx}\} \qquad \text{[6.2-10]}$$

$$\{\varepsilon\} = \{\varepsilon_x \ \varepsilon_y \ \varepsilon_z \ \gamma_{xy} \ \gamma_{yz} \ \gamma_{zx}\}^T$$

Substituting Eqs. (1.4-2) and (1.4-8) into (6.2-8) or (6.2-9) gives

$$u = \frac{1}{2E}\left[\sigma_x^2 + \sigma_y^2 + \sigma_z^2 - 2v(\sigma_x\sigma_y + \sigma_y\sigma_z + \sigma_z\sigma_x) + 2(1 + v)(\tau_{xy}^2 + \tau_{yz}^2 + \tau_{zx}^2)\right]$$

$$\text{[6.2-11]}$$

† See Appendix I for matrix notation.

Equation (6.2-11) can be written in terms of the principal stresses σ_2, σ_2, and σ_3, which is basically the same element transformed to eliminate the shear stresses. Thus, using Eq. (6.2-5), the strain energy per unit volume for the general case can also be expressed as

$$u = \frac{1}{2E}\left[\sigma_1^2 + \sigma_2^2 + \sigma_3^2 - 2v(\sigma_1\sigma_2 + \sigma_2\sigma_3 + \sigma_3\sigma_1)\right] \qquad \textbf{[6.2-12]}$$

6.2.5 PLANE STRESS

For this case (see Fig. 6.2-3) $\sigma_z = \tau_{yz} = \tau_{zx} = 0$, and Eq. (6.2-11) reduces to

$$u = \frac{1}{2E}\left[\sigma_x^2 + \sigma_y^2 - 2v\sigma_x\sigma_y + 2(1 + v)\tau_{xy}^2\right] \qquad \textbf{[6.2-13]}$$

or, in terms of the two in-plane principal stresses, σ_{p1} and σ_{p2},

$$u = \frac{1}{2E}\left(\sigma_{p1}^2 + \sigma_{p2}^2 - 2v\sigma_{p1}\sigma_{p2}\right) \qquad \textbf{[6.2-14]}$$

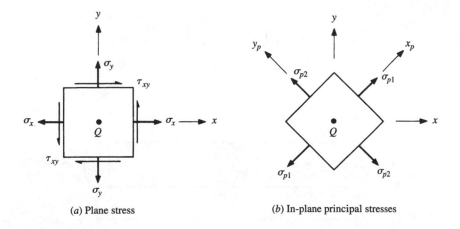

(a) Plane stress

(b) In-plane principal stresses

Figure 6.2-3

6.3   TOTAL STRAIN ENERGY IN BARS WITH SIMPLE LOADING CONDITIONS

6.3.1   AXIAL LOADING

The stress in an elastic bar undergoing axial loading (see Fig. 6.3-1) is $\sigma_x = \pm F/A$ throughout the bar, where the plus sign indicates tension and minus compression. Thus, integrating the strain energy per unit volume [Eq. (6.2-3)] throughout the bar yields the total strain energy. The total strain energy is therefore

$$U_a = \frac{1}{2}\int \frac{1}{E}\left(\frac{F}{A}\right)^2 dV$$

However, $dV = A\,dx$; thus

$$U_a = \frac{1}{2}\int_0^L \frac{F^2 dx}{AE} \qquad\qquad \textbf{[6.3-1]}$$

If F, A, and E are constant, Eq. (6.3-1) reduces to

$$U_a = \frac{1}{2}\frac{F^2 L}{AE} \qquad\qquad \textbf{[6.3-2]}$$

which, as can be seen, is the total work performed by F since $W = 1/2\,(F\delta)$ and $\delta = FL/AE$.

6.3.2   TORSIONAL LOADING OF A SOLID CIRCULAR BAR

The shear stress induced by torsion is a function of radial position, that is, $\tau = Tr/J$. Establish a dV element in which the shear stress is constant (see Fig. 6.3-2). Thus from Eq. (6.2-7)

$$U_t = \int_A \int_0^L \frac{1+v}{E}\left(\frac{Tr}{J}\right)^2 dx\, dA$$

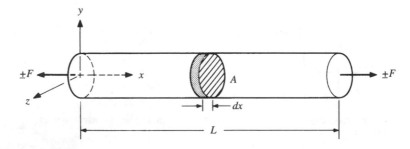

Figure 6.3-1

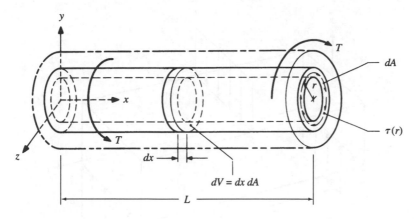

Figure 6.3-2

In general, except for r^2 all the finite terms within the integral will be functions of x. Thus integrating with respect to dA first results in

$$U_t = \int_0^L \left[\frac{1 + v}{E}\left(\frac{T}{J}\right)^2 \int_A r^2 dA \right] dx$$

However, since $\int_A r^2 dA = J$,

$$U_t = \int_0^L \frac{(1 + v)T^2}{JE}\, dx \qquad\qquad \textbf{[6.3-3]}$$

If v, T, E, and J are not functions of x, Eq. (6.3-3) reduces to

$$U_t = (1 + v)\frac{T^2 L}{JE} \qquad\qquad \textbf{[6.3-4]}$$

6.3.3   TRANSVERSE LOADING

Since the normal bending stresses and transverse shear stresses for a beam such as that shown in Fig. 6.3-3(a) vary with respect to x and y in general, a dV element undergoing a constant state of stress is as shown in Fig. 6.3-3(b). Since this volume is undergoing one normal stress

$$\sigma_x = -\frac{M_z y}{I_z}$$

and one shear stress

$$\tau_{xy} = \frac{V_y Q}{I_z b}$$

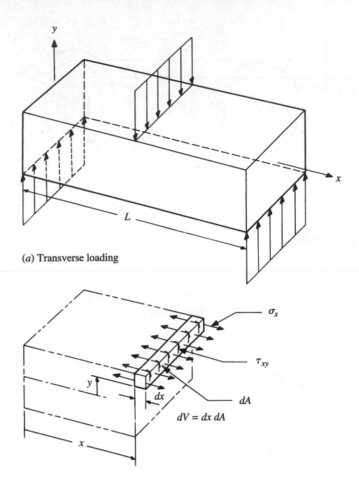

(a) Transverse loading

(b) Volume under constant stress

Figure 6.3-3

from Eq. (6.2-13), with $\sigma_y = 0$, the strain energy will be

$$U = \frac{1}{2} \int_A \int_0^L \frac{1}{E}\left[\left(\frac{M_z y}{I_z} \right)^2 + 2(1 + v)\left(\frac{V_y Q}{I_z b} \right)^2 \right] dx\ dA$$

or

$$U = \frac{1}{2} \int_A \int_0^L \frac{1}{E}\left(\frac{M_z}{I_z} \right)^2 y^2 dx\ dA + \int_A \int_0^L \frac{1 + v}{E} \left(\frac{V_y}{I_z} \right)^2 \left(\frac{Q}{b} \right)^2 dx\ dA$$

The first integral is the part of the total strain energy due to the bending moment; it will be designated U_b. The second integral is the part due to the shear force, designated U_s. Thus

$$U = U_b + U_s \qquad \text{[6.3-5]}$$

Since M_z/I_z is a function of x alone, integration of U_b with respect to dA is performed first. Since $\int y^2 dA = I_z$, the total strain energy due to bending is

$$U_b = \frac{1}{2} \int_0^L \frac{M_z^2}{EI_z} \, dx \qquad \text{[6.3-6]}$$

If E and I_z are constant,

$$U_b = \frac{1}{2EI_z} \int_0^L M_z^2 \, dx \qquad \text{[6.3-7]}$$

An alternate form of Eqs. (6.3-6) and (6.3-7) results from the equation, $M_z = EI_z (d^2 v_c/dx^2)$. Thus the alternate form is

$$U_b = \frac{1}{2} \int_0^L EI_z \left(\frac{d^2 v_c}{dx^2} \right)^2 dx \qquad \text{[6.3-8]}$$

or, if E and I_z are constant,

$$U_b = \frac{1}{2} EI_z \int_0^L \left(\frac{d^2 v_c}{dx^2} \right)^2 dx \qquad \text{[6.3-9]}$$

The second integral in Eq. (6.3-5), U_s, is evaluated in a similar manner. Since V_y/I_z is a function of x alone, integration with respect to dA can be performed first as was done for U_b, but the integral $\int (Q/b)^2 \, dA$ must be evaluated for the particular shape of cross section under investigation.

For example, consider a beam with a rectangular cross section of width b and height h. Recall from Example 3.4-3 that $Q/b = 1/2[(h/2)^2 - y^2]$. Then with $dA = b \, dy$,

$$\int_A \left(\frac{Q}{b} \right)^2 dA = \frac{1}{4} \int_{-\frac{h}{2}}^{\frac{h}{2}} \left(\frac{h^4}{16} - \frac{h^2 y^2}{2} + y^4 \right) b \, dy = \frac{1}{120} bh^5$$

Thus, with $I_z = bh^3/12$ and $A = bh$, the strain energy due to the transverse shear load is given by

$$U_s = \int_A \int_0^L \frac{1 + v}{E} \left(\frac{V_y}{I_z} \right)^2 \left(\frac{Q}{b} \right)^2 dx \, dA$$

$$= \int_0^L \frac{1 + v}{E} \left(\frac{V_y}{bh^3/12} \right)^2 \frac{bh^5}{120} \, dx = 1.2 \int_0^L \frac{(1 + v) V_y^2}{EA} \, dx$$

In general, the energy can be expressed as

$$U_s = k \int_0^L \frac{(1 + v) V_y^2}{EA} \, dx \qquad \text{[6.3-10]}$$

where k is called the *form correction factor for shear.* Table 6.3-1 gives some typical values of the form factor for various cross sections. If the transverse shear stress were uniform at $\tau_{xy} = V_y/A$, the form factor k would be unity. Except for very short beams, the bending energy far surpasses the shear energy. Thus ignoring U_s or using $k = 1$ will normally suffice. If v, E, and A are constant in Eq. (6.3-10),

$$U_s = \frac{(1 + v)k}{EA} \int_0^L V_y^2 \, dx \qquad \qquad \textbf{[6.3-11]}$$

Table 6.3-1   Form correction factor for shear

Beam Cross Section	k
Rectangular	1.2
Circular	$10/9 = 1.11$
Thin-walled circular	2.0
I-section, box section, channel	~1.0

Example 6.3-1   Determine the total strain energy of the steel beam shown in Fig. 6.3-4. The cross section is rectangular, 25 mm wide, and 50 mm deep. $E_s = 205$ GPa; $v = 0.3$.

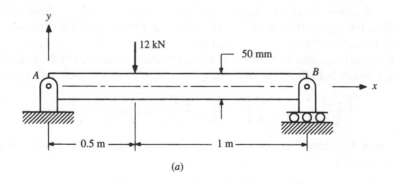

(a)

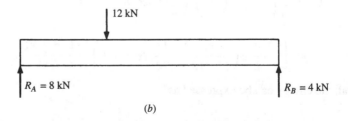

(b)

Figure 6.3-4

Solution:

First, solving for the reactions at A and B gives

$$R_A = 8 \text{ kN} \qquad R_B = 4 \text{ kN}$$

The next step is to determine the shear-force and bending-moment equations:

$$V_y = \begin{cases} -8 \text{ kN} & 0 < x < 0.5 \\ 4 \text{ kN} & 0.5 < x < 1.5 \end{cases}$$

$$M_z = \begin{cases} 8x \text{ kN} \cdot \text{m} & 0 < x < 0.5 \\ 8x - 12(x - 0.5) \text{ kN} \cdot \text{m} & 0.5 < x < 1.5 \end{cases}$$

When Eq. (6.3-7) is used, the bending energy is given by

$$U_b = \frac{1}{2EI_z} \left\{ \int_0^{0.5} (8000x)^2 \, dx + \int_{0.5}^{1.5} [8000x - 12{,}000(x - 0.5)]^2 \, dx \right\}$$

Integration of this and substitution of $E = 205$ GPa and $I_z = 1/12(0.025)(0.05)^3 = 2.604 \times 10^{-7} \text{ m}^4$ results in

$$U_b = 74.93 \text{ N} \cdot \text{m}$$

The strain energy due to the shear force V_y is found using Eq. (6.3-11), which yields

$$U_s = \frac{(1 + 0.3)(1.2)}{(205 \times 10^9)(0.05)(0.025)} \left[\int_0^{0.5} (-8000)^2 \, dx + \int_{0.5}^{1.5} (4000)^2 \, dx \right]$$

$$= 0.292 \text{ N} \cdot \text{m}$$

Thus,

$$U = U_b + U_s = 75.2 \text{ N} \cdot \text{m}$$

Note that the strain energy due to bending is considerably greater than the shear energy. This is usually the case for beams unless they are very short. Thus, in most cases, the shear energy in bending problems can normally be ignored.

6.4   CASTIGLIANO'S FIRST THEOREM[†]

Consider a series of forces P_i gradually applied to a structure ($i = 1, 2, \ldots, n$). At each point of load application i let δ_i be the component of the total deflection of point i along the line of action of P_i. Therefore, the total work applied is

$$W = \sum_{i=1}^{n} \int P_i \, d\delta_i$$

[†] First derived by Alberto Castigliano in 1879.

If energy is conserved, the total work performed on the structure increases the total strain energy of the structure. Thus $U = W$, and

$$U = \sum_{i=1}^{n} \int P_i \, d\delta_i \qquad [6.4\text{-}1]$$

Now, fixing the points of application of all the forces except one specific point i, allow point i to deflect an infinitesimal amount $\Delta\delta_i$ in the direction of P_i by applying an infinitesimal force ΔP_i. The additional work ΔW will cause an infinitesimal change in strain energy of the system ΔU. Thus

$$\Delta W = P_i \, \Delta\delta_i + \int_0^{\Delta\delta_i} \Delta P_i \, d\delta_i = \Delta U^\dagger$$

The work term in the integral is of the order of the product of two infinitesimal terms and thus can be neglected. Therefore, $\Delta U = P_i \, \Delta\delta_i$, or

$$P_i = \frac{\Delta U}{\Delta\delta_i} \qquad [6.4\text{-}2]$$

If $\Delta\delta_i$ is now allowed to approach zero, Eq. (6.4-2) reduces to

$$P_i = \frac{\partial U}{\partial\delta_i} \qquad [6.4\text{-}3]$$

Equation (6.4-3) expresses Castigliano's first theorem and represents n equations since $i = 1, 2, \ldots, n$.

To use Eq. (6.4-3), it is necessary to describe the strain energy as a function of the deflections δ_i, which may be difficult to do as the geometry of deformation is necessary. This is the disadvantage of Eq. (6.4-3).

Equation (6.4-3) is also applicable to problems where the force-deflection relation is nonlinear provided energy is conserved and rotations are small.

Example 6.4-1

Two cables shown in Fig. 6.4-1 are made of a nonlinear material whose stress-strain behavior is $\sigma = E\varepsilon - K\varepsilon^2$, where $E = 30$ Mpsi and $K = 10$ Gpsi. The cross-sectional area A of each cable is 0.2 in², and both have a length L of 5 ft. Determine the vertical deflection δ_B of point B after a load P_B of 2 kips is applied.

Solution:

Thanks to symmetry, the strain energies in each cable are equal. Thus, it is only necessary to evaluate one cable, for example, BC. The total energy is then

$$U = 2 \int F_{BC} \, d\delta_{BC}$$

---

† Here P_i was considered constant through the deflection $\Delta\delta_i$. The force ΔP_i is gradually applied throughout the loading period, resulting in the integral term.

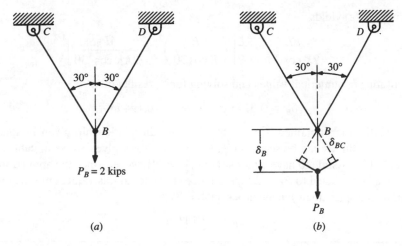

(a)   (b)

Figure 6.4-1

Since $F_{BC} = \sigma A = (E\varepsilon - K\varepsilon^2)_{BC} A$ and $\varepsilon_{BC} = \delta_{BC}/L$, we have

$$F_{BC} = \left[E \frac{\delta_{BC}}{L} - K\left(\frac{\delta_{BC}}{L}\right)^2 \right] A \qquad\qquad [a]$$

$$U = 2A \int_0^{\delta_{BC}} \left[E \frac{\delta_{BC}}{L} - K\left(\frac{\delta_{BC}}{L}\right)^2 \right] d\delta_{BC} = 2AL\left[\frac{E}{2}\left(\frac{\delta_{BC}}{L}\right)^2 - \frac{K}{3}\left(\frac{\delta_{BC}}{L}\right)^3 \right]$$

Now from Eq. (6.4-3)

$$P_B = \frac{\partial U}{\partial \delta_B} = 2A\left[E\left(\frac{\delta_{BC}}{L}\right) - K\left(\frac{\delta_{BC}}{L}\right)^2 \right] \frac{\partial \delta_{BC}}{\partial \delta_B} \qquad\qquad [b]$$

It is now necessary to obtain a relationship between δ_{BC} and δ_B using geometry of deformation. From Fig. 6.4-1(b) it can be seen that

$$\delta_{BC} = \delta_B \cos 30 \qquad\qquad [c]$$

Therefore

$$\frac{\partial \delta_{BC}}{\partial \delta_B} = \cos 30 \qquad\qquad [d]$$

Substituting Eqs. (c) and (d) into Eq. (b) yields

$$P_B = 2A\left[E \frac{\delta_B}{L} - K \cos 30 \left(\frac{\delta_B}{L}\right)^2 \right] \cos^2 30$$

Rearranging results in

$$\left(\frac{\delta_B}{L}\right)^2 - \frac{E}{K \cos 30} \frac{\delta_B}{L} + \frac{P_B}{2AK \cos^3 30} = 0 \qquad\qquad [e]$$

Solving Eq. (e) yields

$$\delta_B = \frac{EL}{2K \cos 30} \pm \frac{L}{2}\left[\left(\frac{E}{K \cos 30}\right)^2 - \frac{4P_B}{2AK \cos^3 30}\right]^{1/2}$$ [f]

Substituting in numerical values and solving for δ_B results in

$$\delta_B = 0.0143 \text{ in} \qquad \text{or} \qquad 0.194 \text{ in}$$

Two solutions are obtained because the stress-strain relationship given is double-valued in strain, as can be seen in Fig. 6.4-2. Thus for a given stress in cables BC and BD, it is possible to have either strain ε_1 or ε_2. However, since the applied force is gradually increased to $P_B = 2$ kips and does not exceed this value, only the lower strain (and displacement) value is acceptable. Thus

$$\delta_B = 0.0143 \text{ in}$$

Note that in Example 6.4-1 the equilibrium force equations were not used in obtaining the solution. Although in this example it would be quite easy to solve for F_{BC} and F_{BD} using the static equilibrium equations, it would not be a simple matter if the problem were statically indeterminate. When Castigliano's first theorem is used, the solution of the forces comes out of the analysis without any difficulty. If δ_B is substituted into Eq. (c),

$$\delta_{BC} = 0.0143 \cos 30 = 0.0124 \text{ in}$$

Substituting δ_{BC} into Eq. (a) yields

$$F_{BC} = \left[(30 \times 10^6)\frac{0.0124}{60} - (10 \times 10^9)\left(\frac{0.0124}{60}\right)^2\right](0.2) = 1155 \text{ lb}$$

The method in evaluating forces in statically indeterminate problems is basically the same as that illustrated in Example 6.4-1.

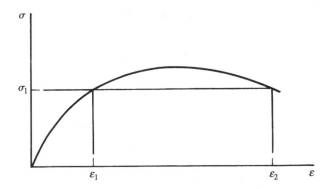

Figure 6.4-2

Example 6.4-2

Use Castigliano's first theorem to determine the cable forces of the symmetric structure shown in Fig. 6.4-3. The length, area, and modulus of cable 1 are L_1, A_1, and E_1 and of cable 2 are L_2, A_2, and E_2; etc. The materials are linear.

Solution:

The total strain energy is

$$U = \int F_1 \, d\delta_1 + 2 \int F_2 \, d\delta_2 + 2 \int F_3 \, d\delta_3 \qquad [a]$$

Since the force-deflection relationship for each cable is

$$F_i = \left(\frac{AE}{L}\right)_i \delta_i \qquad [b]$$

Eq. (a) becomes

$$U = \left(\frac{AE}{L}\right)_1 \frac{\delta_1^2}{2} + \left(\frac{AE}{L}\right)_2 \delta_2^2 + \left(\frac{AE}{L}\right)_3 \delta_3^2 \qquad [c]$$

The deflection of point B is the deflection δ_1 of cable 1. Thus from Eq. (6.4-3)

$$P_B = \frac{\partial U}{\partial \delta_1} = \left(\frac{AE}{L}\right)_1 \delta_1 + 2\left(\frac{AE}{L}\right)_2 \delta_2 \frac{\partial \delta_2}{\partial \delta_1} + 2\left(\frac{AE}{L}\right)_e \delta_e \frac{\partial \delta_3}{\partial \delta_1} \qquad [d]$$

As in Example 6.4-1, a geometric relationship is established:

$$\delta_2 = \delta_1 \cos \alpha_2 \qquad \delta_3 = \delta_1 \cos \alpha_3 \qquad [e]$$

and

$$\frac{\partial \delta_2}{\partial \delta_1} = \cos \alpha_2 \qquad \frac{\partial \delta_3}{\partial \delta_1} = \cos \alpha_3 \qquad [f]$$

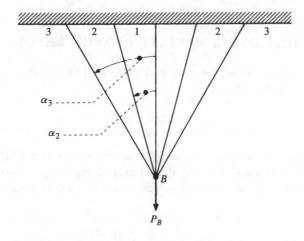

Figure 6.4-3

Substituting Eqs. (e) and (f) into Eq. (d) results in

$$P_B = \left[\left(\frac{AE}{L} \right)_1 + 2\left(\frac{AE}{L} \right)_2 \cos^2\alpha_2 + 2\left(\frac{AE}{L} \right)_3 \cos^2\alpha_3 \right]\delta_1$$

Solving for δ_1 yields

$$\delta_1 = \frac{P_B}{K} \qquad\qquad \text{[g]}$$

where

$$K = \left(\frac{AE}{L} \right)_1 + 2\left(\frac{AE}{L} \right)_2 \cos^2\alpha_2 + 2\left(\frac{AE}{L} \right)_3 \cos^2\alpha_3 \qquad \text{[h]}$$

Substituting δ_1 into Eq. (b) yields the force in cable 1. Thus

$$F_1 = \left(\frac{AE}{L} \right)_1 \frac{P}{K} \qquad\qquad \text{[i]}$$

Substituting δ_1 into Eqs. (e) and then substituting into Eq. (b) yields the forces in cables 2 and 3:

$$F_2 = \left(\frac{AE}{L} \right)_2 \frac{P}{K} \cos\alpha_2 \qquad F_3 = \left(\frac{AE}{L} \right)_3 \frac{P}{K} \cos\alpha_3 \qquad \text{[j]}$$

The approach outlined in Example 6.4-2 can be applied just as easily to any number of cables. The one disadvantage of Castigliano's first theorem comes from the fact that geometry of deformation is necessary, and in many cases some difficulties arise in obtaining the deflection relationships. The complementary-energy theorem circumvents this problem.

6.5   THE COMPLEMENTARY-ENERGY THEOREM

The definition of the complementary work performed by a gradually applied force P undergoing a deflection δ is

$$W_c = \int_0^{P_1} \delta \, dP \qquad\qquad \text{[6.5-1]}$$

where δ is the component of the total deflection in the direction of P. The complementary work is illustrated by the shaded area of Fig. 6.5-1. Defining the complementary energy of the system Φ as being equal to the complementary work, we have

$$\Phi = \int_0^{P_1} \delta \, dP \qquad\qquad \text{[6.5-2]}$$

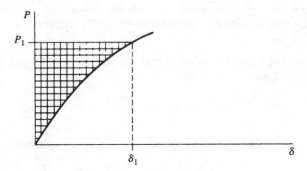

Figure 6.5-1

If a number of forces P_j are applied to a structure and each P_j deflects δ_j, where δ_j is the component of the total deflection of P_j in the direction of the line of action of the force, the total complementary energy of the system is

$$\Phi = \sum_{j=1}^{n} \int \delta_j \, dP_j \qquad \textbf{[6.5-3]}$$

where n is the total number of applied forces. Now, if an infinitesimal force ΔP_i were placed at point i *prior* to the applied loading, the increase in the complementary energy would be

$$\Delta \Phi = \delta_i \, \Delta P_i \qquad \text{or} \qquad \delta_i = \frac{\Delta \Phi}{\Delta P_i}$$

Allowing ΔP_i to approach zero results in

$$\delta_i = \frac{\partial \Phi}{\partial P_i} \qquad \textbf{[6.5-4]}$$

As before with Eq. (6.4-3), Eq. (6.5-4) actually represents n equations since $i = 1, 2, \ldots, n$.

It can be seen in Eq. (6.5-4) that the complementary energy must be written in terms of the applied forces, which are arrived at through the equilibrium equations. Thus, to obtain the deflections, Eq. (6.5-4) is used without resorting to the overall geometry of deformation.

If rotations within a structure are to be considered, it can also be shown that

$$\theta_i = \frac{\partial \Phi}{\partial M_i} \qquad \textbf{[6.5-5]}$$

is a simple extension of Eq. (6.5-4), where θ_i is the rotation point i in the direction of a concentrated moment M_i applied at point i.

Example 6.5-1 | Repeat Example 6.0-1 using the complementary-energy theorem.

Solution:

For a cable made of a linear elastic material, the complementary energy is equal to the strain energy. Thus

$$\Phi = U = \frac{1}{2}\frac{F^2 L}{AE}$$

For the system of two cables in Example 6.0-1, the total complementary energy is

$$\Phi = \frac{1}{2}\frac{F_{BC}^2 L_{BC}}{AE} + \frac{1}{2}\frac{F_{BD}^2 L_{BD}}{AE}$$

But

$$F_{BC} = 0.732P \qquad F_{BD} = 0.518P$$

Therefore

$$\Phi = \frac{0.535P^2 L_{BC}}{2AE} + \frac{0.268P^2 L_{BD}}{2AE}$$

The deflection of point B in the vertical direction is given by Eq. (6.5-4) and is

$$(\delta_B)_V = \frac{\partial \Phi}{\partial P} = \frac{0.536 P L_{BC}}{AE} + \frac{0.268 P L_{BD}}{AE} = \frac{P}{AE}(0.536 L_{BC} + 0.268 L_{BD})$$

Substituting numerical values results in

$$(\delta_B)_V = \frac{2000}{(0.2)(30)(10^6)}[(0.536)(24) + (0.268)(29.4)] = 6.91 \times 10^{-3} \text{ in}$$

To determine the horizontal deflection of point B, a horizontal dummy load H must be applied at point B so that Eq. (6.5-4) can be used (see Fig. 6.5-2).[†] The total forces in cables BC and BD are

$$F_{BC} = 0.732P + 0.732H \qquad F_{BD} = 0.518P - 0.897H$$

The total complementary energy is therefore

$$\Phi = \frac{1}{2AE}(0.732P + 0.732H)^2 L_{BC} + \frac{1}{2AE}(0.518P - 0.897H)^2 L_{BD}$$

Using Eq. (6.5-4) yields[‡]

$$(\delta_B)_H = \frac{\partial U}{\partial H} = \frac{(0.732)(0.732P + 0.732H)L_{BC}}{AE} - \frac{(0.897)(0.518P - 0.897H)L_{BD}}{AE}$$

[†] It is not necessary that H be infinitesimal as in Example 6.0-3. As a matter of fact, if a finite H did exist on the structure, this analysis would yield the correct solution. However, for this example, at the end of the analysis, H will be set equal to zero.

[‡] It is unnecessary to square the terms in Φ before differentiation.

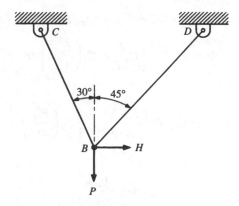

Figure 6.5-2

However, $H = 0$; therefore,

$$(\delta_B)_H = \frac{P}{AE}[(0.732)^2 L_{BC} - (0.897)(0.518)L_{BD}]$$

$$= \frac{2000}{(0.2)(30)(10^6)}[(0.732)^2(24) - (0.897)(0.518)(29.4)]$$

$$= -0.260 \times 10^{-3} \text{ in}^\dagger$$

The complementary-energy theorem is also suitable for nonlinear load-vs.-deflection relations provided rotations are small.

Repeat Example 6.4-1, however, with $\varepsilon = C_1\sigma + C_2\sigma^2$; where C_1 and C_2 are constants.   **Example 6.5-2**

Solution:

By curve fitting, within the range of $0 \leq \sigma \leq 6$ ksi, the constants were found to be $C_1 = 3.309 \, (10^{-8})$ psi^{-1}, and $C_2 = 4.649 \, (10^{-13})$ psi^{-2}.

As in Example 6.4-1, from symmetry the complementary energies of the cables are identical. Thus the total complementary energy can be written in terms of one cable, say, cable BC. Thus

$$\Phi = 2\Phi_{BC} = 2\int_0^{F_{BC}} \delta_{BC} \, dF$$

---

† Again, the negative sign indicates that the deflection is in the direction opposite H.

Since $\delta_{BC} = (\varepsilon L)_{BC}$, where $\varepsilon_{BC} = (C_1\sigma + C_2\sigma^2)_{BC}$ and $\sigma_{BC} = F/A$, we have

$$\Phi = 2L \int_0^{F_{BC}} \left[C_1\frac{F}{A} + C_2\left(\frac{F}{A}\right)^2 \right] dF$$

Integration yields

$$\Phi = 2LA\left[\frac{1}{2}C_1\left(\frac{F_{BC}}{A}\right)^2 + \frac{1}{3}C_2\left(\frac{F_{BC}}{A}\right)^3 \right]$$

from equilibrium

$$F_{BC} = \frac{P_B}{2 \cos 30}$$

Therefore,

$$\Phi = 2L\left(\frac{C_1 P_B^2}{8A \cos^2 30} + \frac{C_2 P_B^3}{24A^2 \cos^3 30} \right)$$

Using Eq. (6.5-4), we see that the deflection of point B is

$$\delta_B = \frac{\partial \Phi}{\partial P_B} = \frac{C_1 P_B L}{2A \cos^2 30} + \frac{C_2 P_B^2 L}{4A^2 \cos^2 30}$$

Substitution of numerical values from Example 6.4-1 results in

$$\delta_B = \frac{(3.309)(10^{-8})(2000)(60)}{(2)(0.2)\cos^2 30} + \frac{(4.649)(10^{-13})(2000)^2(60)}{(4)(0.2)^2\cos^3 30} = 0.0143 \text{ in}$$

which can be seen to agree with the results of Example 6.4-1.

6.6   CASTIGLIANO'S SECOND THEOREM

6.6.1   DEFLECTIONS OF STATICALLY DETERMINATE PROBLEMS

As demonstrated in Example 6.5-1, if the load-displacement relation is *linear,* the complementary energy Φ equals the strain energy U. Thus, Eqs. (6.5-4) and (6.5-5) can be written

$$\delta_i = \frac{\partial U}{\partial P_i} \qquad\qquad \textbf{[6.6-1]}$$

and

$$\theta_i = \frac{\partial U}{\partial M_i} \qquad\qquad \textbf{[6.6-2]}$$

Equations (6.6-1) and (6.6-2) represent Castigliano's second theorem. Many practical engineering problems involve linear load-deflection relations where the forms of the strain energy are known for the basic load cases (see Sec. 6.3). The use of Eqs. (6.6-1) and (6.6-2) provides a simple and straightforward approach to the deflection analysis of a complex collection of simple elements and is commonly referred to as *Castigliano's method*.

Example 6.6-1

For the structure shown in Fig. 6.6-1 use Castigliano's method to determine the vertical deflection of point B. All members have a cross-sectional area A; members 1, 2, 4, and 5 are of length L, and members 3, 6, and 7 are of length $\sqrt{2}L$.

Solution:

The total strain energy is the sum of the strain energies of all the members. Thus

$$U = U_1 + U_2 + U_3 + U_5 + U_6 + U_7$$

Since each member is loaded axially, the energy in the ith member is

$$U_i = \frac{1}{2}\frac{F_i^2 L_i}{AE} \qquad i = 1, 2, \ldots, 7$$

Thus it is necessary to determine the axial load in each member F_i using static analysis. Assuming tension positive for each member, we have

$$F_1 = P \qquad F_2 = P \qquad F_3 = -\sqrt{2}P \qquad F_4 = 0$$

$$F_5 = -2P \qquad F_6 = 0 \qquad F_7 = \sqrt{2}P$$

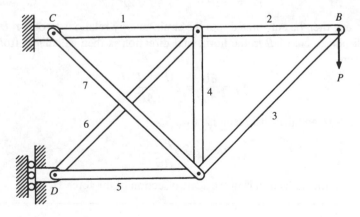

Figure 6.6-1

The total energy is then given by

$$U = \frac{1}{2}\frac{P^2L}{AE} + \frac{1}{2}\frac{P^2L}{AE} + \frac{1}{2}\frac{(-\sqrt{2}P)^2(\sqrt{2}L)}{AE}$$

$$+\ 0 + \frac{1}{2}\frac{(-2P)^2L}{AE} + 0 + \frac{1}{2}\frac{(\sqrt{2}P)^2(\sqrt{2}L)}{AE}$$

Simplifying results in

$$U = \frac{11.66}{2}\frac{P^2L}{AE}$$

From Eq. (6.6-1), the vertical deflection $(\delta_B)_V$ of point B is $\partial U/\partial P$. Thus

$$(\delta_B)_V = \frac{\partial U}{\partial P} = 11.66\frac{PL}{AE}$$

In Example 6.6-1 in order to determine the horizontal deflection of point B using Eq. (6.6-1), it would be necessary to place a dummy force at B in the horizontal direction.

Example 6.6-2 | Using Castigliano's method, determine the horizontal deflection of point B in Example 6.6-1.

Solution:

Place a horizontal dummy force H at point B as shown in Fig. 6.6-2. A static analysis of the forces in the members caused by H alone can be added by superposition to the forces caused by P, as determined in Example 6.6-1. Thus

$$F_1 = F_2 = P + H \qquad F_3 = -\sqrt{2}P \qquad F_4 = F_6 = 0$$

$$F_5 = -2P \qquad F_7 = \sqrt{2}P$$

The forces can then be substituted into the strain-energy relations to obtain U_1, U_2, etc. The displacement of B in the horizontal direction is then calculated from Eq. (6.6-1), where

$$(\delta_B)_H = \frac{\partial U}{\partial H} = 2\frac{(P + H)L}{AE}$$

However, $H = 0$, and the final result is

$$(\delta_B)_H = 2\frac{PL}{AE}$$

Being positive, the deflection is in the same direction of the force H.

The displacement of any point (translational or rotational) in a structure in any direction can be found by (1) placing a dummy force or moment at the point in the

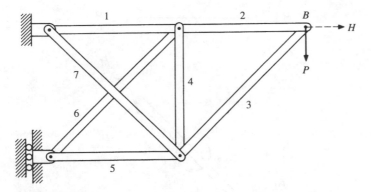

Figure 6.6-2

direction desired, (2) determining the strain energy of the structure due to all applied forces and the dummy load, (3) substituting the total energy into Eq. (6.6-1) or (6.6-2), and (4) equating the dummy load to zero.

Using Castigliano's method, determine the rotation of member 3 in Example 6.6-1. | **Example 6.6-3**

Solution:

In addition to the applied force P, place a counterclockwise moment M *anywhere* on member 3. Perform a static analysis on the structure with only the moment on it. Solving for the reactions at C and D and then performing a joint analysis results in

$$F_1 = F_2 = -M/L \qquad F_4 = F_6 = 0 \qquad F_5 = M/L$$

For member 3 the moment causes transverse forces at the end perpendicular to the member of the magnitude $M/(\sqrt{2}L)$ as shown in Fig. 6.6-3. Since the force in adjoining member 2 must be horizontal, a tensile axial force must develop in member 3 of the value $M/(\sqrt{2}L)$ to cancel the vertical component of the transverse force. The resulting axial forces in the members are

$$F_1 = F_2 = P - M/L \qquad F_3 = -\sqrt{2}P + M/(\sqrt{2}L)$$

$$F_4 = F_6 = 0 \qquad F_5 = -2P + M/L \qquad F_7 = \sqrt{2}P$$

The total energy is then[†]

$$U = 2\left\{\frac{1}{2}\frac{[P - (M/L)]^2 L}{AE}\right\} + \frac{1}{2}\frac{[-\sqrt{2}P + (M/\sqrt{2}L)]^2\sqrt{2}L}{AE}$$

$$+ \frac{1}{2}\frac{[-2P + (M/L)]^2 L}{AE} + \frac{1}{2}\frac{(\sqrt{2}P)^2(\sqrt{2}L)}{AE}$$

[†] The bending energy in member 3 need not be considered, as it is exclusively a function of the dummy moment, which will eventually be set to zero.

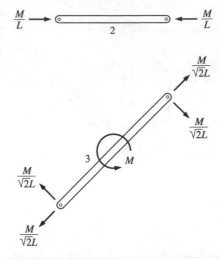

Figure 6.6-3

Differentiate with respect to M and then set M to zero. This gives

$$\theta_3 = -2\frac{P}{AE} - \sqrt{2}\frac{P}{AE} - 2\frac{P}{AE} = -5.414\frac{P}{AE}$$

The minus sign indicates that member 3 rotates opposite M, i.e., clockwise.

The next several examples give additional illustrations of Castigliano's method.

Example 6.6-4 Determine the vertical deflection and slope at point A of the end-loaded cantilever beam shown in Fig. 6.6-4. The stiffness of the beam is EI_z. (Neglect transverse shear stresses.)

Solution:

Since P is applied at point A and in the vertical direction,

$$\delta_A = \frac{\partial U}{\partial P}$$

The strain-energy formulation for a beam in bending is

$$U_b = \frac{1}{2EI}\int_0^L M_z^2 \, dx$$

The bending moment as a function of x is

$$M_z = -Px \qquad 0 < x < L$$

Therefore

$$U_b = \frac{1}{2EI_z} \int_0^L (-Px)^2 \, dx = \frac{P^2 L^3}{6EI_z}$$

Hence

$$\delta_A = \frac{\partial U_b}{\partial P} = \frac{PL^3}{3EI_z}$$

To obtain the slope at point A, it is necessary to apply a dummy moment M_A at A so that Eq. (6.6-2) can be used. This is shown in Fig. 6.6-5. The bending moment as a function of x is then

$$M_z = -Px - M_A \qquad 0 < x < L$$

$$U_b = \frac{1}{2EI_z} \int_0^L (-Px - M_A)^2 \, dx = \frac{1}{2EI_z} \left(\frac{P^2 L^3}{3} + PM_A L^2 + M_A^2 L \right)$$

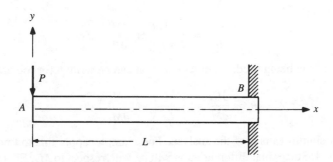

Figure 6.6-4

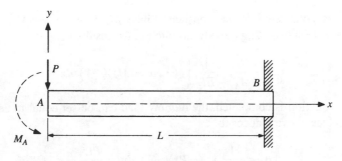

Figure 6.6-5

To find the slope, Eq. (6.6-2) is used:

$$\theta_A = \frac{\partial U_b}{\partial M_A} = \frac{1}{2EI_z}(PL^2 + 2M_AL)$$

However, since M_A is a dummy moment, $M_A = 0$ and

$$\theta_A = \frac{PL^2}{2EI_z}$$

Since θ_A is positive, the beam at point A has rotated in the same direction as M_A.

Note that the applied *concentrated* forces or moments necessary for Eqs. (6.6-1) and (6.6-2) are not functions of x. This means that the partial differentiation with respect to the forces or moments can be performed before integration. That is, since

$$U_b = \frac{1}{2}\int \frac{1}{EI_z} M_z^2 \, dx$$

differentiation with respect to a force, say P_i, when dealing with Eq. (6.6-1) can be written as

$$\delta_i = \frac{\partial U_b}{\partial P_i} = \int \frac{1}{EI_z}\left(M_z \frac{\partial M_z}{\partial P_i}\right) dx \qquad \textbf{[6.6-3]}$$

or if Eq. (6.6-2) is being used, a similar equation can be written for rotations:

$$\theta_i = \frac{\partial U_b}{\partial M_i} = \int \frac{1}{EI_z}\left(M_z \frac{\partial M_z}{\partial M_i}\right) dx \qquad \textbf{[6.6-4]}$$

This greatly simplifies part of the analysis. For example, apply this to finding θ_A of Example 6.6-4. Since final differentiation will be with respect to M_A, Eq. (6.6-4) becomes

$$\theta_A = \frac{\partial U_b}{\partial M_A} = \frac{1}{EI_z}\int M_z \frac{\partial M_z}{\partial M_A} \, dx \qquad \textbf{[a]}$$

where EI_z is considered to be constant. Since $M_z = -Px - M_A$, we have $\partial M_z/\partial M_A = -1$. Substituting these terms into Eq. (a) results in

$$\theta_A = \frac{\partial U_b}{\partial M_A} = \frac{1}{EI_z}\int_0^L (-Px - M_A)(-1) \, dx$$

Once differentiation is completed, the dummy load can be set equal to zero. Thus $M_A = 0$, and the above reduces to

$$\theta_A = \frac{1}{EI_z}\int_0^L (-Px)(-1) \, dx = \frac{PL^2}{2EI_z}$$

The advantages of Eqs. (6.6-3) and (6.6-4) are twofold: (1) when a dummy load is necessary, it can be set equal to zero earlier in the analysis, making the integration easier; (2) the integration of Eqs. (6.6-3) and (6.6-4) is easier than Eqs. (6.6-1) and (6.6-2) when the moment M_z is a function of more than one load.

Using Castigliano's theorem, determine the vertical deflection at point A of the cantilever beam shown in Fig. 6.6-6(a).[†] The stiffness of the beam is $EI_z = 10 \times 10^4$ N·m², $P = 500$ N, $L = 1$ m, and $b = 0.2$ m. (Neglect shear effects.) | **Example 6.6-5**

Solution:

Since there is no force at A, a dummy force Q is introduced. The bending-moment equation is then

$$M_z = \begin{cases} -Qx & 0 < x < b & \text{[a]} \\ -Qx - P(x - b) & b < x < L & \text{[b]} \end{cases}$$

Since

$$\delta_A = \left.\frac{\partial U}{\partial Q}\right|_{Q=0}$$

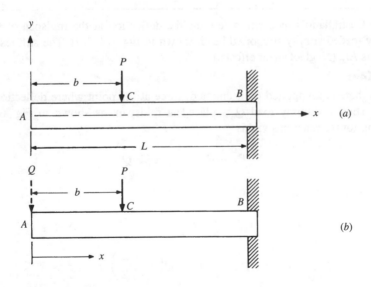

Figure 6.6-6

[†] Note that for cantilever beams the expression for the bending moment is usually slightly simpler if the variable x originates from the free end rather than the wall.

it is necessary to find $\partial M_z / \partial Q$. From Eqs. (a) and (b)

$$\frac{\partial M_z}{\partial Q} = -x \qquad 0 < x < L$$

Thus

$$\delta_A = \frac{\partial U}{\partial Q} = \frac{1}{EI_z} \left\{ \int_0^b (-Qx)(-x) \, dx + \int_b^L [-Qx - P(x-b)](-x) \, dx \right\}$$

However, since $Q = 0$,

$$\delta_A = \frac{\partial U}{\partial Q}\bigg|_{Q=0} = \frac{1}{EI_z} \int_b^L P(x-b)x \, dx$$

and integration yields

$$\delta_A = \frac{1}{EI_z} \left[\frac{P}{3}(L^3 - b^3) - \frac{Pb}{2}(L^2 - b^2) \right]$$

which reduces to

$$\delta_A = \frac{P}{6EI_z}(2L^3 - 3bL^2 + b^3) = \frac{500}{(6)(10 \times 10^4)}[(2)(1^3) - (3)(0.2)(1^2) + (0.2)^3]$$

$$= 1.17 \times 10^{-3} \text{ m} = 1.17 \text{ mm}$$

Example 6.6-6 | Using Castigliano's theorem, determine the deflection at the midspan of the uniformly loaded simply supported beam shown in Fig. 6.6-7(a). The stiffness of the beam is EI_z. (Neglect shear effects.)

Solution:

Again there is no applied (concentrated) force at the point where deflection is desired. Thus, a dummy load Q is placed at midspan, as shown in Fig. 6.6-7(b). Solving for the reactions yields

$$R_A = R_B = \frac{1}{2}(wL + Q)$$

The bending-moment equation is

$$M_z = \begin{cases} \dfrac{1}{2}(wL + Q)x - \dfrac{1}{2}wx^2 & 0 < x < \dfrac{L}{2} \\[2ex] \dfrac{1}{2}(wL + Q)x - \dfrac{1}{2}wx^2 - Q\left(x - \dfrac{L}{2}\right) & \dfrac{L}{2} < x < L \end{cases}$$

However, thanks to symmetry, the energy in the first half of the beam is the same as in the other half. Thus it is only necessary to integrate over half the beam and double the results. Therefore

$$M_z = \frac{1}{2}(wL + Q)x - \frac{1}{2}wx^2 \qquad 0 < x < \frac{L}{2}$$

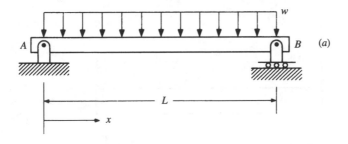

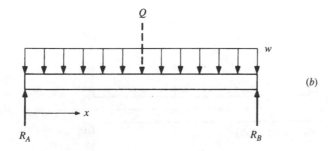

Figure 6.6-7

and

$$\frac{\partial M_z}{\partial Q} = \frac{x}{2}$$

The deflection at midspan is $\delta_Q = \partial U/\partial Q|_{Q=0}$. Thus

$$\delta_Q = \frac{\partial U}{\partial Q}\bigg|_{Q=0} = \frac{2}{EI_z} \int_0^{L/2} \left[\frac{1}{2}(wL + 0)x - \frac{1}{2}wx^2\right]\frac{x}{2}\,dx$$

Integration yields

$$\delta_Q = \frac{5}{384}\frac{wL^A}{EI_z}$$

For the beam shown in Fig. 6.6-8(a), determine the deflection of point A using Castigliano's method. The stiffness of the beam is $EI_z = 20 \times 10^4$ lb·in². | **Example 6.6-7**

Solution:

Substitute P_A and P_B for the loads at points A and B as shown.[†] The deflection at point A is

$$\delta_A = \frac{\partial U}{\partial P_A}$$

[†] Note that although the two applied forces are equal, it is important that they be considered distinctly different by virtue of the fact that they are acting at two different locations.

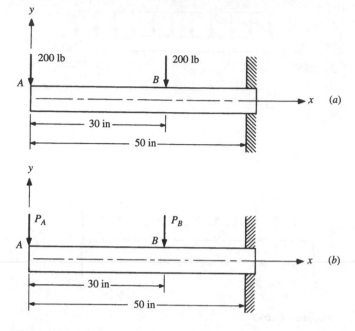

Figure 6.6-8

The bending-moment equations are

$$M_z = \begin{cases} -P_A x & 0 < x < 30 \\ -P_A x - P_B(x - 30) & 30 < x < 50 \end{cases}$$

and

$$\frac{\partial M_z}{\partial P_A} = -x \qquad 0 < x < 50$$

Thus

$$\delta_A = \frac{\partial U}{\partial P_A} = \frac{1}{EI_z}\left\{ \int_0^{30}(-P_A x)\,dx + \int_{30}^{50}[-P_A x - P_B(x - 30)](-x)\,dx\right\}$$

At this point, once differentiation is performed, it is acceptable to write $P_A = P_B = P$. Integration of the above yields

$$\delta_A = 5.033 \times 10^4 \frac{P}{EI_z}$$

Substituting $P = 200$ lb and $EI_z = 20 \times 10^6$ lb·in² results in

$$\delta_A = 0.503 \text{ in}$$

Using Castigliano's theorem, determine the deflection of point A of the step shaft shown in Fig. 6.6-9. The second-area moment of the beam between points A and B is I_1, and from point B to C the second-area moment is $I_2 = 2I_1$. The entire beam is made of a material with a modulus of elasticity of E.

Example 6.6-8

Solution:

The bending moment is

$$M_z = -Px \qquad 0 < x < L$$

Therefore

$$\frac{\partial M_z}{\partial P} = -x$$

Since the second-area moment is a discontinuous function of x, the integration must be divided into two parts. Thus

$$\delta_A = \frac{\partial U}{\partial P} = \frac{1}{EI_1}\int_0^{L/2} (-Px)(-x)\, dx + \frac{1}{EI_2}\int_{L/2}^L (-Px)(-x)\, dx$$

Evaluating the integrals and substituting $I_2 = 2I_1$ yields

$$\delta_A = \frac{3}{16}\frac{PL^3}{EI_1}$$

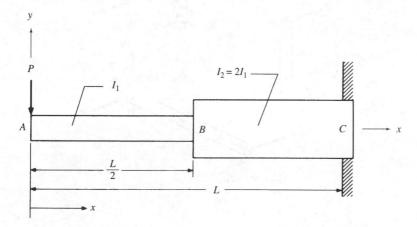

Figure 6.6-9

Example 6.6-9 | For the wire form shown in Fig. 6.6-10(a) determine the deflection of point B in the direction of the applied force P. (Neglect direct shear effects.) The diameter of the wire is d, the modulus of elasticity is E, and Poisson's ratio is v.

Solution:

Using *free-body diagrams,* the reader should verify that the reactions as functions of P in elements BC, CD, and GD are as shown.

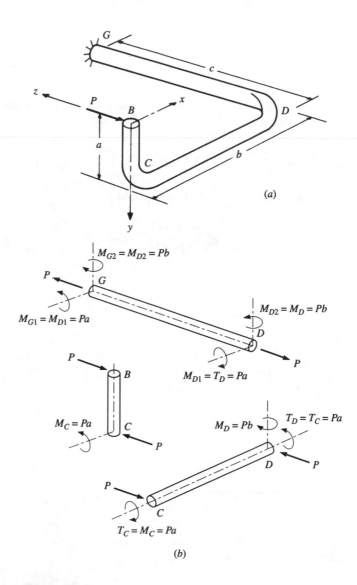

(a)

(b)

Figure 6.6-10

The strain energy in BC is due to bending alone:

$$\frac{\partial U_{BC}}{\partial P} = \frac{1}{EI} \int_0^a (-Py)(-y) \ dy = \frac{Pa^3}{3EI} \qquad \text{[a]}$$

The strain energy in CD is due to torsion and bending.[†]

$$\frac{\partial U_{CD}}{\partial P} = \frac{2(1+v)}{EJ}(Pa)(a)(b) + \frac{1}{EI} \int_0^b (-Px)(-x) \ dx$$

$$= \frac{2(1+v)Pa^2b}{EJ} + \frac{Pb^3}{3EI} \qquad \text{[b]}$$

The strain energy in DG is due to an axial load and bending in two planes:[‡]

$$\frac{\partial U_{DG}}{\partial P} = \frac{Pc}{AE} + \frac{1}{EI} \int_0^c (Pb)(b) \ dz + \frac{1}{EI} \int_0^c (Pa)(a) \ dz$$

$$= \frac{Pc}{AE} + \frac{Pb^2c}{EI} + \frac{Pa^2c}{EI} \qquad \text{[c]}$$

The deflection of point B in the direction of P is

$$(\partial_B)_P = \frac{\partial U}{\partial P} = \frac{\partial U_{BC}}{\partial P} + \frac{\partial U_{CD}}{\partial P} + \frac{\partial U_{DG}}{\partial P}$$

After some rearranging and using the fact that for a circular section $I = \pi d^4/64$ and $J = 2I$, the sum of Eqs. (a), (b), and (c) reduces to

$$(\delta_B)_P = \frac{4}{3\pi} \frac{P}{Ed^4} [16(a^3 + b^3) + 48c(a^2 + b^2) + 48(1+v)a^2b + 3cd^2]$$

The deflections of curved beams can be handled very effectively using Castigliano's theorem, especially when the radius of curvature of the beam is much larger than the lateral dimensions of the beam cross section. When this occurs, it is acceptable to use straight-beam equations for energy, to ignore direct axial and transverse shear forces, and to consider only bending.

[†] Note, for torsion $U_t = [(1+v)/EJ]T^2L$. Therefore,

$$\frac{\partial U_t}{\partial P} = \frac{2(1+v)}{EJ} T \frac{\partial T}{\partial P} L$$

[‡] Likewise, for axial loading

$$U_a = \frac{1}{2} \frac{F^2L}{AE}$$

therefore

$$\frac{\partial U_a}{\partial P} = \frac{F}{AE} \frac{\partial F}{\partial P} L$$

Example 6.6-10   The curved beam shown in Fig. 6.6-11(a) has a radius of curvature R, modulus E, and a sectional second-area moment I_z about an axis out of the page directed through the centroid of an area section. Considering only bending, determine the horizontal and vertical deflections of point A due to the application of the horizontal force P.

Solution:

It is necessary to establish how the bending moment varies throughout the beam. When the variable θ is used, as shown in Fig. 6.6-11(b), the bending moment can be determined as a function of θ. If this is done, an infinitesimal length of beam will be $R\,d\theta$. Substituting $R\,d\theta$ for dx in Eq. (6.6-3) gives

$$\delta_i = \frac{\partial U}{\partial P_i} = \int \frac{M_z}{EI_z} \frac{\partial M_z}{\partial P_i} R\,d\theta \qquad\qquad \textbf{[a]}$$

Summing moments at the center of the break at θ yields

$$M_z = -PR\sin\theta \qquad 0 < \theta < \pi \qquad\qquad \textbf{[b]}$$

The horizontal deflection $(\theta_A)_H$ of point A is given by $\partial U/\partial P$; thus $\partial M/\partial P$ is desired. From Eq. (b)

$$\frac{\partial M_z}{\partial P} = -R\sin\theta \qquad\qquad \textbf{[c]}$$

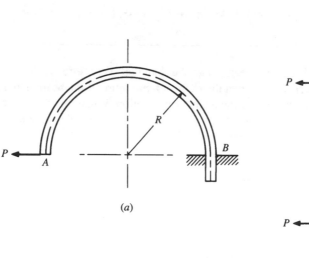

(a)

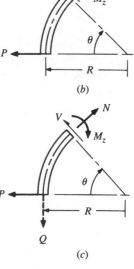

(b)

(c)

Figure 6.6-11

Substitution of Eqs. (b) and (c) into Eq. (a) results in

$$(\delta_A)_H = \frac{\partial U}{\partial P} = \int_0^\pi \frac{1}{EI_z}(-PR \sin \theta)(-R \sin \theta)R \; d\theta$$

Since EI_z, P, and R are constants, the integral simplifies to

$$(\delta_A)_H = \frac{PR^3}{EI_z} \int_0^\pi \sin^2 \theta \; d\theta \qquad \text{[d]}$$

After using the trigonometric identity $\sin^2 \theta = (1 - \cos 2\theta)/2$, integration of Eq. (d) reduces to

$$(\delta_A)_H = \frac{\pi}{2} \frac{PR^3}{EI_z}$$

For the vertical deflection $(\delta_A)_V$ of point A a dummy force Q must be added, as shown in Fig. 6.6-11(c), where

$$(\delta_A)_V = \frac{\partial U}{\partial Q}\bigg|_{Q=0} \qquad \text{[e]}$$

The bending moment as a function of θ is then given by

$$M_z = -PR \sin \theta + QR(1 - \cos \theta) \qquad 0 < \theta < \pi \qquad \text{[f]}$$

and since $\partial M_z/\partial Q$ will be needed,

$$\frac{\partial M_z}{\partial Q} = R(1 - \cos \theta) \qquad \text{[g]}$$

Now that differentiation has been completed, Eqs. (f) and (g) can be substituted into Eq. (e) with $Q = 0$. This results in

$$(\delta_A)_V = \int_0^\pi \frac{1}{EI_z}(-PR \sin \theta)[R(1 - \cos \theta)]R \; d\theta \qquad \text{[h]}$$

Integration of Eq. (h) gives

$$(\delta_A)_V = -2\frac{PR^3}{EI_z}$$

The negative sign indicates that the direction of the vertical displacement is opposite of Q.

Example 6.6-10 neglected the effects of the normal and transverse forces. The effect and order of magnitude of the contribution of these forces can be seen in the following example.

Example 6.6-11 | For Example 6.6-10, determine the horizontal deflection of point A considering the axial and transverse shear forces in addition to the bending moment.

Solution:

From Fig. 6.6-11(*b*) the complete internal reactions are

$$M_z = -PR \sin \theta \qquad \text{[a]}$$

$$N = P \sin \theta \qquad \text{[b]}$$

$$V = -P \cos \theta \qquad \text{[c]}$$

Since N and V are varying, Eqs. (6.3-1) and (6.3-11) are used with $dx = R\, d\theta$. Differentiation with respect to P can be performed before integration, and the resulting deflection equation is

$$(\delta_A)_H = \frac{\partial U}{\partial P} = \frac{1}{EI_z} \int M_z \frac{\partial M_z}{\partial P} R\, d\theta + \frac{1}{EA} \int N \frac{\partial N}{\partial P} R\, d\theta + \frac{2(1+v)}{EA} k \int V \frac{\partial V}{\partial P} R\, d\theta$$

[d]

where k is the form correction factor for shear.

From Eqs. (*a*) to (*c*) the derivative terms are

$$\frac{\partial M_z}{\partial P} = -R \sin \theta \qquad \text{[e]}$$

$$\frac{\partial N}{\partial P} = \sin \theta \qquad \text{[f]}$$

$$\frac{\partial V}{\partial P} = -\cos \theta \qquad \text{[g]}$$

Substituting Eqs. (*a*) to (*c*) and (*e*) to (*g*) into Eq. (*d*) yields

$$(\delta_A)_H = \frac{1}{EI_z} \int_0^\pi (-PR \sin \theta)(-R \sin \theta) R\, d\theta + \frac{1}{EA} \int_0^\pi (P \sin \theta)(\sin \theta) R\, d\theta$$

$$+ \frac{2(1+v)}{EA} k \int_0^\pi (-P \cos \theta)(-\cos \theta) R\, d\theta$$

Integration and simplification results in

$$(\delta_A)_H = \frac{\pi}{2} \frac{PR^3}{EI_z} + \frac{\pi}{2} \frac{PR}{EA} + \pi(1+v)k \frac{PR}{EA}$$

or

$$(\delta_A)_H = \frac{\pi}{2} \frac{PR^3}{EI_z} \left\{ 1 + [1 + 2(1+v)k]\left(\frac{I_z}{AR^2}\right) \right\} \qquad \text{[h]}$$

It can be seen from Eq. (h) of Example 6.6-11 that as the radius of curvature R approaches the radius of gyration $r_g = \sqrt{I_z/A}$, the axial and shear effects become appreciable. In addition, as the radius of curvature approaches the radius of gyration (a thick-walled curved beam), the stress distribution due to bending changes (see Sec. 5.6) and the bending energy must be reformulated. This is done in the next section of this chapter.

6.6.2   DEFLECTIONS DUE TO TEMPERATURE CHANGES

If an axial member that is transmitting a force F undergoes a temperature change ΔT, the strain energy will be

$$U = \frac{1}{2}\frac{F^2 L}{AE} + F\alpha L\,\Delta T \qquad\qquad \textbf{[6.6-5]}$$

where α is the coefficient of linear expansion of the material.

Example 6.6-12

For the structure of Example 6.6-1 (Fig. 6.6-1) let $L = 500$ mm, $P = 2$ kN, $A = 60$ mm^2, and $E = 210$ GPa. If member 5 undergoes an increase of temperature of $100°C$ and the coefficient of expansion is $\alpha = 11.7(10^{-6})\,(°C)^{-1}$, determine the vertical deflection of point B.

Solution:

From Example 6.6-1 the force in member 5 is $-2P$. Thus, from Eq. (6.6-5), the strain energy is

$$U_5 = \frac{1}{2}\frac{(-2P)^2 L}{AE} + (-2P)\alpha L_5\,\Delta T$$

Adding this energy to that of the remaining members gives

$$U = \frac{11.66}{2}\frac{P^2 L}{AE} - 2P\alpha L_5\,\Delta T$$

The vertical deflection of point B in the direction of P is thus

$$(\delta_B)_V = \frac{\partial U}{\partial P} = 11.66\frac{PL}{AE} - 2\alpha L_5\,\Delta T$$

Substitution of values gives

$$(\delta_B)_V = 11.66\frac{(2000)(0.5)}{(60)(10^{-6})(210)(10^9)} - 2(11.7)(10^{-6})(0.5)(100)$$

$$= -2.45(10^{-4})\text{ m} = -0.245\text{ mm}$$

where the minus indicates that point B deflects in the opposite direction to the load P. That is, point B deflects upward.

6.7   DEFLECTIONS OF THICK-WALLED CURVED BEAMS

The bending-stress distribution is given by Eq. (5.6-3), which is

$$\sigma_\theta = -\frac{M(r_n - r)}{Aer} \qquad \text{[6.7-1]}$$

Upon superposing the stress due to the axial force N, the total normal stress across the section becomes

$$\sigma_\theta = \frac{M(r - r_n)}{Aer} + \frac{N}{A} \qquad \text{[6.7-2]}^\dagger$$

In order to formulate the energy, an infinitesimal volume is necessary where the normal stress is constant. Recall Fig. 5.6-1, repeated here as Fig. 6.7-1; a volume element of $r\,d\theta\,dA$ is used. Thus the strain energy due to bending and the axial force is

$$U_b = \frac{1}{2}\int \frac{\sigma_\theta^2}{E}\,dV = \frac{1}{2}\int\int \frac{1}{E}\left[\frac{M(r-r_n)}{Aer} + \frac{N}{A}\right]^2 r\,dA\,d\theta$$

$$= \frac{1}{2}\int \frac{1}{EA^2}\left[\frac{M^2}{e^2}\frac{(r-r_n)^2}{r} + \frac{2MN}{e}(r-r_n) + N^2 r\right]dA\,d\theta \qquad \text{[6.7-3]}^\ddagger$$

Integration will be performed on each term in the square brackets separately and with respect to dA first. For the first term within the brackets only $(r - r_n)^2/r$ is a function of the area A. Thus

$$\int \frac{(r-r_n)^2}{r}\,dA = \int\left(r - 2r_n + \frac{r_n^2}{r}\right)dA = \int r\,dA - 2r_n\int dA + r_n^2\int \frac{dA}{r}$$

On the right-hand side of the equation, the first integral is simply $r_c A$, where r_c is the location of the centroid of the cross section. The second integral is A, and the third integral, recalling Eq. (5.6-2), is A/r_n. Thus

$$\int \frac{(r-r_n)^2}{r}\,dA = (r_c - 2r_n + r_n)A = (r_c - r_n)A = eA$$

Factoring out the first term in the brackets of Eq. (6.7-3) gives

$$\frac{1}{2}\int\int \frac{1}{E}\left(\frac{M_z}{Ae}\right)^2\frac{(r-r_n)^2}{r}\,dA\,d\theta = \frac{1}{2}\int \frac{M_z^2}{EAe}\,d\theta \qquad \text{[6.7-4]}$$

$\dagger$ The strain energies due to M and N are not formulated independently; as will be seen later, they are coupled. This is because the centroidal axis for curved beams has a strain due to bending and thus N has an additional deflection caused by M, giving rise to an additional work term or strain-energy term.

$\ddagger$ This neglects the energy due to the radial stress σ_r, given by Eq. (5.6-10). Normally this energy is extremely small compared with the bending energy.

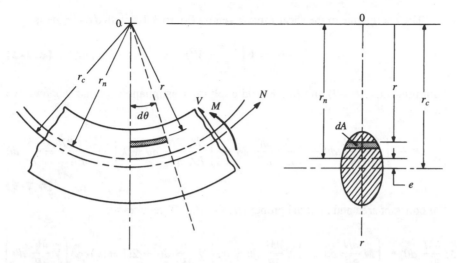

Figure 6.7-1

Returning to Eq. (6.7-3), we see that the second term, $r - r_n$ is a function of A. Thus

$$\int (r - r_n)\, dA = (r_c - r_n)A = eA$$

and the second term in the integral of Eq. (6.7-3) is

$$\frac{1}{2} \int \int \frac{2MN}{EA^2 e}(r - r_n)\, dA\, d\theta = \int \frac{MN}{EA}\, d\theta \qquad \textbf{[6.7-5]}$$

The final term within the [ ] brackets of Eq. (6.7-3) involves the integral of $r\, dA$, which is

$$\int r\, dA = r_c A$$

Thus the last term in the integral of Eq. (6.7-3) is

$$\frac{1}{2} \int \int \frac{N^2 r}{EA^2}\, dA\, d\theta = \frac{1}{2} \int \frac{N^2 r_c}{EA}\, d\theta \qquad \textbf{[6.7-6]}$$

When Eqs. (6.7-4) to (6.7-6) are added, the strain energy due to the bending and axial loading is

$$U_b = \frac{1}{2} \int \frac{M_z^2}{EAe}\, d\theta + \int \frac{M_z N}{EA}\, d\theta + \frac{1}{2} \int \frac{N^2 r_c}{EA}\, d\theta \qquad \textbf{[6.7-7]}$$

The second integral in Eq. (6.7-7) is the additional coupling term between the bending moment and the axial force.

The energy due to the shear force given by Eq. (6.3-10) with $dx = r_c \, d\theta$ is

$$U_s = k \int \frac{1 + \nu}{EA} V^2 r_c \, d\theta \qquad \text{[6.7-8]}$$

Summing Eqs. (6.7-7) and (6.7-8) and applying Castigliano's second theorem, we have

$$\delta_i = \frac{\partial U}{\partial P_i} = \int \frac{M}{EAe} \frac{\partial M}{\partial P_i} \, d\theta + \int \frac{M}{EA} \frac{\partial N}{\partial P_i} \, d\theta + \int \frac{N}{EA} \frac{\partial M}{\partial P_i} \, d\theta + \int \frac{Nr_c}{EA} \frac{\partial N}{\partial P_i} \, d\theta + k \int \frac{2(1 + \nu) r_c V}{EA} \frac{\partial V}{\partial P_i} \, d\theta$$

$$\text{[6.7-9]}$$

For constant area and material properties, Eq. (6.7-9) reduces to[†]

$$\delta_i = \frac{1}{EA} \left[\frac{1}{e} \int M \frac{\partial M}{\partial P_i} \, d\theta + \int M \frac{\partial N}{\partial P_i} \, d\theta + \int N \frac{\partial M}{\partial P_i} \, d\theta + r_c \int N \frac{\partial N}{\partial P_i} \, d\theta + 2(1 + \nu) k r_c \int V \frac{\partial V}{\partial P_i} \, d\theta \right]$$

$$\text{[6.7-10]}$$

Example 6.7-1 | Repeating Example 6.6-11 but for a thick-walled steel ring, as shown in Fig.6.7-2, determine the deflection of point A using Eq. (6.7-10) and compare the results with Eq. (d) of Example 6.6-11.

Solution:

Recall from Example 6.6-11 that the internal reactions and derivatives are

$$M_z = -PR \sin \theta \qquad \frac{\partial M_z}{\partial P} = -R \sin \theta$$

$$N = P \sin \theta \qquad \frac{\partial N}{\partial P} = \sin \theta$$

$$V = -P \cos \theta \qquad \frac{\partial V}{\partial P} = -\cos \theta$$

where $R = r_c$. Substitution of the above terms into Eq. (6.7-10) yields

$$(\delta_A)_H = \frac{PR}{EA} \left(\frac{R}{e} \int_0^\pi \sin^2 \theta \, d\theta - \int_0^\pi \sin^2 \theta \, d\theta - \int_0^\pi \sin^2 \theta \, d\theta + \int_0^\pi \sin^2 \theta \, d\theta + 2(1 + \nu) k \int_0^\pi \cos^2 \theta \, d\theta \right)$$

[†] Care must be exercised when evaluating the sign of M in Eqs. (6.7-9) and (6.7-10). The sign convention for positive M is established in Fig. 6.7-1.

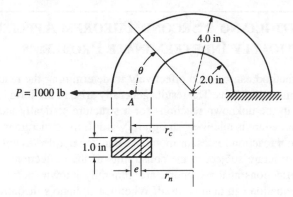

Figure 6.7-2

Evaluating the integrals and simplifying results in

$$(\delta_A)_H = \frac{\pi}{2} \frac{PR}{EA} \left[\frac{R}{e} - 1 + 2(1 + v)k \right] \qquad [a]$$

With the help of Table 5.6-1, the location of the neutral axis in bending r_n is found to be

$$r_n = \frac{(2.0)(1.0)}{(1.0) \ln (4/2)} = 2.8854 \text{ in}$$

and from Fig. 6.7-2 it can be seen that $R = r_c = 3.0$ in. Thus $e = 3.0000 - 2.8854 = 0.1146$ in. For steel, let $E = 30$ Mpsi and $v = 0.3$. For the rectangular section, $k = 1.2$. Substituting the appropriate values into Eq. (a) gives the final result

$$(\delta_A)_H = \frac{\pi}{2} \frac{(1000)(3)}{(30)(10^6)(2)(1)} \left[\frac{3}{0.1146} - 1 + 2(1 + 0.3)(1.2) \right] = 0.00222 \text{ in}$$

For comparison, Eq. (d) of Example 6.6-11 formulates the energy using the straight-beam formulation, where the final result was given by Eq. (h). Thus, with $I_z = (1/12)(1)(2)^3 = 2/3$ in⁴,

$$(\delta_A)_H = \frac{\pi}{2} \frac{(1000)(3)^3}{(30)(10^6)(2/3)} \left\{ 1 + [1 + 2(1 + 0.3)(1.2)] \left[\frac{(2/3)}{(1)(2)(3)^2} \right] \right\}$$

$$= 0.00244 \text{ in}$$

which is 10 percent greater than the solution given by the thick-walled formulation, Eq. (6.7-10).

6.8 CASTIGLIANO'S SECOND THEOREM APPLIED TO STATICALLY INDETERMINATE PROBLEMS

Castigliano's method can be used effectively in determining the reactions of statically indeterminate problems. The results generally arise in the form of n simultaneous equations in the unknown reactions for a structure statically indeterminate of order n. The procedure is relatively straightforward. If the structure is indeterminate of order n, then n reaction forces, or moments, can be regarded as unknown *applied* forces on the structure subject to the constraints of the deflections at the reaction points. The deflections at the supports are generally known (normally zero), and Castigliano's equation can then be used. When Castigliano's theorem is applied to statically indeterminate problems, an extremely important requirement must be understood. Once the n unknown reactions are selected to be considered as applied, the *remaining* unknown reactions must either be solved for in terms of the n unknown reactions and the actual applied forces, or not used in the energy formulation. A number of examples follow to illustrate the technique.

Example 6.8-1

Using Castigliano's theorem, determine the reactions at points A and B of the beam shown in Fig. 6.8-1(a).

Solution:

First, construct the free-body diagram. Since only two equilibrium equations are applicable and there are three unknowns, the structure is indeterminate of the order

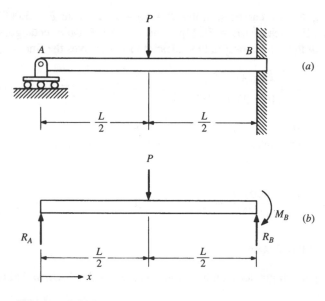

Figure 6.8-1

$n = 1$. This means that one (and *only* one) reaction can be regarded as an applied load and the remaining reactions must be written in terms of P and one unknown applied load. Two solutions will be arrived at by separately considering R_A and M_B as applied.

Solution 1:

Consider R_A as an applied force; then

$$M_z = \begin{cases} R_A x & 0 < x < \dfrac{L}{2} \\[2ex] R_A x - P\left(x - \dfrac{L}{2}\right) & \dfrac{L}{2} < x < L \end{cases}$$

$$\frac{\partial M_z}{\partial R_A} = x \qquad 0 < x < L$$

The vertical deflection of point A is zero; thus[†]

$$(\delta_A)_V = \frac{\partial U}{\partial R_A} = \frac{1}{EI_z}\left\{\int_0^{L/2} (R_A x)x\ dx + \int_{L/2}^L \left[R_A x - P\left(x - \frac{L}{2}\right)\right]x\ dx\right\} = 0$$

Simplifying yields

$$\int_0^{L/2} R_A x^2\ dx - \int_{L/2}^L P\left(x - \frac{L}{2}\right)x\ dx = 0$$

Evaluating the integrals and solving for R_A yields

$$R_A = \frac{5}{16}P$$

Applying the equilibrium equations to Fig. 6.8-1(*b*) results in

$$R_B = \frac{11}{16}P \qquad \text{and} \qquad M_B = \frac{3}{16}PL$$

Solution 2:

In this solution, M_B is considered to be an applied moment. In order to determine how the moment varies within the beam either R_A or R_B must be used, but in this approach P and M_B are applied. Thus R_A or R_B must be written in terms of P and M_B. Summing moments about point B results in

$$R_A = \frac{P}{2} - \frac{M_B}{L}$$

[†] Note that the remaining unknown reactions R_B and M_B are not used in the energy formulation.

and the bending-moment equation is[†]

$$
M_z = \begin{cases} \left(\dfrac{P}{2} - \dfrac{M_B}{L}\right)x & 0 < x < \dfrac{L}{2} \\[2ex] \left(\dfrac{P}{2} - \dfrac{M_B}{L}\right)x - P\left(x - \dfrac{L}{2}\right) & \dfrac{L}{2} < x < L \end{cases}
$$

$$
\frac{\partial M_z}{\partial M_B} = -\frac{x}{L} \qquad 0 < x < L
$$

The slope at point B is zero; thus

$$
\theta_B = \frac{\partial U}{\partial M_B} = \frac{1}{EI_z}\left\{ \int_0^{L/2}\left(\frac{P}{2} - \frac{M_B}{L}\right)(x)\left(-\frac{x}{L}\right)dx + \int_{L/2}^{L}\left[\left(\frac{P}{2} - \frac{M_B}{L}\right)x - P\left(x - \frac{L}{2}\right)\right]\left(-\frac{x}{L}\right)dx \right\} = 0
$$

Simplifying yields

$$
-\int_0^{L}\left(\frac{P}{2} - \frac{M_B}{L}\right)(x)\left(\frac{x}{L}\right)dx + \int_{L/2}^{L} P\left(x - \frac{L}{2}\right)\left(\frac{x}{L}\right)dx = 0
$$

Evaluating the integrals and solving for M_B results in

$$
M_B = \frac{3}{16}PL
$$

and the remaining unknowns, R_A and R_B, are found from equilibrium. Thus both approaches yield the same results, but solution 1 is slightly simpler.

Example 6.8-2　Determine the reactions at points C and D of the structure shown in Fig. 6.8-2(a). All members have equal cross section A and equal modulus of elasticity E; members 1, 2, 4, and 5 are of length L, and members 3, 6, and 7 are of length $\sqrt{2}$.

Solution:

The free-body diagram of the structure is shown in Fig. 6.8-2(b). Summing moments about point C yields $D_x = 2P$, and thus $C_x = 2P$. It can be seen that only one equilibrium equation remains, $C_y + D_y = P$. After using all the pertinent equilibrium equations, there remains only one equation in two unknowns. Therefore, the structure is statically indeterminate of order 1. This means that one unknown can be considered as an applied force. If D_y is selected as the applied force,

$$
C_y = P - D_y
$$

[†] Since in this approach M_B and P are the applied forces, the energy is not formulated using R_A or R_B.

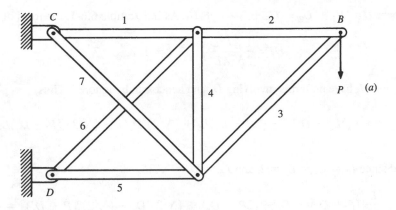

(a)

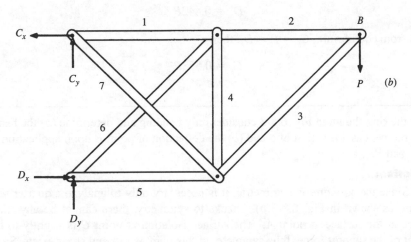

(b)

Figure 6.8-2

Solving for the axial forces in the members in terms of P and D_y only, results in

$$F_1 = P + D_y \qquad F_2 = P \qquad F_3 = -\sqrt{2}P \qquad F_4 = D_y$$

$$F_5 = -(2P - D_y) \qquad F_6 = -\sqrt{2}D_y \qquad F_7 = \sqrt{2}(P - D_y)$$

where positive values indicate tension and negative compression. Since the deflection of point D in the y direction is zero,

$$(\delta_D)_y = \frac{\partial U}{\partial D_y} = 0$$

where $U = U_1 + U_2 + U_3 + \cdots + U_7$. As in Example 6.6-1, the form of U_i is

$$U_i = \frac{1}{2}\left(\frac{F^2 L}{AE}\right)_i \qquad i = 1, 2, \ldots, 7$$

U_2 and U_3 have no terms involving D_y and hence can be ignored. Thus

$$\frac{\partial U}{\partial D_y} = \frac{1}{EA}\left[(P + D_y)L_1 + D_y L_4 - (2P - D_y)L_5 + (\sqrt{2})^2 D_y L_6 - (\sqrt{2})^2 (P - D_y)L_7\right] = 0$$

and since $L_1 = L_4 = L_5 = L$ and $L_6 = L_7 = \sqrt{2}L$,

$$\frac{L}{EA}\left[(P + D_y) + D_y - (2P - D_y) + (\sqrt{2})^3 D_y - (\sqrt{2})^3 (P - D_y)\right] = 0$$

Solving for D_y yields

$$D_y = 0.442P$$

and from equilibrium, $C_y = P - D_y$; then

$$C_y = 0.558P$$

Example 6.8-3   For the ring shown in Fig. 6.8-3 consider only bending and determine (a) the bending moment as a function of θ and (b) the deflection of point B upon application of the load P.

Solution:

(a) Since the structure is symmetric, it is necessary only to analyze a quarter segment, as shown in Fig. 6.8-3(b). Thanks to symmetry, there cannot be any shear force at the surface at point A.[†] The surface isolation at point C is slightly to the right of the support force; thus symmetry at this point is lost and shear exists. Since M_A and M_C are unknown, the problem is statically indeterminate. Thus, consider M_A as an applied load. Thanks to symmetry there is no rotation at point A and using Castigliano's theorem in the form of Eq. (6.6-2) for rotation results in

$$\theta_A = \frac{\partial U}{\partial M_A} = 0$$

Writing the bending moment as a function of θ, we can see from Fig. 6.8-3(c) that

$$M_z = -M_A + \frac{PR}{2}(1 - \cos\theta) \qquad\qquad \textbf{[a]}$$

[†] If this is not clear, refer to Sec. 8.2.

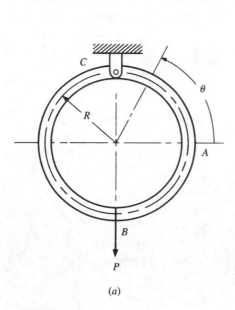

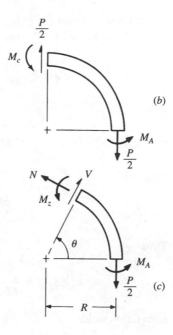

Figure 6.8-3

and

$$\frac{\partial M_z}{\partial M_A} = -1$$

Thus

$$\frac{\partial U}{\partial M_A} = \frac{1}{EI_z} \int_0^{\pi/2} \left[-M_A + \frac{PR}{2}(1 - \cos\theta) \right](-1)R \; d\theta = 0$$

Integration yields

$$M_A = \frac{PR}{2}\left(1 - \frac{2}{\pi}\right)$$

Substituting this into Eq. (a) yields the solution to part (a).

$$M_z = \frac{PR}{2}\left(\frac{2}{\pi} - \cos\theta\right)$$

M_z is greatest at $\theta = 90°$, where $M_z = M_c$. Thus

$$M_{\max} = M_c = \frac{PR}{\pi}$$

(b) Since in part (a) only one-quarter of the ring was analyzed, the total strain energy U_T is 4 times U. The vertical deflection of point B in the direction of P is then

$$(\delta_B)_V = \frac{\partial U_T}{\partial P} = 4\frac{\partial U}{\partial P} = \frac{4}{EI_z}\int_0^{\pi/2} M_z \frac{\partial M_z}{\partial P} R\,d\theta$$

Since

$$M_z = \frac{PR}{2}\left(\frac{2}{\pi} - \cos\theta\right)$$

then

$$\frac{\partial M_z}{\partial P} = \frac{R}{2}\left(\frac{2}{\pi} - \cos\theta\right)$$

Thus

$$(\delta_B)_V = \frac{4}{EI_z}\int_0^{\pi/2} P\left[\frac{R}{2}\left(\frac{2}{\pi} - \cos\theta\right)\right]^2 R\,d\theta$$

Integration yields

$$(\delta_B)_V = \left(\frac{\pi}{4} - \frac{2}{\pi}\right)\frac{PR^3}{EI_z} = 0.1488\frac{PR^3}{EI_z}$$

Thus it can be seen from the examples illustrated that Castigliano's second theorem is extremely powerful in performing deflection analysis of a structure.

6.9   THE VIRTUAL LOAD METHOD

Although not labeled as such, the virtual load method was actually applied in Example 6.0-3. The method is based directly on the complementary energy theorem and is thus applicable to nonlinear load-displacement problems. In the derivation of the complementary energy theorem of Sec. 6.5, an infinitesimal force ΔP_i is placed on a structure at point i prior to the actual loading. The complementary work due to ΔP_i and the deflections caused by the *actual loading* is then equated to the increase in the complementary energy due to ΔP_i, $\Delta\Phi$. That is,

$$\Delta W_C = \delta_i\,\Delta P_i = \Delta\Phi \qquad\qquad \textbf{[6.9-1]}$$

Here δ_i is the deflection of point i on the structure in the direction of ΔP_i due to the actual loading. The final step in obtaining Eq. (6.5-4) of the complementary energy theorem is to take the limit as $\Delta P_i \to 0$. For the *virtual load method*, however, we will basically employ the form of Eq. (6.9-1) and instead of applying ΔP_i and allowing it to approach zero, we will apply a small *imaginary* force or *virtual force*

δP_i to the structure. The increase in the complementary virtual work due to δP_i, δW_c, is then equated to the increase in the complementary virtual energy, $\delta \Phi$. Thus, Eq. (6.9-1) is written as

$$\delta W_c = \delta_i(\delta P_i) = \delta \Phi \qquad\qquad \textbf{[6.9-2]}$$

Determine the vertical deflection of point B of the structure shown in Fig. 6.9-1. Members BC and BD have equal length L, area A, and modulus E.   **Example 6.9-1**

Solution:

Add virtual force δP to P. Thus, the complementary virtual work due to δP is

$$\delta W_C = \delta_B \delta P \qquad\qquad \textbf{[a]}$$

The forces in BC and BD are equal and are found from static analysis to be

$$F_{BC} = F_{BD} = \frac{P}{2 \cos \beta}$$

Thus the deflections of BC and BD due to the internal forces are

$$\delta_{BC} = \delta_{BD} = \frac{PL}{(2 \cos \beta)AE}$$

The forces in members BC and BD due to δP are

$$\delta F_{BC} = \delta F_{BD} = \frac{\delta P}{2 \cos \beta}$$

Hence, the complementary virtual work, say on member BC, due to δP is

$$\delta (W_C)_{BC} = \delta_{BC} \delta F_{BC} = \frac{PL}{(2 \cos \beta)AE} \frac{\delta P}{2 \cos \beta} = \frac{PL}{4AE \cos^2 \beta} \delta P$$

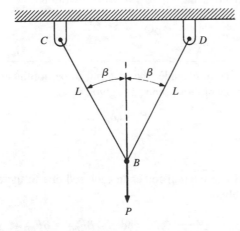

Figure 6.9-1

which is also the increase in the complementary virtual energy of member BC. Since the system is symmetric, the energies of each member are equal, and the total increase in $\delta\Phi$ is double that of BC. Thus

$$\delta\Phi = \frac{PL}{2AE \cos^2 \beta} \, \delta P \qquad\qquad \textbf{[b]}$$

Equating Eqs. (a) and (b), the δP terms cancel, and solving for δ_B yields

$$\delta_B = \frac{PL}{2AE \cos^2 \beta}$$

6.9.1   AXIAL LOADING

For the general case of axial loading considering the longitudinal axis of the member to be in the x direction, let the deflection of an infinitesimal dx element due to the applied loads be $d(\delta_a)$. Consider δF to be the internal force within the element due to a virtual load placed on the structure at a given point; then the work performed by δF only is $\delta F d(\delta_a)$. This is the complementary virtual energy of the dx element. The total complementary virtual energy of the member is determined by integration along the entire length of the member. Thus

$$\delta\Phi_a = \int_0^L \delta F \, d(\delta_a) \qquad\qquad \textbf{[6.9-3]}$$

If the material is linear, the deflection of the dx element is given by $d(\delta_a) = F \, dx/(AE)$, where F is the internal force in the element due to the *applied* loads. Thus for a linear material Eq. (6.9-3) reduces to

$$\delta\Phi_a = \int_0^L \delta F \frac{F}{AE} \, dx \qquad\qquad \textbf{[6.9-4]}$$

If δF, F, A, and E are constant, Eq. (6.9-4) can be written

$$\delta\Phi_a = \delta F \frac{FL}{AE} \qquad\qquad \textbf{[6.9-5]}$$

Example 6.9-2 Repeat Example 6.9-1 but consider the material to be nonlinear, where the tensile load-deflection relationship for each rod is given by

$$F = k\delta_a^{2/3}$$

where k is a constant.

Solution:

From Example 6.9-1 the internal forces in each rod due to the applied and virtual forces are, respectively,

$$F_{BC} = F_{BD} = \frac{P}{2 \cos \beta} \qquad \text{and} \qquad \delta F_{BC} = \delta F_{BD} = \frac{\delta P}{2 \cos \beta}$$

Since δF in each rod is a constant throughout the length, from Eq. (6.9-3) the change in complementary virtual energy for each rod is

$$\delta \Phi_a = \delta F \int_0^L d\delta_a = \delta F \delta_a$$

where δ_a is the deflection of each rod due to the applied forces. Since for the rods, $\delta_a = (F/k)^{3/2}$, the deflection of each member is

$$\delta_a = \left[\frac{P/(2 \cos \beta)}{k} \right]^{3/2}$$

and $\delta \Phi_a$ for each member is

$$\delta \Phi_a = \delta F \delta_a = \frac{\delta P}{2 \cos \beta} \left[\frac{P/(2 \cos \beta)}{k} \right]^{3/2}$$

This complementary virtual work due to δP is $\delta_B \, \delta P$. Equating this to the total complementary virtual energy of the rods yields

$$\delta W_C = \delta \Phi_{BC} + \delta \Phi_{BD} \rightarrow \delta_B \delta P = 2 \left\{ \frac{\delta P}{2 \cos \beta} \left[\frac{P/(2 \cos \beta)}{k} \right]^{3/2} \right\}$$

Solving for δ_B and simplifying yields

$$\delta_B = \left(\frac{P}{2k} \right)^{3/2} (\cos \beta)^{-5/2}$$

6.9.2   TORSIONAL LOADING

In a manner similar to that for axial loading, the complementary virtual energy for a rod in torsion is

$$\delta \Phi_t = \int \delta T \, d\theta \qquad\qquad \textbf{[6.9-6]}$$

where δT is the torque being transmitted at position x in the rod due to a virtual load, and $d\theta$ is the angle of twist of the dx element due to the *applied loading*.

If the material is linear, $d\theta = 2(1 + v)(T/JE) \, dx$, where T is the torque transmitted at x due to the *applied loading*. Thus for a rod of length L made of a linear material, Eq. (6.9-6) reduces to

$$\delta \Phi_t = 2 \int_0^L (1 + v)\delta T \frac{T}{JE} \, dx \qquad\qquad \textbf{[6.9-7]}$$

If δT, T, v, E, and J are constant throughout the rod,

$$\delta \Phi_t = 2(1 + v)\delta T \frac{TL}{JE} \qquad\qquad \textbf{[6.9-8]}$$

6.9.3   BENDING

For beams in bending, the complementary virtual energy due to the bending moment only is

$$\delta\Phi_b = \int \delta M_z \, d\theta \qquad\qquad \textbf{[6.9-9]}$$

where δM_z is the bending moment at x due to a virtual load, and $d\theta$ is the rotational deflection of the dx element due to the *applied loading*.

For a linear material, $d\theta = (M_z / EI_z) \, dx$, where M_z is the internal bending moment due to the applied loads. Thus for a beam of length L made of a linear material, Eq. (6.9-9) reduces to

$$\delta\Phi_b = \int_0^L \delta M_z \frac{M_z}{EI_z} \, dx \qquad\qquad \textbf{[6.9-10]}$$

The complementary virtual energy due to the transverse shear forces is

$$\delta\Phi_s = 2k \int_0^L (1 + v)\delta V_y \frac{V_y}{EA} \, dx \qquad\qquad \textbf{[6.9-11]}$$

where again, k is the form correction factor for shear.

Example 6.9-3 | For the beam shown in Fig. 6.9-2(a), determine the deflection and slope of point A using the virtual load method. Consider the effects of bending only.

Solution:

The bending moment due to the applied load is

$$M_z = -Px \qquad 0 < x < L$$

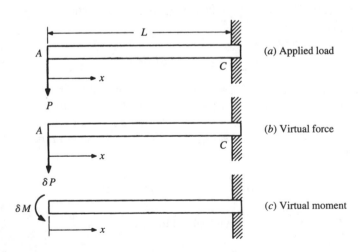

(a) Applied load

(b) Virtual force

(c) Virtual moment

Figure 6.9-2

To find the deflection at point A, apply a virtual force δP, as shown in Fig. 6.9-2(b). Thus

$$\delta M_z = -\delta Px \qquad 0 < x < L$$

The work due to the virtual force is $\delta_A \delta P$. The increase in complementary virtual energy is given by Eq. (6.9-10). Equating the two equations results in

$$\delta_A \delta P = \int_0^L \frac{(Px)(\delta Px)\,dx}{EI_z} = \frac{P\delta PL^3}{3EI_z}$$

Solving for δ_A yields

$$\delta_A = \frac{PL^3}{3EI_z}$$

To find the slope at point A, apply a virtual moment δM at point A, as shown in Fig. 6.9-2(c). Thus

$$\delta M_z = -\delta M \qquad 0 < x < L$$

The complementary virtual work due to δM is $\theta_A \delta M$, where θ_A is the rotational deflection at point A in the direction of δM. Equating this to the increase in complementary virtual energy given by Eq. (6.9-10) results in

$$\theta_A \delta M = \int_0^L \frac{Px}{EI_z}\,\delta M\ dx = \frac{P\delta ML^2}{2EI_z}$$

Solving for θ_A yields

$$\theta_A = \frac{PL^2}{2EI_z}$$

Deflections due to temperature changes can be handled quite effectively by the virtual load method. Equation (6.9-5) can easily be modified by adding on the deflection due to a temperature change. The term FL/AE in Eq. (6.9-5) represents the free deflection in the member due to the internal force. Thus, when a deflection term is added to account for a temperature change, Eq. (6.9-5) becomes

$$\delta\Phi = \left(\frac{FL}{AE} + \alpha\,\Delta TL\right)\delta F$$

or

$$\delta\Phi = \left(\frac{F}{AE} + \alpha\,\Delta T\right)L\delta F \qquad\qquad \textbf{[6.9-12]}$$

where α is the linear temperature coefficient of expansion.

Example 6.9-4

For the structure in Example 6.9-1, determine the deflection of point B if in addition to the applied load P the temperature of the structure increases ΔT. The coefficient of linear expansion of the members is α.

Solution:

The internal forces in each member due to the applied force are equal and are

$$F_{BC} = F_{BD} = \frac{P}{2 \cos \beta} \qquad \text{[a]}$$

When a virtual load δP is applied at point B in the same direction as P, the resulting internal forces are simply

$$\delta F_{BC} = \delta F_{BD} = \frac{P}{2 \cos \beta} \qquad \text{[b]}$$

Thus, from Eq. (6.9-12), the total complementary virtual energy is

$$\delta \Phi = \left(\frac{P}{2 \cos \beta} + \alpha \, \Delta T \right) L \, \frac{\delta P}{\cos \beta} \qquad \text{[c]}$$

Equating this to the complementary virtual work $\delta_B \delta P$ yields

$$\delta_B = \left(\frac{P}{2 \cos \beta} + \alpha \, \Delta T \right) L \, \frac{L}{\cos \beta}$$

6.10 THE VIRTUAL LOAD METHOD APPLIED TO STATICALLY INDETERMINATE PROBLEMS

For statically indeterminate problems, the procedure is similar to that followed in Sec. 6.8. That is, if the structure is indeterminate of order n, then n reactions are treated as applied forces.

Example 6.10-1

The structure shown in Fig. 6.10-1(a) contains a cantilever beam of length L_b, area A_b, modulus E, and second-area moment I_b. In addition, supporting the left end of the beam is a cable of length L_c, area A_c, and modulus E. Determine the reactions at points B and D.

Solution:

The free-body diagram of the structure is shown in Fig. 6.10-1(b). The structure is statically indeterminate of order 1. Thus, consider one unknown reaction as an applied force, say F_D at point D. The force F_D is subjected to the constraint that point D does not deflect. After isolating the two elements of the structure, the free-body

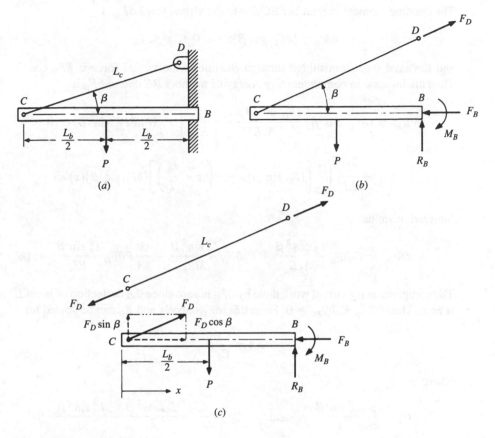

Figure 6.10-1

diagrams are as shown in Fig. 6.10-1(c). The bending moment in member BC can be written

$$
M_z = \begin{cases} (F_D \sin \beta)x & 0 < x < \dfrac{L_b}{2} \\[2ex] (F_D \sin \beta)x - P\left(x - \dfrac{L_b}{2}\right) & \dfrac{L_b}{2} < x < L_b \end{cases}
$$

Apply a virtual force δF_D at point D in the direction of F_D. This is done because the equation for the deflection of point D will be written. The complementary virtual energy of member CD due to δF_D is found from Eq. (6.9-3) and is

$$
\delta \Phi_{CD} = \delta F_D \frac{F_D L_c}{A_c E} \tag{a}
$$

The bending moment in member BC due to the virtual force δF_D is

$$\delta M_z = (\delta F_D \sin \beta)x \qquad 0 < x < L_b$$

and the axial force transmitted through BC due to the virtual force is $\delta F_D \cos \beta$. Thus the increase in complementary energy of member BC due to δF_D is

$$\delta \Phi_{BC} = (\delta F_D \cos \beta) \frac{(F_D \cos \beta)L_b}{A_b E} + \frac{1}{EI_b} \int_0^{L_b/2} (F_D \delta F_D \sin^2 \beta)x^2 \, dx$$

$$+ \frac{1}{EI_b} \int_{L_b/2}^{L_b} \left[(F_D \sin \beta)x - P\left(x - \frac{L_b}{2}\right) \right] (\delta F_D \sin \beta)(x) \, dx$$

Integration yields

$$\delta \Phi_{BC} = F_D \delta F_D \frac{L_b \cos^2 \beta}{A_b E} + F_D \delta F_D \frac{L_b^3 \sin^2 \beta}{3EI_b} - \frac{5}{48} P \delta F_D \frac{L_b^3 \sin \beta}{EI_b} \qquad \textbf{[b]}$$

The complementary virtual work done by δF_D is zero since the deflection of point D is zero. Thus $\delta \Phi_{CD} + \delta \Phi_{BC} = 0$. From this, δF_D cancels and F_D can be solved for

$$F_D = \frac{C_1}{C_2} P$$

where

$$C_1 = \frac{5}{48} \frac{L_b^3 \sin \beta}{I_b} \qquad \text{and} \qquad C_2 = \frac{L_c}{A_c} + \frac{L_b \cos^2 \beta}{A_b} + \frac{L_b^3 \sin^2 \beta}{3I_b}$$

6.11 RAYLEIGH'S METHOD APPLIED TO BEAMS IN BENDING

Rayleigh's method is often employed in applied mechanics when dealing with static or dynamic modes of deformation. Here, we will apply the method to the static deflection of beams in bending. The method can be used for simple beam problems. However, it will be shown in the last section of this chapter that the method can easily be used for beam-column problems where the coupling effect of axial loads with bending can be analyzed. To employ the method, one begins with an assumption of the deflection *shape* using only one unknown coefficient. Based on this shape, the strain energy is formulated and equated to the work done by the applied forces.

Consider a load distribution $q(x)$ applied to the beam as shown in Fig. 6.11-1(a), where the units of $q(x)$ are force per unit length. If a dx element is isolated at position x, the net force on the element is $q(x) \, dx$. If the beam deflects in the positive

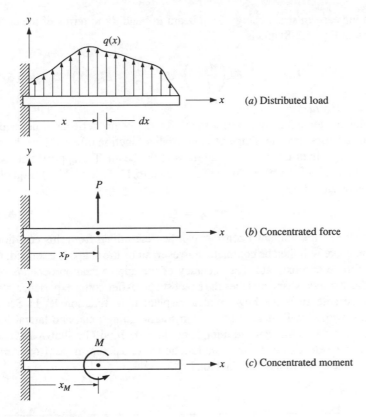

(a) Distributed load

(b) Concentrated force

(c) Concentrated moment

Figure 6.11-1

y direction a distance $v_c(x)$, the net work performed on the element is therefore $1/2 q(x) v_c(x)\ dx$. The total work performed on the beam by the application of $q(x)$ is then

$$W_q = \frac{1}{2} \int_0^L q(x) v_c(x)\ dx \qquad\qquad \textbf{[6.11-1]}$$

For a concentrated force as shown in Fig. 6.11-1(b) the total work would be

$$W_P = \frac{1}{2}\, P(v_c)_{x=x_P} \qquad\qquad \textbf{[6.11-2]}$$

where $v_c(x)$ is evaluated at $x = x_P$.

For a concentrated moment such as that shown in Fig. 6.11-1(c) the total work would be

$$W_M = \frac{1}{2} M \left(\frac{dv_c}{dx} \right)_{x=x_M} \qquad\qquad \textbf{[6.11-3]}$$

where the slope of the beam dv_c/dx is evaluated at $x = x_M$.

The increase in strain energy for a beam in bending in terms of the deflection was given in Eq. (6.3-8) and is

$$U_b = \frac{1}{2} \int_0^L EI_z \left(\frac{d^2 v_c}{dx^2} \right)^2 dx = \frac{1}{2} \int_0^L EI_z (v_c'')^2 \, dx \qquad \textbf{[6.11-4]}$$

where $v_c'' = d^2 v_c / dx^2$.

Rayleigh's technique assumes a shape for the beam deflection, defined up to one unknown constant. The shape is, in general, a function of x and must be consistent with the geometric boundary conditions of the beam. The approximate work W and strain energy U can be evaluated using Eqs. (6.11-1) to (6.11-4). Equating the two terms results in

$$W = U \qquad \textbf{[6.11-5]}$$

and the value of the unknown constant can be determined from this equation. The deflection curve will then be complete; however, since the shape is assumed, the deflection will be approximate. The accuracy of the approximation depends on how closely the assumed shape matches the exact shape. After some experience, the analyst will be able to judge how good a displacement function is. In Sec. 6.13, Rayleigh's method will be applied to beams undergoing axial and lateral loading, where in many cases the exact solution is not known. It will be shown that a simple solution to the selection of an accurate function is to use the shape (from Appendix C) of the same beam minus the axial load.

Example 6.11-1

Determine the deflection of point A of the beam shown in Fig. 6.11-2. Assume that the deflection shape of the beam takes the form[†]

(a)
$$v_c = -\delta \left(1 - \cos \frac{\pi x}{2L} \right)$$

(b)
$$v_c = -\delta \left(\frac{x}{L} \right)^2$$

(c)
$$v_c = \frac{\delta x^2}{2L^3} (x - 3L)$$

[†] Note that every one of the functions given satisfies the geometric boundary conditions that $v_c = dv_c/dx = 0$ at $x = 0$. Also, since the bending moment is directly related to $d^2 v_c/dx^2$, $d^2 v_c/dx^2$ should equal zero at $x = L$. This is not a geometric boundary condition, however, and it is usually not imposed in this type of problem. The functions of part (a) and (c) satisfy this condition, whereas part (b) does not. It will be seen that the function of part (a) is very accurate and the function of part (c) is exact. (The reader should verify the fact that this function is the same shape as given in Appendix C.) The function used in part (b) will be found to be fairly inaccurate. In addition, note that each one of the functions given was normalized to give $v_c = -\delta$ at $x = L$.

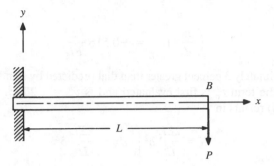

Figure 6.11-2

Solution:
The work due to P is given by Eq. (6.11-2). Since for all three functions, $v_c = -\delta$ at $x = L$, the work is

$$W = -\frac{1}{2}P(-\delta) = \frac{1}{2}P\delta \qquad \text{[a]}$$

(a) The second derivative of v_c with respect to x is

$$v_c'' = -\delta\left(\frac{\pi}{2L}\right)^2 \cos\frac{\pi x}{2L}$$

Substituting this into Eq. (6.11-4) results in

$$U \approx U_b = \frac{EI_z}{2}\delta^2\left(\frac{\pi}{2L}\right)^4 \int_0^L \cos^2\frac{\pi x}{2L}\,dx$$

Integration yields

$$U = \frac{\pi^4 EI_z}{64L^3}\delta^2 \qquad \text{[b]}$$

Substituting Eqs. (a) and (b) into Eq. (6.11-5) yields

$$\frac{1}{2}P\delta = \frac{\pi^4}{64}\frac{EI_z}{L^3}\delta^2$$

and solving for δ results in

$$\delta = \frac{32}{\pi^4}\frac{PL^3}{EI_z} = 0.329\frac{PL^3}{EI_z}$$

which is approximately 2 percent less than that predicted by standard beam theory (see Appendix C). Although not asked for in the problem, the slope at point B can be found by differentiating v_c. Thus

$$\frac{dv_c}{dx} = -\frac{\pi}{2L}\delta\sin\frac{\pi x}{2L} = -0.516\frac{PL^2}{EI}\sin\frac{\pi x}{2L}$$

Therefore

$$\left(\frac{dv_c}{dx}\right)_{x=L} = -0.516 \frac{PL^2}{EI_z}$$

which is approximately 3 percent greater than that predicted by standard beam theory.

(b) Again the term v_c'' is first evaluated and is $v_c'' = -2\delta/L^2$. Substituting this into Eq. (6.11-4) results in

$$U = \frac{2EI_z}{L^4}\delta^2 \int_0^L dx = \frac{2EI_z}{L^3}\delta^2 \qquad \textbf{[c]}$$

Equating Eqs. (a) and (c) gives

$$\frac{1}{2}P\delta = \frac{2EI_z}{L^3}\delta^2$$

and solving for δ yields

$$\delta = 0.25 \frac{PL^3}{EI_z}$$

which is 25 percent lower than the exact solution. Although satisfying the geometric boundary conditions, an x^2 function is not very good.

If a cubic function in x is used, where $v_c = ax^3 + bx^2 + cx + d$, subject to the conditions that $v_c = v_c' = 0$ at $x = 0$ and $v_c'' = 0$ and $v_c = -\delta$ at $x = L$, the function in part (c) is arrived at.

(c) Again, v_c'' is evaluated and is found to be

$$v_c'' = 3\frac{\delta}{L^3}(x - L)$$

Substituting this into Eq. (6.11-4) results in

$$U = \frac{9EI_z\delta^2}{2L^6} \int_0^L (x - L)^2 dx$$

Integration yields

$$U = \frac{3EI_z}{2L^3}\delta^2 \qquad \textbf{[d]}$$

Equating Eqs. (a) and (d) results in

$$\frac{1}{2}P\delta = \frac{3EI_z}{2L^3}\delta^2$$

Solving for δ gives

$$\delta = \frac{1}{3}\frac{PL^3}{EI_z}$$

the exact solution to the problem.

A method of improving the solution obtained by Rayleigh's technique is to select more functions, each with an unknown constant, and by optimization methods
find the best value of each constant. This procedure is most widely known as the
Rayleigh-Ritz method and is covered in the following section.

6.12   THE RAYLEIGH-RITZ TECHNIQUE APPLIED TO BEAMS IN BENDING

Considering beams in bending only, assume that the deflection of the centroidal
axis is given by

$$v_c = \sum_{i=1}^{n} a_i f_i(x) \qquad \textbf{[6.12-1]}$$

where a_i are n unknown constants and $f_i(x)$ are n known (assumed) functions of x,
and v_c satisfies the geometric boundary conditions. When v_c is substituted into Eqs.
(6.11-1) to (6.11-4), the work and strain energy will be functions of a_i only. The
work can then be equated to the strain energy, as is done in the Rayleigh technique.
However, the resulting equation is only one equation with n unknowns a_i. The
Rayleigh-Ritz technique basically performs an optimization of the a_i terms by
slightly varying the a_i, keeping everything else constant. Varying the a_i causes v_c to
change, thus changing the strain energy and the work. However, since $U = W$, then
the changes are equal. That is,

$$dU = dW \qquad \textbf{[6.12-2]}$$

Since only the a_i change, causing a change in v_c, while the applied loads remain
constant, the 1/2 term in the work expression must be omitted. Thus, for a distributed load $q(x)$ [as defined in Eq. (6.11-1)] the change in work is

$$dW_q = d\left(\int_0^L q(x) v_c(x) \, dx \right) \qquad \textbf{[6.12-3]}$$

whereas, for concentrated forces and moments [as defined in Eqs. (6.11-2) and
(6.11-3), respectively], the changes in work are

$$dW_P = d\left[P(v_c)_{x=x_P} \right] \qquad \textbf{[6.12-4]}$$

and

$$dW_M = d\left[M\left(\frac{dv_c}{dx}\right)_{x=x_M} \right] \qquad \textbf{[6.12-5]}$$

The following example serves to illustrate the technique.

Example 6.12-1 In Example 6.11-1 consider only the geometric boundary conditions and determine the deflection of the centroidal axis, assuming that

$$v_c = a_0 + a_1x + a_2x^2 + a_3x^3$$

Solution:

In Example 6.11-1 the geometric boundary conditions are $v_c = v_c' = 0$ at $x = 0$. Therefore, $a_0 = a_1 = 0$ and v_c becomes

$$v_c = a_2x^2 + a_3x^3 \qquad \text{[a]}$$

Evaluating the work term first, we have

$$(v_c)_{x=x_p} = (v_c)_{x=L} = a_2L^2 + a_3L^3 \qquad \text{[b]}$$

Thus from Eq. (6.12-4)

$$dW = dW_p = d[-P(a_2L^2 + a_3L^3)]$$

or

$$dW = -PL^2(da_2 + L\, da_3) \qquad \text{[c]}$$

In order to evaluate the strain energy, v_c'' is obtained from Eq. (a) and is

$$v_c'' = 2a_2 + 6a_3x$$

Thus the strain energy is

$$U = \frac{EI_z}{2}\int_0^L (2a_2 + 6a_3x)^2\, dx$$

Evaluating the integral results in

$$U = 2EI_zL(a_2^2 + 3a_2a_3L + 3a_3^2L^2)$$

Thus

$$dU = 2EI_zL[(2a_2 + 3La_3)\, da_2 + 3L(a_2 + 2La_3)\, da_3] \qquad \text{[d]}$$

Now, Eqs. (c) and (d) are equated, resulting in

$$-PL^2(da_2 + L\, da_3) = 2EI_zL[(2a_2 + 3La_3)\, da_2 + 3L(a_2 + 2La_3)\, da_3]$$

Equating the coefficients of da_2 and da_3 yields

$$-PL^2 = 2EI_zL(2a_2 + 3La_3) \qquad \text{[e]}$$

$$-PL^3 = 6EI_zL^2(a_2 + 2La_3) \qquad \text{[f]}$$

Solving Eqs. (e) and (f) simultaneously results in

$$a_2 = -\frac{PL}{2EI_z} \qquad a_3 = \frac{P}{6EI_z}$$

Substituting a_2 and a_3 back into Eq. (a) and simplifying algebraically yields

$$v_c = -\frac{Px^2}{6EI_z}(3L - x)$$

In Example 6.12-1, the solution converged to the exact solution thanks to the judicious selection of the functions. If higher-order polynomials had been used in that example, the constants associated with these higher-order polynomials would all have converged to zero.

Up to this point, the Rayleigh and the Rayleigh-Ritz techniques have been employed on simple beam problems in which the solution is either known or easily obtained by more standard methods. There are some applications in which the solution to the deflection shape is either difficult or impossible to obtain by the standard beam-deflection methods. In these cases, either the Rayleigh or the Rayleigh-Ritz technique yields relatively straightforward and accurate approximations. Some applications follow in the next section; they will be discussed using the Rayleigh method, but the Rayleigh-Ritz method is just as applicable.

6.13 STRAIGHT BEAMS UNDERGOING THE COMBINED EFFECTS OF AXIAL AND TRANSVERSE LOADING

6.13.1 UNCONSTRAINED BEAMS

Consider a beam loaded by a transverse load w and an axial force N, as shown in Fig. 6.13-1(a). In an elementary strength of materials course, the deflections and stresses caused by these loads are normally considered independently and the results added by superposition. However, if the axial force is large, this is not very accurate, as the axial force affects the lateral deflections as well as the bending stresses. If N is positive (tension), the beam "stiffens" in bending and the lateral deflections and normal stresses are less than predicted by superposition of the individual effects. Thus in terms of design applications, superposition would lead to a conservative result. However, if N is negative (compression), deflections and stresses will be larger than predicted by superposition. If the analysis is to be accurate, or if a beam is constrained axially and the axial force is unknown, the effect of the axial force on bending should be considered. This can only be accomplished by considering deflections in the force analysis.

Making a surface isolation at x, as shown in Fig. 6.13-1(b), and accounting for the deflection due to bending, we can see that the axial force will affect the bending moment. This is where the coupling between axial and transverse loading takes place.

When the force N is known, the problem can be solved by standard beam methods. This requires the solution of a differential equation, and for discontinuous loading, the procedure can be quite involved and time-consuming. For illustration purposes, the following example with continuous loading demonstrates the exact

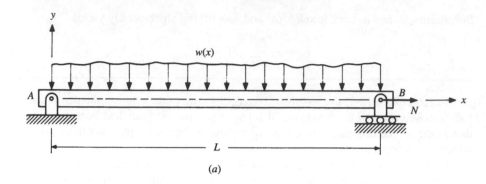

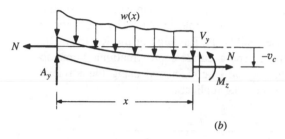

Figure 6.13-1

approach, but since the loading is continuous, much of the difficulty with this approach is alleviated.

Example 6.13-1 For the beam shown in Fig. 6.13-1, determine how the vertical deflection of the centroid v_c varies as a function of x if the load distribution is constant at $w(x) = w_o$.

Solution:

From Fig. 6.13-1 with $w(x) = w_o$ and the reaction $A_y = 1/2 w_o L$, the bending moment is[†]

$$M_z = \frac{1}{2} w_o L x - \frac{1}{2} w_o x^2 + N v_c \qquad \text{[a]}$$

Use of the standard beam equation $M_z = EI_z d^2 v_c/dx^2$ leads to

$$\frac{d^2 v_c}{dx^2} - \frac{N}{EI_z} v_c = \frac{w_o}{2EI_z} x(L - x) \qquad \text{[b]}$$

The solution to this differential equation using the boundary conditions that $v_c = 0$ at $x = 0$ and $x = L$ is

$$v_c = \frac{w_o}{2k^2 N}\left[2(1 - \cosh kx) + 2\frac{\cosh kL - 1}{\sinh kL} \sinh kx - k^2 x(L - x) \right] \qquad \text{[c]}$$

[†] Since the lateral loading is in the downward direction, the vertical deflection is shown negative as $(-v_c)$.

where $k = \sqrt{N/EI_z}$. Note, when $N \leq 0$, Eq. (c) is invalid, as Eq. (b) must be reformulated.

A large problem exists with the method used in Example 6.13-1 when the transverse loading is discontinuous. A differential equation must be written in each zone where M_z changes functionally. Continuity of slope and deflection at the junction of each individual zone provide the necessary conditions for evaluating the two constants of integration for each differential equation. The solution to problems where the loading is discontinuous is a long and tedious process that can be approximated quite easily using the techniques outlined in Secs. 6.11 and 6.12. In addition, although less serious, a problem occurs in numerically calculating deflections when the method of Example 6.13-1 is used. For instance, each term in Eq. (c) is very large compared with the final value of v_c. This means that each term in Eq. (c) must be calculated more accurately than usual. Using either the Rayleigh or the Rayleigh-Ritz method, a reasonably accurate approximate solution can be arrived at through a procedure which is much simpler than solving one or more differential equations. Also, in the case of axially constrained beams where the axial force is unknown, the energy approach is the only practical way to a solution.

Rayleigh's technique is used to equate the work due to the applied loading to the increase in strain energy. The work due to the transverse loading was discussed in Sec. 6.11 and is given by Eqs. (6.11-1) to (6.11-3). The work due to the axial force N is not as obvious as one might expect.

Consider the loading of the beam in Fig. 6.13-1 by first applying the axial force N. Point B will move to B' as shown in Fig. 6.13-2. The deflection δ_{B1} is simply $NL/(AE)$. The work performed in this step is $1/2N\delta_{B1} = 1/2N^2L/(AE)$. Next, with N remaining constant on the beam, application of the lateral load $w(x)$ causes point B' to move in the direction opposite N to point B'' deflecting the distance δ_{B2}. The work done by N during this step is $-N\delta_{B2}$ since during this step N is constant. Thus the net work due to N is

$$W_N = \frac{1}{2}\frac{N^2L}{AE} - N\delta_{B2} \qquad\qquad \textbf{[6.13-1]}$$

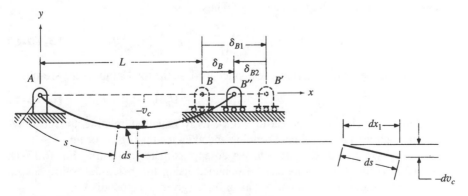

Figure 6.13-2

Assuming that bending does not add any axial stress along the centroidal axis of the beam, the deflection δ_{B2} is simply the difference between the lengths of the arc AB'' and the straight line AB''. Using the curvilinear variable s, one can view a segment of the beam centroid, as shown in Fig. 6.13-2, where $ds^2 = dx_1^2 + dv_c^2$, or

$$ds = dx_1 \left[1 + \left(\frac{dv_c}{dx_1} \right)^2 \right]^{1/2} \qquad [6.13-2]$$

For small slopes, $dv_c/dx_1 \ll 1$, and the term in the [ ] brackets can be expanded using the binomial expansion theorem. When terms $(dv_c/dx_1)^4$ and higher are ignored, Eq. (6.13-2) is approximated by

$$ds \approx dx_1 \left(1 + \frac{1}{2} \left(\frac{dv_c}{dx_1} \right)^2 \right)$$

Integrating each term gives

$$\int ds \approx \int dx_1 + \frac{1}{2} \int \left(\frac{dv_c}{dx_1} \right)^2 dx_1 \qquad [6.13-3]$$

Since $\int ds =$ arc AB'' and $\int dx_1 =$ line AB'', then

$$\delta_{B2} = \int ds - \int dx_1 \approx \frac{1}{2} \int \left(\frac{dv_c}{dx_1} \right)^2 dx_1$$

For small deflections, $dx_1 \approx dx$. Therefore, integrating along the length gives

$$\delta_{B2} \approx \frac{1}{2} \int_0^L \left(\frac{dv_c}{dx} \right)^2 dx = \frac{1}{2} \int_0^L (v_c')^2 \, dx \qquad [6.13-4]$$

where $v_c' = dv_c/dx$.

Substituting Eq. (6.13-4) into Eq. (6.13-1) yields the net work due to N

$$W_N \approx \frac{1}{2} \frac{N^2 L}{AE} - \frac{1}{2} N \int_0^L (v_c')^2 \, dx \qquad [6.13-5]$$

The increase in the strain energy, neglecting shear effects, is the sum of the increase in axial and bending strain energy. Thus

$$U = U_a + U_b = \frac{1}{2} \frac{N^2 L}{AE} + \frac{1}{2} EI_z \int_0^L (v_c'')^2 \, dx \qquad [6.13-6]$$

Equating the total work from the applied loads to the total strain energy results in

$$W_q + W_P + W_M + W_N = U_a + U_b \qquad [6.13-7]$$

Equation (6.13-7) is the basic work-energy equation relating the bending-deflection curve to the combined bending and axial loading of the beam.

If Rayleigh's method is used to get an approximate solution to Eq. (6.13-7), reasonably accurate solutions can be obtained using the beam-deflection shapes found for the same beam without the axial force (given in Appendix C).

(*a*) In Example 6.13-1 the transverse-deflection equation was derived for the centroid of a uniformly loaded simply supported beam with an axial force. Using Eq. (*c*) from that example, determine the deflection at midspan if $L = 1.25$ m, $w_o = 1.6$ kN/m, $I_z = 40 (10^3)$ mm⁴, $E = 70$ GPa, and $N = 3.6$ kN. (*b*) Compare the above results with that obtained by Eq. (6.13-7) using (1) $v_c = -\delta \sin (\pi x/L)$ and (2) the shape given in Appendix C.

Example 6.13-2

Solution:

(*a*) From Example 6.13-1,

$$k = \sqrt{\frac{(3.6)(10^3)}{(70)(10^9)(40)(10^3)(10^{-3})^4}} = 1.1339$$

Substituting k into Eq. (*c*) with $x = 0.625$ m yields

$$(v_c)_{x=L/2} = -0.015083 \text{ m} = -15.1 \text{ mm}^†$$

(*b*) For the approximate solution using Eq. (6.13-7) with $v_c = -\delta \sin (\pi x/L)$,

$$v_c' = -\delta \frac{\pi}{L} \cos \frac{\pi x}{L} \qquad v_c'' = \delta \left(\frac{\pi}{L}\right)^2 \sin \frac{\pi x}{L}$$

For the work due to the uniformly distributed load,

$$W_q = \frac{1}{2} \int_0^L q(x) v_c(x) \, dx = \frac{1}{2} \int_0^L (-w_o) \left(-\delta \sin \frac{\pi x}{L}\right) dx$$

Integration results in

$$W_q = \frac{1}{\pi} w_o L \delta \qquad\qquad \text{[a]}$$

The work due to N, given by Eq. (6.13-5), is

$$W_N = \frac{1}{2} \frac{N^2 L}{AE} - \frac{1}{2} N \int_0^L \left(-\delta \frac{\pi}{L} \cos \frac{\pi x}{L}\right)^2 dx$$

$$= \frac{1}{2} \frac{N^2 L}{AE} - \frac{1}{2} N \left(\delta \frac{\pi}{L}\right)^2 \int_0^L \cos^2 \frac{\pi x}{L} \, dx$$

Integration yields

$$W_N = \frac{1}{2} \frac{N^2 L}{AE} - \frac{\pi^2}{4} \frac{N}{L} \delta^2 \qquad\qquad \text{[b]}$$

---

† Note from Appendix C that for the same beam without the axial force the deflection at midspan is -18.2 mm. Thus it can be seen that there can be quite a difference (21 percent) if N is neglected in the deflection analysis.

The strain energy is given by

$$U = \frac{1}{2}\frac{N^2 L}{AE} + \frac{1}{2}EI_z \int_0^L \left[\delta\left(\frac{\pi}{L}\right)^2 \sin\frac{\pi x}{L}\right]^2 dx$$

$$= \frac{1}{2}\frac{N^2 L}{AE} + \frac{1}{2}EI_z\delta^2\left(\frac{\pi}{L}\right)^4 \int_0^L \sin^2\frac{\pi x}{L}\, dx$$

Integration results in

$$U = \frac{1}{2}\frac{N^2 L}{AE} + \frac{\pi^4}{4}\frac{EI_z}{L^3}\delta^2 \qquad \text{[c]}$$

Substituting Eqs. (a), (b), and (c) into Eq. (6.13-7) results in

$$\frac{1}{\pi}w_o L\delta + \frac{1}{2}\frac{N^2 L}{AE} - \frac{\pi^2}{4}\frac{N}{L}\delta^2 = \frac{1}{2}\frac{N^2 L}{AE} + \frac{\pi^4}{4}\frac{EI_z}{L^3}\delta^2$$

Simplification gives

$$\delta = \frac{4w_o L^4}{\pi^3(NL^2 + \pi^2 EI_z)} \qquad \text{[d]}$$

Substitution of numerical values and simplifying algebraically yields

$$\delta = \frac{4(1.6)(10^3)(1.25)^4}{\pi^3[(3.6)(10^3)(1.25)^2 + \pi^2(70)(10^9)(40)(10^3)(10^{-3})^4]}$$

$$= 0.015151 \text{ m} = 15.2 \text{ mm}$$

which is approximately 0.5 percent higher than the result obtained by the exact method in part (a).

Next we see how the deflection shape of Appendix C is used. Without N, the deflection equation is given by

$$v_c = \frac{wx}{24EI_z}(2Lx^2 - x^3 - L^3)$$

Since only the shape is used, let

$$v_c = Kx(2Lx^2 - x^3 - L^3) \qquad \text{[e]}$$

where K is a constant. Thus

$$v_c' = K(6Lx^2 - 4x^3 - L^3) \qquad v_c'' = 12Kx(L - x)$$

The work due to the distributed load is

$$W_q = -\frac{1}{2}\int_0^L w_o Kx(2Lx^2 - x^3 - L^3)\, dx$$

Integration yields

$$W_q = \frac{1}{10} w_o K L^5 \qquad\qquad [f]$$

The work W_N is

$$W_N = \frac{1}{2} \frac{N^2 L}{AE} - \frac{1}{2} N \int_0^L [K(6Lx^2 - 4x^3 - L^3)]^2 \, dx$$

Integration yields

$$W_N = \frac{1}{2} \frac{N^2 L}{AE} - \frac{17}{70} N K^2 L^7 \qquad\qquad [g]$$

The strain energy is

$$U = \frac{1}{2} \frac{N^2 L}{AE} + \frac{1}{2} \int_0^L EI_z [12Kx(L - x)]^2 \, dx$$

Integration yields

$$U = \frac{1}{2} \frac{N^2 L}{AE} + \frac{12}{5} EI_z K^2 L^5 \qquad\qquad [h]$$

Again, combining Eqs. (f), (g), and (h) with Eq. (6.13-7) results in

$$K = \frac{7w_o}{17NL^2 + 168EI_z} \qquad\qquad [i]$$

Substitution of numerical values yields

$$K = \frac{(7)(1.6)(10^3)}{(17)(3.6)(10^3)(1.25)^2 + (168)(70)(10^9)(40)(10^3)(10^{-3})^4} = 19.787(10^{-3})$$

Substituting K into Eq. (e) gives

$$v_c = 19.787x(2Lx^2 - x^3 - L^3)(10^{-3}) \qquad\qquad [j]$$

The deflection at $x = L/2 = 0.625$ m is

$$(v_c)_{x=L/2} = (19.787)(0.625)[(2)(1.25)(0.625)^2 - (0.625)^3 - (1.25)^3](10^{-3})$$

$$= 0.015096 \text{ m} = -15.1 \text{ mm}$$

Comparing this result with that found in part (a) shows that there is less than 0.1 percent difference.

Determine the maximum deflection of the cantilever beam loaded as shown in Fig. 6.13-3. Approximate the transverse deflection shape using the solution that neglects axial loading.   **Example 6.13-3**

Solution:

The deflection *shape* neglecting the direct compressive force F is found in Appendix C and is

$$v_c = \frac{\delta x^2}{2L^3}(3L - x)$$

Note that v_c was manipulated so that $v_c = -\delta$ at $x = L$, where δ is the maximum deflection. The first derivative of v_c with respect to x is

$$v_c' = -\frac{3\delta x}{2L^3}(2L - x)$$

and the second derivative is

$$v_c'' = -\frac{3\delta}{L^3}(L - x)$$

The work due to P is

$$W_P = \frac{1}{2}P\delta \qquad\qquad\qquad \textbf{[a]}$$

Since $N = -F$, the work due to F is

$$W_F = \frac{1}{2}\frac{F^2 L}{AE} + \frac{1}{2}F\int_0^L \left[-\frac{3\delta x}{2L^3}(2L - x)\right]^2 dx$$

Integration yields

$$W_F = \frac{1}{2}\frac{F^2 L}{AE} + \frac{3}{5}F\frac{\delta^2}{L} \qquad\qquad\qquad \textbf{[b]}$$

The strain energy is

$$U = \frac{1}{2}\frac{F^2 L}{AE} + \frac{1}{2}F\int_0^L EI_z\left[-\frac{3\delta}{L^3}(L - x)\right]^2 dx$$

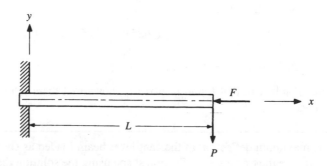

Figure 6.13-3

Integration yields

$$U = \frac{1}{2}\frac{F^2 L}{AE} + \frac{3}{2}EI_z\frac{\delta^2}{L^3}$$ [c]

Substituting Eqs. (*a*) to (*c*) into Eq. (6.13-7) results in

$$\frac{1}{2}P\delta + \frac{1}{2}\frac{F^2 L}{AE} + \frac{3}{5}F\frac{\delta^2}{L} = \frac{1}{2}\frac{F^2 L}{AE} + \frac{3}{2}EI_z\frac{\delta^2}{L^3}$$

Simplifying yields

$$\delta = \frac{5PL^3}{3(5EI_z - 2FL^2)}$$ [d]

If $F = 0$, the maximum deflection agrees with beam C.1 of Appendix C, where $\delta = PL^3/(3EI_z)$, since the shape used agrees with that case. As the compressive force F increases in magnitude, the maximum deflection increases beyond that given by simple beam theory. Further note from Eq. (*d*) that when $F = 2.5EI_z/L^2$, $\delta \to \infty$. From Table 3.10-1, buckling occurs for a fixed-free column when $F = \pi^2 EI_z/4L^2 = 2.467\ EI_z/L^2$.

6.13.2   CONSTRAINED BEAMS

When ground supports cause an axial constraint in a beam loaded by transverse loads, an axial force develops. The problem is similar to the ones previously discussed in this section, but in this case the problem is statically indeterminate, as the axial force N cannot be found from the equilibrium equations. Consider the beam shown in Fig. 6.13-4. The ground constraint is modeled by a simple linear spring of spring constant

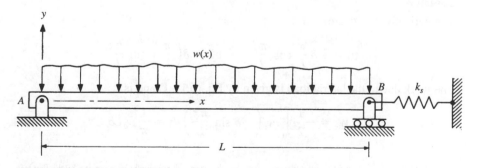

Figure 6.13-4

k_s with units of force per unit length. If the supports are perfectly rigid, $k_s \to \infty$. Recall Fig. 6.13-2; the deflection of point B in the positive x direction is

$$\delta_B = \frac{NL}{AE} - \frac{1}{2} \int_0^L (v_c')^2 \, dx \qquad \textbf{[6.13-8]}$$

As the lateral load $w(x)$ in Fig. 6.13-4 is applied, the spring will go into tension and point B will move to the left. Thus, the deflection of point B in the x direction is

$$\delta_B = -\frac{N}{k_s} \qquad \textbf{[6.13-9]}$$

Equating Eqs. (6.13-8) and (6.13-9) and solving for N yields

$$N = \frac{k_s AE}{2(k_s L + AE)} \int_0^L (v_c')^2 \, dx \qquad \textbf{[6.13-10]}$$

Substituting Eq. (6.13-10) into the second term of Eq. (6.13-5) yields[†]

$$W_N = \frac{1}{2} \frac{N^2 L}{AE} - \frac{k_s AE}{4(k_s L + AE)} \left[\int_0^L (v_c')^2 \, dx \right]^2 \qquad \textbf{[6.13-11]}$$

Thus, if Eq. (6.13-11) is used instead of Eq. (6.13-5) in Eq. (6.13-7), the equation will not contain the unknown N. However, once v_c is found from this equation, N can be determined from Eq. (6.13-10).

If $k_s \to \infty$, Eq. (6.13-11) reduces to

$$W_N = \frac{1}{2} \frac{N^2 L}{AE} - \frac{AE}{4L} \left[\int_0^L (v_c')^2 \, dx \right]^2 \qquad \textbf{[6.13-12]}$$

Example 6.13-4

Figure 6.13-5 shows a uniformly loaded, simply supported beam; however, the axial movement of both supports is zero, as the foundation is very rigid. Determine the maximum deflection and approximate the axial force that develops if $L = 1.25$ m, $w_o = 1.6$ kN/m, $E = 70$ GPa, $I_z = 40 \, (10^3)$ mm^4, and $A = 200$ mm^2. Assume a deflection shape of $v_c = -\delta \sin (\pi x/L)$.

Solution:

The derivatives of v_c are

$$v_c' = -\delta \frac{\pi}{L} \cos \frac{\pi x}{L} \qquad v_c'' = \delta \left(\frac{\pi}{L} \right)^2 \sin \frac{\pi x}{L}$$

The work due to the lateral load is given by Eq. (6.11-1) and is

$$W_q = -\frac{1}{2} \int_0^L w_o \left(-\delta \sin \frac{\pi x}{L} \right) dx = \frac{1}{\pi} w_o L \delta \qquad \textbf{[a]}$$

[†] Substitution of N into the first term is unnecessary, as this term will eventually be canceled by the first term in the strain energy.

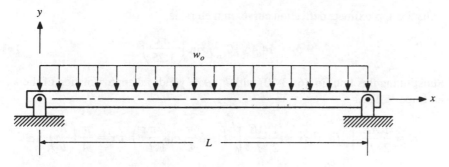

Figure 6.13-5

The work due to N is given by Eq. (6.13-11). Since the supports are to be considered perfectly rigid, $k_s \to \infty$ and

$$W_N = \frac{1}{2} \frac{N^2 L}{AE} - \frac{AE}{4L}\left[\int_0^L \left(-\delta \frac{\pi}{L} \cos \frac{\pi x}{L}\right)^2 dx\right]^2$$

Integration yields

$$W_N = \frac{1}{2} \frac{N^2 L}{AE} - \frac{\pi^4}{16} \frac{AE}{L^3} \delta^4 \qquad \textbf{[b]}$$

The strain energy is

$$U = \frac{1}{2} \frac{N^2 L}{AE} + \frac{1}{2} \int_0^L EI_z \left[\delta \left(\frac{\pi}{L}\right)^2 \sin \frac{\pi x}{L}\right]^2 dx$$

Integration results in

$$U = \frac{1}{2} \frac{N^2 L}{AE} + \frac{\pi^4}{4} \frac{EI_z}{L^3} \delta^2 \qquad \textbf{[c]}$$

Substituting Eqs. (*a*) to (*c*) into Eq. (6.13-7) yields

$$\frac{1}{\pi} w_o L \delta - \frac{\pi^4}{16} \frac{AE}{L^3} \delta^4 = \frac{\pi^4}{4} \frac{EI_z}{L^3} \delta^2$$

Simplifying algebraically gives

$$\delta^3 + 4 \frac{I_z}{A} \delta - \frac{16}{\pi^5} \frac{w_o L^4}{AE} = 0 \qquad \textbf{[d]}$$

Substituting in numerical values yields the cubic equation

$$\delta^3 + 8(10^{-4})\delta - 1.45882(10^{-5}) = 0$$

Solving for δ yields one real root,

$$\delta = 0.014458 \text{ m} = 14.46 \text{ mm}$$

Thus the approximate deflection curve, in meters, is

$$v_c = 14.46(10^{-3}) \sin \left(\frac{\pi x}{1.25} \right) \qquad \text{[o]}$$

Substituting this into Eq. (6.13-10) with $k_s \to \infty$ yields an approximation of the axial force N. Thus

$$N \approx \frac{1}{2} \frac{AE}{L} \int_0^L (v_c')^2 \, dx = \frac{1}{2} \frac{AE}{L} \int_0^L \left(-\delta \frac{\pi}{L} \cos \frac{\pi x}{L} \right)^2 dx = \frac{\pi^2}{4} \left(\frac{\delta}{L} \right)^2 AE$$

Substituting $\delta = 0.014458$ m, $L = 1.25$ m, $A = 200$ mm^2, and $E = 70$ GPa yields

$$N \approx \frac{\pi^2}{4} \left(\frac{0.014458}{1.25} \right)^2 (200)(10^{-3})^2 (70)(10^9) = 4620 \text{ N} = 4.62 \text{ kN}$$

Example 6.13-5 | Compare the maximum tensile stress of Example 6.13-4 with that obtained when one of the supports is free to move axially. Consider the cross section of the beam to be symmetric and the height of the cross section to be 50 mm.

Solution:

When one support has axial freedom, the maximum bending moment occurs at the midspan and is

$$(M_z)_{\max} = \frac{w_o L^2}{8} = \frac{(1.6)(10^3)(1.25)^2}{8} = 312.5 \text{ N·m}$$

and the maximum tensile stress is

$$\sigma_{\max} = \frac{(M_z y)_{\max}}{I_z} = \frac{(312.5)(25)(10^{-3})}{(40)(10^3)(10^{-3})^4} = 195.3(10^6) \text{ N/m}^2 = 195.3 \text{ MPa}$$

For the constrained beam of Example 6.13-4, the maximum bending moment also occurs at midspan, but the bending moment is

$$(M_z)_{\max} = \frac{w_o L^2}{8} - N\delta = 312.5 - (4620)(0.014458) = 245.7 \text{ N·m}$$

The maximum tensile stress is

$$\sigma_{\max} = \frac{(M_z y)_{\max}}{I_z} + \frac{N}{A} = \frac{(245.7)(25)(10^{-3})}{(40)(10^3)(10^{-3})^4} + \frac{4620}{(200)(10^{-3})^2}$$

$$= 176.7(10^6) \text{ N/m}^2 = 176.7 \text{ MPa}$$

By constraining this particular beam, there is a 10 percent reduction in the stress at the midspan. The bearing forces at the beam ends, however, can be shown to be increased by a factor of 4.7.

6.14 PROBLEMS

6.1 At a point the state of stress is given by the stress matrix

$$[\sigma] = \begin{bmatrix} 30 & -15 & 20 \\ -15 & -25 & 10 \\ 20 & 10 & 40 \end{bmatrix} \text{MPa}$$

Determine the strain energy per unit volume if $E = 70$ GPa and $v = 0.33$.

6.2 Determine the strain energy per unit volume for Problem 1.17.

6.3 For the stress matrix shown with $E = 210$ GPa and $v = 0.3$ show that the strain energy per unit volume using Eq. (6.2-11) gives the same result as that using the principal stresses in Eq. (6.2-12).

$$[\sigma] = \begin{bmatrix} 20 & 10 & 10 \\ 10 & 20 & 10 \\ 10 & 10 & 20 \end{bmatrix} \text{MPa}$$

6.4 Substituting Eq. (2.1-30) into Eq. (6.2-14) show that Eq. (6.2-13) results. That is, show that Eqs. (6.2-13) and (6.2-14) are equivalent.

6.5 Determine the total strain energy in the beam shown in Fig. 3.4-1 if $E = 70$ GPa, $v = 0.33$, $I = 50 \, (10^3) \text{ mm}^4$, $A = 150 \text{ mm}^2$, and $k = 1.0$. Compare and comment on the difference between the shear and bending energies. HINT: For the right section of the beam use an isolation of the right section and a variable originating at C and going to the left.

6.6 Show that for a solid circular beam cross section the form correction factor for shear is $k = 10/9 = 1.11$.

6.7 The cables of Problem 3.4 are repeated here. However, let the cable material have a stress-strain relation of

$$\sigma = 200(10^9)\varepsilon \, (1 - 500\varepsilon)$$

where σ is in N/m². Cables AB and BC have an effective area of 125 mm². The lengths of cables AB and BC are 500 and 750 mm, respectively. A vertical force of $P = 8$ kN is applied to joint B. Using Castigliano's first theorem, determine the vertical deflection of point B.

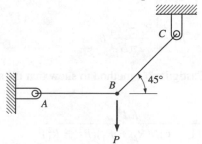

Problem 6.7

6.8 Using the complementary energy theorem, repeat Problem 6.7. Let $\varepsilon = C_1\sigma + C_2\sigma^2$ and using a spreadsheet or mathematics program perform a least-squares fit of the function given in Problem 6.7 in the range $0 \le \sigma \le 100$ MPa.

6.9 Using an energy or work approach, determine the vertical deflection of point C and the deflection of point D of the structure shown. Check your results by using a geometrical approach. Each member has a modulus of elasticity E and area A; the length of members BC and CD is L.

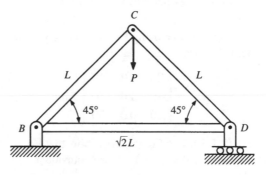

Problem 6.9

6.10 Consider each member of the structure given in Problem 6.9 to have a deflection-load relation of $\delta_i = C_{i1}F_i + C_{i2}F_i^2$, where C_{ij} are constants for each member and F_i is the force transmitted in each member. For members BC and CD, $C_{i1} = C_{11}$ and $C_{i2} = C_{12}$, whereas, for member BD, $C_{i1} = C_{21}$ and $C_{i2} = C_{22}$. Using the complementary energy theorem, determine the vertical deflection of point C.

6.11 The figure shows a compression conical helical coil spring where R_1 and R_2 are the initial and final coil radii, respectively, d is the diameter of the wire cross section, and N is the total number of active coils. The wire cross section primarily transmits a torsional moment which changes with the coil radius. Let the coil radius be given by

$$R = R_1 + \frac{R_2 - R_1}{2\pi N}\theta$$

where θ is in radians. Use Castigliano's method to show that the spring rate is given by

$$k = \frac{P}{\delta_p} = \frac{Ed^4}{32(1 + v)N(R_2 + R_1)(R_2^2 + R_1^2)}$$

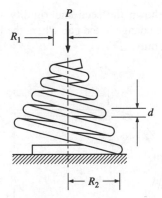

Problem 6.11

6.12 The cantilever beam shown is of constant width, b. The depth of the beam is constant for $0 \leq x \leq a$, and beyond $x = a$ varies gradually and linearly. Using Castigliano's method with the straight beam energy formulation, determine the vertical deflection of the free end.

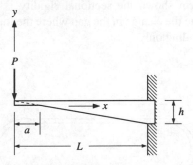

Problem 6.12

6.13 The thin-walled ring shown is loaded by a uniform force per unit length q. Using Castigliano's method, determine the vertical deflection of the free end of the ring. Consider the effects of bending only and let the rigidity of the cross section be EI.

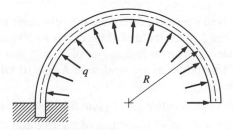

Problem 6.13

6.14 For the wire form shown the sectional rigidity is *EI*. Using Castigliano's method determine the change in the gap where the loads are applied. Consider the effect of bending only.

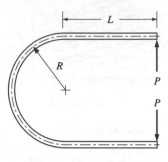

Problem 6.14

6.15 For the wire form shown the sectional rigidity is *EI*. Using Castigliano's method determine the change in the gap where the loads are applied. Consider the effects of bending only.

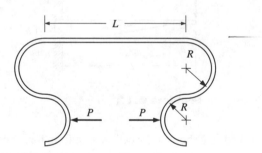

Problem 6.15

6.16 Using Castigliano's method, determine the deflection of the midpoint of the beam in Problem 1.5. Let the area of the cables be 50 mm², and beam *AB* be a circular cylinder with an outer diameter of 40 mm and a wall thickness of 3 mm. The cables and beam are steel with $E = 210$ GPa. Ignore transverse shear in the beam.

6.17 Determine the vertical deflection of point *B* in Problem 6.16.

6.18 Determine the rotational displacement of the free end of the curved beam of Example 6.6-10.

6.19 In Example 3.7-2 use Castigliano's method to determine the equation for the beam deflection $v_c(x)$ in section BC and compare that with Eq. (b) of the example. HINT: Place a dummy load Q anywhere in section BC located at $x = x_Q$. After applying Castigliano's method *and* integration (treating x_Q as constant), set $x_Q = x$.

6.20 Solve part (b) of Problem 3.3 using Castigliano's method.

6.21 Consider the cables of Problem 6.7. However, let the cables be steel with a linear stress-strain relation with $E = 200$ GPa. Cables AB and BC have an effective area of 125 mm². The lengths of cables AB and BC are 500 and 750 mm, respectively. A vertical force of $P = 8$ kN is applied to joint B. Using Castigliano's method, determine the total deflection of point B.

6.22 The tube in Problem 3.32 is made of aluminum with $E = 10$ Mpsi and $v = 0.33$. Using Castigliano's method, determine the deflection of the applied force in the y direction. Ignore transverse shear.

6.23 Let the beam of Problem 3.31 be of steel with $E = 30$ Mpsi and $v = 0.3$. Using Castigliano's method determine the vertical deflection of point A due to bending.

6.24 In Sec. 5.2.1 the angle of twist of a closed thin-walled tube in torsion [Eq. (5.2-6)] is derived using a geometric approach with an averaging assumption. For this problem, derive Eq. (5.2-6) using Castigliano's method and a volumetric element $tL\,ds$. Review the derivation given in Sec. 5.2.1 and comment on the two different approaches in deriving Eq. (5.2-6).

6.25 The truss structure shown contains 15 pin-connected elements each with a cross-sectional area of 200 mm² and a modulus of elasticity of 210 GPa. Determine the vertical deflection of point H using Castigliano's method.

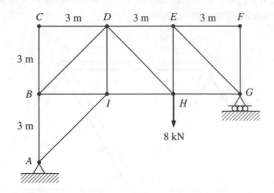

Problem 6.25

6.26 For the structure given in Problem 6.25 determine the horizontal deflection of point G using Castigliano's method.

6.27 For the structure given in Problem 6.25 determine the rotation of member *DH* using Castigliano's method.

6.28 Using Castigliano's method determine the rotation of member 2 in Example 6.6-1.

6.29 For the beam C.3 of Appendix C use Castigliano's method to verify that the slope and deflection at $x = L$ are $-wL^3/(6EI)$ and $-wL^4/(8EI)$, respectively.

6.30 For beam C.4 of Appendix C use Castigliano's method to verify that the slope and deflection at $x = L$ are $ML/(EI)$ and $ML^2/(2EI)$, respectively.

6.31 For beam C.5 of Appendix C with $a = b = L/2$ use Castigliano's method to determine the slope at $x = 0$ and the deflection at $x = L/2$. Compare your results with that given in Appendix C.

6.32 Using Castigliano's method, determine the deflection at the midpoint of the steel step shaft shown. Let $E = 30$ Mpsi.

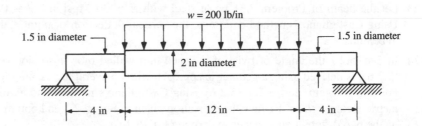

Problem 6.32

6.33 For the truss shown use Castigliano's method to determine the vertical deflection of point *C*. All members are of equal length, area, and modulus of elasticity *L*, *A*, and *E*, respectively.

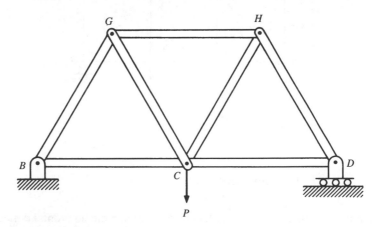

Problem 6.33

6.34 For the truss of Problem 6.33, use Castigliano's method to determine the horizontal deflection of point D.

6.35 For the truss of Problem 6.33, use Castigliano's method to determine the rotation of member BG.

6.36 Using Castigliano's method on the wire form shown determine the vertical deflection of point A considering bending only. The rigidity of the cross section is EI.

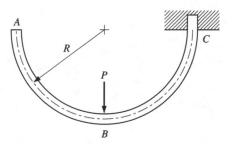

Problem 6.36

6.37 For the wire form shown use Castigliano's method to determine the vertical deflection of point A. Consider bending only and let the bending rigidity of the cross section be EI.

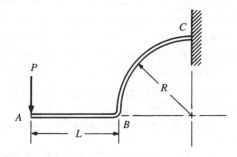

Problem 6.37

6.38 For the wire form shown, use Castigliano's method to determine the deflection of point A in the y direction. Consider the effects of bending and torsion only. Use the straight beam formulation for the bending energy. The wire is steel with $E = 200$ GPa, $v = 0.29$, and has a diameter of 5 mm. Before application of the 200-N force the wire form is in the xz plane where the radius R is 100 mm.

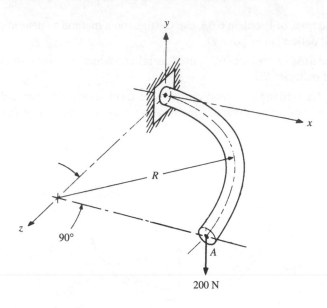

200 N

Problem 6.38

6.39 Using Castigliano's method solve Problem 6.25 with member DH undergoing a temperature increase of 100°C. The temperature coefficient of expansion α is 11.7 (10^{-6})/°C.

6.40 Using Castigliano's method solve Problem 6.33 with member GC undergoing a temperature change of ΔT. The temperature coefficient of expansion is α.

6.41 Using Eq. (6.7-10) for the curved section determine the deflection of the 200-lb load in Problem 5.49. Let $E = 30$ Mpsi, $v = 0.3$, and $k = 1.0$.

6.42 Using Eq. (6.7-10) for the curved section determine the relative displacement of the applied loads in Problem 5.50. Let $E = 210$ GPa, $v = 0.3$, and $k = 1.0$.

6.43 Determine the relative displacement of the applied loads in Problem 5.52. Let $E = 30$ Mpsi, and $v = 0.3$.
   (a) Solve by considering bending only with the straight beam energy formulation.
   (b) Solve using only the first term in Eq. (6.7-10) for the curved section.
   (c) Solve using the entire Eq. (6.7-10) with $k = 1.0$ for the curved section.
   (d) Compare the results of parts (a) to (c).

6.44 Using Eq. (6.7-10) for the curved section determine the relative displacement of the applied forces in Problem 5.51. Let $E = 30$ Mpsi, $v = 0.3$, and $k = 1.2$.

6.45 Repeat Example 6.8-3 except let the cylinder be thick-walled and subject to the equations given in Sec. 6.7.
   (a) Determine the equations for the bending moments at points A and C and the deflection of point B.

(b) Let the cross section be rectangular where the outer and inner radii are 50 and 25 mm, respectively, and the thickness perpendicular to the $r\theta$ plane is 12 mm. If $P = 200$ kN, $E = 210$ GPa, and $\nu = 0.3$, determine M_A, M_C, and δ_B.

(c) Compare the results of part (b) with that determined in Example 6.8-3.

6.46 Using Castigliano's method determine the support reactions for the beam shown.

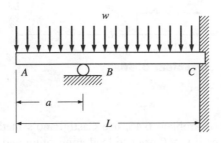

Problem 6.46

6.47 The thin-walled ring shown has three symmetrically positioned forces P. Using Castigliano's method, determine (a) how the bending moment varies in the ring, and (b) the radial deflection of each force considering bending only.

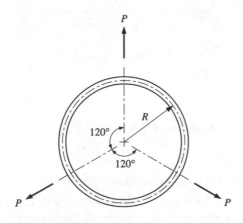

Problem 6.47

6.48 Using Castigliano's method, determine the support reactions if $\epsilon = 0$ prior to load application.

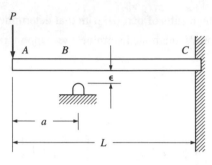

Problem 6.48

6.49 For the figure of Problem 6.48, use Castigliano's method to determine the support reactions if $\epsilon > 0$ prior to load application and P is more than sufficient for the beam to make contact with the support at B.

6.50 Two separate beams are in frictionless contact as shown. Both beams have the same rectangular cross section of constant width and depth and have the same sectional rigidity EI. (1) First explain why that as the end load P is applied, the beams only make contact at B. Then, with this in mind, (2) using Castigliano's method determine the support reactions for each beam.

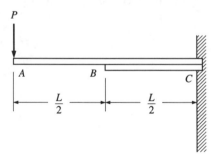

Problem 6.50

6.51 Solve Example 3.8-1 using Castigliano's method.

6.52 Solve Example 3.8-3 using Castigliano's method. Compare the results with that found in the example and comment on the degree of difficulty of the geometric method versus the energy method.

6.53 In Problem 3.44 determine the rotational displacement of the applied torque T_A using Castigliano's method.

6.54 Solve Problem 3.45 using Castigliano's method.

6.55 Solve Problem 3.46 using Castigliano's method.

6.56 The wire form shown models a chain link. Considering bending only and using the straight beam energy formulation, determine (a) how the bending moment varies with respect to position within the wire form and (b) the deflection of the load P. The rigidity of the cross section of the wire is EI.

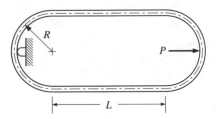

Problem 6.56

6.57 Repeat part (a) of Problem 6.56 considering the link to be thick-walled where the bending term in Eq. (6.7-10) must be used for the curved section. For the curved section let the rigidity be EA, $r_c = R$, and the eccentricity of the neutral axis be e.

6.58 Using Castigliano's method and the straight-beam energy formulation for the wire form shown, determine (a) the reactions at points A and B, (b) how the bending moment varies along the wire, and (c) the deflection of the load P. The bending rigidity of the section is EI.

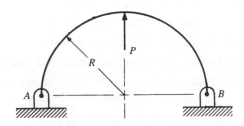

Problem 6.58

6.59 For the wire form shown point A is allowed to slide freely in the z direction. Determine the reactions at points A and B considering bending and torsion in the form. Assume that the cross section is circular with $J = 2I$.

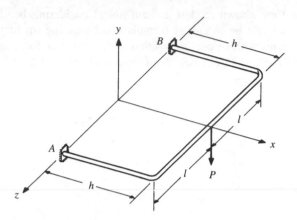

Problem 6.59

6.60 Solve Problem 6.25 with the roller support at point G being replaced by a full pin support.

6.61 Solve Problem 6.13 using the virtual load method.

6.62 Solve Problem 6.16 using the virtual load method.

6.63 Solve Problem 6.18 using the virtual load method.

6.64 Solve Problem 6.21 using the virtual load method.

6.65 Solve Problem 6.39 using the virtual load method.

6.66 Solve Problem 6.40 using the virtual load method.

6.67 Solve Problem 6.48 using the virtual load method.

6.68 Solve Problem 6.49 using the virtual load method.

6.69 Solve Problem 6.52 using the virtual load method.

6.70 Solve Problem 6.53 using the virtual load method.

6.71 Solve Problem 6.54 using the virtual load method.

6.72 Solve part (a) of Problem 6.55 using the virtual load method.

6.73 For the uniformly loaded cantilever beam shown, determine the maximum deflection by Rayleigh's method using the following deflection shapes:

$$v_c(x) = -\delta\left(1 - \cos\frac{\pi x}{2L}\right) \qquad \text{[a]}$$

$$v_c(x) = \frac{x^2}{3L^4} \delta(4Lx - x^2 - 6L^2) \qquad \text{[b]}$$

Compare the results with that given in Appendix C.

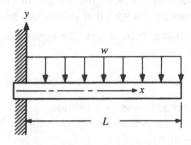

Problem 6.73

6.74 The figure shows a cantilever beam loaded by a moment M. Assuming a deflection shape of $v_c(x) = \delta[1 - \cos(\pi x/2L)]$, approximate δ using Rayleigh's method and compare the result with that given for beam C.4 in Appendix C.

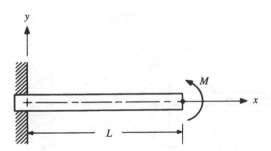

Problem 6.74

6.75 For beam C.6 of Appendix C assume a deflection shape of $v_c(x) = -\delta \sin(\pi x/L)$ and using Rayleigh's method approximate δ. Compare the result with $(v_c)_{max}$ given in Appendix C.

6.76 For beam C.7 of Appendix C with $a = b = L/2$ assume a deflection shape of $v_c(x) = -\delta \sin(2\pi x/L)$ and using Rayleigh's method approximate δ. Compare the result with $(v_c)_{x = L/4}$ using the equations given in Appendix C.

6.77 For the beam given in Problem 6.73, determine the deflection of the centroidal axis using the Rayleigh-Ritz method. Assume the deflection shape to be polynomial in x as given by

$$v_c(x) = a_0 + a_1x + a_2x^2 + a_3x^3 + a_4x^4$$

For boundary conditions, employ only the geometric boundary conditions at $x = 0$. Compare your results with that given in Appendix C.

6.78 For the beam in Problem 6.74 assume the deflection shape to be polynomial in x of the order 3. That is,

$$v_c(x) = a_0 + a_1x + a_2x^2 + a_3x^3$$

Use only the geometric boundary conditions at $x = 0$ and the Rayleigh-Ritz method to determine the deflection equation for the beam. Compare your result with beam C.4 of Appendix C.

6.79 For Example 6.13-3, determine $v_c(x)$ exactly through the development and solution of the appropriate differential equation. If $P = 2$ kN, $F = 3$ kN, $EI_z = 30$ kN·m^2, and $L = 1$ m, determine the maximum deflection and compare this result with that given when substituting the numerical values into Eq. (d) of Example 6.13-3.

6.80 Using Rayleigh's method, estimate the maximum deflection of the steel beam shown, the direct tensile force, and the end reactions using (a) $v_c(x) = -\delta \sin(\pi x/L)$ and (b) the shape given in Appendix C for beam C.5 with $a = b = L/2$.

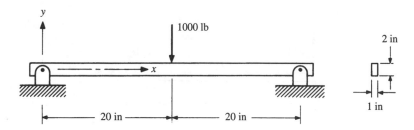

Problem 6.80

6.81 Beam C.14 of Appendix C assumes that the beam is free to slide in the x direction at one or both of the supports. Using Rayleigh's method with the shape given in Appendix C approximate the maximum deflection, the axial force, and the end reactions if the beam is fixed in the x direction at the supports. Let w, L, A, I, and E be 20 kN/m, 1 m, 250 mm^2, 30(10^3) mm^4, and 210 GPa, respectively.

6.82 Consider beam C.3 of Appendix C with a compressive force F applied at the free end.

(a) Determine the equation for the deflection of the centroidal axis $v_c(x)$ by the development and solution of the appropriate differential equation.

(b) Determine the equation for the maximum vertical deflection using Rayleigh's method with the deflection shape

$$v_c(x) = -\delta[1 - \cos{(\pi/2L)}]$$

(c) If $w = 20$ lb/in, $F = 750$ lb, $EI_z = 10(10^5)$ lb·in², and $L = 20$ in, determine the maximum deflection using the results of parts (a) and (b). Compare and comment on the two results.

Computer Problems

6.83 The deflection of the curved beam of Example 6.6-10, considering the straight-beam bending energy formulation only, is

$$\delta = \frac{\pi}{2} \frac{PR^3}{EI_z} \qquad \text{[a]}$$

Example 6.7-1 gives the solution to the same problem using the curved-beam bending energy formulation together with transverse shear and normal force energy formulations. Here the result is given by

$$\delta = \frac{\pi}{2} \frac{PR}{EI} \left[\frac{R}{e} - 1 + 2(1 + v)k \right] \qquad \text{[b]}$$

Use the same cross section as that of Example 6.7.1 (1.0 × 2.0 in), but with varying r_c. Plot the difference, in percent, between Eqs. (a) and (b) as a function of r_c. Let $k = 1.2$ and $1.5 \leq r_c \leq 10$ in. Comment on the results.

6.84 Considering Example 6.5-2, using the numerical values for C_1, C_2, L, and A from Example 6.4-1, plot the complementary strain energy Φ as a function of P_B for $0 \leq P_B \leq 3$ kips. Perform a third-order polynomial fit and then plot $d\Phi/dP_B$ as a function of P_B for $0 \leq P_B \leq 3$ kips. The value of $d\Phi/dP_B$ at $P_B = 2$ kips will yield the deflection for the problem.

SEVEN

STRENGTH, FAILURE MODES, AND DESIGN CONSIDERATIONS

7.0   INTRODUCTION

The intent of this chapter is to provide information related to design applications. The first section is a philosophical discussion of the term *strength*. This is followed by a brief discussion of the design factor or factor of safety. The next four sections introduce various modes of failures such as static failure of ductile and brittle materials, fracture mechanics, fatigue, and structural instability. This is followed by a discussion on the inelastic behavior of ductile materials loaded beyond yield conditions. The final section, based on inelastic behavior, presents methods of approximation which are helpful in reducing statically indeterminate ductile structures to simpler, statically determinate ones.

7.1   "STRENGTH"

One of the earliest topics studied in an introductory mechanics of materials course is that of material strength. The word *strength* is so overworked in the field of stress analysis that its meaning at times becomes vague. For example, strength of materials as a title of a textbook or a course suggests that material behavior per se is the main concern. On the contrary, although material behavior is an important and integral part of a strength of materials textbook or course, the topic is usually considered as an ancillary issue which links stress and strain (via the material properties E and v) and provides some limits on a structural material's capabilities, e.g., yield and/or ultimate strength. Typically, the primary emphasis of a strength of materials textbook or course is on the mechanics of the behavior of structural members under the influence of applied loads (where naturally the members themselves are made of various materials and the properties of the materials are important). A more appropriate name for such a book or course would be something like the *mechanics of deformable bodies, structural analysis,* etc. In any event, at times the word *strength* has mistakenly become synonymous with stress (some textbooks even use the same symbols for stress and strength). An example is a structural element in bending called a beam of *constant strength* (see Problem 3.53). This beam actually is designed geometrically such that the outer fibers along the span have the same (constant) tensile and compressive *stresses*. If the beam were made of a material with differing tensile and compressive strengths, which is possible, the beam would certainly *not* be a beam of constant strength. In order to avoid confusion, it is necessary to divorce the words *stress* and *strength*. Stress has already been defined as the internal-force distribution along a specific surface orientation at a uniquely distinct point within a structural element experiencing applied loads. Strength, on the other hand, denotes some kind of limit and is more difficult to define. Strength is based on the *design intent*. Do we not want the part to yield, buckle, fracture, exceed some deflection criteria, fail under cyclic loading, etc? The answer would constitute the strength of the part to safeguard against such an occurrence. A common practice is to define strength as a maximum allowable stress that can be applied. Consider, for example, a 0.5-in-diameter aluminum rod loaded in pure tension. If the tensile force reaches a maximum value of 13 kips before a complete fracture, it is acceptable to say that the ultimate strength of *that* specific rod loaded in the particular manner stated *was* 13 kips. Note that this definition of strength is restricted by the type of loading and the size, shape, and material capability of the *specific* specimen used. In an attempt to eliminate size and shape from consideration, the force is divided by the original area of the cross section, giving a measure of the maximum tensile stress reached before fracture. For the aluminum rod under discussion, the initial area was $A = \pi(0.5)^2/4 = 0.196$ in^2, and the maximum tensile stress before fracture was $13/0.196 = 66.3$ kpsi.[†] Now, the ultimate tensile strength can be defined by the maximum tensile stress rather than the maximum force obtained before fracture. This value is commonly tabulated in engineering and material handbooks and by definition is the ultimate strength S_U (force per unit area). To avoid confusion with stress, a state property, here the subscripted letter S will be used to denote strength. Other examples include yield strength S_Y and endurance strength S_E used in fatigue analysis.

It is normally assumed that once strength is defined in terms of some maximum obtainable value of stress, size effects are no longer important. Although this is normal practice, in certain cases, it is far from the truth. In general, for a given material the strength in force per unit area decreases as the size increases. This is most dramatic when the size is small. For example, the ultimate strength of music wire for a wire diameter of 0.010 in (0.25 mm) is about 390 kpsi (2.69 GPa). Whereas the

---

[†] Here the definition of stress is based on the initial area differentiating it from the *true stress*, which is based on the final reduced area at fracture. This is standard engineering practice.

same material of diameter 0.25 in (6.35 mm) has an ultimate strength of 240 kpsi (1.65 GPa). However, it is important to note that this effect levels off quickly with increasing size. Thus, in most situations, size effects are neglected, especially when the loading is static. A typical specimen used for obtaining tensile strength properties of metals is shown in Fig. 7.3-1 (Sec. 7.3) where the diameter of the specimen between the threaded ends is nominally 0.505 in (giving a cross-sectional area of 0.2 in^2). The static strength values obtained for the specimen (S_Y and/or S_U) are a measure of the material's performance when in the form of a solid circular rod and loaded gradually in pure tension. But what strength is associated with a structure with a more complex geometric and/or loading condition? An attempt to minimize the amount of experimentation which would be necessary to test each new case comes in the form of *strength theories,* which relate a general state of stress to the strength values obtained from simple tests of the material. Some standard static tests which are performed include pure tension, compression, and torsion. For cyclic loading, a standard test is the bending-fatigue test.

This chapter presents information which will assist the designer or analyst in the awareness and understanding of the many modes of failure and their limits. Material behavior and various static strength theories are presented for ductile and brittle materials. The governing equations are then modified by the *design factor* which results in the design equations typically used in engineering applications. Some rationale associated with the design factor is discussed briefly in the following section. Other topics associated with structural strength are presented at an introductory level. These include fracture mechanics, fatigue, and structural stability. The last section of the chapter, Sec. 7.8, presents two methods commonly used in design to simplify and approximate the strength of statically indeterminate ductile structures. These methods are based, in part, on the inelastic behavior of ductile materials, which is covered in Sec. 7.7.

7.2   THE DESIGN FACTOR

The *design factor* or *factor of safety n* is used to account for the uncertainties in the determination of the strength of a part as well as in the evaluation of the governing parameter (e.g., maximum stress) of the part. Assuming that the governing parameter is the maximum stress, the design factor is defined as

$$n = \frac{\text{Allowable stress (strength)}}{\text{Calculated stress}} = \frac{S}{\sigma} \qquad \textbf{[7.2-1]}$$

Another term that is commonly used is the *margin of safety m,* which is defined as

$$m = n - 1 \qquad \textbf{[7.2-2]}$$

The design factor n is used to account for uncertainties in the determination of the strength of the part as well as uncertainties in the evaluation of the stresses in the part (see Ref. 7.1).

The uncertainties relating to part strength are:

1. The uncertainty of the exact properties of the base material of the part.

2. The uncertainty of the size effect, discussed earlier.

3. The uncertainty in the manufacturing of the part; material forming (casting, forging, drawing, etc.), machining, welding, heat treatment, and surface treatment (such as plating)—all of which have an effect on the part-strength uncertainty.

4. The uncertainty of the effect of the operating environment, e.g., temperature effects and corrosion.

Define a part-strength uncertainty factor f_s such that the "true" minimum strength of the part S_{min} is given by

$$S_{min} = \frac{S}{f_s} \qquad\qquad \textbf{[7.2-3]}$$

where S is the material strength of a specimen (yield and/or ultimate) of a given material obtained from direct test, published data, or the manufacturer's specifications. This number in and by itself may be based on statistical interpretations and uncertainty associated with item 1 above. The value of S may not reflect the true strength of the material of the actual design part and is corrected by the uncertainty factor f_s. The final strength uncertainty factor is the product of each individual uncertainty factor related to items 1 to 4 given above. In general, $f_s > 1$.

The uncertainties associated with the part stresses are:

1. The uncertainties in the analysis of static stresses. The accuracy of the analysis depends on the type of analysis performed and the modeling of the geometry, loading, and boundary conditions. If manual calculations were performed using closed-form equations, how well did the actual part conform to the mathematical model? If a numerical analysis was performed, such as finite elements, was the modeling adequate (mesh density, element type, boundary conditions, etc.)? Were stress concentrations accounted for correctly? Were any critical calculations verified experimentally? Obviously, the list goes on.

2. The uncertainties in the assumptions and the type of calculations made for dynamic loads.

3. The uncertainties in the stress calculations due to manufacturing inaccuracies. Because of manufacturing tolerances, stresses calculated using nominal dimensions may be in error due to cross-sectional dimension tolerances and true load positions relative to the part.

4. The uncertainties in assembly operations. Prior to the service loads, stresses may arise from the methods of fastening one part to another (bolting, riveting, welding, press fitting, etc.), where human error may also be involved. Alignment procedures are important, and if self-alignment devices are used, the uncertainty in stress may be reduced.

5. The uncertainties in the initial stresses even before assembly. Residual stresses may be present because of material forming or fabrication techniques (molding, sheet-metal forming, forging, rolling, drawing, etc.).

Define a stress uncertainty factor f_σ such that the "true" maximum stress σ_{max} is given by

$$\sigma_{max} = \frac{\sigma}{f_\sigma} \qquad \text{[7.2-4]}$$

where σ is the stress determined by analysis. The stress-uncertainty factor is the product of the factors of items 1 to 5 above. In general, $f_\sigma < 1$.

Combining Eqs. (7.2-3) and (7.2-4) gives

$$\frac{S}{\sigma} = \frac{f_s S_{min}}{f_\sigma \sigma_{max}}$$

A strength-reduction factor would not be necessary if S_{min} and σ_{max} were known. Thus, let $S_{min}/\sigma_{max} = 1$, and since the ratio S/σ is the strength-reduction factor n,

$$n = \frac{f_s}{f_\sigma} \qquad \text{[7.2-5]}$$

There is even a good deal of uncertainty associated with obtaining the strength- and stress-uncertainty factors themselves. In many cases, codes dictate the appropriate value of n. Where codes do not apply, the designer must often make an intelligent guess for n based either on intuitive reasoning or past experience. Most of the uncertainty factors must be arrived at through statistical studies, especially those dealing with strength.

Making a guesstimate of the design factor may be the only alternative open to the designer. However, the decision is not to be taken lightly. The following example attempts to illustrate how subtle the selection of a factor of safety can be.

Example 7.2-1 | It is necessary to make a plate of the geometry shown in Fig. 7.2-1. The material to be used has a published yield strength of 50 kpsi. The designer is to specify the maximum acceptable load so that absolutely no yielding of the material will occur. For sake of illustration, the following will be assumed.

1. Everything is ideal, except for the tolerances on the cross section; i.e., the published data on the yield strength are exact, and the material is homogeneous and isotropic; the load is positioned perfectly so that the loading causes no bending; the loading is gradually applied; etc.

2. The designer is unaware of stress-concentration effects.

3. No codes for the establishment of a factor of safety apply to the given design.

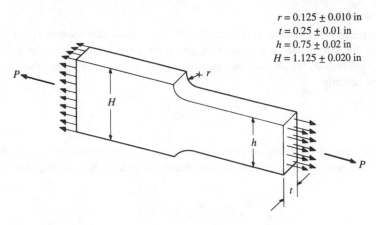

$r = 0.125 \pm 0.010$ in
$t = 0.25 \pm 0.01$ in
$h = 0.75 \pm 0.02$ in
$H = 1.125 \pm 0.020$ in

Figure 7.2-1

Designer's Solution:

It is unfortunate, but assumption 2 applies to some designers. Our designer will probably use nominal dimensions, the normal procedure, finding a stress

$$\sigma = \frac{P_{max}}{A} = \frac{P_{max}}{(0.75)(0.25)} = 5.33 P_{max}$$

He will now equate this to the allowable yield strength divided by the factor of safety. According to the given assumptions, the only place he feels that he is in error is in the determination of the cross-sectional area. The stress-uncertainty factor based on dimensional tolerances is

$$f_\sigma = \frac{\sigma}{\sigma_{max}} = \frac{P_{max}/A_{nom}}{P_{max}/A_{min}} = \frac{A_{min}}{A_{nom}} = \frac{(0.73)(0.24)}{(0.75)(0.25)} = 0.9344$$

Since the strength value is assumed to be exact, $f_s = 1$ and the design factor is

$$n = \frac{f_s}{f_\sigma} = \frac{1}{0.9344} = 1.07$$

To be on the "safe" side, the designer decides to use a design factor of 1.1. To complete the analysis, the stress is equated to the strength divided by the design factor and the result is

$$5.33 P_{max} = \frac{50}{1.1}$$

$$P_{max} = 8.52 \text{ kips}$$

Exact Solution:

The designer was ignorant of the fact that the true stress distribution is not uniform and that the stress near the fillets is larger than the nominal stress he calculated. From Appendix F, the stress-concentration factor is $K_t = 1.8$. Thus, the maximum stress is actually

$$\sigma_{max} = K_t \frac{P_{max}}{A_{min}}$$

If the nominal cross-sectional dimensions are used to calculate the area A, then, as before, a factor of safety of 1.07 should be used to reduce the working stress:

$$1.8 \frac{P_{max}}{(0.75)(0.25)} = \frac{50}{1.07}$$

$$P_{max} = 4.87 \text{ kips}$$

In the above analysis the factor of safety $n = 1.07$ is based on an uncertainty in strength, $f_s = 1$, and in stress, $f_\sigma = 1/1.07$. If in the exact analysis the nominal stress equation is used (the same one used by the designer), a different factor of safety should be used. That is, the uncertainty factor in stress changes to account for the omission of the stress concentration in the nominal stress equation. The uncertainty factor for the stress concentration is $f_\sigma = 1/1.8 = 0.556$, and, as before, the uncertainty factor due to the dimensional tolerances is $1/1.07$. The total uncertainty factor is the product of the individual factors. Thus,

$$f_\sigma = \frac{1}{1.8} \frac{1}{1.07} = 0.5192$$

and the factor of safety using the nominal stress should be

$$n = \frac{f_s}{f_\sigma} = \frac{1}{0.5192} = 1.926 \approx 1.93$$

Thus, a designer who had used this value of n would have obtained the exact solution, i.e.,

$$5.33 P_{max} = \frac{50}{1.926}$$

$$P_{max} = 4.87 \text{ kips}$$

In general, the exact solution is not known as in the overidealized problem just presented. The purpose of the example was to illustrate that a factor of safety is based on how close the analysis reflects the given problem. There will be many cases where a guess is necessary, but one should not be too casual about it.

7.3   STRENGTH THEORIES

7.3.1   BASIS OF THEORIES

The stress-strain curve for a material is obtained by subjecting a test specimen as shown in Fig. 7.3-1 to an axial tensile force. Typical stress-strain curves for various structural materials are shown in Fig. 7.3-2.

Notice the characteristics of the various materials. The initial relationship between stress and strain is linear. Beyond this, the stress reaches the elastic limit, and the material begins to acquire inelastic, or permanent, deformation. The value of stress at this point is called the *yield strength* S_Y. For the carbon steel, the yield point is conspicuous in the sharp dip in its curve. The other materials shown do not exhibit this and their yield strengths are typically defined by the *0.2 percent offset yield stress.* This is obtained by drawing a line parallel to the linear stress-strain line starting at the coordinate (0, 0.002). The stress at the intersection of this line with the material's stress-strain line is defined as the yield strength. If yielding of the material is the criterion for failure, then, all things being equal, the stress in a similar part of the same material should not be allowed to exceed the yield strength *provided* the part is loaded in pure tension.[†] What would be the condition for yielding of a part made of this material if the state of stress was more complicated, e.g.,

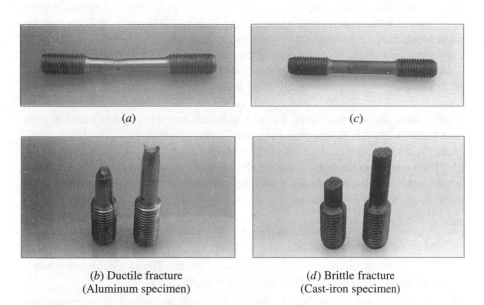

(*a*)

(*c*)

(*b*) Ductile fracture
(Aluminum specimen)

(*d*) Brittle fracture
(Cast-iron specimen)

Figure 7.3-1   Tensile test specimens.

[†] This is neglecting statistical variations in the material as well as size effects.

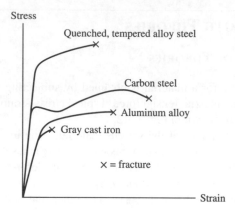

Figure 7.3-2   Tensile stress-strain curves for structural metals.

plane or three-dimensional stress? One logical criterion might be to limit the maximum principal stress to the yield strength of the material. This might be true for one type of material and state of stress, but in general not all materials or stress states have the same mode of failure. For example, if a tensile specimen made of a ductile material is loaded to fracture, the fractures will all be at an angle of approximately 45° from the loading axis [see Fig. 7.3-1(a)]. This would seem to indicate that the shear stress had a more contributing effect to failure than the tensile stress since the maximum shear stress for this load case occurs at 45° (see stress transformations as discussed in Secs. 2.1 and 3.9). To further substantiate this, ductile members tend to yield along 45° "slip" planes at the same pure normal stress, whether in tension or compression. Thus, for ductile materials the *Tresca* (maximum-shear-stress) *theory* is often used. On the other hand, a tensile specimen made of a brittle material *does* tend to fail in tension, as can be seen by the fracture shown in Fig. 7.3-1(b). In compression, a brittle material typically fails in shear.

There are many theories of static failure which can be postulated for which the consequences can be seen in the tensile test. When the tensile specimen begins to yield, the following events occur:

1. The *maximum-principal-stress theory*: the maximum principal stress reaches the tensile yield strength S_Y.

2. The *maximum-shear-stress theory* (also called the *Tresca theory*): the maximum shear stress reaches the shear yield strength, $0.5S_Y$.

3. The *maximum-principal-strain theory*: the maximum principal strain reaches the yield strain S_Y/E.

4. The *maximum-strain-energy theory*: the strain energy per unit volume reaches a maximum of $0.5S_Y^2/E$ (see Sec. 6.2).

5. The *maximum-distortion-energy theory* (also called the *von Mises theory*): the energy causing a change in shape (distortion) reaches $[(1 + v)/(3E)]S_Y^2$ (see Sec. 7.3.3).

6. The *maximum-octahedral-shear-stress theory* (see Problem 2.8): the shear stress acting on each of eight (octahedral) surfaces containing a hydrostatic normal stress $\sigma_{ave} = (\sigma_1 + \sigma_2 + \sigma_3)/3$ reaches a value of $\sqrt{2}\, S_Y/3$. It can be shown that this theory provides the same conditions that the maximum-distortion theory gives; it will not be discussed here (see Problem 7.32).

Theories 3 and 4 will not be discussed in this section since their application to real materials is quite limited. Theory 2 or 5, Tresca or von Mises, respectively, are generally applied when the structural material is ductile. The von Mises theory generally predicts failure more accurately, but the Tresca theory is often used in design because it is simpler to apply and is more conservative. For brittle materials, the Coulomb-Mohr failure theory is quite often used in design. For plane stress, the theory resembles a combination of theories 1 and 2 and is conservative. There are more accurate theories for specific brittle materials, but because of their specialized nature, they will not be discussed here.

Before continuing, consider a general three-dimensional state of stress at a point given by

$$[\sigma] = \begin{bmatrix} \sigma_x & \tau_{xy} & \tau_{zx} \\ \tau_{xy} & \sigma_y & \tau_{yz} \\ \tau_{zx} & \tau_{yz} & \sigma_z \end{bmatrix}$$

It can be shown, by three-dimensional transformations (see Sec. 2.1), that there exists a coordinate system $x'y'z'$ where the state of stress at the same point can be described by the matrix

$$[\sigma] = \begin{bmatrix} \sigma_x' & 0 & 0 \\ 0 & \sigma_y' & 0 \\ 0 & 0 & \sigma_z' \end{bmatrix}$$

These are the principal stresses σ_1, σ_2, and σ_3, where arbitrarily, $\sigma_1 > \sigma_2 > \sigma_3$. The principal stresses are the three roots of Eq. (2.1-25) repeated here for convenience:

$$\sigma_p^3 - (\sigma_x + \sigma_y + \sigma_z)\sigma_p^2 + (\sigma_x\sigma_y + \sigma_y\sigma_z + \sigma_z\sigma_x - \tau_{xy}^2 - \tau_{yz}^2 - \tau_{zx}^2)\sigma_p$$

$$-(\sigma_x\sigma_y\sigma_z + 2\tau_{xy}\tau_{yz}\tau_{zx} - \sigma_x\tau_{yz}^2 - \sigma_y\tau_{zx}^2 - \sigma_z\tau_{xy}^2) = 0 \qquad \textbf{[7.3-1]}$$

For a state of plane stress, $\sigma_z = \tau_{yz} = \tau_{zx} = 0$ and Eq. (7.3-1) reduces to

$$\sigma^3 - (\sigma_x + \sigma_y)\sigma^2 + (\sigma_x\sigma_y - \tau_{xy}^2)\sigma = 0 \qquad \textbf{[7.3-2]}$$

of which the *three* roots are $\sigma = 0$ and

$$\sigma_p = \frac{\sigma_x + \sigma_y}{2} \pm \sqrt{\left(\frac{\sigma_x - \sigma_y}{2}\right)^2 + \tau_{xy}^2} \qquad \textbf{[7.3-3]}$$

Once the principal stresses are determined they can be ordered according to $\sigma_1 > \sigma_2 > \sigma_3$. The maximum shear stress is given by Eq. (2.1-36), which is

$$\tau_{max} = \frac{\sigma_1 - \sigma_3}{2} \qquad \textbf{[7.3-4]}$$

Example 7.3-1 Determine the maximum shear stress on the outer surface of a closed thin-walled circular cylinder internally pressurized at 7 MPa. The mean diameter of the cylinder is 500 mm and the wall thickness is 12 mm.

Solution:

The stresses for a pressurized thin-walled circular cylinder are given in Chap. 3, where the hoop stress σ_θ is

$$\sigma_\theta = \frac{pD}{2t} = \frac{(7)(500)}{(2)(12)} = 145.8 \text{ MPa}$$

and the longitudinal stress σ_z is

$$\sigma_z = \frac{pD}{4t} = \frac{(7)(500)}{(4)(12)} = 72.9 \text{ MPa}$$

The hoop stress is slightly higher at the outer radius of the cylinder and the stress perpendicular to the outer free surface is zero. Since there are no shear stresses on the radial, tangential, and longitudinal surfaces, we have the principal stresses, which are

$$\sigma_1 = 145.8 \text{ MPa} \qquad \sigma_2 = 72.9 \text{ MPa} \qquad \sigma_3 = 0 \text{ MPa}$$

and the maximum shear stress, given by Eq. (7.3-4), is

$$\tau_{max} = \frac{145.8 - 0}{2} = 72.9 \text{ MPa}$$

If a failure were to happen, it would occur along a surface as shown in Fig. 7.3-3(*a*).

Note if one were to restrict themselves to a two-dimensional transformation analysis considering only surfaces perpendicular to the outer free surface, the maximum in-plane shear stress would have been found to be only $(145.8 - 72.9)/2 = 36.4$ MPa. For failure, an erroneous prediction would be that the failure surface is that shown in Fig. 7.3-3(*b*) at a pressure twice that which would cause the actual failure mode of Fig. 7.3-3(*a*).[†] Thus a three-dimensional transformation analysis should always be performed when using the Tresca theory.

[†] The failure surfaces can be 45° from either side of the indicated reference line.

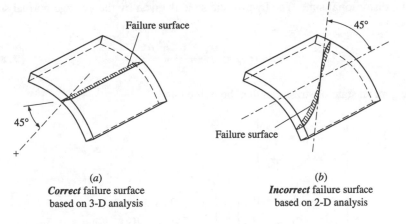

(a)
Correct failure surface
based on 3-D analysis

(b)
Incorrect failure surface
based on 2-D analysis

Figure 7.3-3 Failure surface for cylinder example.

7.3.2 TRESCA (MAXIMUM-SHEAR-STRESS) THEORY FOR DUCTILE MATERIALS

Since the maximum shear stress in an element loaded in pure tension is one-half the maximum tensile stress, the shear yield strength is taken to be $0.5S_Y$. Thus, the yield criterion for the Tresca theory is

$$\tau_{max} = \frac{\sigma_1 - \sigma_3}{2} = 0.5S_Y \qquad \text{or} \qquad \sigma_1 - \sigma_3 = S_Y \qquad \textbf{[7.3-5]}$$

The maximum shear stress is sometimes referred to as the Tresca stress. That is,

$$\tau_{Tresca} = \frac{\sigma_1 - \sigma_3}{2} \qquad \textbf{[7.3-6]}$$

and yielding is said to occur when $2\tau_{Tresca} = S_Y$.

7.3.3 VON MISES (MAXIMUM-ENERGY-OF-DISTORTION) THEORY FOR DUCTILE MATERIALS

This theory relates the distortional energy of a point under a general state of stress to that of the tensile specimen at yielding. A hydrostatic state of stress occurs when all three principal stresses are equal. In this situation, the normal strains in all directions are equal and there is no shear stress. Consequently, no distortion of the stressed element occurs. Any deviation from this state will cause distortion. A general state of stress can be considered as the superposition of a pure hydrostatic state and a distortional state. The hydrostatic state is given by the average normal stress given by

$$\sigma_{ave} = \frac{\sigma_x + \sigma_y + \sigma_z}{3} \qquad \text{[7.3-7]}$$

The general state of stress can then be written as

$$[\boldsymbol{\sigma}] = \begin{bmatrix} \sigma_x & \tau_{xy} & \tau_{zx} \\ \tau_{xy} & \sigma_y & \tau_{yz} \\ \tau_{zx} & \tau_{yz} & \sigma_z \end{bmatrix}$$

$$= \begin{bmatrix} \sigma_{ave} & 0 & 0 \\ 0 & \sigma_{ave} & 0 \\ 0 & 0 & \sigma_{ave} \end{bmatrix} + \begin{bmatrix} \sigma_x - \sigma_{ave} & \tau_{xy} & \tau_{zx} \\ \tau_{xy} & \sigma_y - \sigma_{ave} & \tau_{yz} \\ \tau_{zx} & \tau_{yz} & \sigma_z - \sigma_{ave} \end{bmatrix}$$

On the right-hand side of the equation, the first and second matrices are the hydrostatic and the distortional part of the stress, respectively. This can also be expressed in terms of the principal stresses as

$$[\boldsymbol{\sigma}] = \begin{bmatrix} \sigma_{ave} & 0 & 0 \\ 0 & \sigma_{ave} & 0 \\ 0 & 0 & \sigma_{ave} \end{bmatrix} + \begin{bmatrix} \sigma_1 - \sigma_{ave} & 0 & 0 \\ 0 & \sigma_2 - \sigma_{ave} & 0 \\ 0 & 0 & \sigma_3 - \sigma_{ave} \end{bmatrix}$$

where $\sigma_{ave} = (\sigma_1 + \sigma_2 + \sigma_3)/3$.[†] The energy per unit volume of a stressed element is given in Sec. 6.2. The energy per unit volume of an element with normal stresses is given by Eq. (6.2-12), which is

$$u = \frac{1}{2E} [\sigma_x^2 + \sigma_y^2 + \sigma_z^2 - 2\nu(\sigma_x\sigma_y + \sigma_y\sigma_z + \sigma_z\sigma_x)] \qquad \text{[7.3-8]}$$

To determine the distortional energy, simply substitute $\sigma_1 - \sigma_{ave}, \sigma_2 - \sigma_{ave}$, and $\sigma_3 - \sigma_{ave}$, for σ_x, σ_y, and σ_z into Eq. (7.3-8), respectively. Also, with $\sigma_{ave} = (\sigma_1 + \sigma_2 + \sigma_3)/3$, the distortional energy per unit volume reduces to

$$u_d = \frac{1 + \nu}{6E} [(\sigma_1 - \sigma_2)^2 + (\sigma_2 - \sigma_3)^2 + (\sigma_3 - \sigma_1)^2] \qquad \text{[7.3-9]}$$

For the tensile test, the state of stress at yielding is $\sigma_1 = S_Y$ and $\sigma_2 = \sigma_3 = 0$. Thus at yield, the distortional energy is

$$(u_d)_Y = \frac{1 + \nu}{3E} S_Y^2 \qquad \text{[7.3-10]}$$

[†] Recall from Sec. 2.1.5, Eq. (2.1-26a), that the sum of the normal stresses on any three mutually perpendicular surfaces at a point is an invariant. That is, $\sigma_x + \sigma_y + \sigma_z = \sigma_1 + \sigma_2 + \sigma_3$.

Equating Eqs. (7.3-9) and (7.3-10) and simplifying, the von Mises criterion for yielding of an element under a general state of stress due to distortion is

$$\sqrt{0.5[(\sigma_1 - \sigma_2)^2 + (\sigma_2 - \sigma_3)^2 + (\sigma_3 - \sigma_1)^2]} = S_Y \quad \textbf{[7.3-11]}$$

For yield under a single, uniaxial state of stress, the stress would be equated to S_Y. Thus for yield, a single, uniaxial state of stress *equivalent* to the general state of stress is equated to the left-hand side of Eq. (7.3-11). This equivalent stress is called the *von Mises stress* and is given by

$$\sigma_{\text{von Mises}} = \sqrt{0.5[(\sigma_1 - \sigma_2)^2 + (\sigma_2 - \sigma_3)^2 + (\sigma_3 - \sigma_1)^2]} \quad \textbf{[7.3-12]}$$

Thus, based on the von Mises theory, the condition for yielding is when the equivalent von Mises stress equals the yield strength of the material. That is,

$$\sigma_{\text{von Mises}} = S_Y \quad \textbf{[7.3-13]}$$

Estimate the torque on a 10-mm-diameter steel shaft when yielding begins using (*a*) the Tresca and (*b*) the von Mises theory. The yield strength of the steel is 140 MPa. | **Example 7.3-2**

Solution:

(*a*) The shear yield strength is one-half the tensile yield strength. The maximum shear stress in torsion is thus equated to the shear yield strength, resulting in

$$\tau_{\text{max}} = \frac{16T}{\pi d^3} = 0.5\, S_Y$$

Solving for the torque yields

$$T = \frac{\pi d^3}{32} S_Y = \frac{(\pi)(0.010)^3}{32}\, 140 \times 10^6 = 13.7 \text{ N·m}$$

(*b*) The principal stresses for an element with one set of shear stresses, say $\tau_{xy} = \tau$, are found using Eq. (7.3-3) to be

$$\sigma_1 = \tau \qquad \sigma_2 = 0 \qquad \sigma_3 = -\tau$$

Substituting this into Eq. (7.3-11) yields

$$3\tau^2 = S_Y^2$$

or

$$\tau = 0.577 S_Y$$

Substituting $\tau = 16T/\pi d^3$ and solving for the torque results in

$$T = \frac{(0.577)\pi d^3}{16} S_Y = \frac{(0.577)\pi(0.010)^3}{16}\, 140 \times 10^6 = 15.9 \text{ N·m}$$

Thus, it can be seen that for yielding in pure torsion, the von Mises theory predicts a torque which is 15 percent greater than the prediction of the Tresca theory. Tests on ductile materials have shown that the von Mises theory is much more accurate for predicting yield, but in design work the more conservative answer predicted by the Tresca theory is commonly used.

7.3.4   COMPARISON BETWEEN THE TRESCA AND VON MISES THEORIES (PLANE STRESS)

For plane stress, let $\sigma_z = \tau_{yz} = \tau_{zx} = 0$. Thus one of the principal stresses is zero and is in the z direction. The other two principal stresses σ_{p1} and σ_{p2} are determined from Eq. (7.3-3). To understand the difference between the Tresca and von Mises theories, a graph of σ_{p1} versus σ_{p2} is drawn. First, the Tresca theory will be developed for various cases. The first three cases are for $\sigma_{p1} > \sigma_{p2}$, and are

Case 1 $\sigma_{p1} > \sigma_{p2} > 0$. Here $\sigma_1 = \sigma_{p1}$ and $\sigma_3 = 0$. Thus Eq. (7.3-5) gives $\sigma_{p1} - 0 = S_Y$, or simply

$$\sigma_{p1} = S_Y \qquad \text{[7.3-14a]}$$

Case 2 $\sigma_{p1} > 0 > \sigma_{p2}$. Here $\sigma_1 = \sigma_{p1}$ and $\sigma_3 = \sigma_{p2}$ and Eq. (7.3-5) gives

$$\sigma_{p1} - \sigma_{p2} = S_Y \qquad \text{[7.3-14b]}$$

Case 3 $0 > \sigma_{p1} > \sigma_{p2}$. Here $\sigma_1 = 0$ and $\sigma_3 = \sigma_{p2}$ and Eq. (7.3-5) gives

$$\sigma_{p2} = -S_Y \qquad \text{[7.3-14c]}$$

There are three other cases similar to these, corresponding to $\sigma_{p2} > \sigma_{p1}$. The six cases are plotted in Fig. 7.3-4(a).

For the von Mises theory and plane stress, substitution of σ_{p1}, σ_{p2}, and 0 for the principal stresses into Eq. (7.3-11) gives

$$\sqrt{\sigma_{p1}^2 + \sigma_{p2}^2 - \sigma_{p1}\sigma_{p2}} = S_Y \qquad \text{[7.3-15]}$$

This equation is plotted in Fig. 7.3-4(b). Here the Tresca theory is plotted with dotted lines to show that except for six distinct points Tresca theory is more conservative than the von Mises theory.

The equivalent von Mises stress for plane stress is

$$\sigma_{\text{von Mises}} = \sqrt{\sigma_{p1}^2 + \sigma_{p2}^2 - \sigma_{p1}\sigma_{p2}} \qquad \text{[7.3-16]}$$

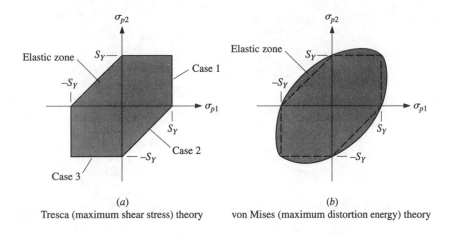

(a)
Tresca (maximum shear stress) theory

(b)
von Mises (maximum distortion energy) theory

Figure 7.3-4 Yield theories for ductile materials in plane stress.

7.3.5 COULOMB-MOHR THEORY FOR BRITTLE MATERIALS

Brittle materials exhibit no distinct point of yielding and fracture very close to when elastic behavior ceases. Thus the strength for failure we will use here will be the ultimate strength at fracture S_U. For some brittle materials, such as hardened tool steel, the ultimate strength in tension S_{UT} is the same as the compressive strength S_{UC} (where S_{UC} is taken as a positive number). For other brittle materials, such as gray cast iron, the ultimate compressive strength can be as much as 3 to 4 times the tensile strength. For these materials the mode of failure is tensile when loaded in tension and shear when loaded in compression. The Coulomb-Mohr theory is based on either considering friction along slip planes or observing the largest of Mohr's circles in three dimensions. Here we will discuss the latter. The dashed-line circles in Fig. 7.3-5 show the Mohr's circles when failure occurs in uniaxial tension or compression. The failure lines are drawn tangent to these circles. The theory bases failure on any Mohr's circle falling on or outside these failure lines. For a general case the largest Mohr's circle connecting σ_1 and σ_3 on the σ axis, shown in solid line in the figure, is indicating that a failure is imminent according to this theory.

For plane stress, similar to the Tresca theory, let us consider some cases.

Case 1 $\quad \sigma_{p1} > \sigma_{p2} > 0$. Here $\sigma_1 = \sigma_{p1}$ and $\sigma_3 = 0$. This is the same as the right dashed circle in Fig. 7.3-5, where

$$\sigma_{p1} = S_{UT} \qquad\qquad \textbf{[7.3-17a]}$$

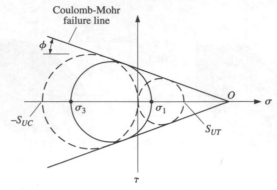

Figure 7.3-5   Coulomb-Mohr theory of failure for brittle materials.

Case 2 $\sigma_{p1} > 0 > \sigma_{p2}$. This case is that shown by the solid-line circle in Fig. 7.3-5. Interpolating the solid-line circle with the dashed-line circles, we have[†]

$$\frac{\dfrac{\sigma_1 - \sigma_3}{2} - \dfrac{S_{UT}}{2}}{\dfrac{S_{UC}}{2} - \dfrac{S_{UT}}{2}} = \frac{\dfrac{S_{UT}}{2} - \dfrac{\sigma_1 + \sigma_3}{2}}{\dfrac{S_{UC}}{2} + \dfrac{S_{UT}}{2}}$$

Cross-multiplying and simplification of the above reduces to

$$\frac{\sigma_1}{S_{UT}} - \frac{\sigma_3}{S_{UC}} = 1$$

Substituting $\sigma_1 = \sigma_{p1}$ and $\sigma_3 = \sigma_{p2}$ yields

$$\frac{\sigma_{p1}}{S_{UT}} - \frac{\sigma_{p2}}{S_{UC}} = 1 \qquad \textbf{[7.3-17b]}$$

Case 3 $0 > \sigma_{p1} > \sigma_{p2}$. Here $\sigma_1 = 0$ and $\sigma_3 = \sigma_{p2}$. This is the same as the left dashed circle in Fig. 7.3-5, where

$$\sigma_{p2} = -S_{UC} \qquad \textbf{[7.3-17c]}$$

Again, as with the Tresca theory, there are three other cases where $\sigma_{p2} > \sigma_{p1}$. A plot of these cases is given in Fig. 7.3-6.

Note that the failure plot for the Coulomb-Mohr theory is quite similar to that found for the Tresca theory. As a matter of fact, for the case of hardened tool steel where $S_{UT} = S_{UC}$, one obtains identical results between the two theories. Although the theories look similar, the modes of failure for most brittle materials are not the same. As stated earlier, when the loading on most brittle materials is primarily ten-

---

[†] The length of a line drawn from the center of a circle to the tangent point of the failure line to the circle is the radius of the circle and is proportional to the distance from point O to the center of the circle.

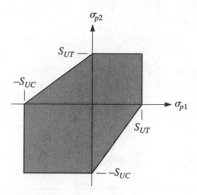

Figure 7.3-6   Coulomb-Mohr theory of failure.

sion, as is case 1, the failure is tension and the Coulomb-Mohr theory agrees with the maximum-principal-stress theory. When the loading is primarily compression, such as case 3, the failure is shear and the Coulomb-Mohr theory does agree with the Tresca theory.

7.3.6   DESIGN EQUATIONS

The equations obtained in the previous section can be modified for design purposes. To obtain a design equation, the strength is reduced by the design factor n. Modifying the commonly used strength theories for plane stress problems results in the following design equations.

Ductile Materials
von Mises theory (maximum-distortion-energy)

$$\sigma_{\text{von Mises}} = \sqrt{\sigma_{p1}^2 + \sigma_{p2}^2 - \sigma_{p1}\sigma_{p2}} \leq \frac{S_Y}{n} \qquad \textbf{[7.3-18]}$$

Tresca theory (maximum-shear-stress) for $\sigma_{p1} \geq \sigma_{p2}$

Case 1 $\sigma_{p1} \geq \sigma_{p2} \geq 0$ $2\tau_{\text{Tresca}} = \sigma_{p1} \leq \dfrac{S_Y}{n}$ $\qquad \textbf{[7.3-19a]}$

Case 2 $\sigma_{p1} \geq 0 \geq \sigma_{p2}$ $2\tau_{\text{Tresca}} = \sigma_{p1} - \sigma_{p2} \leq \dfrac{S_Y}{n}$ $\qquad \textbf{[7.3-19b]}$

Case 3 $0 \geq \sigma_{p1} \geq \sigma_{p2}$ $2\tau_{\text{Tresca}} = -\sigma_{p2} \leq \dfrac{S_Y}{n}$ $\qquad \textbf{[7.3-19c]}$

Brittle Materials
Coulomb-Mohr theory

Case 1 $\sigma_{p1} \geq \sigma_{p2} \geq 0$ $\sigma_{p1} \leq \dfrac{S_{UT}}{n}$ **[7.3-20a]**

Case 2 $\sigma_{p1} \geq 0 \geq \sigma_{p2}$ $\dfrac{\sigma_{p1}}{S_{UT}} - \dfrac{\sigma_{p2}}{S_{UC}} \leq \dfrac{1}{n}$ **[7.3-20b]**

Case 3 $0 \geq \sigma_{p1} \geq \sigma_{p2}$ $|\sigma_{p2}| \leq \dfrac{S_{UC}}{n}$ **[7.3-20c]**

Example 7.3-3

Figure 7.3-7(a) shows a round shaft of diameter 1.5 in loaded by a bending moment $M_z = 5000$ lb·in, a torque $T = 8000$ lb·in, and an axial tensile force $N = 6000$ lb. If the material is ductile with a yield strength $S_Y = 40$ kpsi, determine the design factor corresponding to yield using (a) the Tresca theory and (b) the von Mises theory.

Solution:

Initially, consider each loading state separately. The tensile force N creates a tensile stress σ_x, which is constant over the cross section:

$$(\sigma_x)_N = \frac{N}{A} = \frac{6000}{(0.785)(1.5)^2} = 3400 \text{ psi} = 3.40 \text{ kpsi}$$

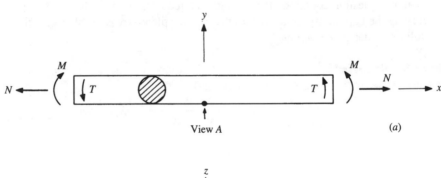

View A   (a)

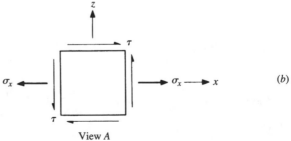

View A   (b)

Figure 7.3-7

The bending moment M gives a normal stress distribution which is linear with respect to y, and the maximum tensile stress due to bending alone occurs at the bottom of the shaft, where

$$(\sigma_x)_M = \frac{Mc}{I} = \frac{(5000)(1.5/2)}{(\pi/64)(1.5)^4} = 15{,}090 \text{ psi} = 15.09 \text{ kpsi}$$

Shear stresses τ arise from the torque T. These stresses are maximum at the outer surface, where

$$\tau_T = \frac{Tr_o}{J} = \frac{(8000)(1.5/2)}{(\pi/32)(1.5)^4} = 12{,}070 \text{ psi} = 12.07 \text{ kpsi}$$

Combining the stresses using superposition, one can see that the worst case will occur on an element located on the bottom of the shaft, where [see Fig. 7.3-7(b)]

$$\sigma_x = (\sigma_x)_N + (\sigma_x)_M = 3.40 + 15.09 = 18.49 \text{ kpsi}$$

$$\tau = \tau_T = 12.07 \text{ kpsi}$$

This is a case of plane stress. Considering only surfaces perpendicular to the plane of the page for Fig. 7.3-7(b), the principal stresses are given by Eq. (7.3-3), where

$$\sigma_{p1} = \frac{18.49 + 0}{2} + \sqrt{\left(\frac{18.49 - 0}{2}\right)^2 + (12.07)^2} = 24.45 \text{ kpsi}$$

$$\sigma_{p2} = \frac{18.49 + 0}{2} - \sqrt{\left(\frac{18.49 - 0}{2}\right)^2 + (12.07)^2} = -5.96 \text{ kpsi}$$

(a) Tresca theory: Since $\sigma_1 = 24.45$ kpsi, $\sigma_2 = 0$ kpsi, and $\sigma_3 = -5.96$ kpsi, this corresponds to case 2, given by the equality in Eq. (7.3-19b). Thus

$$2\tau_{\text{Tresca}} = 24.45 - (-5.96) = \frac{40}{n}$$

Solving for n yields $n = 1.32$.
(b) von Mises theory: Substituting σ_{p1} and σ_{p2} into the equality of Eq. (7.3-18) gives

$$\sigma_{\text{von Mises}} = \sqrt{(24.45)^2 + (-5.96)^2 - (24.45)(-5.96)} = \frac{40}{n}$$

Solving for n yields $n = 1.43$.
Since the von Mises is less conservative than the Tresca theory, it makes sense that a higher factor of safety is obtained in part (b).

7.4 FRACTURE MECHANICS

7.4.1 INTRODUCTION

Failure by fracture is typically thought to be limited by the ultimate strength of the material S_U. Small cracks, however, either present during the fabrication of a part or developed in service, may accelerate fracture of the part at a stress somewhat lower than the ultimate strength or, more surprisingly, may fracture at a stress lower than the *yield strength* of the material. Fractures in this manner resemble the rapid failures of brittle materials and are labeled as brittle fractures. However, this form of failure can even occur in materials normally considered to be ductile.

The study of fracture mechanics is a major subject unto itself, intimately and uniquely intertwined with the infinite combinations of material behavior, service environment, loading conditions, crack orientations, and part geometry. Because of space limitations, the discussion in this section will be limited to an introduction of some of the basic concepts of linear-elastic fracture mechanics (LEFM). For further detailed discussion the reader is urged to consult Refs. 7.2 and 7.3.

The foundation of fracture mechanics was first established by Griffith in 1921 using the stress field calculations for an elliptical flaw in a plate developed by Inglis in 1913. For the infinite plate loaded by an applied uniaxial stress σ (Fig. 7.4-1), the maximum stress occurs at ($\pm a$, 0) and is given by

$$(\sigma_y)_{\max} = \left(1 + 2\frac{a}{b}\right)\sigma \qquad \textbf{[7.4-1]}$$

Note that when $a = b$ the ellipse becomes a circle and Eq. (7.4-1) gives a stress concentration factor of 3. This agrees with the well-known result for an infinite plate

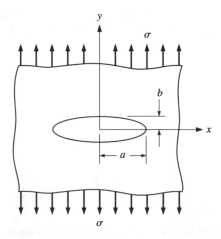

Figure 7.4-1   Elliptical hole in a flat plate loaded in tension.

with a circular hole (see Fig. 5.10-2 with $d/w \to 0$). For a fine crack, $b/a \to 0$, and Eq. (7.4-1) predicts that $(\sigma_y)_{max} \to \infty$. This result, quite similar to the contact stress problem where the contact area, when assumed to be zero, leads one to the conclusion that the stress is infinite and the material would fail under any load no matter how small. On a microscopic level, an infinitely sharp crack is a hypothetical abstraction which is physically impossible, and if any plastic deformation occurs, the stress will be finite at the crack tip.

The *Griffith criterion* is based on the following reasoning. Consider the plate to be of unit thickness and contain a crack such that $b/a \approx 0$. If the plate material is linearly elastic, as the plate is loaded the load-displacement curve would follow line *OAB* as shown in Fig. 7.4-2. Next, assume that when the load reaches P_c the deflection of the load remains fixed and the crack begins to grow such that the half-width of the crack extends from a to $a + da$. If the crack had not changed length, the strain energy in the plate would have been the area *OAC*. However, if the crack extends to $a + da$, and the deflection of the load remains fixed, the cross section of the plate decreases. Hence the stiffness of the plate is reduced and the load must decrease to point D in Fig. 7.4-2. Thus the extension of the crack causes a release of strain energy equal to the area *OAD*. If the plate were loaded at a larger value of P_c when the crack extension da forms, the energy release would be even greater. In order for a crack to propagate, the Griffith criterion requires that the energy released in extending the crack must be sufficient to provide the energy required to create the new surfaces of the propagating crack. Denote the energy or work necessary for crack growth or fracture to be W_f and the strain energy U. Removing the condition of the fixed load during the crack extension da, let the work due to the applied loading be W_P. The condition for crack growth is given by

$$\frac{d}{da}(W_P + U) \ge \frac{dW_f}{da} \qquad \textbf{[7.4-2]}$$

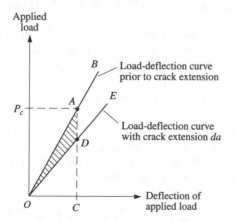

Figure 7.4-2 Load-deflection curve for plate with crack.

Let G be the energy release rate and R be the rate of energy for crack growth (called the *crack resistance force*) such that

$$G = \frac{d}{da}(W_P + U) \qquad R = \frac{dW_f}{da} \qquad \text{[7.4-3]}$$

where the units of G and R are that of force. Thus crack growth occurs when

$$G \geq R \qquad \text{[7.4-4]}$$

Crack growth can be stable or unstable. As shown in Ref. 7.2, the crack resistance R can increase with da, and crack growth can be stable until $dG/da = dR/da$. Thus unstable crack growth occurs when Eq. (7.4-4) is valid *and*

$$\frac{dG}{da} = \frac{dR}{da} \qquad \text{[7.4-5]}$$

From Inglis' stress field equations for a plate with an elliptical crack, Griffith calculated the energy release rate per crack to be

$$G = \frac{\pi\sigma^2 a}{E} \qquad \text{[7.4-6]}$$

Griffith's experimental work was restricted to brittle materials, namely glass, which pretty much confirmed his hypothesis of the energy release upon crack formation exceeding the energy necessary to form the crack surface. However, for ductile materials, the energy needed to perform plastic work at the crack tip is found to be much more crucial than surface energy.

7.4.2 Crack Modes and the Stress Intensity Factor

Three distinct modes of crack propagation exist, as shown in Fig. 7.4-3. A tensile stress field gives rise to mode I, the *opening crack propagation mode,* as shown in Fig. 7.4-3(*a*). This mode is the most common in practice. Mode II is the *sliding mode,* is due to in-plane shear, and can be seen in Fig. 7.4-3(*b*). Mode III is the *tearing mode,* which arises from out-of-plane shear, as shown in Fig. 7.4-3(*c*). Combinations of these modes can also occur. Since mode I is the most common and important mode, the remainder of this section will only consider this mode.

Consider a mode I crack of length $2a$ as shown in Fig. 7.4-4. Using complex stress functions it has been shown that the stress field on a $dx\,dy$ element in the vicinity of the crack tip is given by (see Ref. 7.4)

$$\sigma_x = \sigma\sqrt{\frac{a}{2r}}\cos\frac{\theta}{2}\left(1 - \sin\frac{\theta}{2}\sin\frac{3\theta}{2}\right) \qquad \text{[7.4-7a]}$$

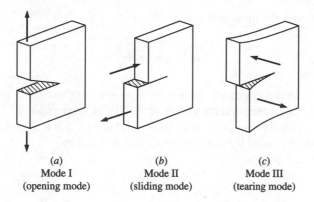

(a)
Mode I
(opening mode)

(b)
Mode II
(sliding mode)

(c)
Mode III
(tearing mode)

Figure 7.4-3 The three modes of crack propagation.

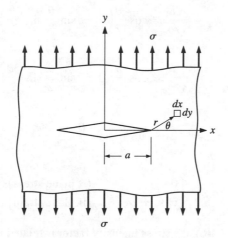

Figure 7.4-4 Mode I crack model.

$$\sigma_y = \sigma\sqrt{\frac{a}{2r}} \, \cos\frac{\theta}{2}\left(1 + \sin\frac{\theta}{2}\sin\frac{3\theta}{2}\right) \qquad \textbf{[7.4-7b]}$$

$$\tau_{xy} = \sigma\sqrt{\frac{a}{2r}} \, \sin\frac{\theta}{2}\cos\frac{\theta}{2}\cos\frac{3\theta}{2} \qquad \textbf{[7.4-7c]}$$

$$\sigma_z = \begin{cases} 0 & \text{(if plane stress)} \\ v(\sigma_x + \sigma_y) & \text{(if plane strain)} \end{cases} \qquad \textbf{[7.4-7d]}$$

The stress σ_y near the tip with $\theta = 0$ is

$$\sigma_y|_{\theta=0} = \sigma\sqrt{\frac{a}{2r}} \qquad\qquad \textbf{[7.4-8]}$$

As with the elliptical crack, we see that $\sigma_y \to \infty$ as $r \to 0$, and again the concept of an infinite stress concentration at the crack tip is inappropriate. The quantity $\sigma_y\sqrt{2r} = \sigma\sqrt{a}$, however, does remain constant as $r \to 0$. It is common practice to define a factor K_I, called the *stress intensity factor*, given by[†]

$$K_I = \sigma\sqrt{\pi a} \qquad\qquad \textbf{[7.4-9]}$$

where the units are MPa·$\sqrt{m}$ or kpsi·$\sqrt{in}$. The stress intensity factor is *not* to be confused with the static stress concentration factor K_t defined in Sec. 5.10.

Thus Eqs. (7.4-7) can be rewritten as

$$\sigma_x = \frac{K_I}{\sqrt{2\pi r}} \cos\frac{\theta}{2}\left(1 - \sin\frac{\theta}{2}\sin\frac{3\theta}{2}\right) \qquad\qquad \textbf{[7.4-10a]}$$

$$\sigma_y = \frac{K_I}{\sqrt{2\pi r}} \cos\frac{\theta}{2}\left(1 + \sin\frac{\theta}{2}\sin\frac{3\theta}{2}\right) \qquad\qquad \textbf{[7.4-10b]}$$

$$\tau_{xy} = \frac{K_I}{\sqrt{2\pi r}} \sin\frac{\theta}{2}\cos\frac{\theta}{2}\cos\frac{3\theta}{2} \qquad\qquad \textbf{[7.4-10c]}$$

$$\sigma_z = \begin{cases} 0 & \text{(if plane stress)} \\ v(\sigma_x + \sigma_y) & \text{(if plane strain)} \end{cases} \qquad\qquad \textbf{[7.4-10d]}$$

Note that from Eq. (7.4-6), the stress intensity factor is related to the Griffith energy release rate G by

$$G = \frac{K_I^2}{E} \qquad\qquad \textbf{[7.4-11]}$$

The stress intensity factor is a function of geometry, size and shape of the crack, and the loading. Tables are available in the literature for basic configurations (see Refs. 7.4 to 7.7). Table 7.4-1 presents a few examples of mode I stress intensity factors.

[†] Here the subscript indicates we are dealing with a mode I crack such as that shown in Fig. 7.4-4. For modes II and III, appropriate subscripts are used.

Table 7.4-1   Mode I stress intensity factors[†]

Case	Intensity Factor with $\eta = a/b$
1. Plate in tension with center crack 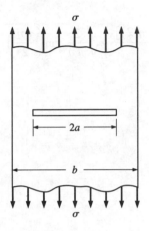	$K_{\mathrm{I}} = C_1 \sigma \sqrt{\pi a}$ $C_1 = (1 - 0.1\eta^2 + 0.96\eta^4)\sqrt{\sec(\pi\eta)}$
2. Plate in tension with double-edge crack 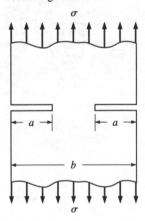	$K_{\mathrm{I}} = C_2 \sigma \sqrt{\pi a}$ $C_2 = \left[1 + 0.122\ \cos^4(\pi\eta)\right]\sqrt{\dfrac{\tan(\pi\eta)}{\pi\eta}}$

[†] This table contains a small sample of the many cases provided in Ref. 7.7. Reproduced and modified with permission of the author and publisher.

Table 7.4-1 (concluded)

Case	Intensity Factor with $\eta = a/b$

3. Plate in tension with single-edge crack

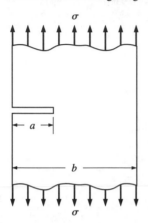

$$K_I = C_3 \sigma \sqrt{\pi a}$$

$$C_3 = \left\{ \frac{0.752 + 2.02\eta + 0.37[1 - \sin(\pi\eta/2)]^3}{\cos(\pi\eta/2)} \right\} \sqrt{\frac{2}{\pi\eta} \tan\left(\frac{\pi\eta}{2}\right)}$$

4. Plate of thickness t in bending with single-edge crack

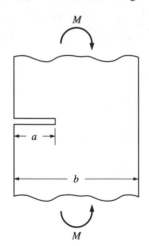

$$K_I = C_4 \sigma \sqrt{\pi a} \qquad \text{where} \qquad \sigma = \frac{6M}{tb^2}$$

$$C_4 = \left\{ \frac{0.923 + 0.199[1 - \sin(\pi\eta/2)]^4}{\cos(\pi\eta/2)} \right\} \sqrt{\frac{2}{\pi\eta} \tan\left(\frac{\pi\eta}{2}\right)}$$

When the mode I stress intensity factor reaches a critical value, K_{Ic}, crack propagation initiates. The *critical stress intensity factor* K_{Ic} is a material constant where this material property can vary with the crack mode, processing of the material, temperature, loading rate, and the state of stress at the crack site (such as plane stress versus plane strain). The critical stress intensity factor K_{Ic} is also called the *fracture toughness* of the material. The fracture toughness for plane strain is normally lower than that for plane stress. For this reason, the term K_{Ic} is typically defined as the *mode I, plane strain fracture toughness.* Table 7.4-2 gives some approximate typical room-temperature values of K_{Ic} for several materials. As previously noted, the fracture toughness is dependent on many factors and the table is only meant to convey some typical magnitudes of K_{Ic}. For an actual application, it is recommended that the material specified for the application be certified using standard test procedures [see the American Society for Testing and Materials (ASTM) standard E399].

Table 7.4-2 Typical K_{Ic} values at room temperature

Material	Yield Strength S_Y (MPa)	Stress Intensity Factor K_{Ic} (MPa· $\sqrt{m}$)
Steel		
Carbon steel	245	>225
Alloy steel		
A533B	500	175
4330V (275°C temper)	1400	85–95
4330V (425°C temper)	1300	103–110
4340 (205°C temper)	1600	44–66
4340 (425°C temper)	1400	79–91
D6AC (540° temper)	1500	100
18 Ni(200) (480°C 6 h)	1450	110
18 Ni(300) (480°C 6 h)	1900	50–64
Titanium alloys		
Ti-6Al-4Zr-2Sn-0.5Mo-0.5V	840	140
Ti-6Al-4V-2Sn	800	110
Aluminum alloys		
2014-T6	440	18–30
7075-T7351	400	26–41
7079-T651	535	19–27

As stated earlier, K_{Ic} is the fracture toughness for plane strain conditions. Plane stress always exists on the free surface perpendicular to the crack surface. However, if the part is thick enough at the crack site, plane strain will dominate. An ASTM recommendation for plane strain conditions is that the thickness, designated as t, be such that

$$t \geq 2.5 \left(\frac{K_{Ic}}{S_Y} \right)^2 \qquad \textbf{[7.4-12]}$$

where S_Y is the yield strength of the material.

This can give rise to severe restrictions on the thickness. For example, for alloy steel A533B in Table 7.4-2, Eq. (7.4-12) requires that $t \geq 0.306$ m $= 306$ mm, a very large value. If the thickness is less than that specified by Eq. (7.4-12), the state of stress approaches plane stress, appreciable yielding may occur at the crack tip, and the actual value of the stress intensity factor may be quite large compared to K_{Ic}.

Using K_{Ic} for these cases might produce a very conservative result. For estimation purposes, the effective fracture toughness can be approximated by (see Ref. 7.8)

$$(K_{Ic})_{eff} = K_{Ic}\sqrt{1 + \frac{1.4}{t^2}\left(\frac{K_{Ic}}{S_Y}\right)^4}$$ [7.4-13]

A similar requirement holds for the crack size, as the plastic zone should be small compared to the length of the crack. The crack size limitation is given by the same requirement for the thickness, i.e., Eq. (7.4-12), (see Ref. 7.2)

$$a \geq 2.5\left(\frac{K_{Ic}}{S_Y}\right)^2$$ [7.4-14]

If the crack size is less than that given by Eq. (7.4-14), the value of K_{Ic} may be larger than the actual value.

7.4.3   THE PLASTIC ZONE CORRECTION

The stress field equations predict a singularity at the crack tip. Ductile metals exhibit a yield stress above which they deform plastically. This means that there will exist a region at the crack tip where plastic deformation occurs and the singularity cannot exist. Based on Eqs. (7.4-10) the shape of the plastic zone is as shown in Fig. 7.4-5. The extent of the plastic zone is given by r_p^*, which is a function of θ. At the free surfaces on both sides of the plate, a state of plane stress exists such that $\sigma_z = 0$. Within the plate, however, strain is constrained, and if the thickness is large enough a state of plane strain exists. In this region, $r_p^*(\theta)$ is different than in the region of plane stress.

The plastic zone correction to the crack length will require an estimate of the extent of the zone by determining r_p^*. First, we will write Eqs. (7.4-10) in terms of the principal stresses. Using Eq. (7.3-3) for σ_1 and σ_2, and $\sigma_3 = 0$ for plane stress or $\sigma_3 = v(\sigma_1 + \sigma_2)$ for the case of plane strain, the principal stresses for the states of stress given by Eqs. (7.4-10) can be shown to be (see Problem 7.38)

$$\sigma_1 = \frac{K_I}{\sqrt{2\pi r}} \cos\frac{\theta}{2}\left(1 + \sin\frac{\theta}{2}\right)$$ [7.4-15a]

$$\sigma_2 = \frac{K_I}{\sqrt{2\pi r}} \cos\frac{\theta}{2}\left(1 - \sin\frac{\theta}{2}\right)$$ [7.4-15b]

$$\sigma_3 = \begin{cases} 0 & \text{(if plane stress)} \\ 2v\dfrac{K_I}{\sqrt{2\pi r}} \cos\dfrac{\theta}{2} & \text{(if plane strain)} \end{cases}$$ [7.4-15c]

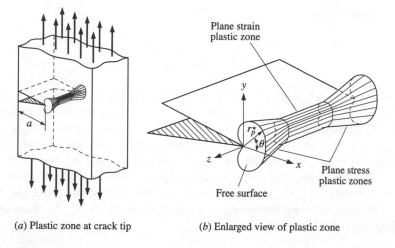

(a) Plastic zone at crack tip   (b) Enlarged view of plastic zone

Figure 7.4-5

Now, to determine r_p^*, let us first consider the case of plane stress. Applying the von Mises yield criteria, plastic behavior occurs according to Eq. (7.3-11). Substituting Eqs. (7.4-15) for the case of plane stress into Eq. (7.3-15) yields

$$\frac{K_I}{\sqrt{2\pi r}}\sqrt{0.5\left(1 + \frac{3}{2}\sin^2\theta + \cos\theta\right)} = S_Y$$

Solving for r gives the extent of the plastic zone as a function of θ, resulting in

$$r_p^* = \frac{1}{4\pi}\left(\frac{K_I}{S_Y}\right)^2\left(1 + \frac{3}{2}\sin^2\theta + \cos\theta\right) \qquad \textbf{[7.4-16]}$$

At $\theta = 0$

$$r_p^* = \frac{1}{2\pi}\left(\frac{K_I}{S_Y}\right)^2 \qquad \textbf{[7.4-17]}$$

This can be repeated for the case of plane strain, which results in

$$r_p^* = \frac{1}{4\pi}\left(\frac{K_I}{S_Y}\right)^2\left[\frac{3}{2}\sin^2\theta + (1 - 2v)^2(1 + \cos\theta)\right] \qquad \textbf{[7.4-18]}$$

At $\theta = 0$ and with $v = 1/3$,

$$r_p^* = \frac{1}{18\pi}\left(\frac{K_I}{S_Y}\right)^2 \qquad\qquad \textbf{[7.4-19]}$$

Note that from Eqs. (7.4-15) for plane strain at $\theta = 0$, $\sigma_2 = \sigma_1$ and $\sigma_3 = 2v\sigma_1$. Substituting σ_1, $\sigma_2 = \sigma_1$, and $\sigma_3 = 2v\sigma_1$ with $v = 1/3$ into Eq. (7.3-11) results in $\sigma_1 = 3S_Y$ when yielding occurs, a rather unexpected result. Actually, plane stress must occur at the free inner surface of the crack where the radius of curvature must be finite, no matter how small. For the case of plane stress, $\sigma_1 = S_Y$ at $\theta = 0$. However, as r increases, the stress does build up very quickly to $3S_Y$. Nevertheless, the average stress in the plastic zone will be something between S_Y and $3S_Y$. Irwin (Ref. 7.9) suggests using the geometric mean. That is, $\sigma_1 = \sqrt{3}S_Y$. Substituting this into Eq. (7.4-17a) results in

$$r_p^* = \frac{1}{6\pi}\left(\frac{K_I}{S_Y}\right)^2 \qquad\qquad \textbf{[7.4-20]}$$

which is the most widely accepted equation for r_p^* for plane strain.

The plastic zone correction replaces a in the determination of the stress intensity factor, K_I, with an effective crack tip length of

$$a_{\text{eff}} = a + r_p^* \qquad\qquad \textbf{[7.4-21]}$$

Example 7.4-1 | Consider a 200-mm-wide and 20-mm-thick plate made of a high-strength steel alloy with the properties $K_{Ic} = 80$ MPa· $\sqrt{m}$ and $S_Y = 1500$ MPa. Using a factor of safety of $n = 2$ determine (a) the maximum allowable tensile force that can be applied to the plate based on yielding and (b) the maximum allowable tensile force that can be applied to the plate if the plate is case 1 of Table 7.4-1 with a crack size $2a = 15$ mm.

Solution:

(a) For simple tension, $P/A = S_Y/n$, or $P = S_Y A/n$, giving

$$P = \frac{1860(10^6)(0.200)(0.020)}{2} = 3.72(10^6)\text{N} = 3.72 \text{ MN}$$

(b) With a factor of safety of 2 against fracture, let $K_I = K_{Ic}/n = 80/2 = 40$ MPa· $\sqrt{m}$. Equation (7.4-20) gives

$$r_p^* = \frac{1}{6\pi}\left(\frac{40}{1500}\right)^2 = 3.77(10^{-5}) \text{ m}$$

Thus

$$a_{\text{eff}} = a + r_p^* = 0.0075 + 0.0000377 = 0.00754 \text{ m}$$

For case 1 of Table 7.4-1 with $\eta = a_{\text{eff}}/b = 0.00754 / 0.200 = 0.0377$, we have

$$C_1 = [1 - 0.1(0.0377)^2 + 0.96(0.0377)^4]\sqrt{\sec(\pi \cdot 0.0377)} = 1.0021$$

$$K_I = 1.0021\sigma\sqrt{\pi \cdot 0.00754} = 0.1542\sigma$$

Setting $K_I = 40$ MPa $\cdot \sqrt{m}$ yields $\sigma = 259.4$ MPa. Solving for P yields

$$P = \sigma A = 259.4(10^6)(0.200)(0.020) = 1.037(10^6) \text{ N} = 1.037 \text{ MN}$$

A check should be made on the conditions of Eqs. (7.4-12) and (7.4-14). The quantity

$$2.5\left(\frac{K_{Ic}}{S_Y}\right)^2 = 2.5\left(\frac{80}{1500}\right)^2 = 0.0071 \text{ m} = 7.1 \text{ mm}$$

Since both t and a exceed this, the solution is acceptable.

Discussion   Note that the force obtained in part (b) is almost 1/4 that of part (a), a substantial difference. Also, with only a factor of safety of 2, note that the maximum allowable stress against fracture is less than 1/5 of the yield strength of the material.

Given a plate made of a titanium alloy such that $K_{Ic} = 110$ MPa$\cdot \sqrt{m}$, $S_Y = 820$ | **Example 7.4-2**
MPa, and width $b = 100$ mm. Determine the largest stable crack size for case 2 of Table 7.4-1 if the applied stress is limited to $0.5S_Y$.

Solution:
The applied stress is to be $\sigma = (0.5)(820) = 410$ MPa. The solution of this problem must be solved by trial and error. This can be done quite easily using a spreadsheet program. The column entries will look as shown below, where C_2 and K_I are solved using the formulas in Table 7.4-1.

a (mm)	$\eta = a/b$	C_2	K_I
18	0.18	1.125047	109.6897
19	0.19	1.127941	112.9853

We see that the solution is between $a = 18$ and 19 mm, as we want K_I to equal 110. After the increments of a are broken down systematically, the solution is found to be 18.09 mm. This can be corrected for the plastic zone if we take this value to be a_{eff}. The depth of the plastic zone is

$$r_p^* = \frac{1}{6\pi}\left(\frac{110}{820}\right)^2 = 0.000955 \text{ m} = 0.955 \text{ mm}$$

Thus the initial crack size is

$$a = a_{\text{eff}} - r_p^* = 18.09 - 0.955 = 17.1 \text{ mm}$$

Checking the crack size requirement of Eq. (7.4-14) we find

$$2.5\left(\frac{K_{\text{Ic}}}{S_Y}\right)^2 = 2.5\left(\frac{110}{820}\right)^2 = 0.0450 \text{ m} = 45 \text{ mm}$$

Since the value of a determined for this example is less than 45 mm, the solution is suspect. However, since a was found to be much larger than the size of the plastic zone, the solution is considered marginally acceptable. Further experimental information is in order here.

Example 7.4-3
The long bar shown in Fig. 7.4-6 is 0.75 in thick and is loaded by a force P offset from center 0.50 in. The material used for the bar is an aluminum alloy with $K_{\text{Ic}} = 30 \text{ MPa} \cdot \sqrt{\text{m}}$ and $S_Y = 500 \text{ MPa}$. Determine the value of P for which the crack will propagate.

Solution:
The offset force causes tension at the crack due to both axial and bending stresses. The net stress intensity factor is determined by the superposition of the axial and bending stress intensity factors. That is,

$$K_{\text{I}} = C_3\sigma_a\sqrt{\pi a} + C_4\sigma_b\sqrt{\pi a} = (C_3\sigma_a + C_4\sigma_b)\sqrt{\pi a}$$

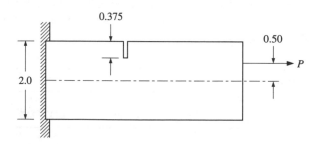

Figure 7.4-6   (Dimensions are given in inches.)

where σ_a and σ_b are the axial and bending stresses, respectively. From Table 7.4-1 with $\eta = a/b = 0.375/2 = 0.1875$,

$$C_3 = \frac{0.752 + (2.02)(0.1875) + 0.37[1 - \sin(0.1875\pi/2)]^3}{\cos(0.1875\pi/2)} \sqrt{\frac{2}{0.1875\pi}} \tan\left(\frac{0.1875\pi}{2}\right) = 1.3395$$

and

$$C_4 = \frac{0.923 + 0.199[1 - \sin(0.1875\pi/2)]^4}{\cos(0.1875\pi/2)} \sqrt{\frac{2}{0.1875\pi}} \tan\left(\frac{0.1875\pi}{2}\right) = 1.0324$$

The stresses are

$$\sigma_a = \frac{P}{A} = \frac{P}{(2.0)(0.75)} = \frac{2}{3}P$$

and with a bending moment, $M = 0.5P$, the bending stress is

$$\sigma_b = \frac{6M}{tb^2} = \frac{6(0.5P)}{(0.75)(2.0)^2} = P$$

Substituting the C_i and the stresses into the equation for K_I yields

$$K_I = \left[(1.3395)\left(\frac{2}{3}P\right) + 1.0324P\right]\sqrt{(\pi)(0.375)} = 2.090P$$

The fracture toughness in U.S. Customary units is

$$K_{Ic} = (30 \text{ MPa·}\sqrt{m})\left(\frac{1 \text{ kpsi}}{6.895 \text{ MPa}}\right)\sqrt{\frac{1 \text{ in}}{0.0254 \text{ m}}} = 27.3 \text{ kpsi·}\sqrt{m}$$

Equating K_I and K_{Ic} yields $P = 13.1$ kips.
    Checking the requirements on a and t

$$2.5\left(\frac{30}{500}\right)^2 = 0.0090 \text{ m} = 0.354 \text{ in}$$

Since both a and t exceed this, the solution is acceptable.

Discussion   The yield strength of the material is

$$S_Y = (500 \text{ MPa})\frac{1 \text{ kpsi}}{6.895 \text{ MPa}} = 72.5 \text{ kpsi}$$

Note that the maximum allowable stress against failure is

$$\sigma_{\max} = \sigma_a + \sigma_b = \frac{2}{3}P + P = \frac{5}{3}(13.06) = 21.8 \text{ kpsi}$$

which is only 30 percent of the yield strength of the material.

7.5 FATIGUE ANALYSIS[†]

7.5.1   FATIGUE STRENGTH AND ENDURANCE LIMIT

Cyclic loading of machine elements occurs often in mechanical systems, and the designer should be aware of some of the characteristics that influence fatigue failures. The governing factor in static design is typically either the yield or ultimate strength in tension or compression. These limiting values are normally obtained from a gradual loading of a standard tensile or compression test specimen. For cyclic loading, the stress intensity factor (as described in the previous section) or the *fatigue strength* S_F is used. One standard test for determining fatigue strength is the R. R. Moore rotating beam shown in Fig. 7.5-1. The rotating specimen has a polished surface to minimize premature failures due to surface finish. The rotating

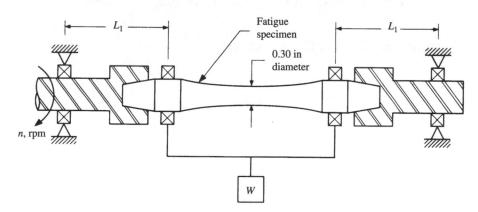

Figure 7.5-1   The R. R. Moore rotating beam test.

[†] For more advanced treatment of fatigue see Refs. 7.10 and 7.11.

beam is loaded by applying a stationary weight symmetrically with respect to the simple supports. Between the load application bearings the shear force is zero and the bending moment is constant at a value of $M = WL_1/2$. The diameter of the specimen is gradually varying to avoid any stress concentration at the midspan of the beam where at any point on the outer surface of the beam, the stress oscillates each revolution between maximum and minimum bending stresses of

$$\sigma_{max,min} = \pm\frac{32M}{\pi d^3} = \pm\frac{32WL_1/2}{\pi(0.30)^3} = \pm188.6WL_1 \qquad \textbf{[7.5-1]}$$

A plot of this stress relative to time is shown in Fig. 7.5-2, where the period of oscillation is $1/n$ and n is the speed of rotation.

The specimen is loaded with a known weight W and cycled until the specimen fractures, and the number of cycles N to fracture is recorded. The value of σ_{max} corresponding to fracture at a specific number of cycles is referred to as S_F, the *fatigue strength*. The test is normally terminated at 10^6 to 10^7 cycles. A number of specimens of a given material are tested using different values of W, and a plot of S_F vs. N can be developed. For most steels, the relationship between S_F and N is linear on a loglog plot for $10^3 \le N \le 10^6$ to 10^7 cycles as shown in Fig. 7.5-3. There is a sharp knee in the curve, which occurs at approximately 10^6 cycles; beyond the knee, failure will not occur no matter how great the number of cycles. The fatigue strength at this point is called the *endurance limit S_E*.

For most steels the relationship between $\log S_F$ and $\log N$ is linear from 10^3 to about 10^6 cycles, whereas from about 10^6 cycles and greater, S_F remains constant at a value S_E. Under 10^3 cycles, the loading can be considered to be quasistatic, and at 10^3 cycles failure occurs at about 0.9 times the ultimate strength of the material S_U.[†] Under these conditions the relationship between S_F and N for 10^3 to 10^6 cycles is

$$S_F = \frac{(0.9S_U)^2}{S_E} N^{-(1/3)\log(0.9S_U/S_E)} \qquad 10^3 \le N \le 10^6 \qquad \textbf{[7.5-2]}$$

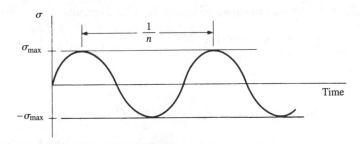

Figure 7.5-2   Stress as a function of time.

[†] This applies to the test specimen which assumes no initial cracks. For an actual application in this range, the principles of fracture mechanics should not be ignored.

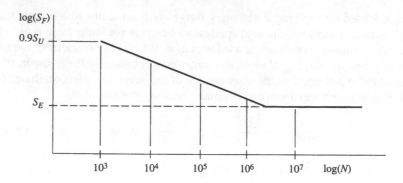

Figure 7.5-3

or (if solving for N)

$$N = \left[\frac{S_F S_E}{(0.9 S_U)^2} \right]^{-3/\log(0.9 S_U / S_E)} \qquad 10^3 \le N \le 10^6 \qquad \textbf{[7.5-3]}$$

The endurance limit for most steels is approximately one-half the ultimate strength. For analysis work, however, it is recommended that the endurance limit be obtained from actual test data. The data will be scattered, and considerable variance will require the analyst to use a statistically acceptable value.

Example 7.5-1

Estimate the life of a steel fatigue specimen with a force $W = 186$ lb and $L_1 = 2.0$ in (see Fig. 7.5-1). The material has an ultimate strength of 100 kpsi and an endurance limit of 48 kpsi.

Solution:

$$\sigma_{max} = 188.6 W L_1 = (188.6)(186)(2.0) = 70{,}160 \text{ psi} = 70.16 \text{ kpsi}$$

Letting $S_F = 70.16$ kpsi and substituting into Eq. (7.5-3) results in

$$N = \left\{ \frac{(70.16)(48)}{[(0.9)(100)]^2} \right\}^{-3/\log[(0.9)(100)/48]} = 15{,}400 \text{ cycles}$$

7.5.2   CYCLIC STRESS WITH A STATIC COMPONENT

Quite often a machine element may have fluctuating stresses which are not purely reversing. For example, if a constant tensile load is placed on the rotating beam, the stress curve appears as shown in Fig. 7.5-4. This case is not purely static, where σ_{max} cannot be directly compared with either S_Y or S_U, nor is it purely reversing, where σ_{max} cannot be compared with S_E. There is no exact analytical approach for

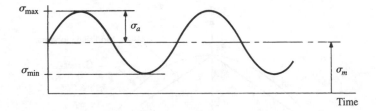

Figure 7.5-4   Cyclic stress with a static component.

this case, and the designer is forced to use an empirical approach. Defining the average or mean stress σ_m (analogous to the static stress component) and the fluctuating stress amplitude σ_a (analogous to the reversing stress component), we have

$$\sigma_m = \frac{\sigma_{max} + \sigma_{min}}{2} \qquad \sigma_a = \frac{\sigma_{max} - \sigma_{min}}{2} \qquad \textbf{[7.5-4]}$$

Up to this point, conditions of failure are known for only two cases of combined values of σ_m and σ_a. That is, when $\sigma_a = 0$, $\sigma_m = S_Y$ or S_U (depending on the criteria); and when $\sigma_m = 0$, $\sigma_a = S_E$. When σ_a and σ_m both have values, some other criterion must be developed.

Possible criteria for failure are shown in Fig 7.5-5. Values of (σ_m, σ_a) falling above the lines shown are deemed to be failure. The Gerber line and the modified Goodman line are used for fracture criteria. For yield criteria, either the Soderberg or the modified Goodman lines are used. The criteria safeguarding against failure for these lines for positive values of σ_m are

Fracture Criteria
Modified Goodman:

$$\frac{\sigma_m}{S_U} + \frac{\sigma_a}{S_E} \leq 1 \qquad \textbf{[7.5-5]}$$

Gerber:

$$\left(\frac{\sigma_m}{S_U}\right)^2 + \frac{\sigma_a}{S_E} \leq 1 \qquad \textbf{[7.5-6]}$$

Yield Criteria
Soderberg:

$$\frac{\sigma_m}{S_Y} + \frac{\sigma_a}{S_E} \leq 1 \qquad \textbf{[7.5-7]}$$

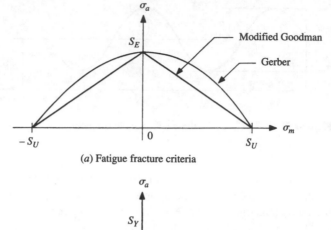

(a) Fatigue fracture criteria

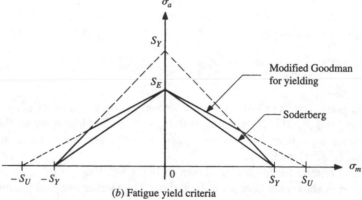

(b) Fatigue yield criteria

Figure 7.5-5   Fatigue failure criteria.

Modified Goodman:

$$\frac{\sigma_m}{S_U} + \frac{\sigma_a}{S_E} \leq 1 \quad \text{for} \quad \frac{\sigma_a}{\sigma_m} \geq \frac{S_E(S_U - S_Y)}{S_U(S_Y - S_E)} \qquad \textbf{[7.5-8a]}$$

$$\frac{\sigma_m + \sigma_a}{S_Y} \leq 1 \quad \text{for} \quad \frac{\sigma_a}{\sigma_m} \leq \frac{S_E(S_U - S_Y)}{S_U(S_Y - S_E)} \qquad \textbf{[7.5-8b]}$$

All other factors being equal, it has been found that a mean tensile stress is more detrimental than a mean compressive stress. Thus all the previously mentioned diagrams are generally much too conservative for negative values of σ_m. Considering only the Goodman yield criteria, the range of permissible values of σ_m and σ_a when σ_m is negative can be increased as shown in Fig. 7.5-6. The criteria safeguarding against failure for negative values of σ_m are

$$\frac{\sigma_a}{S_E} \leq 1 \quad \text{for} \quad \sigma_m \geq S_E - S_Y \qquad \textbf{[7.5-9a]}$$

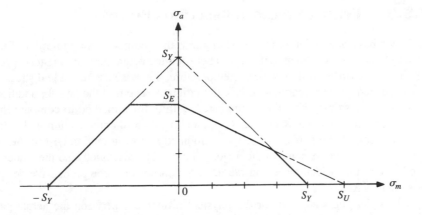

Figure 7.5-6 Modified Goodman diagram with increased range for negative σ_m.

$$\frac{\sigma_a - \sigma_m}{S_Y} \leq 1 \quad \text{for} \quad \sigma_m \leq S_E - S_Y \qquad \textbf{[7.5-9b]}$$

An additional constant tensile load of 11.2 kN is to be applied to a rotating fatigue **Example 7.5-2** specimen where the fluctuating bending stress is 280 MPa. The material values for the specimen are $S_Y = 560$ MPa, $S_U = 700$ MPa, and $S_E = 350$ MPa. Based on the modified Goodman yield criteria, determine whether or not the design criterion of Eq. (7.5-8) is exceeded.

Solution:

The diameter of the specimen is $d = (0.30)(25.4) = 7.62$ mm. The stress amplitude $\sigma_a = 280$ MPa. The mean stress is the axial stress, which is

$$\sigma_m = \frac{11.2(10^3)}{(\pi/4)(0.00762)^2} = 245.6(10^6)\,\text{N/m}^2 = 245.6\,\text{MPa}$$

Since σ_m is positive, we use Eq. (7.5-8). Checking on the ratio σ_a/σ_m we have $280/245.6 = 1.14$. From the strength data

$$\frac{S_E(S_U - S_Y)}{S_U(S_Y - S_E)} = \frac{350(700 - 560)}{700(560 - 350)} = 0.333$$

This means that Eq. (7.5-8a) prevails where

$$\frac{\sigma_m}{S_U} + \frac{\sigma_a}{S_E} = \frac{245.6}{700} + \frac{280}{350} = 1.15$$

Since $1.15 > 1$, this state of stress fails to meet the modified Goodman yield criterion and failure is predicted.

7.5.3 FATIGUE STRENGTH REDUCTION FACTORS

So far we have been relating fatigue strength to the geometric configuration of the rotating beam test specimen. Differences between the actual part and the test specimen in size, surface finish, geometry, temperature, the presence of residual stresses, corrosion, and surface treatment reduce the fatigue strength of the part. In addition, the state of stress may be quite complicated, calling for a method to compare this condition to material values such as S_Y, S_U, and S_E. A third complication is that the cyclic loading of the part may change periodically or even randomly. For the remainder of this section, some of the major points will be discussed, and the reader is urged to consult the references on fatigue cited earlier or a comprehensive design book such as Ref. 7.12.

Many designers account for the physical differences between the actual part and the fatigue specimen of the same material by reducing the allowable fatigue strength or endurance limit through the use of various modifying factors. The main factors, which are the only ones discussed here, are surface-finish imperfections and stress concentrations due to a drastic change in cross section.[†] Figure 7.5-7 illustrates the reduction factor due to surface effects for steel.

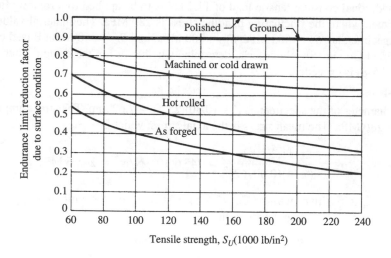

Figure 7.5-7

[†] Many other factors contribute toward reducing the endurance limit of a machine part, e.g., size effect, reliability (the published data on endurance limits are usually based on 50 percent reliability), temperature, residual stresses, corrosion, surface treatment, etc. The standard conservative approach is to consider these effects as separate and independent and reduce the endurance limit by each factor.

Stress concentrations due to a rapid reduction in cross section for static stress applications are discussed in Sec. 5.10. The maximum stress is given by $\sigma_{max} = K_t\sigma_{nom}$, where K_t is called the *static stress concentration factor* and σ_{nom} is the nominal stress given by basic formulations that assume no stress concentration. For fatigue applications, the loading is dynamic and the effective stress concentration factor is called the *fatigue stress concentration factor* K_f and is dependent on the theoretical static stress concentration *and* the material's sensitivity to the *notch radius* r_n at the stress concentration.[†] In general, notch sensitivity decreases with ductility of the material and increases with the notch radius. Peterson (Ref. 5.8) defines a *notch sensitivity factor q* which is given by

$$q = \frac{K_f - 1}{K_t - 1} \qquad [7.5\text{-}10]$$

where q varies between the values of 0 (as $r_n \to 0$) and 1 (as $r_n \to \infty$). Solving for K_f gives

$$K_f = 1 + q(K_t - 1) \qquad [7.5\text{-}11]$$

Based on experimental results, curves for q as a function of r_n have been developed for various grades of steel and aluminum. Work by Neuber (Ref. 7.13) and Kuhn and Hardrath (Ref. 7.14) resulted in the equation

$$q = \frac{1}{1 + \dfrac{\sqrt{\rho'}}{\sqrt{r_n}}} \qquad [7.5\text{-}12]$$

where ρ' is referred to as the *Neuber constant*. Values of $\sqrt{\rho'}$ for various grades of steel and aluminum are given in Tables 7.5-1, 7.5-2, and 7.5-3.[‡] When in doubt concerning a material or notch geometry, a conservative approach is to let $q = 1$, which makes $K_f = K_t$.

Table 7.5-1 Neuber's constant for steel loaded in tension

S_{UT} (kpsi)	50	60	70	80	90	100	120	140	160	180	200	220	240
$\sqrt{\rho'}$ ($\sqrt{\text{in}}$)	0.130	0.108	0.093	0.080	0.070	0.062	0.049	0.039	0.031	0.024	0.018	0.013	0.009

Table 7.5-2 Neuber's constant for annealed aluminum

S_{UT} (kpsi)	10	15	20	25	30	35	40	45
$\sqrt{\rho'}$ ($\sqrt{\text{in}}$)	0.500	0.341	0.264	0.217	0.180	0.152	0.126	0.111

[†] If in shear, $\tau_{max} = K_{ts}\tau_{nom}$ where K_{ts} is the static shear stress concentration factor. For cyclic applications, K_{fs} is the fatigue stress concentration factor for shear.

[‡] The values were constructed from data presented in Ref. 7.15. The tables are reproduced with permission (Ref. 7.12). Note: For steel loaded in torsion, use a Neuber constant for an S_U that is 20 kpsi higher than the given material.

Table 7.5-3   Neuber's constant for hardened aluminum

S_{UT} (kpsi)	15	20	30	40	50	60	70	80	90
$\sqrt{\rho'}$ ($\sqrt{in}$)	0.475	0.380	0.278	0.219	0.186	0.162	0.144	0.131	0.122

The nominal stress can be multiplied by K_f such that

$$\sigma = K_f \sigma_{\text{nom}} \qquad \text{[7.5-13]}$$

or

$$\tau = K_{fs} \tau_{\text{nom}} \qquad \text{[7.5-14]}$$

Rather than doing this, another common design approach is to use the nominal stress as the maximum stress and reduce the fatigue strength or endurance limit by K_f or K_{fs}.

Example 7.5-3

The machined steel shaft shown in Fig. 7.5-8 is in purely reversed bending. The endurance limit of the material, as predicted from a standard rotating-beam fatigue test, is 350 MPa. The ultimate tensile strength of the material is 700 MPa. Estimate the maximum value of the bending moment for which the cyclic life of the shaft is indefinite.

Solution:

Neglecting the stress concentration, we see that the maximum stress at section A-A is

$$\sigma_{\text{max}} = \frac{32M}{\pi d^3} = \frac{32M}{\pi (0.025)^3} (10^{-6}) = 0.6519M$$

where σ_{max} and M are in MPa and N·m, respectively. The ultimate tensile strength is

$$S_U = 700(10^6)6895 = 101.5(10^3) \text{ psi} = 101.5 \text{ kpsi}$$

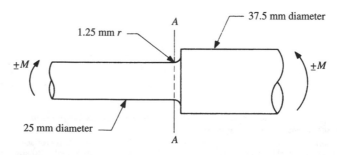

Figure 7.5-8

The surface reduction factor from Fig. 7.5-7 is approximately 0.73. The stress concentration is obtained from Appendix F, Case F.7, where for $D/d = 1.5$ and $r/d = 0.05$, $K_t \approx 1.8$. From Table 7.5-1, $\sqrt{\rho'} \approx 0.062 \sqrt{\text{in}}$. The notch radius is $r_n = 1.25/25.4 = 0.0492$ in. Using Eq. (7.5-12), the notch sensitivity factor is

$$q = \frac{1}{1 + \dfrac{0.062}{\sqrt{0.0492}}} = 0.782$$

Substituting this and $K_t = 1.8$ into Eq. (7.5-11) gives

$$K_f = 1 + 0.782(1.8 - 1) = 1.63$$

The reduced endurance limit of the part S_E' is estimated to be

$$S_E' = \frac{0.73}{1.63}(350) = 156.7 \text{ MPa}$$

Thus

$$0.6519M \approx 156.7$$

$$M \approx 240 \text{ N·m}$$

Remember, however, as stated earlier, other factors can also reduce the endurance limit of this part.

Methods are available to increase the fatigue strength or endurance limit. Inducing a compressive stress at the surface of a part where a fatigue failure is likely will reduce the likelihood of a failure (see Sec. 7.7.5).

7.5.4 EQUIVALENT STRESSES (PLANE STRESS)

Under a general state of stress which is cyclic, it is common practice to use the static failure theories to develop an equivalent one-dimensional stress state. Here, we will demonstrate the method for a state of plane stress. This is the most common stress state of interest since many fatigue failures initiate at free surfaces which are in a state of plane stress.

Combinations of multiple stress states may involve different stress concentration factors for each stress state. Thus, rather than reducing the fatigue strength or endurance limit as was done earlier, we will multiply the nominal stress values by the fatigue stress concentration factors.

If the material is ductile, either the Tresca or von Mises theory is used. The state of stress is separated into its mean and alternating components and the principal stresses are calculated for each component, σ_{1m}, σ_{2m}, and σ_{1a}, σ_{2a}. Next, an equivalent Tresca or von Mises stress is calculated for each component. For example, the equivalent mean and alternating von Mises stresses are

$$(\sigma_m)_{vM} = \sqrt{\sigma_{1m}^2 + \sigma_{2m}^2 - \sigma_{1m}\sigma_{2m}} \qquad \text{[7.5-15a]}$$

$$(\sigma_a)_{vM} = \sqrt{\sigma_{1a}^2 + \sigma_{2a}^2 - \sigma_{1a}\sigma_{2a}} \qquad \text{[7.5-15b]}$$

The final step is to use a σ_m, vs. σ_a criterion such as the modified Goodman criterion to establish the likelihood of failure.

Example 7.5-4

For Example 7.5-3 assume that $S_Y = 560$ MPa, and a purely reversing moment of 100 N·m and a constant torsional moment of 120 N·m are applied. Is the modified Goodman yield criterion exceeded?

Solution:

The mean stress is due to torsion, where

$$\tau_{nom} = \frac{16T}{\pi d^3} = \frac{(16)(120)}{\pi(0.025)^3}(10^{-6}) = 39.1 \text{ MPa}$$

From Appendix F, Case F.8, where $D/d = 1.5$ and $r/d = 0.05$, $K_{ts} \approx 1.7$. The Neuber constant for torsion (for torsion the value of S_U used is 20 kpsi higher than given) is $\sqrt{\rho'} = 0.049$. This gives a notch sensitivity factor of

$$q = \frac{1}{1 + (0.049/\sqrt{0.0492})} = 0.819$$

K_{fs} is then

$$K_{fs} = 1 + q(K_{ts} - 1) = 1 + 0.819(1.7 - 1) = 1.57$$

Thus the maximum shear stress is given by

$$\tau_{max} = K_{fs}\tau_{nom} = 1.57(39.1) = 61.4 \text{ MPa}$$

The principal stresses for this case are $\sigma_{1m} = -\sigma_{2m} = 61.4$ MPa. Then, from Eq. (7.5-15a)

$$(\sigma_m)_{vM} = \sqrt{(61.4)^2 + (-61.4)^2 - (61.4)(-61.4)} = 106.3 \text{ MPa}$$

The alternating stress is due to bending, and together with the fatigue stress concentration factor from Example 7.5-3 is

$$\sigma = \pm K_f \frac{32M}{\pi d^3} = \pm (1.63)\frac{32(100)}{\pi(0.025)^3}(10^{-6}) = \pm 106.3 \text{ MPa}$$

The principal stresses for this case are simply $\sigma_{1a} = 106.3$ MPa and $\sigma_{2a} = 0$ MPa and from Eq. (7.5-15b), $(\sigma_a)_{vM} = 106.3$ MPa.

The modified Goodman yield criterion is given by Eqs. (7.5-8). The ratio of the alternating and mean stresses is $(\sigma_a)_{vM}/(\sigma_m)_{vM} = 106.3/106.3 = 1.0$. From the strength values for the part in Example 7.5-3 we have

$$\frac{S_E(S_U - S_Y)}{S_U(S_Y - S_E)} = \frac{350(700 - 560)}{700(560 - 350)} = 0.333$$

Since $1.0 > 0.333$, we use Eq. (7.5-8a)

$$\frac{\sigma_m}{S_U} + \frac{\sigma_a}{S_E} \le 1$$

Substituting values, we obtain

$$\frac{106.3}{700} + \frac{106.3}{350} = 0.456$$

which is less than 1. Thus the modified Goodman criterion is satisfied with a safety factor over 2.

7.5.5   ESTIMATING LIFE FOR NONREVERSING OR NONREPETITIVE STRESS CYCLES

If the cyclic stress is not purely reversing, that is, $\sigma_m \ne 0$, and a finite life is predicted from the Goodman diagram, the *S-N* diagram cannot be used directly to estimate the number of cycles to failure. However, an estimate of the life can be obtained by first obtaining an *equivalent alternating stress,* which is as damaging as the actual cyclic stress conditions. Consider the case where σ_m and σ_a are outside the Goodman diagram, as shown in Fig. 7.5-9. Under these conditions a finite life is predicted. A line drawn through the points (σ_m, σ_a) and $(0, S_U)$ intersects the σ_a axis

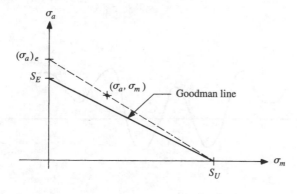

Figure 7.5-9   Equivalent alternating stress.

at $(\sigma_a)_e$, which is said to be the equivalent alternating stress. The equation for this line is

$$\frac{\sigma_a}{(\sigma_a)_e} + \frac{\sigma_m}{S_U} = 1$$

Solving for $(\sigma_a)_e$ results in

$$(\sigma_a)_e = \frac{\sigma_a}{1 - (\sigma_m/S_U)} \qquad \text{[7.5-16]}$$

To obtain an estimate of the number of cycles to failure, $(\sigma_a)_e$ is substituted into Eq. (7.5-3) in place of S_F.

Example 7.5-5 Since in Example 7.5-2, the values of σ_m and σ_a led to a finite life prediction, estimate the number of cycles to failure.

Solution:

From Example 7.5-2

$$\sigma_m = 245.6 \text{ MPa} \qquad \sigma_a = 280 \text{ MPa}$$

and

$$S_E = 350 \text{ MPa} \qquad S_U = 700 \text{ MPa}$$

From Eq. (7.5-16) the equivalent reversing stress is

$$(\sigma_a)_e = \frac{280}{1 - (245.6/700)} = 431.3 \text{ MPa}$$

Substituting this into Eq. (7.5-3) in place of S_F results in

$$N = \left\{ \frac{(431.3)(350)}{[(0.9)(700)]^2} \right\}^{-3/\log[(0.9)(700)/350]} = 85.9(10^3) \text{ cycles}$$

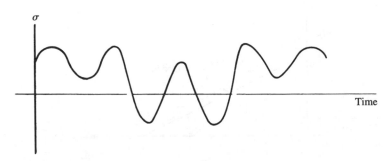

Figure 7.5-10   Nonrepetitive cyclic loading.

If the cyclic loading of a part varies as shown in Fig. 7.5-10, the damaging effect of each individual stress cycle accumulates. An approach which takes this cumulative effect into consideration is Miner's method (see Ref. 7.16).

The basic equation is

$$\sum_{i=1}^{m} \frac{n_i}{N_i} = \frac{n_1}{N_1} + \frac{n_2}{N_2} + \cdots + \frac{n_m}{N_m} \leq 1 \qquad \textbf{[7.5-17]}$$

where n_i is the number of cycles of stress $(\sigma_a)_i$ applied to the part and N_i is the number of cycles for infinite life at $(\sigma_a)_i$. Here equivalent alternating stresses can be used as well.

Example 7.5-6

A rotating-beam fatigue specimen with an endurance limit of 50 kpsi and an ultimate strength of 100 kpsi is cycled 20 percent of the time at $(\sigma_a)_1 = 70$ kpsi, 50 percent at $(\sigma_a)_2 = 55$ kpsi, and 30 percent at $(\sigma_a)_3 = 40$ kpsi. Estimate the number of cycles N to failure.

Solution:

The number of cycles to failure at $(\sigma_a)_i$ is found from Eq. (7.5-3). For $0.2N$ cycles, $(\sigma_a)_1 = 70$ kpsi, where the number of cycles to failure is

$$N_1 = \left\{ \frac{(70)(50)}{[(0.9)(100)]^2} \right\}^{-3/\log[(0.9)(100)/50]} = 19.2(10^3) \text{ cycles}$$

For $0.5N$ cycles, $(\sigma_a)_2 = 55$ kpsi, where the number of cycles to failure is

$$N_2 = \left\{ \frac{(55)(50)}{[(0.9)(100)]^2} \right\}^{-3/\log[(0.9)(100)/50]} = 326(10^3) \text{ cycles}$$

For $0.3N$, $(\sigma_a)_3 = 40$ kpsi, where the number of cycles to failure is $N_3 = \infty$.
Substituting the N_i into Eq. (7.5-17) gives

$$\frac{0.2N}{19.2(10^3)} + \frac{0.5N}{326(10^3)} + \frac{0.3N}{\infty} = 1$$

The third term on the left side of the equation is zero. Solving for N yields

$$N = 84(10^3) \text{ cycles}$$

The discussion on fatigue in this section is far from complete, but the purpose was to provide the reader with some of the concepts and approaches used in fatigue analysis. The methods of analysis and the corresponding results will always be subject to question and debate since there is no universally accepted analytical approach to the exact solution of fatigue problems. For this reason, if the design is critical, extensive reliability tests should be pursued in the final analysis.

7.6 STRUCTURAL STABILITY

Compressive loads and stresses within a long, thin structure can cause structural instabilities (buckling). The compressive stress may be elastic or inelastic and the instability may be global or local. Global instabilities can cause catastrophic failure, whereas local instabilities may cause permanent deformation but not a catastrophic failure. Instabilities can also occur due to shear stresses such as that caused by torsion on an open thin-walled tube. With pure shear, diagonal (45°) compressive stresses are present. This is also the case for the web of a transversely loaded I-beam or plate girder in bending.

In Sec. 3.10, global elastic buckling of columns in compression was presented with a derivation of Euler's equation (3.10-3). Compression of open thin-walled columns can also produce local buckling of the outer flanges as shown by the finite element simulation of the compression of a C-section column shown in Fig. 7.6-1(a). Another collapse mode is shown in Fig. 7.6-1(b), where a column with a + cross section is loaded in compression. From a side view of each flange edge we see the outer fibers of each flange buckle in the same direction and in the same shape as the columns given in Sec. 3.10. This, in turn, causes the column cross section to rotate axially. Hence the collapse mode is again local but one of torsion, not bending.

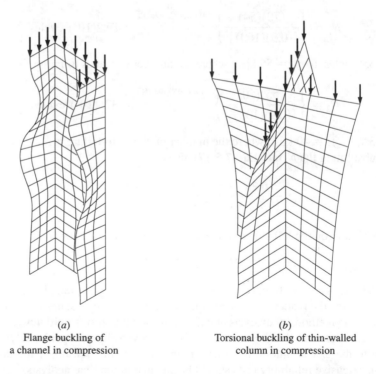

(a)
Flange buckling of
a channel in compression

(b)
Torsional buckling of thin-walled
column in compression

Figure 7.6-1   Buckling modes in thin-walled column (bottom ends fixed).

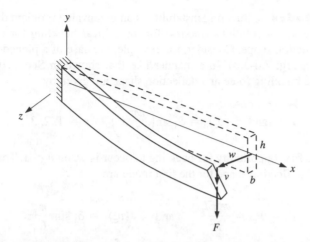

Figure 7.6-2 Torsional buckling of a thin-walled beam in bending.

Combinations of bending and torsion are also possible for columns with cross sections that have L and C shapes.

Thin-walled beams in bending can buckle in a torsional mode, as illustrated in Fig. 7.6-2. Here the cantilever beam is loaded with a lateral force F. As F is increased from zero, the end of the beam will deflect in the negative y direction normally according to the bending equation $v_c|_{x=L} = -FL^3/(3EI_z)$. However, if the beam is long enough and the ratio b/h is sufficiently small, there is a critical value of F for which the beam will collapse in a twisting mode as shown. This is due to the *compression* in the bottom fibers of the beam which causes the fibers to buckle sideways (z direction).

There are a great many other examples of unstable structural behavior such as cylindrical pressure vessels in longitudinal compression or with outer pressure or inner vacuum, arches in compression, frames in compression, shear panels, etc. Owing to the vast array of applications and the complexity of their analyses, this section will only present some basic formulations. The key concept that the designer should be aware of is if any unbraced part of a structural member is thin, long, and in compression, the possibility of buckling should be investigated. For more advanced treatment of stability, the reader is urged to consult Refs. 7.17 and 7.18. For standard construction applications, consult Ref. 7.19. For unique applications the designer may need to revert to a numerical solution such as finite elements. Depending on the application and the finite element code available, a linear analysis can be performed to determine the critical loading (e.g., see Sec. 9.3.7); or a nonlinear analysis can be conducted.

7.6.1 COLUMN BUCKLING

Section 3.10 contains a brief description of Euler buckling. In this section, some additional concepts are discussed such as buckling modes, inelastic buckling, column-beams, and design equations.

Buckling Modes   Buckling instability is an eigenvalue problem that typically has an infinite set of solutions (modes) for the critical buckling force and corresponding deflection shape. Consider, for example, the case of a pinned-pinned column shown in Fig. 7.6-3(a). In a manner like that shown in Sec. 3.10, it can be shown that the buckling force and deflection shape are given by

$$P_n = \left(\frac{n\pi}{L}\right)^2 EI_z \quad \text{and} \quad (v_c)_n = \delta_n \sin\left(\frac{n\pi}{L}x\right) \quad n = 1, 2, 3, \ldots \quad \textbf{[7.6-1]}$$

As before, the first opportunity for buckling to occur is when $n = 1$. Thus the buckling force and deflection shape for the first mode are

$$P_1 = P_{cr} = \frac{\pi^2 EI_z}{L^2} \quad \text{and} \quad (v_c)_1 = \delta_1 \sin\frac{\pi}{L}x \quad \textbf{[7.6-2]}$$

The force for the first mode is called the *critical buckling force*. The deflection shape is shown in Fig. 7.6-3(b). If the center of the column is pinned, this mode of

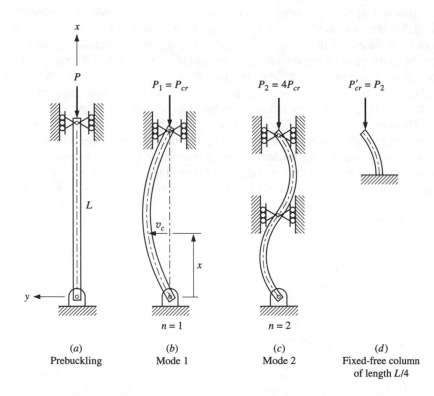

| (a) | (b) | (c) | (d) |
| Prebuckling | Mode 1 | Mode 2 | Fixed-free column of length $L/4$ |

Figure 7.6-3   Buckling modes for a pinned-pinned column.

buckling is suppressed, and buckling will not initiate until $n = 2$, where the buckling force and deflection shape are

$$P_2 = 4P_{cr} = \frac{4\pi^2 EI_z}{L^2} \quad \text{and} \quad (v_c)_2 = \delta_2 \sin\left(\frac{2\pi}{L}x\right) \qquad \textbf{[7.6-3]}$$

The deflection shape is shown in Fig. 7.6-3(c).[†] Note that the buckling force and deflection shape is identical to the fixed-free column of length $L/4$ as shown in Fig. 7.6-3(d), where the critical force will be the same as P_2 of Eq. (7.6-3).

Inelastic Buckling: The Tangent-Modulus Method   As shown in Fig. 7.3-2, some materials such as aluminum and alloy steels do not exhibit a distinct yield point when the proportional limit is reached. Euler's formulation is based on a linear stress-strain relation so that loading beyond the proportional limit must be handled differently. This can be accomplished by the *tangent-modulus method,* which is attributed to Engesser (1889).[‡] Beyond the proportional limit, the modulus of elasticity decreases from E to E_t, where $E_t = d\sigma/d\varepsilon$ is the slope of the stress-strain curve. Thus, for a column with standard geometric conditions, Eq. (3.10-7) becomes

$$\sigma_{cr} = \frac{P_{cr}}{A} = \frac{\pi^2 E_t}{(KL/r_g)^2} \qquad \textbf{[7.6-4]}$$

recalling that $r_g = \sqrt{I/A}$, the radius of gyration relative to the buckling axis of the cross section. The problem with using this equation is that E_t is not known a priori since σ is unknown. A solution can be arrived at by trial and error or, better yet, by curve-fitting the stress-strain diagram.

A 250-mm-long column with pinned-pinned ends has a rectangular cross section $b \times h = 25 \times 35$ mm. The pins are such that bending, if it occurs, will be about the axis of the smaller second-area moment. The column is made of a material for which the compressive stress-strain curve is given by the data in Table 7.6-1 and the corresponding curve shown in Fig. 7.6-4. Determine the critical load for buckling.

Example 7.6-1

Table 7.6-1

$\varepsilon, 10^{-5}$	0	25	50	75	100	125	150	175	200	225	250	275	300	325	350	375	400
σ, MPa	0	50	100	150	200	235	252	263	267	272	276	279	282	285	287	289	290

[†] After buckling initiates, the center pins can be removed without affecting the buckling shape.

[‡] Historically (see Ref. 7.20), in the same year, both F. Engesser and A. G. Considère independently arrived at this method. Shortly afterward, a second method was introduced called the *reduced-modulus* or *double-modulus method.* This method provided for an effective modulus which was lower than what the tangent-modulus theory used. Finally, in 1946, F. R. Shanley developed a theory which explained shortcomings in the two aforementioned theories. However, since the tangent-modulus method is much easier to use and provides slightly more conservative results than Shanley's theory, the tangent-modulus method is preferred in practice. See Ref. 7.19 for further discussion of Shanley's theory.

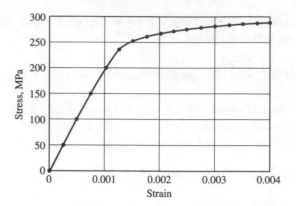

Figure 7.6-4

Solution:

The proportional limit of the curve is reached when $\sigma = 200$ MPa. The radius of gyration about the axis of minimum second-area moment is $(r_g)_{min} = b/\sqrt{12}$. For a pinned-pinned column, $K = 1$. Thus, for elastic buckling

$$\sigma_{cr} = \frac{\pi^2 E}{(KL/r_g)^2} = \left(\frac{\pi r_g}{KL}\right)^2 E = \left(\frac{(\pi)(25/\sqrt{12})}{(1.0)(250)}\right)^2 (200)(10^3) = 1650 \text{ MPa}$$

This far exceeds the proportional limit, thus requiring the use of Eq. (7.6-4). In order to do this, a least-squares curve fit of the curve *above* $\sigma = 200$ MPa is performed. Using a sixth-order polynomial, the curve on a spreadsheet program is approximated by

$$\sigma \approx -1.286(10^{18})\varepsilon^6 + 2.257(10^{16})\varepsilon^5 - 1.636(10^{14})\varepsilon^4$$

$$+ 6.271(10^{11})\varepsilon^3 - 1.343(10^9)\varepsilon^2 + 1.542(10^6)\varepsilon - 483.9 \qquad \textbf{[a]}$$

The slope is then

$$E_t = \frac{d\sigma}{d\varepsilon} \approx -7.714(10^{18})\varepsilon^5 + 1.128(10^{17})\varepsilon^4 - 6.545(10^{14})\varepsilon^3$$

$$+ 1.881(10^{12})\varepsilon^2 - 2.686(10^9)\varepsilon + 1.542(10^6) \qquad \textbf{[b]}$$

Substituting Eqs. (*a*) and (*b*) into Eq. (7.6-4) with $x = \varepsilon(10^3)$ and simplifying gives

$$x^6 - 66.90x^5 + 849.1x^4 - 4675x^3 + 13,080x^2 - 18,380x + 10,240 = 0$$

This polynomial equation can be solved quite easily using a programmable calculator or a math software package. Only two real roots emerge, $x = 1.690$ and 52.28. This results in $\varepsilon = 0.00169$ and 0.0523. Only the first root is within the range of ap-

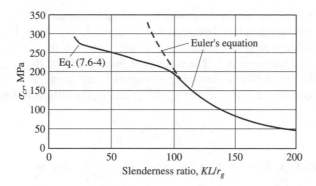

Figure 7.6-5

plication. Substituting $\varepsilon = 0.00169$ into the equations for σ and E_t yields $\sigma = 260$ MPa and $E_t = 31.6$ GPa. Finally, the critical force is given by

$$P_{cr} = \sigma_{cr}A = 260(10^6)(25)(35)(10^{-6}) = 227.5(10^3)N = 227.5 \text{ kN}$$

A plot of σ_{cr} versus the slenderness ratio KL/r can be made for the material used in Example 7.6-1 using Eqs. (7.6-4) and (3.10-7). Equations (a) and (b) of the example are used for Eq. (7.6-4). This is shown in Fig. 7.6-5.

Nonideal Column Behavior   For a structural steel, which is an *EPP* material (see Sec. 7.7), the plot of σ_{cr} versus the slenderness ratio KL/r_g is given in Fig. 3.10-2(b), repeated here as Fig. 7.6-6. However, a large number of tests on hot-rolled wide-flange steel columns have shown that columns of intermediate slenderness ratio tend to buckle at critical stresses significantly lower than S_Y. This behavior can be attributed to either small eccentricities in the loading axis, imperfections in the column, or residual stresses throughout the cross section due to uneven cooling during the manufacturing process.[†] To account for this, an empirical equation called the *parabolic equation* is sometimes used. The parabolic equation is fitted from the Euler equation to the point $(0, S_Y)$. The point on the Euler curve selected is $\sigma_{cr} = S_Y/2$. Substituting this into Eq. (7.6-4) yields the slenderness ratio

$$\left(\frac{KL}{r_g}\right)_c = \sqrt{\frac{2\pi^2 E}{S_Y}} \qquad \textbf{[7.6-5]}$$

The parabolic equation is given by

$$\sigma_{cr} = \left[1 - \frac{1}{2}\frac{(KL/r_g)^2}{(KL/r_g)_c^2}\right]S_Y \qquad (KL/r_g) \le (KL/r_g)_c \qquad \textbf{[7.6-6]}$$

and together with the Euler curve forms the plot shown in Fig. 7.6-7.

[†] See Ref. 7.19 for further discussion of residual stresses.

Known eccentricity can be dealt with using beam-column theory as discussed in Sec. 6.13. Consider the pinned-pinned column shown in Fig. 7.6-8. Due to the eccentricity the column bends immediately upon application of the load P. The vertical reaction at the lower pin is P. The equivalent force-couple at the origin is P and Pe as shown in Fig. 7.6-8(c). Summing moments at the break at x yields $M_z = -Pe - Pv_c$. Recalling that $M_z = EI_z\, d^2 v_c/dx^2$ produces the differential equation

$$\frac{d^2 v_c}{dx^2} + \frac{P}{EI_z} v_c = -\frac{P}{EI_z} e \qquad\qquad \textbf{[7.6-7]}$$

Letting $k^2 = P/(EI_z)$, the solution to Eq. (7.6-7) is

$$v_c = C_1 \sin kx + C_2 \cos kx - e$$

Since $v_c = 0$ at $x = 0$, we find that $C_2 = e$. Also, $v_c = 0$ at $x = L$ gives

$$C_1 = e\,\frac{1 - \cos kL}{\sin kL} = e \tan \frac{kL}{2}$$

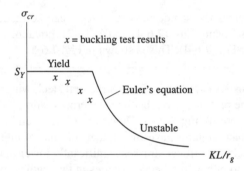

Figure 7.6-6

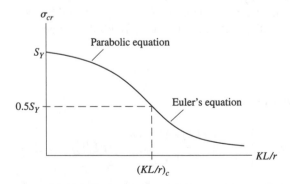

Figure 7.6-7

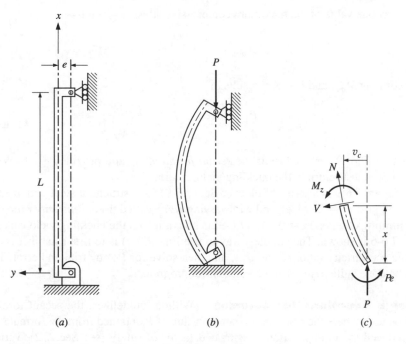

Figure 7.6-8   Eccentrically loaded column.

Thus the equation of the deflection curve is

$$v_c = e\left(\tan \frac{kL}{2} \sin kx + \cos kx - 1\right) \qquad \textbf{[7.6-8]}$$

The maximum deflection and moment occurs at $x = L/2$. The magnitude of the maximum moment is $M_{max} = P(e + v_c|_{x=L/2})$. Substituting Eq. (7.6-8) for $v_c|_{x=L/2}$ gives

$$M_{max} = Pe\left(\tan \frac{kL}{2} \sin \frac{kL}{2} + \cos \frac{kL}{2}\right) = Pe \sec \frac{kL}{2}$$

Substituting $k = \sqrt{P/(EI_z)} = [\sqrt{P/(EA)}]/r_g$, where $I_z = Ar_g^2$ gives

$$M_{max} = Pe \sec\left(\frac{L}{2r}\sqrt{\frac{P}{EA}}\right)$$

The absolute value of the maximum compressive stress is given by

$$\sigma_{max} = \frac{P}{A} + \frac{M_{max}c}{I_z}$$

Substituting M_{max} and $I_z = Ar_g^2$ results in

$$\sigma_{max} = \frac{P}{A}\left[1 + \frac{ec}{r_g^2}\sec\left(\frac{L}{2r_g}\sqrt{\frac{P}{EA}}\right)\right] \qquad \textbf{[7.6-9]}$$

Equation (7.6-9), referred to as the *secant formula,* is valid provided $e > 0$. When $e = 0$, we must return to the buckling/yield criteria.

Figure 7.6-9 is a series of plots of Eq. (7.6-9) for a structural steel with $\sigma_{max} = S_Y = 36$ kpsi, $E = 29$ Mpsi, and with varying $\dot ec/r_g^2$, called the *eccentricity ratio.* For the limiting case, $ec/r_g^2 = 0$, Eq. (7.6-9) is not valid and the buckling/yield curve of Fig. 7.6-6 is shown. The simplest way to use Eq. (7.6-9) is to first establish values of the eccentricity ratio, E, and σ_{max} and then solve for P/A by trial and error. This can be done quite easily using a spreadsheet program.

Design Equations for Columns   Without guidelines, the secant formula can be used where the resulting maximum value of P obtained from the formula can be reduced by an appropriately estimated factor of safety (see Sec. 7.2). Various code formulas are also recommended by various organizations.

For centrally loaded *structural steel* columns, recommendations from the Structural Stability Research Council (SSRC) and the American Institute of Steel

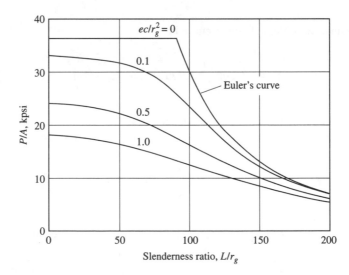

Figure 7.6-9   Eccentrically loaded pinned-pinned column with $\sigma_{max} = S_Y = 36$ kpsi and $E = 29$ Mpsi.

Construction (AISC) specify the use of the parabolic and Euler equations with a recommended factor of safety. These equations are

$$\left(\frac{P}{A}\right)_{allow} = \frac{\left[1 - \frac{(KL/r_g)^2}{2C_c^2}\right]S_Y}{\frac{5}{3} + \frac{3(KL/r_g)}{8C_c} - \frac{(KL/r_g)^3}{8C_c^3}} \qquad \frac{KL}{r_g} \le C_c \qquad \textbf{[7.6-10a]}$$

$$\left(\frac{P}{A}\right)_{allow} = \frac{12\pi^2 E}{23(KL/r_g)^2} \qquad \frac{KL}{r_g} \ge C_c \qquad \textbf{[7.6-10b]}$$

where $C_c = (KL/r_g)_c$ is the critical slenderness ratio given by Eq. (7.6-5). The numerator of Eq. (7.6-10a) is the parabolic equation, whereas the denominator is the recommended factor of safety. Equation (7.6-10b) is the Euler equation with a factor of safety of 23/12.

For *aluminum,* the Aluminum Association recommends equations in three ranges of slenderness ratio: short, intermediate, and long. The equations depend on the particular aluminum alloy. The equations for alloy 2014-T6 often used in building construction and alloy 6061-T6 used in aircraft structures are given below. The units of P/A are in kpsi.

2014-T6

$$\left(\frac{P}{A}\right)_{allow} = 28 \qquad 0 \le \frac{KL}{r_g} \le 12 \qquad \textbf{[7.6-11a]}$$

$$\left(\frac{P}{A}\right)_{allow} = 30.7 - 0.23\frac{KL}{r_g} \qquad 12 \le \frac{KL}{r_g} \le 55 \qquad \textbf{[7.6-11b]}$$

$$\left(\frac{P}{A}\right)_{allow} = \frac{54,000}{(KL/r_g)^2} \qquad 55 \le \frac{KL}{r_g} \qquad \textbf{[7.6-11c]}$$

6061-T6

$$\left(\frac{P}{A}\right)_{allow} = 19 \qquad 0 \le \frac{KL}{r_g} \le 9.5 \qquad \textbf{[7.6-12a]}$$

$$\left(\frac{P}{A}\right)_{allow} = 20.2 - 0.126\frac{KL}{r_g} \qquad 9.5 \le \frac{KL}{r_g} \le 66 \qquad \textbf{[7.6-12b]}$$

$$\left(\frac{P}{A}\right)_{allow} = \frac{51,000}{(KL/r_g)^2} \qquad 66 \le \frac{KL}{r_g} \qquad \textbf{[7.6-12c]}$$

7.6.2   BUCKLING OF PLATES

This section will provide a brief introduction to the buckling of thin rectangular plates. The governing equations for the lateral loading of thin rectangular plates were given in Sec. 5.7.1. The reader is urged to review this before proceeding, as derivations in this section refer to the derivations in Sec. 5.7.1.

In the derivations of Sec. 5.7.1 it is assumed that the lateral deflection is of the order of the plate thickness so that in-plane membrane forces can be neglected. Here, we assume the same; however, we introduce constant in-plane membrane forces as being applied to the plate through external loading. Let the forces per unit length N_x, N_y, V_{xy}, and V_{yx} be applied to the edges of the plate as shown in Fig. 7.6-10. Isolate a $dx\,dy$ element in its deflected state as shown in Fig. 7.6-11. Here, only the constant transmitted forces N_x, N_y, V_{xy}, and V_{yx} are shown. The internal moment and out-of-plane shear force per unit length components are shown in Fig. 5.7-3 in Sec. 5.7.1.

For equilibrium of moment about the z axis, $V_{yx} = V_{xy}$. With this, the net force in the z direction is

$$\left(N_x \frac{\partial^2 w}{\partial x^2} + N_y \frac{\partial^2 w}{\partial y^2} + 2V_{xy} \frac{\partial^2 w}{\partial x\,\partial y} \right) dx\,dy$$

Next, we add this to Eq. (5.7-8) with $p(x, y) = 0$ to obtain the total force in the z direction, which is zero for equilibrium. This results in

$$\frac{\partial V_x}{\partial x} + \frac{\partial V_y}{\partial y} + N_x \frac{\partial^2 w}{\partial x^2} + N_y \frac{\partial^2 w}{\partial y^2} + 2V_{xy} \frac{\partial^2 w}{\partial x\,\partial y} = 0 \qquad \textbf{[7.6-13]}$$

Equation (7.6-13) is analogous to Eq. (5.7-8). Following identically to the derivation of Eqs. (5.7-9) through (5.7-12) we obtain the following equations

$$\frac{\partial^2 M_x}{\partial x^2} - 2\frac{\partial^2 M_{xy}}{\partial x\,\partial y} + \frac{\partial^2 M_y}{\partial y^2} + N_x \frac{\partial^2 w}{\partial x^2} + N_y \frac{\partial^2 w}{\partial y^2} + 2V_{xy} \frac{\partial^2 w}{\partial x\,\partial y} = 0 \qquad \textbf{[7.6-14]}$$

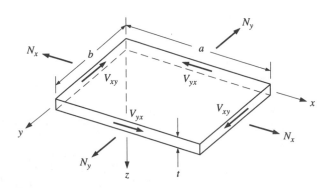

Figure 7.6-10   Rectangular plate with edge forces per unit length.

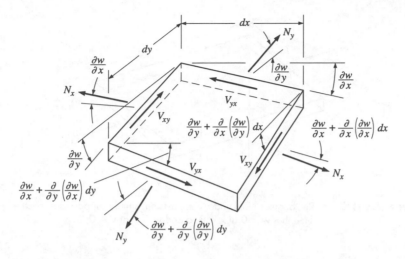

Figure 7.6-11   Differential element with membrane forces only.

$$D\left(\frac{\partial^4 w}{\partial x^4} + 2\frac{\partial^4 w}{\partial x^2 \partial y^2} + \frac{\partial^4 w}{\partial y^4}\right) - N_x \frac{\partial^2 w}{\partial x^2} - N_y \frac{\partial^2 w}{\partial y^2} - 2V_{xy}\frac{\partial^2 w}{\partial x\, \partial y} = 0 \quad \textbf{[7.6-15]}$$

where D is the plate stiffness given by Eq. (5.7-5), repeated here for convenience as

$$D = \frac{Et^3}{12(1 - v^2)} \qquad \textbf{[7.6-16]}$$

Equations (7.6-14) and (7.6-15) represent the governing differential equations for the plate where N_x, N_y, and V_{xy} are uniformly applied forces per unit length.

Consider the case of a plate, simply supported along all edges, with a compressive force per unit length in the x direction P_x as shown in Fig. 7.6-12. Substituting $N_x = -P_x$ and $N_y = V_{xy} = 0$ into Eq. (7.6-15) gives

$$\frac{\partial^4 w}{\partial x^4} + 2\frac{\partial^4 w}{\partial x^2 \partial y^2} + \frac{\partial^4 w}{\partial y^4} + \frac{P_x}{D}\frac{\partial^2 w}{\partial x^2} = 0 \qquad \textbf{[7.6-17]}$$

The following boundary conditions apply to each edge being simply supported.

$$w = 0 \quad \text{at} \quad x = 0, a \qquad \textbf{[7.6-18a]}$$

$$w = 0 \quad \text{at} \quad x = 0, b \qquad \textbf{[7.6-18b]}$$

$$M_x = \frac{\partial^2 w}{\partial x^2} + v\frac{\partial^2 w}{\partial y^2} = 0 \quad \text{at} \quad x = 0, a \qquad \textbf{[7.6-18c]}$$

$$M_y = \frac{\partial^2 w}{\partial y^2} + v\frac{\partial^2 w}{\partial x^2} = 0 \quad \text{at} \quad y = 0, b \qquad \textbf{[7.6-18d]}$$

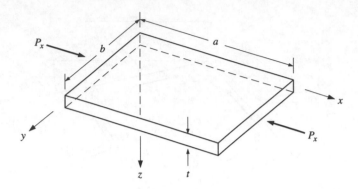

Figure 7.6-12   Rectangular plate under compression, simply supported at each edge such that $w = 0$ at $x = 0$, a and $w = 0$ at $y = 0$, b.

However, since the deflection at the edges at $x = 0$, a cannot vary with respect to y, $\partial^2 w/\partial y^2 = 0$ at $x = 0$, a. Likewise, $\partial^2 w/\partial x^2 = 0$ at $y = 0$, b. Thus the boundary conditions reduce to

$$w = 0 \qquad \text{at} \qquad x = 0, a \qquad\qquad \textbf{[7.6-19a]}$$

$$w = 0 \qquad \text{at} \qquad y = 0, b \qquad\qquad \textbf{[7.6-19b]}$$

$$\frac{\partial^2 w}{\partial x^2} = 0 \qquad \text{at} \qquad x = 0, a \qquad\qquad \textbf{[7.6-19c]}$$

$$\frac{\partial^2 w}{\partial y^2} = 0 \qquad \text{at} \qquad y = 0, b \qquad\qquad \textbf{[7.6-19d]}$$

Equation (7.6-17) with the boundary conditions of Eqs. (7.6-19) is typically solved using the series

$$w = \sum_{m=1}^{\infty} \sum_{n=1}^{\infty} C_{mn} \sin \frac{m\pi x}{a} \sin \frac{n\pi y}{b} \qquad m, n = 1, 2, \ldots \qquad \textbf{[7.6-20]}$$

Note that this expression satisfies the boundary conditions of Eqs. (7.6-19). Substituting w into Eq. (7.6-17) gives

$$\sum_{m=1}^{\infty} \sum_{n=1}^{\infty} C_{mn} \left[\left(\frac{m\pi}{a}\right)^4 + 2\left(\frac{mn\pi^2}{ab}\right)^2 + \left(\frac{n\pi}{b}\right)^4 - \frac{P_x}{D}\left(\frac{m\pi}{a}\right)^2 \right] \sin \frac{m\pi x}{a} \sin \frac{n\pi y}{b} = 0$$

or

$$\sum_{m=1}^{\infty} \sum_{n=1}^{\infty} \pi^2 C_{mn} \left\{ \pi^2 \left[\left(\frac{m}{a}\right)^2 + \left(\frac{n}{b}\right)^2 \right]^2 - \frac{P_x}{D}\left(\frac{m}{a}\right)^2 \right\} \sin \frac{m\pi x}{a} \sin \frac{n\pi y}{b} = 0$$

Since this must be zero for all values of $x(0 \le x \le a)$ and $y(0 \le y \le b)$, then

$$\pi^2 C_{mn} \left\{ \pi^2 \left[\left(\frac{m}{a}\right)^2 + \left(\frac{n}{b}\right)^2 \right]^2 - \frac{P_x}{D}\left(\frac{m}{a}\right)^2 \right\} = 0$$

If $C_{mn} = 0$, we obtain the trivial solution. Setting the term within the { } to zero and solving for P_x yields

$$P_x = D\left(\frac{\pi a}{b}\right)^2 \left[\left(\frac{m}{a}\right)^2 + \left(\frac{n}{b}\right)^2 \right]^2$$

or

$$P_x = \frac{\pi^2 D}{b^2}\left(\frac{mb}{a} + \frac{n^2 a}{mb} \right)^2$$

With respect to n, the lowest value of P_x corresponds to $n = 1$. Thus

$$(P_x)_{n=1} = \frac{\pi^2 D}{b^2}\left(\frac{mb}{a} + \frac{a}{mb} \right)^2 \qquad \textbf{[7.6-21]}$$

To obtain the lowest value of P_x with respect to m we set the derivative $d(P_x)_{n=1}/dm$ to zero. This results in

$$\frac{d(P_x)_{n=1}}{dm} = \frac{2\pi^2 D}{b^2}\left(\frac{mb}{a} + \frac{a}{mb} \right)\left(\frac{b}{a} - \frac{a}{m^2 b} \right) = 0$$

Since mb/a must be a positive quantity, the first term in parentheses cannot be zero. This means that

$$\frac{b}{a} - \frac{a}{m^2 b} = 0$$

or

$$m = \frac{a}{b} \qquad \textbf{[7.6-22]}$$

Since m is an integer, Eq. (7.6-22) is only valid for integer ratios of a/b. For these cases, substituting Eq. (7.6-22) into Eq. (7.6-21) gives the lowest value of P_x, the critical force per unit length

$$P_{cr} = \frac{4\pi^2 D}{b^2} \qquad \textbf{[7.6-23]}$$

For noninteger values of a/b Eq. (7.6-22) is invalid and we must return to Eq. (7.6-21) and review the results for specific integer values of m. To do this, let

$$(P_x)_{n=1} = \frac{k\pi^2 D}{b^2} \qquad \text{[7.6-24]}$$

where from Eq. (7.6-21)

$$k = \left(\frac{mb}{a} + \frac{a}{mb} \right)^2 \qquad \text{[7.6-25]}$$

Plotting k vs. a/b for specific values of m, we obtain the plots shown in Fig. 7.6-13. As the value of a/b increases from zero, the solid lines show the minimum value of k. For $a/b > 1$ the coefficient k does not deviate much from the value of 4. Certainly for $a/b > 4$ the coefficient remains very close to 4.

Similar to the notation used for columns, Eq. (7.6-24) can be expressed in terms of the critical stress σ_{cr}. The critical force per unit length is $P_{cr} = \sigma_{cr}t$, and Eq. (7.6-16) for D when substituted into Eq. (7.6-24) gives

$$\sigma_{cr} = \frac{k\pi^2 E}{12(1 - v^2)(b/t)^2} \qquad \text{[7.6-26]}$$

This expression is quite similar to Euler's formulation of Eq. (3.10-7)

$$\sigma_{cr} = \frac{\pi^2 E}{(KL/r_g)^2}$$

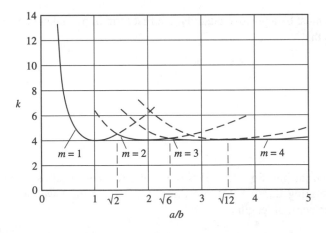

Figure 7.6-13   Buckling coefficient k for a compressed simply supported rectangular plate.

Table 7.6-2 gives the limiting value for k for various side edge boundary conditions with the loading edges being simply supported.[†]

Table 7.6-2   Buckling coefficient for various side edge boundary conditions (loaded edges simply supported)

Edge Conditions	k
Both fixed	6.97
One fixed, one simply supported	5.42
Both simply supported	4.0
One fixed, one free	1.28
One simply supported, one free	0.425

Discussion   It is important to note the difference between plate and column buckling. The buckling results for plates given are based on very small lateral deflections. For postbuckling deflections greater than the plate thickness, Eqs. (7.6-14) and (7.6-15) are still valid; however, the internal membrane forces are no longer constant and the equations become nonlinear. Unlike columns, plates do not collapse when the critical load is reached. Constraints due to the two-dimensional nature of plates allow plates to resist increasing loads and may not fail until the applied loads are considerably higher than the critical loading. For this reason, the nonlinear postbuckling analysis is important. This analysis, however, is beyond the scope of this book, and the reader is urged to consult the references cited earlier.

7.7   INELASTIC BEHAVIOR

7.7.1   EPP MATERIALS

There are situations in design where inelastic behavior can be tolerated. In fact, for structures that are highly statically indeterminate, some standard analyses actually allow the maximum stresses to exceed the elastic limit, and the designer may not be aware of this. As stated in Sec. 5.10, it is sometimes permissible to tolerate plastic behavior in small localized areas undergoing high stresses. As the loading is increased on a member made of a Hookean material, eventually the stress in the highest stressed region will reach the elastic limit of the material. Increasing the loading beyond this point will initiate plastic behavior in this region, and the shape of the stress distribution will change owing to the change in the stress-strain relationship.

[†] See Ref. 7.17 for plots of k vs. a/b for these cases.

Some examples of models of stress-strain relationships are shown in Fig. 7.7-1. The limiting stress for elastic behavior is the yield strength S_Y. For numerical purposes, it is common practice to use piecewise linear functions to approximate the stress-strain relationship for the general elastic-inelastic model. This section will deal only with the elastic, perfectly plastic (*EPP*) material model, which approximates many structural steels.

When the stress in an element within a part made of an *EPP* material reaches the elastic limit, the stress in the element will remain constant as the loading of the structure increases. The stress then is in the inelastic region where the strain can increase, and the element will fracture when $\varepsilon = \varepsilon_F$. The element retains its usefulness in the inelastic region; i.e., it still is carrying part of the load on the structure.

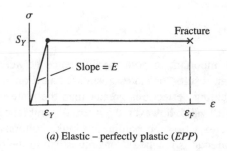

(*a*) Elastic – perfectly plastic (*EPP*)

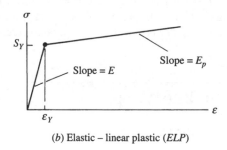

(*b*) Elastic – linear plastic (*ELP*)

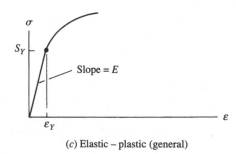

(*c*) Elastic – plastic (general)

Figure 7.7-1   Models of elastic-inelastic behavior.

However, in the inelastic range the element cannot carry any stress greater than S_Y. This means that additional loads on the structure must be carried by other elements of the structure. As an example, consider a plate containing a hole where the plate is loaded in pure tension. As shown in Sec. 5.10, the stress at the edge of the hole is maximum. If the load is increased until the stress at this point reaches S_Y, the resulting stress distribution is the same shape as that shown in Fig. 5.10-1(c) (repeated in Fig. 7.7-2(a)). Let P_Y represent the tensile force on the plate when yielding first occurs. The load can be increased beyond P_Y; however, the stress at the edge of the hole will no longer increase if the material is *EPP*. The stress increases at other points, however, and elements near the edge of the hole increase to S_Y, as shown in Fig. 7.7-2(b). This process continues until the strain at the edge of the hole reaches ε_F, in which case a fracture will occur. A theoretical limit is the point where the

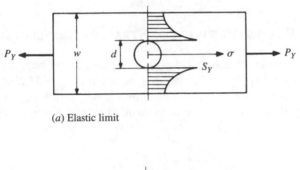

(a) Elastic limit

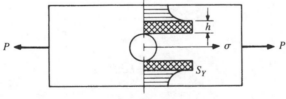

(b) $P > P_Y$, h = plastic zone depth

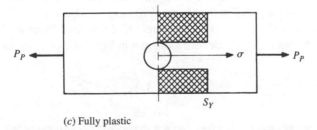

(c) Fully plastic

Figure 7.7-2

entire structure reaches a fully plastic condition, as shown in Fig. 7.7-2(c), where the entire cross section has reached S_Y. The load in which the entire section becomes completely plastic is called P_P.[†] This load can be calculated easily, since at this limit the stress is uniform and the equation $\sigma = P/A$ can be used. Thus the limit load is

$$P_P = S_Y A = S_Y(w - d)t \qquad \textbf{[7.7-1]}$$

where t is the plate thickness.

If the stress concentration at the hole is 2.5, for example, $P_P = 2.5P_Y$. That is, the maximum load carrying capacity of the plate is 2.5 times greater than the yield load. This is quite an increase in usefulness of the plate if inelastic behavior can be tolerated in the design. As mentioned earlier, there are cases where inelastic behavior is not desired, such as cyclic loading, brittle fracture, etc. Also, once the material is loaded beyond the yield point, there will be permanent deformation and residual stresses after the part is unloaded.

7.7.2   PLASTIC BEHAVIOR OF STRAIGHT BEAMS IN BENDING

As another example of plastic behavior, consider the rectangular beam in pure bending shown in Fig. 7.7-3. The bending equation $\sigma_{max} = Mc/I$ is valid until the maximum stress reaches S_Y. Let the bending moment when $\sigma_{max} = S_Y$ be M_Y, the moment when the section first begins to yield. Thus with $I = b(2c)^3/12$,

$$S_Y = \frac{M_Y c}{I} = \frac{M_Y c}{\dfrac{1}{12} b(2c)^3}$$

or

$$M_Y = \frac{2}{3} bc^2 S_Y \qquad \textbf{[7.7-2]}$$

If the bending moment is increased beyond M_Y, plastic behavior begins, as shown in Fig. 7.7-3(c). The depth of the plastic zone h_p will increase as M increases until hypothetically $h_p = c$ [see Fig. 7.7-3(d)]. At this point, the entire section is fully plastic and the section cannot accommodate any further increase in load. The bending moment at this point is called the *plastic limit moment* M_P.[‡]

The plastic limit moment is quite simple to determine. Since each distribution is constant, the equivalent forces are as shown in Fig. 7.7-3(d). The moment due to these forces is

$$M_P = S_Y bc^2 \qquad \textbf{[7.7-3]}$$

[†] This limit is not really possible, as at this point the strain goes to infinity. Thus the true limit would be slightly less than P_P.

[‡] Again, this limit corresponds to infinite strains and will not be reached. Thus the true limit is less than M_P.

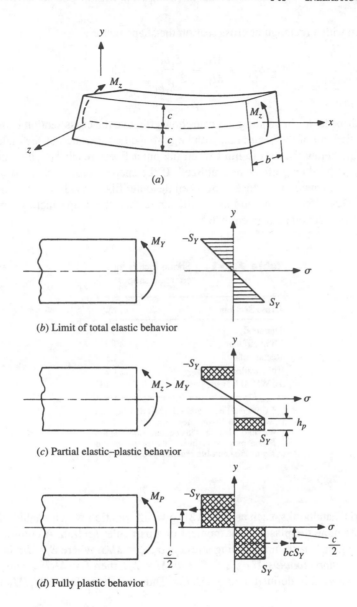

(a)

(b) Limit of total elastic behavior

(c) Partial elastic–plastic behavior

(d) Fully plastic behavior

Figure 7.7-3

Bending Shape Factor   The ratio of the plastic limit moment to the yield moment is defined as the shape factor in bending *SF,* given by

$$SF = \frac{M_P}{M_Y} \qquad\qquad \textbf{[7.7-4]}$$

For a beam with a rectangular cross section the shape factor is

$$SF_\square = \frac{M_P}{M_Y} = \frac{S_Y bc^2}{\frac{2}{3} S_Y bc^2} = 1.5$$

The shape factor in bending indicates how effective the cross section is in terms of the elastic limit of the beam; i.e., with large shape factors much of the material is considerably below the elastic limit when the outer fibers reach S_Y. Thus much of the section is not being effectively utilized. If the shape factor is close to unity, a good deal of the material is close to S_Y when the outer fibers reach S_Y, and hence the section is very effective in bending. To illustrate this, the shape factors for some common cross sections are given in Table 7.7-1.

Table 7.7-1   Shape factors for common shapes

Cross Section	SF
Diamond	2
Solid circular	1.70
Rectangular	1.5
Thin-walled circular	1.2
16 WF 40 I-beam	1.13[†]

[†] Care must be exercised when dealing with wide-flange beam sections, as localized buckling due to the compression in the flange may occur before the plastic or the elastic moments are reached (see Ref. 7.19).

Material handbooks, such as Ref. 7.21, give information which enables the analyst to determine the plastic limit moment of a particular section. A common way of expressing the maximum bending stress is $\sigma_{max} = M/S$, where $S = I/c$ is called the *elastic section modulus*. If $\sigma_{max} = S_Y$ with $M = M_Y$, then $S = M_Y/S_Y$. The *plastic section modulus* Z is defined as $Z = M_P/S_Y$. Thus the shape factor M_P/M_Y is also given by

$$SF = \frac{Z}{S} \qquad\qquad \textbf{[7.7-5]}$$

Neutral Axis Shift   In an elastic analysis of a beam with a cross section that is not symmetric with respect to the bending axis, the maximum tensile and compressive stresses are not equal. Consequently, the neutral axis will shift when the moment exceeds M_Y. For example, consider the section in Fig. 7.7-4(*a*). The cen-

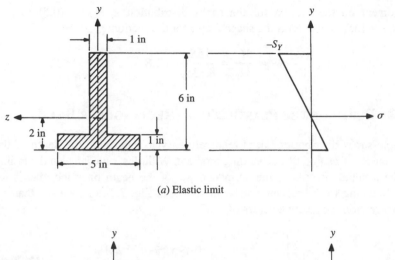

(a) Elastic limit

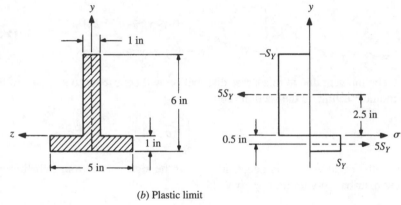

(b) Plastic limit

Figure 7.7-4

troidal axis is shown for the section, and the second area moment is $I = 33.33 \text{ in}^4$. The maximum value for y is 4 in. Thus when the maximum stress reaches the yield strength

$$S_Y = \frac{M_Y(4)}{33.33} \quad \text{or} \quad M_Y = 8.33 S_Y$$

If, in the limit, the section becomes totally plastic where $M = M_P$, then if the neutral axis remains the centroidal axis, the net force above the axis would be $4S_Y$, whereas the net force below the axis would be $6S_Y$. Thus equilibrium is not satisfied unless the neutral axis shifts to balance the forces. To achieve balance, the neutral axis shifts to where the *areas* balance. For the example under discussion, when $M = M_P$, the neutral axis shifts downward to the top of the flange, as shown in Fig. 7.7-4(b). This is where the areas (5 in²) are equal above and below the neutral axis. The limit moment on the section for the stress distribution is $M_P = (0.5)(5S_Y) + (2.5)(5S_Y) = 15S_Y$. This yields the shape factor for the section:

$$SF = \frac{M_P}{M_Y} = \frac{15S_Y}{8.33S_Y} = 1.8$$

7.7.3 DEPTH OF THE PLASTIC ZONE (RECTANGULAR BEAM)

For simple stress distributions, it is relatively easy to determine the depth of the plastic zone as a function of the bending moment. With this, one can also determine either the residual stresses or the permanent set of the beam once the load is released. Returning to the rectangular beam shown in Fig. 7.7-3(c), we see that the stress distribution for positive values of y is

$$\sigma_x = \begin{cases} -\dfrac{S_Y}{c - h_p}y & 0 \le y \le (c - h_p) \\ -S_Y & (c - h_p) \le y \le c \end{cases} \qquad \text{[7.7-6]}$$

The moment due to this stress distribution will contribute to one-half of the total moment acting on the section. Thus,

$$M = -2\int_0^c y\sigma_x \, dA$$

where the negative sign is necessary due to the sign convention. Substituting the stress distribution with $dA = b \, dy$ yields

$$M = -2\left[\int_0^{c-h_p}\left(-\frac{S_Y}{c - h_p}y\right)yb \, dy + \int_{c-h_p}^c - S_Y yb \, dy\right]$$

Integration and simplification yield

$$M = \frac{2}{3}bc^2 S_Y\left[1 + \frac{h_p}{c} - \frac{1}{2}\left(\frac{h_p}{c}\right)^2\right]$$

Since $I = 2/3 \, bc^3$

$$M = \frac{S_Y I}{c}\left[1 + \frac{h_p}{c} - \frac{1}{2}\left(\frac{h_p}{c}\right)^2\right] \qquad \text{[7.7-7]}$$

Note that when $h_p = 0$, $M = S_Y I/c$, or in other words $M = M_Y$. Also, when $h_p = c$, $M = 3/2 S_Y I/c$, or in other words, $M = M_P$.

If M is known and $M_Y < M < M_P$, the depth of the plastic zone can be found by rewriting Eq. (7.7-7) as

$$\left(\frac{h_p}{c}\right)^2 - 2\frac{h_p}{c} + 2\left(\frac{Mc}{S_Y I} - 1\right) = 0 \qquad \text{[7.7-8]}$$

which is a quadratic equation in h_p/c. The one root of this equation which is acceptable is

$$\frac{h_p}{c} = 1 - \sqrt{3 - \frac{2Mc}{S_Y I}} \qquad \textbf{[7.7-9]}$$

Since M (and M_Y) can be coincidentally positive or negative and $M_Y = S_Y I/c$, Eq. (7.7-9) can be written

$$\frac{h_p}{c} = 1 - \sqrt{3 - 2\left|\frac{M}{M_Y}\right|} \qquad \textbf{[7.7-10]}$$

Once h_p is established, the maximum strain for the particular value of M can be determined. The maximum strain occurs at the outer fibers of the beam, but Hooke's law cannot be applied at this point since the outer fibers are in the plastic region. For $|y| < c - h_p$, the beam is still elastic, where

$$\sigma_x = -\frac{S_Y}{c - h_p} y$$

Since Hooke's law is valid in this zone,

$$\varepsilon_x = \frac{\sigma_x}{E} = -\frac{S_Y}{E(c - h_p)} y \qquad 0 \le |y| < c - h_p$$

If it is still assumed that plane surfaces perpendicular to the centroidal axis remain plane up to this point, the linear strain distribution is valid for $|y| > c - h_p$. Thus the strain across the *entire* surface is

$$\varepsilon_x = -\frac{S_Y}{E(c - h_p)} y \qquad \textbf{[7.7-11]}$$

The maximum strain occurs at $y = \pm c$, and at a magnitude of

$$\varepsilon_{max} = \frac{S_Y}{E(c - h_p)} c \qquad \textbf{[7.7-12]}$$

7.7.4 RESIDUAL STRESSES (RECTANGULAR BEAM)

If the load is released after plastic deformation takes place, residual stresses will develop as well as residual strains and deformations. Assume that a rectangular beam is in bending such that $M_Y < M < M_P$. The depth of the plastic zone h_p can be determined from Eq. (7.7-10). The strain in the plastic zone can be calculated from Eq. (7.7-11). Strains greater than ε_Y, however, are in the plastic zone, as shown in Fig 7.7-5. As the beam is unloaded, points in the beam that are in the elastic zone *tend* to return on the linear part of the stress-strain curve. However, points in the beam that are in the plastic zone *tend* to return along a line parallel to the linear curve from the initial position (see Fig. 7.7-5).

Thus the unloading for either point tends to follow a linear path. This is equivalent to superimposing a linear unloading stress distribution over the initial stress distribution, as shown in Fig. 7.7-6.

The moment from the stress distribution, $\sigma_x = ky$, must cancel the initial moment since the net moment on the surface after unloading is zero. Since the loading moment is M, a linear stress distribution for $-M$ is

$$\sigma_x = \frac{My}{I}$$

Therefore, $k = M/I$.

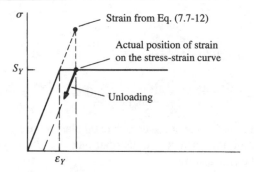

Figure 7.7-5

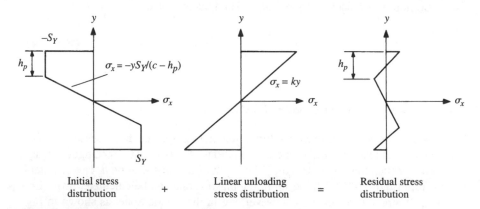

Figure 7.7-6

Adding the distribution of $\sigma_x = ky$ to that in Eq. (7.7-6) yields the residual stress (see Fig. 7.7-6). For positive values of y

$$(\sigma_x)_{\text{res}} = \begin{cases} \left(\dfrac{M}{I} - \dfrac{S_Y}{c - h_p} \right)y & 0 \le y \le (c - h_p) \\ \dfrac{My}{I} - S_Y & (c - h_p) \le y \le c \end{cases} \qquad \textbf{[7.7-13]}$$

The maximum residual stress can occur at either $y = \pm c$ or $\pm(c - h_p)$. It can be shown that for h_p/c less than 0.585, the maximum residual stress occurs at $y = \pm c$, whereas for h_p/c greater than 0.585, the maximum residual stress occurs at $y = \pm(c - h_p)$.

7.7.5 RESIDUAL STRESSES AND FATIGUE (RECTANGULAR BEAM)

Residual stresses can be beneficial when a mechanical element is under the influence of oscillating stresses. Fatigue failures begin at locations of high shear and tensile stresses. High stresses normally occur on the outer surfaces of a mechanical element, and consequently this is where fatigue failures generally begin. The high tensile stresses can be reduced if a compressive residual stress is present before actual loading. One mechanical method of doing this is *shot peening,* in which a rain of metallic shot (such as cast-iron pellets) impinges at high speed on the surface of the mechanical element, inducing a compressive stress on a small outer layer of the element.

If the cyclic loading is primarily in one direction, residual stresses obtained by presetting the element may be more effective.

Example 7.7-1

Consider a beam of rectangular cross section loaded in bending so that the bending moment cycles from zero to $M = 4000$ lb·in (see Fig. 7.7-7(a)). Determine the maximum tensile stress and compare this with the maximum tensile stress at the same point if a preset moment of $M = 1.2M_Y$ is applied before the service load. The elastic limit of the material is 20 kpsi.

Solution:

Without any preset, the maximum tensile stress occurs at the bottom fibers and is

$$(\sigma_x)_{\text{max}} = \frac{(4000)(1)}{0.333} = 12,000 \text{ psi}$$

The complete stress distribution is shown in Fig. 7.7-7(b).

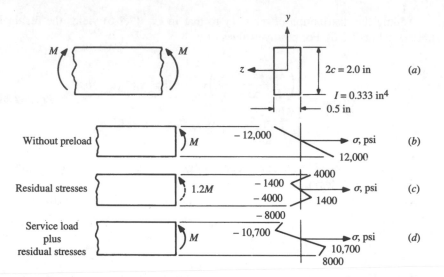

Figure 7.7-7

To reduce the maximum tensile stress by presetting, the preset moment should be applied in the *same* direction as the service moment. From Eq. (7.7-10) the depth of the plastic zone is

$$\frac{h_p}{c} = 1 - \sqrt{3 - \frac{(2)(1.2M_Y)}{M_Y}} = 0.225 \qquad h_p = (0.225)(1.0) = 0.225 \text{ in}$$

The moment necessary to develop h_p is

$$M = 1.2M_Y = (1.2)\left(\frac{2}{3}\right)bc^2S_Y = (1.2)\left(\frac{2}{3}\right)(0.5)(1^2)(20,000) = 8000 \text{ lb·in}$$

Thus the residual stress distribution, given by Eqs. (7.7-13), for positive values of y is

$$\sigma_{res} = \begin{cases} -1810y & 0 \le y \le 0.775 \text{ in} \\ 24,000y - 20,000 & 0.775 \text{ in} \le y \le 1.0 \text{ in} \end{cases}$$

and is shown in Fig. 7.7-7(c).

Adding the service stress to the residual stress results in the maximum stress distribution when $M = 4000$ lb·in. This is illustrated in Fig. 7.7-7(d).

Thus it can be seen that the maximum tensile stress on the outer fiber is reduced from 12,000 to 8000 psi. This will give a drastic improvement in performance with respect to fatigue. Note, however, that at $y = -0.775$ in, the tensile stress has increased to 10,700 psi. This is not as bad as it seems. First, the stress is still lower than the original maximum of 12,000 psi. Second, the maximum stress now occurs internally and not at the surface, where surface effects contribute to fatigue failure (see Sec. 7.5). Note that the preset technique is *not* to be used if the service moment is purely reversing. In this example, if the service moment were purely reversing between 4000 and -4000 lb·in, the preset would cause a tensile stress of 16,000 psi at the top surface when $M = -4000$ lb·in.

7.8 ENGINEERING APPROXIMATIONS USED IN STATICALLY INDETERMINATE PROBLEMS

Although much time is spent on studying statically determinate problems, in reality, structures are normally designed to be highly statically indeterminate. To a large extent this is due to the engineer's basic tendency toward conservatism. Statically indeterminate structures are supported in a redundant fashion. Thus, more than one point will provide safety against collapse. In addition, statically indeterminate structures generally distribute stresses better. Therefore, in a given situation, some stresses can be reduced by adding more restraint. Another basic reason for highly statically indeterminate structures is to reduce the large elastic deflections and rotations which generally arise in statically determinate structures. Although more desirable structurally, statically indeterminate problems are more difficult to analyze manually because the deflections of the structure must be included in the analysis. As the structure becomes more redundant and the number of mechanical elements increases, the exact solution becomes more difficult to obtain. Energy techniques provide the easiest approach to complex statically indeterminate problems; however, there are many times when it is neither possible nor necessary to perform an elaborate and exact analysis of a structure. There are two basic techniques used to make engineering approximations where a statically indeterminate problem is reduced to a statically determinate one. The first is the method of neglecting elastic deformations of members that are much more rigid than others. This technique is used quite extensively with fasteners like bolts and rivets. The second method of approximation is called *limit analysis* (sometimes referred to as limit design or ultimate analysis). Limit analysis allows for a given section under high stress to go completely into the plastic range. Since in this circumstance the stress distribution is known, a deflection equation is no longer necessary and the number of necessary deflection equations for an exact analysis can be reduced to zero.

7.8.1 CONSIDERING DEFLECTIONS OF FLEXIBLE ELEMENTS ONLY

Consider the riveted connection shown in Fig. 7.8-1(*a*). If a break is made isolating the right-hand member by slicing through the rivets, the resulting free-body diagram is as shown in Fig. 7.8-1(*b*).

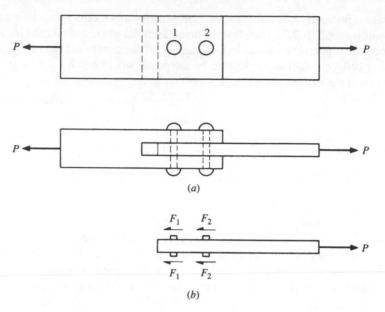

Figure 7.8-1

Thanks to symmetry, the force on the top of each rivet equals the force on the bottom. Summing forces yields $2F_1 + 2F_2 = P$. Since the system is statically inde-terminate, the exact values of F_1 and F_2 would have to be determined by deflection theory, but this is where an approximation is often made. If the deflections of the plates are ignored, the shear deflections of the rivets will be the same. Thus the shear stresses are identical in each rivet. If $A_1 = A_2$, then $F_1 = F_2$. Consequently, ap-plying the equilibrium equation results in $F_1 = F_2 = P/4$.

If the rivets are located as shown in Fig. 7.8-2, twisting of the plates might occur. To avoid twisting, the shear deflections of each rivet should be equal (assum-ing again that the plates are basically rigid). Since shear deflections are proportional to shear stresses, for no twisting the shear stresses in each rivet should be equal. Thus $F_1/A_1 = F_2/A_2$, and using the fact that $2F_1 + 2F_2 = P$ results in

$$F_1 = \frac{A_1}{A_1 + A_2}\frac{P}{2} \qquad F_2 = \frac{A_2}{A_1 + A_2}\frac{P}{2}$$

Summing moments about the center of rivet 2 (keeping in mind that a net force of $2F_1$ is acting on rivet 1, top and bottom surfaces) gives the necessary line of load application for no twist as

$$\bar{y} = \frac{cA_1}{A_1 + A_2}$$

Note that this is the location of the centroid of the shear areas of the rivets. In general, then, loading through the centroid of the shear areas eliminates twisting of the plates.

If the plates are loaded in torsion, as in shaft couplings, or if the load is not through the rivet centroid, twisting will occur and will be centered about the rivet centroid. Considering only the forces which arise from twisting, we see that the line of action of the forces between the rivets or bolts and the plates will be perpendicular to a line drawn between the given rivet and the centroid. An example of the forces applied by a set of rivets to a plate is shown in Fig. 7.8-3, where a torque is being applied to the plate. The forces which arise from the torque must be such that the torque balances for equilibrium and the net force is zero.

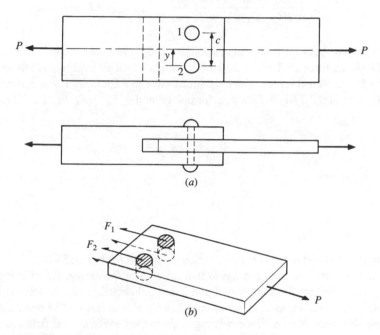

Figure 7.8-2

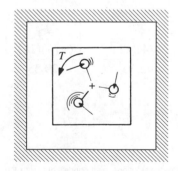

Figure 7.8-3

Example 7.8-1 | The plate shown in Fig. 7.8-4 is mounted to a wall with four bolts. Bolts 1 and 2 each have a diameter of 12 mm, and bolts 3 and 4 each have a diameter of 18 mm. Approximate the shear stress in each bolt if a force of 4 kN is applied to the plate as shown. Neglect any out-of-plane bending.

Solution:

First, the centroid of the bolt areas is found. Thanks to symmetry, $\bar{y} = 25$ mm. The horizontal centroid is[†]

$$\bar{x} = \frac{(2)(50)(\pi/4)(18)^2}{2[(\pi/4)(12)^2 + (\pi/4)(18)^2]} = 34.62 \text{ mm}$$

The net torque about the bolt-area centroid is $(150 - 34.62)(4) = 461.5$ N·m.

The direct force of 4 kN is distributed to each bolt through equal stresses (no twisting for the direct force). Since the areas are equal for bolts 1 and 2, the forces $F_2' = F_1'$ [see Fig. 7.7-4(c)]. Likewise for bolts 3 and 4, $F_4' = F_3'$. For equilibrium of force

$$2F_1' + 2F_3' = 4 \qquad\qquad \textbf{[a]}$$

For equal stresses

$$\frac{F}{A} = \frac{F_1'}{(\pi/4)(12)^2} = \frac{F_3'}{(\pi/4)(18)^2} \qquad\qquad \textbf{[b]}$$

Solving Eqs. (a) and (b) results in $F_1' = 0.615$ kN and $F_3' = 1.385$ kN.

The next step is to find the forces that arise from the torque. Considering only torque, we see that the forces in bolts 1 and 2 are again equal, as are the forces in bolts 3 and 4, because each respective set of bolts has equal area and is at an equal distance from the bolt-area centroid. Since twisting is about the centroid, shear deflections and consequently shear stresses are proportional to the distance from the centroid. Since $r_1 = r_2 = \sqrt{(25)^2 + (34.62)^2} = 42.70$ mm and $r_3 = r_4 = \sqrt{(25)^2 + (15.38)^2} = 29.35$ mm,

$$\frac{\left[\dfrac{F_1''}{(\pi/4)(12)^2}\right]}{42.70} = \frac{\left[\dfrac{F_3''}{(\pi/4)(18)^2}\right]}{29.35}$$

or

$$F_3'' = 1.547 F_1'' \qquad\qquad \textbf{[c]}$$

[†] The $\pi/4$ can be omitted in these calculations, as it cancels out. Thus the centroid of d^2 can be found.

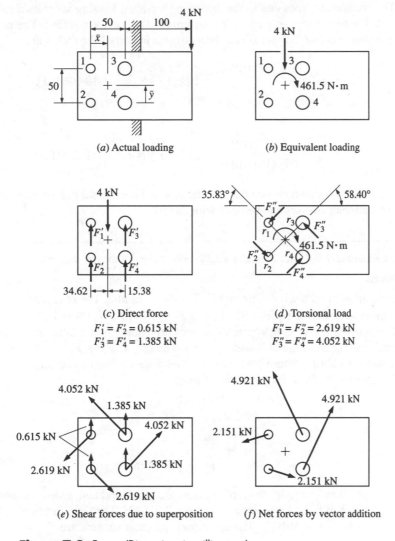

Figure 7.8-4   (Dimensions in millimeters.)

Balancing the torque results in, $T = 2F_1''r_1 + 2F_3''r_3$,   or

$$461.5 = 2F_1''(42.70) + 2F_3''(29.35) \qquad\qquad \textbf{[d]}$$

Substitution of Eq. (c) into (d) results in $F_1'' = 2.619$ kN and $F_3'' = 4.052$ kN.

The combined forces due to the direct and torsional loading are shown in Fig. 7.8-4(e). The total force on each bolt is determined by vector addition. The net results are shown in Fig. 7.8-4(f). The shear stresses in bolts 1 through 4 are

$$\tau_1 = \tau_2 = \frac{2.151(10^3)}{(\pi/4)(0.012)^2} = 19.0(10^6)\,\text{N/m}^2 = 19.0 \text{ MPa}$$

and

$$\tau_3 = \tau_4 = \frac{4.921(10^3)}{(\pi/4)(0.018)^2} = 19.3(10^6)\,\text{N/m}^2 = 19.3 \text{ MPa}$$

When all the bolts have equal areas, the problem becomes much simpler, as the forces rather than stresses can be dealt with directly.

Example 7.8-2 Repeat Example 7.8-1, but consider all the bolts to be 12 mm in diameter.

Solution:

Since the areas of all the bolts are equal, $\bar{x} = \bar{y} = 25$ mm. The reactions from the 4-kN direct force divide equally. Thus $F_1' = F_2' = F_3' = F_4' = 4/4 = 1$ kN.

Considering the torque, again, the forces are equal, as the bolts are of equal area and spacing from the bolt-area centroid. Thus $F_1'' = F_2'' = F_3'' = F_4'' = F''$ and the torque is (4 kN)(125 mm) = 500 N·m. The distance from the centroid to each bolt is 25/(cos 45). Thus, for a balance of torque

$$4F'' \frac{25}{\cos 45} = 500$$

$$F'' = 3.536 \text{ kN}$$

As in the previous example, the force vectors are combined and added vectorially. The final forces in bolts 1 and 2 are $F_1 = F_2 = 2.915$ kN. The final forces in bolts 3 and 4 are $F_3 = F_4 = 4.301$ kN. The corresponding shear stresses are

$$\tau_1 = \tau_2 = \frac{2.915(10^3)}{(\pi/4)(0.012)^2} = 25.8(10^6)\,\text{N/m}^2 = 25.8 \text{ MPa}$$

$$\tau_3 = \tau_4 = \frac{4.301(10^3)}{(\pi/4)(0.012)^2} = 38.0(10^6)\,\text{N/m}^2 = 38.0 \text{ MPa}$$

Comparing the results of Examples 7.8-1 and 7.8-2 shows that increasing the diameter of bolts 3 and 4 equalizes and reduces the stresses in all four bolts. However, the solution with different size bolts is considerably more involved.

7.8.2   LIMIT ANALYSIS[†]

In considering a statically indeterminate structure of order n the approach used in limit analysis is to allow n of the greatest loaded restraints to reach their limit value (totally plastic) and determine the case when the $(n + 1)$th restraint totally yields. At this point, the structure will collapse without bounds. The value of the applied load on the structure when total collapse occurs is called the *limit load*. There are some dangers in using limit analysis which the designer should be aware of. The analysis should be used for statically loaded ductile materials[‡] where the maximum forces are known with confidence. If the structure is undergoing large variation cyclic loading, where the danger of a fatigue failure is present, limit analysis should not be used.

Determine the limit load P_L which will cause the beam shown in Fig. 7.8-5(a) to collapse. The material is *EPP* with an elastic limit of $S_Y = 40$ kpsi.

| **Example 7.8-3** |

Solution:

The structure is statically indeterminate of the first order. Thus, when two restraints reach their limit, the beam will collapse. Since the bending stresses are much greater than the transverse shear stresses, only the bending stresses will be considered. It can be seen that the bending moment will reach maximum values at *two* points, A and B.[§] Thus, when the moment reaches the limit moment M_P at A and B, the structure will collapse. Isolating the beam at A results in the free-body diagram shown in Fig. 7.8-5(b). Summing moments at A yields

$$M_P + 50R_C - 40P_L = 0 \qquad\qquad \textbf{[a]}$$

Isolating section BC just to the right of P_L yields the free-body diagram shown in Fig. 7.8-5(c). Summing moments at B yields

$$R_C = 0.1M_P \qquad\qquad \textbf{[b]}$$

Substituting this into Eq. (a) results in

$$P_L = 0.14M_P \qquad\qquad \textbf{[c]}$$

[†] See Ref. 7.22.

[‡] In this section, discussion will be limited to *EPP* material, as defined in Sec. 7.7.

[§] In limit analysis, it is not important which point reaches the limit first. For example, from Appendix C, an exact analysis would show that point B reaches the limit moment M_P first. When this occurs, it is said that a *hinge* forms at point B. The load can still be increased beyond this point, as section AB still has bending capacity. Since the material is *EPP*, the moment remains constant at the hinge at point B at a value of M_P. Thus the additional bending from the increasing load goes into section AB until the moment at point A reaches the limit M_P. At this point the structure has completely lost its bending capacity and collapses. In the elastic analysis, if the limit moment occurred at point A first, by the same argument, collapse would not occur until the moment at B reached M_P. Thus, in either case the same limit is reached, and consequently the order of the process is immaterial and the results from the elastic analysis are not needed.

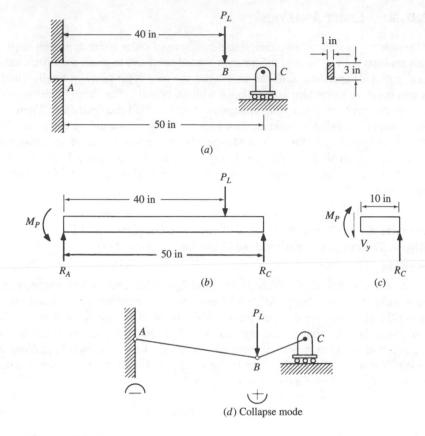

Figure 7.8-5

From Eq. (7.7-3), $M_P = S_Y bc^2$. Thus

$$P_L = (0.15)(40)(1)(1.5)^2 = 13.5 \text{ kips}^\dagger$$

Note that the directions of the limit moments are found by considering hinges forming at the limit points and determining whether the beam is concave up or down at
each point. This is illustrated in Fig. 7.8-5(d).

In some cases, as shown in the next example, more than one collapse mode
must be considered.

Example 7.8-4 | Two forces equal in magnitude are applied to the beam shown in Fig. 7.8-6(a).
Determine the limit value of P_L in terms of M_P.

---

† Note that if an elastic analysis is performed on the beam using the equations for beam C.11 of Appendix
C, when the outer fibers of the beam at point B reach S_Y, the applied force is 8520 lb. Any force greater
than this will cause some permanent deformation. However, if this can be tolerated, the beam has much
more load capacity than the force found in the elastic analysis.

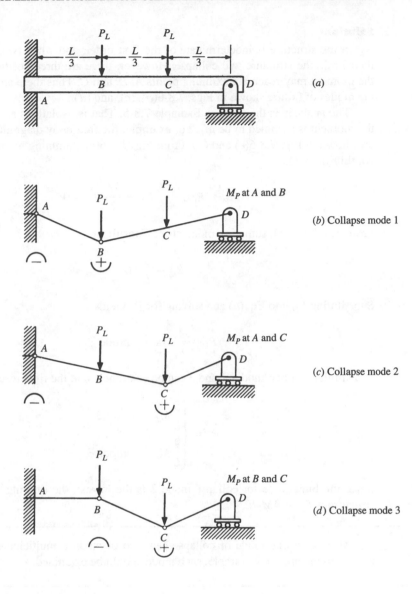

(a)

(b) Collapse mode 1

(c) Collapse mode 2

(d) Collapse mode 3

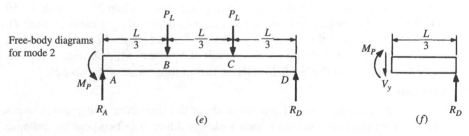

Free-body diagrams for mode 2

(e)

(f)

Figure 7.8-6

Solution:

Again the structure is indeterminate of the first order, and when two points reach their limit, the structure will collapse. However, there are three positions in which the moment may reach a maximum, points A, B, and C. Thus there are three possible modes of failure shown in Figs. 7.8-6(b), (c), and (d).

The analysis is the same as Example 7.8-5. That is, isolations are made where the moment is assumed to be M_P. For example, the free-body diagrams for mode 2 are shown in Fig. 7.8-6(e) and (f). From Fig. 7.8-6(e), summing moments at A for equilibrium yields

$$M_P + R_D L - \frac{L}{3} P_L - \frac{2L}{3} P_L = 0 \qquad \text{[a]}$$

From Fig. 7.8-6(f), summing moments at C results in

$$R_D = \frac{3}{L} M_P$$

Substituting R_D into Eq. (a) and solving for P_L yields

$$P_L = \frac{4}{L} M_P \qquad \text{mode 2}$$

In a similar manner, modes 1 and 3 can be examined, and the final results are

$$P_L = \begin{cases} \dfrac{5}{L} M_P & \text{mode 1} \\[2mm] \dfrac{9}{L} M_P & \text{mode 3} \end{cases}$$

Since the limit force for collapse mode 2 is the lowest, the limiting value for the structure is $P_L = 4M_P/L$.

More than one mode of collapse can also occur in a multielement structure, and, as in the previous example, each mode should be examined.

Example 7.8-5

The beam shown in Fig. 7.8-7(a) has a rectangular cross section 25 mm wide by 50 mm deep and is completely fixed at point B and supported by a cable at point D. The net tensile area of the cable is $A = 12.5$ mm², and both the beam and cable are made of an *EPP* material with an elastic limit of $S_Y = 240$ MPa. A vertical force is applied at midspan of the beam. Determine the limiting value of the load P_L.

Solution:

Since the structure is statically indeterminate of the first order, collapse will occur when two separate sections reach their yield condition. The two possible collapse modes are shown in Figs. 7.8-7(b) and (c). In mode 1, plastic hinges have formed at

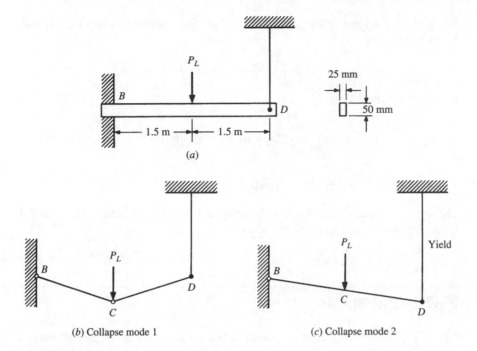

(a)

(b) Collapse mode 1   (c) Collapse mode 2

Figure 7.8-7

the two points where the bending moments are maximum. For mode 2, a hinge has formed at the wall, and the stress in the cable has reached the limit value. A third mode may seem possible, where a hinge forms at the midspan and the cable yields. It can be shown that this is not possible, but the reader should verify through physical reasoning that if this mode occurred, the left-hand side of the beam would still have structural integrity and therefore the structure would not collapse in this mode.

Mode 1 is similar to the case treated in Example 7.8-4. Letting the force in the cable be F_D and using moment summations as was performed for Eqs. (a) and (b) of that example, we find

$$M_P + 3F_D - 1.5P_L = 0 \qquad [a]$$

$$F_D = \frac{2}{3}M_P \qquad [b]$$

Solving for P_L yields $P_L = 2M_P$. Thus, for mode 1,

$$P_L = 2M_P = 2S_Y bc^2 = 2(240)(10^6)(0.25)(0.25)^2$$

$$= 7500 \text{ N} = 7.5 \text{ kN} \qquad \text{mode 1}$$

For mode 2, the force in the cable is $F_D = S_Y A$. Substituting this into Eq. (a) with $M_P = S_Y b c^2$ gives

$$S_Y b c^2 + 3S_Y A - 1.5 P_L = 0$$

Solving for P_L results in

$$P_L = \frac{2}{3} S_Y (bc^2 + 3A) = \frac{2}{3}(240)(10^6)[(0.025)(0.025)^2 + 3(12.5)(10^{-6})]$$

$$= 9500 \text{ N} = 8.5 \text{ kN} \qquad \text{mode 2}$$

The smaller of the collapse loads is the collapse load of the structure. Thus mode 1 is the collapse mode, and the limit force is $P_L = 7.5$ kN.

7.9   PROBLEMS

7.1   Given a horizontal cantilever beam with a vertical force at its free end, make a list of all the stress and strength uncertainties that can exist.

7.2   In a particular mechanism, a connecting rod is used where a tensile force is transmitted through pins located at both ends of the rod. The main portion of the rod will have a circular cross section of diameter d. From past experience it has been found that the following tolerances can be maintained:

1. The diameter can be machined to within a tolerance of ± 0.2 percent.
2. The centerline of the pins can be maintained to within a tolerance of 0.4 percent of the diameter d.
3. The maximum axial force is gradually applied and can be predicted within an accuracy of ± 10 percent.
4. The strength of the material is known within the limits of ± 15 percent.

For analysis purposes assume that the nominal values of dimension, force, and strength will be used with the simple axial stress formulation. Determine each stress and strength uncertainty and the overall factor of safety corresponding to the above stated uncertainties.

7.3   Consider a curved beam with a rectangular cross section of depth h and width b, and a radius of curvature to the centroid of r_c. Develop an equation for the stress uncertainty factor if the maximum stress due to a bending moment M is calculated using straight-beam theory, where

$$\sigma_{max} = \frac{6M}{bh^2}$$

7.4 The stress uncertainty factor due to a suddenly applied force is illustrated in this problem. Consider a weight of magnitude W to be gradually applied to a cable of cross section A. Neglecting the weight of the cable, the static tensile stress in the cable would simply be W/A. Now, consider the weight attached to the cable with the cable being taut but not supporting the weight. At this point, the weight is suddenly released. Owing to the dynamics of the application of the load, the maximum stress in the cable will exceed W/A. To estimate the maximum stress, equate the decrease in potential energy of the weight to the increase in strain energy of the cable and determine the maximum deflection of the cable (this neglects losses due to hysteretic damping of the material). From this, the maximum stress and the stress uncertainty factor based on using W/A can be determined.

7.5 to 7.11 For the plane stress state given

(a) Draw the *three* Mohr's circles corresponding to the three principal stresses.

(b) Determine the surface location and magnitude of the Tresca (maximum shear) stress.

(c) Evaluate the magnitude of the equivalent von Mises stress and compare it with two times the Tresca stress found in part (b).

Problem	σ_x	σ_y	τ_{xy}
7.5	10 kpsi	0 kpsi	0 kpsi
7.6	0 MPa	0 MPa	−20 MPa
7.7	40 MPa	10 MPa	0 MPa
7.8	40 MPa	−10 MPa	0 MPa
7.9	10 kpsi	10 kpsi	0 kpsi
7.10	80 MPa	0 MPa	−30 MPa
7.11	−12 kpsi	−20 kpsi	−6 kpsi

7.12 Determine the maximum shear and von Mises stresses in Problem 3.37.

7.13 Determine the maximum shear and von Mises stresses in Problem 5.51.

7.14 A common problem in design is that of a solid circular shaft of diameter d transmitting a bending moment M and a torsional moment T simultaneously.

(a) Prove that the maximum shear stress is given by

$$\tau_{\max} = \frac{16}{\pi d^3}\sqrt{M^2 + T^2}$$

(b) Develop the equation for the maximum equivalent von Mises stress and compare it with two times the Tresca stress found in part (a).

7.15 Determine the maximum equivalent von Mises stress for Problem 3.32.

7.16 Determine the maximum equivalent von Mises stress for Problem 3.33.

7.17 Determine the maximum equivalent von Mises stress for Problem 3.34.

7.18 Determine the maximum equivalent von Mises stress for Problem 3.35.

7.19 Determine the maximum Tresca and von Mises stresses for Problem 5.71.

7.20 Determine the maximum Tresca and von Mises stresses for Problem 5.72.

7.21 A 20-mm-diameter rod made of a ductile material with a yield strength of 350 MPa is subjected to a torque of 100 N·m and a bending moment of 150 N·m. An axial force is then gradually applied. Determine the value of the force when the rod begins to yield. Solve the problem two ways using the (*a*) Tresca theory and (*b*) von Mises theory.

7.22 A thin-walled cylindrical pressure vessel is made of a ductile material with a yield strength of 53 kpsi. The vessel has an outer diameter of 3 in and a wall thickness of 0.09375 in. What maximum internal pressure can the vessel withstand before yielding occurs?

7.23 In Chapter 5 the locations and magnitudes of the maximum radial and tangential stresses were given for a rotating disk. Determine the locations and magnitudes of the maximum shear and von Mises stresses.

7.24 to 7.27 A ductile material has a yield strength of 280 MPa. Determine the factors of safety corresponding to yielding for the state of plane stress given, according to the (*a*) Tresca theory and (*b*) von Mises theory.

Problem	σ_x	σ_y	τ_{xy}
7.24	140 MPa	140 MPa	0 MPa
7.25	105 MPa	105 MPa	105 MPa
7.26	0 MPa	0 MPa	140 MPa
7.27	−14 MPa	70 MPa	−56 MPa

7.28 to 7.31 A brittle material has the properties S_{UT} = 20 kpsi and S_{UC} = 80 kpsi. Determine the factors of safety corresponding to failure for the state of plane stress given, according to the (*a*) maximum normal stress theory and (*b*) Coulomb-Mohr theory.

Problem	σ_x	σ_y	τ_{xy}
7.28	12 kpsi	0 kpsi	−8 kpsi
7.29	−39 kpsi	−31 kpsi	3 kpsi
7.30	18 kpsi	10 kpsi	0 kpsi
7.31	12 kpsi	−40 kpsi	−15 kpsi

7.32 Problem 2.8 shows an element with octahedral surfaces symmetric to the principal stress axes. The normal stresses on the octahedral surfaces are equal to the average of the principal stresses that form a hydrostatic state of stress with no distortion. The shear stresses on these surfaces, called the *octahedral* *shear stresses,* are said to be the cause of distortion and failure. Problem 2.8 requests the proof that the octahedral shear stresses are given by

$$\tau_{\text{oct}} = \frac{1}{3}\sqrt{(\sigma_1 - \sigma_2)^2 + (\sigma_2 - \sigma_3)^2 + (\sigma_3 - \sigma_1)^2}$$

The *maximum octahedral shear stress theory* is based on equating τ_{oct} to that in a tensile specimen at yield. Show that this results in the same criterion as the von Mises theory.

7.33 to 7.36 Determine the maximum Tresca and von Mises stresses for the given stress matrix. Compare the von Mises stress with two times the Tresca stress.

7.33

$$[\sigma] = \begin{bmatrix} 20 & 10 & 10 \\ 10 & 20 & 10 \\ 10 & 10 & 20 \end{bmatrix} \text{MPa}$$

7.34

$$[\sigma] = \begin{bmatrix} -2 & 3 & -5 \\ 3 & 6 & 2 \\ -5 & 2 & -4 \end{bmatrix} \text{kpsi}$$

7.35

$$[\sigma] = \begin{bmatrix} 10 & 20 & 0 \\ 20 & 0 & -10\sqrt{2} \\ 0 & -10\sqrt{2} & 10 \end{bmatrix} \text{MPa}$$

7.36

$$[\sigma] = \begin{bmatrix} 1 & 2 & -2 \\ 2 & 4 & -4 \\ -2 & -4 & 4 \end{bmatrix} \text{MPa}$$

7.37 For Problem 7.21, let the rod be made of a brittle material with $S_{UT} = 340$ MPa and $S_{UC} = 600$ MPa and determine the value of the axial force when the rod fails using the Coulomb-Mohr theory.

7.38 Show that for the stresses given by Eqs. (7.4-10) the principal stresses are that given by Eqs. (7.4-15).

7.39 Show all the steps in the development of Eq. (7.4-16).

7.40 For a crack tip in a ductile material, show all the steps to prove that $\sigma_1 = 3S_Y$ for the case of plane strain, $v = 1/3$, and near the tip at $\theta = 0$.

7.41 A long bar, 4 in wide and 0.75 in thick, is loaded in tension. The material is steel with $K_{Ic} = 70$ kpsi $\cdot \sqrt{\text{in}}$, $S_Y = 160$ kpsi, and $S_U = 185$ kpsi. If the bar has a crack 0.625 in long on one edge, determine the maximum possible axial load that can be applied.

7.42 A cylindrical pressure vessel with $K_{Ic} = 80$ MPa $\cdot \sqrt{\text{m}}$, $S_Y = 1200$ MPa, and $S_U = 1350$ MPa has an outer diameter of 350 mm and a 25-mm wall thick-

ness. Determine the pressure which will cause a brittle fracture if a longitudinal crack of 25 mm exists through the wall thickness.

7.43 A beam of the configuration of case 4 of Table 7.4-1 is such that $a = 15$ mm, $b = 100$ mm, and is 25 mm thick. For the material, $K_{Ic} = 60$ MPa$\cdot\sqrt{m}$ and $S_U = 1200$ MPa. What value of M will cause fracture? What value of M will cause fracture if the direction is reversed?

7.44 When a crack appears in service, a technique used to arrest the crack is a *stop hole* drilled at the tip of the crack. Consider a crack of length a in a plate such as case 3 of Table 7.4-1, where the radius of the crack tip is almost zero. A stop hole of diameter d is then drilled, centered at the crack tip. Assuming the crack with the stop hole to be an ellipse, calculate the theoretical static stress concentration factor before and after drilling the stop hole. Determine d for a stress concentration of $K_t = 5$ if the crack is 25 mm long. HINT: The minimum radius of curvature of the ellipse of Fig. 7.4-1 must be determined.

7.45 Determine the fracture toughness of a material if it was tested in the form of case 1 of Table 7.4-1 with $a = 1$ in, $b = 20$ in, and the thickness $t = 0.75$ in. The load for which the plate failed was 250 kips. The yield strength of the material is 75 kpsi. Check for yielding, plane strain, and the size of the plastic zone at failure.

7.46 The bar shown is machined from steel with $S_Y = 420$ MPa and $S_U = 560$ MPa. The axial force F is completely reversing. Estimate the value of the force amplitude which will cause a failure at 100,000 cycles.

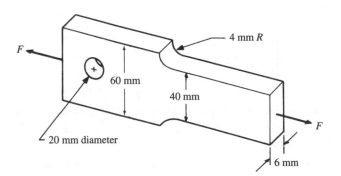

Problem 7.46

7.47 Estimate the number of cycles to failure if the magnitude of the force in Problem 7.46 is 50 kN.

7.48 If the axial force in Problem 7.46 is cycling from zero to a maximum value F_{max}, estimate the value of F_{max} for a life of 100,000 cycles.

7.49 The rotating shaft in Problem 3.33 is machined from steel, where $S_Y = 60$ kpsi and $S_U = 80$ kpsi. If the fatigue stress concentration factor near the pulleys is 1.8, determine the factor of safety of the shaft for an indefinite life based on (*a*) a static yield analysis and (*b*) a modified Goodman fatigue yield analysis.

7.50 A mechanical part is made of steel with the properties $S_U = 560$ MPa, $S_Y = 490$ MPa, and $S_E = 210$ MPa. The part is subjected to a bending stress that alternates between 100 and 200 MPa. Determine the equivalent reversing stress and evaluate the factor of safety corresponding to a life of 500,000 cycles.

7.51 Prove Eqs. (7.5-2) and (7.5-3).

7.52 A grooved steel shaft such as that shown in cases F.10 and F.11 in Appendix F is loaded by a purely reversing bending moment $M = 500$ N·m and a constant torsional moment of $T = 750$ N·m. Given $S_Y = 420$ MPa, $S_U = 560$ MPa, $D = 60$ mm, $d = 40$ mm, and $r = 4$ mm, determine the factor of safety for indefinite life based on (a) the Soderberg criterion and (b) the modified Goodman fatigue yield criterion.

7.53 Consider a column with fixed-free end conditions. If a compressive axial force is applied at the free end with an eccentricity e, determine the resulting lateral deflection and maximum compressive stress equations.

7.54 A long slender bar of rigidity EI and length L is pinned at each end to a very rigid foundation. If the coefficient of thermal expansion of the bar is α, determine the increase in temperature ΔT which will cause the bar to buckle.

7.55 A 500-mm-long column with pinned-pinned ends has a rectangular cross section $b \times h = 30 \times 50$ mm. The pins go through the 30-mm dimension. Consider that the column is fixed-fixed relative to buckling about the *weak* axis. The column is made of a material for which the compressive stress-strain curve is given by the data given. The data are linear to $\sigma = 294$ MPa. Determine the critical load for buckling.

ε (10^{-3})	0	1.1	1.2	1.3	1.4	1.5	1.6	1.8	2.0	2.2	2.5	2.8	3.2	3.6	4.0
σ, MPa	0	231	252	273	294	314.3	333.4	367.7	397.3	422.6	453	475.7	495.5	506	510

7.56 For Problem 7.55, similar to Fig. 7.6-5, plot Eq. (7.6-4) and Euler's equation for buckling in the pinned-pinned mode. Let $25 \le KL/r_g \le 200$.

7.57 A pinned-pinned column 50 in long is made of steel with $S_Y = 60$ kpsi and $E = 29$ Mpsi. The area of the cross section is 0.85 in² and the radius of gyration about the buckling axis is $r_g = 0.65$ in. For this case, plot Euler's and the parabolic equations, and determine the critical buckling force.

7.58 For Problem 7.57, determine the maximum load that can be applied if for the section $c = 1.2$ in and there is an eccentricity of $e = 0.5$ in.

7.59 For Problem 7.57, determine the maximum allowable load using Eq. (7.6-10).

7.60 Consider the plate shown in Fig. 7.6-10 to be square with $b = a$. Furthermore, let each edge be simply supported and loaded with a compressive load per unit length P. That is, $N_x = N_y = -P$, and $V_{xy} = V_{yx} = 0$. Determine the critical value of P.

7.61 Consider a thin plate loaded in tension and containing a centrally located hole [see Fig. 7.7-2(a)]. For the plate let $w = 100$ mm, $d = 20$ mm, $t = 10$ mm, and

$S_Y = 300$ MPa. Determine (a) the maximum axial force P_Y which can be applied if no yielding is allowed and (b) the limit load P_P if the plate material is *EPP*.

7.62 A solid circular shaft with a radius r_o is made of an *EPP* material and transmits a torsional moment T. Within the elastic range, the shear stress distribution is linear with respect to the radial position r ($\tau = Tr/J$). If the torsion is increased such that the outer fibers of the shaft yield, the stress distribution is no longer linear and behaves according to

$$\tau = \begin{cases} \dfrac{S_{YS}}{r_o - h_p}\, r & 0 \le r \le (r_o - h_p) \\ S_{YS} & (r_o - h_p) \le r \le r_o \end{cases}$$

where S_{YS} is the *yield strength in shear* and h_p is the depth of the plastic zone.
(a) Show that the depth of the plastic zone is given by

$$h_p = r_o - \sqrt[3]{4r_o^3 - \frac{6T}{\pi S_{YS}}}$$

(b) If $T = 6500$ lb·in, $r_o = 0.5$ in, and $S_{YS} = 28.85$ kpsi determine the residual shear stress distribution after the torsional moment is removed.

7.63 A square thin-walled box section with outer dimensions $b \times b$ has a constant wall thickness of $0.1b$. Determine the shape factor of the section if the bending axis is parallel to a side of the box. Is this value of the shape factor a "good" value? Explain your answer.

7.64 The cross section shown is transmitting a bending moment about the z axis. If the beam material is an *EPP* type with a yield strength of $S_Y = 40$ kpsi, determine (a) the maximum moment which can be transmitted by the cross section such that the bending stresses do not exceed the yield strength and (b) the shape factor of the cross section.

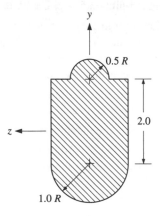

Problem 7.64   (All dimensions are in inches.)

7.65 Using a steel handbook (such as Ref. 7.21) select several WF beam sections and show that the bending shape factor $SF \approx 1.13$ for all.

7.66 Prove that the shape factor in bending is 1.27 for thin-wall circular sections.

7.67 Determine the shape factor in torsion for a solid circular bar in torsion.

7.68 Determine the shape factor in torsion for a thin-walled circular bar in torsion.

7.69 Determine the bending shape factor for Problem 5.51. Why is the SF larger than a straight rectangular beam in bending?

7.70 Given the cross section of Problem 3.10, with the thickness of each wall being 5 mm, determine the shape factor in bending.

7.71 Prove that the shape factor in bending of a beam with a circular cross section is 1.7.

7.72 Prove that the shape factor in bending of a beam with a diamond cross section, such as in Problem 3.15, is 2.0.

7.73 The yield strength of the *EPP* material of the section shown is $S_Y = 350$ MPa. Considering bending about the horizontal axis, determine the limit bending moment M_P and the shape factor.

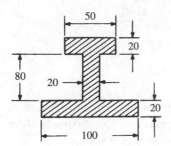

Problem 7.73   (All dimensions are in millimeters.)

7.74 A cantilever beam of rectangular cross section of depth $2c$ and width b is loaded such that $PL = 1.2M_Y$. If the material of the beam is *EPP* and the yield strength is $S_Y = 50$ kpsi, determine (*a*) the maximum theoretical depth of the plastic zone and (*b*) the minimum value for x for which the depth of the plastic zone is zero.

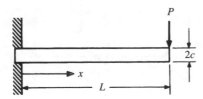

Problem 7.74

7.75 A steel beam with a rectangular cross section $b = 2$ in wide and $2c = 4$ in deep is loaded such that the bending moment is $1.3M_Y$. The material is *EPP* with $S_Y = 60$ kpsi. The load is then released. Determine (*a*) the stress distribution before the load is released, (*b*) the maximum strain before the load is released, and (*c*) the residual stress distribution after the load is released.

7.76 Four rivets each with a cross-sectional area of 250 mm² hold the plate as shown. Determine the maximum shear stress in each rivet.

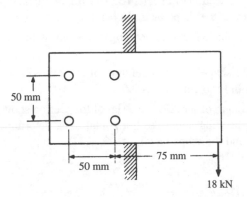

50 mm

50 mm

75 mm

18 kN

Problem 7.76

7.77 A plate is bolted to two walls as shown. Considering the moment and vertical force reactions at the walls, determine the magnitude of the maximum shear stresses in the bolts if they have a cross-sectional area of 0.4 in².

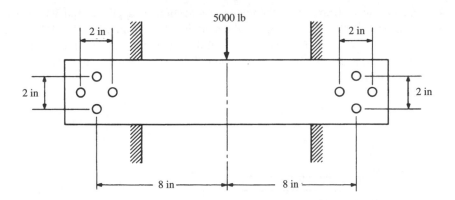

2 in

5000 lb

2 in

2 in

2 in

8 in

8 in

Problem 7.77

7.78 A gusset plate is welded to a base plate as shown. The base plate is secured to the foundation by four bolts, each with an effective cross-sectional area of 0.2 in^2. Before application of the 1000-lb force, the bolts were torqued down so that a preload of 5000 lb tension was developed in each bolt. Determine the maximum tensile stress that develops upon the application of the 1000-lb force. HINT: Assume that the plate is rigid and rotates about corner A.

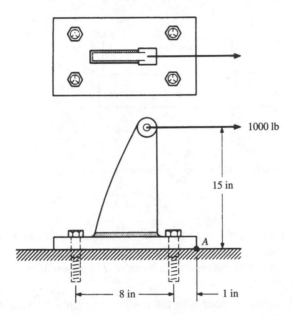

Problem 7.78

7.79 The shafts shown are transmitting 200 hp at 800 rpm. The shafts are connected by a coupling with eight bolts. The bolt circle diameters are 10 and 20 in.
- (*a*) Determine the minimum nominal shoulder diameter of a bolt if all bolts are to have the same diameter and not exceed a shear stress of 20 kpsi.
- (*b*) Determine the minimum number of bolts of the diameter determined in part (*a*) if the shear stress is not to exceed 20 kpsi and the bolts are to be placed only at the outer bolt circle.

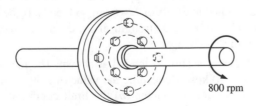

Problem 7.79

7.80 The bar loaded as shown is to be attached to a frame with bolts of equal diameters. The shear stress is not to exceed 120 MPa. Determine the minimum bolt diameter if (a) three bolts are to be used at points A, B, and C, and (b) two bolts are used at points A and C.

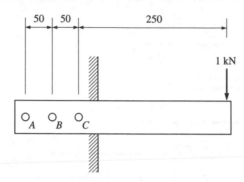

Problem 7.80   (All dimensions are in millimeters.)

7.81 Consider the cables in Problem 3.46 to be of an *EPP* material with $S_Y = 420$ MPa and equal diameters d. Assuming beam *ABCD* to be rigid and using a factor of safety of 2, determine the minimum value of d based on (a) no yielding; (b) a limit analysis.

7.82 For beam C.14 of Appendix C determine the limit value of w for collapse if the bending moment is limited to M_P. Use limit analysis and do not use any equations given in the appendix.

7.83 For beam C.13 of Appendix C, with $a = b = L/2$, determine the limit value of the force F for collapse if the bending moment is limited to M_P. Use limit analysis and do not use any equations given in the appendix.

7.84 For beam C.13 of Appendix C, with $a = b = L/2$, determine the maximum value of the force F such that no yielding occurs. For this, let the bending moment at yield be limited to $M_Y = 2/3 M_P$. Compare your result with that determined in Problem 7.83.

7.85 A beam has a rectangular cross section b wide by 2c deep and a span length L and is cantilevered at one end and simply supported at the other end. Three equal forces P are placed at equal intervals of L/4 from both ends. If the beam is made of an *EPP* material with a yield strength of S_Y, determine the limit value of P and show the resulting collapse mode.

7.86 A clamp with a rectangular cross section is loaded as shown. The material is *EPP* with a yield strength of 40 kpsi. Determine the value of P when (a) the highest stressed element in section a-a yields (b) the limit condition of the cross section is reached. NOTE: The neutral axis due to the combined axial and bending load across section *a-a shifts* when plastic behavior develops. Keep in mind that the stress distribution must always satisfy the state of equilibrium of the structure.

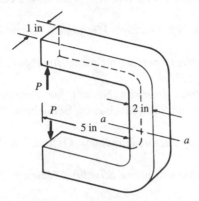

Problem 7.86

7.87 Three steel cables of the same diameter are to be used to support a static load of 100 kips as shown. Assuming the load to be rigid and the cable material to be *EPP* with $S_Y = 40$ kpsi, determine the minimum cross-sectional area of the cable based on (*a*) no yielding and (*b*) a limit analysis.

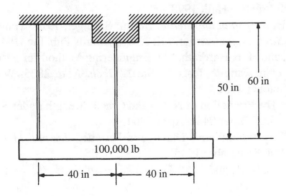

Problem 7.87

Computer Problems

7.88 For a pinned-pinned column, plot $(P/A)_{\text{allow}}$ in kpsi versus the slenderness ratio L/r_g using Eqs. (7.6-10). Let $0 \leq L/r_g \leq 200$, $S_Y = 36$ kpsi, and $E = 29$ Mpsi. Compare the plot with Fig. 7.6-9.

7.89 Using a software package, re-create the plot in Fig. 7.6-13.

7.10   REFERENCES

7.1   Johnson, R. C. "Predicting Part Failures," *Mach. Des.*, vol. 37, no. 1, pp. 137–42, January 1965; no. 2, pp. 157–62, January 1965.

7.2   Broek, D. *Elementary Engineering Fracture Mechanics*, 4th ed. London: Martinus Nijhoff, 1985.

7.3   Broek, D. *The Practical Use of Fracture Mechanics*, London: Kluwar Academic Pub., 1988.

7.4   Tada, H., and P. C. Paris. *The Stress Analysis of Cracks Handbook*, 2nd ed. St. Louis, MO: Paris Productions, 1985.

7.5   Sih, G. C. *Handbook of Stress Intensity Factors for Researchers and Engineers*, Institute of Fracture and Solid Mechanics. Bethlehem, PA: Lehigh University, 1973.

7.6   Murakami, Y., ed. *Stress Intensity Factors Handbook*. Oxford, UK: Pergamon Press, 1987.

7.7   Pilkey, W. D. *Formulas for Stress, Strain, and Structural Matrices*. New York: Wiley, 1994.

7.8   Irwin, G. R. "Fracture Mode Transition for a Crack Traversing a Plate." *J. Basic Eng.,* vol. 82, 1960, pp. 417–25.

7.9   Irwin, G. R. "Plastic Zone Near a Crack and Fracture Toughness." *Sagamore Research Conference Proceedings,* vol. 4, 1961.

7.10   Juvinall, R. C. *Stress Strain and Strength*. New York: McGraw-Hill, 1967.

7.11   Osgood, C. C. *Fatigue Design*. New York: Wiley, 1970.

7.12   Norton, R. L. *Machine Design: An Integrated Approach*, 2nd ed. Englewood Cliffs, NJ: Prentice Hall, 1998.

7.13   Neuber, H. *Kerbspannungslehre*. Berlin: Springer, 1937. Translation: *Theory of Notch Stresses*. Ann Arbor, MI: J. W. Edwards Pub. Inc, 1946.

7.14   Kuhn, P., and H. F. Hardrath. "An Engineering Method for Estimating Notch-Size Effect in Fatigue Tests on Steel." *Tech. Note* 2805, Washington, DC: NACA, October 1952.

7.15   Kuhn, P. "The Prediction of Notch and Crack Strength under Static or Fatigue Loading." *SAE Paper* 843c, April 1964.

7.16   Miner, M. A. "Cumulative Damage in Fatigue." *J. Appl. Mech.,* vol. 12, pp. A159–A164, September 1945.

7.17   Timoshenko, S. P., and J. M. Gere. *Theory of Elastic Stability*, 2nd ed. New York: McGraw-Hill, 1961.

7.18   Bazant, Z. P., and L. Cedolin. *Stability of Structures*. New York: Oxford University Press, 1991.

7.19   Salmon, C. G., and J. E. Johnson. *Steel Structures: Design and Behavior*, 4th ed. New York: Harper Collins, 1996.

7.20   Johnson, B. G. "Column Buckling Theory: Historical Highlights." *Journal of Structural Engineering*, Structural Division, American Society of Civil Engineers, vol. 109, no. 9, pp. 2086–96, September 1983.

7.21   *Manual of Steel Construction: Load and Resistance Factor Design*, vol. 1, 2nd ed. Chicago: American Institute of Steel Construction, 1994.

7.22   Van den Broek, J. A. *Theory of Limit Design*. New York: Wiley, 1948.

EIGHT

EXPERIMENTAL STRESS ANALYSIS

8.0   INTRODUCTION

Owing to the advent and prolific use of the numerical finite element method, the application of experimental techniques has diminished. However, when dealing with a complex stress analysis problem in which a complete theoretical solution may prove impractical with respect to time, cost, or degree of difficulty, experimental techniques are often employed. In the design phase, scale models can be analyzed using experimental methods, and in many cases design changes can be scrutinized quite effectively. If the structure is already in existence and its effectiveness or behavior under a change in loading specifications is to be determined, the structure can easily be studied by experimental methods. Even if a given problem is analyzed by a theoretical or numerical approach, it may be necessary to verify the results, in which case some experimentation may be required. Thus a knowledge of experimental methods is an essential part of an analyst's background.

Experimental stress analysis is a misnomer, as the techniques employed in this field generally involve the measurement of deflection or strain. Stress, then, is implicitly measured. The most widely used experimental stress analysis method employs the electrical-resistance *strain gage* and its associated instrumentation. However, other methods include transmission and reflection *photoelasticity, brittle coating, Moiré gratings, Moiré interferometry, x-ray diffraction, holographic* and *laser speckle interferometry,* and *thermoelastic stress analysis.* Because of space limitations, this chapter covers only the basics of strain gages and photoelasticity.[†] For details on the other techniques mentioned, consult Refs. 8.1 and 8.2.

---

† In the previous edition of this book, the *brittle coating method* was also described. However, owing to the problems associated with the materials used and environmental issues, the materials are no longer readily available.

Before strain gages and photoelasticity are discussed, dimensional analysis and analysis techniques are presented.

8.1   DIMENSIONAL ANALYSIS

When attempting to predict the stresses (or any state property) of a prototype design, it sometimes is necessary to construct and test a scale model, which may or may not be of the same material. The prototype may be so large that cost and practicality make testing of the actual prototype out of the question. Conversely, the prototype may be so small that test data are difficult to obtain. In addition, the prototype may be planned to be of a costly or unmanageable material. In either case, a different material for the model would be advantageous. Furthermore the method of testing may require special materials for the model, as in transmission photoelasticity a change in materials between the prototype and model is usually necessary.

Thus the questions arise: How to design the model and how does the prototype behave with respect to the results obtained from the testing of the model? These questions are resolved through the method of dimensional analysis.

For a structure which obeys Hooke's law it is only necessary to prescribe one state property, the stress matrix σ, and two material properties, E and v. With this, strain and displacements are derivable. The independent spatial variables are x, y, and z. The loads can be prescribed by one of the applied loads, say F, and the other forces by simply determining their ratios to F. Likewise, dimensions can be related by ratios to one of the dimensions of the structure, say L. Thus, if all the ratios are known, the fundamental quantities are σ, E, v, x, y, z, F, and L. The units of all of these quantities can be obtained from the units of force and length. These quantities can also be put in dimensionless form using F and L:

$$\frac{\sigma}{F/L^2} \qquad \frac{E}{F/L^2} \qquad v \qquad \frac{x}{L} \qquad \frac{y}{L} \qquad \frac{z}{L} \qquad \frac{F}{F} \qquad \frac{L}{L}$$

Since the last two quantities are obviously trivial, the independent dimensionless quantities are

$$\frac{L^2\sigma}{F} \qquad \frac{L^2E}{F} \qquad v \qquad \frac{x}{L} \qquad \frac{y}{L} \qquad \frac{z}{L}$$

For a given prototype, each quantity will have some value based on geometry, material, and the applied loading. For example, consider the quantity

$$\frac{L_p^2\sigma_p}{F_p} = K$$

where the subscript p denotes the prototype values and K is some dimensionless constant. In order to maintain similarity between the prototype and model, this quantity should be the same for the model. That is,

$$\frac{L_m^2 \sigma_m}{F_m} = K = \frac{L_p^2 \sigma_p}{F_p}$$

where the subscript m denotes the model values. Thus, in order to predict the prototype stress values from the results obtained from the model,

$$\sigma_p = \left(\frac{L_m}{L_p}\right)^2 \left(\frac{F_p}{F_m}\right) \sigma_m \qquad \textbf{[8.1-1]}$$

Let σ be the stress scale factor;[†] then

$$\sigma_p = \sigma \sigma_m$$

or

$$\sigma = \frac{\sigma_p}{\sigma_m} \qquad \textbf{[8.1-2]}$$

Likewise, let the force scale factor be

$$F = \frac{F_p}{F_m} \qquad \textbf{[8.1-3]}$$

and the length scale factor be

$$L = \frac{L_p}{L_m} \qquad \textbf{[8.1-4]}$$

Thus, it can be seen from Eqs. (8.1-1) to (8.1-4) that the stress factor is related to the force and length scale factors

$$\sigma = \frac{F}{L^2} \qquad \textbf{[8.1-5]}$$

A number of scale factors can be created which (except for v) can be described by the scale factors F and L; they are presented in Table 8.1-1.

---

† Here, the scale factor is defined differently than when normally referring to "models." Here we will define the scale factor as the factor which one multiplies the experimentally obtained value from the model to obtain the predicted prototype value.

Table 8.1-1

Scale Factor	Prototype Model	Force and Length Dependence
Force scale F	$F = \dfrac{F_p}{F_m}$	$F = F$
Length scale L	$L = \dfrac{L_p}{L_m}$	$L = L$
Stress scale σ	$\sigma = \dfrac{\sigma_p}{\sigma_m}$	$\sigma = \dfrac{F}{L^2}$
Modulus scale E	$E = \dfrac{E_p}{E_m}$	Function of materials
Moment scale M	$M = \dfrac{M_p}{M_m}$	$M = FL$
Pressure scale P	$P = \dfrac{P_p}{P_m}$	$P = \dfrac{F}{L^2}$
Strain scale ε	$\varepsilon = \dfrac{\varepsilon_p}{\varepsilon_m}$	$\varepsilon = \dfrac{\sigma}{E} = \dfrac{F}{L^2 E}$
Displacement scale δ	$\delta = \dfrac{\delta_p}{\delta_m}$	$\delta = \varepsilon L = \dfrac{F}{LE}$

Example 8.1-1 A half-scale photoelastic model is made to represent a prototype design. A single force of 8 kN will be applied to the proposed prototype.

(a) The model is loaded to 800 N, and the maximum stress is found to be 7 MPa. Determine the maximum stress in the prototype when loaded to 8 kN.

(b) If the modulus of elasticity of the photoelastic model is 3.5 GPa and the modulus of the prototype is 70 MPa, estimate the maximum deflection of the prototype if the maximum deflection of the model is 5 mm.

Solution:

(a) The length and force scale factors are

$$L = \frac{L_p}{L_m} = \frac{L_p}{0.5L_p} = 2 \qquad F = \frac{F_p}{F_m} = \frac{8(10^3)}{800} = 10$$

The stress scale is thus

$$\sigma = \frac{F}{L^2} = \frac{10}{2^2} = 2.5$$

and since $\sigma = \sigma_p/\sigma_m,$ then

$$\sigma_p = (2.5)(7) = 17.5 \text{ MPa}$$

(*b*) The modulus scale factor is

$$E = \frac{E_p}{E_m} = \frac{70}{3.5} = 20$$

This gives a deflection scale factor of

$$\delta = \frac{F}{LE} = \frac{10}{(2)(20)} = 0.25$$

and since $\delta = \delta_p/\delta_m$, then

$$\delta_p = \delta\delta_m = (0.25)(5) = 1.25 \text{ mm}$$

8.2 ANALYSIS TECHNIQUES

Stress in a solid cannot be measured directly and must be deduced from strain measurements and the stress-strain relations. In general, assuming cross shears to be equal, to determine the complete state of stress or strain at a point, six independent quantities must be determined. In free-surface measurements where the z axis is perpendicular to the surface, $\sigma_z = \tau_{yz} = \tau_{zx} = 0$, and the state of stress is that of plane stress. In this case only *three* independent measurements are necessary to establish the complete state of stress at a given point on the surface. Important problems and experimental techniques are not restricted to the state of plane stress. However most problems of interest involve the measurement of the plane stress state since to a large extent maximum-stress states tend to develop at free surfaces.

Considering only cases of plane stress from this point on, a further reduction in the number of independent measured quantities necessary to establish the state of stress can be made if additional information is known. If the directions of the axes of principal strains or stresses are known, it is only necessary to determine the *two* principal strains or stresses. Then, through the use of transformation equations, the state of stress or strain can be determined in any direction. The direction of the axes of the principal stresses can be found:

1. Along lines of symmetry.
2. When a surface perpendicular to the free surface exists without shear stress.
3. From a photoelastic analysis (see Sec. 8.7).

8.2.1 SYMMETRY

The shear stress in the direction of a line of complete symmetry is zero, and thus the line and its perpendicular establish the principal axes. For example, consider the ring loaded as shown in Fig. 8.2-1(*a*). The lines *a-a* and *b-b* are lines of symmetry. Thus $\tau_{xy} = 0$ along lines *a-a* and *b-b*.

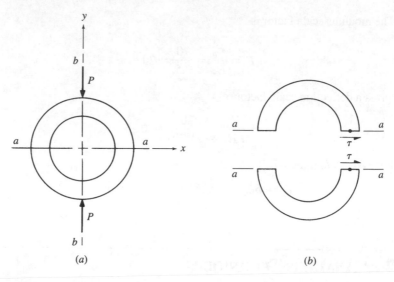

Figure 8.2-1

To understand why this is true, separate the structure along line *a-a*. If a shear stress is present on the top element at a given point on line *a-a*, then thanks to symmetry the shear stress will be present on the bottom element and in the same direction (see Fig. 8.2-1(*b*)). This disobeys Newton's law of action and reaction unless $\tau = 0$.

The same argument holds for line *b-b*, but because of the loading along the line, a sharp shear-stress gradient occurs and shear stresses develop quickly from line *b-b*.

8.2.2 WHEN A SURFACE PERPENDICULAR TO THE FREE SURFACE EXISTS WITHOUT SHEAR STRESS

This occurs at boundaries of plane-stress elastic problems (see Chap. 4) where thin platelike structures undergo in-plane loading. Consider Fig. 8.2-1 again; if the thickness is small, $\sigma_z = \tau_{yz} = \tau_{zx} = 0$ throughout. All surfaces along the boundary, except where loads exist, are also free surfaces where the normal and shear stresses are zero. Consider local *xy* coordinate systems established at points on a free surface such as $x_A\,y_A$ and $x_B\,y_B$ for points *A* and *B*, respectively, as shown in Fig. 8.2-2. The *y* axis for each point is positioned perpendicular to the free surface at the point. Since there is no shear stress on the free surface of an element at *A* and *B*, the $x_A\,y_A$ and $x_B\,y_B$ are axes of principal stress. Furthermore, since the normal stresses on the free surfaces are zero, only one stress at each point remains unknown; the normal stress in the local *x* direction. Thus at *A* and *B*, only one piece of information at each point needs to be determined.

If an external surface is exposed to a known pressure distribution, again only one piece of information is needed. For example, assume that the inner surface of the ring is experiencing a uniform normal pressure along the circumference of a value p. Thus $(\sigma_y)_B = -p$, and the element appears as shown in Fig. 8.2-3; since there is no shear stress on the inner surface, x_B and y_B remain principal axes. In

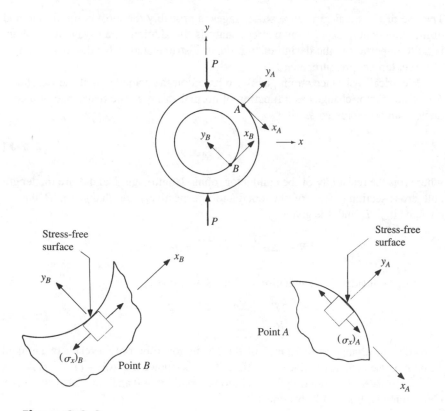

Figure 8.2-2

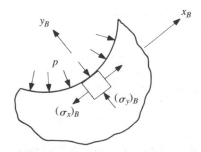

Figure 8.2-3

order to determine the state of stress at point B, again only one additional piece of information is needed, $(\sigma_x)_B$.

8.3   STRAIN GAGES, GENERAL

The use of electrical-resistance strain gages is probably the most common method of measurement in experimental stress analysis. In addition, strain-gage technology is quite important in the design of transducer instrumentation for the measurement of force, torque, pressure, etc.

Electrical-resistance strain gages are based on the principle that the resistance R of a conductor changes as a function of normal strain ε. The resistance of a conductor can be expressed as

$$R = \rho \frac{L}{A} \qquad \text{[8.3-1]}$$

where ρ is the resistivity of the conductor (ohms-length), and L and A are the length and cross-sectional area of the conductor, respectively. A change in R due to changes in ρ, L, and A is given by

$$\Delta R = \Delta \rho \frac{L}{A} + \rho \frac{\Delta L}{A} - \rho \frac{L}{A^2} \Delta A$$

Dividing both sides of the equation by Eq. (8.3-1) gives

$$\frac{\Delta R}{R} = \frac{\Delta \rho}{\rho} + \frac{\Delta L}{L} - \frac{\Delta A}{A} \qquad \text{[8.3-2]}$$

Consider the change in resistance of the conductor to be only a result of an axial strain ε on the conductor. Since $\varepsilon = \Delta L/L$, it can be shown that $\Delta A = (1 - v\varepsilon)^2 A - A = v^2\varepsilon^2 A - 2v\varepsilon A$. Considering the strain to be small, $\varepsilon^2 \approx 0$ and thus $\Delta A/A \approx -2v\varepsilon$. Consequently, Eq. (8.3-2) becomes

$$\frac{\Delta R}{R} = (1 + 2v)\varepsilon + \frac{\Delta \rho}{\rho} \qquad \text{[8.3-3]}$$

If the change in the resistance of the conductor is due only to the applied strain, Eq. (8.3-3) can be written as

$$\frac{\Delta R}{R} = S_A \varepsilon \qquad \text{[8.3-4]}$$

where

$$S_A = 1 + 2v + \frac{\Delta \rho/\rho}{\varepsilon} \qquad \text{[8.3-5]}$$

S_A is the *sensitivity* of the conductor to strain. The first two terms come directly from changes in dimension of the conductor where for most metals the quantity

$1 + 2\nu$ varies from 1.4 to 1.7. The last term in Eq. (8.3-5) is called the change in specific resistance, and for some metals can account for much of the sensitivity to strain. The most commonly used material for strain gages is a copper-nickel alloy called Constantan, which has a strain sensitivity of 2.1. Other alloys used for strain gage applications are modified Karma, Nichrome V, and Isoelastic, which have sensitivities of 2.0, 2.2, and 3.6, respectively. The primary advantages of Constantan are:

1. The strain sensitivity S_A is linear over a wide range of strain and does not change significantly as the material goes plastic.

2. The thermal stability of the material is excellent and is not greatly influenced by temperature changes when used on common structural materials.

3. The metallurgical properties of Constantan and modified Karma are such that they can be processed to minimize the error induced due to the mismatch in the thermal expansion coefficients of the gage and the structure to which it is adhered over a wide range of temperature. For further discussion of this, see Sec. 8.6.3.

Isoelastic, with a higher sensitivity, is used for dynamic applications. Semiconductor gages are also available and can reach sensitivities as high as 175. However, care must be exercised with respect to the poor thermal stability of these piezoresistive gages.

Most gages have a nominal resistance of 120-ohm or 350-ohm. Considering a 120-ohm Constantan gage, to obtain a measurement of strain within an accuracy of $\pm 5\ \mu$, it would be necessary to measure a change in resistance within ± 1.2 m-ohm. To measure these small changes in resistance accurately, the Wheatstone bridge is used (see Sec. 8.5).

Metallic alloy electrical-resistance strain gages used in experimental stress analysis come in two basic types, bonded-wire and bonded-foil (see Fig. 8.3-1). Today, bonded-foil gages are by far the more prevalent. The resistivity of

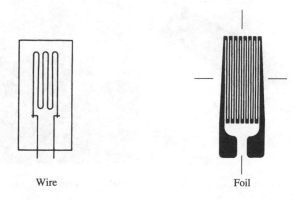

Wire   Foil

Figure 8.3-1

Constantan is approximately 49 μ-ohm $\cdot$ cm. Thus if a strain gage is to be fabricated using a wire 0.025 mm in diameter and is to have a resistance of 120 ohm, the gage would require a wire approximately 120 mm long. To make the gage more compact over a shorter active length, the gage is constructed with many loops as shown in Fig. 8.3-1. Typical commercially available bonded-foil gage lengths vary from 0.20 mm (0.008 in) to 101.6 mm (4.000 in). For normal applications, bonded-foil gages either come mounted on a very thin polyimide film carrier (backing) or are encapsulated between two thin films of polyimide. Other carrier materials are available for special applications.

The most widely used adhesive for bonding a strain gage to a test structure is the pressure-curing methyl-2-cyanoacrylate cement. Other adhesives include epoxy, polyester, and ceramic cements. Extreme care must be exercised when installing a gage as a good bond and an electrically insulated gage are necessary. The installation procedures can be obtained from technical instruction bulletins supplied by the manufacturer. Installations with and without strain-relief loops are shown in Fig. 8.3-2. Once a gage is correctly mounted, wired, resistance tested for continuity and insulation from the test structure, and waterproofed (if appropriate), it is ready for instrumentation and testing.

The remainder of the part of this chapter dedicated to strain gages provides information on strain gage configurations including strain gage rosettes, strain gage instrumentation, and characteristics of strain gage measurements. Appendix G provides equations to determine the principal strains and stresses and their orientation from strains obtained from three-element strain gage rosettes. Appendix H gives correction equations to account for the effects of strains perpendicular (transverse) to the measurement axis of strain gage rosettes (see also Sec. 8.6.2).

At the time of this writing, useful information, as well as *calculators* which perform many of the calculations described in this chapter and the aforementioned

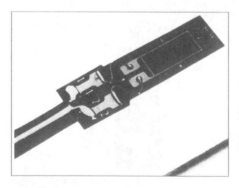

(*a*) With strain relief loops   (*b*) Without strain relief loops

Figure 8.3-2   Examples of strain gage installations. (Courtesy of *Micro-Measurements Division of Measurements Group, Inc.,* Raleigh, NC.)

appendixes, are available on the Internet courtesy of *Micro-Measurements Group, Inc.*, of Raleigh, North Carolina, and the Web location is

www.measurementsgroup.com/guide/index.htm

8.4 STRAIN-GAGE CONFIGURATIONS

In both wire or foil gages many configurations and sizes are available. In wire form, gage lengths ranging from 1/16 to 8 in can be obtained, whereas foil gage lengths range from 0.008 to 4 in. Strain gages come in many forms for transducer or stress-analysis applications. The fundamental configurations for stress-analysis work are shown in Fig. 8.4-1.

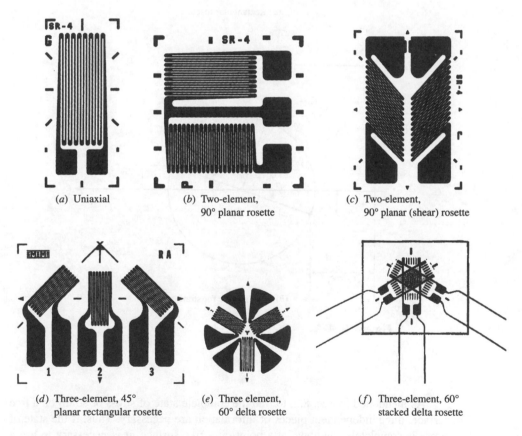

(*a*) Uniaxial

(*b*) Two-element, 90° planar rosette

(*c*) Two-element, 90° planar (shear) rosette

(*d*) Three-element, 45° planar rectangular rosette

(*e*) Three element, 60° delta rosette

(*f*) Three-element, 60° stacked delta rosette

Figure 8.4-1 Examples of commonly used strain gage configurations (SOURCE: Figures *a–c* courtesy of *BLH Electronics, Inc.*, Canton, MA. Figures *d–f* courtesy of *Micro-Measurements Division of Measurements Group, Inc.*, Raleigh, NC.) NOTE: The letters *SR-4* on the *BLH* gages are in honor of E. E. Simmons and Arthur C. Ruge and their two assistants (a total of four individuals), who in 1937–1938, independently produced the first bonded-wire resistance strain gage.

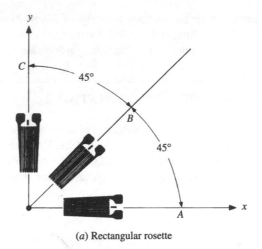

(*a*) Rectangular rosette

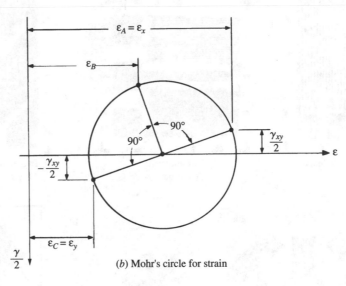

(*b*) Mohr's circle for strain

Figure 8.4-2

As mentioned in Sec. 8.2, to find the complete state of stress or strain on a free surface, three independent pieces of information are necessary. Thus, if the state of stress is completely unknown at a point on a free surface, it is necessary to use a three-element rectangular or delta rosette since *each element* provides only one piece of information, the indicated normal strain at the point in the direction of the gage.

The three-element rectangular rosette provides normal-strain components in three directions spaced at angles of 45°, as shown in Fig. 8.4-2(*a*). If an *xy* coordinate system is assumed to coincide with gages A and C, then $\varepsilon_x = \varepsilon_A$ and $\varepsilon_y = \varepsilon_C$. Gage B in conjunction with gages A and C provides information necessary to determine γ_{xy}. Using the strain-transformation equation (3.9-11) with $\theta = 45°$, we can show that

$$\varepsilon_B = \frac{1}{2}(\varepsilon_x + \varepsilon_y + \gamma_{xy}) = \frac{1}{2}(\varepsilon_A + \varepsilon_C + \gamma_{xy})$$

Thus

$$\gamma_{xy} = 2\varepsilon_B - \varepsilon_A - \varepsilon_C$$

Once ε_x, ε_y, and γ_{xy} are known, we can use Hooke's law [Eqs. (1.4-5) and (1.4-8a)] to determine the stresses σ_x, σ_y, and τ_{xy}.

The relationship between ε_A, ε_B, and ε_C can be seen from the Mohr's circle of strain corresponding to the strain state at the point under investigation (see Fig. 8.4-2(b)).

A three-element rectangular rosette strain gage is mounted on a steel specimen. For a particular state of loading of the structure the strain gage readings are[†]

Example 8.4-1

$$\varepsilon_A = 200\ \mu \qquad \varepsilon_B = 900\ \mu \qquad \varepsilon_C = 1000\ \mu$$

Determine the values and orientations of the principal stresses and the value of the maximum shear stress at the point. Let $E = 200$ GPa and $v = 0.285$.

Solution:

$$\varepsilon_x = \varepsilon_A = 200\ \mu \qquad \varepsilon_y = \varepsilon_C = 1000\ \mu$$

$$\gamma_{xy} = 2\varepsilon_B - \varepsilon_A - \varepsilon_C = (2)(900) - 200 - 1000 = 600\ \mu$$

The stresses can be determined using Eqs. (1.4-5) and (1.4-8a), giving

$$\sigma_x = \frac{E}{1-v^2}(\varepsilon_x + v\varepsilon_y) = \frac{200(10^9)}{1-(0.285)^2}[200 + (0.285)(1000)](10^{-6})$$

$$= 105.6(10^6)\ \text{N/m}^2 = 105.6\ \text{MPa}$$

$$\sigma_y = \frac{E}{1-v^2}(\varepsilon_y + v\varepsilon_x) = \frac{200(10^9)}{1-(0.285)^2}[1000 + (0.285)(200)](10^{-6})$$

$$= 230.1(10^6)\ \text{N/m}^2 = 230.1\ \text{MPa}$$

[†] The strain gage readings are typically corrected owing to the effect of transverse strains on each gage (see Sec. 8.6.2 and Appendix H).

$$\tau_{xy} = \frac{E}{2(1+v)} \gamma_{xy} = \frac{200(10^9)}{2(1+0.285)} 600(10^{-6}) = 46.7(10^6) \text{ N/m}^2 = 46.7 \text{ MPa}$$

For the orientation of the principal stress axes we use Eq. (3.9-7), which is

$$\theta_p = \frac{1}{2} \tan^{-1}\left(\frac{2\tau_{xy}}{\sigma_x - \sigma_y}\right) = \frac{1}{2} \tan^{-1}\left(\frac{2(46.7)}{105.6 - 230.1}\right) = -18.4°, \ 71.6°$$

Substituting $\theta_p = -18.4°$ into Eq. (3.9-4) gives

$$\sigma = \frac{\sigma_x + \sigma_y}{2} + \left(\frac{\sigma_x - \sigma_y}{2}\right) \cos 2\theta_p + \tau_{xy} \sin 2\theta_p$$

$$= \frac{105.6 + 230.1}{2} + \left(\frac{105.6 - 230.1}{2}\right) \cos[2(-18.4)]$$

$$+ (46.7)\sin[2(-18.4)]$$

$$= 90.0 \text{ MPa}$$

Substituting $\theta_p = 71.6°$ into Eq. (3.9-4) yields $\sigma = 245.7$ MPa. For plane stress, the third principal stress is zero. Since the third principal stress normal to the gage is zero, for $\sigma_1 > \sigma_2 > \sigma_3$, then $\sigma_1 = 245.7$ MPa oriented $71.6°$ counterclockwise from the A gage, $\sigma_2 = 90$ MPa oriented $18.4°$ clockwise from the A gage, and $\sigma_3 = 0$ normal to the gages.

The maximum shear stress is given by Eq. (7.3-4) and is

$$\tau_{max} = \frac{\sigma_1 - \sigma_3}{2} = \frac{245.7 - 0}{2} = 122.8 \text{ MPa}$$

and exists on surfaces with normals oriented $\pm45°$ from the normals of the surfaces containing σ_1 and σ_3.

Appendix G provides equations to determine the principal strains and stresses and their orientation directly from strains obtained from three-element strain gage rosettes.

8.5 STRAIN-GAGE INSTRUMENTATION

8.5.1 THE WHEATSTONE BRIDGE

As mentioned earlier, the Wheatstone bridge is the primary circuit used with strain gage measurements. Other circuits are sometimes used such as the potentiometer circuit in special dynamic applications. Only the basic operation of the Wheatstone bridge will be discussed in this section. For more detail on specialized applications refer to the references cited earlier.

The Wheatstone bridge, a circuit sensitive to small resistance changes, is shown in Fig. 8.5-1. A dc voltage V is applied across contacts ac. It can be shown that the resulting voltage across contacts bd is given by

$$E = \frac{R_1R_3 - R_2R_4}{(R_1 + R_2)(R_3 + R_4)} V \qquad \textbf{[8.5-1]}$$

When measuring changes in the output voltage or current, the procedure is to balance the initial resistances so that the nominal output voltage E is zero. Thus, for initial balance

$$R_1R_3 = R_2R_4 \qquad \textbf{[8.5-2]}$$

With the circuit balanced, small changes in the resistances will cause a corresponding change in the output voltage. Substituting $R_1 + \Delta R_1$, $R_2 + \Delta R_2$, etc., into Eq. (8.5-1) will yield the change in voltage ΔE. After some algebraic manipulation if all but the first-order terms in $\Delta R/R$ are neglected (since $\Delta R/R$ is quite small), it can be shown that (see Problem 8.20)

$$\Delta E \approx Vr\left(\frac{\Delta R_1}{R_1} - \frac{\Delta R_2}{R_2} + \frac{\Delta R_3}{R_3} - \frac{\Delta R_4}{R_4}\right) \qquad \textbf{[8.5-3]}$$

where

$$r = \frac{R_1R_2}{(R_1 + R_2)^2} = \frac{R_3R_4}{(R_3 + R_4)^2}$$

For specific applications where maximum sensitivity is desired, there is some flexibility in adjusting the ratios of R_1/R_2, and the analyst can design a specific Wheatstone bridge circuit.

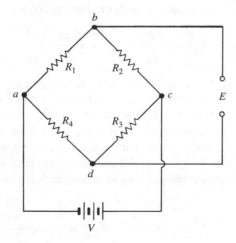

Figure 8.5-1

From Eq. (8.5-3) it can be seen that the output voltage is proportional to resistance change in the bridge circuit. If only one gage of resistance R_g is used in the bridge so that $R_1 = R_g$, then $\Delta R_1 = \Delta R_g$ and $\Delta R_2 = \Delta R_3 = \Delta R_4 = 0$. Thus the voltage change is given by

$$\Delta E = Vr\frac{\Delta R_g}{R_g}$$

If the strain sensitivity of the gage is S_A, then[†]

$$\frac{\Delta R_g}{R_g} = S_A\varepsilon \qquad\qquad \textbf{[8.5-4]}$$

it follows that

$$\Delta E = VrS_A\varepsilon \qquad\qquad \textbf{[8.5-5]}$$

Thus the output voltage is directly proportional to the strain on the gage, and the strain indicator can be calibrated directly to strain.

Each resistance position of the Wheatstone bridge circuit is called an *arm*. Equation (8.5-3) can be used to advantage by employing more than one arm of the bridge for strain gages. Using one gage on one arm is called a *quarter-bridge* circuit, two gages on adjacent arms is called a *half-bridge* circuit, and four gages on all arms is a *full-bridge* circuit. The half- or full-bridge circuits are typically used for temperature compensation or for transducer applications.

As will be discussed in the next section, temperature changes on a strain gage installation can cause a change in resistance. Thus if the temperature changes occur during a strain gage measurement, erroneous readings will be obtained if not corrected. Consider the R_1 resistance in the bridge to be a strain gage of resistance R_g installed on a specimen. If a temperature change ΔT occurs during the measurement of strain ε, the unit change in resistance R_1 can be expressed as

$$\frac{\Delta R_1}{R_1} = \frac{\Delta R_g}{R_g} = S_A\varepsilon + \left(\frac{\Delta R_g}{R_g}\right)_{\Delta T} \qquad\qquad \textbf{[8.5-6]}$$

where the last term in the equation is the unit change in resistance due to the temperature change (see Sec. 8.6.3). The effect of the last term in Eq. (8.5-6) on the output voltage of Eq. (8.5-3) can be canceled by mounting an identical gage on a block of material identical to the specimen, placing this installation in the same temperature environment as the specimen (including an identical lead-wire routing), and attaching this second gage to the R_2 (or R_4) position of the circuit. This forms a half-bridge circuit, with the unit change in the R_2 resistance being

$$\frac{\Delta R_2}{R_2} = \left(\frac{\Delta R_g}{R_g}\right)_{\Delta T} \qquad\qquad \textbf{[8.5-7]}$$

[†] The sensitivity of a strain gage is actually represented by a term called the gage factor S_g. This is discussed in detail in Sec. 8.6.2.

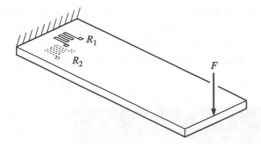

Figure 8.5-2   Strain gage transducer application.

Substituting Eqs. (8.5-6) and (8.5-7) into Eq. (8.5-3) with $\Delta R_3 = \Delta R_4 = 0$ results in

$$\Delta E = VrS_A\varepsilon$$

which is the same as Eq. (8.5-5). Thus it can be seen that the effect of the temperature change has been canceled (compensated). The R_1 gage is called the *active* gage, whereas the R_2 gage is called the inactive *compensating,* or *dummy* gage.

Half- or full-bridge arrangements can be used in transducer applications with temperature compensation where all the gages are active. Consider, for example, a cantilever beam with a force applied to the free end with two identical strain gages one directly below the other as shown in Fig. 8.5-2. Attach the top gage to the R_1 position of the bridge and the bottom gage to the R_2 (or R_4) position of the bridge. As before, since the gages are on adjacent arms of the bridge, temperature compensation is achieved. Furthermore, since the strain on the bottom gage is equal and opposite in sign to the strain on the top gage, then from Eq. (8.5-3) $\Delta E = 2VrS_A\varepsilon$. This yields a gain of 2 in the output of the circuit. Similarly, if an additional gage is mounted adjacent to the R_1 gage and placed in the R_3 position in the circuit, and an additional gage is mounted adjacent to the R_2 gage and placed in the R_4 position in the circuit, the full-bridge circuit will be temperature compensated and yield an output gain of 4. Many other arrangements of strain gages in the Wheatstone bridge can provide unique transducer applications. However, since this text is concerned with the use of strain gages for stress analysis purposes, the reader is urged to consult a reference on mechanical measurements for more detail on transducer applications.

8.5.2   COMMERCIAL STRAIN GAGE INDICATOR SYSTEMS

Early commercial strain gage indicators for static strain measurements employed the Wheatstone bridge using a *null-balance* method. Since no resistors are perfect, additional circuitry was required in order to adjust ΔE to zero (null) prior to loading the specimen. As the strain changed during testing it was necessary to readjust the circuit to null ΔE each time a reading was made. The circuit was designed such that the null adjustment dial readout was directly in units of strain. This allowed for maximum sensitivity and linearity of the circuit as the measurements were taken. Today, more convenient direct-reading indicators are produced, many of which utilize the "unbalanced" Wheatstone bridge system. The circuit description of the various strain gage indicator systems available is beyond the scope of this section.

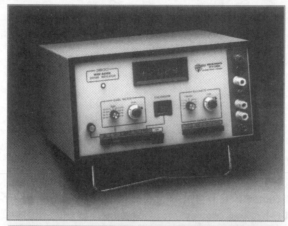

(*a*) Measurements Group wide-range, laboratory-type Model 3800

(*b*) Measurements Group portable Model P-3500

Figure 8.5-3   Strain indicators. (Courtesy of *Measurements Group, Inc.,* Raleigh, NC.)

However, what is important is that the Wheatstone bridge is only linear for sufficiently small strains, and corrections in strain measurements must be made when exceeding the small strain limitations. For strain measurement corrections of commercial Wheatstone bridge systems, see the manufacturer's specifications.

Figure 8.5-3 shows strain indicators produced by a major producer of strain-gage instrumentation. *Switch and balance* units are also available for multiple strain-gage installations and can be seen in Fig. 8.5-4. Usually these units come with 10 independent strain gage channels, which can be individually balanced prior to testing, and the channel can be directed to the strain indicator through manual switching.

Large-scale computer-based data acquisition systems are also available where up to 1200 channels can be scanned at intervals as short as 0.1 second. Figure 8.5-5 shows Measurements Group, Inc., System 5000 Stress Analysis Data System. The software is capable of automatically making the many necessary corrections due to bridge nonlinearity and the additional factors which are discussed in Sec. 8.6.

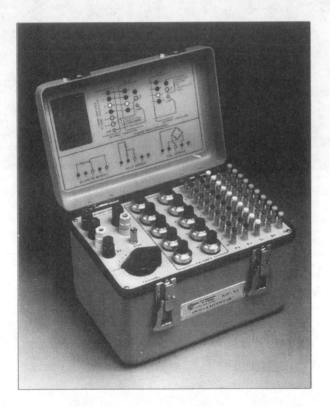

Figure 8.5-4   Switch and balance unit. Measurements Group Model SB-10. (Courtesy of *Measurements Group, Inc.*, Raleigh, NC.)

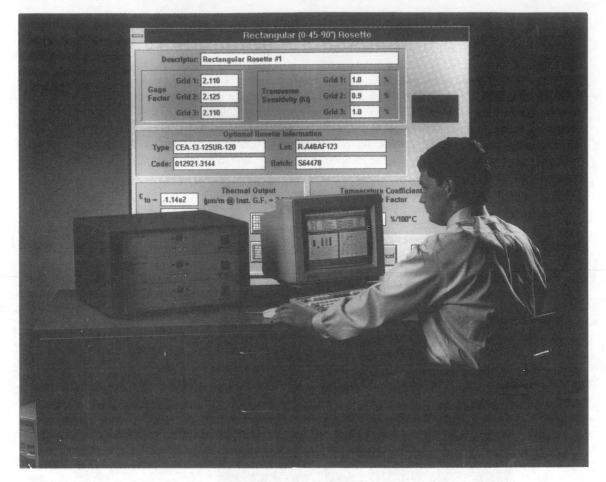

Figure 8.5-5   Measurements Group System 5000 Stress Analysis Data System. (Courtesy of *Measurements Group, Inc.*, Raleigh, NC.)

8.6   CHARACTERISTICS OF STRAIN-GAGE MEASUREMENTS

Some of the fundamental measurement characteristics which influence the output of a strain gage are discussed in this section. The following effects are considered:

1. Linearity of the gage material.
2. Transverse sensitivity of the gage.
3. Temperature effects.
4. Lead-wire connection.

5. Strain gradient.
6. Zero shift and hysteresis effects.
7. Dynamic response.
8. Gage-current heating effects.
9. Noise from electric and/or magnetic fields.

8.6.1 LINEARITY OF THE GRID MATERIAL

As mentioned earlier, Constantan is linear over a wide range of strain. However, if higher sensitivity is desired, as in dynamic measurements, the isoelastic material is commonly used. Isoelastic is linear at a value of $S_A = 3.6$ for strains up to about 7500 μ. Beyond this point S_A drops from 3.6 to 2.5. This must be accounted for if testing goes beyond 7500 μ.

8.6.2 TRANSVERSE SENSITIVITY OF THE GAGE

The strain sensitivity of a single straight uniform length of conductor in a uniform *uniaxial strain field* ε in the longitudinal direction of the conductor is defined as

$$S_A = \frac{\Delta R/R}{\varepsilon}$$ [8.6-1]

In a general strain field there will be strains perpendicular to the longitudinal axis of the conductor *(transverse strains)*. Owing to the width of the conductor and the geometric configuration of the conductor in the gage pattern, the transverse strains will also effect a change in resistance in the conductor. This is not desirable, as only the effect of the strain in the direction of the gage length is expected.

To further complicate things, the sensitivity of the strain gage provided by the gage manufacturer is *not* based on a uniaxial strain field, but on a uniaxial *stress* field from a tensile test specimen. For a uniaxial stress field let the axial and transverse strains be ε_a and ε_t, respectively. The sensitivity provided by the gage manufacturer, called the *gage factor* S_g, is defined as $S_g = (\Delta R/R)/\varepsilon_a$, where under a uniaxial stress field $\varepsilon_t = -v_0\varepsilon_a$. Thus

$$\frac{\Delta R}{R} = S_g\varepsilon_a \quad \text{with} \quad \varepsilon_t = -v_0\varepsilon_a$$ [8.6-2]

The term v_0 is Poisson's ratio of the material on which the manufacturer's gage factor was measured and is normally taken to be 0.285. If the gage is used under conditions where the transverse strain is $\varepsilon_t = -v_0\varepsilon_a$, then the equation $\Delta R/R = S_g\varepsilon_a$ would yield exact results. If $\varepsilon_t \neq -v_0\varepsilon_a$, some error will occur. This error depends on the sensitivity of the gage to transverse strain and the deviation of the ratio of $\varepsilon_t/\varepsilon_a$ from $-v_0$.

Consider a gage installed in a general biaxial strain field ε_a, ε_t. Let the sensitivities of the gage in the axial and transverse directions be S_a and S_t, respectively, where the change in resistance is given by

$$\frac{\Delta R}{R} = S_a \varepsilon_a + S_t \varepsilon_t$$

The ratio S_t/S_a is defined as the *transverse sensitivity coefficient* K_t. Thus

$$\frac{\Delta R}{R} = S_a(\varepsilon_a + K_t \varepsilon_t) \qquad \text{[8.6-3]}$$

In addition to the gage factor, the strain gage manufacturer generally supplies the value of K_t for each gage element.

Substituting $\varepsilon_t = -v_0 \varepsilon_a$ into Eq. (8.6-3) yields

$$\frac{\Delta R}{R} = S_a(1 - v_0 K_t)\varepsilon_a \qquad \text{with} \qquad \varepsilon_t = -v_0 \varepsilon_a \qquad \text{[8.6-4]}$$

Comparing Eqs. (8.6-2) and (8.6-4) we see that

$$S_g = S_a(1 - v_0 K_t) \qquad \text{[8.6-5]}$$

If the gage is not mounted in a uniaxial stress field and Eq. (8.6-2) is used, an error will develop. To see this, let ε_a and ε_t be the actual strains and $\hat{\varepsilon}_a$ be the strain measurement based on Eq. (8.6-2). Equating Eqs. (8.6-2) and (8.6-3) yields

$$S_g \hat{\varepsilon}_a = S_a(\varepsilon_a + K_t \varepsilon_t) \qquad \text{[8.6-6]}$$

Substituting Eq. (8.6-5) and solving for $\hat{\varepsilon}_a$ gives

$$\hat{\varepsilon}_a = \frac{\varepsilon_a + K_t \varepsilon_t}{1 - v_0 K_t} \qquad \text{[8.6-7]}$$

Let the error in the strain measurement based on $\hat{\varepsilon}_a$ be $e = \hat{\varepsilon}_a - \varepsilon_a$. Then from Eq. (8.6-7) the error is

$$e = \frac{K_t}{1 - v_0 K_t}(v_0 \varepsilon_a + \varepsilon_t) \qquad \text{[8.6-8]}$$

We see that the error is zero if K_t is zero or $\varepsilon_t/\varepsilon_a = -v_0$.

Example 8.6-1 | A strain gage with a transverse sensitivity factor of 5 percent is mounted on a specimen which at the strain gage location has a strain of $\varepsilon_a = 500\ \mu$ and $\varepsilon_t = 2500\ \mu$. Determine the indicated strain $\hat{\varepsilon}_a$.

Solution:

With $K_t = 0.05$, Eq. (8.6-7) gives

$$\hat{\varepsilon}_a = \frac{500 + (0.05)2500}{1 - (0.285)(0.05)} = 634 \ \mu$$

Thus we see an error of $+134 \ \mu$, or $+27$ percent.

In order to determine the error and correct a strain gage reading at a point, at least two independent strain measurements must be taken. The method of obtaining the correction equations is given in Appendix H as well as the correction equations for commonly used strain gage rosettes. For example, consider the two-element $90°$ strain gage rosette shown in Fig. 8.6-1 mounted on a test specimen. From Appendix H the equations for the corrected strains are

$$\varepsilon_x = \frac{(1 - v_0 K_{tA})\hat{\varepsilon}_A - (1 - v_0 K_{tB})K_{tA}\hat{\varepsilon}_B}{1 - K_{tA}K_{tB}} \qquad \textbf{[8.6-9a]}$$

$$\varepsilon_y = \frac{(1 - v_0 K_{tB})\hat{\varepsilon}_B - (1 - v_0 K_{tA})K_{tB}\hat{\varepsilon}_A}{1 - K_{tA}K_{tB}} \qquad \textbf{[8.6-9b]}$$

where ε_x, ε_y are the corrected strains; $\hat{\varepsilon}_A$, $\hat{\varepsilon}_B$ are the measurement strains from gages A and B, respectively; K_{tA}, K_{tB} are the transverse sensitivity factors of gages A and B, respectively; and v_0 is normally taken as 0.285.

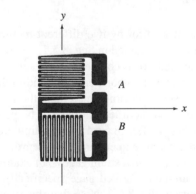

Figure 8.6-1

8.6.3   TEMPERATURE EFFECTS

Temperature changes on an installed strain gage effect a change in resistance which is due to a mismatch in the thermal expansion coefficients of the gage and the specimen, a change in the resistivity of the gage material, and a change in the gage factor S_g.

The strain on the gage due to a mismatch in the thermal expansion coefficients of the gage and the specimen is given by

$$\varepsilon_{\Delta(\alpha,T)} = (\alpha_s - \alpha_g)\,\Delta T \qquad\qquad \textbf{[8.6-10]}$$

where α_s and α_g are the thermal coefficients of expansion of the specimen and gage materials, respectively. This applies to both the axial and transverse strains. Substituting Eq. (8.6-10) for ε_a and ε_t into Eq. (8.6-3) with $S_a = S_g/(1 - v_0 K_t)$ yields

$$\left(\frac{\Delta R}{R}\right)_{\Delta(\alpha,\Delta T)} = S_g\,\frac{1 + K_t}{1 - v_0 K_t}(\alpha_s - \alpha_g)\,\Delta T \qquad\qquad \textbf{[8.6-11]}$$

As stated, a temperature change will also cause a change in resistivity. Let this be given by

$$\left(\frac{\Delta R}{R}\right)_{\Delta(\rho,\Delta T)} = \gamma\,\Delta T \qquad\qquad \textbf{[8.6-12]}$$

where γ is the temperature coefficient of resistivity of the gage material. The combined effects of these changes produce the overall change in resistance due to temperature change and is

$$\left(\frac{\Delta R}{R}\right)_{\Delta T} = \left[S_g\,\frac{1 + K_t}{1 - v_0 K_t}(\alpha_s - \alpha_g) + \gamma\right]\Delta T \qquad\qquad \textbf{[8.6-13]}$$

where the apparent strain due to a change in temperature is

$$\varepsilon_{\Delta T} = \left[\frac{1 + K_t}{1 - v_0 K_t}(\alpha_s - \alpha_g) + \frac{\gamma}{S_g}\right]\Delta T \qquad\qquad \textbf{[8.6-14]}$$

This effect can be compensated for by two different methods. The first method of temperature compensation is discussed in Sec. 8.5.1, where temperature compensation is achieved using an additional compensating gage in the Wheatstone bridge circuit. The second method involves compensation within the gage itself.

For application on specific specimen materials, the metallurgical properties of alloys such as Constantan and modified Karma can be processed to minimize the term within the [ ] brackets of Eq. (8.6-14) over a limited range of temperatures, somewhat centered about room temperature. Gages processed in this manner are called *self-temperature-compensated* strain gages. An example of the characteristics of a BLH self-temperature-compensated gage specifically processed for use on a low-carbon steel is shown in Fig. 8.6-2. Note that the apparent strain is zero at 22 and 45°C and approximately zero in the vicinity of these temperatures. For temperatures beyond this region, compensation can be achieved by monitoring the temper-

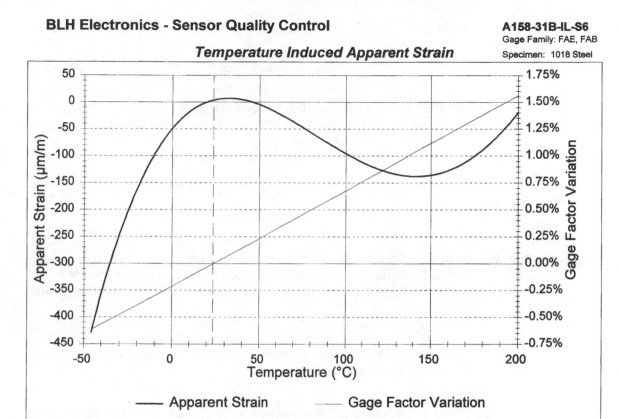

BLH Electronics - Sensor Quality Control

A158-31B-IL-S6
Gage Family: FAE, FAB

Temperature Induced Apparent Strain

Specimen: 1018 Steel

— Apparent Strain   — Gage Factor Variation

Apparent Strain = -48.85 + 3.86 T - 7.85E-02 T² + 4.05E-04 T³ - 5.28E-07 T4   04/24/97

Figure 8.6-2   Strain-gage temperature characteristics. (SOURCE: Data sheet courtesy of *BLH Electronics, Inc., Canton, MA.*)

ature at the strain gage site. Then, using either the curve from the data sheet or the fitted polynomial equation, the strain readings can be corrected numerically. Note, however, the curve and the polynomial equation given on the data sheet are based on a gage factor of 2.0. If corrections are anticipated, the gage factor adjustment of the strain indicator should be set to 2.0. An example which demonstrates this correction is given at the end of this section.

The gage factor variation with temperature is also presented in the data sheet of Fig. 8.6-2. If the strain gage indicator is initially set at $(S_g)_i$, the actual gage factor at temperature T is $(S_g)_T$, and the indicator registers a strain measurement of $\varepsilon_{\text{reading}}$, the corrected strain is

$$\varepsilon_{\text{actual}} = \frac{(S_g)_i}{(S_g)_T} \varepsilon_{\text{reading}} \qquad \textbf{[8.6-15]}$$

where

$$(S_g)_T = \left[1 + \frac{\Delta \dot{S}_g(\%)}{100}\right](S_g)_i \qquad \textbf{[8.6-16]}$$

and $\Delta S_g(\%)$ is the percent variation in gage factor given in Fig. 8.6-2.

If a simultaneous correction for apparent strain and gage factor variation is necessary, the corrected strain is given by

$$\varepsilon_{\text{actual}} = \frac{(S_g)_i}{(S_g)_T}(\varepsilon_{\text{reading}} - \varepsilon_{\text{apparent}}) \qquad \textbf{[8.6-17]}$$

Example 8.6-2 | A strain gage with the characteristics of Fig. 8.6-2 has a room temperature gage factor of 2.1 and is mounted on a 1018 steel specimen. A strain measurement of -1800 μ is recorded during the test when the temperature is 150°C. Determine the value of actual test strain if

(a) the gage is in a half-bridge circuit with a dummy temperature compensating gage and prior to testing the indicator is zeroed with the gage factor set at 2.1.

(b) the gage is the only gage in a quarter-bridge circuit and prior to testing the indicator is zeroed with the gage factor set at 2.0.

Solution:

From Fig. 8.6-2, the gage factor variation at 150°C is $\Delta S_g(\%) = 1.13$ percent. Thus from Eq. (8.6-16) the gage factor at the test temperature is

$$(S_g)_T = \left(1 + \frac{1.13}{100}\right)(2.1) = 2.124$$

(a) Since in this part a dummy gage is present which cancels the apparent strain, the only correction that is necessary is due to the change in the gage factor. From Eq. (8.6-15)

$$\varepsilon_{\text{actual}} = \frac{2.1}{2.124}(-1800) = -1780 \ \mu$$

which we see is a minor correction.

(b) In this part we must use Eq. (8.6-17). Using the equation given in Fig. 8.6-2 the apparent strain at the test temperature is

$$\varepsilon_{\text{apparent}} = -48.85 + (3.86)(150) - (7.85E\text{-}02)(150)^2$$

$$+ (4.05E\text{-}04)(150)^3 - (5.28E\text{-}07)(150)^4 = -136.5 \ \mu$$

Substituting this into Eq. (8.6-17), with $(S_g)_i = 2.0$, gives

$$\varepsilon_{actual} = \frac{2.0}{2.124}[-1800 - (-136.5)] = -1566\mu$$

which is *not* a minor correction.

8.6.4   LEAD-WIRE CONNECTION

Figure 8.3-2 shows examples of strain gage installations. Notice that a three-wire connection is used where two wires are soldered to one of the tabs on the strain gage. For single gage installations, the three-wire quarter-bridge circuit is recommended. Figure 8.6-3 shows a two- and three-wire quarter-bridge circuit where the lead-wire resistance is represented by R_L. The two-wire circuit is inherently unbalanced. If small enough, this imbalance can be offset by the strain indicator. However, the additional resistance in the arm produces a loss in sensitivity and the linear output range. A more serious consequence, however, may result if the lead wires experience temperature changes during the measurement process. The change in resistance for copper lead wires is approximately 22 percent for a 100°F (55°C) temperature change. For example, if a 120-ohm strain gage is installed with

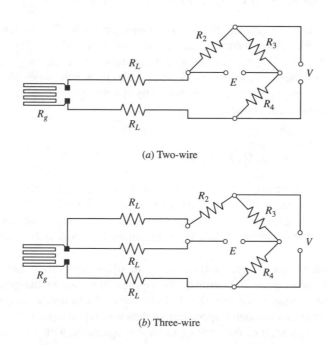

(a) Two-wire

(b) Three-wire

Figure 8.6-3   Two- and three-wire quarter-bridge circuits.

a pair of 20 ft (6 m) AWG 26 (0.4 mm diameter) copper lead wires, and the temperature changes by 10°F (5.5°C), an apparent strain of approximately 150 μ will be measured. The use of a 350-ohm gage will reduce these effects but will not eliminate them.

The three-wire quarter-bridge circuit shown in Fig. 8.6-3(*b*) is inherently balanced and will not cause any apparent strain due to temperature changes provided the wires experience identical temperature variations. There still will be a loss in sensitivity in the circuit; however, the loss will be 50 percent of that obtained in the two-wire circuit.

8.6.5   STRAIN GRADIENT

Since the gage is of finite length, the change in resistance is due to the *average* strain along the gage and not the center strain in general. If the strain along the gage is constant or linearly varying, the average strain is the center strain. However, for any other case, the average strain will differ from the center strain. If the strain gradient or change in strain along the gage is small, the error will be small, but if there is a large change in strain over a small distance, the gage should be as small as possible to minimize the error.

8.6.6   ZERO SHIFT AND HYSTERESIS EFFECTS

When a new gage goes through the first reversal-strain cycle, the resistance will not return to its initial value, thereby causing a zero shift as well as a deviation in linearity, or hysteresis. These factors can cause significant errors in interpreting the results. This is primarily due to cold working of the gage, which causes changes in resistivity. These effects are reduced significantly if a new gage installation is cycled about five times before taking data.

8.6.7   DYNAMIC RESPONSE

In dynamic-strain measurements there is a small lag time before the gage responds to a strain wave; this is due to the intermediate material, namely, the adhesive and carrier. The response time is of the order of 0.1 μs, but this valve has not been verified because of the inherent difficulty in measuring this extremely small response time.

A second, and more important, characteristic of dynamic-strain measurement is the fact that since a gage is measuring the average strain across the grid, there are measurement distortions of the actual strain wave. To illustrate this, consider a strain pulse traveling through a specimen as shown in Fig. 8.6-4(*a*). The magnitude of the pulse is ε_p with time duration t_p, and it is propagating with a velocity c. Thus the effective length of the pulse is ct_p.

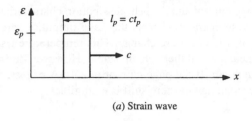

(a) Strain wave

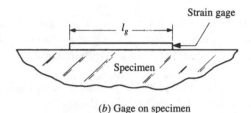

(b) Gage on specimen

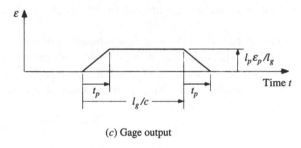

(c) Gage output

Figure 8.6-4

As the wave enters the gage area, the gage output is zero. It begins to increase linearly (since averaging is basically an integrating method) until the full wave is within the gage length, where the average strain would be $(l_p/l_g)\varepsilon_p$. The time this takes is t_p. The average strain across the gage remains constant at this value until the wave begins to leave the gage area. The total time this takes is l_g/c. As the wave leaves the gage area, the gage output again drops to zero in a linear fashion at the same rate of entry. Thus the output shape, magnitude, and time duration are distorted due to the wavelength vs. gage length. The smaller the gage, the less distortion in the output will be observed.

8.6.8 GAGE-CURRENT HEATING EFFECTS

Although this effect is primarily related to the instrumentation circuitry, the gage current alters the gage behavior. The gage is a resistor R, and since the gage current I produces heat according to the I^2R law, there is a temperature rise on the gage which depends on the gage current and the heat-dissipation characteristics of the installation. Since the gage temperature affects the gage output, a zero shift of the gage will occur if there is a temperature change. This temperature rise reaches an equilibrium point, and the gage will then remain stable. However, during the warm-up time, large errors can occur. If the gage current is 25 mA or less, the warm-up time, in general, is quite small and the zero shift is negligible.

8.6.9   NOISE FROM ELECTRIC AND/OR MAGNETIC FIELDS

Electric and/or magnetic fields can superimpose electrical noise on strain gage measurement signals. Some sources of noise are in close proximity to ac power lines, motors, transformers, relays, fluorescent lamps, radio transmitters, and atmospheric electrical disturbances.

Electrostatic noise can be reduced by eliminating *ground loops* (more than one connection of the system to ground) and shielding the lead wires of the strain gages. Special braided wire or conductive foil cable is available for this purpose. The shielded wire should be grounded only through the bridge ground to avoid a ground loop.

Magnetic noise can be reduced by *equalizing* the effect of the magnetic field on the gage and lead wires. This can be accomplished through the use of twisted-conductor cables, gages with closely spaced tabs, and direct soldering of the lead wires to the gage tabs as shown in Fig. 8.3-2(*b*). Flat ribbon cable should be avoided when magnetic fields are present.

8.7   THE THEORY OF PHOTOELASTICITY

Photoelastic analysis is a very powerful tool for both educational purposes and applications. Beyond understanding the basic photoelastic phenomenon the method of analysis is quite straightforward. However, one can become so fully engrossed in the phenomenon when first introducing the theory that the methods of actually using the technique become unclear and often misunderstood. To avoid becoming bogged down in a complex mathematical development of the phenomenon, a rather loose (but physically accurate) presentation of the photoelastic effect will be given in this section.

To begin to understand photoelasticity, it is necessary to review the properties of light, as based on Maxwell's electromagnetic wave theory, and the polarization and refraction of light.

8.7.1   ELECTROMAGNETIC WAVE REPRESENTATION OF LIGHT

A monochromatic light source emits light rays of one particular wavelength λ, which propagates at the speed of light c. Assuming the general case, this emission of light rays propagates in many directions from the source. Consider each ray to be made up of a series of waves, where each wave can be thought of as a vector varying sinusoidally with time and position. One such wave, shown in Fig. 8.7-1, has wavelength λ and amplitude A_1 and is propagating at speed c in the z direction. This wave can be described by

$$a_1(t, z) = A_1 \sin \frac{2\pi}{\lambda}(z - ct) \qquad \textbf{[8.7-1]}$$

If the observation of the wave is fixed at some point along the z axis, the wave can be described as

$$a_1(t) = A_1 \sin \frac{2\pi}{\lambda} ct \qquad \textbf{[8.7-2]}$$

Thus the light amplitude at some point along the z axis can be thought of as a vector whose magnitude varies with time between the values of $\pm A_1$, as shown in Fig. 8.7-2. Since the period of a complete cycle is λ/c, the frequency of the wave is c/λ.

Because each wave, a_1, a_2, etc., can be thought of as a vector, each wave can be divided into two perpendicular components. For example, the wave a_1 of Fig. 8.7-2 has components in the x and y directions, which are, respectively,

$$a_{1x}(t) = A_1 \cos \alpha \sin \frac{2\pi}{\lambda} ct \qquad \textbf{[8.7-3a]}$$

$$a_{1y}(t) = A_1 \sin \alpha \sin \frac{2\pi}{\lambda} ct \qquad \textbf{[8.7-3b]}$$

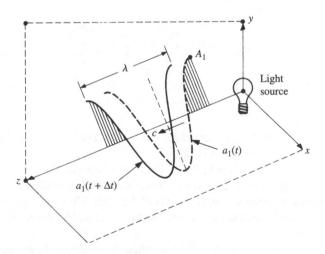

Figure 8.7-1

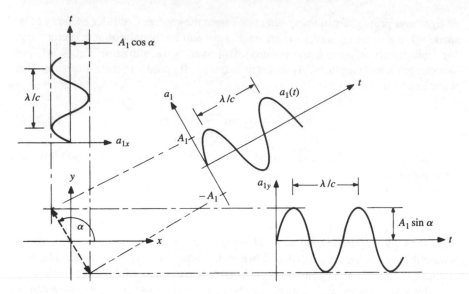

Figure 8.7-2

Thus the wave $a_1(t)$ can be considered as the vector sum of two waves oscillating in phase in the x and y directions propagating in the z direction.

8.7.2   POLARIZATION

Consider an arbitrary wave $a(t)$. The components of $a(t)$ in the x and y direction can be written

$$a_x(t) = A_x \sin \frac{2\pi}{\lambda} ct \qquad\qquad \text{[8.7-4a]}$$

$$a_y(t) = A_y \sin \frac{2\pi}{\lambda} ct \qquad\qquad \text{[8.7-4b]}$$

where A_x and A_y are the magnitudes of the components of the wave in the x and y directions, respectively (see Fig. 8.7-3).

One of the resultant waves in Fig. 8.7-3 can be canceled by placing a polarization filter in the light path. A polarization filter can almost completely absorb a light wave perpendicular to the polarization axis of the filter. Thus if the polarization axis is arbitrarily aligned with the y direction, most of $A_x(t)$ will be absorbed and most of $A_y(t)$ will be transmitted.

The filter which polarizes the light is called the *polarizer P*. If a second polarization filter is placed in the light path, the light polarized by the first filter will be completely absorbed when the polarization axes of the two filters are perpendicular to each other (crossed), as shown in Fig. 8.7-4.

The second filter is called the *analyzer A*. If the analyzer axis is at an angle less than 90° with respect to the polarizer axis, some light will be transmitted. That is, the component of the polarized wave along the analyzer axis will be transmitted. If the analyzer and polarizer axis are aligned (parallel), most of the polarized light will be transmitted through the analyzer.

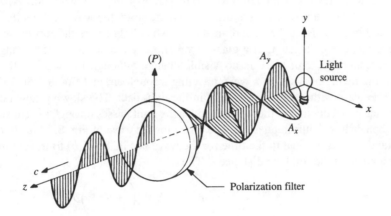

Figure 8.7-3

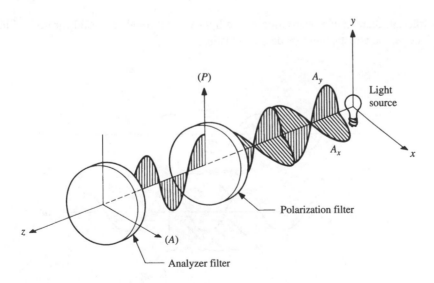

Figure 8.7-4

8.7.3   REFRACTION

When light travels from one medium into another, the speed of wave propagation changes. The relative index of refraction n_r, of a substance is defined as the index of refraction of the substance relative to surrounding medium, e.g., air, and is the ratio of the speed c_a in the surrounding medium to the speed c_s in the substance. Thus

$$n_r = \frac{c_a}{c_s} \qquad\qquad \text{[8.7-5]}$$

When a wave travels from air into a transparent solid, the wave slows down. However, when the wave reenters the air from the solid, the wave regains its original speed in air. This can be seen dramatically when light enters the substance at an oblique angle, e.g., a pencil in a glass of water. On the other hand, if the angle of light incidence is zero, there is no visible change although the wave still slows down. Figure 8.7-5(a) shows a wave traveling through air, and Fig. 8.7-5(b) shows the same wave with a transparent solid in the light path. The slowing down of the wave causes a reduction in the wavelength as the light passes through the substance. This then causes a shift in phase Δ. The time t for the wave of Fig. 8.7-5(a) to travel a distance $d + \Delta$ is equal to the time for the wave of Fig. 8.7-5(b) to travel through the thickness of the solid the distance d. Thus

$$t = \frac{d + \Delta}{c_a} = \frac{d}{c_s} \qquad \text{or} \qquad \Delta = \left(\frac{c_a}{c_s} - 1\right)d$$

Since $n_r = c_a/c_s$,

$$\Delta = (n_r - 1)d \qquad\qquad \text{[8.7-6]}$$

Although there is this retardation of the light wave through the solid for normal incidence, the effect cannot be detected visibly.

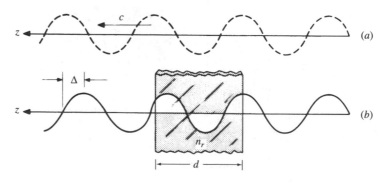

Figure 8.7-5

8.7.4 BIREFRINGENCE

For some transparent materials, the index of refraction varies with respect to the in-plane angle for which polarized light transmits through the material. The principal axes of minimum and maximum refraction are perpendicular and are called the *fast* and *slow* axes, respectively. Materials which behave in this manner are called *doubly refracting* or *birefringent*. Figure 8.7-6 shows a disk made of a birefringent material, where (n_1, c_1) and (n_2, c_2) are the (relative index of refraction, speed of light transmission) along principal axes 1 and 2, respectively. If $n_2 > n_1$, then $c_1 > c_2$ where axis 1 is the fast axis and axis 2 is the slow axis. If light enters the birefringent plate where the light is polarized in the direction of either the fast or slow axis as shown in Fig. 8.7-6, the light will emerge from the plate polarized as when enter-

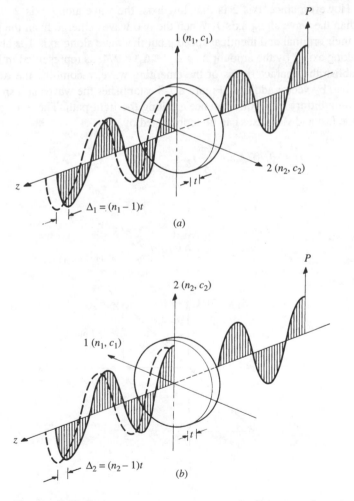

Figure 8.7-6

ing but retarded by either $\Delta_1 = (n_1 - 1)t$ or $\Delta_2 = (n_2 - 1)t$, where t is now used to represent the plate thickness, not time. If $n_2 > n_1$, the phase shifts are such that $\Delta_2 > \Delta_1$. If the polarization axis of the entering light is at an angle relative to the 1,2 axes, the light will have components along each axis and the form of the emerging light will be complex as the emerging components will be out of phase by $\Delta_2 - \Delta_1$.

As a special case, assume that the birefringent plate is such that $\Delta_2 - \Delta_1 = \lambda/4$. This means that there is a quarter of a wavelength difference in retardation between the slow and fast axes, and the plate is called a *quarter-wave plate*. Also assume that the wave is polarized at an angle $\beta = 45°$ with respect to the fast and slow axes. The light entering the plate is illustrated in Fig. 8.7-7.

As before, the wave of magnitude A can be divided into components A_1 and A_2 along axes 1 and 2, respectively. This results in two waves each of magnitude $0.707A$ and each traveling in phase. As the waves enter the plate, they both slow down. However, since axis 2 is the slow axis, the wave along axis 2 slows down more than the wave along axis 1. When the two waves emerge from the plate, they regain their original and identical speed c but the wave along axis 1 is ahead of the wave along axis 2 by the amount $\Delta = \Delta_2 - \Delta_1 = \lambda/4$, as represented in Fig. 8.7-8. To establish the characteristics of the emerging wave, recombine the two component waves by vector addition. Figure 8.7-8 establishes the waves at a specific time and at an arbitrary point of reference z_1 along the light path. The emerging waves along the fast and slow axes can be represented by

$$a_1 = 0.707A \sin \frac{2\pi}{\lambda} z_1 \qquad\qquad \textbf{[8.7-7a]}$$

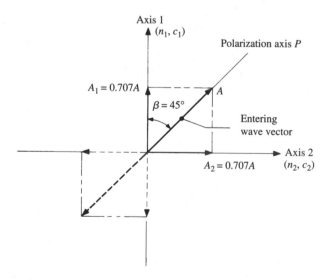

Figure 8.7-7

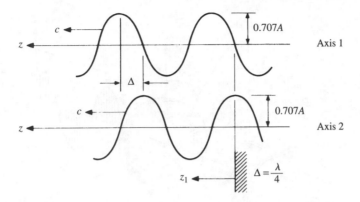

Figure 8.7-8   Light wave vector components emerging from the quarter-wave plate.

$$a_2 = 0.707A \sin \frac{2\pi}{\lambda}\left(z_1 + \frac{\lambda}{4}\right) = 0.707A \cos \frac{2\pi}{\lambda} z_1 \qquad \textbf{[8.7-7b]}$$

The final wave is the vector addition of the waves of Eqs. (8.7-7). The amplitude of the emerging wave is

$$A_f = \sqrt{a_1^2 + a_2^2} = \sqrt{\left(0.707A \sin \frac{2\pi}{\lambda} z_1\right)^2 + \left(0.707A \cos \frac{2\pi}{\lambda} z_1\right)^2}$$

$$= 0.707\,A \qquad\qquad \textbf{[8.7-8]}$$

which is constant. Let the angle that the final wave vector makes with axis 2 be ϕ, where $\phi = \tan^{-1}(a_1/a_2)$. Thus

$$\phi = \tan^{-1}\left[\frac{0.707A \, \sin(2\pi/\lambda)z_1}{0.707A \, \cos(2\pi/\lambda)z_1}\right] = \frac{2\pi}{\lambda} z_1 \qquad \textbf{[8.7-9]}$$

This means that the wave vector emerging from the quarter-wave plate has a constant magnitude of $0.707A$ and is rotating one complete revolution (2π) per wavelength. A representation of the emerging wave vector is shown in Fig. 8.7-9. The light emerging from the quarter-wave plate is called *circularly polarized light*.

If the analyzer filter A is placed in the light path with its polarization axis perpendicular to the polarizer P, the components of the circularly polarized light will be transmitted and observed through the analyzer. Without the quarter-wave plate in the light path, virtually no light is observed through the analyzer when it is perpendicular to the polarizer. Thus this special case of a doubly refracting material illustrates how light can be transmitted through crossed polarization filters.

If the relative wave shift of a birefringent plate Δ is an integral multiple of λ, that is, $\Delta = \lambda, 2\lambda, 3\lambda, \ldots$, the component waves leaving the plate will be in phase

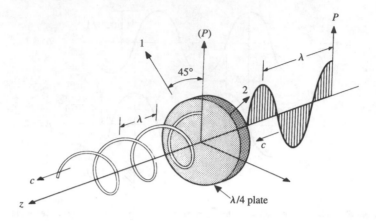

Figure 8.7-9   Circularly polarized light.

regardless of the value of β and their vector combination will produce a wave identical to the wave entering the birefringent plates. Thus, in this case, with the polarization filters crossed, no light will be observed since the birefringent plate does not change the light characteristics. If Δ is a half order of λ, that is, $\Delta = \lambda/2, 3/2\lambda, 5/2\lambda,$ ..., it can be shown that the emerging wave is identical to the entering wave except that it is perpendicular to the polarization axis. Then if the analyzer is perpendicular to the polarizer, the full light wave will be transmitted through the analyzer (minus slight transmission losses).

Returning to the quarter-wave plate, if the angle β between the polarizer axis and axis 1 of the quarter-wave plate is different from 45°, the waves along axis 1 and 2 will have different amplitudes A_1 and A_2. The emerging wave is rotating as before, but instead of being circular the wave will be elliptical with a value of A_1 along axis 1 and A_2 along axis 2, where $A_1 \neq A_2$. If the angle β between the polarizer and axis 1 is 0 or 90°, the wave will not have components along one of the retardation axes regardless of the value of Δ. Hence the emerging wave remains polarized in the same direction as when it entered, and if the analyzer is crossed with the polarizer, again no light will be transmitted. Thus, when $\beta = 0$ or 90° or when $\Delta = N\lambda$ ($N = 0, 1, 2, 3, \ldots$), no light will be transmitted through the analyzer. These two cases are the most important concepts in understanding photoelasticity.

8.7.5   STRESS AND BIREFRINGENCE

When stressed, certain transparent materials behave in a birefractive manner, where the values of the principal indices of refraction n_1 and n_2 are directly related to the principal strains. However, since the principal strains are directly related to the principal stresses, we will relate the refraction behavior directly to the stresses. For two-dimensional photoelasticity, we will consider the birefractive material to be a thin plate of constant thickness t and in a state of plane stress where at a point the in-plane principal stresses are σ_1 and σ_2, and the third principal stress perpendicular to the plate is zero.[†] The difference in the indices of refraction n_1 and n_2 is directly re-lated to the difference in the principal stresses. That is,

$$n_1 - n_2 = k(\sigma_1 - \sigma_2) \qquad\qquad \textbf{[8.7-10]}$$

where axes 1 and 2 refer to the principal-stress axes and k is called the relative stress-optic coefficient, a material property. If t is the thickness of a specimen made of this type of material, then from Eq. (8.7-6)

$$n_1 = \frac{\Delta_1}{t} + 1 \qquad\qquad \textbf{[8.7-11a]}$$

and

$$n_2 = \frac{\Delta_2}{t} + 1 \qquad\qquad \textbf{[8.7-11b]}$$

Therefore

$$n_1 - n_2 = \frac{\Delta_1 - \Delta_2}{t} = \frac{\Delta}{t} \qquad\qquad \textbf{[8.7-12]}$$

where Δ is the relative shift in phase between the waves emerging from the speci-men along axes 1 and 2. Combining Eqs. (8.7-10) and (8.7-12) yields

$$\sigma_1 - \sigma_2 = \frac{\Delta}{kt} \qquad\qquad \textbf{[8.7-13]}$$

When $\Delta = N\lambda$, where $N = 0, 1, 2, 3, \ldots$, there is a full-wave shift which theoreti-cally results in no net change in the wave. Thus, when the model is between two crossed polarization filters (called a *plane polariscope,* see Fig. 8.7-10) and $\Delta = N\lambda$ for integer values of N, no light is transmitted. Therefore, one condition for the ex-tinction of light is

$$\sigma_1 - \sigma_2 = \frac{N\lambda}{kt} \qquad N = 0, 1, 2, 3, \ldots \qquad \textbf{[8.7-14]}$$

Equation (8.7-14) is the basic photoelastic equation which enables one to quantify the stress difference $\sigma_1 - \sigma_2$ at points where light is not transmitted. Notice that the extinction of light is also a function of the wavelength λ except when $N = 0$. This ex-plains why whenever white light is directed through a photoelastic model, various colored bands or fringes will be seen. Each specific color represents the extinction of a particular wavelength. Due to the relative-retardation effect, only the $N = 0$ fringe

---

[†] Here we are ignoring the ordering $\sigma_1 > \sigma_2 > \sigma_3$.

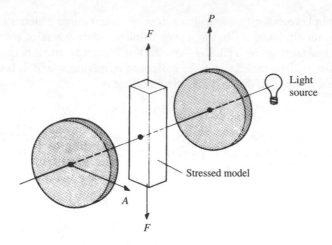

Figure 8.7-10   Plane polariscope.

will be black since it is independent of λ (all wavelengths extinguish for $N = 0$). The remainder of these fringes will be like colors in the rainbow. Thus the entire collection of fringes (including $N = 0$) is called *isochromatic fringes*. The technique for counting fringe orders will be discussed shortly.

There is a second condition for the extinction of light. Recall from the discussion of the quarter-wave plate that when axes 1 or 2 align with the polarizer and the analyzer is crossed, no light is transmitted. This applies to the stressed material as well. Thus, when the principal stresses *align* with the polarizer and the analyzer is crossed, light is extinguished. Again, bands or fringes appear where light is extinguished. These bands, however, are always black regardless of whether or not a white light source is used. In addition, the sensitivity of this effect is relatively small, so that the fringes are usually wider than the isochromatic fringes. The fringes which develop due to the alignment of either principal stress with the polarizer are called *isoclinic fringes* since all points on a specific fringe have principal stresses in the same direction of the polarizer. If the polarizer and analyzer are rotated together relative to the stressed model and their crossed orientation is maintained, the isoclinic fringes change since only points whose stresses align with the polarizer extinguish the light. Thus, to determine the directions of the principal stresses at a point, it is a simple task then to rotate the polarizer/analyzer filter set until an isoclinic fringe intersects the point and record the angular orientation of the filters.

To recapitulate; when the polarizer and analyzer filters are crossed and a thin stressed birefringent material is inserted between the filters, two sets of fringes will be observed:

1. *Isochromatic fringes*   These are dependent on the stress-optic effect. Whenever a zero or multiple of a full-wave shift of a particular wavelength occurs, the light wave is extinguished. This behavior is according to Eq. (8.7-14). If white light is di-

rected through the model, only the zero fringe ($N = 0$) is black and the remaining isochromatic fringes are colored. Orientation of the crossed filter set is independent of this effect.

2. *Isoclinic fringes* These fringes depend on the orientation of the crossed filter set. For a particular angular position of the crossed filter set, a series of wide black fringes appear on the model. The principal stresses for every point within the fringes are roughly in the same direction as the polarizer and analyzer. This is approximate, as there is no sharp cutoff of light. Normally the center of the band is used when recording an isoclinic fringe. These fringes are always black, regardless of the wavelength of the light source. The fringes move as the filter set is rotated relative to the model. If all fringes go through the same point regardless of the orientation of the filters, the point is an *isotropic point* in the plane of the specimen. That is, Mohr's circle of stress at this location is a point, and the state of the stress in any direction in the plane of the specimen is the same, totally lacking any shear stress on any internal surface perpendicular to the plane of the specimen.

Since the isoclinic fringes are wide and black, they tend to mask the isochromatic fringes, making it difficult to analyze them. There is a method using *two quarter-wave plates* which eliminates the isoclinic fringes. Before it is described, however, the analysis of the isoclinic fringes will be discussed.

8.7.6   ISOCLINIC FRINGE ANALYSIS

Consider a disk loaded in compression by two diametrically opposing nearly concentrated forces located on the top and bottom of the disk. The corresponding fringe patterns are shown in Fig. 8.7-11. The faint fringes are the isochromatic fringes and will be discussed later.[†] The wide, dark fringes are the isoclinic fringes. The white dotted lines were hand-drawn on the photographs to show the location of the centerline of each isoclinic fringe. Each photograph shows a different angular orientation of the filter set relative to the model. It can be seen that a different set of isoclinics occur for each orientation. The cases of 0 and 90° are actually the same.

Figure 8.7-12 illustrates the relative positions of all the isoclinics shown in Fig. 8.7-11. Knowledge of the location of the principal axes will make it easier to analyze the isochromatic fringes, as will be seen later. The isoclinic data alone can be quite helpful as from this the *stress trajectories* can be constructed graphically. Stress trajectories depict the "flow" of the principal stresses (recall the hydrodynamic analogy described in Sec. 5.10) and can also help in design.

One method of constructing the stress trajectories is shown in Fig. 8.7-13(*a*). For illustration, consider one quarter of the disk. Mark off a uniform grid along the horizontal diameter. At each point in the grid (here, for illustration, point *a*) draw a vertical line *ab* to the next isoclinic. Since the horizontal line is a 0°, 90° isoclinic,

---

[†] The photographs were made by directing white light through the model, and a time exposure longer than optimum was used to wash out the isochromatics. For this model, the black zero isochromatic fringe occurs only along the outer edge. Thus, the black line completely encircling the outer edge of the disk is not an isoclinic fringe.

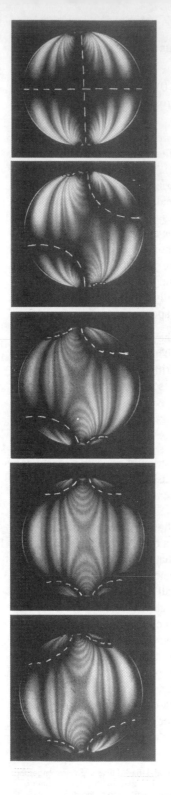

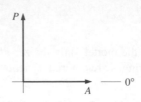

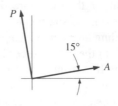

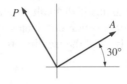

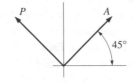

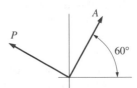

Figure 8.7-11   Isoclinic fringes.

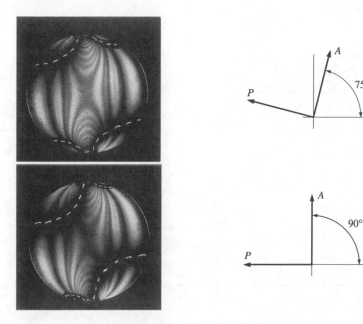

Figure 8.7-11 *(concluded)*

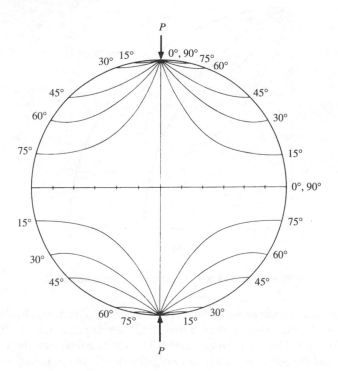

Figure 8.7-12

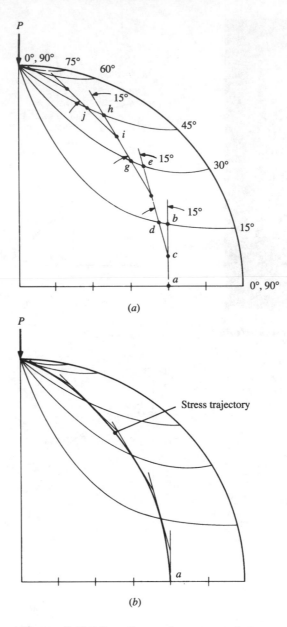

Figure 8.7-13 Constructing a stress trajectory.

one of the principal stresses at point a is in the vertical direction. Divide line ab in half at point c. From point c, draw a line 15° in the direction that the filter set rotated when the 15° isoclinic fringe formed. The 15° line intersects the 15° isoclinic at point d and the 30° isoclinic at point e. Again, divide line de in half, now at point f. From point f draw a line 30° from the vertical. Continue the process as shown until the 90° isoclinic is reached. Then using a french curve, draw a continuous line

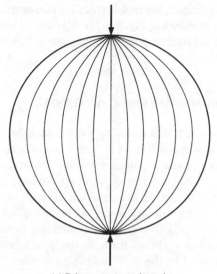

(a) Primary stress trajectories

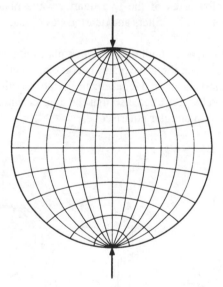

(b) Primary and secondary stress trajectories

Figure 8.7-14

tangent to each of the construction lines as shown in Fig. 8.7-13(b). This presents a good approximation of one stress trajectory. The process is continued for each initial horizontal grid point. Thanks to the symmetry of this problem, the remaining three quarters can be drawn. Thus the first set of the stress trajectories (primary) is constructed as shown in Fig. 8.7-14(a). In a similar fashion, the secondary stress trajectories can be constructed, but in this case, the construction is begun along the vertical-diameter line, as shown in Fig. 8.7-14(*b*). It can be seen from Fig. 8.7-14(*b*) that the primary principal stresses provide an indication of how the flow of stress is transmitting through the disk.

8.7.7   ISOCHROMATIC FRINGE ANALYSIS

There is an optical method which completely eliminates the isoclinic fringes when an analysis of the isochromatic fringes is performed. Recall that when a quarter-wave plate is positioned after the polarizer and the fast and slow axes of the plate are at 45° with respect to the polarizer axis, the wave which emerges from the plate is circularly polarized. If the model is inserted in the path of this circular polarized light, the relative refractions still take place, but since the light is spiraling through the model, the light vector aligns only with the principal stress axes at infinitesimal distances. Thus the isoclinics will not form. As the light leaves the model, however, the spiraling effect induced by the quarter-wave plate must be canceled; otherwise the isochromatic fringes will be affected. This is done by placing a second quarter-wave plate between the model and the analyzer filter, as shown in Fig. 8.7-15, where the fast and slow axes of the two quarter-wave plates are crossed. This arrangement of the polarization filters and quarter-wave plates is referred to as a *circular polariscope*.

   If the fast or slow axes of both quarter-wave plates are rotated 45° so that the axes align with the polarizer, the wave plates become ineffective, the polariscope reverts to a plane polariscope, and consequently the isoclinic fringes reappear. Polariscopes are generally constructed so that the quarter-wave plate can be rotated to form either a plane or a circular polariscope.

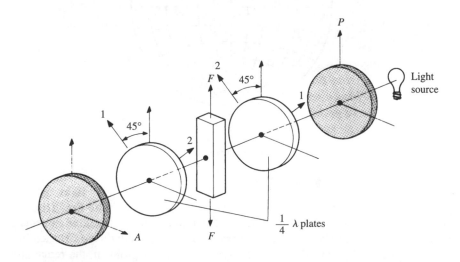

Figure 8.7-15   Circular polariscope.

Consider the compressed disk again. If the disk is placed in a circular polariscope, only the isochromatic fringes will be present. A schematic representation of the fringes and their color ordering is presented in Fig. 8.7-16 (because of high cost a color photograph cannot be reproduced). In white light, the black isochromatic fringe is the zero-order fringe ($N = 0$). On the compressed disk, a zero-order fringe occurs everywhere along the outer edge except near the points of load application. The horizontal dotted line shown in Fig. 8.7-16 was drawn to illustrate the method of fringe counting.

Starting at point A, the fringe is black; thus $N = 0$ at this point. Now, as one moves from point A to the right and along the dotted line, the color makes the following changes: black to yellow, yellow to orange, orange to red, red to blue, blue to green, and finally green back to yellow. This is due to the fact that each wavelength of the light refracts differently. From Eq. (8.7-14), $\sigma_1 - \sigma_2 = N\lambda/kt$, it can be seen that lower wavelengths extinguish first since less stress difference is re-

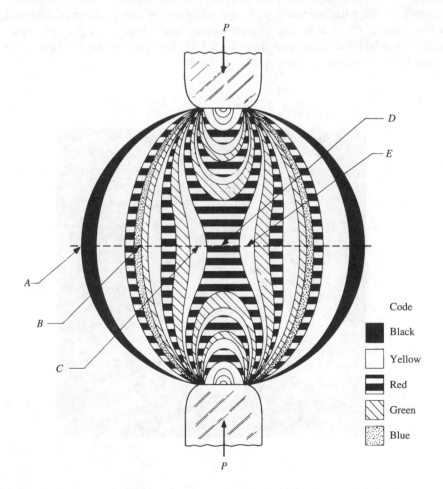

Figure 8.7-16 Isochromatic fringes using a white light source in a circular polariscope.

quired. Since the wavelength of blue or violet is the shortest in the visible spectrum, it extinguishes first, leaving a yellowish light. As the higher wavelengths extinguish, the transmission of light goes from yellow to red to green (the progression of a traffic light). The first fringe zone begins from the zero-order black fringe starting with the yellowish area and extending to the end of green. This band of colors constitutes the first fringe. Quantitative work, however, should be done using monochromatic light since Eq. (8.7-14) pertains to a particular wavelength. Also, the fringes will be narrow and distinct for a given wavelength.

As the fringe order increases, the order of progression of yellow, red, green continues. The order of progression changes if the fringe order decreases. For example, if we observe the color order going from point C to point E in the model shown in Fig. 8.7-16, we see that the order begins to reverse at point D, where the order changes to green, red, yellow, indicating that the fringe order is decreasing. This makes sense since for the model being discussed the loading is symmetric. Using white light initially makes it easier to count the fringes later with monochromatic light since the ascending and descending orders can be distinguished with white light. If the disk is now observed using monochromatic light, only black fringes will be observed as shown in Fig. 8.7-17. However, through the use of the white light, the orders are determined quite easily.

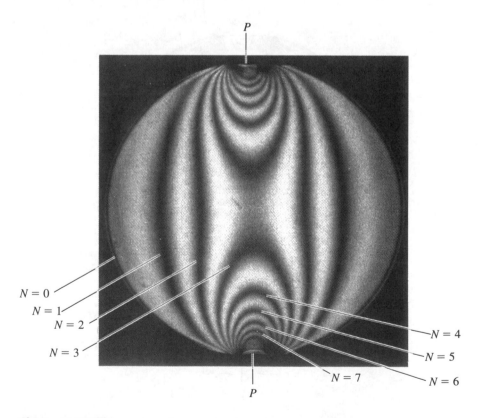

Figure 8.7-17

Using white light is not the only method of obtaining the fringe order. If the formation of the fringes is observed, one can keep account of the fringes as they develop. Another method is to use a *Babinet compensator,* covered in detail in Sec. 8.8.2.

Another characteristic of isochromatic fringes in white light is that a blue zone occurs between the red and green zones and is very predominant in the first-order fringe ($N = 1$); in higher-order fringes the blue zone is extremely fine and almost indiscernible. The wavelength corresponding to this color is 5770 Angstrom units. Thus a monochromatic filter of this wavelength is preferred for producing narrow fringes. For approximate work using white light, this red-to-green transition (called the *tint of passage*) can be used to establish fringe orders.

If a black isochromatic fringe is present ($N = 0$), fringe counting should start at this point. If no black fringe is present, a fringe present with a distinct blue between red and green is a first-order fringe ($N = 1$) and counting should start from this point. If neither the zero nor first-order fringes are present in the model, fringe development up to the final loading should be watched carefully, or the Babinet compensator can be used.

8.8   TECHNIQUES USED IN PHOTOELASTIC APPLICATIONS

Photoelastic techniques are primarily applied in practice to:

1. Two-dimensional models using transmission photoelasticity
2. Photoelastic coatings on actual components using reflection photoelastic techniques
3. Three-dimensional models using the stress-freezing method

Since the scope of this chapter is introductory in nature, only a discussion of the basic procedures used in the above techniques is presented; further development can be found in the references.

As shown in the previous section, two types of fringe patterns develop in a photoelastic model, isoclinic and isochromatic fringes. The isoclinic fringes provide full-field information on the directions of the principal stress axes, and the isochromatic fringes give full-field information on the principal stress difference, $\sigma_1 - \sigma_2$, according to Eq. (8.7-14).

8.8.1   PHOTOELASTIC MATERIAL CALIBRATION

Before testing a photoelastic model we must calibrate the photoelastic material to determine the value of k used in Eq. (8.7-14). There are various standard specimens used to calibrate a photoelastic material. Here we will demonstrate the calibration using a tensile test as shown in Fig. 8.8-1. In practice, the model and calibration tests are performed for a given wavelength of light λ (whether the tests are performed in monochromatic light or using the tint of passage under white light) and thus λ/k is constant and is evaluated from the tensile test. If we let $\lambda/k = C$, Eq. (8.7-14) can be written

$$\sigma_1 - \sigma_2 = \frac{CN}{t} \qquad \text{[8.8-1]}$$

For the tensile test, $\sigma_1 = P/A$ and $\sigma_2 = 0$, and if the thickness and width of the tensile specimen are t and w, respectively, $A = wt$. Thus

$$\sigma_1 - \sigma_2 = \frac{P}{wt}$$

and from Eq. (8.8-1) $CN/t = P/wt$, or

$$\frac{P}{N} = Cw \qquad \text{[8.8-2]}$$

The procedure is to load the specimen until $N = 1$ in the area where P/A is valid and record the load P. Continuing this for $N = 2, 3$, etc., and plotting P vs. N yields a curve, as shown in Fig. 8.8-1. From Eq. (8.8-2), the slope of the curve is Cw. Table 8.8-1 summarizes some of the mechanical properties and the photoelastic constants of various photoelastic materials.

Table 8.8-1

Material	Tensile Strength, psi	E, psi	v	C, lb/(in) (fringe)[†]
CR-39	3000	300,000	0.42	100
PSM-5 epoxy[‡]	—	450,000	0.36	60
PSM-1 polycarbonate[‡]	—	340,000	0.38	40
PSM-4 polyurethane[‡]	—	1,000	0.50	3–5

[†] $\lambda = 5770$ Angstrom units.
[‡] Supplied by Measurements Group, Inc., Raleigh, NC.

For a complete quantification of the value of the principal stresses at a given point in the model two questions remain unanswered: If the point does not fall on an isochromatic fringe, how is the fractional fringe order established? Since Eq. (8.8-1) represents only one equation in two unknowns, σ_1 and σ_2, how does one obtain the individual values of σ_1 and σ_2?

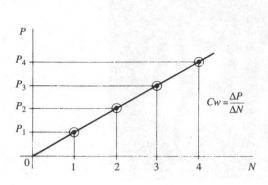

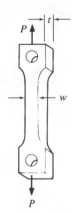

$$Cw = \frac{\Delta P}{\Delta N}$$

Figure 8.8-1

8.8.2 FRACTIONAL FRINGE ORDERS

In the previous section it was shown how to determine the fringe orders at points in the model where light extinction (isochromatic fringes) occurs for a particular wavelength. The fringe orders for these points have integer values, that is, $N = 0, 1, 2, 3, \ldots$. Points not on the fringe lines have fractional values of N. There are two basic methods for determining the fractional fringe order at a point, the *Tardy method* and the *Babinet-Soleil method* of compensation. The Tardy method can be used with a standard polariscope, but the second requires an additional piece of equipment called a Babinet-Soleil compensator.

Tardy Method   Consider a point in the model to be between fringes of the Nth and $(N + 1)$st orders; the following steps are taken when using the Tardy method:

1. With the quarter-wave plates in the plane-polariscope position align the filters with the principal stress directions at the point in question by rotating the crossed polarizer and analyzer filters until an isoclinic fringe intersects the point.

2. Moving only the quarter-wave plates, rotate the plates in proper position to convert to a circular polariscope.

3. Rotate only the analyzer filter until an isochromatic fringe coincides with the point. Determine the angle γ through which the analyzer filter was rotated.

4. If the $(N + 1)$st order fringe "moves" to the point as the analyzer rotates through the angle γ, the fringe order at the point N_Q is

$$N_Q = (N + 1) - \frac{\gamma}{180} \qquad \textbf{[8.8-3]}$$

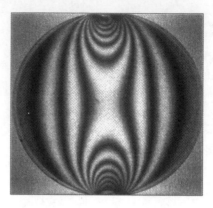

Figure 8.8-2   Light field transmission. Half-order isochromatic fringes ($N = \frac{1}{2}, \frac{3}{2}, \ldots$).

If the Nth order fringe "moves" to the point as the analyzer rotates through the angle γ, the fringe order at the point is

$$N_Q = N + \frac{\gamma}{180} \qquad [8.8\text{-}4]$$

When the analyzer is rotated 90°, half-order fringes appear. In this case, the filters can be oriented with respect to the model in any direction. In addition, the light field around the model is no longer dark, and maximum light is being transmitted. This is called *light-field transmission,* as opposed to dark-field transmission, where $\gamma = 0$. An example of light field is shown in Fig. 8.8-2.

Babinet-Soleil Method   This method requires an additional optical device which has, in effect, an adjustable birefringence. Assume that one has a material which is doubly refracting in its free state, e.g., the material used in wave plates, and that the refraction indices of axes 1 and 2 are n_1 and n_2, respectively. The phase-shift equation, given by Eq. (8.7-12), can be written as

$$\Delta = (n_1 - n_2)t \qquad [8.8\text{-}5]$$

The phase shift can be adjusted by changing the thickness t. This is accomplished by making two wedges of the same material, as shown in Fig. 8.8-3(a). By moving the wedges as shown, the thickness t_a can be made to vary. For practicality in designing the compensator, a third element is used. A plate of thickness t_b of the same material as the wedges is introduced in the optical path so that the axes are crossed with respect to the wedges, as shown in Fig. 8.8-3(b). When $t_a = t_b$, there is no phase shift through the three-element compensator. If the wedges are adjusted so that $t_a > t_b$, there is a phase shift of

$$\Delta = (n_1 - n_2)(t_a - t_b) \qquad [8.8\text{-}6]$$

If the compensator is placed in the optical path of the polariscope after the light is transmitted through the model so that the axes of the compensator line up with the

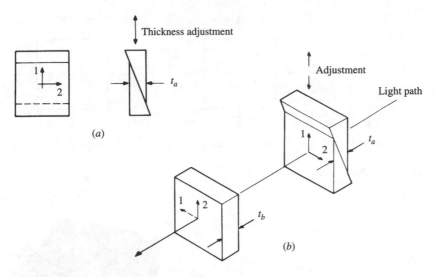

Thickness adjustment

t_a

(a)

Adjustment

Light path

t_a

t_b

(b)

Figure 8.8-3

principal axes, a shift of the final wavefront entering the analyzer can be accomplished simply by the wedge adjustment. The wedge adjustment can be calibrated to a particular wavelength (usually 5770 Angstrom units) in terms of percentage of full-wave shifts in the compensator. Thus the optical effect of using the compensator is as if an additional stress field were superposed on the model stress field. The effective stress difference of the compensator will be proportional to the thickness change, and therefore

$$(\sigma_1 - \sigma_2)_c = K(t_a - t_b) \qquad \qquad \textbf{[8.8-7]}$$

where K is a constant dependent on wavelength and material. The wedge thickness of the compensator can then be increased so that the sum of the model stresses and compensator "stresses" yields equal stresses along axes 1 and 2 (see Fig. 8.8-4). When $\sigma_1 - \sigma_2 = (\sigma_1)_c - (\sigma_2)_c$, the combined optical effect yields equal stresses σ, where

$$\sigma = (\sigma_1)_c + \sigma_2 = (\sigma_2)_c + \sigma_1$$

Thus the net effect at the point as seen through the analyzer is a zero-order fringe which is black when white light is used. The compensator is calibrated in terms of fringe orders. The compensator shown in Fig. 8.8-5 is calibrated to x units per fringe. The adjustment screw is moved until a lower-order fringe intersects the point in question. If, for example, $x = 52$ and the screw is moved 42 units such that a fringe of order 2 moves to align some point Q, the fringe order at the point is $N_Q = 2 + 42/52 = 2.81$. If the screw was turned such that the black zero-order fringe intersected the point, the indicator would read $(2.81)(52) = 146$ units.

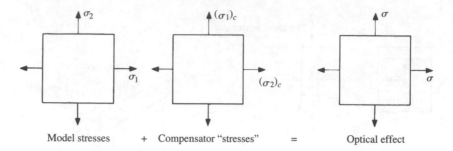

Figure 8.8-4

Figure 8.8-5   Babinet-Soleil compensator. (Courtesy of *Measurements Group, Inc.,* Raleigh, NC.)

Note if positive wedge adjustment causes $(\sigma_1)_c$ to increase relative to $(\sigma_2)_c$, then to obtain a net zero-order fringe with the compensator it must be aligned so that $(\sigma_1)_c$ is aligned in the direction of the lower principal stress σ_2 in the model.

If the analyzer is constructed such that positive wedge adjustment direction causes the relative compensator stresses to decrease in the direction of the wedge adjustment, then the compensator must be aligned in the direction of σ_1 in the model. It is easy to test this on a known stress state.

8.8.3   SEPARATION OF THE PRINCIPAL STRESSES, σ_1 AND σ_2

For a general state of plane stress, three independent pieces of information are necessary. Isochromatic and isoclinic fringes provide full-field visual information which supplies only *two* pieces of information at any point. The third is usually arrived at by another independent equation in the principal stresses σ_1 and σ_2, since in the general case, Eq. (8.8-1) contains these two unknowns. However, in most plane stress problems, the maximum stress state occurs at a boundary, where either σ_1 or

σ_2 are zero (see Sec. 8.2.2). Thus, in these cases, isochromatic information is sufficient for the determination of the one unknown principal stress. For interior points of the model, however, only the stress difference and principal-axis directions are determined from the two sets of fringes. An additional relationship between σ_1 and σ_2 must be found in order to separate the principal stresses from Eq. (8.8-1). Some of the common methods are:

1. Measuring the out-of-plane strain ε_z, where for a thin model undergoing biaxial stress

$$\varepsilon_z = -\frac{v}{E}(\sigma_x + \sigma_z) \quad \text{or} \quad \varepsilon_z = -\frac{v}{E}(\sigma_1 + \sigma_2)$$

This can be done very accurately using interferometry methods (see Ref. 8.1).

2. Directing the light at an oblique angle to the model, causing new isochromatic data (called the *oblique-incidence method*).

3. Using a technique called the *shear-difference method,* where the finite-difference method is applied to the equilibrium equations [Eqs. (2.4-2) repeated]

$$\frac{\partial \sigma_x}{\partial x} + \frac{\partial \tau_{xy}}{\partial y} + \overline{F}_x = 0 \qquad \textbf{[8.8-8a]}$$

$$\frac{\partial \tau_{xy}}{\partial x} + \frac{\partial \sigma_y}{\partial y} + \overline{F}_y = 0 \qquad \textbf{[8.8-8b]}$$

In practice the most common techniques are the oblique-incidence and the shear-difference methods.

The Oblique-Incidence Method The oblique-incidence method can be performed by rotating the model about the σ_1 axis corresponding to the point in question or by placing a set of prisms aligned along the σ_1 axis, as shown in Fig. 8.8-6, where σ_1 is perpendicular to the plane of the page.

The isochromatic fringe order observed at normal incidence N_0 will generally differ from that of oblique incidence N_θ. From Eq. (8.8-1)

$$\sigma_1 - \sigma_2 = \frac{CN_0}{t} \qquad \textbf{[8.8-9]}$$

and for the oblique ray

$$\sigma_1 - \sigma_2' = \frac{CN_\theta}{t/(\cos \theta)} \qquad \textbf{[8.8-10]}$$

where $t/(\cos \theta)$ is the distance the oblique ray travels through the model. From Eq. (3.9-1)

$$\sigma_2' = \sigma_2 \cos^2\theta$$

Combining this with Eqs. (8.8-9) and (8.8-10) and solving for σ_1 and σ_2 yields

$$\sigma_1 = \frac{C \cos \theta}{t \sin^2\theta} (N_\theta - N_0 \cos \theta) \qquad \textbf{[8.8-11a]}$$

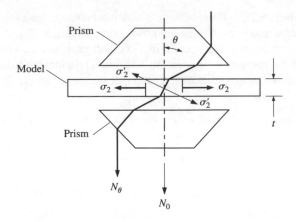

Figure 8.8-6   Oblique incidence using prisms.

$$\sigma_2 = \frac{C}{t \sin^2\theta}(N_\theta \cos\theta - N_0) \qquad \textbf{[8.8-11}\textit{b}\textbf{]}$$

The Shear-Difference Method   The shear-difference method is based on the equilibrium equations. Assuming the absence of body forces, we can write Eq. (8.8-8a) as

$$\frac{\partial \sigma_x}{\partial x} + \frac{\partial \tau_{xy}}{\partial y} = 0$$

For an approximation, this equation can be expressed by finite differences as

$$\frac{\Delta\sigma_x}{\Delta x}\bigg|_{y=\text{const}} + \frac{\Delta\tau_{xy}}{\Delta y}\bigg|_{x=\text{const}} = 0$$

or

$$\Delta\sigma_x = -\frac{\Delta\tau_{xy}}{\Delta y}\bigg|_{x=\text{const}} \Delta x \bigg|_{y=\text{const}} \qquad \textbf{[8.8-12]}$$

where Δx and Δy are made as small as practical. Starting at a free surface (or any surface where the state of stress is known), construct a grid as shown in Fig. 8.8-7. Assuming that σ_x at point 0 is known, $(\sigma_x)_0$, we see from Eq. (8.8-12) that the change in σ_x from point 0 to point 1 is

$$\Delta\sigma_x = (\sigma_x)_1 - (\sigma_x)_0 = -\frac{(\tau_{xy})_a - (\tau_{xy})_b}{\Delta y}\Delta x$$

Thus the value of σ_x at point 1 is

$$(\sigma_x)_1 = (\sigma_x)_0 - \frac{\Delta x}{\Delta y}[(\tau_{xy})_a - (\tau_{xy})_b] \qquad \textbf{[8.8-13]}$$

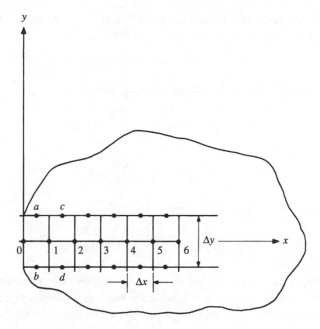

Figure 8.8-7

The shear stress τ_{xy} can be determined from the isoclinic and isochromatic information. To see this, consider the transformation equation, Eq. (3.9-5)

$$\tau_{x'y'} = -\left(\frac{\sigma_x - \sigma_y}{2}\right)\sin 2\theta + \tau_{xy}\cos 2\theta$$

Starting from the principal stress state (with zero shear), transform in the clockwise direction $(-\theta)$ to determine τ_{xy}. That is

$$\tau_{xy} = -\left(\frac{\sigma_1 - \sigma_2}{2}\right)\sin[2(-\theta)] + (0)\cos[2(-\theta)]$$

or

$$\tau_{xy} = \frac{\sigma_1 - \sigma_2}{2}\sin 2\theta \qquad \textbf{[8.8-14]}$$

where $\sigma_1 - \sigma_2$ can be determined from isochromatic fringe data via Eq. (8.8-1) and θ is the angle between the x axis and the direction of σ_1, which is determined from the *isoclinic* fringe data at a specific point a.[†] Thus the shear stresses $(\tau_{xy})_a$ and $(\tau_{xy})_b$ can be determined from the photoelastic fringes. Once $(\sigma_x)_1$ is calculated from

---

[†] The direction of σ_1 can easily be found using the Babinet-Soleil compensator.

Eq. (8.8-13), the principal stresses for point 1 can be determined. Transforming *from* the principal stress state (with zero shear) Eq. (3.9-1) gives

$$(\sigma_x)_i = (\sigma_1)_i \cos^2\theta_i + (\sigma_2)_i \sin^2\theta_i \qquad \textbf{[8.8-15]}$$

Solving Eqs. (8.8-15) and (8.8-1) simultaneously yields

$$(\sigma_1)_i = (\sigma_x)_i + \frac{CN}{t} \sin^2\theta_i \qquad \textbf{[8.8-16a]}$$

$$(\sigma_2)_i = (\sigma_x)_i - \frac{CN}{t} \cos^2\theta_i \qquad \textbf{[8.8-16b]}$$

The procedure is then repeated for point 2. That is,

$$(\sigma_x)_2 = (\sigma_x)_1 - \frac{\Delta x}{\Delta y}[(\tau_{xy})_c - (\tau_{xy})_d]$$

etc. The technique can be continued throughout the interior of the model to obtain the stress values at any interior point. Since the method is based on numerical integration, errors are cumulative, and the photoelastic data should be taken carefully.

8.8.4   REFLECTION PHOTOELASTICITY

Reflection techniques can be used on a model of any material. Photoelastic coatings are cemented onto areas of investigation, or the entire part can be coated (see Fig. 8.8-8). In principle, as can be seen in Fig. 8.8-9, the technique is similar to transmission photoelasticity.

Since light transmits through the coating twice,

$$\sigma_1 - \sigma_2 = \frac{CN}{2t}$$

(*a*) Totally coated part   (*b*) Coating sections

Figure 8.8-8   Photoelastic coating of a machine component. (Courtesy of *Measurements Group, Inc.*, Raleigh, NC.)

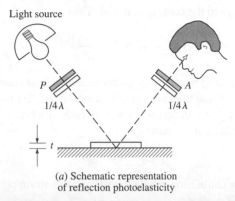

Light source

P $\quad$ A

$1/4\lambda$ $\quad$ $1/4\lambda$

t

(*a*) Schematic representation
of reflection photoelasticity

(*b*) Measurements Group Model 031 Reflection Polariscope with Model 632 Digital Strain
Indicator/Printer coupled to an optical Babinet-Soleil compensator transducer

Figure 8.8-9 $\quad$ Reflection photoelasticity. (Courtesy of *Measurements Group, Inc.*, Raleigh, NC.)

However, since the coating is used on a specimen which may be of a different material, it is preferable to deal with strains. From Hooke's law, Eq. (1.4-5), it can be shown that

$$\sigma_1 - \sigma_2 = \frac{E_c}{1 + v_c}(\varepsilon_1 - \varepsilon_2)$$

where E_c and v_c apply to the coating material. Thus

$$\varepsilon_1 - \varepsilon_2 = \frac{(1 + v_c)CN}{2E_c t}$$ **[8.8-17]**

Calibration of the coating is generally performed on a beam test specimen using the same coating material and thickness as that used on the test model. Thus $(1 + v_c) \cdot C/(2E_c t) = f$, where f is a constant for a particular thickness of coating material. Equation (8.8-17) can then be written as

$$\varepsilon_1 - \varepsilon_2 = fN$$ **[8.8-18]**

The constant f is the calibration constant with units of strain per fringe.

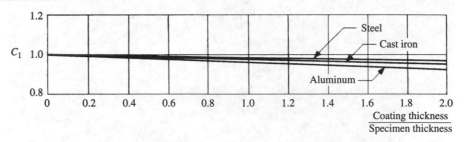

(a) Correction factors for plane-stress problems

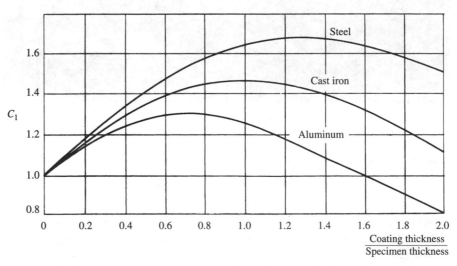

(b) Correction factors for plates or beams in bending, perpendicular to plane of coating

Figure 8.8-10

If the test specimen is thin, the coating itself may carry an appreciable portion of the loading and a correction must be made. The corrective equation is approximated by

$$\varepsilon_1 - \varepsilon_2 = \frac{fN}{C_1}$$ **[8.8-19]**

where C_1 is given in Fig. 8.8-10(a) and (b) (see Prob. 5.45).

Example 8.8-1

The cantilever beam provides a very reliable method for the calibration of a coating. Consider a cantilever beam made of aluminum, 1/4 in thick and 1 in wide, and coated with a strip of 0.080-in-thick photoelastic plastic bonded to the top surface of the beam, as shown in Fig. 8.8-11. With a Babinet-Soleil compensator, the fringe value 6 in from the point of loading is found to be $N = 1.54$ after a 20-lb weight is applied to the end of the beam. Determine the fringe constant f for the coating. The material constants for the aluminum are $E_a = 10.3 \times 10^6$ psi and $v_a = 0.33$.

Solution:

The stress on the beam (if uncoated) at the measurement point is

$$\sigma_1 = \frac{Mc}{I} = \frac{(6)(20)(0.25/2)}{(\frac{1}{12})(1)(0.25)^3} = 11,500 \, \text{psi} \qquad \sigma_2 = 0$$

Since $\varepsilon_1 - \varepsilon_2 = (1 + v)(\sigma_1 - \sigma_2)/E$,

$$\varepsilon_1 - \varepsilon_2 = \frac{1 + 0.33}{10.3 \times 10^6}(11,500 - 0) = 1490 \, \mu$$

The thickness ratio is

$$\frac{t_p}{t_s} = \frac{0.08}{0.25} = 0.32$$

Therefore, from Fig. 8.8-10(b), $C_1 = 1.2$. Assuming that the strain in the coating is the same as the beam, from Eq. (8.8-19)

$$f = \frac{C_1}{N}(\varepsilon_1 - \varepsilon_2) = \frac{1.2}{1.54}(1490) = 1160 \, \mu/\text{fringe}$$

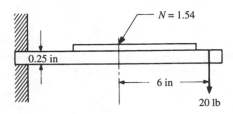

Figure 8.8-11

If the material constants for the coating are $E_c = 340$ kpsi and $v_c = 0.38$, the C value for the material will be

$$C = \frac{2Etf}{1 + v} = \frac{(2)(340,000)(0.8)}{1 + 0.38} \ 1160 \times 10^{-6} = 45.7 \ \text{lb}/(\text{in})\cdot(\text{fringe})$$

The principal strains can be separated using either an oblique incidence adapter or special strain separator gages for photoelastic applications.[†] A schematic of the oblique incidence adapter is shown in Fig. 8.8-12. Light emerging from the polarizer is directed through an aperture in the adapter and reflected by a mirror which directs the light to the coating at an angle of incidence θ_i. The detail at the test point in Fig. 8.8-12 shows the light transmitting through the coating at an angle of θ from normal where the change in angle is due to refraction. The return path of the light to the analyzer is basically the same as the path from the polarizer to the coating.

If the reflection polariscope is aligned such that ε_1 is perpendicular to the page of Fig. 8.8-12, the strains in the plane of the page are ε_2 and ε_3, parallel and perpendicu-

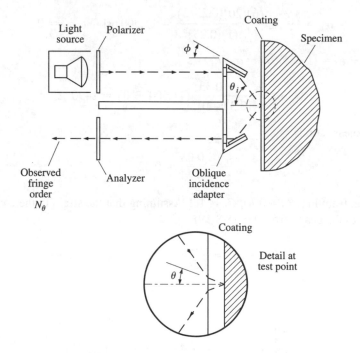

Figure 8.8-12   Schematic representation of Model 033 oblique-incidence attachment manufactured by Measurements Group, Inc., Raleigh, NC.

[†] Strain separator gages are available from Measurements Group, Inc., Raleigh, NC, and are called PhotoStress® separator gages.

lar to the coating, respectively. Using the strain transformation equation with no shear strain, the strain in the plane of the page perpendicular to the oblique light ray will be

$$\varepsilon_2' = \varepsilon_2 \cos^2\theta + \varepsilon_3 \sin^2\theta \qquad \text{[8.8-20]}$$

The strain-optic equation for the oblique fringe order N_θ will be

$$\varepsilon_1 - \varepsilon_2' = \frac{(1 + v_c)CN_\theta}{2E_c t/\cos\theta} = fN_\theta \cos\theta \qquad \text{[8.8-21]}$$

where the oblique ray has passed through a thickness of $2t/\cos\theta$. Before substituting ε_2' from Eq. (8.8-20) into Eq. (8.8-21), ε_3 must be determined. Since the photoelastic coating is thin and assumed to be in a state of plane stress, $\varepsilon_3 = -v_c(\sigma_1 + \sigma_2)/E_c$. However, since

$$\sigma_1 = \frac{E_c}{1 - v_c^2}(\varepsilon_1 + v_c\varepsilon_2) \qquad \sigma_2 = \frac{E_c}{1 - v_c^2}(\varepsilon_2 + v_c\varepsilon_1)$$

it can be shown that

$$\varepsilon_3 = -\frac{v_c}{1 - v_c}(\varepsilon_1 + \varepsilon_2) \qquad \text{[8.8-22]}$$

Substitute Eq. (8.8-22) into Eq. (8.8-20). Substitute the result of this into Eq. (8.8-21) and simplify. This results in

$$\left(1 + \frac{v_c}{1 - v_c}\sin^2\theta\right)\varepsilon_1 - \left(\cos^2\theta - \frac{v_c}{1 - v_c}\sin^2\theta\right)\varepsilon_2 = fN_\theta \cos\theta \qquad \text{[8.8-23]}$$

This, together with Eq. (8.8-18) with $N = N_0$ for normal incidence, gives two equations for the unknowns ε_1 and ε_2. Solving these equations simultaneously yields

$$\varepsilon_1 = \frac{f[(1 - v_c)\cos\theta N_\theta - (\cos^2\theta - v_c)N_0]}{(1 + v_c)\sin^2\theta} \qquad \text{[8.8-24a]}$$

$$\varepsilon_2 = \frac{f[(1 - v_c)\cos\theta N_\theta - (1 - v_c\cos^2\theta)N_0]}{(1 + v_c)\sin^2\theta} \qquad \text{[8.8-24b]}$$

For the the oblique incidence attachment shown in Fig. 8.8-12, $\phi = 28°$. Thus the incidence angle is $\theta_i = 2(28) = 56°$. For light entering a refractive medium from air, $\sin\theta_i = n_c \sin\theta$, where for the coating n_c is the index of refraction relative to air and θ is the angle of the refracted wave in the coating (see Fig. 8.8-12). Photoelastic's PSM-1 material is a commonly used material for reflection work, where $n_c = 1.58$ and $v_c = 0.38$. Thus $\theta = \sin^{-1}[(\sin 56°)/1.58] = 30.6°$. Substituting $\theta = 30.6°$ and $v_c = 0.38$ into Eq. (8.8-24) yields

$$\varepsilon_1 = f(1.492 N_\theta - 1.009 N_0) \qquad \text{[8.8-25a]}$$

$$\varepsilon_2 = f(1.492 N_\theta - 2.009 N_0) \qquad \text{[8.8-25b]}$$

which can be approximated by

$$\varepsilon_1 = f(1.5N_\theta - N_0) \qquad\qquad \textbf{[8.8-26a]}$$

$$\varepsilon_2 = f(1.5N_\theta - 2N_0) \qquad\qquad \textbf{[8.8-26b]}$$

8.8.5   STRESS FREEZING IN THREE-DIMENSIONAL PHOTOELASTICITY

There is a method for locking or freezing the strain field in certain photoelastic materials. Once the strain field is frozen in the model, segments can be cut off the model, and if this is done carefully, the *strain field* is left undisturbed. In this way, three-dimensional investigations can be performed. The technique is called *stress freezing*, but this is a misnomer. The analysis procedure is beyond the scope of this book; a detailed description can be found in Ref. 8.1. In this section, a simple explanation of the mechanics of stress freezing is presented.

Many polymers exhibit diphase behavior; i.e., these materials undergo drastic changes in their mechanical properties at a critical temperature. As an example, the characteristics of PLM-4B epoxy are given in Table 8.8-2.

Table 8.8-2   Characteristics of PLM-4B epoxy

	E, kpsi	C, lb/(in)(fringe)	Tensile Strength, kpsi	ν
Room temperature	450	60	9	0.36
Critical temperature, 240–260°F	2.5	2.2	>2.5	0.50

| SOURCE: Data supplied from Measurements Groups, Inc., Raleigh, NC.

The diphase behavior is due to the nature of the molecular bonds in the polymer. Above the critical temperature the weak though numerous secondary bonds break down, causing extreme changes in the mechanical properties of the material. To understand how this applies to stress freezing a simplification will suffice. Neglecting the strain induced by the coefficient of expansion of the solid, we have the stress-strain curve shown in Fig. 8.8-13 for the material at room temperature and at the critical temperature. The slope of the curve for room temperature is $E_{RT} = 450$ kpsi, and at the critical temperature is $E_{CT} = 2.5$ kpsi. Assuming a simple case such as simple tension, the model is first loaded at room temperature to

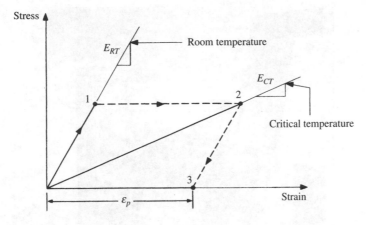

Figure 8.8-13

point 1. With the load *constant,* the temperature is increased gradually to the critical temperature and held there for 2 to 4 h so that the temperature throughout the model is uniform. At this point the secondary bonds have broken and the strain has increased to a steady value established by point 2. The next step is to gradually decrease the temperature back down to room temperature. During this phase, the secondary bonds re-form but in the deformed state established at point 2. Thus, for all practical purposes, when room temperature is reached, the state of the model is still basically at point 2. At this point, the loading is removed, and since the room-temperature behavior of the material is regained, the slope of the unloading curve is E_{RT} and the final state is given by point 3. The unloaded model can now be viewed through a polariscope, and photoelastic fringe patterns will be present, due to the residual *strain* field ε_p. Since no residual stresses are present, the three-dimensional model can be sliced and the photoelastic analysis can be performed on two-dimensional slices using a transmission polariscope. For the method of stress freezing, care must be taken for the following:

1. The heating and cooling cycles should be performed slowly so that temperature gradients are minimized, thus reducing the possibility of thermally induced stresses.

2. Attempts should be made to simulate loading with dead weights so that the loading will not change with deflection. An example of a loading arrangement is shown in Fig. 8.8-14.

3. To avoid machining stresses sharp tools, air cooling, and light cuts at high cutting speeds should be used when slicing the model for analysis.

Figure 8.8-14   Dead weight loading used in stress freezing. (Courtesy of *Measurements Group, Inc.*, Raleigh, NC.)

8.9 PROBLEMS

8.1 A model scaled to ten times the size of a prototype is tested under a loading of five times greater than the actual prototype loading. The model material has a modulus of elasticity which is 10 percent less than that for the prototype material. The maximum stress and deflection of the model were found to be 1.2 kpsi and 0.025 in, respectively. Determine the expected values for the prototype.

8.2 A simply supported steel beam 3 m long of rectangular cross section 150 mm deep and 75 mm wide is to carry a uniform load of 6 kN/m along a span (for steel let $E_s = 200$ GPa). A photoelastic scale model 500 mm long is to be constructed of a material with a modulus of elasticity of 20 GPa.

(a) Determine the cross section of the model.

(b) The prototype is to be loaded with a uniformly distributed force at a particular location. To assure that the stresses in the model do not exceed the linear range of the material, it was decided to apply a uniform load of 240 N/m on the model. Determine all scale factors.

(c) The model test revealed a maximum bending stress and deflection of 1.4 MPa and 9.50 mm, respectively. Determine the expected prototype values.

8.3   Figure 8.2-2 shows a thick-walled disk loaded by diametrically opposing concentrated forces. Consider this to be a test model with a material such that $E_m = 71$ GPa and $v_m = 0.334$. The disk is also relatively thin in the axial direction. Upon testing the model, the tangential strain at point A was found to be 300 μ after a load of $P = 80$ kN/m was applied to the disk.

(a)   Consider the prototype to be four times larger than the model with $E_p = 200$ GPa and $v_p = 0.29$. Determine the tangential stress at point A in the prototype if the applied load on the prototype is to be 800 kN/m.

(b)   Consider the radial dimensions of the prototype to be four times larger than the model with $E_p = 200$ GPa and $v_p = 0.29$. Also consider the prototype to be very long in the axial direction. Determine the tangential stress at point A in the prototype if the applied load on the prototype is to be 800 kN/m.

8.4   The structure shown is of constant thickness, 0.25 in. Assuming that the state of stress throughout the entire structure is that of plane stress, show the state of stress where one or more of the three possible stresses are known. For each specifically different area show the stress element properly oriented and indicate the known and unknown stresses.

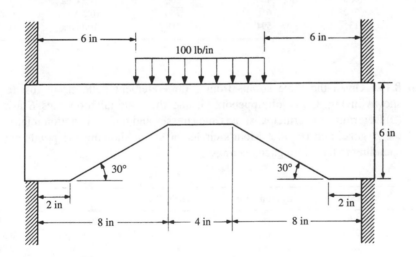

Problem 8.4

8.5   Determine the shear stress distribution along section A-A of Problem 5.49.

8.6   Determine the shear stress distribution along section A-A of Problem 5.51.

8.7   Show the complete development of Eqs. (G.1-1) and (G.1-2) given in Appendix G for the three-element rectangular strain gage rosette.

8.8 Develop the single-valued equation, Eq. (G.1-3), for the counterclockwise angle between gage A of the three-element rectangular strain gage rosette of Fig. G.1-1 and the greater of the two principal stresses in the plane of analysis σ_{p1}.

8.9 Develop Eqs. (G.2-1) and (G.2-2) given in Appendix G for the three-element delta strain gage rosette.

8.10 Develop the single-valued equation, Eq. (G.2-3), for the counterclockwise angle between gage A of the three-element delta strain gage rosette of Fig. G.2-1 and the greater of the two principal stresses in the plane of analysis σ_{p1}.

8.11–8.15 Given the three strains from a three-element rectangular strain rosette as shown in Fig. G.1-1 of Appendix G, and the material properties E and v, (a) Determine the principal strains and stresses and their orientation relative to the A gage, and (b) Plot Mohr's circle for strain showing the points corresponding to the strain gage readings.

	ε_A (μStrain)	ε_B (μStrain)	ε_C (μStrain)	E	v
8.11	0	500	0	71 GPa	0.34
8.12	220	500	780	29 Mpsi	0.285
8.13	−400	−400	200	105 GPa	0.324
8.14	−250	−616	750	10.3 Mpsi	0.334
8.15	131	594	269	100 GPa	0.211

8.16–8.19 Given the three strains from a three-element delta strain rosette as shown in Fig. G.2-1 of Appendix G, and the material properties E and v, (a) Determine the principal strains and stresses and their orientation relative to the A gage, and (b) Plot Mohr's circle for strain showing the points corresponding to the strain gage readings.

	ε_A (μStrain)	ε_B (μStrain)	ε_C (μStrain)	E	v
8.16	−150	−150	300	200 GPa	0.285
8.17	220	100	340	10.3 Mpsi	0.334
8.18	−215	442	440	71 GPa	0.34
8.19	56	−125	−89	29 Mpsi	0.285

8.20 Using Eqs. (8.5-1) and (8.5-2) derive Eq. (8.5-3) for small changes in resistance.

8.21 Consider a strain gage with sensitivity $S_A = 2$ to be in a quarter-arm bridge circuit in the R_1 position of Fig. 8.5-1 with $R_2 = R_1$. Using Eqs. (8.5-1), (8.5-2), and (8.5-4) derive the exact equation for $\Delta E/(Vr)$ considering change only in the gage resistance due to strain ε (i.e., $\Delta R_2 = \Delta R_3 = \Delta R_4 = 0$). Plot

$\Delta E/(Vr)$ as a function of ε for $0 \le \varepsilon \le 0.1$. Plot Eq. (8.5-5) on the same graph. Also, tabulate the two equations in increments of 0.001 for ε giving the percent error in Eq. (8.5-5). For what value of ε is the error of the linear Eq. (8.5-5) less than 5 percent?

8.22 A single strain gage in a quarter-bridge circuit is mounted on the outer surface of a solid circular shaft of diameter d which is loaded by a torsional moment T. Show the orientation of the gage which will provide the maximum tensile strain on the gage. What is the strain output of the gage in terms of T, d, and the shaft material properties E and v?

8.23 A two-element rectangular strain gage rosette is mounted on the outer surface of a solid circular shaft of diameter d which is loaded by a torsional moment T. Show the orientation of the gages which will provide the maximum tensile and compressive strains on the gages. If the tensile gage and compressive gages are connected to adjacent arms of a half-bridge circuit, what is the indicated output strain in terms of T, d, and the shaft material properties E and v? Does this arrangement provide temperature compensation?

8.24 A two-element *shear* strain gage is shown below where the three terminal tabs 1 to 3 can be connected to the bridge circuit shown.

(a) Indicate which positions (a, b, c, d) of the bridge the gage terminals (1, 2, and 3) should be connected to such that the output of the circuit is directly related to the shear strain only.

(b) Considering that the output of the bridge gives a direct reading of the strain if a single gage is connected across terminals ab, determine the output of the circuit if the arrangement of the shear gage as you have indicated in part (a) is employed. Assume that the gage is in a general strain field (ε_x, ε_y, γ_{xy}).

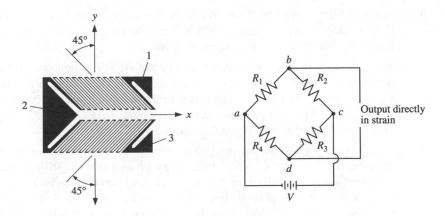

Problem 8.24

8.25–8.29 Let the strains given in Problems 8.11 to 8.15 be indicated values. Based on the transverse sensitivities given determine the corrected strain values, the principal strains and stresses, and their orientation relative to the A gage.

	Make Correction to Problem	K_{tA}	K_{tB}	K_{tC}
8.25	8.11	0.05	0.035	0.05
8.26	8.12	7%	5%	7%
8.27	8.13	5%	5%	5%
8.28	8.14	0.08	0.08	0.08
8.29	8.15	8%	5%	8%

8.30–8.33 Let the strains given in Problems 8.16 to 8.19 be indicated values. Based on the transverse sensitivities given determine the corrected strain values, the principal strains and stresses, and their orientation relative to the A gage.

	Make Correction to Problem	K_{tA}	K_{tB}	K_{tC}
8.30	8.16	0.05	0.035	0.05
8.31	8.17	7%	5%	7%
8.32	8.18	5%	5%	5%
8.33	8.19	0.08	0.08	0.08

8.34 Develop the equations given in Sec. H.3 of Appendix H for the transverse sensitivity corrections of a three-element delta strain gage rosette (see Sec. H.2 for the development of the correction equations for a three-element rectangular strain gage rosette).

8.35 A strain gage is mounted perpendicular to the direction of a uniaxial stress field. Determine the percentage error indicated by the gage if for the stressed material $v = 0.29$ and for the gage $K_t = 0.08$.

8.36 For each strain field given, a strain gage of length 0.5 in is mounted such that the gage is oriented in the x direction and centered at $x = 0$. Ignoring transverse sensitivity, determine the indicated strain from the gage and compare it with the value of the strain at the center of the gage. (a) $\varepsilon_x = 200 \ \mu$, (b) $\varepsilon_x = 200(1 - 2x) \ \mu$, (c) $\varepsilon_x = 200(1 - 2x - 3x^2) \ \mu$.

8.37 A large thin plate is loaded in tension where the plate contains a 40-mm-diameter hole (see Fig. 4.2-4). A strain gage with an active length of 10 mm is cemented along the wall of the hole in the circumferential direction and centered at $\theta = 90°$. A second gage with an active length of 2 mm is also mounted along the wall 180° from the outer gage. If $\sigma = 10$ MPa, determine the theoretical stress at the edge of the hole for $\theta = \pm 90°$. Based on the theoretical stress field given by Eqs. (4.2-27) determine the stress that would result if calculated from the strain output of each strain gage and compare these values with the theoretical result.

8.38 Determine the magnitude and speed of the strain pulse for the output strain shown if the active length of the gage is 1.0 in.

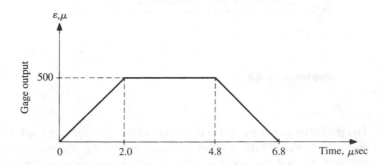

Problem 8.38

8.39 Determine the output of a strain gage of active length ℓ_g if subjected to a triangular strain pulse of magnitude ε and total pulse time t_p. Consider the pulse to be symmetric with time and $l_g > ct_p$.

8.40 A strain gage with the temperature characteristics given in Fig. 8.6-2 with a room temperature gage factor of 2.1 is mounted on a 1018 steel specimen. A strain measurement of 800 μ is recorded during a test when the temperature of the installation is $-30°C$. Determine the value of the actual test strain if the gage is in a quarter-bridge circuit and prior to testing, the indicator was zeroed at room temperature with the gage factor set at 2.0.

8.41 Another specimen used to calibrate photoelastic materials is the compressed disk shown in Fig. 8.7-17. If the disk has a radius and thickness of 1.5 and 0.25 in, respectively, and $P = 200$ lb, estimate the fringe constant C in Eq. (8.8-1).

8.42 Another specimen used to calibrate photoelastic materials is a beam in pure bending as shown. The loading fixture is simple to construct and the specimens are much simpler to make than the tensile specimen. If the beam shown has a rectangular cross section with a thickness 6.35 mm, $h = 25$ mm, $e = 20$

mm, and $P = 80$ N, determine the fringe constant C in Eq. (8.8-1). Assume that the specimen is in a dark-field polariscope.

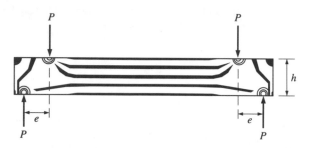

Problem 8.42

8.43 The photoelastic model shown is a thick-walled circular ring loaded by two opposing forces. If the photoelastic constant of the material is $C = 60$ lb/(in·fringe) and the model is 0.25 in thick, determine σ_1 and σ_2 at points A, B, and C.

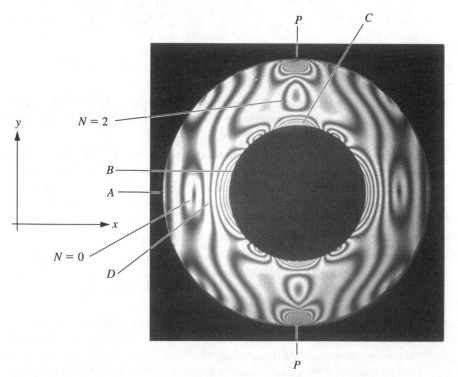

Problem 8.43

8.44 For Problem 8.43, an oblique incidence adapter is used with an angle of transmission of 30° and the fringe order at point D is found to be $N_\theta = 1.64$. Determine the values of σ_1 and σ_2 at this point.

8.45 A step beam in a dark-field polariscope is loaded by a pure moment of $M = 100$ lb·in. If the thickness of the beam is 0.25 in and $h = 2.0$ in, determine (a) the fringe constant C, (b) the maximum stress at the fillet, (c) the static stress concentration factor *without* using the fringe constant, and (d) all locations where the fringe order $N = 0$.

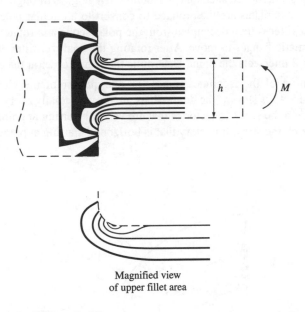

Magnified view
of upper fillet area

Problem 8.45

8.46 A part with a photoelastic coating is analyzed under loading. Using a reflection polariscope (in circular mode) with a Babinet-Soleil compensator, the fringe order under normal incidence at a specific location is found to be 1.07. The quarter-wave plates are then rotated establishing a plane polariscope. Defining an xy coordinate system on the part's surface centered at the measured point, the filter set is rotated 53° counterclockwise relative to the x axis before an isoclinic fringe intersects the measurement point. Again using the Babinet-Soleil compensator, it is established that σ_1 is 143° counterclockwise relative to the x axis. The quarter-wave plates are then rotated re-establishing a circular polariscope, and using an oblique incidence adapter, the fringe order at the same point is found to be 1.02. Assume that Eqs. (8.8-26) apply to the adaptor. The fringe constant of the coating is $f = 1015$ μ/fringe and for the part material, $E = 200$ GPa and $v = 0.29$. For the part, determine the principal stresses and σ_x, σ_y, and τ_{xy} at the measurement point. Draw the principal stress element properly oriented relative to the xy coordinate system. Also, draw the element containing σ_x, σ_y, and τ_{xy} showing the true directions of the stresses.

8.47 A point in a photoelastic part is between the fringe orders 2 and 3. With the axes of the quarter-wave plates aligned with the polarizer and analyzer filters, the filter set is rotated such that an isoclinic fringe goes through the point. The quarter-wave plates are then rotated to convert to a circular polariscope. The analyzer filter is then decoupled from the polariscope and rotated causing the isochromatic fringes to move. After rotating the analyzer filter 30° the fringe of order 2 intersects the point. Determine the fringe order at the point.

8.48 A mapping of the isoclinic fringes for one-quarter of the circular ring of Problem 8.43 is shown (the dotted lines are drawn radially every 15°).
(a) Sketch the stress trajectories that are vertical starting at points a and b.
(b) Sketch the stress trajectory that is horizontal starting at point c.

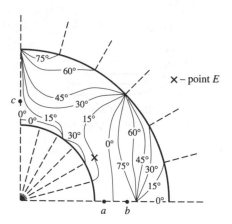

Problem 8.48

8.49 From the figures of Problems 8.43 and 8.48, estimate τ_{xy} for point E.

8.50 Consider the uniformly loaded rectangular cantilever beam shown. In order to determine how σ_y varies using a photoelastic model, data were taken at $x = 5.0$ in and $x = 5.2$ in, thus making $\Delta x = 0.2$ in. With a Δy of 0.2 in the following data were obtained ($C = 60$ lb/(in·fringe).

y, in	x = 5.0 in		x = 5.2 in	
	N	θ°	N	θ°
−0.9	5.63	1	6.09	1
−0.7	4.41	4	4.77	4
−0.5	3.24	8	3.49	8
−0.3	2.15	16	2.31	15
−0.1	1.36	33	1.42	32
0.1	1.43	60	1.50	61
0.3	2.30	75	2.45	76
0.5	3.40	82	3.65	82
0.7	4.58	86	4.93	86
0.9	5.80	89	6.25	89

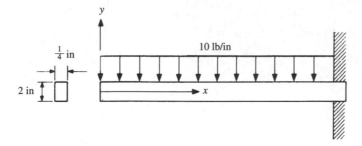

Problem 8.50

Computer Problems

8.51 Write a program, either directly or using a spreadsheet program, which uses the equations in Sec. G.1 of Appendix G and H.2 of Appendix H. The input to the program is to be the transverse sensitivities and the indicated output of a three-element rectangular strain gage rosette. The program should then output the corrected gage strains, the principal strains and stresses, and the counterclockwise angle the greater principal stress makes with gage *A*.

8.52 Write a program, either directly or using a spreadsheet program, which uses the equations in Sec. G.2 of Appendix G and H.3 of Appendix H. The input to the program is to be the transverse sensitivities and the indicated output of a three-element delta strain gage rosette. The program should then output the corrected gage strains, the principal strains and stresses, and the counterclockwise angle the greater principal stress makes with gage *A*.

8.53 Write a program, either directly or using a spreadsheet program, which accepts data such as that given in Problem 8.50 and performs the necessary calculations.

8.10   REFERENCES

8.1   Dally, J. W., and W. F. Riley. *Experimental Stress Analysis,* 3rd ed. New York: McGraw-Hill, 1991.

8.2   Kobayashi, A. S., editor. *Handbook on Experimental Mechanics,* 2nd ed. Bethel, CT: Society for Experimental Mechanics, 1993, VCH Publishers, Inc. New York.

chapter NINE

INTRODUCTION TO THE FINITE ELEMENT METHOD

9.0 INTRODUCTION

The purpose of this chapter is to expose the reader to some of the fundamental concepts of the finite element method (FEM), and therefore the coverage is only introductory in nature. For further detail, the reader is urged to consult the references cited at the end of this chapter.

The finite element method is a numerical technique, ideally suited to digital computers, in which a continuous elastic structure (continuum) is divided (discretized) into small but finite well-defined substructures (*elements*). Using matrices, the continuous elastic behavior of each element is categorized in terms of the element's material and geometric properties, the distribution of loading (static, dynamic, and thermal) within the element, and the loads and displacements at the *nodes* of the element. The element's nodes are the fundamental governing entities of the element, as it is the node where the element connects to other elements, where elastic properties of the element are established, where boundary conditions are assigned, and where forces (contact or body) are ultimately applied. A node possesses *degrees of freedom* (dof's). Degrees of freedom are the translational and rotational motion that can exist at a node. At most, a node can possess three translational and three rotational degrees of freedom. Once each element within a structure is defined locally in matrix form, the elements are then globally assembled (attached) through their common nodes (dof's) into an overall system matrix. Applied loads and boundary conditions are then specified, and through matrix operations the values of all unknown displacement degrees of freedom are determined. Once this is done, it is a simple matter to use these displacements to determine strains and stresses through the constitutive equations of elasticity. Since matrix notation and algebra are extensively used in this chapter, it is recommended that the reader fully understand the treatment of matrices presented in Appendix I before continuing beyond this point.

Many geometric shapes of elements are used in finite element analysis for specific applications. The various elements used in a general-purpose commercial FEM software code constitute what is referred to as the *element library* of the code. Elements can be placed in the following categories: *line elements, surface elements, solid elements,* and *special-purpose elements.* Table 9.0-1 provides some, but not all, of the types of elements available for finite element analysis.

Since the finite element method is a numerical technique which discretizes the domain of a continuous structure, errors are inevitable. These errors are:

1. *Computational errors.*   These are due to round-off errors from the computer floating-point calculations and the formulations of the numerical integration schemes that are employed. Most commercial finite element codes concentrate on reducing these errors[†] and consequently the analyst generally is concerned with discretization factors.

Table 9.0-1   Sample finite element library

Element Type	None	Shape	Number of Nodes	Applications
Line	Truss		2	Pin-ended bar in tension or compression
	Beam		2	Bending
	Frame		2	Axial, torsional, and bending. With or without load stiffening.
Surface	4-noded quadri-lateral		4	Plane stress or strain, axisymmetry, shear panel, thin flat plate in bending
	8-noded quadri-lateral		8	Plane stress or strain, thin plate or shell in bending
	3-noded triangular		3	Plane stress or strain, axisymmetry, shear panel, thin flat plate in bending. Prefer quad where possible. Used for transitions of quads.

[†] For example, most commercial programs use automatic bandwidth optimization, which can significantly reduce numerical error.

Table 9.0-1 *(concluded)*

Element Type	None	Shape	Number of Nodes	Applications
Surface	6-noded Triangular	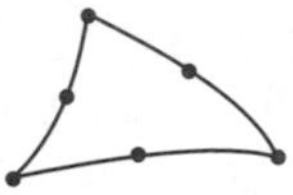	6	Plane stress or strain, axisymmetry, thin plate or shell in bending. Prefer quad where possible. Used for transitions of quads.
Solid[†]	8-noded hexagonal (brick)	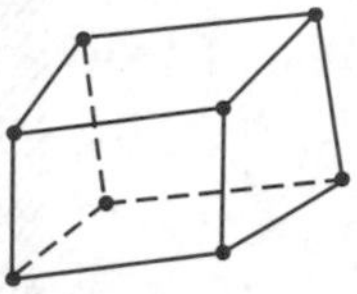	8	Solid, thick plate (using midside nodes)
	6-noded pentagonal (wedge)	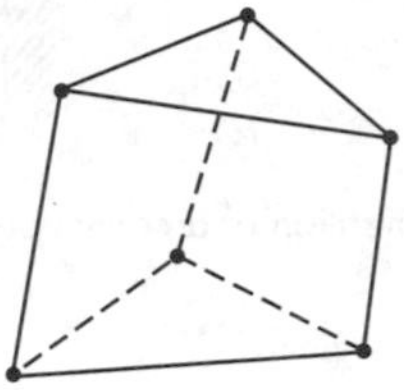	6	Solid, thick plate (using midside nodes). Used for transitions.
	4-noded tetrahedron (tet)	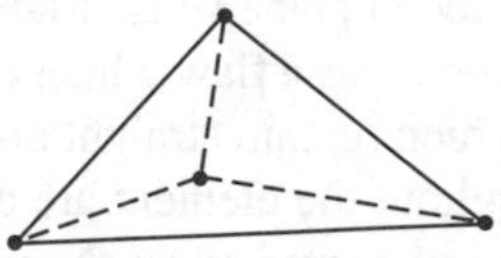	4	Solid, thick plate (using midside nodes). Used for transitions.
Special purpose	Gap		2	Free displacement for prescribed compressive gap
	Hook	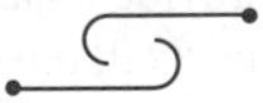	2	Free displacement for prescribed extension gap
	Rigid		Variable	Rigid constraints between nodes

[†] These elements are also available with midside nodes.

2. *Discretization errors.* The geometry and the displacement distribution of a true structure continuously varies. Using a finite number of elements to model the structure introduces errors in matching geometry and the displacement distribution due to the inherent limitations of the elements. For example, consider the thin plate structure shown in Fig. 9.0-1(*a*). Figure 9.0-1(*b*) shows a finite element model of

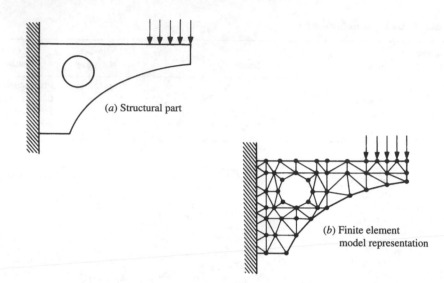

(a) Structural part

(b) Finite element
model representation

Figure 9.0-1   Discretization of a continuous structure.

the structure where three-noded plane stress triangular elements are employed. The plane stress triangular element has a flaw which creates two basic problems. The element has straight sides which remain straight after deformation. As will be seen in Sec. 9.4, the strains throughout the element are constant. The first problem, a geometric one, is the modeling of curved edges. Note that the surface of the model with a large curvature appears reasonably modeled, whereas the surface of the hole is very poorly modeled. The second problem, which is much more severe, is that the strains in various regions of the actual structure are changing rapidly, and the constant strain element will only provide an approximation of the average strain at the center of the element. So, in a nutshell, the results predicted using this model will be relatively poor. The results can be improved by significantly increasing the number of elements used (increased mesh density). Alternatively, using a better element, such as an eight-noded quadrilateral, which is more suited to the application will provide improved results. This element can model curved edges and provides for a higher-order strain distribution.

This chapter is intended to provide a basic understanding of element matrices, and the assembly and solution process. Entire books as well as multivolume books are dedicated to the finite element method. These books typically develop the equations and processes for many of the elements and applications shown in Table 9.0-1, as well as devote many pages to the numerical computational techniques associated with finite element software programs. Obviously, here in one chapter we can only try to provide a glimpse of the theory of the method and how it is applied. Toward this purpose, this chapter goes into some depth for the line elements (truss, beam, and frame) where the formulations do not require any involved numerical operations. Later, simple membrane elements are introduced to provide some insight into the differences between line and spatial elements in the formulation of the element equations.

Chapter 10 presents some modeling techniques employed when using commercial finite element codes. Hopefully, with these two chapters, some insight can be gained so as to eliminate some of the pitfalls that can arise from the ignorant use of an unintelligent computer program.[†]

9.1 NODE AND ELEMENT SUBSCRIPT NOTATION

In succeeding sections, the following subscript notation will be used. For the initial definition of a specific element type, nodes will be defined by the letters $i, j, k, \ldots$. Nodal displacements and loads will be subscripted by these letters. For example, the nodal force and translational displacement of node i in the x direction are f_{xi} and u_i, respectively. When dealing with a specific application, numbers rather than letters will be assigned to nodes. For example, the nodal force and translational displacement of node 2 in the y direction are f_{y2} and v_2, respectively.

For element notation, the subscript e will be used for the initial definition, whereas for applications, the elements will be assigned numbers. For example, the cross-sectional area of element 5 will be designated as A_5. It should be obvious as to when a subscript is referring to a node or to an element.

Note that since node i belongs to element e perhaps the force and deflection of the node, say in the x direction, should be written as $(f_{xi})_e$ and $(u_i)_e$, respectively. However, to avoid an overly cumbersome use of subscripts in the initial derivation of the element equations, we will write the force and deflection as f_{xi} and u_i, respectively. In the final key matrix equations we will insert the e subscript, as will be seen later.

9.2 THE TRUSS ELEMENT

In this section we will demonstrate the finite element technique on the simplest elastic element, the truss element. First we will develop the element in one-dimensional space, and then through coordinate transformations we will cast the element into two- and three-dimensional space. The one-dimensional element equations will first be developed using the *direct stiffness method*. This method is very easy to

---

[†] This is not meant to belittle any specific commercial finite element code. Certainly, it takes a great deal of intelligence to create the capabilities that current finite element software packages possess. Some of the commercial codes provide excellent mesh generation techniques, optimized numerical procedures, and comprehensive error tracking capabilities. However, it takes an intelligent analyst to continuously scrutinize modeling techniques and their corresponding results. Intelligent use of computer software minimizes the "gigo" (garbage in = garbage out) factor.

understand physically. However, the method becomes inadequate as more abstract conditions develop. For this purpose the variational energy procedure, the Rayleigh-Ritz method (discussed in Chap. 6), will be presented.

9.2.1 THE ONE-DIMENSIONAL TRUSS ELEMENT— DIRECT STIFFNESS METHOD

The definition of the one-dimensional element is provided by Fig. 9.2-1 (for clarity, the deflected element is shown offset in the y direction). Once the element is defined, a displacement field $u(x)$ is assumed. As will be seen later, we want $u(x)$ to be a function of x and the nodal parameters. Since there are two nodes, we write $u(x)$ in terms of two unknown constants. The simplest function is linear given by

$$u(x) = a_1 + a_2 x \qquad \text{[9.2-1]}$$

where a_1 and a_2 are constants. Since $u_i = u(x_i)$ and $u_j = u(x_j)$

$$u_i = a_1 + a_2 x_i \qquad u_j = a_1 + a_2 x_j$$

Solving the two equations simultaneously yields

$$a_1 = \frac{u_i x_j - u_j x_i}{x_j - x_i} \qquad a_2 = \frac{u_j - u_i}{x_j - x_i}$$

Substituting a_1 and a_2 back into Eq. (9.2-1) with $x_j - x_i = L_e$ gives

$$u(x) = \frac{1}{L_e}\left[(u_i x_j - u_j x_i) + (u_j - u_i)x\right] \qquad \text{[9.2-2]}$$

Equation (9.2-2) can be rewritten in the form

$$u(x) = \frac{x_j - x}{L_e} u_i + \frac{x - x_i}{L_e} u_j \qquad \text{[9.2-3]}$$

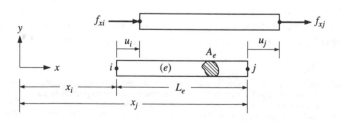

Figure 9.2-1   Definition of the one-dimensional axial element.

or

$$u(x) = N_i(x)u_i + N_j(x)u_j \qquad \text{[9.2-4]}$$

where

$$N_i(x) = \frac{x_j - x}{L_e} \qquad N_j(x) = \frac{x - x_i}{L_e} \qquad \text{[9.2-5]}$$

The $N_i(x)$ and $N_j(x)$ are called *shape functions* or *interpolation functions*.

Instead of writing $u(x)$ as a function of the two meaningless unknowns a_1 and a_2, we now have $u(x)$ expressed as a function of the two unknowns u_i and u_j, which are the unknown displacements of the nodes.

Next, the strain on the element is given by $\varepsilon_x = \partial u/\partial x$. Thus, from Eq. (9.2-2) or (9.2-3)

$$\varepsilon_x = \frac{1}{L_e}(u_j - u_i) \qquad \text{[9.2-6]}$$

which should be obvious. Equation (9.2-6) can be written in matrix form as

$$\varepsilon_x = \frac{1}{L_e}\{-1 \ \ 1\}\begin{Bmatrix} u_i \\ u_j \end{Bmatrix}_e \qquad \text{[9.2-7]}$$

where the last matrix is the element nodal displacement vector. The matrix which relates the displacement to strain is commonly referred to as the [**B**] matrix, which is a function of the element geometry. Thus, for the element

$$[\mathbf{B}]_e = \frac{1}{L_e}\{-1 \ \ 1\} \qquad \text{[9.2-8]}$$

For a one-dimensional axial element, the stress is given by $\sigma_x = E_e\varepsilon_x$, where E_e is the modulus of elasticity of the element material. Thus, from Eq. (9.2-7), the element stress is

$$(\sigma_x)_e = \left(\frac{E}{L}\right)_e \{-1 \ \ 1\}\begin{Bmatrix} u_i \\ u_j \end{Bmatrix}_e \qquad \text{[9.2-9]}$$

The last step in formulating the element's governing equations is to bring in the nodal forces. From Fig. 9.2-1 the forces are related to stress by

$$f_{xi} = -(\sigma_x A)_e \qquad f_{xj} = (\sigma_x A)_e$$

where A_e is the cross-sectional area of the element. These two equations can be written in matrix form as

$$\begin{Bmatrix} f_{xi} \\ f_{xj} \end{Bmatrix}_e = A_e \begin{Bmatrix} -1 \\ 1 \end{Bmatrix}(\sigma_x)_e \qquad \text{[9.2-10]}$$

Substituting Eq. (9.2-9) into this yields

$$\begin{Bmatrix} f_{xi} \\ f_{xj} \end{Bmatrix}_e = \left(\frac{AE}{L} \right)_e \begin{Bmatrix} -1 \\ 1 \end{Bmatrix} \{-1 \ 1\} \begin{Bmatrix} u_i \\ u_j \end{Bmatrix}_e \qquad \textbf{[9.2-11]}$$

Multiplying the first two matrices on the right-hand side of Eq. (9.2-11) gives

$$\begin{Bmatrix} f_{xi} \\ f_{xj} \end{Bmatrix}_e = \left(\frac{AE}{L} \right)_e \begin{bmatrix} 1 & -1 \\ -1 & 1 \end{bmatrix} \begin{Bmatrix} u_i \\ u_j \end{Bmatrix}_e \qquad \textbf{[9.2-12]}$$

The matrix which relates the nodal displacements to the nodal forces on the element is called the *element stiffness matrix* $[\mathbf{k}]_e$. Thus

$$[\mathbf{k}]_e = \left(\frac{AE}{L} \right)_e \begin{bmatrix} 1 & -1 \\ -1 & 1 \end{bmatrix} \qquad \textbf{[9.2-13]}$$

Equation (9.2-12) can also be written as

$$\begin{Bmatrix} f_{xi} \\ f_{xj} \end{Bmatrix}_e = k_e \begin{bmatrix} 1 & -1 \\ -1 & 1 \end{bmatrix} \begin{Bmatrix} u_i \\ u_j \end{Bmatrix}_e \qquad \textbf{[9.2-14]}$$

where

$$k_e = \left(\frac{AE}{L} \right)_e \qquad \textbf{[9.2-15]}$$

is the spring constant of an axial element, and the stiffness matrix is

$$[\mathbf{k}]_e = k_e \begin{bmatrix} 1 & -1 \\ -1 & 1 \end{bmatrix} \qquad \textbf{[9.2-16]}$$

This completes the derivation of the governing equations for the axial element. Before proceeding to a system of axial elements, let us recap the process in deriving the element equations. The steps for the direct stiffness formulation are:

1. Define the element. Define the displacements and loads associated with the degrees of freedom of the nodes, and the geometric and material properties of the element.

2. Assume a deflection field of the element using as many unknown constants as there are degrees of freedom.

3. Determine the unknown constants of step 2 in terms of the nodal positions and displacements. Substitute these constants back into the displacement field equation.

4. Differentiate the displacement field equation to obtain the strain as a function of nodal displacements.

5. Substitute the stress-strain relations to write the stress as a function of nodal displacements.

6. Determine the relationship between the nodal forces and stress on the element. This step becomes more difficult as the complexity of the element increases.

7. Substitute the results of step 5 into the relationship of step 6 and simplify.

9.2.2 THE ONE-DIMENSIONAL TRUSS ELEMENT— THE RAYLEIGH-RITZ METHOD

The Rayleigh-Ritz method is applied to beams in Sec. 6.13. The method is based on the principle of minimum potential energy using variational techniques. To employ the procedure, the displacement field of a structure is expressed by a summation of r functions, ϕ_i ($i = 1, 2, \ldots, r$), where each function satisfies the geometric boundary conditions of the structure, and each function is multiplied by an unknown constant a_i ($i = 1, 2, \ldots, r$).

The total potential energy of an elastic system in static equilibrium Π is the sum of the internal strain energy U and the potential energy of the external forces, called the *work potential* W_p. That is,

$$\Pi = U + W_p \qquad \text{[9.2-17]}$$

The strain energy and work potential can be written in terms of the displacement field. Thus the potential energy is a function of the constants a_i. The Rayleigh-Ritz method then minimizes the potential energy with respect to these constants. Thus

$$\frac{\partial \Pi}{\partial a_i} = 0 \qquad (i = 1, 2, \ldots, r) \qquad \text{[9.2-18]}$$

represents r equations in the r unknowns a_i ($i = 1, 2, \ldots, r$). For the finite element, rather than a_i, the unknown constants will be the nodal displacements u_i [recall Eq. (9.2-3) for the axial element]. Thus

$$\frac{\partial \Pi}{\partial u_i} = 0 \qquad (i = 1, 2, \ldots, r) \qquad \text{[9.2-19]}$$

In general, the strain energy per unit volume is given by Eq. (6.2-9) $u = 1/2\{\sigma\}^T\{\varepsilon\}$. Integrating this through the volume gives

$$U = \frac{1}{2} \int_V \{\sigma\}^T\{\varepsilon\} \, dV \qquad \text{[9.2-20]}$$

For the axial element in a uniaxial stress field we can express the stress vector as $\{\sigma\}^T = \{\sigma_x\}^T$. Taking the transpose of Eq. (9.2-9) gives

$$\{\sigma\}^T = \{(\sigma_x)_e\}^T = \{u_i \quad u_j\}_e \left(\frac{E}{L}\right)_e \begin{Bmatrix} -1 \\ 1 \end{Bmatrix}$$

Substituting this and Eq. (9.2-7) into Eq. (9.2-20) with $dV = A_e dx$ yields the strain energy of the axial element

$$U_e = \frac{1}{2} \int_{L_e} \{u_i \quad u_j\}_e \left(\frac{E}{L}\right)_e \begin{Bmatrix} -1 \\ 1 \end{Bmatrix} \frac{1}{L_e} \{-1 \quad 1\} \begin{Bmatrix} u_i \\ u_j \end{Bmatrix}_e A_e \, dx$$

No term within the integral is a function of x. Rearranging the scalar terms, simplifying, and multiplying the inner matrices yields

$$U_e = \frac{1}{2} \{u_i \quad u_j\}_e \left(\frac{AE}{L}\right)_e \begin{bmatrix} 1 & -1 \\ -1 & 1 \end{bmatrix} \begin{Bmatrix} u_i \\ u_j \end{Bmatrix}_e \qquad \textbf{[9.2-21]}$$

The general form for the strain energy of an element is

$$U_e = \frac{1}{2} \{\mathbf{u}\}_e^T [\mathbf{k}]_e \{\mathbf{u}\}_e \qquad \textbf{[9.2-22]}$$

where $\{\mathbf{u}\}_e$ is the nodal displacement vector of the element and $[\mathbf{k}]_e$ is the element stiffness matrix. Comparing Eqs. (9.2-21) and (9.2-22), the element stiffness matrix is

$$[\mathbf{k}]_e = \left(\frac{AE}{L}\right)_e \begin{bmatrix} 1 & -1 \\ -1 & 1 \end{bmatrix} \qquad \textbf{[9.2-23]}$$

which agrees with that developed in Sec. 9.2.1.

For the element with only nodal forces at the nodes, the work potential is

$$W_p = -u_i f_{xi} - u_j f_{xj} \qquad \textbf{[9.2-24]}$$

The terms are negative since the work potential is opposite in sign to the work performed *on* the system. The total potential energy of the element is thus

$$\Pi_e = \frac{1}{2} \{u_i \quad u_j\}_e \left(\frac{AE}{L}\right)_e \begin{bmatrix} 1 & -1 \\ -1 & 1 \end{bmatrix} \begin{Bmatrix} u_i \\ u_j \end{Bmatrix}_e - u_i f_{xi} - u_j f_{xj} \qquad \textbf{[9.2-25]}$$

Finally, using Eq. (9.2-19) we take the derivative of Π_e with respect to the unknown constants u_i and u_j separately and set to zero. This yields the equations

$$\left(\frac{AE}{L}\right)_e \begin{bmatrix} 1 & -1 \\ -1 & 1 \end{bmatrix} \begin{Bmatrix} u_i \\ u_j \end{Bmatrix}_e - \begin{Bmatrix} f_{xi} \\ f_{xj} \end{Bmatrix}_e = \begin{Bmatrix} 0 \\ 0 \end{Bmatrix}$$

or

$$\begin{Bmatrix} f_{xi} \\ f_{xj} \end{Bmatrix}_e = \left(\frac{AE}{L}\right)_e \begin{bmatrix} 1 & -1 \\ -1 & 1 \end{bmatrix} \begin{Bmatrix} u_i \\ u_j \end{Bmatrix}_e \qquad \textbf{[9.2-26]}$$

which agrees with Eq. (9.2-12) found by the direct stiffness method.

9.2.3   THE ASSEMBLY PROCESS

Consider a structure with two axial elements as shown in Fig. 9.2-2. The structure is fixed at the left end and a force is applied to the right end. The nodes and elements are numbered and the cross-sectional areas of elements (1) and (2) are A_1 and A_2, respectively. Figure 9.2-3 shows the individual elements with their nodal forces defined. Note that the nodal forces are subscripted for the element also. The matrix equations for each element can be written.

Element (1):

$$\begin{Bmatrix} f_{x1} \\ f_{x2} \end{Bmatrix}_1 = \begin{bmatrix} k_1 & -k_1 \\ -k_1 & k_1 \end{bmatrix} \begin{Bmatrix} u_1 \\ u_2 \end{Bmatrix} \qquad \text{[9.2-27]}$$

Element (2):

$$\begin{Bmatrix} f_{x2} \\ f_{x3} \end{Bmatrix}_2 = \begin{bmatrix} k_2 & -k_2 \\ -k_2 & k_2 \end{bmatrix} \begin{Bmatrix} u_2 \\ u_3 \end{Bmatrix} \qquad \text{[9.2-28]}$$

where $k_1 = (AE/L)_1$ and $k_2 = (AE/L)_2$. The forces shown in Fig. 9.2-3 are the element forces. Consider the total forces applied to the nodes to be F_1, F_2, and F_3 as

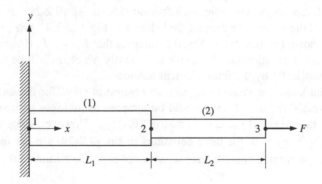

Figure 9.2-2   A two-element axial structure.

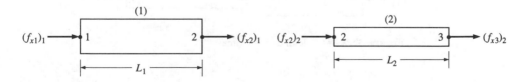

Figure 9.2-3   The individual elements of Fig. 9.2-2.

shown in Fig. 9.2-4. Superposing the forces of Fig. 9.2-3, we see that $F_1 = (f_{x1})_1$, $F_2 = (f_{x2})_1 + (f_{x2})_2$, and $F_3 = (f_{x3})_2$. Expressing these forces using Eqs. (9.2-27) and (9.2-28) gives

$$\begin{Bmatrix} F_1 \\ F_2 \\ F_3 \end{Bmatrix} = \begin{bmatrix} k_1 & -k_1 & 0 \\ -k_1 & (k_1 + k_2) & -k_2 \\ 0 & -k_2 & k_2 \end{bmatrix} \begin{Bmatrix} u_1 \\ u_2 \\ u_3 \end{Bmatrix}$$ [9.2-29]

or

$$\{F\}_{system} = [K]_{system}\{u\}_{system}$$ [9.2-30]

where $\{F\}_{system}$, $[K]_{system}$, and $\{u\}_{system}$ are the system nodal force vector, stiffness matrix, and displacement vectors, respectively, given by

$$\{F\}_{system} = \begin{Bmatrix} F_1 \\ F_2 \\ F_3 \end{Bmatrix} \quad [K]_{system} = \begin{bmatrix} k_1 & -k_1 & 0 \\ -k_1 & (k_1 + k_2) & -k_2 \\ 0 & -k_2 & k_2 \end{bmatrix} \quad \{u\}_{system} = \begin{Bmatrix} u_1 \\ u_2 \\ u_3 \end{Bmatrix}$$

[9.2-31]

Notice in the system stiffness matrix how the element matrices "join" (the $k_1 + k_2$ term) along their common diagonal term associated with their common node 2.

The F_i forces in the vector in the left-hand side of Eq. (9.2-29) are the net applied forces at the nodes. For the structure shown in Fig. 9.2-2, $F_3 = F$, $F_2 = 0$, and F_1 is the unknown reaction force. Yes, it is obvious that $F_1 = -F$. However, this requires us to solve the statics of the problem manually which, as will be seen shortly, is done automatically by the finite element method.

If we did know the value of F_1 and we desired to solve for the unknown displacement vector $\{u_1 \ u_2 \ u_3\}^T$, we would be tempted to multiply both sides of Eq. (9.2-29) by the inverse of the system matrix $[K]_{system}^{-1}$.[†] However, since for equilibrium $F_1 + F_2 + F_3 = 0$, the three equations in Eq. (9.2-29) are not independent. Thus $[K]_{system}$ is *singular* and its inverse does not exist. To eliminate this difficulty,

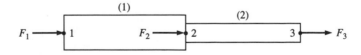

Figure 9.2-4   Total applied forces.

[†] It is known that $u_1 = 0$. This is the key to reducing the equations which will be employed momentarily.

we *partition* out the equation associated with the force for which we know the displacement, namely, the F_1 equation. Thus we have

$$F_1 = k_1(u_1 - u_2) \qquad \text{[9.2-32]}$$

and

$$\begin{Bmatrix} F_2 \\ F_3 \end{Bmatrix} = \begin{bmatrix} -k_1 & (k_1 + k_2) & -k_2 \\ 0 & -k_2 & k_2 \end{bmatrix} \begin{Bmatrix} u_1 \\ u_2 \\ u_3 \end{Bmatrix} \qquad \text{[9.2-33]}$$

Next, since we know the value of u_1, we move u_1 together with its coefficients from the 2×3 matrix to the left-hand side of Eq. (9.2-33) with the other known terms F_2 and F_3. That is, we move u_1 and the column associated with it to the other side of the equation. Thus

$$\begin{Bmatrix} F_2 + k_1 u_1 \\ F_3 - 0u_1 \end{Bmatrix} = \begin{bmatrix} (k_1 + k_2) & -k_2 \\ -k_2 & k_2 \end{bmatrix} \begin{Bmatrix} u_2 \\ u_3 \end{Bmatrix}$$

In this case, since $u_1 = 0$, we do not need to continue to carry the term. However, in some cases the boundary condition will be nonzero and we must perform the above step correctly. Setting $u_1 = 0$ gives

$$\begin{Bmatrix} F_2 \\ F_3 \end{Bmatrix} = \begin{bmatrix} (k_1 + k_2) & -k_2 \\ -k_2 & k_2 \end{bmatrix} \begin{Bmatrix} u_2 \\ u_3 \end{Bmatrix} \qquad \text{[9.2-34]}$$

The 2×2 matrix is called the *reduced stiffness matrix* $[\mathbf{K}]_{\text{red}}$, which has an inverse. The inverse is

$$\begin{bmatrix} (k_1 + k_2) & -k_2 \\ -k_2 & k_2 \end{bmatrix}^{-1} = \frac{\begin{bmatrix} k_2 & k_2 \\ k_2 & (k_1 + k_2) \end{bmatrix}}{(k_1 + k_2)(k_2) - k_2^2} = \frac{1}{k_1 k_2} \begin{bmatrix} k_2 & k_2 \\ k_2 & (k_1 + k_2) \end{bmatrix}$$

Thus the inverse of Eq. (9.2-34) is

$$\begin{Bmatrix} u_2 \\ u_3 \end{Bmatrix} = \frac{1}{k_1 k_2} \begin{bmatrix} k_2 & k_2 \\ k_2 & (k_1 + k_2) \end{bmatrix} \begin{Bmatrix} F_2 \\ F_3 \end{Bmatrix} \qquad \text{[9.2-35]}$$

Writing the equations individually with $F_2 = 0$ and $F_3 = F$ gives

$$u_2 = \frac{1}{k_1} F \qquad \text{[9.2-36a]}$$

$$u_3 = \frac{k_1 + k_2}{k_1 k_2} F \qquad \text{[9.2-36b]}$$

Based on the deflection of springs in series, the reader should verify that this is correct. The deflections can then be substituted into Eq. (9.3-32) to determine the reaction force F_1. This is found to be

$$F_1 = k_1(u_1 - u_2) = k_1\left(0 - \frac{F}{k_1}\right) = -F \qquad \textbf{[9.2-37]}$$

which confirms our earlier observation. Through the partitioning process of the finite element method, the reaction force is determined. However, it is always advisable to check your results by other methods even if your check is only an estimate.

If desired, the element stress, strain, or nodal forces can now be obtained quite easily from Eqs. (9.2-9), (9.2-6), or (9.2-14), respectively. Say, for example, we desire the nodal forces and stress of element 1. The nodal forces on element 1 are

$$\begin{Bmatrix} f_{x1} \\ f_{x2} \end{Bmatrix}_1 = \begin{bmatrix} k_1 & -k_1 \\ -k_1 & k_1 \end{bmatrix} \begin{Bmatrix} u_1 \\ u_2 \end{Bmatrix} = \begin{bmatrix} k_1 & -k_1 \\ -k_1 & k_1 \end{bmatrix} \begin{Bmatrix} 0 \\ F/k_1 \end{Bmatrix} = \begin{Bmatrix} -F \\ F \end{Bmatrix}$$

which the reader should verify are correct. For stress, let $k_1 = (AE/L)_1$. Then from Eq. (9.2-36a), $u_2 = F/(AE/L)_1$, and the stress in element (1) is

$$(\sigma_x)_1 = \left(\frac{E}{L}\right)_1 \{-1 \quad 1\} \begin{Bmatrix} u_1 \\ u_2 \end{Bmatrix} = \left(\frac{E}{L}\right)_1 \{-1 \quad 1\} \begin{Bmatrix} 0 \\ \dfrac{F}{(AE/L)_1} \end{Bmatrix} = \frac{F}{A_1}$$

which is obviously correct.

Example 9.2-1 | For the structure shown in Fig. 9.2-5 consider the walls to be rigid. Using the finite element method determine the deflections of all nodes; the wall reactions; and the deflection equations, stresses, and nodal forces for each element. Plot the deflection

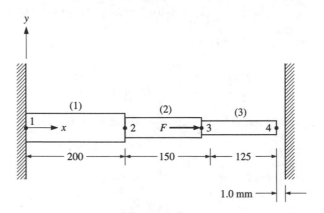

Figure 9.2-5

and stress as a function of x for the entire structure. The applied force is $F = 20$ kN. The areas for the elements are $A_1 = 100$, $A_2 = 75$, and $A_3 = 50$ mm². The elements have the same modulus of elasticity, $E = 50$ GPa.

Solution:

First we will determine whether or not node 4 hits the wall. Thus we will ignore the wall and determine the deflection of node 4. If the deflection is less than 1 mm, the solution is valid. If it is greater than 1 mm, we must reformulate the problem with a specified deflection of 1 mm.[†]

Using $k = AE/L$, the stiffness of each element is

$$k_1 = \frac{(100)(10^{-6})(50)(10^9)}{200(10^{-3})} = 25(10^6) \, \text{N/m}$$

$$k_2 = \frac{(75)(10^{-6})(50)(10^9)}{150(10^{-3})} = 25(10^6) \, \text{N/m}$$

$$k_3 = \frac{(50)(10^{-6})(50)(10^9)}{125(10^{-3})} = 20(10^6) \, \text{N/m}$$

The nodal force-deflection equations can be written for each element and assembled as with the two-element mode. When assembled, with the total nodal force vector being $\{F_1 \, F_2 \, F_3 \, F_4\}^T$, the system force-deflection equations are found to be

$$\begin{Bmatrix} F_1 \\ F_2 \\ F_3 \\ F_4 \end{Bmatrix} = \begin{bmatrix} k_1 & -k_1 & 0 & 0 \\ -k_1 & (k_1 + k_2) & -k_2 & 0 \\ 0 & -k_2 & (k_2 + k_3) & -k_3 \\ 0 & 0 & -k_3 & k_3 \end{bmatrix} \begin{Bmatrix} u_1 \\ u_2 \\ u_3 \\ u_4 \end{Bmatrix}$$

Note again, the stiffness matrices of each element *join* (add) at their common degree of freedom. Substituting the k_i values yields

$$\begin{Bmatrix} F_1 \\ F_2 \\ F_3 \\ F_4 \end{Bmatrix} = (10^6) \begin{bmatrix} 25 & -25 & 0 & 0 \\ -25 & 50 & -25 & 0 \\ 0 & -25 & 45 & -20 \\ 0 & 0 & -20 & 20 \end{bmatrix} \begin{Bmatrix} u_1 \\ u_2 \\ u_3 \\ u_4 \end{Bmatrix}$$

As before, the overall system stiffness matrix is singular and we must apply the boundary conditions and partition the equations. Since we are first checking to see how much node 4 deflects without considering the right wall, the only boundary

---

[†] This, in essence, is what the special-purpose gap element shown in Table 9.0-1 accomplishes.

condition is $u_1 = 0$. The force associated with this degree of freedom is the unknown reaction force F_1, so we partition the F_1 equation. This yields

$$F_1 = 25(10^6)(u_1 - u_2) \qquad \textbf{[a]}$$

and the remaining equations are

$$\begin{Bmatrix} F_2 \\ F_3 \\ F_4 \end{Bmatrix} = (10^6) \begin{bmatrix} -25 & 50 & -25 & 0 \\ 0 & -25 & 45 & -20 \\ 0 & 0 & -20 & 20 \end{bmatrix} \begin{Bmatrix} u_1 \\ u_2 \\ u_3 \\ u_4 \end{Bmatrix}$$

However, the first column of the 3×4 matrix multiplies the known u_1 and can be moved with u_1 to the left-hand side of the equation. But since $u_1 = 0$, the left-hand side of the equation does not change. This gives

$$\begin{Bmatrix} F_2 \\ F_3 \\ F_4 \end{Bmatrix} = (10^6) \begin{bmatrix} 50 & -25 & 0 \\ -25 & 45 & -20 \\ 0 & -20 & 20 \end{bmatrix} \begin{Bmatrix} u_2 \\ u_3 \\ u_4 \end{Bmatrix} \qquad \textbf{[b]}$$

Solving Eq. (b) for $\{u_2 \ u_3 \ u_4\}^T$ by determining the inverse of the reduced modified 3×3 system matrix yields

$$\begin{Bmatrix} u_2 \\ u_3 \\ u_4 \end{Bmatrix} = 4(10^{-8}) \begin{bmatrix} 1 & 1 & 1 \\ 1 & 2 & 2 \\ 1 & 2 & 3.25 \end{bmatrix} \begin{Bmatrix} F_2 \\ F_3 \\ F_4 \end{Bmatrix} \qquad \textbf{[c]}$$

Solving for u_4 with $F_2 = F_4 = 0$, and $F_3 = 20$ kN yields

$$u_4 = 4(10^{-8})[(1)(0) + (2)(20)(10^3) + (3.25)(0)] = 1.6(10^{-3}) \, \text{m} = 1.6 \, \text{mm}$$

Since 1.6 mm > 1 mm, node 4 hits the wall and the problem must be reformulated.

Hitting the wall means that we now have two boundary conditions, $u_1 = 0$ and $u_4 = 1$ mm. In the first solution we had already partitioned out the first boundary condition, $u_1 = 0$, yielding Eqs. (a) and (b). All that is necessary now is to apply the second boundary condition, $u_4 = 1$ mm, to Eq. (b). Since the force at the degree of freedom associated with u_4 is now the unknown F_4, we partition out the last equation in Eq. (b). This results in

$$F_4 = 20(10^6)(u_4 - u_3) \qquad \textbf{[d]}$$

and

$$\begin{Bmatrix} F_2 \\ F_3 \end{Bmatrix} = (10^6) \begin{bmatrix} 50 & -25 & 0 \\ -25 & 45 & -20 \end{bmatrix} \begin{Bmatrix} u_2 \\ u_3 \\ u_4 \end{Bmatrix}$$

Again, since u_4 is known, we move the last column of the 2 × 3 matrix, with u_4, to the left-hand side of the equation. However, since u_4 is *not* zero, the force vector *is* modified this time. Thus

$$\left\{ \begin{array}{c} F_2 \\ F_3 + 20(10^6)u_4 \end{array} \right\} = (10^6) \begin{bmatrix} 50 & -25 \\ -25 & 45 \end{bmatrix} \left\{ \begin{array}{c} u_2 \\ u_3 \end{array} \right\} \qquad [e]$$

Solving for $\{u_2 \ u_3\}^T$, obtaining the inverse of the reduced modified 2 × 2 stiffness matrix yields

$$\left\{ \begin{array}{c} u_2 \\ u_3 \end{array} \right\} = (10^{-8}) \begin{bmatrix} 2.7692 & 1.5385 \\ 1.5385 & 3.0769 \end{bmatrix} \left\{ \begin{array}{c} F_2 \\ F_3 + 20(10^6)u_4 \end{array} \right\} \qquad [f]$$

Substitution of $F_2 = 0$, $F_3 = 20$ kN, and $u_4 = 1$ mm yields

$$u_2 = (10^{-8})\{(2.7692)(0) + (1.5385)[20(10^3) + 20(10^6)(1)(10^{-3})]\}$$

$$= 6.154(10^{-4}) \text{ m} = 0.6154 \text{ mm}$$

and

$$u_3 = (10^{-8})\{(1.5385)(0) + (3.0769)[20(10^3) + 20(10^6)(1)(10^{-3})]\}$$

$$= 1.2308(10^{-3}) \text{ m} = 1.2308 \text{ mm}$$

The reactions are determined by substituting the deflections into Eqs. (*a*) and (*d*), giving

$$F_1 = 25(10^6)[0 - 0.6154(10^{-3})] = -15.38(10^3) \text{ N} = -15.38 \text{ kN}$$

$$F_4 = 20(10^6)[1(10^{-3}) - 1.2308(10^{-3})] = -4.615(10^3) \text{ N} = -4.62 \text{ kN}$$

The deflection equations, stresses, and nodal forces for each element are given by Eqs. (9.2-2), (9.2-9), and (9.2-14), respectively. For the deflection equations let the $u_1(x)$ and x_i be expressed in millimeters.

Element (1): $0 < x < 200$ mm

$$u(x) = \frac{1}{L_1}[(u_1 x_2 - u_2 x_1) + (u_2 - u_1)x]$$

$$= \frac{1}{200}[(0)(200) - (0.6154)(0) + (0.6154 - 0)x] = 3.077(10^{-3})x$$

$$(\sigma_x)_1 = \left(\frac{E}{L}\right)_1 \{-1 \quad 1\} \begin{Bmatrix} u_1 \\ u_2 \end{Bmatrix} = \frac{50(10^9)}{200(10^{-3})} \{-1 \quad 1\} \begin{Bmatrix} 0 \\ 0.6154(10^{-3}) \end{Bmatrix}$$

$$= 153.8(10^6) \, \text{N/m}^2 = 153.8 \, \text{MPa}$$

$$\begin{Bmatrix} f_{x1} \\ f_{x2} \end{Bmatrix}_1 = k_1 \begin{bmatrix} 1 & -1 \\ -1 & 1 \end{bmatrix} \begin{Bmatrix} u_1 \\ u_2 \end{Bmatrix} = 25(10^6) \begin{bmatrix} 1 & -1 \\ -1 & 1 \end{bmatrix} \begin{Bmatrix} 0 \\ 0.6154(10^{-3}) \end{Bmatrix}$$

$$= \begin{Bmatrix} -15.38 \\ 15.38 \end{Bmatrix} (10^3) \, \text{N} = \begin{Bmatrix} -15.38 \\ 15.38 \end{Bmatrix} \text{kN}$$

Element (2): $200 < x < 350$ mm

$$u(x) = \frac{1}{L_2}[(u_2 x_3 - u_3 x_2) + (u_3 - u_2)x]$$

$$= \frac{1}{150}[(0.6154)(350) - (1.2308)(200) + (1.2308 - 0.6154)x]$$

$$= 0.2051 + 4.103(10^{-3})x$$

$$(\sigma_x)_2 = \left(\frac{E}{L}\right)_2 \{-1 \quad 1\} \begin{Bmatrix} u_2 \\ u_3 \end{Bmatrix} = \frac{50(10^9)}{150(10^{-3})} \{-1 \quad 1\} \begin{Bmatrix} 0.6154 \\ 1.2308 \end{Bmatrix} (10^{-3})$$

$$= 205.1(10^6) \, \text{N/m}^2 = 205.1 \, \text{MPa}$$

$$\begin{Bmatrix} f_{x2} \\ f_{x3} \end{Bmatrix}_2 = k_2 \begin{bmatrix} 1 & -1 \\ -1 & 1 \end{bmatrix} \begin{Bmatrix} u_2 \\ u_3 \end{Bmatrix} = 25(10^6) \begin{bmatrix} 1 & -1 \\ -1 & 1 \end{bmatrix} \begin{Bmatrix} 0.6154 \\ 1.2308 \end{Bmatrix} (10^{-3})$$

$$= \begin{Bmatrix} -15.38 \\ 15.38 \end{Bmatrix} (10^3) \, \text{N} = \begin{Bmatrix} -15.38 \\ 15.38 \end{Bmatrix} \text{kN}$$

Element (3): $350 < x < 475$ mm

$$u(x) = \frac{1}{L_3}[(u_3 x_4 - u_4 x_3) + (u_4 - u_3)x]$$

$$= \frac{1}{125}[(1.2308)(475) - (1)(350) + (1 - 1.2308)x]$$

$$= 1.8769 - 1.8462(10^{-3})x$$

$$(\sigma_x)_3 = \left(\frac{E}{L}\right)_3 \{-1 \quad 1\} \begin{Bmatrix} u_3 \\ u_4 \end{Bmatrix} = \frac{50(10^9)}{125(10^{-3})} \{-1 \quad 1\} \begin{Bmatrix} 1.2308 \\ 1 \end{Bmatrix} (10^{-3})$$

$$= -92.3(10^6) \, \text{N/m}^2 = -92.3 \, \text{MPa}$$

$$\begin{Bmatrix} f_{x3} \\ f_{x4} \end{Bmatrix}_3 = k_3 \begin{bmatrix} 1 & -1 \\ -1 & 1 \end{bmatrix} \begin{Bmatrix} u_3 \\ u_4 \end{Bmatrix} = 20(10^6) \begin{bmatrix} 1 & -1 \\ -1 & 1 \end{bmatrix} \begin{Bmatrix} 1.2308(10^{-3}) \\ 1(10^{-3}) \end{Bmatrix}$$

$$= \begin{Bmatrix} 4.62 \\ -4.62 \end{Bmatrix} (10^3) \, \text{N} = \begin{Bmatrix} 4.62 \\ -4.62 \end{Bmatrix} \, \text{kN}$$

Figure 9.2-6 depicts the displacement and stress plots of the structure. Note that the simple axial element that we are using is a constant strain, constant stress element. This can produce discontinuities in strain and stress at the nodes. The discontinuities in stress can be seen in the stress plot in Fig. 9.9-6, whereas the discontinuity in strain can be seen in the slope discontinuities of the displacement plot.

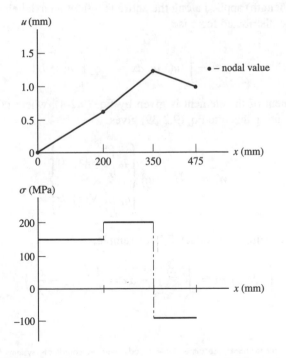

Figure 9.2-6   Displacement and stress as functions of x for Example 9.2-1.

9.2.4   DISTRIBUTED LOADS

Distributed loads can be handled quite easily using the energy approach. Consider a system where the nodes can have all six degrees of freedom, three translation, and three rotation. The work potential arises from the displacement of the loads on the structure. The loads can be body forces per unit volume $\{\overline{\mathbf{F}}\} = \{\overline{F}_x \ \overline{F}_y \ \overline{F}_z\}$ surface force distributions $\{\mathbf{q}\} = \{q_x \ q_y \ q_z\}$, concentrated forces $\{\mathbf{P}\} = \{P_x \ P_y \ P_z\}$, and concentrated moments $\{\mathbf{M}\} = \{M_x \ M_y \ M_z\}$. The form of the work potential can then be expressed as

$$W_p = -\int_V \{\mathbf{u}\}^T\{\overline{\mathbf{F}}\} \, dV - \int_A \{\mathbf{u}\}^T\{\mathbf{q}\} \, dA - \sum_i \{\mathbf{u}_i\}^T\{\mathbf{P}_i\} - \sum_j \{\mathbf{\theta}_j\}^T\{\mathbf{M}_j\}$$

[9.2-38]

where $\{\mathbf{u}\}^T = \{u(x,y,z) \ v(x,y,z) \ w(x,y,z)\}$ is the transpose of the vector of the translational displacement fields, $\{\mathbf{u}_i\}^T = \{u \ v \ w\}_i$ is the transpose of the displacement vector at the site of the force vector $\{\mathbf{P}_i\}$, and $\{\mathbf{\theta}_j\}^T = \{\theta_x \ \theta_y \ \theta_z\}_j$ is the transpose of the rotational displacement vector at the site of the moment vector $\{\mathbf{M}_j\}$. The i, j subscripts correspond to a set of discrete loads.

For the one-dimensional axial element there is no rotational displacement vector, and $\{\mathbf{u}\}^T = \{u(x)\} = u(x)$. Consider a uniformly distributed constant surface load q_o (force/length) applied along the entire length of an axial element. The work potential of the distributed force is

$$W_p = -\int_{L_e} u(x)q_o \, dx = -q_o \int_{L_e} u(x) \, dx$$

[9.2-39]

The displacement of the element is given by Eq. (9.2-4), where $u(x) = N_i(x)u_i + N_j(x)u_j$. Substituting this into Eq. (9.2-39) gives

$$W_p = -\{u_i \quad u_j\} \begin{Bmatrix} q_o \int_{x_i}^{x_j} N_i(x) \, dx \\[2mm] q_o \int_{x_i}^{x_j} N_j(x) \, dx \end{Bmatrix}$$

[9.2-40]

The integrals are simple to evaluate.[†] For example,

$$\int_{x_i}^{x_j} N_i(x) \, dx = \int_{x_i}^{x_j} \frac{x_j - x}{L_e} \, dx = \frac{1}{L_e}\left(x_j x - \frac{x^2}{2}\right)\Big|_{x_i}^{x_j}$$

[†] Most finite element textbooks use dimensionless body-centered coordinate systems for elements, which makes this integration even simpler.

$$\int_{x_i}^{x_j} N_i(x)\,dx = \frac{1}{L_e}\left[x_j^2 - \frac{x_j^2}{2} - \left(x_j x_i - \frac{x_i^2}{2}\right)\right]$$

$$= \frac{1}{2L_e}(x_j^2 - 2x_j x_i + x_i^2) = \frac{(x_j - x_i)^2}{2L_e} = \frac{L_e}{2}$$

The other integral can also be shown to be $L_e/2$. Thus Eq. (9.2-40) reduces to

$$W_p = -\{u_i \quad u_j\}\begin{Bmatrix} \dfrac{q_o L_e}{2} \\[2mm] \dfrac{q_o L_e}{2} \end{Bmatrix} \qquad\qquad \textbf{[9.2-41]}$$

If concentrated forces $\{F_i \quad F_j\}^T$ were placed at the nodes, the work potential would be

$$W_p = -\{u_i \quad u_j\}\begin{Bmatrix} F_i \\ F_j \end{Bmatrix} \qquad\qquad \textbf{[9.2-42]}$$

Thus the *work equivalent nodal forces* for the uniform load are both $q_o L_e/2$. This makes sense as the total force on the element is $q_o L_e$, and dividing it equally to each node would be statically equivalent. The work equivalent nodal forces are not always so obvious, as we will see later for the beam element. The equivalent nodal forces can be added to the nodal forces in the element equation. Thus Eq. (9.2-12) can be rewritten as

$$\begin{Bmatrix} f_{xi} \\ f_{xj} \end{Bmatrix}_e + \frac{q_o L_e}{2}\begin{Bmatrix} 1 \\ 1 \end{Bmatrix} = \left(\frac{AE}{L}\right)_e \begin{bmatrix} 1 & -1 \\ -1 & 1 \end{bmatrix}\begin{Bmatrix} u_i \\ u_j \end{Bmatrix}_e \qquad \textbf{[9.2-43]}$$

The stress in the element with a distributed force will no longer be constant. Isolating the element at location x as shown in Fig. 9.2-7, the stress is given by

$$(\sigma_x)_e = -\left[\frac{f_{xi} + q_o(x - x_i)}{A}\right]_e \qquad\qquad \textbf{[9.2-44]}$$

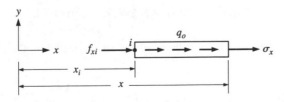

Figure 9.2-7

From the first equation of Eq. (9.2-43) we have

$$f_{xi} = \left(\frac{AE}{L}\right)_e \{1 \quad -1\} \begin{Bmatrix} u_i \\ u_j \end{Bmatrix}_e - \frac{q_o L_e}{2} \qquad \textbf{[9.2-45]}$$

Substituting this into Eq. (9.2-44) yields

$$(\sigma_x)_e = \left(\frac{E}{L}\right)_e \{1 \quad -1\} \begin{Bmatrix} u_i \\ u_j \end{Bmatrix}_e + \left(\frac{q_o}{2A}\right)_e (L_e - 2x + 2x_i) \qquad \textbf{[9.2-46]}$$

We will see an example of how to apply these equations after we develop the equations to account for thermal effects.

9.2.5   THERMAL STRESS

If a stressed axial element first undergoes an unconstrained temperature change, the total strain is given by

$$\varepsilon_x = \alpha \ \Delta T + (\varepsilon_x)_s = \alpha \ \Delta T + \frac{\sigma_x}{E} \qquad \textbf{[9.2-47]}$$

where α is the thermal expansion coefficient, ΔT is the change in temperature, and $(\varepsilon_x)_s$ is the *structural strain*. Consequently the stress is

$$\sigma_x = E(\varepsilon_x - \alpha \ \Delta T) \qquad \textbf{[9.2-48]}$$

Recall that in terms of the nodal displacements, the strain is given by Eq. (9.2-7), which is

$$\varepsilon_x = \frac{1}{L_e}\{1 \quad -1\} \begin{Bmatrix} u_i \\ u_j \end{Bmatrix}_e$$

Substituting this into Eq. (9.2-48) gives

$$(\sigma_x)_e = \left(\frac{E}{L}\right)_e \{1 \quad -1\} \begin{Bmatrix} u_i \\ u_j \end{Bmatrix}_e - (E\alpha \ \Delta T)_e \qquad \textbf{[9.2-49]}$$

The transpose of this is

$$\{(\sigma_x)_e\}^T = \frac{1}{L_e}\{u_i \ u_j\}_e \begin{Bmatrix} -1 \\ 1 \end{Bmatrix} - (E\alpha \ \Delta T)_e \qquad \textbf{[9.2-50]}$$

The part of the strain which contributes to the strain energy is the structural part, namely,

$$\{(\varepsilon_x)_s\} = \varepsilon_x - \alpha \ \Delta T = \frac{1}{L_e}\{1 \quad -1\} \begin{Bmatrix} u_i \\ u_j \end{Bmatrix}_e - \alpha \ \Delta T \qquad \textbf{[9.2-51]}$$

Substituting this and Eq. (9.2-50) into Eq. (9.2-20) gives the strain energy

$$U_e = \frac{1}{2} \int_{L_e} \left(\{u_i \ u_j\}_e \left(\frac{E}{L}\right)_e \left\{ \begin{matrix} -1 \\ 1 \end{matrix} \right\} - (E\alpha \ \Delta T)_e \right) \left(\frac{1}{L_e} \{-1 \ \ 1\} \left\{ \begin{matrix} u_i \\ u_j \end{matrix} \right\}_e - (\alpha \Delta T)_e \right) A_e \, dx$$

This can be simplified to

$$U_e = \frac{1}{2} \{u_i \ \ u_j\}_e \left(\frac{AE}{L}\right)_e \begin{bmatrix} 1 & -1 \\ -1 & 1 \end{bmatrix} \left\{ \begin{matrix} u_i \\ u_j \end{matrix} \right\}_e + (EA\alpha \ \Delta T)_e (u_i - u_j)_e + \frac{1}{2} [EAL(\alpha \ \Delta T)^2]_e$$

[9.2-52]

When this together with the work potential associated with the element nodal forces are included in the total potential energy and minimized relative to u_i and u_j, the following results (see Sec. 9.2.2):

$$\left(\frac{AE}{L}\right)_e \begin{bmatrix} 1 & -1 \\ -1 & 1 \end{bmatrix} \left\{ \begin{matrix} u_i \\ u_j \end{matrix} \right\}_e + \left\{ \begin{matrix} EA\alpha \ \Delta T \\ -EA\alpha \ \Delta T \end{matrix} \right\}_e - \left\{ \begin{matrix} f_{xi} \\ f_{xj} \end{matrix} \right\}_e = \left\{ \begin{matrix} 0 \\ 0 \end{matrix} \right\}$$ [9.2-53]

which can be rewritten as

$$\left\{ \begin{matrix} f_{xi} \\ f_{xj} \end{matrix} \right\}_e = \left(\frac{AE}{L}\right)_e \begin{bmatrix} 1 & -1 \\ -1 & 1 \end{bmatrix} \left\{ \begin{matrix} u_i \\ u_j \end{matrix} \right\}_e + \left\{ \begin{matrix} EA\alpha \ \Delta T \\ -EA\alpha \ \Delta T \end{matrix} \right\}_e$$ [9.2-54]

It is assumed that the temperature changes are known. Thus, when assembling the individual elements into the system equations, a better form of this equation is

$$\left\{ \begin{matrix} f_{xi} \\ f_{xj} \end{matrix} \right\}_e + \left\{ \begin{matrix} -EA\alpha \ \Delta T \\ EA\alpha \ \Delta T \end{matrix} \right\}_e = \left(\frac{AE}{L}\right)_e \begin{bmatrix} 1 & -1 \\ -1 & 1 \end{bmatrix} \left\{ \begin{matrix} u_i \\ u_j \end{matrix} \right\}_e$$ [9.2-55]

The second vector on the left-hand side of Eq. (9.2-55) is called the thermal nodal force vector, $\{f\}_{\Delta T}$ where

$$\left\{ \begin{matrix} f_{xi} \\ f_{xj} \end{matrix} \right\}_{\Delta T} = \left\{ \begin{matrix} -EA\alpha \ \Delta T \\ EA\alpha \ \Delta T \end{matrix} \right\}_e$$ [9.2-56]

Example 9.2-2

For Example 9.2-1 let the entire structure undergo an increase in temperature of 50°C. Let $\alpha = 12 \ \mu/°C$. Also, in addition to the force of 20 kN on node 3, consider that element 1 has a uniformly distributed load of 25 kN/m in the positive x direction.

Solution:

Under the conditions of Example 9.2-1, node 4 hits the wall. An increase in temperature and the additional distributed force in the positive x direction will not change that boundary condition. Also, in Example 9.2-1 the stiffness of each element is calculated. Incorporating Eqs. (9.2-43) and (9.2-55) the finite element equations for each element can be written.

For element (1), the uniform force is divided to each node with $1/2 q_o L_1 = 1/2(25)(200) = 2500$ N. The element thermal force is $(EA\alpha\,\Delta T)_1 = (50)(10^9)(100)(10^{-6})(12)(10^{-6})(50) = 3000$ N. Thus

Element (1):

$$\left\{\begin{matrix} f_{x1} \\ f_{x2} \end{matrix}\right\}_1 + \left\{\begin{matrix} 2500 \\ 2500 \end{matrix}\right\} + \left\{\begin{matrix} -3000 \\ 3000 \end{matrix}\right\} = 25(10^6)\begin{bmatrix} 1 & -1 \\ -1 & 1 \end{bmatrix}\left\{\begin{matrix} u_1 \\ u_2 \end{matrix}\right\}$$

or

$$\left\{\begin{matrix} f_{x1} \\ f_{x2} \end{matrix}\right\}_1 + \left\{\begin{matrix} -500 \\ 5500 \end{matrix}\right\} = 25(10^6)\begin{bmatrix} 1 & -1 \\ -1 & 1 \end{bmatrix}\left\{\begin{matrix} u_1 \\ u_2 \end{matrix}\right\} \qquad \textbf{[a]}$$

Elements (2) and (3) only have thermal forces which are, respectively,

$$(EA\alpha\,\Delta T)_2 = (50)(10^9)(75)(10^{-6})(12)(10^{-6})(50) = 2250 \text{ N}$$

$$(EA\alpha\,\Delta T)_3 = (50)(10^9)(50)(10^{-6})(12)(10^{-6})(50) = 1500 \text{ N}$$

The equations for these elements are

Element (2):

$$\left\{\begin{matrix} f_{x2} \\ f_{x3} \end{matrix}\right\}_2 + \left\{\begin{matrix} -2250 \\ 2250 \end{matrix}\right\} = 25(10^6)\begin{bmatrix} 1 & -1 \\ -1 & 1 \end{bmatrix}\left\{\begin{matrix} u_2 \\ u_3 \end{matrix}\right\} \qquad \textbf{[b]}$$

Element (3):

$$\left\{\begin{matrix} f_{x3} \\ f_{x4} \end{matrix}\right\}_3 + \left\{\begin{matrix} -1500 \\ 1500 \end{matrix}\right\} = 25(10^6)\begin{bmatrix} 1 & -1 \\ -1 & 1 \end{bmatrix}\left\{\begin{matrix} u_3 \\ u_4 \end{matrix}\right\} \qquad \textbf{[c]}$$

Next, we assemble the matrices according to the total nodal forces where $F_1 = (f_{x1})_1$, $F_2 = (f_{x2})_1 + (f_{x2})_2$, $F_3 = (f_{x3})_2 + (f_{x3})_3$, and $F_4 = (f_{x4})_3$. The final result is

$$\left\{\begin{matrix} F_1 - 500 \\ F_2 + 5500 - 2250 \\ F_3 + 2250 - 1500 \\ F_4 + 1500 \end{matrix}\right\} = (10^6)\begin{bmatrix} 25 & -25 & 0 & 0 \\ -25 & 50 & -25 & 0 \\ 0 & -25 & 45 & -20 \\ 0 & 0 & -20 & 20 \end{bmatrix}\left\{\begin{matrix} u_1 \\ u_2 \\ u_3 \\ u_4 \end{matrix}\right\}$$

or

$$\begin{Bmatrix} F_1 - 500 \\ F_2 + 3250 \\ F_3 + 750 \\ F_4 + 1500 \end{Bmatrix} = (10^6) \begin{bmatrix} 25 & -25 & 0 & 0 \\ -25 & 50 & -25 & 0 \\ 0 & -25 & 45 & -20 \\ 0 & 0 & -20 & 20 \end{bmatrix} \begin{Bmatrix} u_1 \\ u_2 \\ u_3 \\ u_4 \end{Bmatrix} \qquad \textbf{[d]}$$

The boundary conditions are $u_1 = 0$ and $u_4 = 1(10^{-3})$ m. Thus we partition out the first and last equations in (d). This gives

$$F_1 = 25(10^6)(u_1 - u_2) + 500 \qquad \textbf{[e]}$$

$$F_4 = 20(10^6)(u_4 - u_3) - 1500 \qquad \textbf{[f]}$$

and

$$\begin{Bmatrix} F_2 + 3250 \\ F_3 + 750 + 20(10^6)u_4 \end{Bmatrix} = (10^6) \begin{bmatrix} 50 & -25 \\ -25 & 45 \end{bmatrix} \begin{Bmatrix} u_2 \\ u_3 \end{Bmatrix}$$

Substituting in $u_4 = 1(10^{-3})$ m gives

$$\begin{Bmatrix} F_2 + 3.25(10^3) \\ F_3 + 20.75(10^3) \end{Bmatrix} = (10^6) \begin{bmatrix} 50 & -25 \\ -25 & 45 \end{bmatrix} \begin{Bmatrix} u_2 \\ u_3 \end{Bmatrix} \qquad \textbf{[g]}$$

With $F_2 = 0$ and $F_3 = 20(10^3)$ N, inverting the equation and solving for the nodal displacements yields

$$\begin{Bmatrix} u_2 \\ u_3 \end{Bmatrix} = (10^{-8}) \begin{bmatrix} 2.7692 & 1.5385 \\ 1.5385 & 3.0769 \end{bmatrix} \begin{Bmatrix} 3.25(10^3) \\ 40.75(10^3) \end{Bmatrix}$$

$$= \begin{Bmatrix} 0.7169 \\ 1.3038 \end{Bmatrix} (10^{-3}) \, \text{m} = \begin{Bmatrix} 0.7169 \\ 1.3038 \end{Bmatrix} \text{mm} \qquad \textbf{[h]}$$

Once the nodal displacements are determined, the reaction forces and the element displacement function, strains, and stresses can be evaluated. This is done in the same manner as in Example 9.2-1. Here, we will only show the calculations for the reaction forces and stress in element (1). The reaction forces are given by Eqs. (e) and (f), which result in

$$F_1 = 25(10^6)[0 - 0.7169(10^{-3})] + 500 = -17.42(10^3) \, \text{N} = -17.42 \, \text{kN}$$

$$F_4 = 20(10^6)[1(10^{-3}) - 1.3038(10^{-3})] - 1500 = -7.58(10^3) \, \text{N} = -7.58 \, \text{kN}$$

Note that this balances the applied loads of $F_2 + (q_oL)_1 = [20 + 25(0.200)] = 25$ kN in the positive x direction.

The stress in element (1) is given by combining Eq. (9.2-46) and the thermal part of Eq. (9.2-49), which yields

$$(\sigma_x)_1 = \left(\frac{E}{L}\right)_1 \{-1 \quad 1\} \begin{Bmatrix} u_1 \\ u_2 \end{Bmatrix} + \left(\frac{q_0}{2A_1}\right)_1 (L_1 - 2x + 2x_1) - (E\alpha \ \Delta T)_1$$

$$= \frac{50(10^9)}{0.200}[0.7169(10^{-3}) - 0] + \frac{25(10^3)}{2(100)(10^{-6})}[0.200 - 2x + 2(0)]$$

$$- 50(10^9)(12)(10^{-6})(50)$$

$$= (174.23 - 250x)(10^6) \, \text{N/m}^2 = 174.23 - 250x \, \text{MPa}$$

where x is in meters.

9.2.6   THE TWO-DIMENSIONAL TRUSS ELEMENT

The one-dimensional axial element can now be placed into two- or three-dimensional space through the use of coordinate transformations. Consider first a simple two-dimensional transformation as shown in Fig. 9.2-8. The xyz coordinate system is referred to as the *global coordinate system* where the nodal positions are defined, (x_i, y_i) and (x_j, y_j). The $x'y'$ system is referred to as the element *local coordinate system* where the local one-dimensional element matrices are defined.[†] The angles ϕ_x and ϕ_y are defined *from* the global coordinate system *to* the local coordinate x' axis. The angles are obtained from the nodal coordinates (x_i, y_i) and (x_j, y_j) and are

$$\phi_x = \tan^{-1}\left(\frac{y_j - y_i}{x_j - x_i}\right)$$

[9.2-57]

$$\phi_y = \tan^{-1}\left(\frac{x_j - x_i}{y_j - y_i}\right)$$

The local deflection can be written in terms of the global deflections as [see Fig. 9.2-8(b)]

$$u_i' = u_i \cos \phi_x + v_i \cos \phi_y$$

[9.2-58]

$$u_j' = u_j \cos \phi_x + v_j \cos \phi_y$$

[†] Note that if one desired the displacement field of an element or had stresses due to distributed loading on an element, the corresponding Eqs. (9.2-2) and (9.2-46) would need to be reformulated with respect to the local coordinate x'.

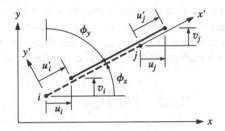

(a) Local and global coordinates

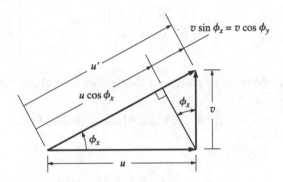

(b) General relationship between
local and global deflections at a node

Figure 9.2-8   Two-dimensional transformation of displacements.

The following shorthand notation expresses the directional cosines for the element as

$$\ell = \cos \phi_x \qquad m = \cos \phi_y \qquad \textbf{[9.2-59]}$$

Thus Eq. (9.2-58) can be rewritten as

$$u_i' = u_i \ell + v_i m$$

$$\textbf{[9.2-60]}$$

$$u_j' = u_j \ell + v_j m$$

Equations (9.2-60) can be expressed in matrix form as

$$\{\mathbf{u}\}_{\text{local}} = [\mathbf{T}]\{\mathbf{u}\}_{\text{global}} \qquad \textbf{[9.2-61]}$$

where

$$\{\mathbf{u}\}_{\text{local}} = \begin{Bmatrix} u_i' \\ u_j' \end{Bmatrix} \qquad [\mathbf{T}] = \begin{bmatrix} \ell & m & 0 & 0 \\ 0 & 0 & \ell & m \end{bmatrix} \qquad \{\mathbf{u}\}_{\text{global}} = \begin{Bmatrix} u_i \\ v_i \\ u_j \\ v_j \end{Bmatrix} \qquad \textbf{[9.2-62]}$$

[T] is referred to as the *global transformation matrix* which transforms vectors in global coordinates to local coordinates. Forces can be transformed in the same manner. Consider the global x and y components of the forces at the nodes being $\{f_{xi}\ f_{yi}\ f_{xj}\ f_{yj}\}$. This can be transformed to the local components of the forces at the nodes, $\{f_{x'i}\ f_{x'j}\}$ by

$$\{f\}_{\text{local}} = [T]\{f\}_{\text{global}} \qquad \text{[9.2-63]}$$

where

$$\{f\}_{\text{local}} = \begin{Bmatrix} f_{x'i} \\ f_{x'j} \end{Bmatrix} \qquad [T] = \begin{bmatrix} \ell & m & 0 & 0 \\ 0 & 0 & \ell & m \end{bmatrix} \qquad \{f\}_{\text{global}} = \begin{Bmatrix} f_{xi} \\ f_{yi} \\ f_{xj} \\ f_{yj} \end{Bmatrix} \qquad \text{[9.2-64]}$$

The form of the potential energy for the element in local coordinates is given by

$$\Pi_e = \frac{1}{2}\{u\}_{\text{local}}^T[k]_{\text{local}}\{u\}_{\text{local}} - \{u\}_{\text{local}}^T\{f\}_{\text{local}} \qquad \text{[9.2-65]}$$

where

$$[k]_{\text{local}} = \left(\frac{AE}{L}\right)_e \begin{bmatrix} 1 & -1 \\ -1 & 1 \end{bmatrix} \qquad \text{[9.2-66]}$$

The form of the potential energy for the element in global coordinates will be

$$\Pi_e = \frac{1}{2}\{u\}_{\text{global}}^T[k]_{\text{global}}\{u\}_{\text{global}} - \{u\}_{\text{global}}^T\{f\}_{\text{global}} \qquad \text{[9.2-67]}$$

If we transform the local displacements in Eq. (9.2-65) using Eq. (9.2-61) we obtain

$$\Pi_e = \frac{1}{2}\{u\}_{\text{global}}^T[T]^T[k]_{\text{local}}[T]\{u\}_{\text{global}} - \{u\}_{\text{global}}^T[T]^T\{f\}_{\text{local}} \qquad \text{[9.2-68]}$$

Comparing Eqs. (9.2-67) and (9.2-68) we see that

$$[k]_{\text{global}} = [T]^T[k]_{\text{local}}[T] \qquad \text{[9.2-69]}$$

and

$$\{f\}_{\text{global}} = [T]^T\{f\}_{\text{local}} \qquad \text{[9.2-70]}$$

Note that Eq. (9.2-70) is consistent with Eq. (9.2-63) since for orthogonal transformations, $[T]^T = [T]^{-1}$.

Minimizing Π_e in Eq. (9.2-67) will give

$$\{f\}_{\text{global}} = [k]_{\text{global}}\{u\}_{\text{global}} \qquad \text{[9.2-71]}$$

which is the element equation in the global coordinate system. The element global stiffness matrix can be determined from Eq. (9.2-69). Substituting Eqs. (9.2-62) and (9.2-66) into (9.2-69) results in

$$[\mathbf{k}]_{\text{global}} = \begin{bmatrix} \ell & 0 \\ m & 0 \\ 0 & \ell \\ 0 & m \end{bmatrix} \left(\frac{AE}{L}\right)_e \begin{bmatrix} 1 & -1 \\ -1 & 1 \end{bmatrix} \begin{bmatrix} \ell & m & 0 & 0 \\ 0 & 0 & \ell & m \end{bmatrix}$$

Performing the matrix multiplication gives

$$[\mathbf{k}]_{\text{global}} = \left(\frac{AE}{L}\right)_e \begin{bmatrix} \ell^2 & \ell m & -\ell^2 & -\ell m \\ \ell m & m^2 & -\ell m & -m^2 \\ -\ell^2 & -\ell m & \ell^2 & \ell m \\ -\ell m & -m^2 & \ell m & m^2 \end{bmatrix}_e \qquad \textbf{[9.2-72]}$$

In truss structures the nodes do not follow each other in a serial manner, as in the case of the one-dimensional systems. The process of defining the element stiffness matrix and assembling it into the system matrix equations now becomes a little more complicated. To ease the process we will order the displacement vector according to the node number and the degree of freedom. That is, for n nodes, the system displacement vector will be defined as

$$\{\mathbf{u}\}_{\text{system}} = \begin{Bmatrix} u_1 \\ v_1 \\ u_2 \\ v_2 \\ \cdot \\ \cdot \\ \cdot \\ u_n \\ v_n \end{Bmatrix} \qquad \textbf{[9.2-73]}$$

To further reduce the complexity, two procedures will be followed. First, for a given element, *always* define the lower node number as the i node and the higher node number as the j node. Second, after the element equations are formulated with respect to the element degrees of freedom, expand the element equations with respect to the system degrees of freedom. This will make the assembly process simply a summation process. Let us look at an example.

Example 9.2-3

For the two-dimensional truss structure shown in Fig. 9.2-9, determine the nodal deflections and the stress and nodal forces of each element. The support reaction forces R_{1x}, R_{1y}, and R_{3x} are also indicated in the figure next to the supports. Members (1) and (2) each have a length of 2 m. Each member has a cross-sectional

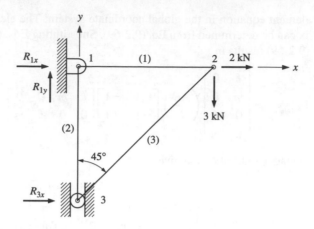

Figure 9.2-9

area of 80 mm^2 and a modulus of elasticity of 200 GPa. Ignore the possibility of buckling in compression members.[†]

Solution:

Element (1): $k_1 = (AE/L)_1 = 80(10^{-6})(200)(10^9)/2 = 8(10^6)$ N/m. For the node numbers, let $i_1 = 1$ and $j_1 = 2$. Thus, $(\phi_x)_1 = 0$ and $\ell_1 = \cos(0) = 1$ and $m_1 = \sin(0) = 0$. From Eq. (9.2-72) the global stiffness matrix is

$$[k_1]_{global} = 8(10^6) \begin{array}{c} \begin{array}{cccc} u_1 & v_1 & u_2 & v_2 \end{array} \\ \begin{bmatrix} 1 & 0 & -1 & 0 \\ 0 & 0 & 0 & 0 \\ -1 & 0 & 1 & 0 \\ 0 & 0 & 0 & 0 \end{bmatrix} \begin{array}{c} u_1 \\ v_1 \\ u_2 \\ v_2 \end{array} \end{array}$$

The nodal displacements are shown to indicate the associativity of the rows and columns of the stiffness matrix to the nodal degrees of freedom. The stiffness matrix in terms of the system degrees of freedom u_1, v_1, u_2, v_2, u_3, and v_3 is

$$[k_1]_{global} = 8(10^6) \begin{array}{c} \begin{array}{cccccc} u_1 & v_1 & u_2 & v_2 & u_3 & v_3 \end{array} \\ \begin{bmatrix} 1 & 0 & -1 & 0 & 0 & 0 \\ 0 & 0 & 0 & 0 & 0 & 0 \\ -1 & 0 & 1 & 0 & 0 & 0 \\ 0 & 0 & 0 & 0 & 0 & 0 \\ 0 & 0 & 0 & 0 & 0 & 0 \\ 0 & 0 & 0 & 0 & 0 & 0 \end{bmatrix} \begin{array}{c} u_1 \\ v_1 \\ u_2 \\ v_2 \\ u_3 \\ v_3 \end{array} \end{array} \qquad [a]$$

where the last two rows and columns of zeros are added.

[†] Buckling of planar elements is covered in Sec. 9.3.7.

Element (2): $k_2 = (AE/L)_2 = 80(10^{-6})(200)(10^9)/2 = 8(10^6)$ N/m. For the node numbers 1 and 3 let i_2 be the lower number 1 and j_2 be the higher number 3. That is, $i_2 = 1$ and $j_2 = 3$. Thus, $(\phi_x)_2 = 270°$ and $\ell_2 = \cos(270) = 0$ and $m_2 = \sin(270) = -1$. Then from Eq. (9.2-72), the global stiffness matrix is

$$[k_2]_{global} = 8(10^6) \begin{array}{cccc} u_1 & v_1 & u_3 & v_3 \\ \begin{bmatrix} 0 & 0 & 0 & 0 \\ 0 & 1 & 0 & -1 \\ 0 & 0 & 0 & 0 \\ 0 & -1 & 0 & 1 \end{bmatrix} & \begin{array}{c} u_1 \\ v_1 \\ u_3 \\ v_3 \end{array} \end{array}$$

where again the degrees of freedom associated with each row and column are shown. In terms of the system degrees of freedom the stiffness matrix is

$$[k_2]_{global} = 8(10^6) \begin{array}{cccccc} u_1 & v_1 & u_2 & v_2 & u_3 & v_3 \\ \begin{bmatrix} 0 & 0 & 0 & 0 & 0 & 0 \\ 0 & 1 & 0 & 0 & 0 & -1 \\ 0 & 0 & 1 & 0 & 0 & 0 \\ 0 & 0 & 0 & 0 & 0 & 0 \\ 0 & 0 & 0 & 0 & 0 & 0 \\ 0 & -1 & 0 & 0 & 0 & 1 \end{bmatrix} & \begin{array}{c} u_1 \\ v_1 \\ u_2 \\ v_2 \\ u_3 \\ v_3 \end{array} \end{array}$$ **[b]**

where the third and fourth rows and columns of zeros were added.

Element (3): $k_3 = (AE/L)_3 = 80(10^{-6})(200)(10^9)/(2\sqrt{2}) = 4\sqrt{2}(10^6)$ N/m. For the node numbers, let $i_3 = 2$ and $j_3 = 3$. Thus, $(\phi_x)_3 = 225°$ and $\ell_3 = \cos(225) = -\sqrt{2}/2$ and $m_3 = \sin(225) = -\sqrt{2}/2$. The global stiffness matrix is thus

$$[k_3]_{global} = 2\sqrt{2}(10^6) \begin{array}{cccc} u_2 & v_2 & u_3 & v_3 \\ \begin{bmatrix} 1 & 1 & -1 & -1 \\ 1 & 1 & -1 & -1 \\ -1 & -1 & 1 & 1 \\ -1 & -1 & 1 & 1 \end{bmatrix} & \begin{array}{c} u_2 \\ v_2 \\ u_3 \\ v_3 \end{array} \end{array}$$

In terms of the system degrees of freedom, the stiffness matrix is

$$[k_3]_{global} = 2\sqrt{2}(10^6) \begin{array}{cccccc} u_1 & v_1 & u_2 & v_2 & u_3 & v_3 \\ \begin{bmatrix} 0 & 0 & 0 & 0 & 0 & 0 \\ 0 & 0 & 0 & 0 & 0 & 0 \\ 0 & 0 & 1 & 1 & -1 & -1 \\ 0 & 0 & 1 & 1 & -1 & -1 \\ 0 & 0 & -1 & -1 & 1 & 1 \\ 0 & 0 & -1 & -1 & 1 & 1 \end{bmatrix} & \begin{array}{c} u_1 \\ v_1 \\ u_2 \\ v_2 \\ u_3 \\ v_3 \end{array} \end{array}$$ **[c]**

Using a common factor of $2(10^6)$, adding the stiffness matrices of Eqs. (a), (b), and (c) results in

$$[K]_{system} = 2(10^6) \begin{array}{c} \begin{array}{cccccc} u_1 & v_1 & u_2 & v_2 & u_3 & v_3 \end{array} \\ \begin{bmatrix} 4 & 0 & -4 & 0 & 0 & 0 \\ 0 & 4 & 0 & 0 & 0 & -4 \\ -4 & 0 & (4+\sqrt{2}) & \sqrt{2} & -\sqrt{2} & -\sqrt{2} \\ 0 & 0 & \sqrt{2} & \sqrt{2} & -\sqrt{2} & -\sqrt{2} \\ 0 & 0 & -\sqrt{2} & -\sqrt{2} & \sqrt{2} & \sqrt{2} \\ 0 & -4 & -\sqrt{2} & -\sqrt{2} & \sqrt{2} & (4+\sqrt{2}) \end{bmatrix} \end{array} \begin{array}{c} u_1 \\ v_1 \\ u_2 \\ v_2 \\ u_3 \\ v_3 \end{array} \qquad [d]$$

The system force vector is $\{F_{1x}\, F_{1y}\, F_{2x}\, F_{2y}\, F_{3x}\, F_{3y}\}^T = \{R_{1x}\, R_{1y}\, 2(10^3)\, -3(10^3)$ $R_{3x}\, 0\}^T$ N, where R_{1x}, R_{1y}, and R_{3x} are the unknown reaction forces at nodes 1 and 3. The system equation is thus

$$\begin{Bmatrix} R_{1x} \\ R_{1y} \\ 2(10^3) \\ -3(10^3) \\ R_{3x} \\ 0 \end{Bmatrix} = 2(10^6) \begin{bmatrix} 4 & 0 & -4 & 0 & 0 & 0 \\ 0 & 4 & 0 & 0 & 0 & -4 \\ -4 & 0 & (4+\sqrt{2}) & \sqrt{2} & -\sqrt{2} & -\sqrt{2} \\ 0 & 0 & \sqrt{2} & \sqrt{2} & -\sqrt{2} & -\sqrt{2} \\ 0 & 0 & -\sqrt{2} & -\sqrt{2} & \sqrt{2} & \sqrt{2} \\ 0 & -4 & -\sqrt{2} & -\sqrt{2} & \sqrt{2} & (4+\sqrt{2}) \end{bmatrix} \begin{Bmatrix} u_1 \\ v_1 \\ u_2 \\ v_2 \\ u_3 \\ v_3 \end{Bmatrix} \qquad [e]$$

The boundary conditions are $u_1 = v_1 = u_3 = 0$. This means that we partition out the first, second, and fifth equations in Eq. (e). This gives

$$R_{1x} = 8(10^6)(u_1 - u_2)$$

$$R_{1y} = 8(10^6)(v_1 - v_3) \qquad [f]$$

$$R_{3x} = 2\sqrt{2}(10^6)(-u_2 - v_2 + u_3 + v_3)$$

and the remaining equations are

$$\begin{Bmatrix} 2(10^3) \\ -3(10^3) \\ 0 \end{Bmatrix} = 2(10^6) \begin{bmatrix} -4 & 0 & (4+\sqrt{2}) & \sqrt{2} & -\sqrt{2} & -\sqrt{2} \\ 0 & 0 & \sqrt{2} & \sqrt{2} & -\sqrt{2} & -\sqrt{2} \\ 0 & -4 & -\sqrt{2} & -\sqrt{2} & \sqrt{2} & (4+\sqrt{2}) \end{bmatrix} \begin{Bmatrix} u_1 \\ v_1 \\ u_2 \\ v_2 \\ u_3 \\ v_3 \end{Bmatrix} \qquad [g]$$

The first, second, and fifth columns together with the known values of u_1, v_1, and u_3 can be moved to the left-hand side of the equations. However, since $u_1 = v_1 = u_3 = 0$, the left-hand side of the equations remains unchanged. Thus

$$\begin{Bmatrix} 2(10^3) \\ -3(10^3) \\ 0 \end{Bmatrix} = 2(10^6) \begin{bmatrix} (4 + \sqrt{2}) & \sqrt{2} & -\sqrt{2} \\ \sqrt{2} & \sqrt{2} & -\sqrt{2} \\ -\sqrt{2} & -\sqrt{2} & (4 + \sqrt{2}) \end{bmatrix} \begin{Bmatrix} u_2 \\ v_2 \\ v_3 \end{Bmatrix} \qquad \textbf{[h]}$$

Obtaining the inverse of the modified stiffness matrix yields

$$\begin{Bmatrix} u_2 \\ v_2 \\ v_3 \end{Bmatrix} = 1.25(10^{-7}) \begin{bmatrix} 1 & -1 & 0 \\ -1 & 4.828 & 1 \\ 0 & 1 & 1 \end{bmatrix} \begin{Bmatrix} 2(10^3) \\ -3(10^3) \\ 0 \end{Bmatrix} = \begin{Bmatrix} 6.250 \\ -20.61 \\ -3.750 \end{Bmatrix} (10^{-4}) \, \text{m}$$

With the nodal displacements determined, the calculations of the reactions and element stresses and local nodal forces are straightforward. The reactions, given by Eqs. (f), are

$$R_{1x} = 8(10^6)[0 - 6.25(10^{-3})] = -5000 \, \text{N} = -5 \, \text{kN}$$

$$R_{1y} = 8(10^6)[0 - (-3.75)(10^{-3})] = 3000 \, \text{N} = 3 \, \text{kN}$$

$$R_{3x} = 2\sqrt{2}(10^6)[-6.25 - (-20.61) + 0 + (-3.75)](10^{-4}) = 3000 = 3 \, \text{kN}$$

Using basic statics, the reader should verify that these are correct.

For the element stresses and local nodal forces, we use Eqs. (9.2-9) and (9.2-14), respectively. These equations are written in terms of the local displacements. However, since $\{u\}_{\text{local}} = [T]\{u\}_{\text{global}}$, where $[T]$ is given in Eq. (9.2-62), these equations can be rewritten as

$$(\sigma_x)_e = \left(\frac{E}{L} \right)_e \{-1 \quad 1\}[T] \begin{Bmatrix} u_i \\ v_i \\ u_j \\ v_j \end{Bmatrix}_e \qquad \textbf{[9.2-74]}$$

and

$$\begin{Bmatrix} f_{x'i} \\ f_{x'j} \end{Bmatrix}_e = k_e \begin{bmatrix} 1 & -1 \\ -1 & 1 \end{bmatrix} [\mathbf{T}] \begin{Bmatrix} u_i \\ v_i \\ u_j \\ v_j \end{Bmatrix} \qquad \textbf{[9.2-75]}$$

Element (1): $\{u_i\, v_i\, u_j\, v_j\}^T = \{u_1\, v_1\, u_2\, v_2\}^T$, $\ell_1 = 1$, and $m_1 = 0$. This gives

$$(\sigma_{x'})_1 = \frac{200(10^9)}{2}\{-1 \quad 1\}\begin{bmatrix} 1 & 0 & 0 & 0 \\ 0 & 0 & 1 & 0 \end{bmatrix}\begin{Bmatrix} 0 \\ 0 \\ 6.25 \\ -20.61 \end{Bmatrix}(10^{-4})$$

$$= 62.5(10^6) \text{ N/m}^2 = 62.5 \text{ MPa}$$

$$\begin{Bmatrix} f_{x'1} \\ f_{x'2} \end{Bmatrix}_1 = 8(10^6)\begin{bmatrix} 1 & -1 \\ -1 & 1 \end{bmatrix}\begin{bmatrix} 1 & 0 & 0 & 0 \\ 0 & 0 & 1 & 0 \end{bmatrix}\begin{Bmatrix} 0 \\ 0 \\ 6.25 \\ -20.61 \end{Bmatrix}(10^{-4})$$

$$= \begin{Bmatrix} -5000 \\ 5000 \end{Bmatrix} \text{N} = \begin{Bmatrix} -5 \\ 5 \end{Bmatrix} \text{kN}$$

Element (2): $\{u_i\, v_i\, u_j\, v_j\}^T = \{u_1\, v_1\, u_3\, v_3\}^T$, $\ell_2 = 0$, and $m_2 = -1$. This gives

$$(\sigma_{x'})_2 = \frac{200(10^9)}{2}\{-1 \quad 1\}\begin{bmatrix} 0 & -1 & 0 & 0 \\ 0 & 0 & 0 & -1 \end{bmatrix}\begin{Bmatrix} 0 \\ 0 \\ 0 \\ -3.75 \end{Bmatrix}(10^{-4})$$

$$= 37.5(10^6) \text{ N/m}^2 = 37.5 \text{ MPa}$$

$$\begin{Bmatrix} f_{x'1} \\ f_{x'3} \end{Bmatrix}_2 = 8(10^6)\begin{bmatrix} 1 & -1 \\ -1 & 1 \end{bmatrix}\begin{bmatrix} 0 & -1 & 0 & 0 \\ 0 & 0 & 0 & -1 \end{bmatrix}\begin{Bmatrix} 0 \\ 0 \\ 0 \\ -3.75 \end{Bmatrix}(10^{-4})$$

$$= \begin{Bmatrix} -3000 \\ 3000 \end{Bmatrix} \text{N} = \begin{Bmatrix} -3 \\ 3 \end{Bmatrix} \text{kN}$$

Element (3): $\{u_i \; v_i \; u_j \; v_j\}^T = \{u_2 \; v_2 \; u_3 \; v_3\}^T$, $\ell_3 = -\sqrt{2}/2$, and $m_3 = -\sqrt{2}/2$.
This gives

$$(\sigma_{x'})_3 = \frac{200(10^9)}{2\sqrt{2}}\{-1 \;\; 1\}\left(\frac{-\sqrt{2}}{2}\right)\begin{bmatrix} 1 & 1 & 0 & 0 \\ 0 & 0 & 1 & 1 \end{bmatrix}\begin{Bmatrix} 6.25 \\ -20.61 \\ 0 \\ -3.75 \end{Bmatrix}(10^{-4})$$

$$= -53.03(10^6) \, \text{N/m}^2 = -53.03 \, \text{MPa}$$

$$\begin{Bmatrix} f_{x'2} \\ f_{x'3} \end{Bmatrix}_3 = \frac{8}{\sqrt{2}}(10^6)\begin{bmatrix} 1 & -1 \\ -1 & 1 \end{bmatrix}\left(\frac{-\sqrt{2}}{2}\right)\begin{bmatrix} 1 & 1 & 0 & 0 \\ 0 & 0 & 1 & 1 \end{bmatrix}\begin{Bmatrix} 6.25 \\ -20.61 \\ 0 \\ -3.75 \end{Bmatrix}(10^{-4})$$

$$= \begin{Bmatrix} 4243 \\ -4243 \end{Bmatrix} \text{N} = \begin{Bmatrix} 4.243 \\ -4.243 \end{Bmatrix} \text{kN}$$

Based on the finite element results, Fig. 9.2-10 shows the nodal forces for each element together with the local coordinate systems for clarity.

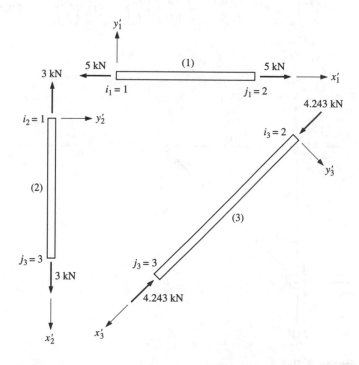

Figure 9.2-10   Local element nodal forces and coordinate systems.

9.2.7   SKEW SUPPORTS

The support boundary conditions may not necessarily align with the global coordinate system as shown in Fig. 9.2-11 where the translational freedom of node 2 is in neither the x nor the y direction. There are three basic methods of dealing with this situation (for the first two methods see Ref. 9.2): (1) coordinate transformation of the nodal degrees of freedom, (2) imposing constraint equations, and (3) utilizing a stiff spring element in the direction of the constraint (referred to as *boundary elements*: see Ref. 9.3). Here we will only apply the first method as this is the most straightforward.

Consider the coordinate system of the constrained and free directions (x_c, y_c) to be rotated the angle β counterclockwise from the global coordinates. Let the translational displacements of node i relative to the x_c, y_c coordinate be u_{ic}, v_{ic} respectively. The global displacements can be written as

$$\begin{Bmatrix} u_i \\ v_i \end{Bmatrix}_{global} = \begin{bmatrix} \cos\beta & -\sin\beta \\ \sin\beta & \cos\beta \end{bmatrix}_c \begin{Bmatrix} u_i \\ v_i \end{Bmatrix}_c = [\mathbf{t}]_c \begin{Bmatrix} u_i \\ v_i \end{Bmatrix}_c \qquad \textbf{[9.2-76]}$$

where the $[\mathbf{t}]_c$ is the transformation matrix for node i which transforms vectors in the constraint coordinate system to the global coordinate system. Likewise, the forces are

$$\begin{Bmatrix} f_{xi} \\ f_{yi} \end{Bmatrix}_{global} = \begin{bmatrix} \cos\beta & -\sin\beta \\ \sin\beta & \cos\beta \end{bmatrix}_c \begin{Bmatrix} f_{xi} \\ f_{yi} \end{Bmatrix}_c = [\mathbf{t}]_c \begin{Bmatrix} f_{xi} \\ f_{yi} \end{Bmatrix}_c \qquad \textbf{[9.2-77]}$$

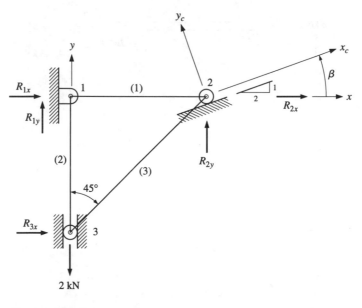

Figure 9.2-11

Assuming there was only one such constraint, the transformation matrix for the entire system of nodes would be

$$[\mathbf{T}]_c = \begin{bmatrix} 1 & & & & & & & \\ & 1 & & & & & & \\ & & 1 & & & & & \\ & & & \cdot & & & & \\ & & & & \cdot & & & \\ & & & \cos\beta & -\sin\beta & & & \\ & & & \sin\beta & \cos\beta & & & \\ & & & & & \cdot & & \\ & & & & & & 1 & \\ & & & & & & & 1 \end{bmatrix}$$ [9.2-78]

where blank row and column terms are zero. The single nodal transformation matrix is inserted in the rows and columns pertaining to node i. This can be written as

$$[\mathbf{T}]_c = \begin{bmatrix} [\mathbf{I}] & & \\ & [\mathbf{t}]_c & \\ & & [\mathbf{I}] \end{bmatrix}$$ [9.2-79]

where $[\mathbf{I}]$ is the identity matrix.

Similar to the transformation from the element local coordinate system to the global coordinate system, the transformed system stiffness matrix is

$$[\mathbf{K}]_c = [\mathbf{T}]_c^T [\mathbf{K}]_{\text{global}} [\mathbf{T}]_c$$ [9.2-80]

and the transformed force vector is

$$\{\mathbf{F}\}_c = [\mathbf{T}]_c^T \{\mathbf{F}\}_{\text{global}}$$ [9.2-81]

Determine the displacements and support reaction forces for the truss system shown in Fig. 9.2-11 using the length, cross-sectional area, and material property of each member given in Example 9.2-3.

Example 9.2-4

Solution:

The structure is identical to that solved in Example 9.2-3 with the exception of the boundary conditions on node 2 and the applied forces. The partitioning of the equations up to Eq. (h) of that example is still applicable. The forces are different, however. Rewriting Eq. (h) we have

$$\begin{Bmatrix} R_{2x} \\ R_{2y} \\ -2(10^3) \end{Bmatrix} = 2(10^6) \begin{bmatrix} (4+\sqrt{2}) & \sqrt{2} & -\sqrt{2} \\ \sqrt{2} & \sqrt{2} & -\sqrt{2} \\ -\sqrt{2} & -\sqrt{2} & (4+\sqrt{2}) \end{bmatrix} \begin{Bmatrix} u_2 \\ v_2 \\ v_3 \end{Bmatrix}$$ [a]

We can now apply the transformation on node 2 to account for the skew constraint. From Fig. 9.2-11 the angle β is given by $\beta = \tan^{-1}(1/2) = 26.57°$, resulting in $\cos \beta = 0.8944$ and $\sin \beta = 0.4472$. From Eq. (9.2-79), the transformation matrix is

$$[\mathbf{T}]_c = \begin{bmatrix} 0.8944 & -0.4472 & 0 \\ 0.4472 & 0.8944 & 0 \\ 0 & 0 & 1 \end{bmatrix} \qquad [b]$$

Applying Eq. (9.2-80) by pre- and postmultiplying the stiffness matrix of Eq. (a) by $[\mathbf{T}]_c^T$ and $[\mathbf{T}]_c$, respectively, yields

$$\begin{bmatrix} 0.8944 & 0.4472 & 0 \\ -0.4472 & 0.8944 & 0 \\ 0 & 0 & 1 \end{bmatrix} 2(10^6) \begin{bmatrix} (4 + \sqrt{2}) & \sqrt{2} & -\sqrt{2} \\ \sqrt{2} & \sqrt{2} & -\sqrt{2} \\ -\sqrt{2} & -\sqrt{2} & (4 + \sqrt{2}) \end{bmatrix} \begin{bmatrix} 0.8944 & -0.4472 & 0 \\ 0.4472 & 0.8944 & 0 \\ 0 & 0 & 1 \end{bmatrix}$$

$$= 1(10^6) \begin{bmatrix} 11.491 & -1.5029 & -3.7947 \\ -1.5029 & 2.1657 & -1.2649 \\ -3.7947 & -1.2649 & 10.828 \end{bmatrix}$$

The transformed force vector is $\{(R_{2x})_c\ (R_{2y})_c\ -2000\}^T$, where $(R_{2x})_c$ and $(R_{2y})_c$ are the components of the support reaction forces in the directions of the free and constrained directions, respectively. The transformed force deflection equation is then

$$\begin{Bmatrix} (R_{2x})_c \\ (R_{2y})_c \\ -2000 \end{Bmatrix} = 1(10^6) \begin{bmatrix} 11.491 & -1.5029 & -3.7947 \\ -1.5029 & 2.1657 & -1.2649 \\ -3.7947 & -1.2649 & 10.828 \end{bmatrix} \begin{Bmatrix} u_{2c} \\ v_{2c} \\ v_3 \end{Bmatrix} \qquad [c]$$

The component of the support reaction force in the free direction $(R_{2x})_c = 0$, whereas $(R_{2y})_c$ is unknown and $v_{2c} = 0$. This means that we partition the second equation (with $v_{2c} = 0$), yielding

$$(R_{2y})_c = (-1.5029 u_{2c} - 1.2649 v_3)(10^6) \qquad [d]$$

$$\begin{Bmatrix} 0 \\ -2000 \end{Bmatrix} = 1(10^6) \begin{bmatrix} 11.491 & -3.7947 \\ -3.7947 & 10.828 \end{bmatrix} \begin{Bmatrix} u_{2c} \\ v_3 \end{Bmatrix} \qquad [e]$$

Solving the inverse of Eq. (e) yields

$$\begin{Bmatrix} u_{2c} \\ v_3 \end{Bmatrix} = 1(10^{-7}) \begin{bmatrix} 98.41 & 34.49 \\ 34.49 & 1.0444 \end{bmatrix} \begin{Bmatrix} 0 \\ -2000 \end{Bmatrix} = \begin{Bmatrix} -6.898 \\ -20.887 \end{Bmatrix} (10^{-5})\ \text{m}$$

With the displacements known, the support reactions can be determined. For R_{1x}, R_{1y}, and R_{3x}, Eqs. (f) from Example 9.2-3 can be used. However, these equations are written in terms of global displacements. We would need to transform the deflections of node 2, known in terms of the free and constraint components to global displacements. This is easily accomplished using Eq. (9.2-76). Thus

$$\begin{Bmatrix} u_2 \\ v_2 \end{Bmatrix}_{global} = \begin{bmatrix} 0.8944 & -0.4472 \\ 0.4472 & 0.8944 \end{bmatrix}_c \begin{Bmatrix} -6.898(10^{-5}) \\ 0 \end{Bmatrix}_c = \begin{Bmatrix} -6.169 \\ -3.085 \end{Bmatrix} (10^{-5}) \text{ m}$$

From Example 9.2-3, Eqs. (f),

$$R_{1x} = 8(10^6)[0 - (-6.169)(10^{-5})] = 493.5 \text{ N}$$

$$R_{1y} = 8(10^6)[0 - (-2.089)(10^{-4})] = 1671.0 \text{ N}$$

$$R_{3x} = 2\sqrt{2}(10^6)[-(-6.169) - (-3.085) + 0 + (-20.89)](10^{-5})$$

$$= -329.0 \text{ N}$$

We can calculate the forces on node 2 in terms of either coordinate system using either Eq. (a) or (c). In terms of the global coordinates we use the first two equations of Eq. (a), giving

$$R_{2x} = 2(10^6)[(4 + \sqrt{2})(-6.169) + \sqrt{2}(-3.085) - \sqrt{2}(-20.89)](10^{-5})$$

$$= -164.5 \text{ N}$$

$$R_{2y} = 2(10^6)[\sqrt{2}(-6.169) + \sqrt{2}(-3.085) - \sqrt{2}(-20.89)](10^{-5})$$

$$= -329.0 \text{ N}$$

The reader is urged to check these results using basic statics.

9.2.8   THE THREE-DIMENSIONAL TRUSS ELEMENT

The truss element in three-dimensional space is handled quite similarly to the two-dimensional procedure. Consider the axial element shown in Fig. 9.2-12 where the general global coordinate system is given by the xyz axes and the element local coordinate system is given by the $x'y'z'$ axes. The coordinates of the i and j nodes are given by (x_i, y_i, z_i) and (x_j, y_j, z_j), respectively. The length of the element is $L_e = [(x_j - x_i)^2 + (y_j - y_i)^2 + (z_j - z_i)^2]^{1/2}$. The directional cosines are

$$\ell = \cos \theta_x = \frac{x_j - x_i}{L_e}$$

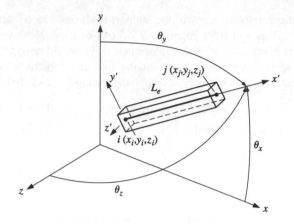

Figure 9.2-12

$$m = \cos \theta_y = \frac{y_j - y_i}{L_e} \qquad \textbf{[9.2-82]}$$

$$n = \cos \theta_z = \frac{z_j - z_i}{L_e}$$

In a manner identical to the development of Eq. (9.2-61) the local axial displacement of the nodes can be written in terms of the global displacements as

$$u_i' = u_i \ell + v_i m + w_i n$$

$$u_j' = u_j \ell + v_j m + w_j n \qquad \textbf{[9.2-83]}$$

which can be expressed in matrix form as

$$\left\{ \begin{array}{c} u_i' \\ u_j' \end{array} \right\} = \begin{bmatrix} \ell & m & n & 0 & 0 & 0 \\ 0 & 0 & 0 & \ell & m & n \end{bmatrix} \left\{ \begin{array}{c} u_i \\ v_i \\ w_i \\ u_j \\ v_j \\ w_j \end{array} \right\} \qquad \textbf{[9.2-84]}$$

As before, let

$$\{\mathbf{u}\}_{local} = \left\{ \begin{array}{c} u_i' \\ u_j' \end{array} \right\} \qquad [\mathbf{T}] = \begin{bmatrix} \ell & m & n & 0 & 0 & 0 \\ 0 & 0 & 0 & \ell & m & n \end{bmatrix} \qquad \{\mathbf{u}\}_{global} = \left\{ \begin{array}{c} u_i \\ v_i \\ w_i \\ u_j \\ v_j \\ w_j \end{array} \right\} \qquad \textbf{[9.2-85]}$$

where $[\mathbf{T}]$ is the three-dimensional transformation matrix.

Every equation in the previous section involving the two-dimensional transformation matrix can be utilized using the three-dimensional transformation matrix. For example, Eq. (9.2-69) gives the global stiffness matrix. Using the three-dimensional transformation matrix, we obtain

$$[\mathbf{k}]_{\text{global}} = [\mathbf{T}]^T[\mathbf{k}]_{\text{local}}[\mathbf{T}]$$

$$= \begin{bmatrix} \ell & 0 \\ m & 0 \\ n & 0 \\ 0 & \ell \\ 0 & m \\ 0 & n \end{bmatrix} \left(\frac{AE}{L}\right)_e \begin{bmatrix} 1 & -1 \\ -1 & 1 \end{bmatrix} \begin{bmatrix} \ell & m & n & 0 & 0 & 0 \\ 0 & 0 & 0 & \ell & m & n \end{bmatrix}$$

Performing the matrix multiplication results in

$$[\mathbf{k}]_{\text{global}} = \left(\frac{AE}{L}\right)_e \begin{bmatrix} \ell^2 & \ell m & \ell n & -\ell^2 & -\ell m & -\ell n \\ \ell m & m^2 & mn & -\ell m & -m^2 & -mn \\ \ell n & mn & n^2 & -\ell n & -mn & -n^2 \\ -\ell^2 & -\ell m & -\ell n & \ell^2 & \ell m & \ell n \\ -\ell m & -m^2 & -mn & \ell m & m^2 & mn \\ -\ell n & -mn & -n^2 & \ell n & mn & n^2 \end{bmatrix} \qquad \textbf{[9.2-86]}$$

All joints of the space truss shown in Fig. 9.2-13 are to be considered ball joints. **Example 9.2-5** Each member has a cross-sectional area of 0.2 in² and a modulus of elasticity of 30 Mpsi. The coordinates of the nodes are given in inches. Determine the deflection of node 2 and the stress in each member. Ignore the possibility of buckling.

Solution:

Element (1): The length is obviously $L_1 = 40$ in. With $i_1 = 1$ and $j_1 = 2$ the directional cosines are $\ell_1 = 1$, and $m_1 = n_1 = 0$. The stiffness is $k_1 = (AE/L)_1 = 0.2(30)(10^6)/40 = 1.5(10^5)$ lb/in. Thus the global stiffness matrix is

$$[\mathbf{k}]_{\text{global}} = 1.5(10^5) \begin{array}{c} \begin{array}{cccccc} u_1 & v_1 & w_1 & u_2 & v_2 & w_2 \end{array} \\ \begin{bmatrix} 1 & 0 & 0 & -1 & 0 & 0 \\ 0 & 0 & 0 & 0 & 0 & 0 \\ 0 & 0 & 0 & 0 & 0 & 0 \\ -1 & 0 & 0 & 1 & 0 & 0 \\ 0 & 0 & 0 & 0 & 0 & 0 \\ 0 & 0 & 0 & 0 & 0 & 0 \end{bmatrix} \begin{array}{c} u_1 \\ v_1 \\ w_1 \\ u_2 \\ v_2 \\ w_2 \end{array} \end{array}$$

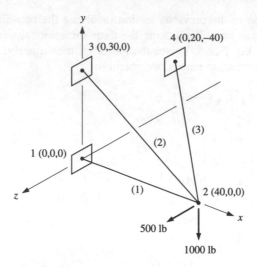

Figure 9.2-13

Element (2): Let $i_2 = 2$ and $j_2 = 3$. The length is

$$L_2 = [(0 - 40)^2 + (30 - 0)^2 + (0 - 0)^2]^{1/2} = 50 \text{ in}$$

and the directional cosines are

$$\ell_2 = \frac{0 - 40}{50} = -0.8 \qquad m_2 = \frac{30 - 0}{50} = 0.6 \qquad n_2 = \frac{0 - 0}{50} = 0$$

The stiffness is $k_2 = 0.2(30)(10^6)/50 = 1.2(10^5)$ lb/in. Thus the stiffness matrix is

$$[\mathbf{k}_2]_{\text{global}} = 1.2(10^5)
\begin{array}{c}
\begin{array}{cccccc}
\quad u_2 \quad & v_2 \quad & w_2 \quad & u_3 \quad & v_3 \quad & w_3
\end{array} \\
\left[
\begin{array}{cccccc}
0.64 & -0.48 & 0 & -0.64 & 0.48 & 0 \\
-0.48 & 0.36 & 0 & 0.48 & -0.36 & 0 \\
0 & 0 & 0 & 0 & 0 & 0 \\
-0.64 & 0.48 & 0 & 0.64 & -0.48 & 0 \\
0.48 & -0.36 & 0 & -0.48 & 0.36 & 0 \\
0 & 0 & 0 & 0 & 0 & 0
\end{array}
\right]
\begin{array}{c}
u_2 \\ v_2 \\ w_2 \\ u_3 \\ v_3 \\ w_3
\end{array}
\end{array}$$

Element (3): Let $i_3 = 2$ and $j_3 = 4$. The length is

$$L_3 = [(0 - 40)^2 + (20 - 0)^2 + (-40 - 0)^2]^{1/2} = 60 \text{ in}$$

and the directional cosines are

$$\ell_3 = \frac{0 - 40}{60} = -\frac{2}{3} \qquad m_3 = \frac{20 - 0}{60} = \frac{1}{3} \qquad n_3 = \frac{-40 - 0}{50} = -\frac{2}{3}$$

The stiffness is $k_3 = 0.2(30)(10^6)/60 = 1.0(10^5)$ lb/in. Thus the stiffness matrix is

$$[\mathbf{k}_3]_{\text{global}} = 1.0(10^5)\left(\frac{1}{3^2}\right)
\begin{array}{c}
\begin{array}{cccccc} u_2 & v_2 & w_2 & u_4 & v_4 & w_4 \end{array} \\
\begin{bmatrix}
4 & -2 & 4 & -4 & 2 & -4 \\
-2 & 1 & -2 & 2 & -1 & 2 \\
4 & -2 & 4 & -4 & 2 & -4 \\
-4 & 2 & -4 & 4 & -2 & 4 \\
2 & -1 & 2 & -2 & 1 & -2 \\
-4 & 2 & -4 & 4 & -2 & 4
\end{bmatrix}
\begin{array}{c} u_2 \\ v_2 \\ w_2 \\ u_4 \\ v_4 \\ w_4 \end{array}
\end{array}$$

Adding the matrices according to the degree of freedom yields the system stiffness matrix[†]

$$[\mathbf{K}]_{\text{system}} = 1.0(10^5)$$

	u_1	v_1	w_1	u_2	v_2	w_2	u_3	v_3	w_3	u_4	v_4	w_4	
	1.5	0	0	−1.5	0	0	0	0	0	0	0	0	u_1
	0	0	0	0	0	0	0	0	0	0	0	0	v_1
	0	0	0	0	0	0	0	0	0	0	0	0	w_1
	−1.5	0	0	2.7124	−0.7982	0.4444	−0.768	0.576	0	−0.4444	0.2222	−0.4444	u_2
	0	0	0	−0.7982	0.5431	−02222	0.576	−0.432	0	0.2222	−0.1111	0.2222	v_2
	0	0	0	0.4444	−0.2222	0.4444	0	0	0	−0.4444	0.2222	−0.4444	w_2
	0	0	0	−0.768	0.576	0	0.768	−0.576	0	0	0	0	u_3
	0	0	0	0.576	−0.432	0	−0.576	0.432	0	0	0	0	v_3
	0	0	0	0	0	0	0	0	0	0	0	0	w_3
	0	0	0	−0.4444	0.2222	−0.4444	0	0	0	0.4444	−0.2222	0.4444	u_4
	0	0	0	0.2222	−0.1111	0.2222	0	0	0	−0.2222	0.1111	−0.2222	v_4
	0	0	0	−0.4444	0.2222	−0.4444	0	0	0	0.4444	−0.2222	0.4444	w_4

The system force vector is $\{F_{1x}\ F_{1y}\ F_{1z}\ F_{2x}\ F_{2y}\ F_{2z}\ F_{3x}\ F_{3y}\ F_{3z}\ F_{4x}\ F_{4y}\ F_{4z}\}^T = \{R_{1x}\ R_{1y}\ R_{1z}\ 0 -1000\ 500\ R_{3x}\ R_{3y}\ R_{3z}\ R_{4x}\ R_{4y}\ R_{4z}\}^T$ where the R's are the x, y, and z components of the unknown support reactions at nodes 1, 3, and 4. This means that the first three and the last six force equations partition out of the system equations. These are (with $u_1 = v_1 = w_1 = u_3 = v_3 = w_3 = u_4 = v_4 = w_4 = 0$)

$$R_{1x} = 1.5(10^5)(-u_2)$$

$$R_{1y} = 0$$

$$R_{1z} = 0$$

$$R_{3x} = 1(10^5)(-0.768u_2 + 0.576v_2)$$

$$R_{3y} = 1(10^5)(0.576u_2 - 0.432v_2)$$

[†] If the reader is still unsure of how the matrices add, it is recommended that he or she expand each matrix in terms of the system degrees of freedom $\{u_1\ v_1\ w_1\ u_2\ v_2\ w_2\ u_3\ v_3\ w_3\ u_4\ v_4\ w_4\}^T$ before adding.

$$R_{3z} = 0$$

$$R_{4x} = 1(10^5)(-0.4444u_2 + 0.2222v_2 - 0.4444w_2)$$

$$R_{4y} = 1(10^5)(0.2222u_2 - 0.1111v_2 + 0.2222w_2)$$

$$R_{4z} = 1(10^5)(-0.4444u_2 + 0.2222v_2 - 0.4444w_2)$$

Applying the boundary conditions to the remainder of the system equations yields

$$\begin{Bmatrix} 0 \\ -1000 \\ 500 \end{Bmatrix} = 1(10^5) \begin{bmatrix} 2.7124 & -0.7982 & 0.4444 \\ -0.7982 & 0.5431 & -0.2222 \\ 0.4444 & -0.2222 & 0.4444 \end{bmatrix} \begin{Bmatrix} u_2 \\ v_2 \\ w_2 \end{Bmatrix}$$

Solving for the inverse yields

$$\begin{Bmatrix} u_2 \\ v_2 \\ w_2 \end{Bmatrix} = 1(10^{-1}) \begin{bmatrix} 0.6667 & 0.8889 & -0.2222 \\ 0.8889 & 3.5 & 0.8611 \\ -0.2222 & 0.8611 & 2.9028 \end{bmatrix} \begin{Bmatrix} 0 \\ -1000 \\ 500 \end{Bmatrix} = \begin{Bmatrix} -0.01 \\ -0.03069 \\ 0.005903 \end{Bmatrix} \text{ in}$$

Substituting the deflections back into the equations for the support reaction forces gives

$$R_{1x} = 1.5(10^5)[-(-0.01)] = 1500 \text{ lb}$$

$$R_{1y} = 0$$

$$R_{1z} = 0$$

$$R_{3x} = 1(10^5)[-0.768(-0.01) + 0.576(-0.03069)] = -1000 \text{ lb}$$

$$R_{3y} = 1(10^5)[0.576(-0.01) - 0.432(-0.03069)] = 750 \text{ lb}$$

$$R_{3z} = 0$$

$$R_{4x} = 1(10^5)[-0.4444(-0.01) + 0.2222(-0.03069) - 0.4444(0.005903)]$$

$$= -500 \text{ lb}$$

$$R_{4y} = 1(10^5)[0.2222(-0.01) - 0.1111(-0.03069) + 0.2222(0.005903)]$$

$$= 250 \text{ lb}$$

$$R_{4z} = 1(10^5)[-0.4444(-0.01) + 0.2222(-0.03069) - 0.4444(0.005903)]$$

$$= -500 \text{ lb}$$

The stresses are given by Eq. (9.2-9) with Eq. (9.2-84), which is

$$(\sigma_{x'})_e = \left(\frac{E}{L}\right)_e \{-1 \quad 1\} \begin{bmatrix} \ell & m & n & 0 & 0 & 0 \\ 0 & 0 & 0 & \ell & m & n \end{bmatrix}_e \begin{Bmatrix} u_i \\ v_i \\ w_i \\ u_j \\ v_j \\ w_j \end{Bmatrix} \qquad \textbf{[9.2-87]}$$

For the three elements of the example, the stresses are

$$(\sigma_{x'})_1 = \frac{30(10^6)}{40} \{-1 \quad 1\} \begin{bmatrix} 1 & 0 & 0 & 0 & 0 & 0 \\ 0 & 0 & 0 & 1 & 0 & 0 \end{bmatrix} \begin{Bmatrix} 0 \\ 0 \\ 0 \\ -0.01 \\ -0.03069 \\ 0.005903 \end{Bmatrix} = -7500 \text{ psi}$$

$$(\sigma_{x'})_2 = \frac{30(10^6)}{50} \{-1 \quad 1\} \begin{bmatrix} -0.8 & 0.6 & 0 & 0 & 0 & 0 \\ 0 & 0 & 0 & -0.8 & 0.6 & 0 \end{bmatrix} \begin{Bmatrix} -0.01 \\ -0.03069 \\ 0.005903 \\ 0 \\ 0 \\ 0 \end{Bmatrix}$$

$$= 6250 \text{ psi}$$

$$(\sigma_{x'})_3 = \frac{30(10^6)}{60} \{-1 \quad 1\} \left(\frac{1}{3}\right) \begin{bmatrix} -2 & 1 & -2 & 0 & 0 & 0 \\ 0 & 0 & 0 & -2 & 1 & -2 \end{bmatrix} \begin{Bmatrix} -0.01 \\ -0.03069 \\ 0.005903 \\ 0 \\ 0 \\ 0 \end{Bmatrix}$$

$$= 3750 \text{ psi}$$

9.3   BEAM AND FRAME ELEMENTS[†]

In this section we will first develop the governing equations of the planar beam element. Distributed forces and hinges will also be covered. This will be followed by the two-plane bending element. The frame element will then be presented, which incorporates two-plane bending, axial loading, and torsional loading. For general applications, the frame element will then be transformed into three-dimensional space. In the last part of this section, the effects of load stiffening and buckling will be introduced.

9.3.1   THE PLANAR BEAM ELEMENT

Similar to the truss element, the beam element has two nodes. Figure 9.3-1 shows the positive definitions of the nodal loads and deflections. For the planar element the nodal shear forces and bending moments are defined positive according to the coordinate system and *not* according to the mechanics of materials convention established in Sec. 3.4.1. Also note that we are defining the *xyz* coordinate system as a local system originating at node *i*. In Sec. 9.3.6 we will use the *xyz* as a global system and when we refer to the local system we will use the notation *x′y′z′* for the local coordinate system. Let the modulus of elasticity of the element be E_e and the second-area moment about the bending (z) axis be I_e.

As before, once the element is defined, we assume a displacement field. Since each node has *two* displacement degrees of freedom, v and $\theta = dv/dx$, at each of

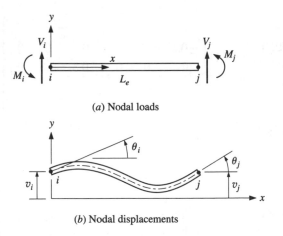

(*a*) Nodal loads

(*b*) Nodal displacements

Figure 9.3-1   The planar beam element.

[†] In this section we will ignore the effects of the transverse shear on the deflections of a beam. This is acceptable for most beam problems with the exception of short beams.

two nodes, we can assume a displacement field for $v(x)$ using *four* unknown constants. That is,

$$v(x) = a_1 + a_2x + a_3x^2 + a_4x^3 \qquad \textbf{[9.3-1]}$$

Consequently, the slope of the beam is

$$\theta = \frac{dv}{dx} = a_2 + 2a_3x + 3a_4x^2 \qquad \textbf{[9.3-2]}$$

As with the axial element, we want to express the field in terms of the nodal displacements, not the a_i. Thus substitute the nodal displacements

$$v(0) = v_i \qquad \frac{dv}{dx}(0) = \theta_i \qquad v(L_e) = v_j \qquad \frac{dv}{dx}(L_e) = \theta_j$$

These four conditions yield

$$v_i = a_1$$

$$\theta_i = a_2$$

$$v_j = a_1 + a_2L_e + a_3L_e^2 + a_4L_e^3$$

$$\theta_j = a_2 + 2a_3L_e + 3a_4L_e^2$$

The first two equations yield a_1 and a_2 directly. Substituting these into the last two equations and solving simultaneously for a_3 and a_4 results in

$$a_1 = v_i$$

$$a_2 = \theta_i$$

$$a_3 = \frac{3}{L_e^2}(v_j - v_i) - \frac{1}{L_e}(2\theta_i + \theta_j)$$

$$a_4 = \frac{2}{L_e^3}(v_i - v_j) + \frac{1}{L_e^2}(\theta_i + \theta_j) \qquad \textbf{[9.3-3]}$$

Substitute Eqs. (9.3-3) into (9.3-1) and rearrange in terms of the nodal displacements. This yields

$$v(x) = \left[1 - 3\left(\frac{x}{L_e}\right)^2 + 2\left(\frac{x}{L_e}\right)^3\right]v_i + \left[x - 2\frac{x^2}{L_e} + \frac{x^3}{L_e^2}\right]\theta_i$$

$$+ \left[3\left(\frac{x}{L_e}\right)^2 - 2\left(\frac{x}{L_e}\right)^3\right]v_j + \left[-\frac{x^2}{L_e} + \frac{x^3}{L_e^2}\right]\theta_j \qquad \textbf{[9.3-4]}$$

Equation (9.3-4) can be expressed as

$$v(x) = (N_v)_i v_i + (N_\theta)_i \theta_i + (N_v)_j v_j + (N_\theta)_j \theta_j \qquad \textbf{[9.3-5]}$$

where

$$(N_v)_i = 1 - 3\left(\frac{x}{L_e}\right)^2 + 2\left(\frac{x}{L_e}\right)^3$$

$$(N_\theta)_i = x - 2\frac{x^2}{L_e} + \frac{x^3}{L_e^2}$$

$$(N_v)_j = 3\left(\frac{x}{L_e}\right)^2 - 2\left(\frac{x}{L_e}\right)^3$$

$$(N_\theta)_j = -\frac{x^2}{L_e} + \frac{x^3}{L_e^2} \qquad \textbf{[9.3-6]}$$

are the *shape functions* for the beam element. Equation (9.3-5) can be written in matrix form as $v(x) = \{N\}\{u\}$, where $\{N\}$ is the shape function row vector and $\{u\}$ is the nodal displacement vector. That is,

$$v(x) = \{(N_v)_i \quad (N_\theta)_i \quad (N_v)_j \quad (N_\theta)_j\} \begin{Bmatrix} v_i \\ \theta_i \\ v_j \\ \theta_j \end{Bmatrix} \qquad \textbf{[9.3-7]}$$

Using the Rayleigh-Ritz method, the next step is to develop the strain energy in terms of the nodal displacements. Recall from Sec. 6.3, Eq. (6.3-8), the strain energy for a beam in bending is given by

$$U = \frac{1}{2} \int_{L_e} EI \left(\frac{d^2v}{dx^2}\right)^2 dx \qquad \textbf{[9.3-8]}$$

The second derivative can be determined from Eq. (9.3-7) and expressed as

$$\frac{d^2v}{dx^2} = \left\{\frac{d^2(N_v)_i}{dx^2} \quad \frac{d^2(N_\theta)_i}{dx^2} \quad \frac{d^2(N_v)_j}{dx^2} \quad \frac{d^2(N_\theta)_j}{dx^2}\right\} \begin{Bmatrix} v_i \\ \theta_i \\ v_j \\ \theta_j \end{Bmatrix}$$

$$= \{B_1 \quad B_2 \quad B_3 \quad B_4\} \begin{Bmatrix} v_i \\ \theta_i \\ v_j \\ \theta_j \end{Bmatrix} \qquad \textbf{[9.3-9]}$$

where

$$B_1 = \frac{d^2(N_v)_i}{dx^2} = -\frac{6}{L_e^2} + 12\frac{x}{L_e^3}$$

$$B_2 = \frac{d^2(N_\theta)_i}{dx^2} = -\frac{4}{L_e} + 6\frac{x}{L_e^2}$$

$$B_3 = \frac{d^2(N_v)_j}{dx^2} = \frac{6}{L_e^2} - 12\frac{x}{L_e^3}$$

$$B_4 = \frac{d^2(N_\theta)_j}{dx^2} = -\frac{2}{L_e} + 6\frac{x}{L_e^2} \qquad \textbf{[9.3-10]}$$

The form of Eq. (9.3-9) is $d^2v/dx^2 = \{\mathbf{B}\}\{\mathbf{u}\}$ where $\{\mathbf{B}\} = \{B_1\ B_2\ B_3\ B_4\}$ and the displacement vector is $\{\mathbf{u}\} = \{v_i\ \theta_i\ v_j\ \theta_j\}^T$. The square of the product of two matrices, $[\mathbf{C}] = [\mathbf{A}]\,[\mathbf{B}]$, is given by $[\mathbf{C}]^2 = [\mathbf{B}]^T\,[\mathbf{A}]^T\,[\mathbf{A}]\,[\mathbf{B}]$. Thus $(d^2v/dx^2)^2 = \{\mathbf{u}\}^T\{\mathbf{B}\}^T\,\{\mathbf{B}\}\{\mathbf{u}\}$. Substituting this into the strain energy, and assuming $(EI)_e$ is constant for the element, yields the form

$$U = \frac{1}{2}(EI)_e \int_{L_e} \{\mathbf{u}\}^T\{\mathbf{B}\}^T\{\mathbf{B}\}\{\mathbf{u}\}\ dx \qquad \textbf{[9.3-11]}$$

Since $\{\mathbf{u}\}$ is not a function of x, the above can be written as

$$U = \frac{1}{2}\{\mathbf{u}\}^T\left[(EI)_e \int_{L_e} \{\mathbf{B}\}^T\{\mathbf{B}\}\ dx\right]\{\mathbf{u}\} \qquad \textbf{[9.3-12]}$$

Recall that the form of the strain energy is $U = 1/2\{\mathbf{u}\}_e^T\,[\mathbf{k}]_e\,\{\mathbf{u}\}_e$, where $[\mathbf{k}]_e$ is the element stiffness matrix. Thus, from Eq. (9.3-12), we see that for the beam element the stiffness matrix is

$$[\mathbf{k}]_e = (EI)_e \int_{L_e} \{\mathbf{B}\}^T\{\mathbf{B}\}\ dx \qquad \textbf{[9.3-13]}$$

or

$$[\mathbf{k}]_e = (EI)_e \int_{L_e} \begin{Bmatrix} B_1 \\ B_2 \\ B_3 \\ B_4 \end{Bmatrix} \{B_1\ B_2\ B_3\ B_4\}_e\ dx \qquad \textbf{[9.3-14]}$$

Performing the matrix multiplication results in

$$[\mathbf{k}]_e = (EI)_e \int_{L_e} \begin{bmatrix} B_1^2 & B_1B_2 & B_1B_3 & B_1B_4 \\ B_1B_2 & B_2^2 & B_2B_3 & B_2B_4 \\ B_1B_3 & B_2B_3 & B_3^2 & B_3B_4 \\ B_1B_4 & B_2B_4 & B_3B_4 & B_4^2 \end{bmatrix}_e dx \qquad \textbf{[9.3-15]}$$

Since the B_i terms are functions of x, all of the terms within the matrix must be integrated. Integration yields

$$[\mathbf{k}]_e = \left(\frac{EI}{L^3}\right)_e \begin{bmatrix} 12 & 6L & -12 & 6L \\ 6L & 4L^2 & -6L & 2L^2 \\ -12 & -6L & 12 & -6L \\ 6L & 2L^2 & -6L & 4L^2 \end{bmatrix}_e$$ **[9.3-16]**

For an example of an integration, consider the B_1B_2 term. That is,

$$(EI)_e \int_0^{L_e} B_1 B_2 \ dx = (EI)_e \int_0^{L_e} \left(-\frac{6}{L_e^2} + 12\frac{x}{L_e^3}\right)\left(-\frac{4}{L_e} + 6\frac{x}{L_e^2}\right) dx$$

$$= (EI)_e \int_0^{L_e} \left(\frac{24}{L_e^3} - 84\frac{x}{L_e^4} + 72\frac{x^2}{L_e^5}\right) dx$$

$$= (EI)_e \left(24\frac{x}{L_e^3} - 42\frac{x^2}{L_e^4} + 24\frac{x^3}{L_e^5}\right)\Big|_0^{L_e} = 6\left(\frac{EI}{L^2}\right)_e = \left(\frac{EI}{L^3}\right)_e (6L_e)$$

Note that this agrees with the k_{12} and k_{21} terms of the matrix in Eq. (9.3-16).

The forms of loading of a beam are shown in Fig. 9.3-2, which include force distributions $q(x)$ (normally given in force per unit length), concentrated forces P_r, acting at specific points r, and concentrated moments M_s acting at specific points s. The total potential energy is the strain energy plus the work potential, giving

$$\Pi = \frac{1}{2}\{\mathbf{u}\}^T[\mathbf{k}]_e\{\mathbf{u}\} - \int_{L_e} q(x)v(x)\,dx - \sum_r P_r v_r - \sum_s M_s \theta_s$$ **[9.3-17]**

where $\{\mathbf{u}\} = \{v_i\,\theta_i\,v_j\,\theta_j\}^T$ and $[\mathbf{k}]_e$ is given by Eq. (9.3-16). The summations refer to the collections of concentrated forces and moments acting on the beam. In the next section we will describe the procedure for dealing with distributed forces. For now, let us only consider the nodal loads shown in Fig. 9.3-1(a).

The total potential energy of the beam shown in Fig. 9.3-1(a) is

$$\Pi_e = \frac{1}{2}\{\mathbf{u}\}^T[\mathbf{k}]_e\{\mathbf{u}\} - V_i v_i - V_j v_j - M_i \theta_i - M_j \theta_j$$ **[9.3-18]**

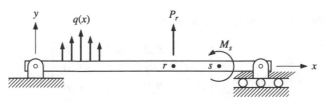

Figure 9.3-2

For the Rayleigh-Ritz method we differentiate this with respect to the nodal displacements $\{u\} = \{v_i\ \theta_i\ v_j\ \theta_j\}^T$; and substituting Eq. (9.3-16) for $[k]_e$ we obtain

$$\left(\frac{EI}{L^3}\right)_e \begin{bmatrix} 12 & 6L & -12 & 6L \\ 6L & 4L^2 & -6L & 2L^2 \\ -12 & -6L & 12 & -6L \\ 6L & 2L^2 & -6L & 4L^2 \end{bmatrix}_e \begin{Bmatrix} v_i \\ \theta_i \\ v_j \\ \theta_j \end{Bmatrix}_e - \begin{Bmatrix} V_i \\ M_i \\ V_j \\ M_j \end{Bmatrix}_e = \begin{Bmatrix} 0 \\ 0 \\ 0 \\ 0 \end{Bmatrix}$$

or

$$\begin{Bmatrix} V_i \\ M_i \\ V_j \\ M_j \end{Bmatrix}_e = \left(\frac{EI}{L^3}\right)_e \begin{bmatrix} 12 & 6L & -12 & 6L \\ 6L & 4L^2 & -6L & 2L^2 \\ -12 & -6L & 12 & -6L \\ 6L & 2L^2 & -6L & 4L^2 \end{bmatrix}_e \begin{Bmatrix} v_i \\ \theta_i \\ v_j \\ \theta_j \end{Bmatrix}_e \qquad \textbf{[9.3-19]}$$

Example 9.3-1

For the step beam shown in Fig. 9.3-3 the second-area moment and section modulus of element (1) are $I_1 = 40(10^{-6})$ m^4 and $S_1 = 35(10^{-4})$ m^3, respectively. For element (2), $I_2 = 10(10^{-6})$ m^4 and $S_2 = 15(10^{-4})$ m^3. Determine the deflections of the three nodes, the support reactions, the element loads, and the maximum bending stress in each beam. The beam material is steel with $E = 200$ GPa.

Solution:

Element (1): $(EI/L^3)_1 = 200(10^9)(40)(10^{-6})/2^3 = 1.0(10^6)$ N/m. The stiffness matrix for the element is

$$[k]_1 = 1.0(10^6) \begin{bmatrix} 12 & (6)(2) & -12 & (6)(2) \\ (6)(2) & (4)(2^2) & -(6)(2) & (2)(2^2) \\ -12 & -(6)(2) & 12 & -(6)(2) \\ (6)(2) & (2)(2^2) & -(6)(2) & (4)(2^2) \end{bmatrix}$$

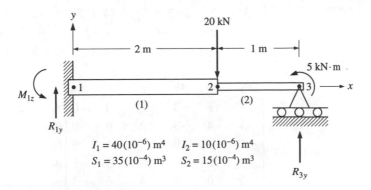

$$I_1 = 40(10^{-6})\ \text{m}^4 \qquad I_2 = 10(10^{-6})\ \text{m}^4$$
$$S_1 = 35(10^{-4})\ \text{m}^3 \qquad S_2 = 15(10^{-4})\ \text{m}^3$$

Figure 9.3-3

$$[\mathbf{k}]_1 = 4.0(10^6) \begin{array}{cccc} v_1 & \theta_1 & v_2 & \theta_2 \\ \begin{bmatrix} 3 & 3 & -3 & 3 \\ 3 & 4 & -3 & 2 \\ -3 & -3 & 3 & -3 \\ 3 & 2 & -3 & 4 \end{bmatrix} \begin{array}{c} v_1 \\ \theta_1 \\ v_2 \\ \theta_2 \end{array} \end{array} \ \text{N/m}$$

where in the final matrix, the degrees of freedom associated with the rows and columns are shown.

Element (2): $(EI/L^3)_2 = 200(10^9)(10)(10^{-6})/1^3 = 2.0(10^6)$ N/m. The stiffness matrix for the element is

$$[\mathbf{k}]_2 = 2.0(10^6) \begin{bmatrix} 12 & (6)(1) & -12 & (6)(1) \\ (6)(1) & (4)(1^2) & -(6)(1) & (2)(1^2) \\ -12 & -(6)(1) & 12 & -(6)(1) \\ (6)(1) & (2)(1^2) & -(6)(1) & (4)(1^2) \end{bmatrix}$$

$$= 4.0(10^6) \begin{array}{cccc} v_2 & \theta_2 & v_3 & \theta_3 \\ \begin{bmatrix} 6 & 3 & -6 & 3 \\ 3 & 2 & -3 & 1 \\ -6 & -3 & 6 & -3 \\ 3 & 1 & -3 & 2 \end{bmatrix} \begin{array}{c} v_2 \\ \theta_2 \\ v_3 \\ \theta_3 \end{array} \end{array} \ \text{N/m}$$

Assembling the matrices into the system stiffness matrix we see that the matrices add at their common rows and columns associated with the degrees of freedom v_2 and θ_2. That is,

$$[\mathbf{K}]_{\text{system}} = 4.0(10^6) \begin{bmatrix} 3 & 3 & -3 & 3 & 0 & 0 \\ 3 & 4 & -3 & 2 & 0 & 0 \\ -3 & -3 & (3+6) & (-3+3) & -6 & 3 \\ 3 & 2 & (-3+3) & (4+2) & -3 & 1 \\ 0 & 0 & -6 & -3 & 6 & -3 \\ 0 & 0 & 3 & 1 & -3 & 2 \end{bmatrix}$$

$$= 4.0(10^6) \begin{array}{cccccc} v_1 & \theta_1 & v_2 & \theta_2 & v_3 & \theta_3 \\ \begin{bmatrix} 3 & 3 & -3 & 3 & 0 & 0 \\ 3 & 4 & -3 & 2 & 0 & 0 \\ -3 & -3 & 9 & 0 & -6 & 3 \\ 3 & 2 & 0 & 6 & -3 & 1 \\ 0 & 0 & -6 & -3 & 6 & -3 \\ 0 & 0 & 3 & 1 & -3 & 2 \end{bmatrix} \begin{array}{c} v_1 \\ \theta_1 \\ v_2 \\ \theta_2 \\ v_3 \\ \theta_3 \end{array} \end{array}$$

The applied load vector is $\{R_{1y}, M_{1z} -20(10^3)\ 0\ R_{3y}, 5(10^3)\}^T$ where R_{1y} and R_{3y} are the vertical force support reactions at nodes 1 and 3, respectively, and M_{1z} is the moment support reaction at node 1; $-20(10^3)$ N is the applied vertical force at node 2; and 0 and $5(10^3)$ N·m are the applied moments at nodes 2 and 3, respectively. Thus the system equation is

$$\begin{Bmatrix} R_{1y} \\ M_{1z} \\ -20(10^3) \\ 0 \\ R_{3y} \\ 5(10^3) \end{Bmatrix} = 4.0(10^6) \begin{bmatrix} 3 & 3 & -3 & 3 & 0 & 0 \\ 3 & 4 & -3 & 2 & 0 & 0 \\ -3 & -3 & 9 & 0 & -6 & 3 \\ 3 & 2 & 0 & 6 & -3 & 1 \\ 0 & 0 & -6 & -3 & 6 & -3 \\ 0 & 0 & 3 & 1 & -3 & 2 \end{bmatrix} \begin{Bmatrix} v_1 \\ \theta_1 \\ v_2 \\ \theta_2 \\ v_3 \\ \theta_3 \end{Bmatrix}$$

The boundary conditions are $v_1 = \theta_1 = v_3 = 0$. Thus the first, second, and fifth equations are partitioned out, with $v_1 = \theta_1 = v_3 = 0$, resulting in

$$R_{1y} = 12(10^6)(-v_2 + \theta_2)$$

$$M_{1z} = 4(10^6)(-3v_2 + 2\theta_2)$$

$$R_{3y} = -12(10^6)(2v_2 + \theta_2 + \theta_3)$$

The first, second, and fifth columns of the remaining stiffness matrix can be placed into the load vector. But since $v_1 = \theta_1 = v_3 = 0$, the load vector is unchanged. The result of this is

$$\begin{Bmatrix} -20(10^3) \\ 0 \\ 5(10^3) \end{Bmatrix} = 4.0(10^6) \begin{bmatrix} 9 & 0 & 3 \\ 0 & 6 & 1 \\ 3 & 1 & 2 \end{bmatrix} \begin{Bmatrix} v_2 \\ \theta_2 \\ \theta_3 \end{Bmatrix}$$

Taking the inverse of this gives

$$\begin{Bmatrix} v_2 \\ \theta_2 \\ \theta_3 \end{Bmatrix} = \frac{1}{4}(10^{-6})\left(\frac{1}{90}\right) \begin{bmatrix} 22 & 6 & -36 \\ 6 & 18 & -18 \\ -36 & -18 & 108 \end{bmatrix} \begin{Bmatrix} -20(10^3) \\ 0 \\ 5(10^3) \end{Bmatrix}$$

$$= \begin{Bmatrix} -1.722(10^{-3})\ \text{m} \\ -5.833(10^{-4})\ \text{rad} \\ 3.500(10^{-3})\ \text{rad} \end{Bmatrix}$$

Substituting the deflections into the support reaction force equations yields

$$R_{1y} = 12(10^6)[-(-1.722)(10^{-3}) + (-5.833)(10^{-4})] = 13.667(10^3)\ \text{N}$$

$$= 13.67\ \text{kN}$$

$$M_{1z} = 4(10^6)[-3(-1.722)(10^{-3}) + 2(-5.833)(10^{-4})] = 16.0(10^3) \text{ N·m}$$

$$= 16.0 \text{ kN·m}$$

$$R_{3y} = -12(10^6)[2(-1.722)(10^{-3}) + (-5.833)(10^{-4}) + 3.50(10^{-3})]$$

$$= 6.333(10^3) \text{ N} = 6.33 \text{ kN}$$

The element nodal loads are given by Eq. (9.3-19).

Element (1):

$$\begin{Bmatrix} V_1 \\ M_1 \\ V_2 \\ M_2 \end{Bmatrix}_1 = 4.0(10^6) \begin{bmatrix} 3 & 3 & -3 & 3 \\ 3 & 4 & -3 & 2 \\ -3 & -3 & 3 & -3 \\ 3 & 2 & -3 & 4 \end{bmatrix} \begin{Bmatrix} v_1 \\ \theta_1 \\ v_2 \\ \theta_2 \end{Bmatrix}$$

$$= 4.0(10^6) \begin{bmatrix} 3 & 3 & -3 & 3 \\ 3 & 4 & -3 & 2 \\ -3 & -3 & 3 & -3 \\ 3 & 2 & -3 & 4 \end{bmatrix} \begin{Bmatrix} 0 \\ 0 \\ -1.722(10^{-3}) \\ -5.833(10^{-4}) \end{Bmatrix}$$

$$= \begin{Bmatrix} 13.667(10^3) \text{ N} \\ 16.0(10^3) \text{ N·m} \\ -13.667(10^3) \text{ N} \\ 11.333(10^3) \text{ N·m} \end{Bmatrix} = \begin{Bmatrix} 13.67 \text{ kN} \\ 16.0 \text{ kN·m} \\ -13.67 \text{ kN} \\ 11.33 \text{ kN·m} \end{Bmatrix}$$

Element (2):

$$\begin{Bmatrix} V_2 \\ M_2 \\ V_3 \\ M_3 \end{Bmatrix}_2 = 4.0(10^6) \begin{bmatrix} 6 & 3 & -6 & 3 \\ 3 & 2 & -3 & 1 \\ -6 & -3 & 6 & -3 \\ 3 & 1 & -3 & 2 \end{bmatrix} \begin{Bmatrix} v_2 \\ \theta_2 \\ v_3 \\ \theta_3 \end{Bmatrix}$$

$$= 4.0(10^6) \begin{bmatrix} 6 & 3 & -6 & 3 \\ 3 & 2 & -3 & 1 \\ -6 & -3 & 6 & -3 \\ 3 & 1 & -3 & 2 \end{bmatrix} \begin{Bmatrix} -1.722(10^{-3}) \\ -5.833(10^{-4}) \\ 0 \\ 3.500(10^{-3}) \end{Bmatrix}$$

$$= \begin{Bmatrix} -6.333(10^3) \text{ N} \\ -11.333(10^3) \text{ N·m} \\ 6.333(10^3) \text{ N} \\ 5.0(10^3) \text{ N·m} \end{Bmatrix} = \begin{Bmatrix} -6.33 \text{ kN} \\ -11.33 \text{ kN·m} \\ 6.33 \text{ kN} \\ 5.0 \text{ kN·m} \end{Bmatrix}$$

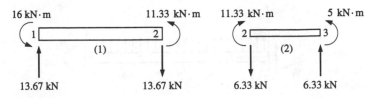

Figure 9.3-4

Figure 9.3-4 shows the element nodal loads. The reader is urged to verify the results using basic statics and mechanics of materials. Note that the nodal forces at node 2 add up to the applied 20-kN force in the negative y direction.

The maximum bending moments in a beam structure with only concentrated loads occur where the loads are applied. Since we are applying loads only at nodes, the maximum bending moments can occur only at the nodes. This will not be the case with distributed loading, as will be seen in the next section.

Thus, for element (1) the maximum bending moment magnitude is 16 kN·m and the maximum bending stress is

$$(\sigma_{max})_1 = \frac{(M_{max})_1}{S_1} = \frac{16(10^3)}{35(10^{-4})} = 4.57(10^3)\, \text{N/m}^2 = 4.57\, \text{MPa}$$

For element (2), the maximum bending moment magnitude is 11.33 kN·m. Thus

$$(\sigma_{max})_2 = \frac{(M_{max})_2}{S_2} = \frac{11.33(10^3)}{15(10^{-4})} = 7.55(10^6)\, \text{N/m}^2 = 7.55\, \text{MPa}$$

9.3.2 DISTRIBUTED LOADING

Consider the uniformly distributed force/length q_o shown in Fig. 9.3-5. With the finite element method, all loads are eventually reduced to nodal loads. Following the procedure used for the axial element, an equivalent nodal force vector for the distributed force can be represented by the *static* equivalent vector

$$\begin{Bmatrix} V_i \\ M_i \\ V_j \\ M_j \end{Bmatrix}_{q_o} = \begin{Bmatrix} \dfrac{q_o L_e}{2} \\ 0 \\ \dfrac{q_o L_e}{2} \\ 0 \end{Bmatrix} \qquad \textbf{[9.3-20]}$$

This is referred to as the "lumped" static equivalent load vector. However, this does not lead to the best results. A better equivalent model is that which is static and *work* equivalent.

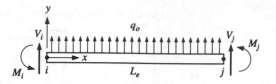

Figure 9.3-5

Consider the work potential of the distributed load

$$W_p = -\int_{L_e} q(x)v(x)\, dx = -q_o \int_{L_e} v(x)\, dx \qquad \textbf{[9.3-21]}$$

The displacement $v(x)$ is given in terms of the shape functions by Eq. (9.3-4). That is,

$$W_p = -q_o \int_0^{L_e} \left\{ \left[1 - 3\left(\frac{x}{L_e}\right)^2 + 2\left(\frac{x}{L_e}\right)^3 \right]v_i + \left[x - 2\frac{x^2}{L_e} + \frac{x^3}{L_e^2} \right]\theta_i \right.$$

$$\left. + \left[3\left(\frac{x}{L_e}\right)^2 - 2\left(\frac{x}{L_e}\right)^3 \right]v_j + \left[-\frac{x^2}{L_e} + \frac{x^3}{L_e^2} \right]\theta_j \right\} dx$$

which integrates to

$$W_p = -q_o \left[\left(x - \frac{x^3}{L_e^2} + \frac{1}{2}\frac{x^4}{L_e^3} \right)v_i + \left(\frac{x^2}{2} - \frac{2}{3}\frac{x^3}{L_e} + \frac{1}{4}\frac{x^4}{L_e^2} \right)\theta_i \right.$$

$$\left. + \left(\frac{x^3}{L_e^2} - \frac{1}{2}\frac{x^4}{L_e^3} \right)v_j + \left(-\frac{1}{3}\frac{x^3}{L_e} + \frac{1}{4}\frac{x^4}{L_e^2} \right)\theta_j \right]_0^{L_e}$$

Substituting the limits results in

$$W_p = -q_o \left(\frac{L_e}{2}v_i + \frac{L_e^2}{12}\theta_i + \frac{L_e}{2}v_j - \frac{L_e^2}{12}\theta_j \right) \qquad \textbf{[9.3-22]}$$

or

$$W_p = -\{v_i \ \theta_i \ v_j \ \theta_j\} \begin{Bmatrix} \dfrac{q_o L_e}{2} \\[2mm] \dfrac{q_o L_e^2}{12} \\[2mm] \dfrac{q_o L_e}{2} \\[2mm] -\dfrac{q_o L_e^2}{12} \end{Bmatrix} \qquad \textbf{[9.3-23]}$$

From the Rayleigh-Ritz method the nodal forces come from the derivatives of W_p relative to the nodal displacements. Thus the equivalent nodal load vector for the uniform distributed load is

$$\begin{Bmatrix} V_i \\ M_i \\ V_j \\ M_j \end{Bmatrix}_{q_o} = \begin{Bmatrix} \dfrac{q_o L_e}{2} \\[2ex] \dfrac{q_o L_e^2}{12} \\[2ex] \dfrac{q_o L_e}{2} \\[2ex] -\dfrac{q_o L_e^2}{12} \end{Bmatrix} \qquad \textbf{[9.3-24]}$$

This equivalent nodal load vector is called the *static* and *work equivalent* or *consistent* load vector. Figure 9.3-6 shows the static equivalent (lumped) and consistent models of the uniformly distributed loading.

The consistent load model will yield exact displacement results no matter how few elements are used. The lumped load model yields higher displacements than the exact solution. Consider, for example, that the beam shown in Fig. 9.3-6(*a*) is fixed at the left end and a single beam element is to be used to model the cantilever beam with a uniform load. From the tables in Appendix C the vertical deflection and rotation of the right end would be found to be $q_o L_e^4/(8EI)$ and $q_o L_e^3/(6EI)$, respectively. A single-element lumped-load beam model of Fig. 9.3-6(*b*) would yield values of $q_o L_e^4/(6EI)$ and $q_o L_e^3/(4EI)$ which are 33 and 50 percent higher than the exact results. A single-element consistent-load model of Fig. 9.3-6(*b*) would give values of $q_o L_e^4/(8EI)$ and $q_o L_e^3/(6EI)$ which are the exact results. In order to obtain reasonably

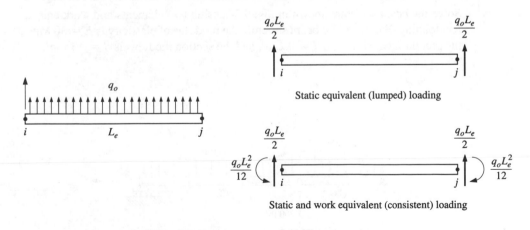

Static equivalent (lumped) loading

Static and work equivalent (consistent) loading

(*a*) Uniformly distributed loading

(*b*) Equivalent load models

Figure 9.3-6

accurate results using the lumped equivalent load model, one could use more than one element. If the elements were of equal length, it would take more than five elements to obtain an accuracy of 5 percent or better.

For the lumped, static equivalent model of the element shown in Fig 9.3-5, the finite element equations are

$$\begin{Bmatrix} V_i \\ M_i \\ V_j \\ M_j \end{Bmatrix}_e + \begin{Bmatrix} \dfrac{q_oL_e}{2} \\ 0 \\ \dfrac{q_oL_e}{2} \\ 0 \end{Bmatrix}_e = \left(\dfrac{EI}{L^3}\right)_e \begin{bmatrix} 12 & 6L & -12 & 6L \\ 6L & 4L^2 & -6L & 2L^2 \\ -12 & -6L & 12 & -6L \\ 6L & 2L^2 & -6L & 4L^2 \end{bmatrix}_e \begin{Bmatrix} v_i \\ \theta_i \\ v_j \\ \theta_j \end{Bmatrix}_e \qquad \textbf{[9.3-25]}$$

For the consistent model, the finite element equations are

$$\begin{Bmatrix} V_i \\ M_i \\ V_j \\ M_j \end{Bmatrix}_e + \begin{Bmatrix} \dfrac{q_oL_e}{2} \\ \dfrac{q_oL_e^2}{12} \\ \dfrac{q_oL_e}{2} \\ -\dfrac{q_oL_e^2}{12} \end{Bmatrix} = \left(\dfrac{EI}{L^3}\right)_e \begin{bmatrix} 12 & 6L & -12 & 6L \\ 6L & 4L^2 & -6L & 2L^2 \\ -12 & -6L & 12 & -6L \\ 6L & 2L^2 & -6L & 4L^2 \end{bmatrix}_e \begin{Bmatrix} v_i \\ \theta_i \\ v_j \\ \theta_j \end{Bmatrix}_e \qquad \textbf{[9.3-26]}$$

Example 9.3-2 | Solve the beam structure shown in Fig. 9.3-7 using two elements and work equivalent loading. For the entire beam structure the modulus of elasticity is $E = 30$ Mpsi, the second-area moment is $I = 0.2$ in^4, and the section modulus is $S = 0.15$ in^3.

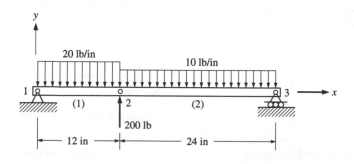

Figure 9.3-7

Solution:

Element (1): The stiffness matrix is

$$[\mathbf{k}]_1 = \frac{30(10^6)(0.2)}{(12)^3} \begin{bmatrix} 12 & (6)(12) & -12 & (6)(12) \\ (6)(12) & (4)(12^2) & -(6)(12) & (2)(12^2) \\ -12 & -(6)(12) & 12 & -(6)(12) \\ (6)(12) & (2)(12^2) & -(6)(12) & (4)(12^2) \end{bmatrix}$$

$$= 3472 \begin{bmatrix} 12 & 72 & -12 & 72 \\ 72 & 576 & -72 & 288 \\ -12 & -72 & 12 & -72 \\ 72 & 288 & -72 & 576 \end{bmatrix}$$

$$= 5208 \begin{matrix} \quad v_1 \quad\quad \theta_1 \quad\quad v_2 \quad\quad \theta_2 \\ \begin{bmatrix} 8 & 48 & -8 & 48 \\ 48 & 384 & -48 & 192 \\ -8 & -48 & 8 & -48 \\ 48 & 192 & -48 & 384 \end{bmatrix} \begin{matrix} v_1 \\ \theta_1 \\ v_2 \\ \theta_2 \end{matrix} \end{matrix}$$

It will become clear why we factored out the (5208) term as we did when we develop the equations for element (2).

From Eq. (9.3-24), the consistent nodal load vector for the -20 lb/in load on element (1) is

$$\left\{ \begin{matrix} \dfrac{(-20)(12)}{2} \\ \dfrac{(-20)(12^2)}{12} \\ \dfrac{(-20)(12)}{2} \\ \dfrac{-(-20)(12^2)}{12} \end{matrix} \right\} = \left\{ \begin{matrix} -120 \\ -240 \\ -120 \\ 240 \end{matrix} \right\}$$

Using Eq. (9.3-26), the element equations for element (1) are

$$\left\{ \begin{matrix} V_1 \\ M_1 \\ V_2 \\ M_2 \end{matrix} \right\}_1 + \left\{ \begin{matrix} -120 \\ -240 \\ -120 \\ 240 \end{matrix} \right\} = 5208 \begin{bmatrix} 8 & 48 & -8 & 48 \\ 48 & 384 & -48 & 192 \\ -8 & -48 & 8 & -48 \\ 48 & 192 & -48 & 384 \end{bmatrix} \left\{ \begin{matrix} v_1 \\ \theta_1 \\ v_2 \\ \theta_2 \end{matrix} \right\} \qquad [a]$$

Element (2):

$$[\mathbf{k}]_2 = \frac{30(10^6)(0.2)}{(24)^3}\begin{bmatrix} 12 & (6)(24) & -12 & (6)(24) \\ (6)(24) & (4)(24^2) & -(6)(24) & (2)(24^2) \\ -12 & -(6)(24) & 12 & -(6)(24) \\ (6)(24) & (2)(24^2) & -(6)(24) & (4)(24^2) \end{bmatrix}$$

$$= 434 \begin{bmatrix} 12 & 144 & -12 & 144 \\ 144 & 2304 & -144 & 1152 \\ -12 & -144 & 12 & -144 \\ 144 & 1152 & -144 & 2304 \end{bmatrix}$$

$$= 5208 \begin{array}{c} \begin{array}{cccc} v_2 & \theta_2 & v_3 & \theta_3 \end{array} \\ \begin{bmatrix} 1 & 12 & -1 & 12 \\ 12 & 192 & -12 & 96 \\ -1 & -12 & 1 & -12 \\ 12 & 96 & -12 & 192 \end{bmatrix} \begin{array}{c} v_2 \\ \theta_2 \\ v_3 \\ \theta_3 \end{array} \end{array}$$

We have created a common factor (5208) for the two elements to ease the assembly process.

The consistent nodal load vector for element (2) with -10 lb/in is

$$\begin{Bmatrix} \dfrac{(-10)(24)}{2} \\ \dfrac{(-10)(24^2)}{12} \\ \dfrac{(-10)(24)}{2} \\ -\dfrac{(-10)(24^2)}{12} \end{Bmatrix} = \begin{Bmatrix} -120 \\ -480 \\ -120 \\ 480 \end{Bmatrix}$$

The element equations for element (2) are

$$\begin{Bmatrix} V_2 \\ M_2 \\ V_3 \\ M_3 \end{Bmatrix}_2 + \begin{Bmatrix} -120 \\ -480 \\ -120 \\ 480 \end{Bmatrix} = 5208 \begin{bmatrix} 1 & 12 & -1 & 12 \\ 12 & 192 & -12 & 96 \\ -1 & -12 & 1 & -12 \\ 12 & 96 & -12 & 192 \end{bmatrix} \begin{Bmatrix} v_2 \\ \theta_2 \\ v_3 \\ \theta_3 \end{Bmatrix} \qquad \textbf{[b]}$$

Assembling the stiffness matrices of Eqs. (*a*) and (*b*) where the matrices have the common degrees of freedom v_2 and θ_2 results in

$$[\mathbf{K}]_{\text{system}} = 5208 \begin{array}{cc} & \begin{array}{cccccc} v_1 & \theta_1 & v_2 & \theta_2 & v_3 & \theta_3 \end{array} \\ & \begin{bmatrix} 8 & 48 & -8 & 48 & 0 & 0 \\ 48 & 384 & -48 & 192 & 0 & 0 \\ -8 & -48 & (8+1) & (-48+12) & -1 & 12 \\ 48 & 192 & (-48+12) & (384+192) & -12 & 96 \\ 0 & 0 & -1 & -12 & 1 & -12 \\ 0 & 0 & -12 & 96 & -12 & 192 \end{bmatrix} \begin{array}{c} v_1 \\ \theta_1 \\ v_2 \\ \theta_2 \\ v_3 \\ \theta_3 \end{array} \end{array}$$

This can be simplified to

$$[\mathbf{K}]_{\text{system}} = 5208 \begin{bmatrix} 8 & 48 & -8 & 48 & 0 & 0 \\ 48 & 384 & -48 & 192 & 0 & 0 \\ -8 & -48 & 9 & -36 & -1 & 12 \\ 48 & 192 & -36 & 576 & -12 & 96 \\ 0 & 0 & -1 & -12 & 1 & -12 \\ 0 & 0 & -12 & 96 & -12 & 192 \end{bmatrix} \qquad [c]$$

The applied load vector is

$$\begin{Bmatrix} R_{1y} \\ 0 \\ 200 \\ 0 \\ R_{3y} \\ 0 \end{Bmatrix} + \begin{Bmatrix} -120 \\ -240 \\ -120 \\ 240 \\ 0 \\ 0 \end{Bmatrix} + \begin{Bmatrix} 0 \\ 0 \\ -120 \\ -480 \\ -120 \\ 480 \end{Bmatrix} = \begin{Bmatrix} R_{1y} - 120 \\ -240 \\ -40 \\ -240 \\ R_{3y} - 120 \\ 480 \end{Bmatrix} \qquad [d]$$

where the first vector on the left-hand side of the equation contains the externally applied concentrated loads with R_{1y} and R_{3y} being the unknown support reaction forces. The second and third vectors are the consistent nodal loads on elements (1) and (2), respectively, from the externally applied distributed forces.

From Eqs. (c) and (d) the system equations are

$$\begin{Bmatrix} R_{1y} - 120 \\ -240 \\ -40 \\ -240 \\ R_{3y} - 120 \\ 480 \end{Bmatrix} = 5208 \begin{bmatrix} 8 & 48 & -8 & 48 & 0 & 0 \\ 48 & 384 & -48 & 192 & 0 & 0 \\ -8 & -48 & 9 & -36 & -1 & 12 \\ 48 & 192 & -36 & 576 & -12 & 96 \\ 0 & 0 & -1 & -12 & 1 & -12 \\ 0 & 0 & -12 & 96 & -12 & 192 \end{bmatrix} \begin{Bmatrix} v_1 \\ \theta_1 \\ v_2 \\ \theta_2 \\ v_3 \\ \theta_3 \end{Bmatrix}$$

The boundary conditions are $v_1 = v_3 = 0$. Thus the first and fifth equations are partitioned out. This gives (with $v_1 = v_3 = 0$)

$$R_{1y} = 120 + 5208(48\theta_1 - 8v_2 + 48\theta_2)$$

$$R_{3y} = 120 - 5208(v_2 + 12\theta_2 + 12\theta_3) \qquad [e]$$

Placing the first and fifth columns of the stiffness matrix with $v_1 = v_3 = 0$ over to the force vector creates no change. The modified system equations become

$$\begin{Bmatrix} -240 \\ -40 \\ -240 \\ 480 \end{Bmatrix} = 5208 \begin{bmatrix} 384 & -48 & 192 & 0 \\ -48 & 9 & -36 & 12 \\ 192 & -36 & 576 & 96 \\ 0 & -12 & 96 & 192 \end{bmatrix} \begin{Bmatrix} \theta_1 \\ v_2 \\ \theta_2 \\ \theta_3 \end{Bmatrix}$$

Solving the inverse yields

$$\begin{Bmatrix} \theta_1 \\ v_2 \\ \theta_2 \\ \theta_3 \end{Bmatrix} = \frac{1}{5208}(10^{-3}) \begin{bmatrix} 10.42 & 69.44 & 1.736 & -5.208 \\ 69.44 & 666.7 & 27.78 & -55.56 \\ 1.736 & 27.78 & 3.472 & -3.472 \\ -5.208 & -55.56 & -3.472 & 10.42 \end{bmatrix} \begin{Bmatrix} -240 \\ -40 \\ -240 \\ 480 \end{Bmatrix}$$

$$= \begin{Bmatrix} -1.573(10^{-3})\ \text{rad} \\ -1.472(10^{-2})\ \text{in} \\ -7.734(10^{-4})\ \text{rad} \\ 1.787(10^{-3})\ \text{rad} \end{Bmatrix}$$

Substituting the displacements into the support reaction force equations gives

$$R_{1y} = 120 + 5208[48(-1.573)(10^{-3}) - 8(-1.472)(10^{-2}) + 48(-7.734)(10^{-4})]$$

$$= 146.67\ \text{lb}$$

$$R_{3y} = 120 - 5208[(-1.472)(10^{-2}) + 12(-7.734)(10^{-4}) + 12(1.787)(10^{-3})]$$

$$= 133.33\ \text{lb}$$

which balances the net applied force down of 280 lb.

The element loads are determined from Eqs. (a) and (b).

Element (1):

$$\begin{Bmatrix} V_1 \\ M_1 \\ V_2 \\ M_2 \end{Bmatrix}_1 = \begin{Bmatrix} 120 \\ 240 \\ 120 \\ -240 \end{Bmatrix} + 3472 \begin{bmatrix} 12 & 72 & -12 & 72 \\ 72 & 576 & -72 & 288 \\ -12 & -72 & 12 & -72 \\ 72 & 288 & -72 & 576 \end{bmatrix} \begin{Bmatrix} 0 \\ -1.573(10^{-3}) \\ -1.472(10^{-2}) \\ -7.734(10^{-4}) \end{Bmatrix}$$

$$= \begin{Bmatrix} 146.67\ \text{lb} \\ 0\ \text{lb·in} \\ 93.33\ \text{lb} \\ 320\ \text{lb·in} \end{Bmatrix}$$

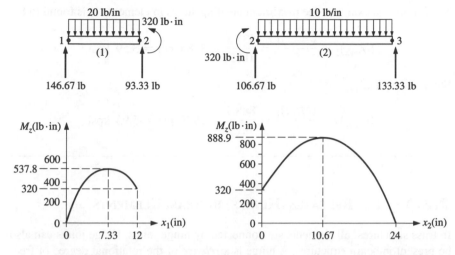

Figure 9.3-8   Element nodal loads and bending moment diagrams.

Element (2):

$$
\begin{Bmatrix} V_2 \\ M_2 \\ V_3 \\ M_3 \end{Bmatrix}_2 = \begin{Bmatrix} 120 \\ 480 \\ 120 \\ -480 \end{Bmatrix} + 434 \begin{bmatrix} 12 & 144 & -12 & 144 \\ 144 & 2304 & -144 & 1152 \\ -12 & -144 & 12 & -144 \\ 144 & 1152 & -144 & 2304 \end{bmatrix} \begin{Bmatrix} -1.472(10^{-2}) \\ -7.734(10^{-4}) \\ 0 \\ 1.787(10^{-3}) \end{Bmatrix}
$$

$$
= \begin{Bmatrix} 106.67\ \text{lb} \\ -320\ \text{lb·in} \\ 133.33\ \text{lb} \\ 0\ \text{lb·in} \end{Bmatrix}
$$

A plot of the individual elements and their loads is shown in Fig. 9.3-8. In addition, the bending moment diagram for each beam element is drawn using the element's local coordinate system along the x axis. The sign convention used for the bending moment in each diagram is according to the mechanics of materials convention established in Sec. 3.4.1 and *not* according to vector mechanics. The reader is urged to verify the results using basic mechanics.

It is important to note that in the case of the beam element with distributed loading, the maximum bending moment does not necessarily occur at a node, which is the case in this example. Thus some interpretation of the nodal results is necessary. This is a very key observation about the finite element method and will be emphasized in the next chapter. The finite element method is not to be taken as a panacea which always provides the complete picture of the analysis. The user of finite elements and commercial finite element programs must still have a knowledge of the fundamental principles of mechanics.

So for this example, the maximum bending stress in element (1) is found to be

$$(\sigma_{\max})_1 = \frac{(M_{\max})_1}{S_1} = \frac{537.8}{0.15} = 3585 \text{ psi} = 3.59 \text{ kpsi}$$

For element (2),

$$(\sigma_{\max})_2 = \frac{(M_{\max})_2}{S_2} = \frac{888.9}{0.15} = 5926 \text{ psi} = 5.93 \text{ kpsi}$$

9.3.3   PIN RELEASES (HINGES) IN BEAM ELEMENTS

In truss structures, all elements are connected by hinge joints. Hinge joints can also be present in beam structures. A hinge is a *release* of the rotational degree of freedom between two connecting elements.[†] Hence the term *pin release*. To see how the release is affected, consider the beam structure shown in Fig. 9.3-9 where a hinge exists at node 2. If the hinge were not present, the rotational deflection of node 2 for elements (1) and (2) would be identical and a bending moment would be transmitted through the joint. However, with the hinge, the rotations will not be equal [as shown in Fig. 9.3-9(*b*)] and the transmitted bending moment at the joint is zero. With the moment at the joint being a known quantity, a relationship exists between

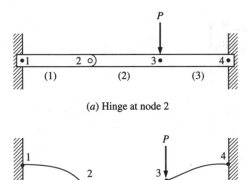

(*a*) Hinge at node 2

(*b*) Deflection (exaggerated)

Figure 9.3-9   Beam with a hinge joint.

[†] Translational (sliding) releases can also be applied in the same manner as described here.

the nodal displacements. Since only one relationship can be established, the pin release must be specified on only *one* of the two connecting elements. Consider, for example, that the pin release is specified on node 2 for element (1). From Eq. (9.3-19) the equations for element (1) can be expressed as

$$
\left\{ \begin{matrix} V_1 \\ M_1 \\ V_2 \\ M_2 \end{matrix} \right\}_1 = \begin{bmatrix} k_{11} & k_{12} & k_{13} & k_{14} \\ k_{12} & k_{22} & k_{23} & k_{24} \\ k_{13} & k_{23} & k_{33} & k_{34} \\ k_{14} & k_{24} & k_{34} & k_{44} \end{bmatrix}_1 \left\{ \begin{matrix} v_1 \\ \theta_1 \\ v_2 \\ (\theta_2)_1 \end{matrix} \right\}
\qquad \textbf{[9.3-27]}
$$

where the k_{ij} are the coefficients of the symmetric element stiffness matrix. Note that the displacement θ_2 is subscripted by the element number since θ_2 of elements (1) and (2) are not equal. That is, $(\theta_2)_1 \neq (\theta_2)_2$. Now, with the pin release, $(M_2)_1 = 0$. Partitioning the last equation out of Eq. (9.3-27) and solving for $(\theta_2)_1$ results in

$$
(\theta_2)_1 = -\frac{k_{14}}{k_{44}} v_1 - \frac{k_{24}}{k_{44}} \theta_1 - \frac{k_{34}}{k_{44}} v_2
\qquad \textbf{[9.3-28]}
$$

and

$$
\left\{ \begin{matrix} V_1 \\ M_1 \\ V_2 \end{matrix} \right\}_1 = \begin{bmatrix} k_{11} & k_{12} & k_{13} & k_{14} \\ k_{12} & k_{22} & k_{23} & k_{24} \\ k_{13} & k_{23} & k_{33} & k_{34} \end{bmatrix}_1 \left\{ \begin{matrix} v_1 \\ \theta_1 \\ v_2 \\ (\theta_2)_1 \end{matrix} \right\}
\qquad \textbf{[9.3-29]}
$$

Substituting Eq. (9.3-28) into (9.3-29) and rearranging results in

$$
\left\{ \begin{matrix} V_1 \\ M_1 \\ V_2 \end{matrix} \right\}_1 = \begin{bmatrix} \left(k_{11} - \dfrac{k_{14}^2}{k_{44}} \right) & \left(k_{12} - \dfrac{k_{14}k_{24}}{k_{44}} \right) & \left(k_{13} - \dfrac{k_{14}k_{34}}{k_{44}} \right) \\[2ex] \left(k_{12} - \dfrac{k_{14}k_{24}}{k_{44}} \right) & \left(k_{22} - \dfrac{k_{24}^2}{k_{44}} \right) & \left(k_{23} - \dfrac{k_{24}k_{34}}{k_{44}} \right) \\[2ex] \left(k_{13} - \dfrac{k_{14}k_{34}}{k_{44}} \right) & \left(k_{23} - \dfrac{k_{24}k_{34}}{k_{44}} \right) & \left(k_{33} - \dfrac{k_{34}^2}{k_{44}} \right) \end{bmatrix}_1 \left\{ \begin{matrix} v_1 \\ \theta_1 \\ v_2 \end{matrix} \right\}
$$

$$
\textbf{[9.3-30]}
$$

When assembling this matrix into the system matrix there will be no contribution due to the rotational degree of freedom of node 2 for this element, $(\theta_2)_1$. The finite element solution of the system will yield $(\theta_2)_2$. The value of $(\theta_2)_1$ must be determined from Eq. (9.3-28).

Example 9.3-3 | In Fig. 9.3-9 let $P = 20$ kN and each element be the same as element (2) of Example 9.3-1 where for each element

$$[\mathbf{k}] = 4.0(10^6) \begin{bmatrix} 6 & 3 & -6 & 3 \\ 3 & 2 & -3 & 1 \\ -6 & -3 & 6 & -3 \\ 3 & 1 & -3 & 2 \end{bmatrix}$$

Determine the displacement vector for each element and the nodal loads for elements (1) and (2).

Solution:

For the system stiffness matrix the above stiffness matrix applies to elements (2) and (3). For element (1), Eq. (9.3-30) yields

$$[\mathbf{k}]_1 = 4.0(10^6) \begin{bmatrix} [6-(3)^2/2] & [3-(3)(1)/2] & [-6-(3)(-3)/2] \\ [3-(3)(1)/2] & [2-(1)^2/2] & [-3-(1)(-3)/2] \\ [-6-(3)(-3)/2] & [-3-(1)(-3)/2] & [6-(-3)^2/2] \end{bmatrix}$$

$$= 4.0(10^6) \begin{bmatrix} 1.5 & 1.5 & -1.5 \\ 1.5 & 1.5 & -1.5 \\ -1.5 & -1.5 & 1.5 \end{bmatrix}$$

Assembling the stiffness matrices yields

$$[\mathbf{K}]_{\text{system}} = 4.0(10^6) \begin{array}{c} \begin{array}{cccccccc} v_1 & \theta_1 & v_2 & (\theta_2)_2 & v_3 & \theta_3 & v_4 & \theta_4 \end{array} \\ \begin{bmatrix} 1.5 & 1.5 & -1.5 & 0 & 0 & 0 & 0 & 0 \\ 1.5 & 1.5 & -1.5 & 0 & 0 & 0 & 0 & 0 \\ -1.5 & -1.5 & (1.5+6) & 3 & -6 & 3 & 0 & 0 \\ 0 & 0 & 3 & 2 & -3 & 1 & 0 & 0 \\ 0 & 0 & -6 & -3 & (6+6) & (-3+3) & -6 & 3 \\ 0 & 0 & 3 & 1 & (-3+3) & (2+2) & -3 & 1 \\ 0 & 0 & 0 & 0 & -6 & -3 & 6 & -3 \\ 0 & 0 & 0 & 0 & 3 & 1 & -3 & 2 \end{bmatrix} \begin{array}{c} v_1 \\ \theta_1 \\ v_2 \\ (\theta_2)_2 \\ v_3 \\ \theta_3 \\ v_4 \\ \theta_4 \end{array} \end{array}$$

Note that elements (1) and (2) only combine in the column and row term associated with the v_2 degree of freedom. The applied load vector is $\{R_{1y}\ M_{1z}\ 0\ 0\ -20(10^3)$ $0\ R_{4y}\ M_{4z}\}^T$, where the R_{iy} and M_{iz} terms are the force and moment reactions supplied by the walls. The system equations are then

$$
\begin{Bmatrix}
R_{1y} \\
M_{1z} \\
0 \\
0 \\
-20(10^3) \\
0 \\
R_{4y} \\
M_{4z}
\end{Bmatrix}
= 4.0(10^6)
\begin{bmatrix}
1.5 & 1.5 & -1.5 & 0 & 0 & 0 & 0 & 0 \\
1.5 & 1.5 & -1.5 & 0 & 0 & 0 & 0 & 0 \\
-1.5 & -1.5 & 7.5 & 3 & -6 & 3 & 0 & 0 \\
0 & 0 & 3 & 2 & -3 & 1 & 0 & 0 \\
0 & 0 & -6 & -3 & 12 & 0 & -6 & 3 \\
0 & 0 & 3 & 1 & 0 & 4 & -3 & 1 \\
0 & 0 & 0 & 0 & -6 & -3 & 6 & -3 \\
0 & 0 & 0 & 0 & 3 & 1 & -3 & 2
\end{bmatrix}
\begin{Bmatrix}
v_1 \\
\theta_1 \\
v_2 \\
(\theta_2)_2 \\
v_3 \\
\theta_3 \\
v_4 \\
\theta_4
\end{Bmatrix}
$$

Partitioning the first, second, seventh, and eighth equations, and applying the boundary conditions $v_1 = \theta_1 = v_4 = \theta_4 = 0$, the system equations reduce to

$$
\begin{Bmatrix}
0 \\
0 \\
-20(10^3) \\
0
\end{Bmatrix}
= 4.0(10^6)
\begin{bmatrix}
7.5 & 3 & -6 & 3 \\
3 & 2 & -3 & 1 \\
-6 & -3 & 12 & 0 \\
3 & 1 & 0 & 4
\end{bmatrix}
\begin{Bmatrix}
v_2 \\
(\theta_2)_2 \\
v_3 \\
\theta_3
\end{Bmatrix}
$$

Solving the inverse equation yields

$$
\begin{Bmatrix}
v_2 \\
(\theta_2)_2 \\
v_3 \\
\theta_3
\end{Bmatrix}
= \frac{1}{4(10^6)}\left(\frac{1}{54}\right)
\begin{bmatrix}
32 & -24 & 10 & -18 \\
-24 & 72 & 6 & 0 \\
10 & 6 & 11 & -9 \\
-18 & 0 & -9 & 27
\end{bmatrix}
\begin{Bmatrix}
0 \\
0 \\
-20(10^3) \\
0
\end{Bmatrix}
$$

$$
=
\begin{Bmatrix}
-9.259(10^{-4})\ \text{m} \\
-5.556(10^{-4})\ \text{rad} \\
-1.019(10^{-3})\ \text{m} \\
8.333(10^{-4})\ \text{rad}
\end{Bmatrix}
$$

The displacement $(\theta_2)_1$ determined from Eq. (9.3-28) with $v_1 = \theta_1 = 0$, $v_2 = -9.259(10^{-4})$, and $k_{34}/k_{44} = (-3)/2$ is

$$(\theta_2)_1 = -\frac{(-3)}{2}(-9.258)(10^{-4}) = -1.389(10^{-3}) \text{ rad}$$

The element loads are given by Eq. (9.3-19)

Element (1):

$$\begin{Bmatrix} V_1 \\ M_1 \\ V_2 \\ M_2 \end{Bmatrix}_1 = 4.0(10^6) \begin{bmatrix} 6 & 3 & -6 & 3 \\ 3 & 2 & -3 & 1 \\ -6 & -3 & 6 & -3 \\ 3 & 1 & -3 & 2 \end{bmatrix} \begin{Bmatrix} v_1 \\ \theta_1 \\ v_2 \\ (\theta_2)_1 \end{Bmatrix}$$

$$= 4.0(10^6) \begin{bmatrix} 6 & 3 & -6 & 3 \\ 3 & 2 & -3 & 1 \\ -6 & -3 & 6 & -3 \\ 3 & 1 & -3 & 2 \end{bmatrix} \begin{Bmatrix} 0 \\ 0 \\ -9.259(10^{-4}) \\ -1.389(10^{-3}) \end{Bmatrix}$$

$$= \begin{Bmatrix} 5.556(10^3) \text{ N} \\ 5.556(10^3) \text{ N·m} \\ -5.556(10^3) \text{ N} \\ 0 \text{ N·m} \end{Bmatrix} = \begin{Bmatrix} 5.56 \text{ kN} \\ 5.56 \text{ kN·m} \\ -5.56 \text{ kN} \\ 0 \text{ N·m} \end{Bmatrix}$$

Element (2):

$$\begin{Bmatrix} V_2 \\ M_2 \\ V_3 \\ M_3 \end{Bmatrix}_2 = 4.0(10^6) \begin{bmatrix} 6 & 3 & -6 & 3 \\ 3 & 2 & -3 & 1 \\ -6 & -3 & 6 & -3 \\ 3 & 1 & -3 & 2 \end{bmatrix} \begin{Bmatrix} v_2 \\ (\theta_2)_2 \\ v_3 \\ \theta_3 \end{Bmatrix}$$

$$= 4.0(10^6) \begin{bmatrix} 6 & 3 & -6 & 3 \\ 3 & 2 & -3 & 1 \\ -6 & -3 & 6 & -3 \\ 3 & 1 & -3 & 2 \end{bmatrix} \begin{Bmatrix} -9.259(10^{-4}) \\ -5.556(10^{-4}) \\ -1.019(10^{-3}) \\ 8.333(10^{-4}) \end{Bmatrix}$$

$$= \begin{Bmatrix} 5.556(10^3) \text{ N} \\ 0 \text{ N·m} \\ -5.556(10^3) \text{ N} \\ 5.556(10^3) \text{ N·m} \end{Bmatrix} = \begin{Bmatrix} 5.56 \text{ kN} \\ 0 \text{ kN·m} \\ -5.56 \text{ kN} \\ 5.56 \text{ kN·m} \end{Bmatrix}$$

We see that the moment at node 2, M_2, is zero for both elements, confirming that no moment about the z axis is being transmitted through the hinged joint.

9.3.4 BEAMS IN TWO-PLANE BENDING

In the previous sections, bending was limited to the local xy plane. With two-plane bending, bending can also occur in the xz plane. The notation needs to reflect this, and consequently we add subscripts for this purpose. Thus, for bending in the xy plane we redefine Eq. (9.3-19) to be

$$
\begin{Bmatrix} (V_y)_i \\ (M_z)_i \\ (V_y)_j \\ (M_z)_j \end{Bmatrix}_e = \left(\frac{EI_z}{L^3} \right)_e \begin{bmatrix} 12 & 6L & -12 & 6L \\ 6L & 4L^2 & -6L & 2L^2 \\ -12 & -6L & 12 & -6L \\ 6L & 2L^2 & -6L & 4L^2 \end{bmatrix}_e \begin{Bmatrix} v_i \\ (\theta_z)_i \\ v_j \\ (\theta_z)_j \end{Bmatrix}_e \qquad \textbf{[9.3-31]}
$$

For bending in the xz plane, the stiffness matrix is quite similar. Forces and moments are still defined positive according to the right-handed coordinate system. However, owing to the nature of the cartesian coordinate system, some of the signs of the coefficients will be different from those in Eq. (9.3-31). Here, we will simply give the results (see Problem 9.30). For bending in the xz plane the load-deflection relation is

$$
\begin{Bmatrix} (V_z)_i \\ (M_y)_i \\ (V_z)_j \\ (M_y)_j \end{Bmatrix}_e = \left(\frac{EI_y}{L^3} \right)_e \begin{bmatrix} 12 & -6L & -12 & -6L \\ -6L & 4L^2 & 6L & 2L^2 \\ -12 & 6L & 12 & 6L \\ -6L & 2L^2 & 6L & 4L^2 \end{bmatrix}_e \begin{Bmatrix} w_i \\ (\theta_y)_i \\ w_j \\ (\theta_y)_j \end{Bmatrix}_e \qquad \textbf{[9.3-32]}
$$

For the two-plane bending element we combine Eqs. (9.3-31) and (9.3-32) and order the degrees of freedom for each node according to translation and then rotation. That is, the displacement vector is $\{v_i \ w_i \ (\theta_y)_i \ (\theta_z)_i \ v_j \ w_j \ (\theta_y)_j \ (\theta_z)_j\}^T$. After some rearranging, the resulting matrix equation is

$$
\begin{Bmatrix}
(V_y)_i \\
(V_z)_i \\
(M_y)_i \\
(M_z)_i \\
(V_y)_j \\
(V_z)_j \\
(M_y)_j \\
(M_z)_j
\end{Bmatrix}_e
=
\left(\frac{E}{L^3}\right)_e
\begin{bmatrix}
12I_z & 0 & 0 & 6I_zL & -12I_z & 0 & 0 & 6I_zL \\
0 & 12I_y & -6I_yL & 0 & 0 & -12I_y & -6I_yL & 0 \\
0 & -6I_yL & 4I_yL^2 & 0 & 0 & 6I_yL & 2I_yL^2 & 0 \\
6I_zL & 0 & 0 & 4I_zL^2 & -6I_zL & 0 & 0 & 2I_zL^2 \\
-12I_z & 0 & 0 & -6I_zL & 12I_z & 0 & 0 & -6I_zL \\
0 & -12I_y & 6I_yL & 0 & 0 & 12I_y & 6I_yL & 0 \\
0 & -6I_yL & 2I_yL^2 & 0 & 0 & 6I_yL & 4I_yL^2 & 0 \\
6I_zL & 0 & 0 & 2I_zL^2 & -6I_zL & 0 & 0 & 4I_zL^2
\end{bmatrix}_e
\begin{Bmatrix}
v_i \\
w_i \\
(\theta_y)_i \\
(\theta_z)_i \\
v_j \\
w_j \\
(\theta_y)_j \\
(\theta_z)_j
\end{Bmatrix}
$$

$$[9.3\text{-}33]$$

The procedure in using this equation is identical to that provided in earlier sections. However, owing to the size of the matrix equations for even one element, manual examples become laborious.

9.3.5   THE FRAME ELEMENT

This element is what most commercial finite element software refer to as the beam element. Basically, the axial element and a similar torsional element (discussed next) are simply added to the two-plane bending element. It is very important to understand that with the simple frame element axial, torsional, and bending loads are completely uncoupled and no stiffening effects due to interaction between the loading are considered. Stiffening can be incorporated and is discussed briefly in Sec. 9.3.7 for single-plane bending of a beam.

For *torsion,* the load-displacement relationship is very similar to axial loading. The stiffness of an axial member is EA/L, whereas for a torsional member the stiffness is $GJ/L = EJ/[2(1 + v)L]$. Recall, the definition of the local axial element equations given by Eq. (9.2-12) repeated here as

$$
\begin{Bmatrix}
(N_x)_i \\
(N_x)_j
\end{Bmatrix}_e
=
\left(\frac{AE}{L}\right)_e
\begin{bmatrix}
1 & -1 \\
-1 & 1
\end{bmatrix}_e
\begin{Bmatrix}
u_i \\
u_j
\end{Bmatrix}_e
\qquad [9.3\text{-}34]
$$

where the notation for the nodal axial forces has been changed to $(N_x)_i$ and $(N_x)_j$ to reflect the notation for normal force used in earlier chapters. If torsion about the local x axis is defined as T_x and the angle of twist as θ_x, then similar to Eq. (9.3-34) the equation for an element undergoing torsion is

$$
\begin{Bmatrix}
(T_x)_i \\
(T_x)_j
\end{Bmatrix}_e
=
\left(\frac{EJ}{2(1 + v)L}\right)_e
\begin{bmatrix}
1 & -1 \\
-1 & 1
\end{bmatrix}_e
\begin{Bmatrix}
(\theta_x)_i \\
(\theta_x)_j
\end{Bmatrix}
\qquad [9.3\text{-}35]
$$

The second area polar moment for a beam J can be found in either Secs. 3.3.1, 4.3.2, 5.2 or Ref. 9.1.

The frame element simply places Eqs. (9.3-33), (9.3-34), and (9.3-35) into one equation, which is

$$
\begin{Bmatrix}
(N_x)_i \\
(V_y)_i \\
(V_z)_i \\
(T_x)_i \\
(M_y)_i \\
(M_z)_i \\
(N_x)_j \\
(V_y)_j \\
(V_z)_j \\
(T_x)_j \\
(M_y)_j \\
(M_z)_j
\end{Bmatrix}_e
=
\left(\frac{E}{L}\right)_e
\begin{bmatrix}
A & 0 & 0 & 0 & 0 & 0 & -A & 0 & 0 & 0 & 0 & 0 \\
0 & 12I_z/L^2 & 0 & 0 & 0 & 6I_z/L & 0 & -12I_z/L^2 & 0 & 0 & 0 & 6I_z/L \\
0 & 0 & 12I_y/L^2 & 0 & -6I_y/L & 0 & 0 & 0 & -12I_y/L^2 & 0 & -6I_y/L & 0 \\
0 & 0 & 0 & J/[2(1+v)] & 0 & 0 & 0 & 0 & 0 & -J/[2(1+v)] & 0 & 0 \\
0 & 0 & -6I_y/L & 0 & 4I_y & 0 & 0 & 0 & 6I_y/L & 0 & 2I_y & 0 \\
0 & 6I_z/L & 0 & 0 & 0 & 4I_z & 0 & -6I_z/L & 0 & 0 & 0 & 2I_z \\
-A & 0 & 0 & 0 & 0 & 0 & A & 0 & 0 & 0 & 0 & 0 \\
0 & -12I_z/L & 0 & 0 & 0 & -6I_z/L & 0 & 12I_z/L^2 & 0 & 0 & 0 & -6I_z/L \\
0 & 0 & -12I_y/L^2 & 0 & 6I_y/L & 0 & 0 & 0 & 12I_y/L^2 & 0 & 6I_y/L & 0 \\
0 & 0 & 0 & -J/[2(1+v)] & 0 & 0 & 0 & 0 & 0 & J/[2(1+v)] & 0 & 0 \\
0 & 0 & -6I_y/L & 0 & 2I_y & 0 & 0 & 0 & 6I_y/L & 0 & 4I_y & 0 \\
0 & 6I_z/L & 0 & 0 & 0 & 2I_z & 0 & -6I_z/L & 0 & 0 & 0 & 4I_z
\end{bmatrix}_e
\begin{Bmatrix}
u_i \\
v_i \\
w_i \\
(\theta_x)_i \\
(\theta_y)_i \\
(\theta_z)_i \\
u_j \\
v_j \\
w_j \\
(\theta_x)_j \\
(\theta_y)_j \\
(\theta_z)_j
\end{Bmatrix}_e
$$

[9.3-36]

9.3.6 THREE-DIMENSIONAL TRANSFORMATION OF THE FRAME ELEMENT

The equations developed in the previous sections for the beam and frame elements are all based on the local xyz coordinate system of the beam. For a frame element in three-dimensional space, refer back to Sec. 9.2.8 where the axial element is oriented in three-dimensional space (see Fig. 9.2-12). The local $x'y'z'$ coordinate system originates at node i and the x' direction is from node i to node j. In the case of the axial element, the $y'z'$ coordinate axes, other than being perpendicular to the x' axis, are arbitrary. For the frame element, it is necessary that the $y'z'$ axes *align with the axes of the principal second area moments*. To orient the $y'z'$ axes in three-dimensional space, it is customary practice to position a third point or node k in addition to nodes i and j. In practice, point k is *defined* to lie in the local $x'y'$ plane. If it is a node, the degrees of freedom are completely fixed so that they are automatically partitioned out of the system stiffness matrix. Consider the third point or node k shown in Fig. 9.3-10 where the shaded area is in the $x'y'$ plane. In this figure, for simplicity and mathematics, the beam is represented by the vector **L**. The vector **L** is given by

$$\mathbf{L} = (x_j - x_i)\mathbf{i} + (y_j - y_i)\mathbf{j} + (z_j - z_i)\mathbf{k} \qquad \textbf{[9.3-37]}$$

where **i**, **j**, and **k** are unit vectors in the global x, y, and z directions, respectively. The directional cosines for the x' axis relative to the xyz coordinate system are given by

$$\ell_{x'} = \frac{x_j - x_i}{L}$$

$$m_{x'} = \frac{y_j - y_i}{L} \qquad \textbf{[9.3-38]}$$

$$n_{x'} = \frac{z_j - z_i}{L}$$

where $L = [(x_j - x_i)^2 + (y_j - y_i)^2 + (z_j - z_i)^2]^{1/2}$.

Define the vector **K** from node i to node k as shown in Fig. 9.3-10, where

$$\mathbf{K} = (x_k - x_i)\mathbf{i} + (y_k - y_i)\mathbf{j} + (z_k - z_i)\mathbf{k} \qquad \textbf{[9.3-39]}$$

A vector can now be established in the z' direction **Z'**, by taking the *vector cross product* of **L** and **K**. That is,

$$\mathbf{Z'} = Z'_x \mathbf{i} + Z'_y \mathbf{j} + Z'_z \mathbf{k} = \mathbf{L} \times \mathbf{K} \qquad \textbf{[9.3-40]}$$

The components of **Z'** establish the directional cosines of axis z' relative to the global xyz coordinate system as

$$\ell_{z'} = \frac{Z'_x}{|Z'|}$$

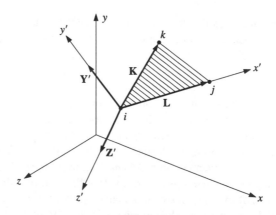

Figure 9.3-10   Orienting the local $x'y'z'$ coordinate system of a beam in space.

$$m_{z'} = \frac{Z'_y}{|Z'|} \qquad\qquad [9.3-41]$$

$$n_{z'} = \frac{Z'_z}{|Z'|}$$

where $|Z'| = [Z'^2_x + Z'^2_y + Z'^2_z]^{1/2}$ is the magnitude of $\mathbf{Z}'$. Finally, the y' axis is located by the vector $\mathbf{Y}'$, defined by the vector cross product $\mathbf{Z}' \times \mathbf{L}$. That is,

$$\mathbf{Y}' = Y'_x \mathbf{i} + Y'_y \mathbf{j} + Y'_z \mathbf{k} = \mathbf{Z}' \times \mathbf{L} \qquad\qquad [9.3-42]$$

and the directional cosines of the y' axis relative to the global xyz coordinate system are

$$\ell_{y'} = \frac{Y'_x}{|Y'|}$$

$$m_{y'} = \frac{Y'_y}{|Y'|} \qquad\qquad [9.3-43]$$

$$n_{y'} = \frac{Y'_z}{|Y'|}$$

where $|Y'| = [Y'^2_x + Y'^2_y + Y'^2_z]^{1/2}$ is the magnitude of $\mathbf{Y}'$.

To transform a single vector $\mathbf{V}$ represented by global x, y, and z components V_x, V_y, and V_z, respectively, to one represented by the local x', y', and z' components $V_{x'}$, $V_{y'}$, and $V_{z'}$, respectively, one performs a coordinate transformation

$$\begin{Bmatrix} V_{x'} \\ V_{y'} \\ V_{z'} \end{Bmatrix} = [\mathbf{T}] \begin{Bmatrix} V_x \\ V_y \\ V_z \end{Bmatrix}$$

or

$$\begin{Bmatrix} V_x \\ V_y \\ V_z \end{Bmatrix}_{\text{local}} = [\mathbf{T}] \begin{Bmatrix} V_x \\ V_y \\ V_z \end{Bmatrix}_{\text{global}} \qquad\qquad [9.3-44]$$

where

$$[\mathbf{T}] = \begin{bmatrix} \ell_{x'} & m_{x'} & n_{x'} \\ \ell_{y'} & m_{y'} & n_{y'} \\ \ell_{z'} & m_{z'} & n_{z'} \end{bmatrix} \qquad\qquad [9.3-45]$$

and the directional cosines are defined by Eqs. (9.3-38), (9.3-41), and (9.3-43).

Recall, from Sec. 9.2.6, the global stiffness matrix is given by

$$[\mathbf{k}]_{\text{global}} = [\mathbf{T}]^T [\mathbf{k}]_{\text{local}} [\mathbf{T}] \qquad\qquad [9.3-46]$$

where $[\mathbf{T}]$ is the transformation matrix converting the global degrees of freedom to the local degrees of freedom. There are 12 degrees of freedom for the frame element, two sets of vectors (translation and rotation) for each node $\{u_i\ v_i\ w_i\ (\theta_x)_i\ (\theta_y)_i$ $(\theta_z)_i\ u_j\ v_j\ w_j\ (\theta_x)_j\ (\theta_y)_j\ (\theta_z)_j\}^{T}.^{\dagger}$ The transformation matrix for the element is given by

$$[\mathbf{T}] = \begin{bmatrix}
\ell_{x'} & m_{x'} & n_{x'} & 0 & 0 & 0 & 0 & 0 & 0 & 0 & 0 & 0 \\
\ell_{y'} & m_{y'} & n_{y'} & 0 & 0 & 0 & 0 & 0 & 0 & 0 & 0 & 0 \\
\ell_{z'} & m_{z'} & n_{z'} & 0 & 0 & 0 & 0 & 0 & 0 & 0 & 0 & 0 \\
0 & 0 & 0 & \ell_{x'} & m_{x'} & n_{x'} & 0 & 0 & 0 & 0 & 0 & 0 \\
0 & 0 & 0 & \ell_{y'} & m_{y'} & n_{y'} & 0 & 0 & 0 & 0 & 0 & 0 \\
0 & 0 & 0 & \ell_{z'} & m_{z'} & n_{z'} & 0 & 0 & 0 & 0 & 0 & 0 \\
0 & 0 & 0 & 0 & 0 & 0 & \ell_{x'} & m_{x'} & n_{x'} & 0 & 0 & 0 \\
0 & 0 & 0 & 0 & 0 & 0 & \ell_{y'} & m_{y'} & n_{y'} & 0 & 0 & 0 \\
0 & 0 & 0 & 0 & 0 & 0 & \ell_{z'} & m_{z'} & n_{z'} & 0 & 0 & 0 \\
0 & 0 & 0 & 0 & 0 & 0 & 0 & 0 & 0 & \ell_{x'} & m_{x'} & n_{x'} \\
0 & 0 & 0 & 0 & 0 & 0 & 0 & 0 & 0 & \ell_{y'} & m_{y'} & n_{y'} \\
0 & 0 & 0 & 0 & 0 & 0 & 0 & 0 & 0 & \ell_{z'} & m_{z'} & n_{z'}
\end{bmatrix}$$

$$[\textbf{9.3-47}]$$

and the local stiffness matrix, given in Eq. (9.3-36), is

$$[\mathbf{k}]_{\text{local}} = \left(\frac{E}{L}\right)_e \begin{bmatrix}
A & 0 & 0 & 0 & 0 & 0 & -A & 0 & 0 & 0 & 0 & 0 \\
0 & 12I_z/L^2 & 0 & 0 & 0 & 6I_z/L & 0 & -12I_z/L^2 & 0 & 0 & 0 & 6I_z/L \\
0 & 0 & 12I_y/L^2 & 0 & -6I_y/L & 0 & 0 & 0 & -12I_y/L^2 & 0 & -6I_y/L & 0 \\
0 & 0 & 0 & J/[2(1+v)] & 0 & 0 & 0 & 0 & 0 & -J/[2(1+v)] & 0 & 0 \\
0 & 0 & -6I_y/L & 0 & 4I_y & 0 & 0 & 0 & 6I_y/L & 0 & 2I_y & 0 \\
0 & 6I_z/L & 0 & 0 & 0 & 4I_z & 0 & -6I_z/L & 0 & 0 & 0 & 2I_z \\
-A & 0 & 0 & 0 & 0 & 0 & A & 0 & 0 & 0 & 0 & 0 \\
0 & -12I_z/L & 0 & 0 & 0 & -6I_z/L & 0 & 12I_z/L^2 & 0 & 0 & 0 & -6I_z/L \\
0 & 0 & -12I_y/L^2 & 0 & 6I_y/L & 0 & 0 & 0 & 12I_y/L^2 & 0 & 6I_y/L & 0 \\
0 & 0 & 0 & -J/[2(1+v)] & 0 & 0 & 0 & 0 & 0 & J/[2(1+v)] & 0 & 0 \\
0 & 0 & -6I_y/L & 0 & 2I_y & 0 & 0 & 0 & 6I_y/L & 0 & 4I_y & 0 \\
0 & 6I_z/L & 0 & 0 & 0 & 2I_z & 0 & -6I_z/L & 0 & 0 & 0 & 4I_z
\end{bmatrix}_e$$

$$[\textbf{9.3-48}]$$

The procedures are similar to that outlined in Sec. 9.2.8. That is, once the properties of the beam and the coordinates of nodes i, j, and k are defined, the local stiffness matrix is computed according to Eq. (9.3-48), the transformation matrix is computed according to Eq. (9.3-47), the global stiffness is then computed according to Eq. (9.3-46), and then the element global stiffness matrix is assembled into the system stiffness matrix. Applying loads, boundary conditions, partitioning, and solving the inverse of the modified equations for the unknown global displacements

---

$\dagger$ The vector components of the translational displacement are the u, v, and w for each node. The components $(\theta_x)_i$, $(\theta_y)_i$, and $(\theta_z)_i$ can be approximated to be the components of a rotational vector provided they are very small.

follows. Once the global displacements are determined, the element loads can be computed from the equation

$$\{\mathbf{F}\}_{\text{local}} = [\mathbf{k}]_{\text{local}}[\mathbf{T}]\{\mathbf{u}\}_{\text{global}} \qquad \textbf{[9.3-49]}$$

where $\{\mathbf{F}\}_{\text{local}}$ is the local load vector $\{(N_{x'})_i\ (V_{y'})_i\ (V_{z'})_i\ (T_{x'})_i\ (M_{y'})_i\ (M_{z'})_i\ (N_{x'})_j\ (V_{y'})_j$ $(V_{z'})_j\ (T_{x'})_j\ (M_{y'})_j\ (M_{z'})_j\}^T$, and $[\mathbf{k}]_{\text{local}}$ and $[\mathbf{T}]$ are given by Eqs. (9.3-48) and (9.3-47), respectively, and $\{\mathbf{u}\}_{\text{global}}$ is the global displacement vector $\{u_i\ v_i\ w_i\ (\theta_x)_i\ (\theta_y)_i$ $(\theta_z)_i\ u_j\ v_j\ w_j\ (\theta_x)_j\ (\theta_y)_j\ (\theta_z)_j\}^T_{\text{global}}$. Stresses are then determined from the local load vector according to the individual states of axial, torsional, and bending loading.

Obviously, owing to the size of all the matrices involved, examples involving manual calculations in this book would take many pages to convey. However, small problems can be solved fairly easily using spreadsheet software.

9.3.7   Load Stiffening and Buckling of Beams

Straight beams undergoing the combined effects of axial and transverse loading are covered in Sec. 6.13. Two classes of problems are discussed: unconstrained beams where the axial force is known, and constrained beams where the axial force is unknown. In this section, only the first class of problem will be presented.

In addition to bending, assume that the planar beam of Fig. 9.3-1 is also transmitting a known tensile axial force N. In Sec. 6.13, the work due to N is given by Eq. (6.13-5) and is repeated here as

$$W_N \approx \frac{1}{2}\frac{N^2 L}{AE} - \frac{1}{2}N\int_0^L \left(\frac{dv}{dx}\right)^2 dx \qquad \textbf{[9.3-50]}$$

The strain energy due to axial and bending effects is given by Eq. (6.13-6) and is repeated here as

$$U = \frac{1}{2}\frac{N^2 L}{AE} + \frac{1}{2}EI_z\int_0^L \left(\frac{dv^2}{dx^2}\right)^2 dx \qquad \textbf{[9.3-51]}$$

where it is assumed that EI_z is constant throughout the beam. The work potential is the negative of the work performed. That is, $W_p = -W_N$. Thus the potential energy $\Pi = U - W_N$ is given by

$$\Pi = \frac{1}{2}EI_z\int_0^L \left(\frac{dv^2}{dx^2}\right)^2 dx + \frac{1}{2}N\int_0^L \left(\frac{dv}{dx}\right)^2 dx \qquad \textbf{[9.3-52]}$$

In Sec 9.3.1 the first integral of Eq. (9.3-52) led to the stiffness matrix given by Eq. (9.3-16). The second integral can be developed in a similar fashion. Recall that

$v(x)$ is expressed in terms of the shape functions by Eqs. (9.3-5) and (9.3-6). The first derivative can be written as

$$\frac{dv}{dx} = \left\{ \frac{d(N_v)_i}{dx} \quad \frac{d(N_\theta)_i}{dx} \quad \frac{d(N_v)_j}{dx} \quad \frac{d(N_\theta)_j}{dx} \right\} \begin{Bmatrix} v_i \\ \theta_i \\ v_j \\ \theta_j \end{Bmatrix}$$

$$= \left\{ C_1 \, C_2 \, C_3 \, C_4 \right\} \begin{Bmatrix} v_i \\ \theta_i \\ v_j \\ \theta_j \end{Bmatrix} \qquad \text{[9.3-53]}$$

where

$$C_1 = \frac{d(N_v)_i}{dx} = -6\frac{x}{L_e^2} + 6\frac{x^2}{L_e^3}$$

$$C_2 = \frac{d(N_\theta)_i}{dx} = 1 - 4\frac{x}{L_e} + 3\frac{x^2}{L_e^2}$$

$$C_3 = \frac{d(N_v)_j}{dx} = 6\frac{x}{L_e^2} - 6\frac{x^2}{L_e^3}$$

$$C_4 = \frac{d(N_\theta)_i}{dx} = -2\frac{x}{L_e} + 3\frac{x^2}{L_e^2} \qquad \text{[9.3-54]}$$

The form of Eq. (9.3-53) is $dv/dx = \{C\}\{u\}$ where $\{C\} = \{C_1 \, C_2 \, C_3 \, C_4\}$ and the displacement vector is $\{u\} = \{v_i \, \theta_i \, v_j \, \theta_j\}^T$. The square of dv/dx is of the form $(dv/dx)^2 = \{u\}^T\{C\}^T\{C\}\{u\}$. Thus the second integral in Eq. (9.3-52) is of the form

$$\frac{1}{2} N \int_{L_e} \{u\}^T\{C\}^T\{C\}\{u\} \, dx \qquad \text{[9.3-55]}$$

Since $\{u\}$ is not a function of x, the above can be rewritten as

$$\frac{1}{2} \{u\}^T \left[N \int_{L_e} \{C\}^T\{C\} \, dx \right] \{u\} \qquad \text{[9.3-56]}$$

which is in the familiar form of $1/2\{\mathbf{u}\}^T[\mathbf{k}]\{\mathbf{u}\}$. Thus the term within the [ ] brackets of Eq. (9.3-56) is the additional element stiffness due to the axial force that we add to the basic beam stiffness matrix of Eq. (9.3-16). Referring to the term in the [ ] brackets as $[\mathbf{k}_N]_e$ and expanding $\{\mathbf{C}\}$ we obtain

$$[\mathbf{k}_N]_e = N \int_{L_e} \begin{Bmatrix} C_1 \\ C_2 \\ C_3 \\ C_4 \end{Bmatrix} \{C_1\ C_2\ C_3\ C_4\}\, dx \qquad \textbf{[9.3-57]}$$

Performing the matrix multiplication yields

$$[\mathbf{k}_N]_e = N \int_{L_e} \begin{bmatrix} C_1^2 & C_1C_2 & C_1C_3 & C_1C_4 \\ C_1C_2 & C_2^2 & C_2C_3 & C_2C_4 \\ C_1C_3 & C_2C_3 & C_3^2 & C_3C_4 \\ C_1C_4 & C_2C_4 & C_3C_4 & C_4^2 \end{bmatrix} dx \qquad \textbf{[9.3-58]}$$

Substituting Eqs. (9.3-54) and integrating from 0 to L_e results in

$$[\mathbf{k}_N]_e = \left(\frac{N}{30L}\right)_e \begin{bmatrix} 36 & 3L & -36 & 3L \\ 3L & 4L^2 & -3L & -L^2 \\ -36 & -3L & 36 & -3L \\ 3L & -L^2 & -3L & 4L^2 \end{bmatrix}_e \qquad \textbf{[9.3-59]}$$

Adding this to Eq. (9.3-16) yields the stiffness matrix for the beam, which is stiffened by the tensile axial force N. That is,

$$[\mathbf{k}]_e = \left(\frac{EI}{L^3}\right)_e \begin{bmatrix} 12 & 6L & -12 & 6L \\ 6L & 4L^2 & -6L & 2L^2 \\ -12 & -6L & 12 & -6L \\ 6L & 2L^2 & -6L & 4L^2 \end{bmatrix}_e + \left(\frac{N}{30L}\right)_e \begin{bmatrix} 36 & 3L & -36 & 3L \\ 3L & 4L^2 & -3L & -L^2 \\ -36 & -3L & 36 & -3L \\ 3L & -L^2 & -3L & 4L^2 \end{bmatrix}_e$$

$$\textbf{[9.3-60]}$$

Example 9.3-4

Examples 6.13-1 and 6.13-2 involve a uniformly loaded, simply supported beam structure with an axial tensile force where the length of the beam is 1.25 m, the uniform load is 1.6 kN/m in the negative y direction, $I_z = 40(10^3)$ mm^4, $E = 70$ GPa, and $N = 3.6$ kN. Using a two-element model, where each element is $1.25/2 = 0.625$ m long, determine the maximum tensile stress and lateral deflection at the midpoint of the beam structure. For the stress calculation let the cross-sectional area and section modulus of the beam be 700 mm^2 and $2(10^3)$ mm^3, respectively.

Solution:

Since each of the two elements are identical, the stiffness matrix for each element is

$$[\mathbf{k}]_{1,2} = \left(\frac{70(10^9)(40)(10^3)(10^{-3})^4}{(0.625)^3}\right)\begin{bmatrix} 12 & 6(0.625) & -12 & 6(0.625) \\ 6(0.625) & 4(0.625)^2 & -6(0.625) & 2(0.625)^2 \\ -12 & -6(0.625) & 12 & -6(0.625) \\ 6(0.625) & 2(0.625)^2 & -6(0.625) & 4(0.625)^2 \end{bmatrix}$$

$$+ \left(\frac{3.6(10^3)}{30(0.625)}\right)\begin{bmatrix} 36 & 3(0.625) & -36 & 3(0.625) \\ 3(0.625) & 4(0.625)^2 & -3(0.625) & -(0.625)^2 \\ -36 & -3(0.625) & 36 & -3(0.625) \\ 3(0.625) & -(0.625)^2 & -3(0.625) & 4(0.625)^2 \end{bmatrix}$$

$$= \begin{bmatrix} 144.538 & 43.368 & -144.538 & 43.368 \\ 43.368 & 18.220 & -43.368 & 8.885 \\ -144.538 & -43.368 & 144.538 & -43.368 \\ 43.368 & 8.885 & -43.368 & 18.220 \end{bmatrix}(10^3)\,\text{N/m} \qquad\qquad \textbf{[a]}$$

In terms of the system degrees of freedom $\{v_1\ \theta_1\ v_2\ \theta_2\ v_3\ \theta_3\}^T$ the individual element stiffness matrices are

$$[\mathbf{k}]_1 = \begin{bmatrix} 144.538 & 43.368 & -144.538 & 43.368 & 0 & 0 \\ 43.368 & 18.220 & -43.368 & 8.885 & 0 & 0 \\ -144.538 & -43.368 & 144.538 & -43.368 & 0 & 0 \\ 43.368 & 8.885 & -43.368 & 18.220 & 0 & 0 \\ 0 & 0 & 0 & 0 & 0 & 0 \\ 0 & 0 & 0 & 0 & 0 & 0 \end{bmatrix}(10^3)\,\text{N/m}$$

and

$$[\mathbf{k}]_2 = \begin{bmatrix} 0 & 0 & 0 & 0 & 0 & 0 \\ 0 & 0 & 0 & 0 & 0 & 0 \\ 0 & 0 & 144.538 & 43.368 & -144.538 & 43.368 \\ 0 & 0 & 43.368 & 18.220 & -43.368 & 8.885 \\ 0 & 0 & -144.538 & -43.368 & 144.538 & -43.368 \\ 0 & 0 & 43.368 & 8.885 & -43.368 & 18.220 \end{bmatrix}(10^3)\,\text{N/m}$$

Adding the two matrices yields the system stiffness matrix

$$[\mathbf{K}]_{system} = \begin{bmatrix} 144.538 & 43.368 & -144.538 & 43.368 & 0 & 0 \\ 43.368 & 18.220 & -43.368 & 8.885 & 0 & 0 \\ -144.538 & -43.368 & 289.075 & 0 & -144.538 & 43.368 \\ 43.368 & 8.885 & 0 & 36.440 & -43.368 & 8.885 \\ 0 & 0 & -144.538 & -43.368 & 144.538 & -43.368 \\ 0 & 0 & 43.368 & 8.885 & -43.368 & 18.220 \end{bmatrix} (10^3) \ N/m$$

The beam structure is pinned at both ends. Thus the boundary conditions are $v_1 = v_3 = 0$. Thus the first and fifth rows and columns are partitioned out. This leaves the reduced system stiffness matrix

$$[\mathbf{K}]_{red} = \begin{bmatrix} 18.220 & -43.368 & 8.885 & 0 \\ -43.368 & 289.075 & 0 & 43.368 \\ 8.885 & 0 & 36.440 & 8.885 \\ 0 & 43.368 & 8.885 & 18.220 \end{bmatrix} (10^3) \ N/m$$

The inverse of this is

$$[\mathbf{K}]_{red}^{-1} = \begin{bmatrix} 13.202 & 2.881 & -1.756 & -6.001 \\ 2.881 & 1.210 & 0 & -2.881 \\ -1.756 & 0 & 3.600 & -1.756 \\ -6.001 & -2.881 & -1.756 & 13.202 \end{bmatrix} (10^{-5}) \ N/m \qquad \textbf{[b]}$$

The uniform load on each beam element is handled by Eq. (9.3-24) with $q_o = -w_o = -1600$ N/m, and $L_e = 0.625$ m. Thus for each element the nodal loads due to the distributed force, written in terms of the system degrees of freedom $\{v_1 \ \theta_1 \ v_2 \ \theta_2 \ v_3 \ \theta_3\}^T$, are

Element (1):

$$\begin{Bmatrix} \dfrac{-1600(0.625)}{2} \\[2ex] \dfrac{-1600(0.625)^2}{12} \\[2ex] \dfrac{-1600(0.625)}{2} \\[2ex] -\left(\dfrac{-1600(0.625)^2}{12}\right) \\[2ex] 0 \\[1ex] 0 \end{Bmatrix} = \begin{Bmatrix} -500 \\ -52.083 \\ -500 \\ 52.083 \\ 0 \\ 0 \end{Bmatrix}$$

Element (2):

$$\left\{ \begin{array}{c} 0 \\ 0 \\ \dfrac{-1600(0.625)}{2} \\ \dfrac{-1600(0.625)^2}{12} \\ \dfrac{-1600(0.625)}{2} \\ -\left(\dfrac{-1600(0.625)^2}{12}\right) \end{array} \right\} = \left\{ \begin{array}{c} 0 \\ 0 \\ -500 \\ -52.083 \\ -500 \\ 52.083 \end{array} \right\}$$

Adding the vectors and including the unknown support reactions R_{1y} and R_{3y} at nodes 1 and 3, respectively, we obtain the load vector

$$\left\{ \begin{array}{c} R_{1y} - 500 \\ -52.083 \\ -1000 \\ 0 \\ R_{3y} - 500 \\ 52.083 \end{array} \right\}$$

However, since the first and fifth load-deflection equations are partitioned, the reaction rows of the applied load vector partition out and the remaining load vector is $\{-52.083 \quad -1000 \quad 0 \quad 52.083\}^T$. Multiplying this by the inverse stiffness matrix of Eq. (*b*) gives the deflections

$$\left\{ \begin{array}{c} \theta_1 \\ v_2 \\ \theta_2 \\ \theta_3 \end{array} \right\} = \left[\begin{array}{cccc} 13.202 & 2.881 & -1.756 & -6.001 \\ 2.881 & 1.210 & 0 & -2.881 \\ -1.756 & 0 & 3.600 & -1.756 \\ -6.001 & -2.881 & -1.756 & 13.202 \end{array} \right] (10^{-5}) \left\{ \begin{array}{c} -52.083 \\ -1000 \\ 0 \\ 52.083 \end{array} \right\}$$

$$= \left\{ \begin{array}{c} -3.881(10^{-2}) \text{ rad} \\ -1.5104(10^{-2}) \text{ m} \\ 0 \\ -3.881(10^{-2}) \text{ rad} \end{array} \right\}$$

Thus the vertical deflection of the beam structure at midspan is $v_2 = -15.1$ mm, which agrees with the results given for Example 6.13-2. In order to determine the stress at node 2 we use Eq. (9.3-26) on element (1). *However,* we replace the stiffness term in this equation with the stiffness given by Eq. (*a*).

$$
\begin{Bmatrix} V_1 \\ M_1 \\ V_2 \\ M_2 \end{Bmatrix}_1 =
\begin{bmatrix}
144.538 & 43.368 & -144.538 & 43.368 \\
43.368 & 18.220 & -43.368 & 8.885 \\
-144.538 & -43.368 & 144.538 & -43.368 \\
43.368 & 8.885 & -43.368 & 18.220
\end{bmatrix}(10^3)
\begin{Bmatrix} 0 \\ -3.881(10^{-2}) \\ -1.5104(10^{-2}) \\ 0 \end{Bmatrix}
-
\begin{Bmatrix} -500 \\ -52.083 \\ -500 \\ 52.083 \end{Bmatrix}
$$

$$
= \begin{Bmatrix} 1000 \text{ N} \\ 0 \text{ N·m} \\ 0 \text{ N} \\ 258.13 \text{ N·m} \end{Bmatrix}
$$

Thus the bending moment at midspan is 258.13 N·m. The maximum tensile stress is thus

$$
\sigma_{max} = \frac{M_2}{S} + \frac{N}{A} = \frac{258.13}{2(10^3)(10^{-3})^3} + \frac{3.6(10^3)}{700(10^{-3})^2}
$$

$$
= 134.2(10^6) \text{ N/m} = 134.2 \text{ MPa}
$$

If the axial force is compressive, the effect is the opposite of stiffening and the beam will have larger deflections and bending stresses. Here N is negative, and it is also possible for the determinant of the modified system matrix to go to zero. This, in turn, would make the modified system matrix singular and the free displacements tend toward infinity. This is the condition for buckling instability.

Repeat Example 3.10-1 for the case when $EI_z = 1.2$ kN·m^2, $L = 750$ mm, and $a = 500$ mm. Figure 3.10-3(*a*) of the example is repeated here for convenience as Fig. 9.3-11. | **Example 9.3-5**

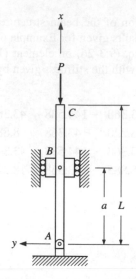

Figure 9.3-11   Column of Example 3.10-1.

Solution:

We will model the column with two elements where AB is element (1) and BC is element (2), and points A, B, and C are nodes 1, 2, and 3, respectively. Using Eq. (9.3-60), the stiffness matrices are (with $N = -P$)

Element (1):

$$
[\mathbf{k}]_1 = \left(\frac{1.2(10^3)}{(0.5)^3} \right)
\begin{bmatrix}
12 & 3 & -12 & 3 \\
3 & 1 & -3 & 0.5 \\
-12 & -3 & 12 & -3 \\
3 & 0.5 & -3 & 1
\end{bmatrix}
- \frac{P}{15}
\begin{bmatrix}
36 & 1.5 & -36 & 1.5 \\
1.5 & 1 & -1.5 & -0.25 \\
-36 & -1.5 & 36 & -1.5 \\
1.5 & -0.25 & -1.5 & 1
\end{bmatrix}
$$

$$
=
\begin{bmatrix}
115.2 & 28.8 & -115.2 & 28.8 \\
28.8 & 9.6 & -28.8 & 4.8 \\
-115.2 & -28.8 & 115.2 & -28.8 \\
28.8 & 4.8 & -28.8 & 9.6
\end{bmatrix}
(10^3) - P
\begin{bmatrix}
2.4 & 0.1 & -2.4 & 0.1 \\
0.1 & 1/15 & -0.1 & -1/60 \\
-2.4 & -0.1 & 2.4 & -0.1 \\
0.1 & -1/60 & -0.1 & 1/15
\end{bmatrix}
$$

In terms of the system degrees of freedom the stiffness matrix is

$$
[\mathbf{k}]_1 =
\begin{bmatrix}
115.2 & 28.8 & -115.2 & 28.8 & 0 & 0 \\
28.8 & 9.6 & -28.8 & 4.8 & 0 & 0 \\
-115.2 & -28.8 & 115.2 & -28.8 & 0 & 0 \\
28.8 & 4.8 & -28.8 & 9.6 & 0 & 0 \\
0 & 0 & 0 & 0 & 0 & 0 \\
0 & 0 & 0 & 0 & 0 & 0
\end{bmatrix}(10^3) - P
\begin{bmatrix}
2.4 & 0.1 & -2.4 & 0.1 & 0 & 0 \\
0.1 & 1/15 & -0.1 & -1/60 & 0 & 0 \\
-2.4 & -0.1 & 2.4 & -0.1 & 0 & 0 \\
0.1 & -1/60 & -0.1 & 1/15 & 0 & 0 \\
0 & 0 & 0 & 0 & 0 & 0 \\
0 & 0 & 0 & 0 & 0 & 0
\end{bmatrix}
$$

[a]

Element (2):

$$
[\mathbf{k}]_2 = \left(\frac{1.2(10^3)}{(0.25)^3}\right)
\begin{bmatrix}
12 & 1.5 & -12 & 1.5 \\
1.5 & 0.25 & -1.5 & 0.125 \\
-12 & -1.5 & 12 & -1.5 \\
1.5 & 0.125 & -1.5 & 0.25
\end{bmatrix}
- \frac{2P}{15}
\begin{bmatrix}
36 & 0.75 & -36 & 0.75 \\
0.75 & 0.25 & -0.75 & -0.0625 \\
-36 & -0.75 & 36 & -0.75 \\
0.75 & -0.0625 & -0.75 & 0.25
\end{bmatrix}
$$

$$
=
\begin{bmatrix}
921.6 & 115.2 & -921.6 & 115.2 \\
115.2 & 19.2 & -115.2 & 9.6 \\
-921.6 & -115.2 & 921.6 & -115.2 \\
115.2 & 9.6 & -115.2 & 19.2
\end{bmatrix}(10^3) - P
\begin{bmatrix}
4.8 & 0.1 & -4.8 & 0.1 \\
0.1 & 1/30 & -0.1 & -1/120 \\
-4.8 & -0.1 & 4.8 & -0.1 \\
0.1 & -1/120 & -0.1 & 1/30
\end{bmatrix}
$$

In terms of the system degrees of freedom the stiffness matrix is

$$[k]_2 = \begin{bmatrix} 0 & 0 & 0 & 0 & 0 & 0 \\ 0 & 0 & 0 & 0 & 0 & 0 \\ 0 & 0 & 921.6 & 115.2 & -921.6 & 115.2 \\ 0 & 0 & 115.2 & 19.2 & -115.2 & 9.6 \\ 0 & 0 & -921.6 & -115.2 & 921.6 & -115.2 \\ 0 & 0 & 115.2 & 9.6 & -115.2 & 19.2 \end{bmatrix}(10^3)$$

$$- P \begin{bmatrix} 0 & 0 & 0 & 0 & 0 & 0 \\ 0 & 0 & 0 & 0 & 0 & 0 \\ 0 & 0 & 4.8 & 0.1 & -4.8 & 0.1 \\ 0 & 0 & 0.1 & 1/30 & -0.1 & -1/120 \\ 0 & 0 & -4.8 & -0.1 & 4.8 & -0.1 \\ 0 & 0 & 0.1 & -1/120 & -0.1 & 1/30 \end{bmatrix} \qquad \textbf{[b]}$$

Adding Eqs. (a) and (b) yields the system stiffness matrix

$$[K]_{\text{system}} = \begin{bmatrix} 115.2 & 28.8 & -115.2 & 28.8 & 0 & 0 \\ 28.8 & 9.6 & -28.8 & 4.8 & 0 & 0 \\ -115.2 & -28.8 & 1036.8 & 86.4 & -921.6 & 115.2 \\ 28.8 & 4.8 & 86.4 & 28.8 & -115.2 & 9.6 \\ 0 & 0 & -921.6 & -115.2 & 921.6 & -115.2 \\ 0 & 0 & 115.2 & 9.6 & -115.2 & 19.2 \end{bmatrix}(10^3)$$

$$- P \begin{bmatrix} 2.4 & 0.1 & -2.4 & 0.1 & 0 & 0 \\ 0.1 & 1/15 & -0.1 & -1/60 & 0 & 0 \\ -2.4 & -0.1 & 7.2 & 0 & -4.8 & 0.1 \\ 0.1 & -1/60 & 0 & 0.1 & -0.1 & -1/120 \\ 0 & 0 & -4.8 & -0.1 & 4.8 & -0.1 \\ 0 & 0 & 0.1 & -1/120 & -0.1 & 1/30 \end{bmatrix}$$

The boundary conditions are $v_1 = v_2 = 0$. Thus the first and third rows and columns are partitioned from the system stiffness matrix, yielding the reduced system stiffness matrix

$$[K]_{\text{red}} = \begin{bmatrix} 9.6 & 4.8 & 0 & 0 \\ 4.8 & 28.8 & -115.2 & 9.6 \\ 0 & -115.2 & 921.6 & -115.2 \\ 0 & 9.6 & -115.2 & 19.2 \end{bmatrix}(10^3) - P \begin{bmatrix} 1/15 & -1/60 & 0 & 0 \\ -1/60 & 0.1 & -0.1 & -1/120 \\ 0 & -0.1 & 4.8 & -0.1 \\ 0 & -1/120 & -0.1 & 1/30 \end{bmatrix}$$

When the determinant of this matrix goes to zero, the matrix becomes singular and buckling occurs.

Because of the large difference of the order of the two matrices in $[K]_{red}$ and for convenience and exactness of expression, define a term p such that

$$P = 12(10^4)p \qquad \qquad \text{[c]}$$

Substitute Eq. (c) into $[K]_{red}$ and factor (10^3) from both matrices. Setting the determinant of the resulting matrix equal to zero provides the condition for buckling. This results in

$$(10^3)\begin{vmatrix} (9.6 - 8p) & (4.8 + 2p) & 0 & 0 \\ (4.8 + 2p) & (28.8 - 12p) & (-115.2 + 12p) & (9.6 + p) \\ 0 & (-115.2 + 12p) & (921.6 - 576p) & (-115.2 + 12p) \\ 0 & (9.6 + p) & (-115.2 + 12p) & (19.2 - 4p) \end{vmatrix} = 0$$

Canceling the (10^3) term, and using a mathematics software package to evaluate the determinant symbolically, results in

$$(1.8720p^4 - 18.248p^3 + 43.429p^2 - 28.665p + 3.0576)(10^5) = 0$$

dividing by $1.8720(10^5)$ gives

$$p^4 - 9.748p^3 + 23.200p^2 - 15.313p + 1.633 = 0$$

Using a calculator or software which solves polynomial equations, the roots of the above equation are found to be $p = 0.1314, 0.86549, 2.18863$, and 6.56217. The lowest value of p produces the lowest value of P, which is the critical force P_{cr}. Thus $p_{cr} = 0.1314$ and from Eq. (c)

$$P_{cr} = 12(10^4)p_{cr} = 12(10^4)(0.1314) = 15.77(10^3)\,\text{N} = 15.77\,\text{kN}$$

which is only 1.1 percent higher than the exact solution given in Example 3.10-1.

9.4 TWO-DIMENSIONAL ELASTIC ELEMENTS

A brief introduction to the simplest spatial elements, two-dimensional elastic elements, will be given here. For more detail on these elements, as well as the other elements shown in Table 9.0-1, the reader is urged to consult with the references given at the end of this chapter. In this section we will briefly discuss the three-noded triangular and four-noded quadrilateral two-dimensional elastic elements.

Two-dimensional elastic elements can be used for plane elastic and axisymmetric problems. The initial development of the triangular element will be based on a plane stress assumption. However, converting the equations to apply to plane strain or axisymmetric problems is a simple task. Plane strain is covered at the end of the next section, whereas for axisymmetric problems the reader is referred to the references cited at the end of the chapter.

9.4.1 THE TWO-DIMENSIONAL CONSTANT STRAIN TRIANGLE (CST) ELEMENT

Consider the triangular element shown in Fig. 9.4-1. It is assumed that the element is continuous, is thin in the z direction of constant thickness t, and has three nodes i, j, and k ordered *counterclockwise* in the xy plane. Each node has two degrees of freedom, translation in the x and y directions u and v, respectively. Since there are three nodes, the element has *six degrees of freedom*. As with earlier elements, this allows us to assume up to six unknown constants for the displacement field $u(x, y)$ and $v(x, y)$. Thus, let

$$u(x, y) = a_1 + a_2 x + a_3 y \qquad \text{[9.4-1a]}$$

$$v(x, y) = b_1 + b_2 x + b_3 y \qquad \text{[9.4-1b]}$$

which is a linear displacement field. Again, as before, to determine the constants we substitute the nodal displacements u_i, v_i, u_j, v_j, u_k, and v_k together with the nodal coordinates into the displacement equations and solve for the constants in terms of the nodal displacements and coordinates. That is,

$$u_i = a_1 + a_2 x_i + a_3 y_i \qquad v_i = b_1 + b_2 x_i + b_3 y_i$$

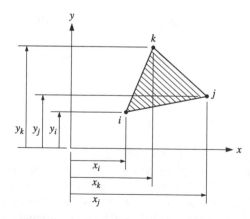

Figure 9.4-1   The constant strain triangle (CST) element.

$$u_j = a_1 + a_2 x_j + a_3 y_j \qquad v_j = b_1 + b_2 x_j + b_3 y_j$$

$$u_k = a_1 + a_2 x_k + a_3 y_k \qquad v_k = b_1 + b_2 x_k + b_3 y_k$$

When the above equations are solved simultaneously, the constants a_i, b_i are determined in terms of the nodal displacements and coordinates. The results are

$$a_1 = \frac{\alpha_i u_i + \alpha_j u_j + \alpha_k u_k}{2A} \qquad \text{[9.4-2a]}$$

$$a_2 = \frac{\beta_i u_i + \beta_j u_j + \beta_k u_k}{2A} \qquad \text{[9.4-2b]}$$

$$a_3 = \frac{\gamma_i u_i + \gamma_j u_j + \gamma_k u_k}{2A} \qquad \text{[9.4-2c]}$$

$$b_1 = \frac{\alpha_i v_i + \alpha_j v_j + \alpha_k v_k}{2A} \qquad \text{[9.4-2d]}$$

$$b_2 = \frac{\beta_i v_i + \beta_j v_j + \beta_k v_k}{2A} \qquad \text{[9.4-2e]}$$

$$b_3 = \frac{\gamma_i v_i + \gamma_j v_j + \gamma_k v_k}{2A} \qquad \text{[9.4-2f]}$$

where

$$\alpha_i = x_j y_k - x_k y_j \qquad \beta_i = y_j - y_k \qquad \gamma_i = x_k - x_j$$

$$\alpha_j = x_k y_i - x_i y_k \qquad \beta_j = y_k - y_i \qquad \gamma_j = x_i - x_k$$

$$\alpha_k = x_i y_j - x_j y_i \qquad \beta_k = y_i - y_j \qquad \gamma_k = x_j - x_i$$

$$2A = \alpha_i + \alpha_j + \alpha_k \qquad \text{[9.4-3]}$$

It can be shown that A is the area of the triangle in the xy plane.

Substituting Eqs. (9.4-2) back into Eqs. (9.4-1) and with some rearranging yields

$$u(x, y) = \frac{1}{2A}[(\alpha_i + \beta_i x + \gamma_i y)u_i + (\alpha_j + \beta_j x + \gamma_j y)u_j + (\alpha_k + \beta_k x + \gamma_k y)u_k]$$

$$\text{[9.4-4a]}$$

$$v(x, y) = \frac{1}{2A}[(\alpha_i + \beta_i x + \gamma_i y)v_i + (\alpha_j + \beta_j x + \gamma_j y)v_j + (\alpha_k + \beta_k x + \gamma_k y)v_k]$$

$$\text{[9.4-4b]}$$

or

$$u(x, y) = N_i(x, y)u_i + N_j(x, y)u_j + N_k(x, y)u_k \qquad \textbf{[9.4-5a]}$$

$$v(x, y) = N_i(x, y)v_i + N_j(x, y)v_j + N_k(x, y)v_k \qquad \textbf{[9.4-5b]}$$

where $N_i(x, y)$, $N_j(x, y)$, and $N_k(x, y)$ are the shape functions given by

$$N_i(x, y) = \frac{1}{2A}(\alpha_i + \beta_i x + \gamma_i y) \qquad \textbf{[9.4-6a]}$$

$$N_j(x, y) = \frac{1}{2A}(\alpha_j + \beta_j x + \gamma_j y) \qquad \textbf{[9.4-6b]}$$

$$N_k(x, y) = \frac{1}{2A}(\alpha_k + \beta_k x + \gamma_k y) \qquad \textbf{[9.4-6c]}$$

Next, the strains are related to the nodal deflections. The strains in the plane are

$$\varepsilon_x = \frac{\partial u}{\partial x} = a_2 = \frac{1}{2A}(\beta_i u_i + \beta_j u_j + \beta_k u_k) \qquad \textbf{[9.4-7a]}$$

$$\varepsilon_y = \frac{\partial v}{\partial y} = b_3 = \frac{1}{2A}(\gamma_i v_i + \gamma_j v_j + \gamma_k v_k) \qquad \textbf{[9.4-7b]}$$

$$\gamma_{xy} = \frac{\partial v}{\partial x} + \frac{\partial u}{\partial y} = b_2 + a_3$$

$$= \frac{1}{2A}(\gamma_i u_i + \gamma_j u_j + \gamma_k u_k + \beta_i v_i + \beta_j v_j + \beta_k v_k) \qquad \textbf{[9.4-7c]}$$

Equations (9.4-7) can be placed in matrix form as

$$\begin{Bmatrix} \varepsilon_x \\ \varepsilon_y \\ \gamma_{xy} \end{Bmatrix} = \frac{1}{2A} \begin{bmatrix} \beta_i & 0 & \beta_j & 0 & \beta_k & 0 \\ 0 & \gamma_i & 0 & \gamma_j & 0 & \gamma_k \\ \gamma_i & \beta_i & \gamma_j & \beta_j & \gamma_k & \beta_k \end{bmatrix} \begin{Bmatrix} u_i \\ v_i \\ u_j \\ v_j \\ u_k \\ v_k \end{Bmatrix} \qquad \textbf{[9.4-8]}$$

or

$$\{\varepsilon\} = [\mathbf{B}]\{\mathbf{u}\} \qquad \textbf{[9.4-9]}$$

where

$$\{\varepsilon\} = \begin{Bmatrix} \varepsilon_x \\ \varepsilon_y \\ \gamma_{xy} \end{Bmatrix} \qquad \{u\} = \begin{Bmatrix} u_i \\ v_i \\ u_j \\ v_j \\ u_k \\ v_k \end{Bmatrix}$$

and

$$[\mathbf{B}] = \frac{1}{2A} \begin{bmatrix} \beta_\iota & 0 & \beta_\xi & 0 & \beta_\kappa & 0 \\ 0 & \gamma_\iota & 0 & \gamma_\xi & 0 & \gamma_\kappa \\ \gamma_\iota & \beta_\iota & \gamma_\xi & \beta_\xi & \gamma_\kappa & \beta_\kappa \end{bmatrix} \qquad \textbf{[9.4-10]}$$

We see from the strain equations that the three strains are constant within a given triangular element. Hence the name constant strain triangle (CST). This is the element's serious limitation. Within an elastic body undergoing large strain gradients, extremely small CST elements would be necessary to model these strain fields. This would require a vast number of elements to accurately model even the most simple problems. The quadrilateral element provides for strain variation within the element and is preferred over the CST element. The CST element can be used for transitions within fields of quadrilateral elements. The quadrilateral element is introduced in the next section.

Plane Stress   The stress-strain relations for a homogeneous, isotropic, plane-stress element are given by Eqs. (1.4-5) and (1.4-8a), which in matrix form is written as

$$\begin{Bmatrix} \sigma_x \\ \sigma_y \\ \tau_{xy} \end{Bmatrix} = \frac{E}{2(1-v^2)} \begin{bmatrix} 2 & 2v & 0 \\ 2v & 2 & 0 \\ 0 & 0 & (1-v) \end{bmatrix} \begin{Bmatrix} \varepsilon_x \\ \varepsilon_y \\ \gamma_{xy} \end{Bmatrix} \qquad \textbf{[9.4-11]}$$

or as

$$\{\sigma\} = [\mathbf{D}]\{\varepsilon\} \qquad \textbf{[9.4-12]}$$

where

$$\{\sigma\} = \begin{Bmatrix} \sigma_x \\ \sigma_y \\ \tau_{xy} \end{Bmatrix}$$

and

$$[\mathbf{D}] = \frac{E}{2(1-v^2)} \begin{bmatrix} 2 & 2v & 0 \\ 2v & 2 & 0 \\ 0 & 0 & (1-v) \end{bmatrix} \qquad \textbf{[9.4-13]}$$

Substituting Eq. (9.4-9) into (9.4-12) gives

$$\{\sigma\} = [D][B]\{u\} \qquad\qquad \textbf{[9.4-14]}$$

The strain energy is given by

$$U = \frac{1}{2}\int_V \{\sigma\}^T\{\varepsilon\}\,dV = \frac{1}{2}\int_A \{\sigma\}^T\{\varepsilon\}\,t\,dA$$

From Eq. (9.4-14) $\{\sigma\}^T = \{u\}^T[B]^T[D]^T = \{u\}^T[B]^T[D]$, since $[D]$ is symmetric. Substituting this and Eq. (9.4-9) into the strain energy expression results in

$$U = \frac{1}{2}\int_A \{u\}^T[B]^T[D][B]\{u\}\,t\,dA$$

Everything within the integral except for dA contains constants. Thus

$$U = \frac{1}{2}\{u\}^T At[B]^T[D][B]\{u\} \qquad\qquad \textbf{[9.4-15]}$$

As shown in earlier sections, the form of the strain energy is $1/2\{u\}_e^T[k]_e\{u\}_e$, where $[k]_e$ is the element stiffness matrix. Therefore, for the CST element, the stiffness matrix is given by

$$[k]_e = (At[B]^T[D][B])_e \qquad\qquad \textbf{[9.4-16]}$$

When determining $[B]$, care must be taken in assigning the i, j, and k nodes. The derivation of the element equations was based on the i, j, and k nodes being in a *counterclockwise* order.

Example 9.4-1 For a CST element in a finite element model the nodal coordinates for the i, j, and k nodes are $(15, -8)$ and $(10, 5)$ and $(2, 0)$ mm, respectively. The element is 2 mm thick and is of a material with properties $E = 70$ GPa and $v = 0.3$. Upon loading of the model the deflections of the given element were found to be $u_i = 100\ \mu\text{m}$, $v_i = -50\ \mu\text{m}$, $u_j = 75\ \mu\text{m}$, $v_j = -40\ \mu\text{m}$, $u_k = 80\ \mu\text{m}$, and $v_k = -45\ \mu\text{m}$. Determine (*a*) the element stiffness matrix, (*b*) the nodal force vector, and (*c*) the stress in the element.

Solution:

(*a*) First we determine the terms in Eqs. (9.4-3)

$$\alpha_i = (10)(0) - (2)(5) = -10\ \text{mm}^2 \qquad \beta_i = 5 - 0 = 5\ \text{mm}$$

$$\gamma_i = 2 - 10 = -8\ \text{mm}$$

$$\alpha_j = (2)(-8) - (15)(0) = -16\,\text{mm}^2 \qquad \beta_j = 0 - (-8) = 8\,\text{mm}$$

$$\gamma_j = 15 - 2 = 13\,\text{mm}$$

$$\alpha_k = (15)(5) - (10)(-8) = 155\ \text{mm}^2 \qquad \beta_k = -8 - 5 = -13\ \text{mm}$$

$$\gamma_k = 10 - 15 = -5\,\text{mm}$$

$$2A = -10 + (-16) + 155 = 129\,\text{mm}^2$$

The [B] matrix, given by Eq. (9.4-10), is

$$[\mathbf{B}] = \frac{1}{129(10^{-3})^2}\begin{bmatrix} 5 & 0 & 8 & 0 & -13 & 0 \\ 0 & -8 & 0 & 13 & 0 & -5 \\ -8 & 5 & 13 & 8 & -5 & -13 \end{bmatrix}(10^{-3})\,\text{m}^{-1} \qquad [\textbf{\textit{a}}]$$

The transpose of the [B] matrix is

$$[\mathbf{B}]^T = \frac{1}{129(10^{-3})^2}\begin{bmatrix} 5 & 0 & -8 \\ 0 & -8 & 5 \\ 8 & 0 & 13 \\ 0 & 13 & 8 \\ -13 & 0 & -5 \\ 0 & -5 & -13 \end{bmatrix}(10^{-3})\ \text{m}^{-1} \qquad [\textbf{\textit{b}}]$$

The [D] matrix is

$$[\mathbf{D}] = \frac{70(10^9)}{2(1 - 0.3^2)}\begin{bmatrix} 2 & 2(0.3) & 0 \\ 2(0.3) & 2 & 0 \\ 0 & 0 & (1 - 0.3) \end{bmatrix}$$

$$= 38.46(10^9)\begin{bmatrix} 2 & 0.6 & 0 \\ 0.6 & 2 & 0 \\ 0 & 0 & 0.7 \end{bmatrix}\text{N/m}^2 \qquad [\textbf{\textit{c}}]$$

From Eq. (9.4-16) the stiffness matrix is

$$[\mathbf{k}]_e = At[\mathbf{B}]^T[\mathbf{D}][\mathbf{B}]$$

Substitution of Eqs. (*a*), (*b*), and (*c*) matrix multiplication yields

$$
[\mathbf{k}]_e =
\begin{bmatrix}
28.265 & -15.504 & 2.147 & -1.729 & -30.411 & 17.233 \\
-15.504 & 43.381 & 2.117 & -53.667 & 13.387 & 10.286 \\
2.147 & 2.117 & 73.435 & 40.310 & -75.581 & -42.427 \\
-1.729 & -53.667 & 40.310 & 114.132 & -38.581 & -60.465 \\
-30.411 & 13.387 & -75.581 & -38.581 & 105.993 & 25.194 \\
17.233 & 10.286 & -42.427 & -60.465 & 25.194 & 50.179
\end{bmatrix}
(10^6) \ \text{N/m}
$$

(*b*) The displacement vector is $\{\mathbf{u}\}_e = \{100 \ -50 \ 75 \ -40 \ 80 \ -45\}^T (10^{-6})$ m. To obtain the nodal force vector we premultiply $\{\mathbf{u}\}_e$ by $[\mathbf{k}]_e$. The resulting force vector is $\{f_{xi} \ f_{yi} \ f_{xj} \ f_{yj} \ f_{xk} \ f_{yk}\} = \{623.4 \ -805.9 \ -133.3 \ 602.9 \ -490.2 \ 203.0\}^T$ N.

(*c*) The stress is determined from Eq. (9.4-14), $\{\boldsymbol{\sigma}\}_e = [\mathbf{D}][\mathbf{B}]\{\mathbf{u}\}_e$. This gives

$$
\{\boldsymbol{\sigma}\}_e = 38.46(10^9)
\begin{bmatrix}
2 & 0.6 & 0 \\
0.6 & 2 & 0 \\
0 & 0 & 0.7
\end{bmatrix}
\left(\frac{1}{129(10^{-3})^2}\right)
\begin{bmatrix}
5 & 0 & 8 & 0 & -13 & 0 \\
0 & -8 & 0 & 13 & 0 & -5 \\
-8 & 5 & 13 & 8 & -5 & -13
\end{bmatrix}
(10^{-3})
\begin{Bmatrix}
100 \\ -50 \\ 75 \\ -40 \\ 80 \\ -45
\end{Bmatrix}
(10^{-6})
$$

$$
=
\begin{Bmatrix}
54.56 \\ 73.35 \\ -43.83
\end{Bmatrix}
(10^6) \ \text{N/m}^2 =
\begin{Bmatrix}
54.56 \\ 73.35 \\ -43.83
\end{Bmatrix}
\text{MPa}
$$

When subdividing a continuous structure into a system of triangular elements such as shown in Fig. 9.0-1, the assembly, partitioning, and solution processes are the same as that performed in previous sections. However, when solving small problems manually, care must be taken in defining the element's degrees of freedom and assembling them correctly in the system stiffness matrix. Automating the process in software is very simple however.

Example 9.4-2 | Show how the system stiffness matrix is formed for the thin plate loaded as shown in Fig. 9.4-2(*a*) using the two-element model shown in Fig. 9.4-2(*b*).

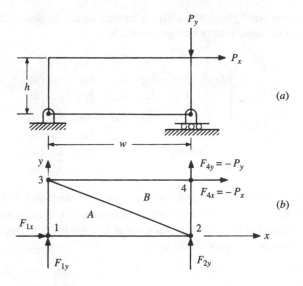

Figure 9.4-2

Solution:

For element A let the $i, j,$ and k nodes be nodes 1, 2, and 3 with nodal coordinates (0, 0), $(w, 0)$, and $(0, h)$, respectively. If the nodal coordinates and material properties were given, the stiffness matrix would be determined using Eq. (9.4-16) as done in Example 9.4-1. Let the stiffness matrix be represented by

$$[\mathbf{k}]_A = \begin{array}{c} \begin{array}{cccccc} u_1 & v_1 & u_2 & v_2 & u_3 & v_3 \end{array} \\ \begin{bmatrix} k_{11} & k_{12} & k_{13} & k_{14} & k_{15} & k_{16} \\ & k_{22} & k_{23} & k_{24} & k_{25} & k_{26} \\ & & k_{33} & k_{34} & k_{35} & k_{36} \\ & & & k_{44} & k_{45} & k_{46} \\ & \text{Sym} & & & k_{55} & k_{56} \\ & & & & & k_{66} \end{bmatrix}_A \end{array} \begin{array}{c} u_1 \\ v_1 \\ u_2 \\ v_2 \\ u_3 \\ v_3 \end{array}$$

where the degrees of freedom are shown for the rows and columns.

For element B, the $i, j,$ and k nodes *cannot* be ordered as 2, 3, and 4 as that would be in clockwise order. The order can be either 2, 4, and 3 or 4, 3, and 2 or 3, 2, and 4. Here, we will use the order 2, 4, and 3. Again, if the nodal coordinates and material properties were given, the stiffness matrix would be determined using Eq. (9.4-16). Let the stiffness matrix be represented by

$$
[\mathbf{k}]_B =
\begin{array}{c}
\begin{array}{cccccc} u_2 & v_2 & u_4 & v_4 & u_3 & v_3 \end{array} \\
\begin{bmatrix}
k_{11} & k_{12} & k_{13} & k_{14} & k_{15} & k_{16} \\
 & k_{22} & k_{23} & k_{24} & k_{25} & k_{26} \\
 & & k_{33} & k_{34} & k_{35} & k_{36} \\
 & & & k_{44} & k_{45} & k_{46} \\
 & \text{Sym} & & & k_{55} & k_{56} \\
 & & & & & k_{66}
\end{bmatrix}
\begin{array}{l} u_2 \\ v_2 \\ u_4 \\ v_4 \\ u_3 \\ v_3 \end{array}
\end{array}
$$

The system matrix will be 8×8 and we must assemble the matrices for elements A and B. Placing element A into the system stiffness matrix is quite straightforward owing to the ordered node sequence. Element B is a bit harder. The reader should verify that the system stiffness matrix is

$$
[\mathbf{K}]_{\text{system}} =
\begin{array}{c}
\begin{array}{cccccccc} u_1 & v_1 & u_2 & v_2 & u_3 & v_3 & u_4 & v_4 \end{array} \\
\begin{bmatrix}
(k_{11})_A & (k_{12})_A & (k_{13})_A & (k_{14})_A & (k_{15})_A & (k_{16})_A & 0 & 0 \\
 & (k_{22})_A & (k_{23})_A & (k_{24})_A & (k_{25})_A & (k_{26})_A & 0 & 0 \\
 & & [(k_{33})_A + (k_{11})_B] & [(k_{34})_A + (k_{12})_B] & [(k_{35})_A + (k_{15})_B] & [(k_{36})_A + (k_{16})_B] & (k_{13})_B & (k_{14})_B \\
 & & & [(k_{44})_A + (k_{22})_B] & [(k_{45})_A + (k_{25})_B] & [(k_{46})_A + (k_{26})_B] & (k_{23})_B & (k_{24})_B \\
 & \text{Sym} & & & [(k_{55})_A + (k_{55})_B] & [(k_{56})_A + (k_{56})_B] & (k_{35})_B & (k_{45})_B \\
 & & & & & [(k_{66})_A + (k_{66})_B] & (k_{36})_B & (k_{46})_B \\
 & & & & & & (k_{33})_B & (k_{34})_B \\
 & & & & & & & (k_{44})_B
\end{bmatrix}
\begin{array}{l} u_1 \\ v_1 \\ u_2 \\ v_2 \\ u_3 \\ v_3 \\ u_4 \\ v_4 \end{array}
\end{array}
$$

where the degrees of freedom are indicated.

Plane Strain   For plane elastic problems where the thickness t is very large in the z direction, the assumption is that $\varepsilon_z = 0$. Substituting this into Eqs. (1.4-3a) and (1.4-3b) gives

$$
\sigma_x = \frac{E}{(1 + v)(1 - 2v)}[(1 - v)\varepsilon_x + v\varepsilon_y] \qquad \textbf{[9.4-17a]}
$$

$$
\sigma_y = \frac{E}{(1 + v)(1 - 2v)}[(1 - v)\varepsilon_y + v\varepsilon_x] \qquad \textbf{[9.4-17b]}
$$

This together with Eq. (1.4-8a) for shear gives a new stress-strain relation matrix equation of

$$
\begin{Bmatrix} \sigma_x \\ \sigma_y \\ \tau_{xy} \end{Bmatrix} = \frac{E}{2(1 + v)(1 - 2v)}
\begin{bmatrix}
2(1 - v) & 2v & 0 \\
2v & 2(1 - v) & 0 \\
0 & 0 & (1 - 2v)
\end{bmatrix}
\begin{Bmatrix} \varepsilon_x \\ \varepsilon_y \\ \gamma_{xy} \end{Bmatrix}
$$

Thus, for plane strain we can replace the **[D]** matrix in Eqs. (9.4-12) to (9.4-16) with

$$[\tilde{\mathbf{D}}] = \frac{E}{2(1+v)(1-2v)} \begin{bmatrix} 2(1-v) & 2v & 0 \\ 2v & 2(1-v) & 0 \\ 0 & 0 & (1-2v) \end{bmatrix} \qquad \textbf{[9.4-18]}$$

This is all that needs to be changed for a plane strain problem. In some commercial codes, the thickness of the element for plane strain is assumed to be unity so that it is unnecessary to input the thickness in Eq. (9.4-16) when evaluating the stiffness matrix. However, the forces must be input as force per unit thickness.

9.4.2   THE TWO-DIMENSIONAL ISOPARAMETRIC QUADRILATERAL ELEMENT

The four-node, four-sided, two-dimensional quadrilateral element shown in Fig. 9.4-3(*a*) is a drastic improvement over the constant strain triangle. With four nodes the displacement field can be written as

$$u(x, y) = a_1 + a_2 x + a_3 y + a_4 xy \qquad \textbf{[9.4-19a]}$$

$$v(x, y) = b_1 + b_2 x + b_3 y + b_4 xy \qquad \textbf{[9.4-19b]}$$

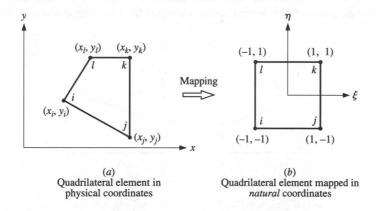

(a)
Quadrilateral element in physical coordinates

(b)
Quadrilateral element mapped in *natural* coordinates

Figure 9.4-3   Two-dimensional plane elastic quadrilateral element.

and with the xy term in the displacement fields the strains will not be constant, thus allowing the strains to vary through the element.

The drawback with the additional term is that the [**B**] matrix will no longer be made up of constant terms as was the case with the simple CST element. This creates a problem in the integration, especially when it comes to the limits of integration with an arbitrarily shaped quadrilateral. The use of the *isoparametric* element reduces this difficulty. Here, the element is mapped from physical space to a space (ξ, η) where the element is a well-defined square of sides $\xi = \pm 1$ and $\eta = \pm 1$ as shown in Fig. 9.4-3(*b*). The (ξ, η) are called *natural coordinates*. If the mapping of the physical coordinates uses the same shape (interpolation) functions that are used for the displacement shape functions, the element is said to be *isoparametric*. If the coordinate shape functions are higher-ordered than the displacement shape functions, the element is said to be *superparametric*. If the displacement shape functions are higher-ordered than the coordinate shape functions, the element is said to be *subparametric*.

With the mapping, the displacements can be expressed in terms of the natural coordinates by

$$u(\xi, \eta) = a_1 + a_2\xi + a_3\eta + a_4\xi\eta \qquad \textbf{[9.4-20a]}$$

$$v(\xi, \eta) = b_1 + b_2\xi + b_3\eta + b_4\xi\eta \qquad \textbf{[9.4-20b]}$$

We will determine the shape functions using the natural coordinates. We could do this using the same approach as the previous sections by substituting the u and v displacements and coordinate positions for each node into Eqs. (9.4-20) and solving the simultaneous equations for the a and b coefficients. This would require some effort, and since there is a simpler method, we will describe it here.

The interpolation or shape functions were introduced for the simple axial element in Sec. 9.2.1. Let us return to this case briefly to demonstrate some simple properties of an interpolation function. Figure 9.4-4 shows how each shape function, through simple linear interpolation, provides the two components of the inter-

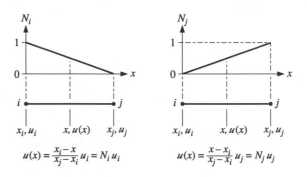

Figure 9.4-4   Interpolation functions for the simple axial element.

polated value of $u(x) = N_i u_i + N_j u_j$. From this, the properties of the interpolation functions are observed to be

1. $N_i = 1$ at $x = x_i$ $\qquad N_j = 1$ at $x = x_j$
   $\quad = 0$ at $x = x_j$ $\qquad\quad\ = 0$ at $x = x_i$
2. $N_i + N_j = 1$ for any value of x.

Using these properties in conjunction with Fig. 9.4-3(b), let us determine the shape function $N_i(\xi, \eta)$. If $N_i = 0$ at nodes j, k, and l, then $N_i = 0$ when $\xi = \eta = 1$. The equation

$$N_i(\xi, \eta) = C(1 - \xi)(1 - \eta) \qquad\qquad \textbf{[9.4-21]}$$

where C is a constant that satisfies these conditions. Since $N_i = 1$ at node i, where $\xi = -1$ and $\eta = 1$, then $C = 1/4$ and Eq. (9.4-21) becomes

$$N_i(\xi, \eta) = \frac{1}{4}(1 - \xi)(1 - \eta) \qquad\qquad \textbf{[9.4-22]}$$

The remaining shape functions can be determined in a similar manner. The complete set of shape functions are

$$N_i(\xi, \eta) = \frac{1}{4}(1 - \xi)(1 - \eta) \qquad\qquad \textbf{[9.4-23a]}$$

$$N_j(\xi, \eta) = \frac{1}{4}(1 + \xi)(1 - \eta) \qquad\qquad \textbf{[9.4-23b]}$$

$$N_k(\xi, \eta) = \frac{1}{4}(1 + \xi)(1 + \eta) \qquad\qquad \textbf{[9.4-23c]}$$

$$N_l(\xi, \eta) = \frac{1}{4}(1 - \xi)(1 + \eta) \qquad\qquad \textbf{[9.4-23d]}$$

We can now express the deflections as

$$u(\xi, \eta) = N_i u_i + N_j u_j + N_k u_k + N_l u_l \qquad\qquad \textbf{[9.4-24a]}$$

$$v(\xi, \eta) = N_i v_i + N_j v_j + N_k v_k + N_l v_l \qquad\qquad \textbf{[9.4-24b]}$$

Using the isoparametric formulation, any position within the element can also be expressed with the same shape functions. That is,

$$x(\xi, \eta) = N_i x_i + N_j x_j + N_k x_k + N_l x_l \qquad\qquad \textbf{[9.4-25a]}$$

$$y(\xi, \eta) = N_i y_i + N_j y_j + N_k y_k + N_l y_l \qquad\qquad \textbf{[9.4-25b]}$$

For the strains we need to express the derivatives of u and v in terms of x and y, but now that x and y are functions of ξ and η, we will also need the derivatives relative to these coordinates. Using the chain rule for differentiation, we have

$$\frac{\partial u}{\partial \xi} = \frac{\partial u}{\partial x}\frac{\partial x}{\partial \xi} + \frac{\partial u}{\partial y}\frac{\partial y}{\partial \xi} \qquad \text{[9.4-26a]}$$

$$\frac{\partial u}{\partial \eta} = \frac{\partial u}{\partial x}\frac{\partial x}{\partial \eta} + \frac{\partial u}{\partial y}\frac{\partial y}{\partial \eta} \qquad \text{[9.4-26b]}$$

which written in matrix form is

$$\begin{Bmatrix} \dfrac{\partial u}{\partial \xi} \\[2mm] \dfrac{\partial u}{\partial \eta} \end{Bmatrix} = \begin{bmatrix} \dfrac{\partial x}{\partial \xi} & \dfrac{\partial y}{\partial \xi} \\[2mm] \dfrac{\partial x}{\partial \eta} & \dfrac{\partial y}{\partial \eta} \end{bmatrix} \begin{Bmatrix} \dfrac{\partial u}{\partial x} \\[2mm] \dfrac{\partial u}{\partial y} \end{Bmatrix} = [\mathbf{J}] \begin{Bmatrix} \dfrac{\partial u}{\partial x} \\[2mm] \dfrac{\partial u}{\partial y} \end{Bmatrix} \qquad \text{[9.4-27]}$$

The square matrix is the transformation matrix which transforms derivatives in the physical coordinates to derivatives in the natural coordinates and is called the *Jacobian matrix* [**J**], where

$$[\mathbf{J}] = \begin{bmatrix} J_{11} & J_{12} \\ J_{21} & J_{22} \end{bmatrix} = \begin{bmatrix} \dfrac{\partial x}{\partial \xi} & \dfrac{\partial y}{\partial \xi} \\[2mm] \dfrac{\partial x}{\partial \eta} & \dfrac{\partial y}{\partial \eta} \end{bmatrix} \qquad \text{[9.4-28]}$$

The derivatives of x and y with respect to ξ and η are determined from Eqs. (9.4-25) and (9.4-23) and are

$$J_{11} = \frac{1}{4}[(1 - \eta)(x_j - x_i) + (1 + \eta)(x_k - x_l)]$$

$$J_{12} = \frac{1}{4}[(1 - \eta)(y_j - y_i) + (1 + \eta)(y_k - y_l)]$$

$$J_{21} = \frac{1}{4}[(1 - \xi)(x_l - x_i) + (1 + \xi)(x_k - x_j)]$$

$$J_{22} = \frac{1}{4}[(1 - \xi)(y_l - y_i) + (1 + \xi)(y_k - y_j)] \qquad \text{[9.4-29]}$$

For the strains we will need the derivatives relative to the physical coordinates. Thus we invert Eq. (9.4-27), which gives

$$\begin{Bmatrix} \dfrac{\partial u}{\partial x} \\[2mm] \dfrac{\partial u}{\partial y} \end{Bmatrix} = [\mathbf{J}]^{-1} \begin{Bmatrix} \dfrac{\partial u}{\partial \xi} \\[2mm] \dfrac{\partial u}{\partial \eta} \end{Bmatrix} = \frac{1}{|J|} \begin{bmatrix} J_{22} & -J_{12} \\ -J_{21} & J_{11} \end{bmatrix} \begin{Bmatrix} \dfrac{\partial u}{\partial \xi} \\[2mm] \dfrac{\partial u}{\partial \eta} \end{Bmatrix} \qquad \text{[9.4-30]}$$

where $|J|$ is the determinant of the Jacobian matrix. Likewise, derivatives of v can be expressed as

$$\left\{\begin{array}{c} \dfrac{\partial v}{\partial x} \\[2mm] \dfrac{\partial v}{\partial y} \end{array}\right\} = \frac{1}{|J|}\begin{bmatrix} J_{22} & -J_{12} \\ -J_{21} & J_{11} \end{bmatrix}\left\{\begin{array}{c} \dfrac{\partial v}{\partial \xi} \\[2mm] \dfrac{\partial v}{\partial \eta} \end{array}\right\} \qquad \textbf{[9.4-31]}$$

The strains are given by $\varepsilon_x = \partial u/\partial x$, $\varepsilon_y = \partial v/\partial y$, and $\gamma_{xy} = \partial v/\partial x + \partial u/\partial y$. From Eqs. (9.4-30) and (9.4-31) this gives

$$\{\boldsymbol{\varepsilon}\} = \left\{\begin{array}{c} \varepsilon_x \\ \varepsilon_y \\ \gamma_{xy} \end{array}\right\} = \left\{\begin{array}{c} \dfrac{\partial u}{\partial x} \\[2mm] \dfrac{\partial v}{\partial y} \\[2mm] \dfrac{\partial v}{\partial x} + \dfrac{\partial u}{\partial y} \end{array}\right\} = \frac{1}{|J|}\begin{bmatrix} J_{22} & -J_{12} & 0 & 0 \\ 0 & 0 & -J_{21} & J_{11} \\ -J_{21} & J_{11} & J_{22} & -J_{12} \end{bmatrix}\left\{\begin{array}{c} \dfrac{\partial u}{\partial \xi} \\[2mm] \dfrac{\partial u}{\partial \eta} \\[2mm] \dfrac{\partial v}{\partial \xi} \\[2mm] \dfrac{\partial v}{\partial \eta} \end{array}\right\}$$

$$\textbf{[9.4-32]}$$

For shorthand notation let us refer to Eq. (9.4-32) as

$$\{\boldsymbol{\varepsilon}\} = [\mathbf{G}]\{\mathbf{u'}\} \qquad \textbf{[9.4-33]}$$

where $\{\mathbf{u'}\} = \{\partial u/\partial \xi \;\; \partial u/\partial \eta \;\; \partial v/\partial \xi \;\; \partial v/\partial \eta\}^T$ and

$$[\mathbf{G}] = \frac{1}{|J|}\begin{bmatrix} J_{22} & -J_{12} & 0 & 0 \\ 0 & 0 & -J_{21} & J_{11} \\ -J_{21} & J_{11} & J_{22} & -J_{12} \end{bmatrix} \qquad \textbf{[9.4-34]}$$

The derivatives of u and v with respect to ξ and η are determined directly from Eqs. (9.4-24) and (9.4-23) and are

$$\left\{\begin{array}{c} \dfrac{\partial u}{\partial \xi} \\[2mm] \dfrac{\partial u}{\partial \eta} \\[2mm] \dfrac{\partial v}{\partial \xi} \\[2mm] \dfrac{\partial v}{\partial \eta} \end{array}\right\} = \frac{1}{4}\begin{bmatrix} -(1-\eta) & 0 & (1-\eta) & 0 & (1+\eta) & 0 & -(1+\eta) & 0 \\ -(1-\xi) & 0 & -(1+\xi) & 0 & (1+\xi) & 0 & (1-\xi) & 0 \\ 0 & -(1-\eta) & 0 & (1-\eta) & 0 & (1+\eta) & 0 & -(1+\eta) \\ 0 & -(1-\xi) & 0 & -(1+\xi) & 0 & (1+\xi) & 0 & (1-\xi) \end{bmatrix}\left\{\begin{array}{c} u_i \\ v_i \\ u_j \\ v_j \\ u_k \\ v_k \\ u_l \\ v_l \end{array}\right\}$$

$$\textbf{[9.4-35]}$$

which can be expressed as

$$\{u'\} = [\mathbf{H}]\{u\} \tag{9.4-36}$$

where $\{\mathbf{u}\} = \{u_i \, v_i \, u_j \, v_j \, u_k \, v_k \, u_l \, v_l\}^T$ and

$$[\mathbf{H}] = \frac{1}{4}\begin{bmatrix} -(1-\eta) & 0 & (1-\eta) & 0 & (1+\eta) & 0 & -(1+\eta) & 0 \\ -(1-\xi) & 0 & -(1+\xi) & 0 & (1+\xi) & 0 & (1-\xi) & 0 \\ 0 & -(1-\eta) & 0 & (1-\eta) & 0 & (1+\eta) & 0 & -(1+\eta) \\ 0 & -(1-\xi) & 0 & -(1+\xi) & 0 & (1+\xi) & 0 & (1-\xi) \end{bmatrix}$$

$$\tag{9.4-37}$$

Substituting Eq. (9.4-36) into (9.4-33) gives

$$\{\boldsymbol{\varepsilon}\} = [\mathbf{G}][\mathbf{H}]\{\mathbf{u}\} = [\mathbf{B}]\{\mathbf{u}\} \tag{9.4-38}$$

where the familiar [**B**] matrix is given by

$$[\mathbf{B}] = [\mathbf{G}][\mathbf{H}] \tag{9.4-39}$$

The stiffness matrix is given by

$$[\mathbf{k}]_e = \left(\int_A [\mathbf{B}]^T[\mathbf{D}][\mathbf{B}]\, t \, dx \, dy\right)_e \tag{9.4-40}$$

The matrix [**B**], however, is written exclusively as a function of ξ and η. It can be shown that (see Appendix A of Ref. 9.4)

$$dx \, dy = |J| \, d\xi \, d\eta \tag{9.4-41}$$

Thus $[\mathbf{k}]_e$ is given by

$$[\mathbf{k}]_e = \int_{-1}^{1}\int_{-1}^{1} [\mathbf{B}]^T[\mathbf{D}][\mathbf{B}]t|J| \, d\xi \, d\eta \tag{9.4-42}$$

Although the limits of integration of Eq. (9.4-42) are uncomplicated, the terms within the integral are very involved functions of ξ and η. For this reason, the integral is performed by numerical techniques. Here, we will introduce some basic concepts. For more detail consult the finite element references.

Numerical Integration of [k].   The numerical method normally employed is called *Gauss-Legendre quadrature*. In this method, the integration of a function is approximated using weighted sampling points, called *Gauss points*. Consider, for example, the integral

$$\int_{-1}^{1} f(x)\, dx \tag{9.4-43}$$

This can be approximated by

$$\int_{-1}^{1} f(x)\, dx \approx w_1 f(x_1) + w_2 f(x_2) + \cdots + w_n f(x_n) = \sum_{i=1}^{n} w_i f(x_i) \qquad \textbf{[9.4-44]}$$

The x_i are the Gauss sampling points with corresponding weights w_i.

Considering the integration of some simple polynomials, the conditions for exact results can be explored. If $f(x)$ were a linear function, only one $(n = 1)$ sampling point would be necessary. To see this, let $f(x) = a_0 + a_1 x$. The integral is

$$\int_{-1}^{1} (a_0 + a_1 x)\, dx = 2a_0$$

Setting this equal to $w_1 f(x_1)$ yields

$$w_1(a_0 + a_1 x_1) = 2a_0$$

Since a_0 and a_1 are arbitrary, we will obtain exact results if

$$w_1 = 2 \qquad x_1 = 0 \qquad \textbf{[9.4-45]}$$

Thus exact results of the integration of a linear function between the limits of -1 and 1 are obtained by simply evaluating the function at $x = 0$ and multiplying the value by 2.

With a one-point sampling, a polynomial with two coefficients can be integrated exactly since a one-point sampling can adjust two parameters w_1 and x_1. With a two-point sampling, we have four parameters w_1, x_1, w_2, and x_2. Thus we can expect exact results for a polynomial with four coefficients. With this, consider the integration of $f(x) = a_0 + a_1 x + a_2 x^2 + a_3 x^3$

$$\int_{-1}^{1} (a_0 + a_1 x + a_2 x^2 + a_3 x^3)\, dx = 2a_0 + \frac{2}{3} a_2$$

With two-point quadrature we set $w_1 f(x_1) + w_2 f(x_2)$ equal to the integral, giving

$$w_1(a_0 + a_1 x_1 + a_2 x_1^2 + a_3 x_1^3) + w_2(a_0 + a_1 x_2 + a_2 x_2^2 + a_3 x_2^3) = 2a_0 + \frac{2}{3} a_2$$

Equating the coefficients of the a_i terms gives

$$w_1 + w_2 = 2$$

$$w_1 x_1 + w_2 x_2 = 0$$

$$w_1 x_1^2 + w_2 x_2^2 = \frac{2}{3}$$

$$w_1 x_1^3 + w_2 x_2^3 = 0$$

The solutions of these nonlinear equations are

$$w_1 = w_2 = 1 \qquad x_1 = -\frac{1}{\sqrt{3}} \qquad x_2 = \frac{1}{\sqrt{3}} \qquad \textbf{[9.4-46]}$$

In evaluating x_1 and x_2, high accuracy is required and double precision should be used. That is, $1/\sqrt{3} = 0.5773502691.\ldots$ For higher-order quadratures, the finite element references provide tables for the locations and weights of the sampling points.

Two-dimensional integrals are handled in a similar fashion, integrating with respect to one of the variables first, followed by integration with respect to the second variable. That is,

$$\int_{-1}^{1}\int_{-1}^{1} f(x, y)\, dx\, dy = \sum_{i=1}^{n}\sum_{j=1}^{n} w_i w_j f(x_i, y_i) \qquad \textbf{[9.4-47]}$$

For a 2×2 quadrature

$$\int_{-1}^{1}\int_{-1}^{1} f(x, y)\, dx\, dy = w_1^2 f(x_1, y_1) + w_1 w_2 f(x_1, y_2)$$

$$+ w_2 w_1 f(x_2, y_1) + w_2^2 f(x_2, y_2) \qquad \textbf{[9.4-48]}$$

where $w_1 = w_2 = 1$, $x_1 = y_1 = 1/\sqrt{3}$, and $x_2 = y_2 = -1/\sqrt{3}$. Thus

$$\int_{-1}^{1}\int_{-1}^{1} f(x, y)\, dx\, dy = f(x_1, y_1) + f(x_1, y_2) + f(x_2, y_1) + f(x_2, y_2) \qquad \textbf{[9.4-49]}$$

For the stiffness of the quadrilateral element, the integral in Eq. (9.4-42) is desired, where the dependent variables are the natural coordinates ξ and η. Let the matrix within the integral be

$$[\Phi(\xi, \eta)]_e = ([\mathbf{B}]^T[\mathbf{D}][\mathbf{B}]t|J|)_e \qquad \textbf{[9.4-50]}$$

where the $[\mathbf{B}]$ matrix and $|J|$ are functions of ξ and η. Based on Eq. (9.4-49) the integral can be written as

$$[\mathbf{k}]_e = [\Phi(\xi_1, \eta_1)] + [\Phi(\xi_1, \eta_2)] + [\Phi(\xi_2, \eta_1)] + [\Phi(\xi_2, \eta_2)] \qquad \textbf{[9.4-51]}$$

where $\xi_1 = \eta_1 = 1/\sqrt{3}$, and $\xi_2 = \eta_2 = -1/\sqrt{3}$. The Gauss points in natural coordinates are shown in Fig. 9.4-5.

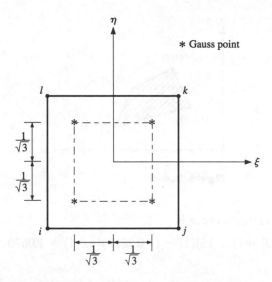

Figure 9.4-5 Gauss points for a 2 × 2 quadrature integration.

The coordinates for the quadrilateral element shown in Fig. 9.4-6 are in millimeters. **Example 9.4-3**
Using a 2 × 2 quadrature determine the first quadrature matrix, $[\Phi(\xi_1, \eta_1)]$, which
forms part of the element stiffness matrix. The element thickness is 2 mm, $E = 200$
GPa, and $v = 0.3$.

Solution:[†]

For convenience we will use the units of N and mm for calculations. For the first
Gauss point, $\xi_1 = \eta_1 = 1/\sqrt{3}$. From Eq. (9.4-29) the terms in the Jacobean matrix are

$$J_{11} = \frac{1}{4}[(1 - 1/\sqrt{3})(7 - 3) + (1 + 1/\sqrt{3})(6 - 4)] = 1.2113$$

$$J_{12} = \frac{1}{4}[(1 - 1/\sqrt{3})(2 - 3) + (1 + 1/\sqrt{3})(4 - 5)] = -0.5$$

$$J_{21} = \frac{1}{4}[(1 - 1/\sqrt{3})(4 - 3) + (1 + 1/\sqrt{3})(6 - 7)] = -0.2887$$

$$J_{22} = \frac{1}{4}[(1 - 1/\sqrt{3})(5 - 3) + (1 + 1/\sqrt{3})(4 - 2)] = 1$$

[†] In practice, double precision is used. However, to depict the calculations in this example, the numbers will
be abbreviated to save space.

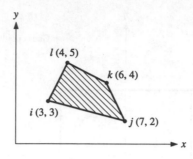

Figure 9.4-6

The determinant of the Jacobean is

$$|J| = (1.2113)(1) - (-0.2887)(-0.5) = 1.0670 \qquad \textbf{[a]}$$

From Eq. (9.4-32)

$$[G] = \frac{1}{1.0670}\begin{bmatrix} 1 & 0.5 & 0 & 0 \\ 0 & 0 & 0.2887 & 1.2113 \\ 0.2887 & 1.2113 & 1 & 0.5 \end{bmatrix} \qquad \textbf{[b]}$$

For the [H] matrix we need

$$1 + \xi_1 = 1.5774 \qquad 1 - \xi_1 = 0.4226 \qquad 1 + \eta_1 = 1.5774 \qquad 1 - \eta_1 = 0.4226$$

Thus, from Eq. (9.4-37) we have

$$[H] = \frac{1}{4}\begin{bmatrix} -0.4226 & 0 & 0.4226 & 0 & 1.5774 & 0 & -1.5774 & 0 \\ -0.4226 & 0 & -1.5774 & 0 & 1.5774 & 0 & 0.4226 & 0 \\ 0 & -0.4226 & 0 & 0.4226 & 0 & 1.5774 & 0 & -1.5774 \\ 0 & -0.4226 & 0 & -1.5774 & 0 & 1.5774 & 0 & 0.4226 \end{bmatrix} \qquad \textbf{[c]}$$

Since $[B] = [G][H]$, matrix multiplication yields

$$[B] = \begin{bmatrix} -0.14854 & 0 & -0.08576 & 0 & 0.5544 & 0 & -0.3201 & 0 \\ 0 & -0.14854 & 0 & -0.4191 & 0 & 0.5544 & 0 & 0.013267 \\ -0.14854 & -0.14854 & -0.4191 & -0.08576 & 0.5544 & 0.5544 & 0.013267 & -0.3201 \end{bmatrix} \qquad \textbf{[d]}$$

The transpose is

$$[B]^T = \begin{bmatrix} -0.14854 & 0 & -0.14854 \\ 0 & -0.14854 & -0.14854 \\ -0.08576 & 0 & -0.4191 \\ 0 & -0.4191 & -0.08576 \\ 0.5544 & 0 & 0.5544 \\ 0 & 0.5544 & 0.5544 \\ -0.3201 & 0 & 0.013267 \\ 0 & 0.013267 & -0.3201 \end{bmatrix} \qquad \textbf{[e]}$$

Using $E = 200 \times 10^3$ N/mm^2, the [**D**] matrix, given by Eq. (9.4-13), is

$$[\mathbf{D}] = \frac{200(10^3)}{2(1 - 0.3^2)} \begin{bmatrix} 2 & 0.6 & 0 \\ 0.6 & 2 & 0 \\ 0 & 0 & 0.7 \end{bmatrix}$$

$$= \begin{bmatrix} 21.978 & 6.593 & 0 \\ 6.593 & 21.978 & 0 \\ 0 & 0 & 7.692 \end{bmatrix} (10^4) \, \text{N/mm}^2 \qquad [\boldsymbol{f}]$$

Performing the matrix multiplication $[\mathbf{B}]^T \, [\mathbf{D}] \, [\mathbf{B}]$ and multiplying the result by $t \, |J|$ (with $t = 2$ mm) yields $[\Phi(\xi_1, \eta_1)]$. That is,

$$[\Phi(\xi_1, \eta_1)] = \begin{bmatrix} 13.971 & 6.727 & 16.194 & 10.850 & -52.139 & -25.104 & 21.975 & 7.527 \\ 6.727 & 13.971 & 12.011 & 31.288 & -25.104 & -52.139 & 6.366 & 6.880 \\ 16.194 & 12.011 & 32.281 & 10.957 & -60.436 & -44.828 & 11.961 & 21.859 \\ 10.850 & 31.288 & 10.957 & 83.584 & -40.494 & -116.770 & 18.687 & 1.898 \\ -52.139 & -25.104 & -60.436 & -40.494 & 194.586 & 93.690 & -82.011 & -28.092 \\ -25.104 & -52.139 & -44.828 & -116.770 & 93.690 & 194.586 & -23.758 & -25.677 \\ 21.975 & 6.366 & 11.961 & 18.687 & -82.011 & -23.758 & 48.075 & -1.295 \\ 7.527 & 6.880 & 21.859 & 1.898 & -28.092 & -25.677 & -1.295 & 16.899 \end{bmatrix} (10^3) \, \text{N/mm}$$

To complete the stiffness matrix the remaining matrices $[\Phi(\xi_1, \eta_2)]$, $[\Phi(\xi_2, \eta_1)]$, and $[\Phi(\xi_2, \eta_2)]$ would be calculated in an identical fashion. A spreadsheet program would make this an easy task.

If one were to compare the final $[\mathbf{k}]_e$ matrix in the preceding example with that determined in a commercial finite element program, the matrices would not be the same. The isoparametric element as defined in this section is too rigid when in an in-plane bending mode. This is caused by an effect called *parasitic shear*. An improved quadrilateral element with incompatible modes is discussed in Ref. 9.2 and is referred to as the QM6 element.

9.5   HIGHER-ORDER AND THREE-DIMENSIONAL ELASTIC ELEMENTS

This section offers only some concluding remarks about the more advanced elastic elements that are beyond the scope of this chapter. For detailed discussion of the higher-order and three-dimensional elements, consult the references cited at the end of the chapter.

Adding midside nodes to the triangular element permits a higher-order polynomial to portray the displacement field and, in the case of an isoparametric element, model curved sides of the element using the higher-ordered shape functions for the nodal coordinates. This, in turn, will produce an element with much more capability of modeling a structure and producing much better results than the CST element. Likewise, a quadrilateral element with midside nodes permits a higher-order polynomial than a quadrilateral element without midside nodes. Interior nodes can be added to the elements. The better performance comes with a cost, however. The elements are much more complicated numerically. There are computational trade-offs between using more lower-order elements versus fewer higher-order elements.

The three-dimensional elements consist of the surface triangular and quadrilateral thin-plate elements that are capable of bending and twisting, and the solid brick, wedge, and tetrahedron elements shown in Table 9.0-1. Higher-order versions of these elements are also available.

The thin-plate elements typically support all three translation degrees of freedom at each node; the in-plane *membrane* deflections and the transverse deflections due to bending. In addition, at a minimum, the plate element also provides two rotational degrees of freedom at each node for plate bending and twisting moments as discussed in Sec. 5.7.1. Thus, for a thin-plate quadrilateral element with four nodes, at a minimum the element supports $4 \times 5 = 20$ degrees of freedom.

The solid elements are similar to the two-dimensional plane elastic elements as they only support translational degrees of freedom at each node. For the three-dimensional elements, all three translational degrees of freedom are supported at each node. Thus, for a brick element with eight nodes, the element supports $8 \times 3 = 24$ degrees of freedom. The lack of rotational degrees of freedom implies that rotations within the structure only result from the translations of the nodes; and furthermore, concentrated moments *cannot* be applied or transmitted at a node. This has additional implications when attempting to connect these elements to an element that *does* support rotational degrees of freedom (such as the beam, frame, and thin-plate elements). More will be said about this in the next chapter.

9.6   PROBLEMS

9.1   Solve Problem 3.41 using the finite element method.

9.2   For the figure of Problem 3.41 let $F = 0$ and let element AB be undergoing a uniformly distributed load of 65 kN/m in the positive x direction. Using the finite element method, determine the deflections of all nodes, the wall reactions, and the element stresses and nodal forces.

9.3   Solve Problem 3.42 using the finite element method.

9.4   A tapered plate 0.125 in thick is uniformly loaded in tension as shown. Assume that the taper is gradual enough such that $\sigma_x = P/A$ is still valid (where A is the area of the cross section of the plate and is a function of x. The material of the plate is steel with $E = 30$ Mpsi and $v = 0.3$.

(*a*)   Using the conventional mechanics of materials approach, solve for and separately graph σ_x and the displacement u as functions of x.

(b) Using the two-element model shown in Fig. (b), evaluate and graph σ_x and u as functions of x on the same plots of part (a).

(c) Repeat part (b) using the three-element model of Fig. (c).

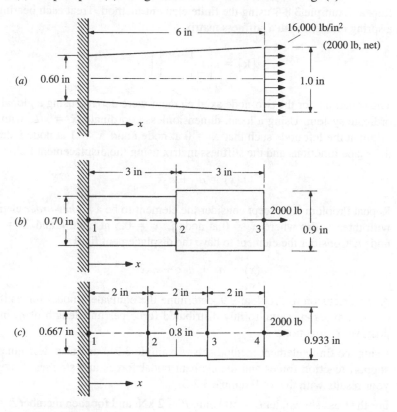

Problem 9.4

9.5 Example 3.2-2 pertains to a rod hanging under its own weight. Return to the solution of this example and in separate graphs, plot the displacement u and stress σ_x as functions of x. Then, using the finite element method, solve the problem using (a) a one-element model, (b) a two-element model, (c) a three-element model, and (d) a four-element model. On the same graphs discussed earlier, plot the resulting element displacements and stresses as functions of x.

9.6 Using three finite elements, solve Problem 3.3c by the finite element method. Show that the net "dynamic force" on each element in the positive x direction is

$$F_d = \rho A_e L_e \omega^2 (x_{cm})_e$$

where $(x_{cm})_e$ is the location of the center of mass of the element. As an approximation, assume the distribution to be uniform and divide half the force in each element to each node.

Solve Problem 3.3c and plot $u(x)$ and the finite element displacement fields on the same graph. On a separate graph, as functions of x, plot the stress fields from the two solutions.

9.7 Repeat Example 3.8-1 using the finite element method. Treat each bearing as a spring element with a stiffness matrix.

$$[\mathbf{k}] = k\begin{bmatrix} 1 & -1 \\ -1 & 1 \end{bmatrix}$$

9.8 The equations for the two-node axial element were derived using a global coordinate system. Using a local, dimensionless coordinate $\bar{x} = x/L_e$ with the origin at the left node such that $\bar{x} = 0$ at node i and $\bar{x} = 1$ at node j, derive the shape functions and the stiffness matrix using the displacement field

$$u(x) = a_1 + a_2\bar{x}$$

9.9 Repeat Problem 9.8 except consider the element to be a *higher-order* element with three nodes where $\bar{x} = 0$ at node i, $\bar{x} = 0.5$ at node j, and $\bar{x} = 1$ at node k. Consider the element to have the displacement field

$$u(x) = a_1 + a_2\bar{x} + a_3\bar{x}^2$$

9.10 As an extension to Problem 9.9, determine the equivalent nodal forces if the element experiences uniformly distributed force per unit length of q_o in the positive $\bar{x}$ direction.

9.11 Using the finite element method solve Example 3.2-3 and also determine the support reaction forces and the element nodal forces and stresses. Compare your results with that of Example 3.2-3.

9.12 For the truss shown, let $L = 500$ mm, $P = 2$ kN, and for each member $E = 70$ GPa, and the cross-section area $A = 950$ mm^2.

(a) Using the finite element method, determine the deflections of the pins, the reaction forces, the nodal forces, and the element stresses.

(b) Verify the results of part (a) using standard methods of mechanics.

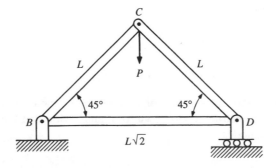

Problem 9.12

9.13 For the truss shown all members have a cross-sectional area of 500 mm² and a modulus of elasticity of 210 GPa. Other than member (5) all members have a length of 400 mm. Using the finite element method determine the global displacements of each node, all support force reactions, and the element nodal forces and stress in element (5).

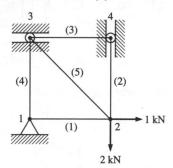

Problem 9.13

9.14 Solve Problem 9.13 with the slot at node 3 rotated 30° counterclockwise.

9.15 The i and j nodes of an element in a truss structure have the coordinates $(1, -1, 0)$ and $(2, 1, -2)$ in. The element has an area of 0.2 in² and the modulus of elasticity is 10 Mpsi. The global displacements are found to be

$$\{u_i \, v_i \, w_i \, u_j \, v_j \, w_j\}^T = \{-2.5 \ 1.8 \ 2.1 \ -1.5 \ 1.2 \ 2.8\}^T \ (10^{-2}) \text{ in}$$

Using the finite element equations for each calculation, determine the local nodal forces and stress for the element.

9.16 For the space truss shown let the cross-sectional area and modulus of elasticity of each element be 600 mm² and 70 GPa, respectively. Using the finite element method exclusively, determine the displacement vector of node 2 and the local nodal forces and stress in element 2. Verify your results using standard methods of mechanics.

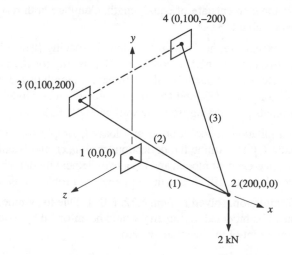

Problem 9.16

9.17 Determine the slope and deflections at nodes 2 and 3 for the beam shown. The modulus of elasticity of the beam is E and $I_1 = 2I_2$. HINT: When dealing with beam systems without numerical parameters the element stiffness equation can be put into the form

$$\left\{\begin{array}{c} V_i \\ M_i/L \\ V_j \\ M_j/L \end{array}\right\}_e = \left(\frac{EI}{L^3}\right)_e \begin{bmatrix} 12 & 6 & -12 & 6 \\ 6 & 4 & -6 & 2 \\ -12 & -6 & 12 & -6 \\ 6 & 2 & -6 & 4 \end{bmatrix}_e \left\{\begin{array}{c} v_i \\ \theta_i L \\ v_j \\ \theta_j L \end{array}\right\}_e$$

Care must be exercised, however, when applying a moment to a node (see Problem 9.18).

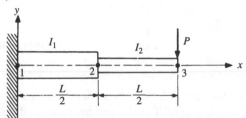

Problem 9.17

9.18 Repeat Problem 9.17; however, replace the force P with a clockwise moment M on node 3.

9.19 Solve Problem 3.39 using the finite element method.

9.20 Solve Problem 3.43 using the finite element method.

9.21 Determine the nodal translational and rotational displacements of the uniformly loaded cantilever beam given by Table C.3 of Appendix C. Let $L = 1.2$ m, $w = 2$ kN/m, and $EI = 320$ kN·m². Use lumped equivalent loading and (a) a two-element model where the elements are of equal length, and (b) a three-element model where the elements are of equal length. Compare both results with that obtained from Table C.3.

9.22 Repeat Problem 9.21 except use the following arithmetic spacing (where the change in length of adjacent elements is constant). That is, (a) for the two-element model let the element connected to the wall have a length of $L/3$ and the next element have a length $2L/3$, and (b) for the three-element model let the lengths of the elements progressing from the wall be $2L/9$, $3L/9$, and $4L/9$.

9.23 Using a consistent formulation, model a uniformly loaded, simply supported beam (C.6 of Appendix C). Determine the deflections $v(x)$, $\theta(x)$, and bending moment $M(x)$ for (a) a one-element model, and (b) a two-element model where the elements are of equal length. HINT: See the hint given in Problem 9.17.

9.24 Using a consistent formulation, solve Problem 6.32. HINT: Due to symmetry, only half the beam need be modeled. Symmetry would be enforced by requiring the slope of the midpoint of the beam to be zero.

9.25 Using a two-element model with consistent formulation for the beam shown, determine the nodal translational and rotational displacements, the support reactions, the element nodal loads, and the maximum bending stress. Let $a = 500$ mm, $L = 1.5$ m, $w = 2$ kN/m, $E = 210$ GPa, $I = 1.5(10^6)$ mm^4, and the section modulus $S = 3.5(10^4)$ mm^3.

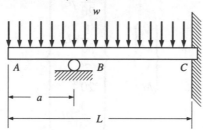

Problem 9.25

9.26 Model the beam shown using two elements. Given $a = 2$ in, $L = 10$ in, $\epsilon = 0.125$ in, $EI = 2.8\,(10^5)$ lb·in^2, determine the value of P required to cause contact with the support at B.

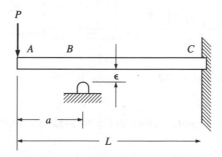

Problem 9.26

9.27 For the beam shown, $EI = 160$ kN·m^2, and the section modulus is $20(10^3)$ mm^3. A hinge exists at point B. Using a two-element model with consistent formulation, determine all nodal translational and rotational displacements, the support reactions, the element nodal loads, and the maximum bending stresses in each element.

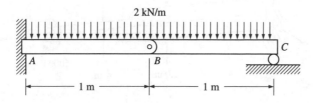

Problem 9.27

9.28 Solve Problem 9.27 with the uniform load applied only across span BC.

9.29 For the beam with the linearly varying distributed load shown, show that the consistent equivalent nodal loads are

$$\begin{Bmatrix} V_i \\ M_i \\ V_j \\ M_j \end{Bmatrix}_e = \begin{Bmatrix} \dfrac{3}{20}qL_e \\ \dfrac{1}{30}qL_e^2 \\ \dfrac{7}{20}qL_e \\ -\dfrac{1}{20}qL_e^2 \end{Bmatrix}$$

Show, also, that the consistent equivalent nodal loads are equivalent to the actual loading.

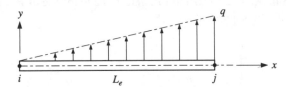

Problem 9.29

9.30 Derive the stiffness matrix given in Eq. (9.3-32) for a beam bending in the xz plane.

9.31 The step shaft shown is loaded in torsion. Using a two-element finite element model, determine the angular displacements of each node. For the material $E = 30$ Mpsi and $v = 0.3$.

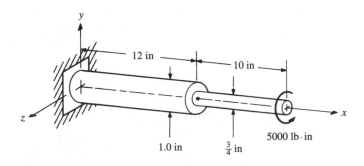

Problem 9.31

9.32 For the shaft in Problem 9.31, apply the torsional moment of 5000 lb·in at the step and fix the right end of the shaft. Using the finite element method determine the resulting rotational deflections, the support reactions, and the element nodal loads. Evaluate the shear stress in each element.

9.33 A frame element has a rectangular cross-section where the lengths of the sides parallel to the local y' and z' axes are 50 mm and 20 mm, respectively. The coordinates of the i, j, and k nodes are (500, 200, 0), (800, 500, −300), and (500, 1000, 100) mm, respectively. The material properties of the element are $E = 70$ GPa and $v = 0.33$. Determine:

(a) the local stiffness matrix of the element. Note the effective polar second-area moment for a rectangular cross section is given by (see Table 20, Ref. 9.1)

$$J = \left[1 - 0.63\left(\frac{b}{h}\right)\left(1 - \frac{b^4}{12h^4}\right)\right]\left(\frac{hb^3}{3}\right)$$

where $h > b$.

(b) the global transformation matrix [**T**].

(c) the global stiffness matrix.

9.34 Solve Example 6.13-3 using the finite element method and (a) a one-element model, and (b) a two-element model where the elements are of equal length. Let $P = 2$ kN, $F = 9$ kN, $L = 1$ m, and $EI = 16$ kN·m². Compare your results with Example 6.13-3.

9.35 Consider a simply supported beam with an intermediate load such as shown in Table C.5 of Appendix C. Let $a = 20$ in, $L = 40$ in, $EI = 1(10^6)$ lb·in², and $P = 400$ lb. Consider that the beam is also transmitting a tensile axial force of 600 lb. Using a two-element model determine the lateral deflection of the midpoint of the beam and the maximum normal stress. The area of the cross section is 0.85 in² and the section modulus is 1.85 in³.

9.36 Consider the beam shown in Table C.10 of Appendix C with $a = 1$ m, $L = 2$ m, $w = 2$ kN/m, and $EI = 12$ kN·m². Consider that the beam is also transmitting a tensile axial force of 1.2 kN. Using a two-element model determine the lateral deflection of the midpoint of the beam and the maximum normal stress. The area of the cross section is 300 mm² and the section modulus is $32(10^3)$ mm³.

9.37 Model the fixed-fixed column shown in Table 3.10-1 using two elements of equal length. The length of the column is L and the minimum bending rigidity is EI. Determine the critical buckling force and compare your results with Eq. (3.10-5). HINT: See the hint for Problem 9.17.

9.38 Model the fixed-pinned column shown in Table 3.10-1 using two elements of equal length. The length of the column is L and the minimum bending rigidity is EI. Determine the critical buckling force and compare your results with Eq. (3.10-5). HINT: See the hint for Problem 9.17.

9.39 Model the fixed-free column shown in Table 3.10-1 using two elements of equal length. The length of the column is $L = 500$ mm and the minimum bending rigidity is $EI = 5$ kN·m². Determine the critical buckling force and compare your results with Eq. (3.10-5).

9.40 A CST element has the nodal coordinates $(0, 0)$, $(3, -2)$, $(3, 2)$ in for the i, j, and k nodes, respectively. The element is to be considered in plane stress with a thickness of 0.5 in and is of a material with properties $E = 10$ Mpsi and $v = 0.33$. Upon loading the model the nodal deflections were found to be

$$\{u_i\, v_i\, u_j\, v_j\, u_k\, v_k\}^T = \{-5.45\ \ 8.24\ \ -3.15\ \ 6.10\ \ 1.21\ 5.56\}^T(10^{-3})\ \text{in}$$

(a) Determine the element stiffness matrix.
(b) Determine the nodal force vector.
(c) Determine the element stresses.

9.41 The plate in Problem 9.4 is modeled very crudely as shown using CST elements.[†] Thanks to symmetry, the model can be simplified to that of Fig. (b). For elements A, B, and C, determine all displacements and stresses.

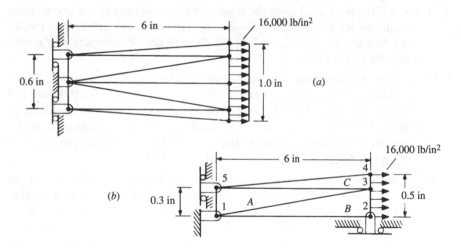

Problem 9.41

---

† The intent of this problem is only to provide an exercise in applying the mechanics of the finite element method. The model is not meant to convey good modeling practice.

9.42 The thin plate structure shown is to be modeled with the four elements shown. The four elements with corresponding nodes are to be labeled accordingly.

Element	i Node	j Node	k Node
A	1	4	3
B	3	4	5
C	2	3	5
D	1	3	2

In a manner similar to Example 9.4-2, determine the system stiffness of the structure.

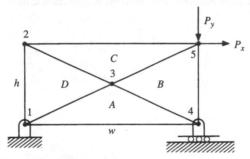

Problem 9.42

Computer Problems

9.43 Develop a program or spreadsheet program for a three-dimensional frame element where the input includes the global coordinates of nodes i and j, the geometric and the material properties of the beam, and the output is the global stiffness matrix.

9.44 Using the program developed in Problem 9.43 determine the reactions and deflection of the beams in Problem 3.45, using three frame elements.

9.45 Determine the second quadrature matrix $[\Phi(\xi_1, \eta_2)]$ for Example 9.4-3.

9.46 Determine the remaining quadrature matrices for Example 9.4-3. With these and the quadrature matrix determined in Example 9.5-1 evaluate the complete stiffness matrix for the element. Finally, applying pin supports to nodes i and l and a force of 100 N in the y direction to node j, determine the resulting deflections of the element.

9.7   REFERENCES

9.1   Young, W. *Roark's Formulas for Stress and Strain*, 6th ed. New York: McGraw-Hill, 1989.

9.2   Cook, R. D., D. S. Malkus, and M. E. Plesha. *Concepts and Applications of Finite Element Analysis*, 3rd ed. New York: Wiley, 1989.

9.3   Logan, D. L. *A First Course in the Finite Element Method Using Algor*. Boston, MA: PWS Publishing Company, 1997.

9.4   Chandrupatla, T. R., and A. D. Belegundu. *Introduction to Finite Elements in Engineering*, 2nd ed. Upper Saddle River, NJ: Prentice Hall, 1997.

Additional Uncited References in Finite Elements

9.5   Zienkiewicz, O. C., and R. L. Taylor. *The Finite Element Method*, 4th ed., vols. 1 and 2. New York: McGraw-Hill, 1989, 1991.

9.6   Reddy, J. N. *An Introduction to the Finite Element Method*, 2nd ed. New York: McGraw-Hill, 1984.

9.7   Bathe, K. J. *Finite Element Procedures*. Englewood Cliffs, NJ: Prentice Hall, 1996.

chapter

TEN

FINITE ELEMENT MODELING TECHNIQUES

10.0 INTRODUCTION

Chapter 9 provides an introduction to the finite element analysis (FEA) method, which describes how some basic elements are developed; how elements are assembled to model some very simple structures; how forces and boundary conditions are applied; and how the displacement solutions are arrived at and manipulated to obtain other results such as element forces and stresses. Except for the computational process, these procedures are basically the same even when performing a linear static analysis on a very complex structural problem using a commercial finite element program. Nonetheless, there are a great many modeling strategies to be learned and experienced before becoming proficient with a finite element software package. There is no substitute for experience; however, one must start somewhere. The intent of this chapter is to present a brief introduction to some basic finite element modeling considerations and practices employed when using a commercial software package. Owing to space limitations much of the discussion here is qualitative and given without complete justification. For more detail the reader is urged to consult the more advanced finite element textbooks cited at the end of this chapter and/or the technical literature supplied by the manufacturer of the commercial finite element program that the reader will be using. For a more extensive discussion of finite element modeling see Ref. 10.1.

A prerequisite to becoming a capable finite element modeler of stress analysis problems is a good background in theoretical stress analysis. Even as one becomes more proficient and experienced with finite element modeling, one should continue to expand one's theoretical base. Anyone can be "trained" to build finite element models. However, only one knowledgeable in structural analysis can understand how the elements behave and be able to effectively judge the results of a finite element analysis. After spending a good deal of time modeling a complex problem the model may look very impressive and it is quite tempting to simply blindly accept the results of the analysis. Also, the graphics capabilities of commercial software are generally quite extraordinary, and even totally incorrect results can appear quite convincing. One should *always* question and scrutinize the results of any numerical solution such as a finite element analysis.

Using a specific commercial finite element software package adds another component to the learning process. In addition to a comprehensive understanding of theoretical stress analysis and finite element theory, the analyst must now learn how to use FEA software effectively. This is not a trivial part of the total knowledge base. Certainly, one must learn the mechanical steps of entering the necessary information to create the model, process it, and analyze the results. The software vendors typically provide volumes of user manuals just for this purpose (if readable). However, one must also develop good modeling practices, understand the use and limitations of the particular elements used in the software, study the verification models that the software vendor furnishes, and comprehend the various status, warning, and error messages that the software generates. Not a simple task.

Before discussing the modeling process, a single but very important piece of advice is given here. Students tend to ignore this from time to time, but take it from one who has "graduated from the school of hard knocks": *keep good records.* This means to keep an engineering logbook which contains records of all activities engaged in during each stage of modeling a specific problem, a listing of file names and what they contain, and the results of various procedures tested during the modeling process. There is nothing more frustrating than having to repeat previously performed work because of lost or forgotten information.

In a broad sense, the complete finite element modeling process consists of three basic stages:

1. *Preprocessing.* Using graphical and text input, this stage is where the model is created and the following are defined: geometry; elements and nodes; loads, constraints, and boundary conditions; and element geometric and material parameters. The final output of this step is files (text and/or binary) or data bases which contain the information pertaining to the entire model and the instructions necessary for the finite element analysis processor to run the problem.

2. *Processing.* In this stage, the finite element analysis processor solves the problem and creates additional files (text and/or binary) or appends the data base with the information on nodal deflections, element stresses, etc.

3. *Postprocessing.* The output information generated by the processor enables the analyst to review the results in graphical and text mode. In the graphical mode the analyst can view exaggerated scaled deflection plots of the structure either alone or superimposed on the undeflected structure. View control enables the analyst to change views so as to assist in the visualization of the three-dimensional structure on the two-dimensional computer monitor. Contour or dithered color carpet fringe plots allow the analyst to easily visualize the change of stresses, strains, deflections, etc., throughout the structure. In the text

mode the analyst can very quickly review the "echo" of the input data to ascertain whether any mistakes were made when creating the model. Typically, displacement and stress output is also available in text format. Furthermore, the analyst can review status messages or, if requested, the details on the element and system matrices.

The remainder of this chapter is organized to follow the above process and expand upon the various considerations the analyst must confront.

10.1 PLANNING AND CREATING THE FINITE ELEMENT MODEL (PREPROCESSING)

This and the next four sections are devoted to the considerations made in the preprocessing stage of the finite element process. When planning an analysis, several questions should be posed in the very beginning. The questions should be focused on what information is being sought from the analysis: displacements and/or stresses, accuracy, detail, etc. The answers to these questions will control the complexity of the model(s) necessary to complete the analysis.[†] Probably the most difficult process the beginning modeler must learn is how to *simplify* a complex structure. Invariably, within the limitations of a finite element analysis (due to real and time and/or cost constraints) a real structure cannot be modeled exactly. The modeler must learn how to arrive at the simplest model which effectively and efficiently provides the answers the modeler searches for. Building the model in stages of complexity may prove worthwhile. The beginning modeler is normally impatient, desires an immediate solution, and goes for a complicated model from the very start. Actually, this approach generally wastes much modeling time going down "blind alleys." The results from a sequence of models, starting with a simpler representation of the structure, can be used to continuously improve the model. Often, too, modelers are faced with attempting to simulate the behavior of some part of a structure using an approach that they are unfamiliar with. Another key suggestion here is to use the "kiss" method—keep it simple, silly! That is, do not attempt to try something new on a very complicated structural system. Create a simple model which can test your hypotheses of this *new* approach. It may even be necessary to experimentally validate the results of this simple model test. When confident that the behavior is understood, only then should you incorporate it in the actual model.

In the initial stage of modeling, one must very carefully study the structure to be analyzed and attempt to visualize its behavior. The stages of planning and creating the model include:

- Selection of the element type(s) and mesh strategy (Sec. 10.2).

- Load simulation (Sec. 10.3).

[†] Yes, more than one model may be necessary in a structural analysis. See, for example, Sec. 10.2.6 on submodeling.

- Constraint simulation (Sec. 10.4).
- Preprocessing checks (Sec. 10.5).

Certainly, the first three items on the list are interrelated and cannot be performed independently. However, load and constraint simulation cannot be finalized until the selection of the element type(s) is made.

Probably one of the first decisions made in the modeling process concerns what units to use. The analysis processors are capable of using any units desired. However, be completely consistent with whatever units you do use. Your work environment may dictate the units such as that utilized in the design phase of the structure. For example, if you are building the model from a CAD file in which the design dimensional units are given in millimeters, it is unnecessary to change the system of units or to scale the model to units of meters. However, if the input forces are in Newtons in this example, then the output stresses will be in N/mm^2 or MPa. If applied moments are to be specified in this example, the units should be N-mm. For deflections in this example, the modulus of elasticity E should also be specified in MPa and the output deflections will be in millimeters. The preprocessor software may use default values for specific entities. For example, the gravity constant may be preset to a default value of $g = 386.4$ in/s^2 to accommodate gravity loading of a structure. If SI units are being used, the default will need to be changed to $g = 9.81$ m/s^2 or 9.81 (10^3) mm/s^2 if gravity loading is to be specified. Other default settings may be present in the preprocessor, and it is the user's responsibility to be aware of them.

10.2   ELEMENT SELECTION AND MESH STRATEGY

10.2.1   INTRODUCTORY REMARKS

By nature, each region of a structural system is three-dimensional with boundary surfaces of various geometric forms (planar, cylindrical, quadric, etc.). The finite elements that are used to model these regions must provide a *reasonable* representation of these boundaries and the continuously varying deflection, strain, and stress fields. If only one element type could be used in a finite element analysis, the tetrahedron would be the only element capable of meshing a general three-dimensional structure.[†] The simple tetrahedron element (see Table 9.0-1) has only translational degrees of freedom in three directions at each of four nodes and is the three-dimensional equivalent of the two-dimensional CST element. That is, the element is a constant strain element with flat sides. As was the case of the CST element, the simple tetrahedron element is not a particularly good element. To adequately model the geometry and stress fields of a general three-dimensional structure utilizing simple tetrahedrons, an enormous number of elements is normally necessary. Higher-

---

[†] The first truly automeshing software was developed to mesh solid structures with tetrahedrons.

order tetrahedrons with midside nodes could be used where the sides and shape functions would be capable of parabolic behavior. Fewer elements would be necessary to model a given geometry, but limiting oneself to one type of element severely restricts the capability of modeling a given structural problem efficiently. Thus a full understanding of the structural problem matched with a proper element selection from the code's element library will result in the optimum usage of FEA.

As indicated in Table 9.0-1, there are three basic geometric categories of elastic element that can be used in modeling a structure. They are *line elements* (truss, beam, frame—where area properties are specified in text format), *surface elements* (plane elastic, membrane, plate, shell—where thickness is specified in text format[†]), and *solid elements* (brick, wedge, tetrahedron, thick shell—where the geometry is completely specified by the node locations). A structure can be modeled using one type of element exclusively or different element types can be employed in the same model provided certain rules concerning the degrees of freedom (*dof*) are followed. These rules are discussed in Sec. 10.2.2.

Consider, for example, a simple beam structure such as that shown in Fig. 10.2-1. Modeling this type of structure using tetrahedrons would require an unnecessarily large number of tetrahedron elements where only two beam elements might do the job. Two beam elements might be adequate if one were concerned about nominal stresses and the deflections of the centroidal axis of the beam. However, if one sought more detail on the stresses and deflections throughout the beam structure such as that due to the stress concentration at the step at point B or the beam/support interaction at point C, different elements would be necessary and more modeling detail would need to be provided.

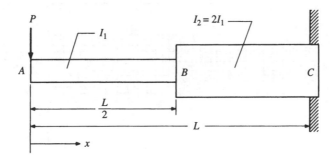

Figure 10.2-1   Step beam.

[†] The two-dimensional triangular and quadrilateral elastic element is used to model plane stress, plane strain, and axisymmetric problems. The thickness is only specified for plane stress applications. The software typically uses unity for thickness if plane strain is specified. For axisymmetric applications, the finite element equations are based on rotating the triangular or quadrilateral about the axis of symmetry 360° (see any textbook on finite elements for a derivation of the element equations for the axisymmetric element).

For example, consider that the stresses in the vicinity of the step at point B are to be investigated. If the beam is thin of constant thickness, two-dimensional quadrilateral plane-stress elements could be used effectively.[†] Furthermore, it is necessary to model a fillet at the step. Without this, the maximum stress predicted by the analysis at the internal corners of the step would continually increase as the mesh density (number of elements used) is increased. That is, the solution does not converge. This is due to a theoretically infinite stress condition when a fillet is not present. In reality, a fillet must exist, no matter how small, and must be included if detail concerning the stresses at the step is required. The minimum element size would be based on the size of the fillet. A fillet is unnecessary at external corners since for the plane stress condition the stresses are zero at external corners. A reasonable mesh of quadrilateral elements is shown in Fig. 10.2-2, where 388 elements are used. Note the following characteristics of the mesh pattern:

1. All elements are properly connected.[‡] That is, the nodes of adjacent elements must be common to both elements.
2. The mesh is uniform throughout the model.
3. The elements gradually reduce in size to a minimum where the stress concentration occurs at the base of the fillet.
4. In the vicinity of the stress concentration the elements are small, nearly square, and uniformly follow around the contour of the fillet.

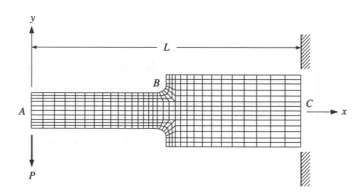

Figure 10.2-2   A two-dimensional quadrilateral mesh of the step beam.

[†] Brick elements or plate elements which support out-of-plane bending could also be used. However, the additional dof that these elements support will not change the results obtained using the 2-D quad element. Thus we use the simplest element that will do the job.

[‡] See item 1 of the last paragraph of this section.

The quadrilateral element behaves best numerically when it is square. Squareness implies:

1. The length of the sides are near equal. That is, the *aspect ratio* (the ratio of the length of the long side to the length of the short side) is near unity. Element performance deteriorates as the aspect ratio increases. Some software packages can monitor the aspect ratio.

2. The interior angles of the quadrilateral are near 90°. A measure of this is called *skew*. Some software packages can also monitor this characteristic.

Near the ends of the beam in Fig. 10.2-2 the aspect ratio of the elements is of the order of 4. More elements can be used to reduce the aspect ratio if better accuracy is required at these locations. However, it was stated earlier that detailed stress information at the stress concentration location was the goal. The inaccuracy of the stresses determined at points A and C will not affect the results at the step provided points A and C are at reasonable distances from the step.[†]

If the beam had a circular cross section, solid elements would be necessary. Rather than using tetrahedron, parallelepiped brick elements would be preferred (see Table 9.0-1). The mesh would be similar to Fig. 10.2-2; however, the additional sides of the brick elements would be obtained by rotating the vertices shown in Fig. 10.2-2 about the x axis. The top and bottom rows adjacent to the x axis would need to be wedge elements with the "peak" of the wedge lying on the x axis. If the rotational increment of the vertices of Fig. 10.2-2 was 15°, a total of 9312 elements would result. Using symmetry, this model could be cut in 1/4, resulting in 2328 elements. However, specific boundary conditions at the surfaces of symmetry would be necessary. Symmetry is discussed in Sec. 10.2.7.

If the beam had a circular cross section but was loaded axially in the x direction at A, the same mesh used for the two-dimensional plane stress elements (Fig. 10.2-2) can be utilized. This is because the problem is now axisymmetric, and instead of the plane-stress element, the *axisymmetric* element can be specified. Using a cylindrical coordinate system, the x axis in Fig. 10.2-2 becomes the (longitudinal) axis of symmetry, the y axis becomes the radial axis, and the axis perpendicular to the plane to the page is the tangential axis. Mathematically, the finite element equations for the axisymmetric element are based on rotating the quadrilateral about the longitudinal axis 360° (see Sec. 10.2.7 or any textbook on finite elements for a derivation of the equations for the axisymmetric element).

Modeling the forces and boundary conditions in the above examples is discussed in Secs. 10.3 and 10.4, respectively.

Finally, a brief comment concerning element connectivity is given here. Further discussion is given in Sec. 10.2.2. Connecting elements properly can be a subtle process. Two basic situations that arise should be kept in mind:

1. *Connecting the same element type.* Nodes of connecting elements should match. For example, the four-node two-dimensional plane elastic quadrilateral elements

---

[†] It is very important to note here that Saint-Venant's principle also applies to element size and distortion inaccuracies the model itself may produce.

(Q4) shown in Fig. 10.2-3 are to be connected along line *A-A*. The purpose might be to increase the mesh density going from left to right. However node 3′ would be left "hanging" and consequently the edges 2′-3′ and 4′-3′ would only be connected to the left element at nodes 2 and 4. In order to increase the mesh density, a *transition region* would be necessary (see Sec. 10.2.5).

2.   *Connecting different element types.* There are two basic instances to consider:

(*a*) The elements have different defining shape functions. Consider the connection shown in Fig. 10.2-3 along line *B-B* where the Q4 elements are to be connected with the higher-order eight-node quadrilateral element (Q8). In this case the nodes of the connecting elements do match. However, when deformed, the edges of the Q4 elements remain linear whereas the Q8 element is defined by a parabolic shape function. Thus the deformations of the connecting elements are incompatible. A fatal error would not occur in this case, but the results are affected in this area and could be misleading if this connection was in a region of concern.

(*b*) The elements have different supported dof. In Fig. 10.2-3, a beam element is being connected to a Q4 element at node 1. The shape functions are certainly different for the two element types, but no problem occurs since there is no edge connection. However, the two elements support different dof at the connected node. The Q4 element only supports in-plane translation whereas the beam supports translation and rotation with respect to three mutually perpendicular axes at the connection node. If not handled correctly this could cause rigid-body motion and a singularity in the system stiffness matrix (see Secs. 10.2.2 and 10.4).

10.2.2 ELEMENT SELECTION

What is offered in this section are some general rules for element selection based on the geometric aspects of the structure. As was indicated in the examples of the pre-

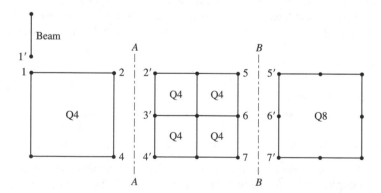

Figure 10.2-3   Unacceptable element connectivity.

vious section, loading and boundary conditions, and the information being pursued will also play a role in element selection. For linear problems, the matching of the basic elements to a structure exhibiting simple characteristics is given in Table 10.2-1. With some software, higher-order elements with midside nodes are available. Use of these elements in modeling is beyond the scope of this chapter. If these elements are available, the reader is urged to consult the vendor's technical literature and verification models and the advanced finite element textbooks cited at the end of the chapter, and to experiment with the element(s) using examples with known closed-form solutions.

Table 10.2-1   Elements suited to structural characteristics

Structure Characteristics	Preferred Basic Element for Analysis	Number of Nodes	DOF at Each Node
Two- or three-dimensional truss with pin joints	Truss	2	(3) Translation in 3 directions
Two- or three-dimensional frame [a] with fixed joints [b]	Beam [c]	2	(6) Translation and rotation in 3 directions
Two-dimensional plane stress or strain	2-D elastic quadrilateral [d]	4	(2) In-plane translation in 2 directions
Axisymmetric [e]	Quadrilateral [d]	4	(2) Translation in 2 directions [f]
Thin-walled three-dimensional structure	Quadrilateral plate [d,g]	4	(5) Translation in 3 directions and 2 rotations about axes in the plane of the element [h]
Three-dimensional solid structure	Brick [i]	8	(3) Translation in 3 directions

[a] A frame structure is a structure consisting of elements where the longitudinal dimensions are much larger than the lateral dimensions.

[b] Some joints may be hinged by using pin releases (see Sec. 9.3.3).

[c] Most commercial software packages refer to the frame element (see Secs. 9.3.5 and 9.3.6) as the *beam* element.

[d] The triangular element can be used for interior transition regions.

[e] Geometry, loading, and boundary conditions must be completely symmetric to a longitudinal axis.

[f] The element geometry, loading, and boundary conditions are defined in the plane of the radial and longitudinal axes. Stresses are reported in three directions.

[g] The nodes of the quadrilateral plate element are defined at the midplane of the wall of the structure. The element can be used to model curved surfaces. For single curvature, such as a thin-walled cylinder, the element can be formed in the tangential and longitudinal directions so that the nodes of the element lie in a single plane. For double-curvature, the nodes may not lie in a single plane. A measure of this is called *warp*. Warp affects accuracy. By proper node placement and element size, warp should be kept to a minimum (ideally zero).

[h] Rotation perpendicular to the plane of the element may be handled differently by the various software products. Refer to the software vendor's technical literature.

[i] The four-noded tetrahedron and six-noded wedge elements can be used for interior transition regions.

Typically, a structure may exhibit mixed characteristics where the combination of more than one element type is advantageous. The basic concern when connecting nodes of dissimilar elements is the incompatibility of the degrees of freedom that each element supports. For example, consider the two structures shown in Fig. 10.2-4(a) and (b) each containing a beam and cable(s). The cables can be modeled by one of two methods. The first method would be to model the cables as beams with rotational pin releases[†] about the z axis at the connection with the actual beam and appropriate boundary conditions at the ground supports. Here, a second-area moment would need to be entered for the cable. However, it really doesn't matter what the value is because the pin release and pinned boundary condition will eliminate bending of the cable.

The second method models the cables as truss elements. This automatically eliminates bending of the cables. However, connecting a truss element to a beam element results in a mismatch in the dof supported by the elements at the node. With this approach, consider Fig. 10.2-4(a), where the structure is modeled using two beam elements each of length $L_B/2$ for member BC and one truss element for member CD. For the fixed support of the beam at the wall all six (three translation and three rotation) dof would be set to zero at point B, whereas for the truss element all three translation dof at point D would be set to zero. The fact that there is a mis-

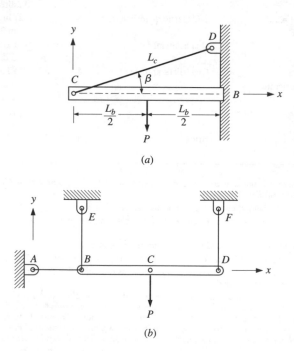

(a)

(b)

Figure 10.2-4   Beam/cable examples.

† See Sec. 9.3.3.

match of supported dof for the beam and truss at the connection point C causes no problem since the beams are fully constrained through the support at point B.

Now consider Fig. 10.2-4(b), with the three cables modeled by truss elements. Here the support constraints would be to fix the three translations of the nodes at points A, E, and F. If nothing else is prescribed, a fatal error will occur owing to improper restraint of the beam's dof. The trusses are incapable of restraining the rigid body freedom of the beam's rotation about the x and y axes and translation in the z direction. To remedy this, place a set of boundary conditions on any node of the beam (B, C, or D) fixing z translation, and rotation about axes parallel to the x and y directions.[†]

Another example of incompatibility of dof when connecting different element types is shown in Fig. 10.2-5. Consider the structure in Fig. 10.2-5(a) to be in a state of plane stress. The structure shows two geometric characteristics, a large rectangular section with a low aspect ratio and an appendage, where the force P is applied, with a very high aspect ratio. The large rectangular section must be discretized in the x and y directions. Based on the state of plane stress, this calls for the use of the quadrilateral element as shown in Fig. 10.2-5(b). The long, thin appendage could be

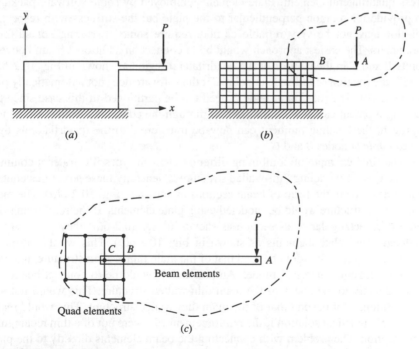

Figure 10.2-5

---

† Some software packages have features which will automatically correct the problem and will partition all unconstrained rigid-body dof. Care must be taken, however, when using this feature as an unconstrained rigid-body dof may be due to a modeling error.

modeled using quadrilateral elements; however, this would lead to a similar situation as in Fig. 10.2-2 where a fillet would need to be added if stress detail at the connection of the appendage and the block was required. This, in turn, would result in an extremely large number of elements. If accurate deflection information was desired, with less stress detail, a simpler approach would be to model the appendage using one or more beam elements. This is indicated by line AB in Fig. 10.2-5(b). Thus the connection at node B would be between a beam element with all six dof and a quadrilateral element with only two translation dof (TX and $TY^\dagger$) at the node. The bending moment necessary at node B of the beam cannot be transmitted to the quadrilateral. This, in turn, creates a mechanism joint which would lead to dynamic rigid-body motion, a singular system stiffness matrix, and a fatal error. If the software automatically partitions out the rigid-body dof, the rotation at node B would be set to zero. This means that only the force P would be transmitted to the structure modeled with quadrilaterals. An answer would result, but it may be quite conservative depending on the length of the appendage. Another alternative would be to model the rectangular region with quadrilateral plate elements. Table 10.2-1 indicates, however, that rotation perpendicular to the plane of the plate is not supported. This will result with the same problem as that experienced when using the plane stress quadrilateral element. Plate elements employed by some software packages do provide for rotation perpendicular to the plate but the stiffness with respect to this dof may not be very reliable or may require some "tweaking" of stiffness parameters. The easiest approach would be to connect an additional beam between points B and C in the interior of the quadrilateral region as shown in Fig. 10.2-5(c) and fix the rotation of the node at point C if the software does not automatically partition the dof. The rotation at node B need not be partitioned in this case since the bending moment can now be transmitted through the connected beams. Using this approach, the bending moment can now be transferred to the quadrilaterals by a force *couple* at nodes B and C.

The final example of combining different element types illustrates a common practice used in modeling thin-walled structures. Generally these structures contain reinforcing ribs in the form of beam sections as shown in Fig. 10.2-6(a). The main thin-walled structure would be modeled using plate elements. If the reinforcing ribs had a thin rectangular cross section as shown in Fig. 10.2-6(a) they too could be modeled using plate elements as shown in Fig. 10.2-6(b). This would introduce many additional nodes (dof) beyond that of the main thin-walled structure and consequently a large numerical model. An alternative would be to connect beam elements directly to the nodes of the main thin-walled structure. This would not add any additional dof beyond that of the main thin-walled structure. This would really be the only practical solution if the reinforcement ribs were not of a thin rectangular cross section. The problem with connecting the beam elements directly to the plate nodes is that the centroid of the beam would lie on the midplane of the plate structure. The centroids of the reinforcement ribs are offset from the plate structure in the

---

† For simplicity, the notation TX, TY, and TZ can represent the translation dof of a node in the global x, y, and z directions, respectively. Likewise, RX, RY, and RZ can represent the rotation dof about axes parallel to the global x, y, and z directions, respectively.

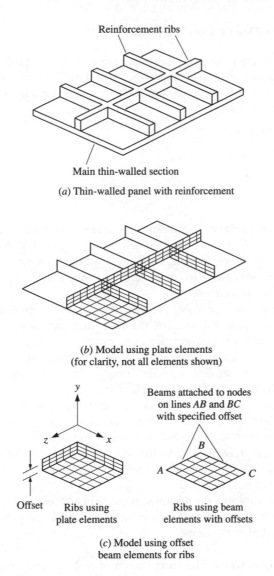

Reinforcement ribs

Main thin-walled section

(a) Thin-walled panel with reinforcement

(b) Model using plate elements
(for clarity, not all elements shown)

Beams attached to nodes
on lines AB and BC
with specified offset

Offset   Ribs using
plate elements

Ribs using beam
elements with offsets

(c) Model using offset
beam elements for ribs

Figure 10.2-6   Modeling a thin-walled panel containing reinforcement ribs.

positive y direction as shown in Fig. 10.2-6(c). To solve this problem, the beam element as supplied by most software has the capability of offsetting the centroidal axis of the beam from each node a prescribed distance in any direction. Thus, for the case of the rectangular rib given, each beam connected to each node along lines AB and BC of Fig. 10.2-6(c) would need to have offsets specified in the y direction equal to 1/2 the height of the rib. Since the beams are interconnected, the mismatch in supported dof will not be a problem.

10.2.3 ELEMENT INPUT INFORMATION

Each element type has its unique input requirements in the preprocessing stage. The element information is entered either by "pull-down" menus in the software's graphic preprocessing, by a spreadsheet form, or in a text mode. Generally, the categories of information are:

1. Geometric properties.
2. Material properties.
3. Element loading.

Item 3 will be discussed in Sec. 10.3.2.

Geometric Properties   Spatially, an element is defined by its nodes. Additional information which completes the geometric definition of the element might be necessary. For example, the longitudinal axis of the truss element is established by the location of two nodes. To complete the geometric definition of the truss element, the area must be entered.

The beam element is probably the most complicated when dealing with specifying additional element information. The local longitudinal axis is defined from node i to node j. Additional information is necessary to define the transverse principal axes of the second-area moments. Normally this is done by defining a third point or node in space (see Sec. 9.3.6 for further elaboration). Once this is done the various area properties can be entered. These properties include the axial area, the effective shear area,[†] the principal second-area moments relative to the axes defined as described in Sec. 9.3.6, the effective axial second-area moment,[‡] and either the section moduli relative to the transverse axes or location points in the transverse plane for stress output. Depending on the application, further information might include pin releases relative to any of six dof at a node (see Sec. 9.3.3) and/or nodal offsets as discussed in Sec. 10.2.2.

Beyond node location, the thickness is the only additional geometric definition to specify for the surface elements such as the two-dimensional plane elastic and the three-dimensional plate elements.

Material Properties   The specification of material properties, using proper units, is dependent on the element type and the element load factors. At a minimum, the modulus of elasticity E must be specified for an element (the truss element). If the material is isotropic, Poisson's ratio v *or* the shear modulus G must be specified for the remaining element types.[§] If the software is capable of handling an orthotropic material for a specific element type, the orientation of the principal axes

---

[†] This is to account for shear deflection in the case of beams with lateral dimensions of the order of the axial dimension. The shear area is given by A/k, where A is the cross-sectional area and k, the form correction factor, is given in Table 6.3-1.

[‡] For J_{eff} see Secs. 3.3.1, 4.3.2, 5.2, or Tables 20 and 21 of Ref. 10.2.

[§] Recall, for isotropic materials, there is a relationship between E, v, and G. Typically, the software will only allow you to enter E and v or G.

of the material for the element must be specified as well as the values of E, G, and ν relative to each axis.

Element loads such as gravity and thermal loading necessitate additional material properties of mass (or weight) density ρ (or γ) and the thermal expansion coefficient α, respectively. If the software is capable of handling an orthotropic material for a specific element type, the orientation of the principal axes of the material for the element must be specified as well as the values of α relative to each axis.

10.2.4 MESH GENERATION

The network of elements and nodes that discretize a region is referred to as a *mesh*. The *mesh density* increases as more elements are placed within a given region. *Mesh refinement* is when the mesh is modified from one analysis of a model to the next analysis to yield improved results. Results generally improve when the mesh density is increased in areas of high stress gradients and/or when geometric transition zones are meshed smoothly. Generally, but not always, the FEA results converge toward the exact results as the mesh is continuously refined. To assess improvement, there are methods to estimate the order of error in a given analysis. These methods range from evaluating the maximum stress differences between predicted values at the common nodes of adjacent elements relative to some maximum global stress value to estimating errors in element and global strain energy using the element-by-element and averaged (smoothed) stress fields (see Ref. 10.3, chap. 14). The idea is to minimize these differences.

Thus, as one can plainly see, mesh generation and regeneration are by far the most time-consuming facets of the preprocessing phase. There are three basic ways to generate an element mesh, manually, semiautomatically, or fully automated.

Manually Editing an Input Text File This is how the element mesh was created in the early days of the finite element method. This is a very labor-intensive method of creating the mesh, and except for some quick modifications of a model it is rarely done. Note: Care must be exercised when editing an input text file. With some FEA software, other files such as the preprocessor binary graphics file may not change. Consequently, the files may no longer be compatible with each other.

Semiautomatic Mesh Generation Over the years computer algorithms have been developed which enable the modeler to automatically mesh regions of the structure that have been divided up using well-defined boundaries. Since the modeler has to define these regions, the technique is deemed *semiautomatic*. Owing to space limitations, further discussion of this will be confined to only one well-known technique which uses shape functions to mesh two-dimensional quadratic regions. The development of the many computer algorithms for mesh generation emanates from the field of computer graphics. Readers who desire more information on this subject should review the literature available in this field.

Consider the four-sided region shown in Fig. 10.2-7. Let each boundary be defined by a parabolic function which can be represented by three points, called the

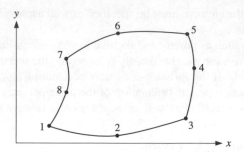

Figure 10.2-7   Four-sided quadratic region with boundary edge points.

boundary edge points. We will map this region into the same ξ,η space as was done for the quadrilateral element of Fig. 9.4-3. Similar to the development of the shape functions for the isoparametric quadrilateral discussed in Sec. 9.4.2, the shape functions for the quadratic region of Fig. 10.2-7 in ξ,η space are

$$N_1 = -\frac{1}{4}(1 - \xi)(1 - \eta)(\xi + \eta + 1) \qquad N_2 = \frac{1}{2}(1 - \xi^2)(1 - \eta)$$

$$N_3 = \frac{1}{4}(1 + \xi)(1 - \eta)(\xi - \eta - 1) \qquad N_4 = \frac{1}{2}(1 - \eta^2)(1 + \xi)$$

$$N_5 = \frac{1}{4}(1 + \xi)(1 + \eta)(\xi + \eta - 1) \qquad N_6 = \frac{1}{2}(1 - \xi^2)(1 + \eta)$$

$$N_7 = -\frac{1}{4}(1 - \xi)(1 + \eta)(\xi - \eta + 1) \qquad N_8 = \frac{1}{2}(1 - \eta^2)(1 - \xi) \quad \textbf{[10.2-1]}$$

Similar to Eqs. (9.4-25), to determine points within the *xy* region of Fig. 10.2-7 in terms of ξ and η, we use the equations

$$x(\xi,\eta) = \sum_{i=1}^{8} N_i x_i \qquad\qquad \textbf{[10.2-2a]}$$

$$y(\xi,\eta) = \sum_{i=1}^{8} N_i y_i \qquad\qquad \textbf{[10.2-2b]}$$

where the N_i are the shape functions given by Eqs. (10.2-1) and the (x_i, y_i) are the coordinates of the boundary edge points.

To create a uniform mesh, we divide the region in ξ, η space uniformly obtaining a set of values of (ξ, η). The shape functions for each (ξ, η) are determined from Eqs. (10.2-1) and then substituted into Eq. (10.2-2) to obtain the corresponding values of (x, y). Geometric spacing also can be employed in ξ, η space. This was used in the end regions of Fig. 10.2-2 to provide a gradual change in element size.

Example 10.2-1

The coordinates of the boundary edge points of the four-sided region of Fig. (10.2-7) are

$$(x_1, y_1) = (1, 2) \quad (x_2, y_2) = (3, 1) \quad (x_3, y_3) = (5, 3) \quad (x_4, y_4) = (5.2, 9)$$

$$(x_5, y_5) = (5, 13) \quad (x_6, y_6) = (3, 13) \quad (x_7, y_7) = (1.5, 10) \quad (x_8, y_8) = (1.5, 6)$$

Mesh the region uniformly with four-noded quadrilaterals, four elements along sides 1-2-3 and 5-6-7, and three elements along sides 1-8-7 and 3-4-5.

Solution:

The grid for $-1 \leq \xi \leq 1$ will be divided by 4. Thus ξ will take on the values of -1, $-0.5, 0, 0.5$, and 1. The grid for $-1 \leq \eta \leq 1$ will be divided by 3. Thus η will take on the values of $-1, -1/3, 1/3$, and 1. The grid mesh in ξ, η space is shown in Fig. 10.2-8(a). Table 10.2-2 gives the grid coordinates of all points in the mesh. Substitute a specific set of (ξ, η) coordinates into Eq. (10.2-1) to determine the corresponding shape functions. Then substitute these shape functions and the given boundary edge point coordinates into Eq. (10.2-2). The result is the (x, y) coordinates associated with the (ξ, η) coordinates. A sample calculation is given for the grid point $(\xi, \eta) = (0.5, 1/3)$.

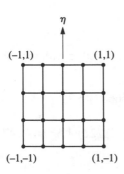

(a) Transform ξ, η space

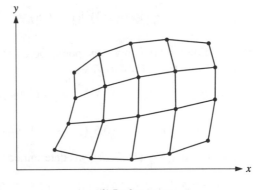

(b) Real xy space

Figure 10.2-8 Mesh generation within a quadratic region.

Table 10.2-2   Mesh (ξ, η) coordinates

ξ	-1	-0.5	0	0.5	1	-1	-0.5	0	0.5	1
η	-1	-1	-1	-1	-1	$-1/3$	$-1/3$	$-1/3$	$-1/3$	$-1/3$

ξ	-1	-0.5	0	0.5	1	-1	-0.5	0	0.5	1
η	$1/3$	$1/3$	$1/3$	$1/3$	$1/3$	1	1	1	1	1

$$N_1 = -\frac{1}{4}(1 - 0.5)(1 - 1/3)(0.5 + 1/3 + 1) = -0.15278$$

$$N_2 = \frac{1}{2}(1 - 0.5^2)(1 - 1/3) = 0.25$$

$$N_3 = \frac{1}{4}(1 + 0.5)(1 - 1/3)(0.5 - 1/3 - 1) = -0.20833$$

$$N_4 = \frac{1}{2}[1 - (1/3)^2](1 + 0.5) = 0.66667$$

$$N_5 = \frac{1}{4}(1 + 0.5)(1 + 1/3)(0.5 + 1/3 - 1) = -0.08333$$

$$N_6 = \frac{1}{2}(1 - 0.5^2)(1 + 1/3) = 0.5$$

$$N_7 = -\frac{1}{4}(1 - 0.5)(1 + 1/3)(0.5 - 1/3 + 1) = -0.19444$$

$$N_8 = \frac{1}{2}[1 - (1/3)^2](1 - 0.5) = 0.22222$$

Substituting the boundary edge point coordinates given for x_i together with the values of N_i into Eq. (10.2-2a) gives

$$x = (-0.15278)(1) + (0.25)(3) + (-0.20833)(5) + (0.66667)(5.2)$$

$$+ (-0.08333)(5) + (0.5)(3) + (-0.19444)(1.5) + (0.22222)(1.5) = 4.147$$

Likewise, using Eq. (10.2-2b), y is determined to be

$$y = (-0.15278)(2) + (0.25)(1) + (-0.20833)(3) + (0.66667)(9)$$

$$+ (-0.08333)(13) + (0.5)(13) + (-0.19444)(10) + (0.22222)(6) = 10.125$$

Table 10.2-3 gives the (x, y) coordinates which correspond to the (ξ, η) coordinates in Table 10.2-2. Figure 10.2-8(b) shows the resulting mesh in real xy space. Note that since quadrilaterals are straight-sided, once the grid coordinates are established, they are connected by straight lines.

Table 10.2-3 Mesh (x, y) coordinates

x	1	2	3	4	5	1.388	2.273	3.2	4.168	5.177
y	2	1.125	1	1.625	3	4.667	4.931	5.444	6.208	7.222

x	1.556	2.336	3.2	4.147	5.178	1.5	2.188	3	3.938	5
y	7.333	8.514	9.444	10.125	10.556	10	11.875	13	13.375	13

Triangular regions can be meshed in an identical fashion simply by the degeneration of a side into a common vertex. For example, the upper side of the region shown in Fig. 10.2-7 can be eliminated by making the coordinates of points 5, 6, and 7 equal (see Problem 10.2).

Fully Automated Mesh Generation   Many software vendors are concentrating their efforts on developing fully automatic mesh generation and, in some instances, automatic *self-adaptive* mesh refinement. The obvious goal is to significantly reduce the modeler's preprocessing time and effort to arrive at a final well-constructed FEA mesh. Once the complete boundary of the structure is defined, without subdivisions as in semiautomatic mesh generation and with a minimum of user intervention, schemes such as *advancing front (paving)*[†] or *recursive spatial decomposition (octree)*[‡] techniques are employed to discretize the region with *one element type*. For plane elastic problems the boundary is defined by a series of internal and external geometric lines and the element type to be automeshed would be the plane elastic element. For thin-walled structures, the geometry would be defined by three-dimensional surface representations and the automeshed element type would be the three-dimensional plate element. For solid structures, the boundary could be constructed using *constructive solid geometry* (CSG) or *boundary representation* (B-rep) techniques. In the case of three-dimensional structures the finite element types for automeshing would be the brick and/or tetrahedron.

Automatic self-adaptive mesh refinement programs estimate the error of the FEA solution, as discussed earlier. Based on the error, the mesh is automatically revised and reanalyzed. The process is repeated until some convergence or termination criterion is satisfied.

Modeling with fully automated software is highly dependent on the specific code used. Since the primary intent of this chapter is to concentrate on basic modeling techniques, further discussion of automatic mesh generation will not be pursued. It is highly recommended that the beginning modeler concentrate on the fun-

---

† See, for example, Ref. 10.4.
‡ See, for example, Ref. 10.5.

damentals and become comfortable with the semiautomatic meshing techniques before applying fully automated procedures. As stated earlier, the automatic mesh generation schemes are used with one element type and may work fine for specialized problems. However, in many engineering applications, the full model cannot be completed with automated procedures alone. In these cases, it is important that the analyst have the skills necessary to develop the final model.

10.2.5 Two-Dimensional Meshing Strategies

In this section we will briefly describe some basic techniques employing semiautomatic mesh generation to mesh transitions throughout a general two-dimensional plane elastic region. The philosophy described can also be extended somewhat to three-dimensional modeling.

The author once asked an experienced FEA modeler how he judged whether a mesh was good or bad. The answer was "the mesh is good when it *looks* good." Well, what makes a mesh look good? Two basic observations concerning the relation between a finite element representation of a continuous structure are (1) a finite element mesh is generally more rigid than the continuous model, and (2) the maximum stress for most structural problems occurs at boundaries. The more that an element mesh conforms smoothly to the boundaries of the model, and the elements uniformly reduce in size and are closer to being square, the less rigid the model becomes and the results more closely approximate the continuous model. Rapid changes in mesh density of the model create "stress raisers" that may not exist in the real structure. If these stress raisers occur in areas of little concern and are far enough away from areas of interest, the analyst should realize it and ignore the results at these locations. So, in a nutshell, a mesh will *look good* if it flows smoothly around boundaries with small elements where the boundaries have small radii of curvature and there are gradual changes in the mesh density. For curvature in two-dimensional applications, elements should be no larger than 15° in arc length.

Note that the characteristics of the mesh for the step plate shown in Fig. 10.2-2 satisfy the criteria of a good mesh. Figure 10.2-9 shows the same step shaft with a mesh generated by an automated program called Supergen™ produced by Algor, Inc. To create the mesh, the boundary was drawn and three parameters were set: minimum angle for the division of arcs = 15°, geometric ratio which defines the maximum approximate size ratio of adjacent elements = 1.25, and the approximate number of elements = 300. The resulting mesh shown, completely generated automatically, contains 508 elements. Again, note the characteristics of the mesh. The program uses a paving technique which first meshes the boundaries and then progresses inward from the boundaries. Notice the symmetry and the well-shaped elements along the boundary. Well within the plate the mesh symmetry disappears and there are a few interior elements that have some skew problems. However, this occurs at locations well removed from areas of concern.

For semiautomatic mesh generation, the key is to reduce the interior of the structures to three- or four-sided well-defined regions where the edges, boundary edge points, and the corresponding mesh density of the regions match. If quadratic regions are used to mesh around holes or arcs, for accuracy, no more than 45° of the arc should be used for a boundary edge. This is because we are representing the equation of a circle by a parabolic equation. Figure 10.2-10(*a*) illustrates how the region around a hole is meshed. First a square box centered with the hole is drawn. The circle is divided by increments of 45° and lines are drawn from the circle to the square. This creates eight quadratic regions. The boundary edge points of the upper left quadratic region are shown in Fig. 10.2-10(*a*). For illustration, the quadratic region is then meshed by a 3 × 4 mesh as shown in Fig. 10.2-10(*b*). If the software is capable of representing a circular edge exactly, then it is only necessary to divide the region into four subregions and meshed as shown in Fig. 10.2-10(*c*) and (*d*). The use of the computer-aided graphic methods of *mirroring* and *copy-rotate* will shorten this procedure. Once the hole is meshed, it is now easier to connect the edges of the box to other boundaries within the structure.

The mesh given in Fig. 10.2-2 was generated by dividing the top half of the step plate into the regions shown in Fig. 10.2-11. The process begins by first identifying

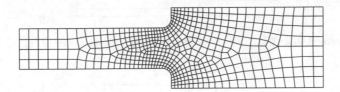

Figure 10.2-9   Step beam modeled with an automated mesh generator Supergen, a product of Algor, Inc., Pittsburgh, PA.

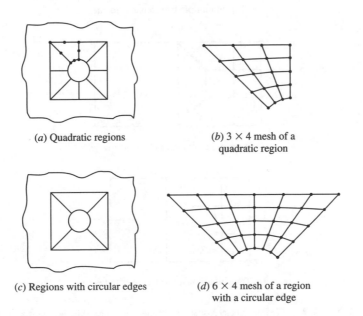

(*a*) Quadratic regions

(*b*) 3 × 4 mesh of a quadratic region

(*c*) Regions with circular edges

(*d*) 6 × 4 mesh of a region with a circular edge

Figure 10.2-10   Meshing a region around a hole.

the curved surfaces. Using a technique similar to that shown in Fig. 10.2-10(*a*), the fillet is "boxed" by straight sides. The fillet in Fig. 10.2-11 is simply 1/4 of the regions shown in Fig. 10.2-10(*a*). The remainder of the straight boundaries together with the box lines for the fillet are joined filling the region. Finally, each quadratic region is automeshed where the number of elements along common edges are identical. Some of the regions in Fig. 10.2-2 were meshed using geometric rather than uniform spacing. After meshing, the entire result was mirror-copied about the *x* axis to create the final mesh. Depending on the software, it may be necessary to eliminate duplicate nodes or edges that may exist from mesh and copy operations.

For mesh density transitions, various techniques can be employed. Figure 10.2-12 shows some common methods.

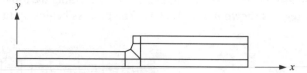

Figure 10.2-11   Quadratic regions to mesh the step plate of Fig. 10.2-2.

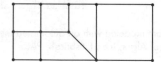

(*a*) Transition from two elements to one

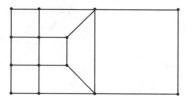

(*b*) Transition from three elements to one

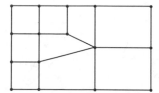

(*c*) Transition from three elements to two

Figure 10.2-12   Mesh density transitions in the horizontal direction (left to right) using quadrilateral elements.

10.2.6 SUBMODELING

When following the mesh strategies outlined in the previous section to model a complex structure, many transitions may occur in various regions with high stress gradients. The model will require much modeling time and inevitably contain a great many elements. Another approach to this, called *submodeling,* employs separate models. The first model is called the global model and is of a coarser mesh than what might have been created using good meshing procedures. The global model should be good enough to model everything satisfactorily other than within the regions of high stress gradients. The remaining models, called the *submodels,* only represent the high stress gradient regions and are meshed more finely than the global model. The boundaries of each submodel should match the boundaries isolated from within the global model with the exception of the mesh density and should be located sufficiently far away from the high stress gradients according to Saint-Venant's principle. The global model is solved first. Then stresses or deflections and boundary conditions consistent with the global model at the isolation boundaries are applied to the submodels. Since, for a given boundary, the mesh densities of the global model and the submodel are different, interpolation of the applied stresses or displacements is necessary. For this reason, all nodes of the global model along the isolation boundaries should be included in the submodel.

For illustration, consider the example of Fig. 10.2-2 modeled with plane-stress quadrilaterals. Using submodeling, the model shown in Fig. 10.2-13(*a*) serves as

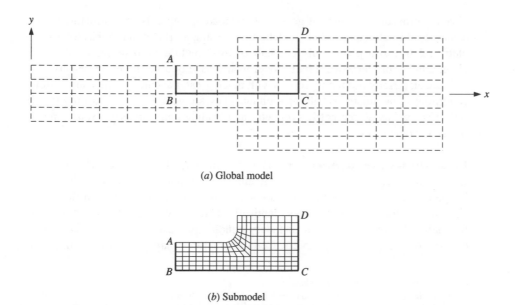

(*a*) Global model

(*b*) Submodel

Figure 10.2-13 Submodeling the step shaft of Fig. 10.2-2.

the global model where the elements are shown in dotted lines.[†] This model contains 96 solid elements as compared with the 388 elements for the model shown in Fig. 10.2-2. Here no effort is made to model the fillet in the global model. The submodel will match the isolation boundaries *AB, BC,* and *CD* of the global model. The submodel is modeled with 147 elements as shown in Fig. 10.2-13(*b*). A number of different approaches can be taken to apply loading and boundary conditions. The simplest approach would be to apply deflections (or relative deflections) along the isolation boundaries using the results obtained from the analysis of the global model. However, since there are more nodes along the isolation boundaries of the refined model as compared to the global model, interpolation of the results from the global model must be performed.

As one can see, submodeling is not as simple to apply as it might seem from a casual description of the technique. This approach is not a remedy for all FEA problems. Much thought and care should be exercised when defining the global model, as it must provide sufficient accuracy to provide good results at the isolation boundaries. The isolation boundaries in the global model should be in regions where displacements or stresses are relatively uniform or linear such that the interpolation needed for the refined model is accurate. Finally, the stresses at the isolation boundary for the refined and global models should be compared. The coarse global model will be stiffer than reality and the deflections and stresses may be underestimated at the isolation boundaries.

10.2.7 SYMMETRY

Various forms of symmetry of geometry and loading can exist in a structure which can be used to advantage when modeling. For example, if a structure has total symmetry relative to a plane, only one-half of the structure need be modeled. For a given number of elements, a half model would have double the mesh density of a full model, thereby producing results with much greater accuracy. There are various conditions of symmetry worth mentioning. They are geometric and loading (total) *reflective symmetry,* geometric symmetry and loading *antisymmetry, cyclic symmetry,* and *axisymmetry.*

Total Reflective Symmetry   The conditions for this are met when a plane exists such that half the structure *and* loading on one side of the plane is the mirror image of the other half of the structure and loading on the other side of the plane. Each half of the structure behaves identically. Thus, only half of the structure need be analyzed. However, certain boundary conditions must be placed on *all nodes* existing on the plane of symmetry. Certain structures may contain more than one

---

[†] Two qualifications should be stated here:
1. Aside from the stress concentration, we have a basic beam in bending and we know what the stress distribution is. Thus we really do not need the global model in this case. Here the global model is provided to demonstrate the technique.
2. The global model used here could be made simpler using symmetry. However, this is not covered until the next section, and we will try not to confuse the issue.

plane of symmetry. For example, the simple stress concentration problem of a rectangular plate in tension containing a centrally located hole has two planes of symmetry. For this case, called *one-quarter symmetry,* it is necessary to model and analyze only one-fourth of the structure. *Axisymmetry* is a special case of reflective symmetry where *all* planes intersecting a single axis are planes of symmetry.

Figure 10.2-14 shows three examples of total reflective symmetry and the corresponding reduced models. If we were modeling the beam example shown with

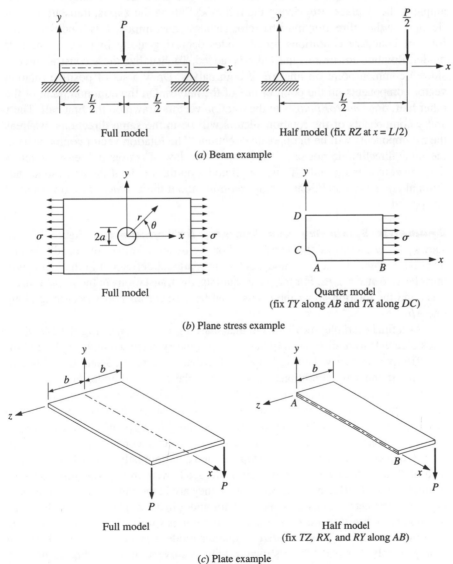

(*a*) Beam example

(*b*) Plane stress example

(*c*) Plate example

Figure 10.2-14   Examples of total reflective symmetry.

beam elements, the half model would need a boundary condition imposed at $x = L/2$ to ensure that the half model deflects the same as the full model. The additional boundary condition would be to set the rotation about an axis parallel to the z axis at $x = L/2$ equal to zero.

The plane stress example shown, as indicated earlier, has two axes of symmetry, the x and y axes. Thus a quarter model is used. In order for the quarter model to behave in a fashion identical to the full model, deflections in the y direction for all nodes on line AB must be fixed, and deflections in the x direction for all nodes on line CD must be fixed.

The plate example is a little more subtle. The plane of symmetry for this example is the xy plane. Reviewing the full model along the x axis, translational deflections in the z direction must be zero. Thus we must impose this condition in the form of boundary conditions for all nodes occurring along line AB of the half model. Rotation constraints must also be added. To visualize these constraints consider a common point on line AB of both halves. Draw a set of positive rotation vector components on the point of one of the halves. On the common point of the other half, draw *mirror images* of the rotation vectors drawn on the first half. The x and y components of the rotation vectors will be in the same directions, whereas the z components will be in opposite directions. The rotation vector components in the same direction do not satisfy Newton's third law of action and reaction unless they are zero in magnitude. Thus, for all nodes on the x axis of the half-model, additional boundary conditions setting *rotations* about the *x and y* axes to zero must be applied.

Geometric Symmetry and Loading Antisymmetry   Figure 10.2-15 shows two examples of this condition. For the beam example, the condition that must be imposed on the half model is the translational deflection in the y direction must be zero at $x = L/2$. For the plate example, the translations in the *x and y* directions and the rotation about the z axis must be zero for all nodes occurring along line AB.

As a final example, recall the circular shaft representation of Fig. 10.2-2 where brick elements were discussed. In that example, the xy plane would be a plane of total reflective symmetry. Thus only half of the shaft need be modeled. In addition to the support boundary conditions, only the z translation for all nodes in the xy plane are fixed to ensure symmetry. Rotation boundary conditions are unnecessary, as the brick element does not support rotational dof. In the earlier discussion it was stated that a quarter model could be used. This is because the xz plane is an axis of geometric symmetry and *loading antisymmetry*. To see this, consider the shaft in Fig. 10.2-2 divided at the xz plane into a top half and a bottom half. On each half place a load of $P/2$ in the negative y direction. If the loads were in opposite directions we would have total reflective symmetry. Since they are in the same direction, the loading is antisymmetric. Thus the conditions for nodes in the xz plane are that the translation in the x direction are zero. Again, no rotation is specified since we are dealing with brick elements. To summarize, a quarter model can be analyzed, where, for example, only that part of the structure with positive x, y, z coordinates need be

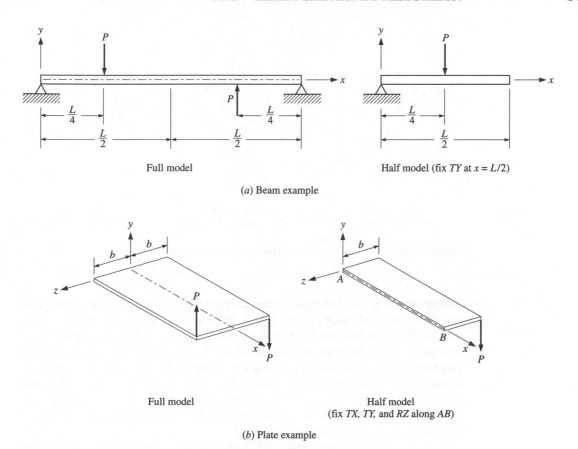

(a) Beam example

(b) Plate example

Figure 10.2-15 Geometric symmetry with loading antisymmetry.

modeled. The net force to apply is $P/4$ at $x = 0$, and other than support constraints for all nodes at $x = L$, translation of all nodes in the xy plane must be fixed in the z direction, and all nodes in the xz plane must be fixed in the x direction.

Cyclic Symmetry   This condition occurs when there is a repetition of geometry and loading. Consider the disk shown in Fig. 10.2-16. The holes and forces are repeated every 120°. Thus the full model can be separated into one-third segments as shown by the dotted lines. To ensure that the third model duplicates the full model, translation in the tangential (θ) direction of all nodes along lines AB and BC must be set to zero. The third model can be reduced in half, as it exhibits reflective symmetry about line BD. Thus the final model is a sixth of the original as shown and loaded by $P/2$. For boundary conditions, translation in the tangential (θ) direction of all nodes along lines AB and the straight-line segments of BD must be set to zero.

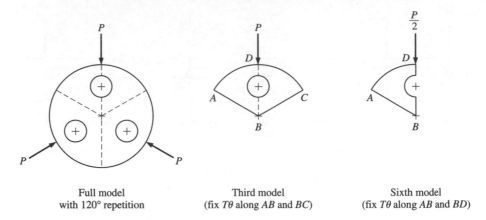

Full model with 120° repetition	Third model (fix $T\theta$ along AB and BC)	Sixth model (fix $T\theta$ along AB and BD)

Figure 10.2-16   Cyclic symmetry (with reflective symmetry).

Axisymmetry   As indicated earlier, axisymmetry is a special case of reflective symmetry when all planes intersecting a single axis are planes of symmetry. The thick-walled cylinders and rotating disks covered in Sec. 5.8 are examples of axisymmetry where the axial direction is the axis of symmetry. The models for these problems can be reduced in one of two ways. One method is to model the problem in the same fashion as with cyclic symmetry. In the example of Fig. 10.2-16, repetition was at 120° intervals so a 120° segment was modeled. In the case of axisymmetry, *any* angle segment will work and the translations in the tangential direction of all nodes on the isolation surfaces are fixed.

The second, most commonly used method, employs special elements called *axisymmetric elements*. These elements are represented graphically in a manner identical to the two-dimensional elements such as the quadrilateral element of Fig. 9.4-3 and the CST element of Fig. 9.4-1. However, the mathematical formulations of the equations for the elements are based on rotating the elements about a specific axis creating a *solid of revolution* as shown in Fig. 10.2-17. In the figure the element is rotated about the *y* axis. For a specific software, consult with the vendor's users manual for proper definition of the axis of revolution.

All node locations, loading, and boundary conditions of the axisymmetric model are defined in the two-dimensional *xy* plane only. Based on the coordinate system of the element shown in solid in Fig. 10.2-17, the *x* axis is the radial direction, the *y* axis is the longitudinal direction, and the *z* direction is the tangential direction. Owing to symmetry, deflection output for the element is restricted to the *xy* plane. However, output for the normal stresses is in all three directions. Owing to symmetry, shear stress is restricted to the *xy* plane only.

There are other ways to exploit symmetry. If the software is capable of applying superposition, then combinations of symmetric and antisymmetric loads can be addressed. For example, consider the structure shown in Fig. 10.2-18(*a*). The loading of the structure can be modeled as the superposition of the two load states

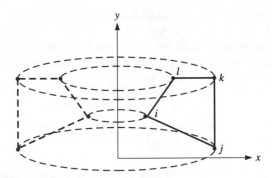

Figure 10.2-17   Axisymmetric element based on a two-dimensional quadrilateral.

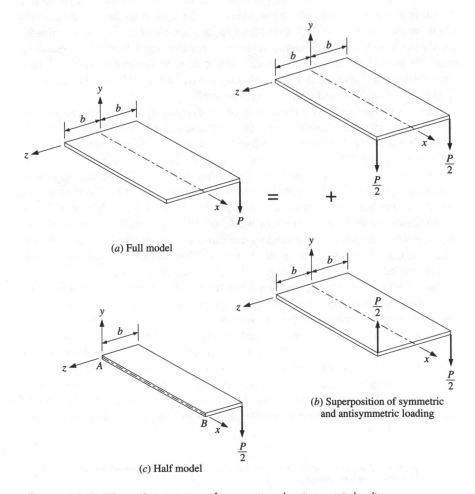

(a) Full model

=   +

(b) Superposition of symmetric
and antisymmetric loading

(c) Half model

Figure 10.2-18   Superposition of symmetric and antisymmetric loading.

shown in Fig. 10.2-18(*b*). The top load state is symmetric and the lower load state is antisymmetric. The half model shown in Fig. 10.2-18(*c*) can be used for both cases. One case would apply symmetric boundary conditions as indicated in Fig. 10.2-14(*c*), and the other case would apply antisymmetric boundary conditions as indicated in Fig. 10.2-15(*b*).

10.3   LOAD APPLICATION

There are two basic forms of specifying loads on a structure, nodal and element loading. However, as explained in Chap. 9, element loads are eventually applied to the nodes using equivalent nodal loads.

More than one set of load configurations can easily be handled in a *single* run of a finite element analysis where the inverse of the system stiffness matrix needs only to be computed *once*. This is referred to as *multiple load cases* where computational time is reduced significantly relative to running each load case separately where the inverse of the system stiffness matrix must be computed with each run. Some software are also capable of combining the results of multiple load cases using superposition. Scaling factors can also be applied.

Some software enables a third form of applying a load on a structure when *specified* displacement boundary conditions are applied. In Chap. 9, specified displacement boundary conditions were handled as part of the partitioning process. That is, the system stiffness matrix was reduced by the specified boundary conditions prior to obtaining the inverse of the stiffness matrix. Using this approach, changes in specified displacement boundary conditions cannot be incorporated as part of multiple load cases since the system stiffness matrix for each case of specified displacement boundary conditions would be different. Another method of handling specified displacement boundary conditions does not reduce the stiffness matrix[†] and is referred to as the *penalty method.* This method uses a special finite element, called the *boundary element,* and is discussed in Sec. 10.3.3.

One final note on load application is related to Saint-Venant's principle. If one is not concerned about the stresses near points of load application, it is not necessary to attempt to distribute the loading very precisely. The net force and/or moment can be applied to a single node provided the element supports the dof associated with the force and/or moment at the node. However, the analyst should not be surprised, or concerned, when reviewing the results and the stresses in the vicinity of the load application point are found to be very large. Concentrated moments can be applied to the nodes of beam and plate elements. However, as stated earlier in this chapter, concentrated moments cannot be applied to truss, two-dimensional plane

---

[†] For large structural problems, reducing the stiffness matrix for specified (nonzero) boundary conditions does not offer significant time savings.

elastic, axisymmetric, or brick elements. A pure moment can be applied to these elements only by using forces in the form of a *couple*. From the mechanics of *statics* a couple can be generated using two or more forces acting in a plane where the net force from the forces is zero and the net moment is a vector perpendicular to the plane and is the summation of the moments from the forces taken about any common point.

10.3.1 NODAL LOADS

As stated earlier, concentrated forces and/or moments can be applied to nodes of an element provided the element supports the dof associated with the load. Force and moment distributions can be simulated manually by equivalent concentrated nodal forces. The software may be capable of doing this. If not, the basic technique is a simple application of the mechanics of statics as shown in Sec. 1.2. For example, consider a force distribution applied along a boundary of a model where the span of the boundary contains the edges and nodes of several elements. First, the equivalent nodal forces on the two nodes of each element edge containing the distributed load are determined. This is done by determining the moment about each node due to the force distribution. It is then a simple task to determine the equivalent nodal forces that provide the same moments. Once the equivalent nodal forces for each edge are established, the forces at each node are summed to obtain the total equivalent nodal forces.

Loads can also be distributed using additional elements. Figure 10.3-1(*a*) shows a plane stress structure with a distributed load which totals to a force *P*. This can be simulated using additional elements and a single concentrated load as shown

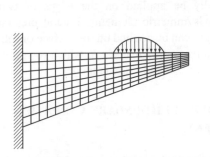

(*a*) Structure with a distributed load

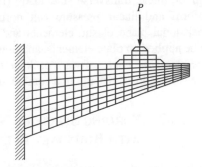

(*b*) Using elements to distribute the loading

Figure 10.3-1

in Fig. 10.3-1(*b*). The shape of the distribution can be controlled by the relative stiffness of the elements above the loading plane to the actual structure by changing the modulus of elasticity *E*.

With problems using axisymmetric elements, the concentrated forces applied to nodes are *line* loads in reality as the load revolves around the axis of revolution. Consult the software manual as to how to specify the value. Some software codes specify the nodal force as the total force/radian. In this case the nodal force is the total force divided by 2π.

10.3.2 ELEMENT LOADS

Element loads include static loads due to gravity (weight), thermal effects, surface loads such as uniform and hydrostatic pressure, and dynamic loads due to constant acceleration and steady-state rotation (centrifugal acceleration). Some of these are discussed in detail in Chap. 9. As stated earlier, element loads are converted by the software to equivalent nodal loads and in the end are treated as concentrated loads applied to nodes.

For gravity loading, the gravity constant in appropriate units and the direction of gravity must be supplied by the modeler. If the model length and force units are inches and lbf, $g = 386.4$. If the model length and force units are meters and newtons, $g = 9.81$. The gravity direction is normally toward the center of the earth.

For thermal loading, the thermal expansion coefficient α must be given for each material, as well as the initial temperature of the structure, and the final nodal temperatures. Most software packages have the capability of first performing a finite element heat transfer analysis on the structure to determine the final nodal temperatures. The temperature results are written to a file to use with the static stress analysis. Here the heat transfer model should have the same nodes and element type the static stress analysis model has.

Surface loading can generally be applied to most elements. For example, uniform or linear transverse line loads (force/length) can be specified on beams. Uniform and linear pressure can normally be applied on the edges of two-dimensional plane elastic elements and axisymmetric elements. Lateral pressure can be applied on plate elements, and pressure can be applied on the surface of solid brick elements. Each software package has its unique manner in which to specify these surface loads usually in a combination of text and graphic modes.

10.3.3 FORCING SPECIFIED NONZERO BOUNDARY CONDITIONS WITH BOUNDARY ELEMENTS

The *boundary element* is simply a linear spring with a large stiffness oriented in the direction of the specified boundary condition. Care must be taken when specifying the size of the stiffness of the element. The stiffness must be large enough to yield the specified boundary condition, but not too large to cause numerical problems (see Sec. 10.4). To demonstrate the method, we will return to Example 9.2-1. Figure 10.3-2 repeats Fig. 9.2-5 of the example where $F = 20$ kN, $A_1 = 100$, $A_2 = 75$,

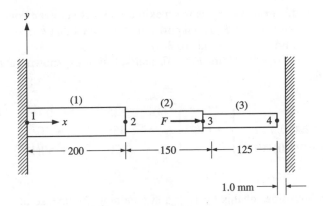

Figure 10.3-2   Step shaft of Example 9.2-1.

$A_3 = 50$ mm^2, and $E = 50$ GPa. Furthermore, it was found in the example that node 4 deflected 1.0 mm and contacted the rigid wall. Thus we will start at this point by stating that the specified boundary condition on node 4 is 1.0 mm.

Node 1 is fixed and is partitioned as usual. The resulting force-deflection equations are given by Eq. (*b*) of Example 9.2-1, given by

$$\begin{Bmatrix} F_2 \\ F_3 \\ F_4 \end{Bmatrix} = \begin{bmatrix} k_{22} & k_{23} & k_{24} \\ k_{32} & k_{33} & k_{34} \\ k_{42} & k_{43} & k_{44} \end{bmatrix} \begin{Bmatrix} u_2 \\ u_3 \\ u_4 \end{Bmatrix} \qquad \textbf{[10.3-1]}$$

where $k_{22} = 50(10^6)$, $k_{23} = k_{32} = -25(10^6)$, $k_{24} = k_{42} = 0$, $k_{33} = 45(10^6)$, $k_{34} = k_{43} = -20(10^6)$, and $k_{44} = 20(10^6)$ N/m.

Imagine now that an additional spring of stiffness k is attached between node 4 and ground. In order for the displacement of node 4 to be a specific value $\delta = 1.0$ mm, the force in the spring must be $k\delta$. This force adds to F_4 in Eq. (10.3-1) and the spring stiffness adds to k_{44} in the stiffness matrix. Thus

$$\begin{Bmatrix} F_2 \\ F_3 \\ F_4 + k\delta \end{Bmatrix} = \begin{bmatrix} k_{22} & k_{23} & k_{24} \\ k_{32} & k_{33} & k_{34} \\ k_{42} & k_{43} & (k_{44} + k) \end{bmatrix} \begin{Bmatrix} u_2 \\ u_3 \\ u_4 \end{Bmatrix} \qquad \textbf{[10.3-2]}$$

Since $u_4 = \delta$, we see that Eqs. (10.3-1) and (10.3-2) are the same equations. However, up to this point, k is unknown. Dividing the last equation of Eq. (10.3-2) by k, the equation is rewritten as

$$\frac{F_4}{k} + \delta = \frac{k_{42}}{k} u_2 + \frac{k_{43}}{k} u_3 + \left(\frac{k_{44}}{k} + 1 \right) u_4 \qquad \textbf{[10.3-3]}$$

Now, if k is made sufficiently large, each term in Eq. (10.3-3) containing k will be small and $u_4 \approx \delta$. A rule of thumb is to make the value of k of the order 10^4 greater than the largest value of k_{ij} of the system stiffness matrix. This is somewhat arbitrary, since for this example, k_{22} is the maximum and is not even contained in Eq. (10.3-3). Since the value of F_4 is unknown, it is assumed that k is large enough to eliminate it from Eqs. (10.3-2) and (10.3-3).

Letting $k = 50(10^{10})$ N/m, $F_4 = 0$, and substituting known values into Eq. (10.3-2) yields

$$\begin{Bmatrix} 0 \\ 20(10^3) \\ 50(10^{10})(1.0)(10^{-3}) \end{Bmatrix} = (10^6) \begin{bmatrix} 50 & -25 & 0 \\ -25 & 45 & -20 \\ 0 & -20 & [20 + 50(10^4)] \end{bmatrix} \begin{Bmatrix} u_2 \\ u_3 \\ u_4 \end{Bmatrix}$$

$$[\textbf{10.3-4}]$$

Multiplying both sides of this by 10^{-6} and obtaining the inverse of the remaining 3×3 matrix gives

$$\begin{Bmatrix} u_2 \\ u_3 \\ u_4 \end{Bmatrix} = \begin{bmatrix} 0.027692 & 0.015385 & 6.15(10^{-7}) \\ 0.015385 & 0.030770 & 1.23(10^{-6}) \\ 6.15(10^{-7}) & 1.23(10^{-6}) & 2.00(10^{-6}) \end{bmatrix} \begin{Bmatrix} 0 \\ 0.02 \\ 500 \end{Bmatrix}$$

$$= \begin{Bmatrix} 0.6154 \\ 1.2308 \\ 1.0000 \end{Bmatrix} (10^{-3}) \, \text{m} = \begin{Bmatrix} 0.6154 \\ 1.2308 \\ 1.0000 \end{Bmatrix} \text{mm}$$

Note that the displacement results agree with that found in Example 9.2-1.

To summarize, the stiffness of the boundary element is *added* to the diagonal of the system stiffness matrix associated with the dof of the specified boundary condition. The stiffness of the boundary element should be of the order 10^4 greater than the maximum stiffness coefficient in the system stiffness matrix. The value of the load associated with the dof of the specified boundary condition is set equal to the product of the boundary element stiffness and the specified boundary condition.

The boundary element can be used when modeling a skew boundary condition such as that on node 2 in Fig. 9.2-11. The approach presented in Sec. 9.2.7 was to transform the coordinate system at the node to align with the skew. This is the most direct and effective manner to deal with skew boundary conditions. However, if the software is not capable of doing this, a boundary element can be used by attaching the element to node 2, ground aligning the element in the direction of the constraint, and applying a specified deflection of zero. Again, the analyst is forewarned that using elements with a large stiffness can cause numerical problems. A rigid element would be a better alternative (see Sec. 10.4).

10.3.4 MULTIPLE LOAD CASES

During the finite element analysis of a structural problem most of the computational effort goes into the inversion of the stiffness matrix. Once this is done, deflections are obtained from a simple matrix multiplication of the inverted reduced and/or modified stiffness matrix and the nodal force vector as indicated by

$$\{\mathbf{u}\} = [\tilde{\mathbf{K}}]^{-1}\{\mathbf{F}\} \qquad \text{[10.3-5]}$$

where $\{\mathbf{u}\}$, $[\tilde{\mathbf{K}}]^{-1}$, and $\{\mathbf{F}\}$ are the system nodal displacement vector, inverse of the reduced and/or modified stiffness matrix, and nodal load vector, respectively. A different load case would simply change the load vector. Rather than rerunning the analysis and repeating the inversion of the stiffness matrix, the force vector in Eq. (10.3-5) can be replaced by a matrix which contains multiple force vectors. Considering n multiple load cases, Eq. (10.3-5) can be rewritten as

$$[\{\mathbf{u}_1\}\{\mathbf{u}_2\}\dots\{\mathbf{u}_n\}] = [\tilde{\mathbf{K}}]^{-1}[\{\mathbf{F}_1\}\{\mathbf{F}_2\}\dots\{\mathbf{F}_n\}] \qquad \text{[10.3-6]}$$

Thus n solutions for $\{\mathbf{u}_i\}$ are obtained from n cases of force vectors $\{\mathbf{F}_i\}$, where it is only necessary to determine $[\tilde{\mathbf{K}}]^{-1}$ once.

10.3.5 LOAD SCALE FACTORS

For versatility in applying various combinations of load cases, most FEA programs enable the modeler to specify a multiplier for each of the various element load states such as surface pressure, thermal, constant acceleration, and specified displacements. If, for example, the analyst wants to double the pressure loads on a structure while keeping the thermal loading the same, it is unnecessary to change all the pressure loads. In this case, only the pressure multiplier needs to be changed from a scale factor of 1.0 to 2.0.

10.4 CONSTRAINTS

The simulation of boundary conditions and other forms of constraint is probably the single most difficult part of the accurate modeling of a structure for a finite element analysis. When specifying constraints it is relatively easy to make mistakes of omission or misrepresentation. It may be necessary for the analyst to test different approaches to model esoteric constraints such as bolted joints, welds, etc., which are not as simple as the idealized pinned or fixed joints. As mentioned earlier, testing should be confined to simple problems and not to a large, complex structure. Sometimes, when uncertain as to the exact nature of a boundary condition, limits of behavior may only be possible. For example, consider beam C.12 of Appendix C. The horizontal beam is uniformly loaded and is fixed on one end and simply supported on the other end. Although not explicitly stated, tables such as these assume that the beams are not restrained in the horizontal direction. If they were restrained, analytical techniques such as shown at the end of Sec. 6.13 can be used. With a finite element analysis, a special element, a beam with stiffening, would be necessary. Assuming no restraint in the horizontal direction, let us focus on the fixed support. This is an idealization, as the wall will have some degree of flexibility. The wall could be included as part of the finite element model, but this could overcomplicate the problem. A simple alternative would be to first solve the model with the support fixed and then resolve the model with the support pinned. The results of both analyses would provide bounds on the stresses and deflections of the actual structure.

As indicated earlier in the chapter, a node can have up to six dof. However, the elements used in a model may or may not support all six dof. If a dof without an associated stiffness exists, the system stiffness matrix will be *singular* and a fatal error will result. Thus all unused dof must be partitioned out of the system stiffness matrix.[†] Some software have automatic methods to do this, and the analyst should be aware of this capability. Without an automatic feature, the analyst would have to apply fixed boundary conditions to all unused dof.

The minimum set of constraints on a model must at most restrain *rigid-body motion.* If not, the system stiffness matrix will be *singular* and a fatal error will occur. Establishing constraint of rigid-body motion is not always so obvious. For example, consider the simple block structure shown in Fig. 10.4-1. Assume that the block is modeled using a number of brick elements. Fixing the block at point Q would not sufficiently restrain rigid-body motion. Brick elements only have translational dof. Fixing point Q would not restrain rotation about the three axes. Rotation of a brick structure can only be restrained by constraining translational dof's. Additional translational constraint of point G in the z direction will have no effect of restraining any rotation as the line of action of the constraint intersects point Q. However, additional constraint of translation of point G in the x and y directions will restrain rigid-body rotation about the y and x axes, respectively. Rigid-body rotation

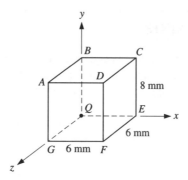

Figure 10.4-1

[†] Each of the two nodes of a beam element support all six dof. However, it is common practice to introduce a *third* node to define the lateral principal second area-moment axes. The dof of this node must be partitioned out, as it plays no role in the structural analysis.

about the z axis could be restrained by fixing the translation of points A, B, C, or D in the x direction or constraining the translation of points C, D, E, or F in the y direction.

For one reason or another, rather than using symmetry, suppose that it was desired to model the entire C-clamp of Fig. 5.6-3(a). In the figure, there are no indications of any boundary conditions, only equal and opposite forces. Boundary conditions *must* be applied to eliminate rigid-body motion. Only one of the forces is required to be applied to load the clamp. That is, one could fix the translation of the point where one of the loads exists in Fig. 5.6-3(a). This would not eliminate rigid-body rotation. This can be eliminated by fixing the translation of other points. For example, fixing the point where the second load is applied in the two directions perpendicular to the load direction would fix rotation about two axes. For the other rotation axis it would only be necessary to fix the translation of any *one* node in a direction perpendicular to the loading axis where the node does not lie along the line of the loading axis.

The constraints used when reducing the size of a symmetric model are discussed in Sec. 10.2.7. The constraints are associated with the plane(s) of symmetry and restrain specific dof's in the plane(s) according to the conditions of symmetry while allowing other dof's in the plane(s) to displace freely. In some cases, the free displacement allows for Poisson expansion or contraction. For example, let us return to the C-clamp shown in Fig. 5.6-3(a). Let the horizontal axis (line a-a) going through the centroid of the cross section be the x direction and the vertical be the y direction. One could model the C-clamp by modeling the clamp above section a-a, a half model, or continue the reduction with a second plane of symmetry through the center of the cross section, a quarter model. The boundary conditions would depend on the element type and which reduced model was used. For simplicity, let us consider the half model, ignore the solid part of the clamp, and model the body of the C-clamp with plate elements as shown in Fig. 10.4-2. One would not apply completely fixed boundary conditions on the nodes along a-a, as that would constrict the horizontal Poisson expansion and contraction that occurs in the plane. The proper boundary conditions for all nodes in the xz plane would be (1) for all nodes

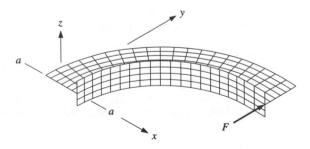

Figure 10.4-2   Half model of C-clamp shown in Fig. 5.6-3(a) modeled with plate elements.

in the xz plane, fix translation in the y direction; and rotations about the x and y axes,[†] (2) for all nodes *on* the x axis, fix the translation in the z direction, (3) for any *one* node on the x axis, fix the translation in the x direction. Condition (1) maintains symmetry in the xz plane and allows for Poisson expansion and contraction in the x and z directions. Condition (2) maintains symmetry relative to the xy plane and allows for Poisson expansion and contraction in the x and z directions. Condition (3) restrains rigid-body translation in the x direction. Since condition (3) restrains only one node from translating in the x direction, Poisson expansion and contraction in the x and z directions is possible.

Multipoint constraint equations are quite often used to model boundary conditions or *rigid* connections between elastic members. When used in the latter form, the equations are acting as elements and are thus referred to as *rigid elements*. In Sec. 9.2.7, skew boundary conditions were discussed. In Fig. 9.2-11, a skew boundary condition on node 2 is shown. The approach presented was to transform the coordinate system for the node to align with the skew. This is the most direct and effective manner to deal with skew boundary conditions. However, if the software is not capable of this, other approaches may be necessary. In Sec. 10.3.3, an alternative of using a boundary element was suggested. Using elements with large stiffness can cause numerical problems. If the software is capable of specifying multipoint constraint equations, another alternative would be to specify the condition that $u_2 = 2v_2$ in Fig. 9.2-11. Using a transformation, the dof associated with u_2 or v_2 is eliminated from the system equations without the possible numerical problems of a boundary element.

Rigid elements perform the same function that multipoint constraint equations do. For example, in Fig. 9.2-11 a rigid truss[‡] can be connected from node 2 to a ground node where the line of the nodes of the link aligns with the constraint. Rigid elements can also be used when a structure contains an element that is very stiff compared to the other elements. As stated earlier, an overly stiff element in a structure can cause numerical problems. As an example of this, return to Example 9.2-3 and with a spreadsheet program continually increase the stiffness of element 3. Eventually [when k_3 goes beyond $1.0(10^{21})$ N/m], the stiffness matrix will approach being singular and the solution will go unstable.

Example 10.4-1 | Replace element 3 of Example 9.2-3 with a rigid element.

Solution:

Returning to Fig. 9.2-9, we will write a constraint equation for element 3 based on its being rigid. Let the displacement of node 3 relative to node 2 in the x and y directions be $u_{3/2}$ and $v_{3/2}$, respectively, where $u_{3/2} = u_3 - u_2$ and $v_{3/2} = v_3 - v_2$. Since the element is rigid, node 3 moves in a circular path relative to node 2. For

---

[†] If the reader is unsure about the rotation boundary conditions, review the discussion concerning the rotation boundary conditions for Fig. 10.2-14(c).

[‡] A rigid truss would have two nodes. Some software also support rigid elements with three or more nodes.

small rotations we see the relationship between the relative deflections, as shown in Fig. 10.4-3, to be

$$u_{3/2} = -v_{3/2} \tan 45 = -v_{3/2} \qquad \qquad [a]$$

or

$$u_3 - u_2 = v_2 - v_3 \qquad \qquad [b]$$

Equation (b) is the multipoint constraint equation for rigid element 3. Arbitrarily, we will select to eliminate v_3. Thus Eq. (b) is rewritten as

$$v_3 = v_2 + u_2 - u_3 \qquad \qquad [c]$$

The system dof can now be written in terms of the independent dof as

$$
\begin{Bmatrix} u_1 \\ v_1 \\ u_2 \\ v_2 \\ u_3 \\ v_3 \end{Bmatrix}
=
\begin{bmatrix}
1 & 0 & 0 & 0 & 0 \\
0 & 1 & 0 & 0 & 0 \\
0 & 0 & 1 & 0 & 0 \\
0 & 0 & 0 & 1 & 0 \\
0 & 0 & 0 & 0 & 1 \\
0 & 0 & 1 & 1 & -1
\end{bmatrix}
\begin{Bmatrix} u_1 \\ v_1 \\ u_2 \\ v_2 \\ u_3 \end{Bmatrix}
\qquad \qquad [d]
$$

or

$$\{u_G\} = [T]\{u_R\} \qquad \qquad [10.4\text{-}1]$$

where $\{u_G\}$, $[T]$, and $\{u_R\}$ are the global dof, the transformation matrix, and the *retained* independent dof respectively.

Equation (e) of Example 9.2-3 represents the overall system stiffness equations prior to partitioning. For this example, the stiffness of element 3 is infinite, since it is a rigid element. However, the transformation process will completely eliminate the stiffness of this element from the system equations. To see how this happens we

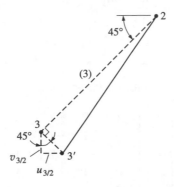

Figure 10.4-3 Rotation of rigid element 3 relative to node 2.

will rewrite Eq. (e) of Example 9.2-3 in terms of the stiffness of each element k_1, k_2, and k_3, where for the example $k_1 = k_2 = k = 8(10^6)$ N/m. Thus

$$
\begin{Bmatrix} R_{1x} \\ R_{1y} \\ 2(10^3) \\ -3(10^3) \\ R_{3x} \\ 0 \end{Bmatrix} = \frac{1}{2} \begin{bmatrix} 2k & 0 & -2k & 0 & 0 & 0 \\ 0 & 2k & 0 & 0 & 0 & -2k \\ -2k & 0 & (2k+k_3) & k_3 & -k_3 & -k_3 \\ 0 & 0 & k_3 & k_3 & -k_3 & -k_3 \\ 0 & 0 & -k_3 & -k_3 & k_3 & k_3 \\ 0 & -2k & -k_3 & -k_3 & k_3 & (2k+k_3) \end{bmatrix} \begin{Bmatrix} u_1 \\ v_1 \\ u_2 \\ v_2 \\ u_3 \\ v_3 \end{Bmatrix} \qquad [e]
$$

or

$$\{F\} = [K]\{u_G\} \qquad\qquad [10.4\text{-}2]$$

To eliminate the dependent dof, we will substitute Eq. (10.4-1) into Eq. (10.4-2) and premultiply both sides of the equation by $[T]^T$. That is,

$$[T]^T \{F\} = [T]^T [K][T]\{u_R\} \qquad\qquad [10.4\text{-}3]$$

Equation (10.4-3) represents the system equations in terms of the retained independent dof. For the example let us perform the transformations according to Eq. (10.4-3) with

$$
[T] = \begin{bmatrix} 1 & 0 & 0 & 0 & 0 \\ 0 & 1 & 0 & 0 & 0 \\ 0 & 0 & 1 & 0 & 0 \\ 0 & 0 & 0 & 1 & 0 \\ 0 & 0 & 0 & 0 & 1 \\ 0 & 0 & 1 & 1 & -1 \end{bmatrix} \qquad \{F\} = \begin{Bmatrix} R_{1x} \\ R_{1y} \\ 2(10^3) \\ -3(10^3) \\ R_{3x} \\ 0 \end{Bmatrix}
$$

$$
[K] = \frac{1}{2} \begin{bmatrix} 2k & 0 & -2k & 0 & 0 & 0 \\ 0 & 2k & 0 & 0 & 0 & -2k \\ -2k & 0 & (2k+k_3) & k_3 & -k_3 & -k_3 \\ 0 & 0 & k_3 & k_3 & -k_3 & -k_3 \\ 0 & 0 & -k_3 & -k_3 & k_3 & k_3 \\ 0 & -2k & -k_3 & -k_3 & k_3 & (2k+k_3) \end{bmatrix}
$$

The reader should confirm that the results are

$$
\begin{Bmatrix} R_{1x} \\ R_{1y} \\ 2(10^3) \\ -3(10^3) \\ R_{3x} \end{Bmatrix} = k \begin{bmatrix} 1 & 0 & -1 & 0 & 0 \\ 0 & 1 & -1 & -1 & 1 \\ -1 & -1 & 2 & 1 & -1 \\ 0 & -1 & 1 & 1 & -1 \\ 0 & 1 & -1 & -1 & 1 \end{bmatrix} \begin{Bmatrix} u_1 \\ v_1 \\ u_2 \\ v_2 \\ u_3 \end{Bmatrix} \qquad [f]
$$

We note that the system equations are totally devoid of k_3. Thus it is not necessary to even specify a value of k_3. Since $u_1 = v_1 = u_3 = 0$, the first two and the last equations of Eq. (f) can be partitioned as usual. The resulting equations, with $u_1 = v_1 = u_3 = 0$ and $k = 8(10^6)$, are

$$\begin{Bmatrix} 2(10^3) \\ -3(10^3) \end{Bmatrix} = 8(10^6)\begin{bmatrix} 2 & 1 \\ 1 & 1 \end{bmatrix}\begin{Bmatrix} u_2 \\ v_2 \end{Bmatrix} \qquad \text{[g]}$$

Solving Eqs.(g) for u_2 and v_2 yields

$$\begin{Bmatrix} u_2 \\ v_2 \end{Bmatrix} = \begin{Bmatrix} 0.625 \\ -1.0 \end{Bmatrix}(10^{-3})\text{ m} = \begin{Bmatrix} 0.625 \\ -1.0 \end{Bmatrix}\text{mm}$$

From Eq. (c)

$$v_3 = v_2 + u_2 - u_3 = -1.0 + 0.625 + 0 = -0.375\text{ mm}$$

10.5 PREPROCESSING CHECKS

There are some checks that can be performed prior to running the model through the FEA processor.

In the graphic preprocessor, various visual checks can be made. Through view control, the model can be viewed virtually at any angle. A single two-dimensional view of a three-dimensional model can be misleading. Most preprocessors provide the capability of displaying element numbers, node numbers, *shrink plots,* shaded or unshaded *hidden-line plots,* multiple load cases, element area property and material identification, element local coordinate identification, element subgrouping or a *hide* feature, etc. Shrink plots of the model are where the size of each element is reduced by a user-prescribed amount. If an element is *missing* in a structure, a shrink plot will show it quite readily. A wire frame graphic representation of a large model and its elements can appear to be quite confusing. This is where shaded or unshaded hidden-line plots with multiple views can reduce confusion. Similarly, element subgrouping or a hide feature allows for the display of smaller localized portions of the model.

Algor FEA software[†] has a unique system where in the preprocessing stage the structure is meshed graphically with lines and boundary conditions and loading are applied. This allows for a direct interface with other CAD software. The nodes and elements are created in a secondary preprocessor called a *decoder* which applies material and area properties and decodes the lines to create the nodes and elements. With Algor, the preprocessing checks noted earlier are actually performed in their postprocessor which views either a decoded model or a processed model.

Some processors will allow the model to be submitted and produce an output, in text form, of the input without running the actual analysis. This part of the output

---

[†] Produced by Algor, Inc., Pittsburgh, PA.

is called the *echo* of the input. The echo can be quite helpful in providing a check on the input in a different format than when specified. Inputting data with a graphic preprocessor on the computer is wondrous when compared to the early days of FEA. However, mistakes of omission and submission will inevitably take place. For large problems, correspondingly, the echo will be large. However, the bulk of the data will be in the form of node and element information most of which is typically checked with the graphic preprocessor. Checks on material properties, loads and load cases, boundary conditions, element properties, etc., can be made quite effectively and efficiently when viewing the echo.

Another check that can made on the model is that the processor can run a weight and center of gravity analysis prior to running the structural analysis. Discrepancies between the model and the actual structure may indicate how to improve the model.

10.6   PROCESSING THE MODEL

The major concerns when submitting the model to run in the processor are the *run-time* options and the computer memory capacity.

Run-time options are generally prescribed either in the preprocessor or at the time the processor is activated. These options are typically associated with what output, in text form, the modeler desires. The default settings for the processor may be such that stresses, deflections, forces, etc., are only generated in binary form by the computer where it is only available for review through the graphic postprocessor. The modeler can change these defaults through the run-time options. Other options could include terminating or suspending the processor after the echo, or stiffness matrices, or weight and c.g., etc., are produced. Other options for text output include the element and system stiffness matrices.

FEA of large structural problems requires a computer with large disk storage and rapid access (RAM) memories. In the early days of FEA, software was exclusively run on large, main-frame computers. Later, with the development of efficient minicomputers, FEA had yet another source of execution. With these types of computer resources, the analyst typically relies on a computer manager to assist in the daily management of computer memory. Today, with the proliferation of powerful stand-alone personal and workstation microcomputers, the analyst in many instances has to deal with all aspects of the analysis including memory management of the computer.

10.7   POSTPROCESSING

There are two basic formats in which the results of a finite element analysis can be reviewed, graphic and text output. In the early days of FEA the only way to review results of an analysis of a large structure was by sifting through a voluminous quantity of computer printout. With today's slick computer graphics capabilities much time is saved by reviewing the results in the graphic mode. However, there are occasions when some portion of the output is better seen in text format.

10.7.1 GRAPHIC OUTPUT

The graphic postprocessor provides an enormous capability to interrogate the results of the finite element analysis of a structural model. Here the analyst can attempt to ascertain whether the results make sense or not, verify expectations, and compare calculations based on closed-form solutions or handbook results if available. With view and section control and subgrouping capabilities, the graphic postprocessor assists analysts as they survey the model. A given screen display in the postprocessor is sometimes referred to as a *plot*.

The first plots that the analyst should review are those of scaled displacements. Blatant errors will most certainly be obvious here. If the deflections are suspect, and since the stresses are less accurate (being calculated from the displacements), one should have little confidence in the stress results. The graphic scaling exaggerates the deflections which if plotted with a 1:1 scale would normally be indiscernible to the eye. The scaling should be large enough to see the deflections adequately but not too large to distort the displacements relative to the dimensions of the model. The displacements can be viewed with or without the undeformed structure (shown in a different color). Numerical values of the displacements can be obtained by selecting specific nodes. Constraint conditions should be verified. No overlaps (interpenetration) between disconnected elements should occur (scaling might show overlaps that do not actually occur). Elements that are supposed to be connected should not separate. If the model is symmetric, deflections should be symmetric as well. The basic elements have straight sides and since deflection plots only depict the translational deflections of the nodes, the sides will remain straight in the deflection plots. For example, if a single beam was used to model an end-loaded cantilever beam, a deflection plot would still show the beam straight. However, the node with the load will be deflected. Occasionally, the model and deflections are too difficult to visualize regardless of the view. Contour plotting, used primarily for stress results, can also be used for deflections.

Contour plots come in two basic forms, line contour plots and fringe carpet (dithered) plots. Figure 10.7-1 shows examples of the two types as applied to the von Mises stress for the step-plate of Fig. 10.2-2. The contour plots are typically displayed in color and the legend values are easily matched with the contour lines or fringes. However, since color reproduction in this book is cost prohibitive they are all shown in black. Instead of colors, letters can be assigned to the legend and the plot for black and white plots.

For surface and solid elements, the raw stress output from a finite element analysis is typically relative to the Gauss points of each element (see Sec. 9.4.2). For contour plots, the stresses at the Gauss points are interpolated to the element's nodes using the shape functions of the specific element. Once the nodal values are

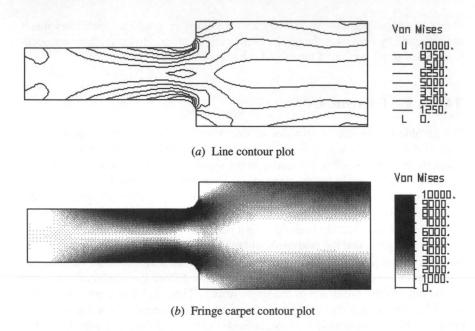

(a) Line contour plot

(b) Fringe carpet contour plot

Figure 10.7-1   Stress contour plots (obtained from Superview™, the postprocessor supplied by Algor, Inc., Pittsburgh, PA).

obtained, the stress at any point within the element can be determined by using the shape functions again. Owing to discretization errors, the value for the stress at a node that is common to two or more elements will generally be different. This may cause the contours between elements to be discontinuous. A technique commonly employed in postprocessors is to *smooth* the results. Smoothing is a technique which averages the conflicting stresses reported for each element connected to a common node. What results is a pleasing, continuous contour plot which may not convey some serious problems with the model and the results. The analyst should *always* view the unsmoothed contour plots first. Highly discontinuous contours between elements in an area of an unsmoothed plot reveal modeling problems, which typically calls for a further refinement of the element mesh in the area.

If the discontinuities in an unsmoothed contour plot are small or are in areas of little consequence, a smoothed contour plot can normally be used for reporting purposes. There are situations, however, when smoothing should *never* be used. If between adjacent elements, the thickness or the material stiffness changes, the stresses will most assuredly be different. When plate elements connect where the plates are not in the same plane, smoothing should not be employed. Problem 10.10 illustrates another, more subtle, case where smoothing should not be used. The problem simulates the press-fit between two mating cylinders. At the interface, the tangential stresses will be different and smoothing will obliterate this.

Normal and shear stresses can be depicted in terms of local or global coordinates. If local, the analyst will need to know the direction of the local coordinates.

Most postprocessors have simple graphical means to display local coordinates. Other stresses can be displayed in contour plots, including the principal stresses, the equivalent *von Mises* stress,[†] or the maximum shear stress.[‡] The stresses perpendicular to a free surface or a surface with a normal pressure p should be zero or $-p$, respectively. All stresses should be zero at an external corner. When comparing FEA results with expected results, the analyst must realize that this is a numerical technique and obtaining some specific exact value will not happen. For example, if a value of zero is expected somewhere, as long as the finite element analysis produces a value that is considerably small when compared with the average stress values determined in the FEA model, the expectation is basically satisfied.

The principal stress trajectories can be displayed, which shows how the stress *flows* through the structure.[§] The stress trajectories should be tangential and perpendicular to free boundaries, boundaries loaded by normal pressures, or a plane of reflective symmetry.

In the case of *plate* elements, the stresses are a superposition of in-plane membrane and bending and twisting shear and normal stresses.[¶] At a given location in the plane of the plate, the membrane stresses are constant through the thickness. The bending and twisting stresses vary linearly across the thickness, being zero at the center. Some software have the capability of viewing the stresses on either side of the plate without having to rotate the view.

Other capabilities may be present in the postprocessor such as displaying contour plots for strains, strain energy, error estimation, shear force and bending moment for beams, and nodal force and moment loading. The nodal loads can be used to check reactions to determine if static equilibrium is satisfied. Care should be exercised here as constraint equations can give rise to fictitious nodal loads. The graphic postprocessor may also have the capability of allowing the analyst to inquire on element characteristics such as aspect ratio, skew, and warp.

The analyst should always be on guard when interpreting the results that the postprocessor displays. As a simple example, for beam problems, Superview, the postprocessor for Algor FEA software in addition to other quantities provides the following stress output for each element: P/A, M_2/S_2, M_3/S_3, and *Worst*. P/A, M_2/S_2, and M_3/S_3 are the axial and bending stresses about the local transverse principal axes of the second-area moment, respectively. The quantity *Worst* is calculated from the formula

$$Worst = (sign\ of\ P/A)[ABS(P/A) + ABS(M_2/S_2) + ABS(M_3/S_3)] \qquad \textbf{[10.7-1]}$$

This formula may seem appropriate to determine the maximum normal stress in a beam. However, the reader should verify that this equation is valid for some but not all cross sections. For example, the equation is not valid for circular cross sections. Because the quantity is not always appropriate does not mean it is of no use. The

---

[†] See Sec. 7.3.3 for the definition of the equivalent von Mises stress.
[‡] Sometimes referred to as the Tresca stress, which is defined in Sec. 7.3.2.
[§] See Sec. 5.10 for a brief discussion on stress trajectories.
[¶] See Sec. 5.7 for a definition of the in-plane bending and twisting (normal and shear) stresses for plates in bending.

point that is being made here is that the analyst should completely understand what an output quantity means if it is going to be reported (such as *von Mises* stress, as another example).

10.7.2 TEXT OUTPUT

As indicated earlier, a text output file offers an alternative form to review various output from the finite element analysis. Quite often, this is where various diagnostic messages are found. Information such as memory allocation, and matrix parameters such as the maximum and relative values of terms along the diagonal of the system stiffness matrix may be communicated. Warning and/or fatal messages may indicate various modeling problems and/or outright errors which exist in the model. The messages may not pinpoint where or what the problems are exactly; however, they alert the analyst as to the type of problem to look for. The analyst may also want to review the echo of the input to confirm that certain properties or conditions were entered correctly. In text form, certain entities in the echo can be scanned quite readily. For more detailed diagnostic information, the analyst can also request various outputs such as the element and system matrices.

10.8 CLOSURE

A great deal more could have been presented in this chapter. Unfortunately, because of space limitations, further discussion must be abandoned. The intent of this book is to provide theory, applications, and insight in the field of stress analysis. The finite element method is only one part of the stress analysis picture. There are a great many commendable textbooks on the finite element method. However, the one element that is generally seriously lacking in these books is how to apply the technique with an FEA software package. The vast majority of stress analysts that use the finite element method for industrial applications use commercial FEA software, and many times the beginning analyst has not a clue as to how to apply the software. There are many subtle aspects of the technique that may be obvious to some but are not to others. Some simple examples are:

Question: Why doesn't the finite element analysis show the failure of the structure when it yields or fractures?

Answer: The fundamental finite element analysis solution is based on a linear relationship between the loads and deflections. The loads on a *model* can be increased *indefinitely* while the resulting deflections, stresses, etc., continue to increase linearly. Some FEA software is capable of alerting the analyst when in the postprocessor that the stresses have exceeded some amount specified by the analyst. For behavior beyond the elastic limit, or applications when the deflections and load are coupled, a nonlinear finite element analysis must be performed.

Question: Will the fundamental finite element static analysis solution predict local or global instabilities (buckling)?

Answer: No (see the answer to the first question). Most producers of FEA software provide independent processor modules which perform finite element buckling analyses.

Hopefully, this chapter has accomplished its goal to provide some insight into the use of commercial FEA software for basic stress analysis applications. Furthermore, it is hoped that readers are eager to continue their development in this area.

10.9 PROBLEMS

Preliminary Remarks

Some of the problems in this section are not necessarily placed in an order relative to the sections of the chapter. Also, if one observes the nature of many of the problems given, it can be seen that most of them come from the examples and problems of past chapters. For a source of additional problems beyond what is given in this section, the reader is urged to revisit the examples and problems of Chaps. 4 to 6.

For problems of beams or tubes modeled with brick or plate elements the hints given generally generate elements with medium-high aspect ratios. This is to reduce the size of the model for processing purposes. Satisfactory results should result, however. If the reader's computer memory is large enough, more elements along the length of the beams or tubes would reduce the aspect ratios.

10.1 Using the semiautomatic strategies shown in Sec. 10.2.5 mesh the structure shown in Fig. (*a*) with quadrilateral elements. HINT: The right external fillet can be modeled by placing a square region inside the fillet as shown in Fig. (*b*).

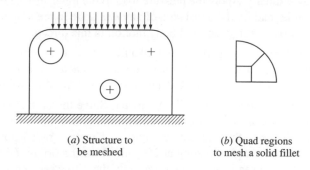

(*a*) Structure to be meshed

(*b*) Quad regions to mesh a solid fillet

Problem 10.1

10.2 Repeat Example 10.2-1, except the region is to be three-sided with the coordinates of points 5 and 6 being the same as point 7 where $(x_7, y_7) = (1.5, 10)$.

10.3   For Fig. 10.2-10(*a*) employ the technique used in Example 10.2-1 to mesh the upper quadratic region with a 3 × 4 mesh as shown in Fig. 10.2-10(*b*). Let the radius of the circle be 2 in and the dimensions of the square be 5 in × 5 in.

10.4   This problem requires the solution of Problem 10.3. Copy-mirror the mesh obtained in Problem 10.3 about the right-hand vertical edge. Next consider the upper quadratic region shown in Fig. 10.2-10(*c*). Mesh this complete region with a 6 × 4 mesh using Eqs. (10.2-1) and (10.2-2). Compare the two meshes created.

10.5   Determine the nodal deflections of Example 9.2-3 if element 1 is rigid.

10.6   Solve Example 9.2-4 using the multipoint constraint equation $u_2 = 2\,v_2$.

Computer Problems

For the remaining problems in this section, use a commercial FEA software package available to you.

10.7   If available, use an automatic mesh generator similar to Algor's Supergen to mesh the structure of Problem 10.1. Compare results if you have solved Problem 10.1.

10.8   Solve Problem 5.73(*b*). The elements for each cylinder should be assigned their unique material properties.

    (*a*) Solve the problem using plane-stress two-dimensional quadrilateral elements with an arbitrary thickness of 10 mm. Use a quarter model and mesh each cylinder uniformly with four elements radially and 18 elements tangentially (5° increments). Apply the pressure load to the inner surface of the inner cylinder and apply boundary conditions consistent with symmetry.

    (*b*) Solve the problem using axisymmetric two-dimensional quadrilateral elements. Arbitrarily, let the longitudinal length be 10 mm. Mesh each cylinder uniformly with four elements radially and two elements longitudinally. Apply the pressure load to the inner surface of the inner cylinder and fix the translation of any *one* node in the longitudinal direction to eliminate rigid-body motion in this direction.

Compare the results of parts (*a*) and (*b*) and the analytical solution of Problem 5.73(*b*). Also determine the interface pressure for each case and compare with what is obtained by the equation determined in part (*a*) of Problem 5.73.

10.9   Solve Problem 10.8 using cyclic symmetry. Solve the problem using plane-stress two-dimensional quadrilateral elements with an arbitrary thickness of 10 mm. Generate the elements as follows. Create a radial horizontal line starting at 100 mm and ending at 200 mm from the origin. Divide the line into eight equal segments. Copy-rotate-join these line segments 5° [this will match the size of the elements used in Problem 10.1, part (*a*)]. For boundary conditions, fix the vertical translation of all nodes on the horizontal line. For all nodes on the inclined 5° line apply either a boundary element, a rigid element, or a multipoint constraint equation which fixes translation perpendicular to this line. The elements for each cylinder should be assigned their unique material properties. Apply the pressure load to the inner surface of the inner cylinder. Compare your results with the solution of Problem 10.8, if obtained.

10.10 Solve the press-fit problem of Example 5.8-2 using the following procedure. Using the plane-stress two-dimensional quadrilateral create a quarter model meshing in the radial and tangential directions. The elements for each cylinder should be assigned their unique material properties. To simulate the press-fit, the inner cylinder will be forced to expand thermally. Assign a coefficient of expansion and temperature increase α and ΔT, respectively, for the inner cylinder according to the relation $\delta_r = \alpha \Delta Tb$, where δ_r and b are the radial interference and the outer radius of the inner member, respectively. Explain why axisymmetric elements will not work.

10.11 Solve Problem 5.18 using brick elements. The mesh regions are shown in Fig. (*a*). Rather than placing a half-square box around the arc, a vertical line is drawn though the origin O as shown. The circle is divided into four segments and lines are connected as shown. Line DC is drawn from the end of the arc to a distance of one-third of line BO. The reason for this is that when we mesh the regions as shown in Fig. (*b*) all line segments along line BE are equal. Finally we do a copy-rotate 180° of the mesh to create the final cross-section mesh of Fig. (*c*). After the duplicate lines are removed, the mesh is then copy-joined in the x direction in increments of 50 mm. This will create 960 elements. Apply the 2-kN load to the nodes along the line $z = 0$ at $x = 750$ mm. For boundary conditions the translation in the x, y, and z directions of all nodes at $x = 0$ should be set to zero.

However, if one desired to simulate the analytical results of Problem 5.18, Poisson expansion and contraction at the wall should be freed. To accomplish this, (1) fix the translation in the x direction of all nodes at $x = 0$, (2) at the origin apply the additional fixation of translation in the y and z directions, and (3) at any *one* node at $x = 0$ not on the origin, add the additional fixation of a translation which would eliminate rigid-body rotation about the x axis.

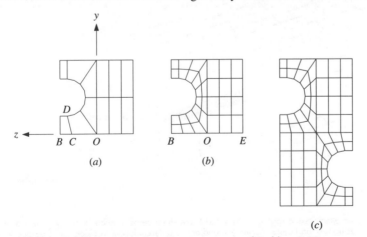

Problem 10.11   Meshing the cross section of Problem 5.18.

10.12 Solve the composite beam problem of Example 5.5-1 using plane-stress two-dimensional quadrilaterals in the xy plane.

10.13 Solve the C-clamp Examples 5.6-3 and 5.6-5 using a half model and plate elements.

10.14 Solve the C-clamp Examples 5.6-3 and 5.6-5 using a quarter model and plate elements.

10.15 Consider an end-loaded cantilever beam with the cross section given in Problem 5.32. You will need to solve Problem 5.32 and locate the shear center before attempting this problem.[†] Let the beam be 2 m long with a concentrated force of 4 kN in the negative y direction applied at the free end. Model the beam with plate elements. Divide the horizontal 50-mm lines, the horizontal 100-mm lines, and the vertical 200-mm lines into two, four, and eight segments respectively. Copy-join these lines in the x direction in increments of 100 mm to a beam length of 2.1 m. The extra length is used to build attachment points to apply the load. The load application can be accomplished by building the plates shaded as shown in the figure for each attachment point. The figure shows the force applied to the shear center. Solve the problem using the following two load cases.

(a) Attach the load on the centroid of the cross section and report on the maximum transverse shear stress in the beam and the angle of twist of the beam. Compare your results with an analytical solution. HINT: The direct transverse shear stress is covered in Sec. 5.4.2. For torsion, the shear stress and angle of twist is given in Sec. 4.3.2 under Thin Rectangular Shapes.

(b) Attach the load at the shear center of the cross section and report on the maximum transverse shear stress in the beam and angle of twist of the beam. Compare the stress to an analytical solution.

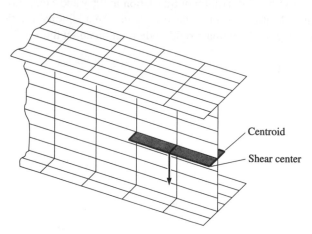

Centroid

Shear center

Problem 10.15

[†] There is an alternative to solving Problem 5.32. The shear center can be found by linear extrapolation of the results using two load cases placing the load at two different locations along a line parallel to the z axis at x = 2 m.

10.16 Solve Problem 3.10 using

(a) Quadrilateral plate elements. Divide the horizontal flange and each vertical web into four sections. Copy-join these segments in the x direction in increments of 25 mm.

(b) Quadrilateral plate elements for the horizontal flange using the same mesh density as in part (a) and offset beam elements for the vertical webs.

Compare the maximum deflections and stresses found in parts (a) and (b).

10.17 Solve Problem 5.50 using a half-model and plate elements.

10.18 Solve Problem 5.3 using plate elements. Draw the centerline of the wall of the section in the yz plane. Divide each quarter arc into four line segments. Copy-join this at 25-mm increments 22 times (we will make the tube 50 mm longer to reduce Saint-Venant's effect at $x = 500$ mm). Fix the nodes at $x = 0$ and apply a torsional moment of 100/48 N·m at each of the 48 nodes at $x = 550$ mm. For stresses, to compare with the theory of Sec. 5.2.1, review the membrane shear stresses on elements at $x = 250$ mm. Distortion will cause the shear stress to vary across the thickness. For deflections, θ_x for each node at $x = 500$ mm can be compared with the results of Problem 5.3. Alternatively, the translation of opposing nodes on the cross section at $x = 500$ mm will provide another method of comparing the angle of twist of the cross section with Problem 5.3.

10.19 Solve Problem 5.7 using plate elements. Draw the centerline of the wall of the section in the yz plane. Divide the arc into eight line segments. Divide each straight line into four segments. Copy-join this at 40-mm increments 22 times (we will make the tube 80 mm longer to reduce Saint-Venant's effect at $x = 800$ mm). Fix the nodes at $x = 0$ and apply a torsional moment of 600/20 N·m at each of the 20 nodes at $x = 880$ mm. For stresses, to compare with the theory of Sec. 5.2.1, review the membrane shear stresses on elements at $x = 400$ mm. Distortion will cause the shear stress to vary across the thickness. For deflections, θ_x for each node at $x = 800$ mm can be compared with the results of Problem 5.7. Alternatively, the translation of opposing nodes on the cross section at $x = 800$ mm will provide another method of comparing the angle of twist of the cross section with Problem 5.7.

10.20 Solve Problem 5.11 using plate elements. Draw the centerline of the wall of the section in the yz plane. Divide each quarter arc into four line segments. Divide each straight line into four segments. Copy-join this at 25-mm increments 22 times (we will make the tube 50 mm longer to reduce Saint-Venant's effect at $x = 500$ mm). Fix the nodes at $x = 0$ and apply a torsional moment of 20/28 kN·m at each of the 28 nodes at $x = 550$ mm. For stresses, to compare with the theory of Sec. 5.2.1, review the membrane shear stresses on elements at $x = 250$ mm. Distortion will cause the shear stress to vary across the thickness. For deflections, θ_x for each node at $x = 500$ mm can be compared with the results of Problem 5.11. Alternatively, the translation of opposing nodes on the cross section at

$x = 800$ mm will provide another method of comparing the angle of twist of the cross section with Problem 5.11.

10.21 Solve Problem 5.60(a) using plate elements. (a) Use a full model with a 20×40 mesh, (b) Use a half model with a 20×40 mesh, and (c) Use a quarter model with a 20×40 mesh. Compare all results with the analytical solution.

10.22 Solve Problem 5.60(d) using plate elements. (a) Use a full model with a 20×40 mesh, (b) Use a half model with a 20×40 mesh, and (c) Use a quarter model with a 20×40 mesh. Compare all results with the approximate solution suggested in Problem 5.60(d).

10.23 Solve Problem 5.67 using plate elements. Mesh the plate with 10 elements radially and 25 elements tangentially. (a) Use a full model with 250 uniformly spaced elements, (b) Use a half model with 250 uniformly spaced elements, (c) Use a quarter model with 250 uniformly spaced elements, and (d) Use a full model with 250 elements uniformly spaced tangentially and geometrically spaced radially to reduce the aspect ratios of the elements. Compare all results with the analytical solution.

10.24 Solve Problem 5.83 with a quarter model using plane-stress two-dimensional quadrilateral elements. After obtaining the solution, double the mesh density and compare results. Compare both results with the reference cited in Problem 5.83.

10.25 Verify the results of Examples 6.6-1, 6.6-2, and 6.6-3 with $L = 200$ mm, $A = 50$ mm^2, $E = 200$ GPa, and $P = 5$ kN.

10.26 Verify the results of Example 6.6-9 with $a = 2$ in, $b = 3$ in, $c = 5$ in, $d = 0.25$ in, $E = 30$ Mpsi, $v = 0.3$, and $P = 15$ lbf.

10.27 Verify the results of Example 6.6-10. Draw the arc and divide it into 20 segments to create 20 beam elements. Let the diameter of the cross section of the beams be 6 mm, $R = 30$ mm, $E = 200$ GPa, and $P = 200$ N.

10.28 Verify the results of Example 6.7-1 using two-dimensional plane-stress quadrilateral elements. Distribute the 1000-lbf force uniformly over the line of nodes on the horizontal line though point A.

10.29 Verify the results of Example 6.8-3 using beam elements. Let the diameter of the beam cross section be 0.25 in, $R = 2$ in, $E = 30$ Mpsi, and $P = 80$ lbf. (a) Solve the full model using 24 elements, (b) Solve the half model using 24 elements, and (c) Solve the quarter model using 24 elements. Compare the results with the equations given in Example 6.8-3 for M_A, M_C, and $(\delta_B)_V$.

10.30 Model the structure shown in Fig. 10.2-4(a). Let $L_b = 500$ mm, $\beta = 30°$, $A_b = 75$ mm^2, $I_b = 600$ mm^4, $A_c = 50$ mm^2, $E_c = E_b = 200$ GPa, and $P = 5$ kN. Determine the vertical deflection of the applied force. Determine the maximum tensile and compressive stresses in each member. Let the section modulus of the beam be $7.5(10^3)$ mm^3. Compare the results with that obtained by analytical methods. (a) Solve the problem using beam elements only, and (b) Model BC with beam elements and CD with a truss element.

10.31 Solve Problem 6.37 using beam elements. Model the arc with 20 elements and the straight section with one. Let the diameter of the beams be 6 mm, $R = 40$ mm, $L = 50$ mm, and $E = 200$ GPa. Compare the results with that obtained by analytical methods.

10.32 Solve Problem 6.47 using beam elements.
(a) Solve the full model using 48 elements.
(b) Use cyclic symmetry with 16 elements. Orient the model such that standard boundary conditions can be applied to one end. On the other end use boundary elements to impose conditions on the translational displacement perpendicular to the radial direction. Rotation boundary conditions must also be applied at both ends.

Compare the results of both parts with that obtained by analytical methods.

10.33 Solve Problem 6.50. Let the cross section of beams AC and BC each be square 10 mm $\times$ 10 mm, $L = 500$ mm, $P = 2$ kN, and $E = 200$ GPa. Model the structure as follows. Model beam AC using three beams, one beam of length $L/2$ from A to B and two beams each of length $L/4$ from B to C. Model beam BC with two beams of length $L/4$, but 5 mm *below* the beams making up beam AC. Connect points B of the upper and lower beams with a beam element of length 5 mm. Release rotation perpendicular to the plane of the figure for *both* ends of the short beam segment. This will allow only a force to transmit between the beams. Dividing both beam segments BC into two beams will allow for the testing as to whether they separate or not. If they do as they should, then the model is correct. Compare the results with that obtained from the analytical solution.

10.34 Solve Problem 6.49 using beam elements. Let $L = 20$ in, $a = 5$ in, $\epsilon = 0.25$ in, $I = 0.05$ in^4, $P = 200$ lbf, and $E = 10$ Mpsi. Compare the results with that obtained from the analytical solution.

10.35 A thin steel disk with an inner diameter of 50 mm and an outer diameter of 500 mm is rotating axially at 2000 rpm. Using axisymmetric elements determine how the radial and tangential stresses vary in the radial direction. Compare the results with Eqs. (5.8-29) and (5.8-30).

10.36 Using brick elements model a rectangular bar with a 25 mm $\times$ 50 mm cross section loaded by a torsional moment $T = 500$ N·m. Fix the bar at one end and apply *forces* at the other end to provide the necessary torque. Make the bar long enough such that Saint-Venant's effect will not affect the results at the midspan of the bar. Compare the shear stress results with Eq. (4.3-11) and/or the distribution given on pp. 309–313 of Ref 5.3 [combine Eqs. (h) and (171)].

10.37 Model the half-plane problem shown in Fig. 5.9-4 with $a = 25$ mm and $p_0 = 1.0$ MPa. Use the plane-stress two-dimensional quadrilateral with an arbitrary thickness of 5 mm. Model the half plane as a square region, 250 mm $\times$ 250 mm. Center the uniform load on the top of the region. Constrain the translation of the bottom point of the half plane at coordinates (250, 0) mm in the x and y directions. Constrain the remaining nodes on the bottom of the plane in the x direction only. Compare the results with that obtained from the equations for σ_x, σ_y, and τ_{xy} given in Example 5.9-1 at points (10, 0), (25, 0), (50, 0), (100, 0), and (25, 25) mm.

10.38 Model an actual "card" table or similar table with a uniform pressure of 0.1 psi over a 10 in $\times$ 10 in area centered on the table top. For modeling purposes, estimate anything that you cannot directly measure.

10.39 Create a structure similar to Fig. 10.2-5 and experiment with creating:

(a) A model using plane-stress two-dimensional quadrilateral elements. Model fillets at the connection with the long narrow appendage and the main rectangular region.

(b) A model where the long narrow appendage is modeled as a beam element. Try different meshes of the block and various interior beam connection schemes and compare the results to part (a).

10.40 Use submodeling to solve Problem 10.24. Compare the results with that problem and/or the reference cited in Problem 5.83.

10.10 REFERENCES

10.1 Cook, R. D. *Finite Element Modeling for Stress Analysis*. New York: Wiley, 1995.

10.2 Young, W. *Roark's Formulas for Stress and Strain*, 6th ed. New York: McGraw-Hill, 1989.

10.3 Zienkiewicz, O. C., and R. L. Taylor. *The Finite Element Method*, 4th ed., vols. 1 and 2. New York: McGraw-Hill, 1989, 1991.

10.4 Löhner, R., and P. Parikh. "Three-Dimensional Grid Generation by the Advancing-Front Method," *Int. J. of Numerical Methods in Fluids*, no. 2, 1998, pp. 1135–1149.

10.5 Perucchio, R.; M. Saxena; and A. Kela. "Automatic Mesh Generation from Solid Models Based on Recursive Spatial Decompositions," *Int. J. of Numerical Methods in Eng.*, vol. 28, 1989, pp. 2469–2501.

Additional Uncited References in Finite Elements

10.6 Bathe, K. J. *Finite Element Procedures*. Englewood Cliffs, NJ: Prentice Hall, 1996.

10.7 Cook, R. D.; D. S. Malkus; and M. E. Plesha. *Concepts and Applications of Finite Element Analysis*, 3rd ed. New York: Wiley, 1989.

APPENDIX

A

SI AND USCU CONVERSIONS†

In today's global industrial climate SI units have become the predominant system of units. However, in the United States, the USCU system is still used in many areas. Thus the analyst is quite often compelled to understand and use both systems. Table A.1 presents the conversions of some common state property units. The most common terms in stress analysis are force, distance, and stress. Conversion approximations are: multiply force in lbf, distance in inches, and stress in psi by 4.5, 25, and 7000 to obtain force in N, distance in mm, and stress in Pa, respectively.

Table A.1 Multiplication factors to convert from USCU units to SI units

To Convert from USCU	To SI	Multiply by
Area:		
ft^2	m^2	$9.29 \, (10^{-2})$
in^2	m^2	$6.452 \, (10^{-4})$
Density:		
$slugs/ft^3 \, (lbf{\cdot}s^2/ft^4)$	kg/m^3	$5.152 \, (10^2)$
$lbf{\cdot}s^2/in^4$	kg/m^3	$2.485 \, (10^{-2})$
Energy, work, moment:		
$ft{\cdot}lbf$ or $lbf{\cdot}ft$	J or N·m	1.356
$in{\cdot}lbf$ or $lbf{\cdot}in$	J or N·m	$1.13 \, (10^{-1})$
Force:		
lbf	N	4.448
Length:		
ft	m	$3.048 \, (10^{-1})$
in	m	$2.54 \, (10^{-2})$

† SI and USCU are abbreviations for the International System of Units (from the French Systéme International d'Unités) and the United States Customary Units, respectively.

Table A.1 (*concluded*)

To Convert from USCU	To SI	Multiply by
Mass:		
slugs (lbf·s^2/ft)	kg	14.59
lbf·s^2/in	kg	1.216
Pressure, stress:		
lbf/ft^2	Pa (N/m^2)	47.88
lbf/in^2	Pa (N/m^2)	6.895 (10^3)
Volume:		
ft^3	m^3	2.832 (10^{-2})
in^3	m^3	1.639 (10^{-5})

PROPERTIES OF CROSS SECTIONS

B.1   TABLES

Section	Area	Centroid	Second-Area Moments
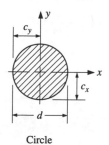 Rectangle	bh	$c_x = \dfrac{h}{2}$ $c_y = \dfrac{b}{2}$	$I_x = \dfrac{bh^3}{12}$ $I_y = \dfrac{hb^3}{12}$
 Circle	$\dfrac{\pi d^2}{4}$	$c_x = \dfrac{d}{2}$ $c_y = \dfrac{d}{2}$	$I_x = \dfrac{\pi d^4}{64}$ $I_y = \dfrac{\pi d^4}{64}$ $I_z = J = \dfrac{\pi d^4}{32}$

Section	Area	Centroid	Second-Area Moments
Thick-walled tube	$\dfrac{\pi}{4}(d_o^2 - d_i^2)$	$c_x = \dfrac{d_o}{2}$ $c_y = \dfrac{d_o}{2}$	$I_x = \dfrac{\pi}{64}(d_o^4 - d_i^4)$ $I_y = \dfrac{\pi}{64}(d_o^4 - d_i^4)$ $I_z = J = \dfrac{\pi}{32}(d_o^4 - d_i^4)$
Thin-walled tube $(\bar{d} \gg t)$	$\pi \bar{d} t$	$c_x = \dfrac{\bar{d}}{2}$ $c_y = \dfrac{\bar{d}}{2}$	$I_x = \dfrac{\pi}{8} t \bar{d}^3$ $I_y = \dfrac{\pi}{8} t \bar{d}^3$ $I_z = J = \dfrac{\pi}{4} t \bar{d}^3$
Circular quadrant	$\dfrac{\pi}{4} r^2$	$c_x = \dfrac{4r}{3\pi}$ $c_y = \dfrac{4r}{3\pi}$	$I_x = \left(\dfrac{\pi}{16} - \dfrac{4}{9\pi}\right) r^4$ $I_y = \left(\dfrac{\pi}{16} - \dfrac{4}{9\pi}\right) r^4$ $P_{xy} = \left(\dfrac{1}{8} - \dfrac{4}{9\pi}\right) r^4$
Triangle	$\dfrac{1}{2} bh$	$c_x = \dfrac{h}{3}$ $c_y = \dfrac{b}{3}$	$I_x = \dfrac{bh^3}{36}$ $I_y = \dfrac{hb^3}{36}$ $P_{xy} = \dfrac{b^2 h^2}{72}$

Section	Area	Centroid	Second-Area Moments
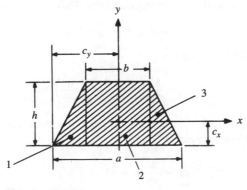Circular sector	αr^2	$c_y = \dfrac{2r}{3\alpha}\sin\alpha$	$I_x = \dfrac{r^4}{8}(2\alpha - \sin 2\alpha)$ $I_y = \dfrac{r^4}{8}(2\alpha + \sin 2\alpha)$

B.2 COMBINATIONS OF SECTIONS

Find the location of the centroids and the second-area moments of the trapezoid section shown in Fig. B.2-1.   **Example B.2-1**

Solution:

By symmetry, $c_y = a/2$. To determine the properties of the cross section, consider the section as one rectangle and two triangles. The area is

$$A = \frac{1}{2}\frac{a-b}{2}h + bh + \frac{1}{2}\frac{a-b}{2}h = \frac{h}{2}(a+b) \qquad \text{[a]}$$

From the base, the location of the centroid of each triangle is $h/3$ and of the rectangle is $h/2$. Thus

$$Ac_x = A_1c_{1x} + A_2c_{2x} + A_3c_{3x} = 2A_1c_{1x} + A_2c_{2x}$$

or

$$\frac{h}{2}(a+b)c_x = 2\left(\frac{1}{2}\frac{a-b}{2}h\right)\frac{h}{3} + bh\frac{h}{2}$$

Figure B.2-1

Simplifying and solving for c_x yields

$$c_x = \frac{h}{3}\left(\frac{a + 2b}{a + b}\right) \qquad [b]$$

To find the second-area moments, the parallel-axis theorem is used. For the rectangular section, the moment of inertia about its own horizontal centroid is $bh^3/12$. The distance from the centroid of the rectangle to the centroid of the trapezoid section is

$$\frac{h}{2} - \frac{h}{3}\left(\frac{a + 2b}{a + b}\right)$$

which simplifies to

$$\frac{h}{6}\left(\frac{a - b}{a + b}\right)$$

Thus, the second-area moment of the rectangle about the x axis is

$$(I_x)_2 = \frac{1}{12}bh^3 + bh\left(\frac{h}{6}\frac{a - b}{a + b}\right)^2 \qquad [c]$$

For each triangular section, the second-area moment about the horizontal centroid is

$$\frac{1}{36}\frac{a - b}{2}h^3$$

and the distance between the horizontal centroid of the triangular section and the centroid of the trapezoidal section is

$$\frac{h}{3}\frac{a + 2b}{a + b} - \frac{h}{3}$$

which simplifies to $hb/3(a + b)$. Thus, the second-area moment of each triangular section is

$$(I_x)_1 = (I_x)_3 = \frac{1}{36}\frac{a - b}{2}h^3 + \frac{1}{2}\frac{a - b}{2}h\left[\frac{hb}{3(a + b)}\right]^2 \qquad [d]$$

From Eqs. (c) and (d) the total second-area moment of the trapezoidal section about its horizontal centroidal axis is obtained by adding each term. Thus

$$I_x = (I_x)_1 + (I_x)_2 + (I_x)_3$$

and simplifying the final results yields

$$I_x = \frac{h^3}{36}\left(\frac{a^2 + 4ab + b^2}{a + b}\right) \qquad [e]$$

Applying the same approach, the reader should verify that the second-area moment of the trapezoidal section about the y axis is

$$I_y = \frac{h}{48}(a + b)(a^2 + b^2) \qquad [f]$$

BEAMS IN BENDING

C.1   Cantilever with End Load.

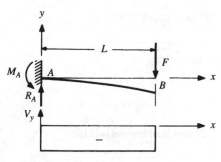

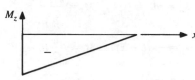

External and internal reactions

$$R_A = -V_y = F$$

$$M_A = FL$$

$$M_z = F(x - L)$$

Deflection

$$v_c = \frac{Fx^2}{6EI}(x - 3L)$$

$$(v_c)_{x=L} = -\frac{FL^3}{3EI}$$

Slope

$$\theta = \frac{dv_c}{dx} = \frac{Fx}{2EI}(x - 2L)$$

C.2   Cantilever with Intermediate Load.

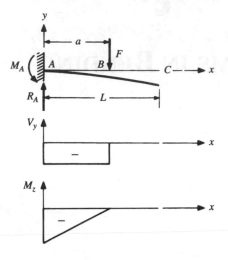

External and internal reactions

$$R_A = -(V_y)_{AB} = F \qquad M_A = Fa$$

$$(M_z)_{AB} = F(x - a) \qquad (M_z)_{AB} = (V_y)_{BC} = 0$$

Deflection

$$(v_c)_{AB} = \frac{Fx^2}{6EI}(x - 3a)$$

$$(v_c)_{BC} = \frac{Fa^2}{6EI}(a - 3x)$$

$$(v_c)_{x=L} = \frac{Fa^2}{6EI}(a - 3L)$$

Slope

$$(\theta)_{AB} = \frac{Fx}{2EI}(x - 2a)$$

$$(\theta)_{BC} = -\frac{Fa^2}{2EI}$$

C.3   Cantilever with Uniform Load.

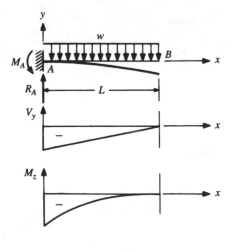

External and internal reactions

$$R_A = wL \qquad\qquad M_A = \frac{wL^2}{2}$$

$$V_y = -w(L - x) \qquad M_z = -\frac{w}{2}(L - x)^2$$

Deflection

$$v_c = \frac{wx^2}{24EI}(4Lx - x^2 - 6L^2)$$

$$(v_c)_{x=L} = -\frac{wL^4}{8EI}$$

Slope

$$\theta = \frac{wx}{6EI}(3Lx - x^2 - 3L^2)$$

C.4 Cantilever with Moment Load.

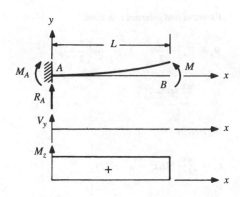

External and internal reactions

$$R_A = V_y = 0 \qquad M_A = M_z = M$$

Deflection

$$v_c = \frac{Mx^2}{2EI} \qquad (v_c)_{x=L} = \frac{ML^2}{2EI}$$

Slope

$$\theta = \frac{Mx}{EI}$$

C.5 Simple Supports with Intermediate Load.

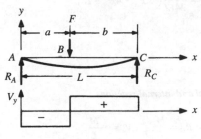

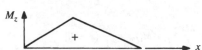

External and internal reactions

$$R_A = \frac{Fb}{L} \qquad R_c = \frac{Fa}{L}$$

$$(V_y)_{AB} = -R_A \qquad (V_y)_{BC} = R_B$$

$$(M_z)_{AB} = \frac{Fbx}{L} \qquad (M_z)_{BC} = \frac{Fa}{L}(L - x)$$

Deflection

$$(v_c)_{AB} = \frac{Fbx}{6EIL}(x^2 + b^2 - L^2)$$

$$(v_c)_{BC} = \frac{Fa(L - x)}{6EIL}(x^2 + a^2 - 2Lx)$$

Slope

$$\theta_{AB} = \frac{Fb}{6EIL}(3x^2 + b^2 - L^2)$$

$$\theta_{BC} = \frac{Fa}{6EIL}(6Lx - 3x^2 - a^2 - 2L^2)$$

C.6   Simple Supports with Uniform Load.

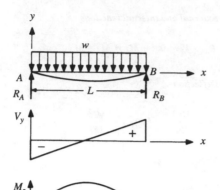

External and internal reactions

$$R_A = R_B = \frac{wL}{2} \qquad V_y = -w\left(\frac{L}{2} - x\right)$$

$$M_z = \frac{wx}{2}(L - x)$$

Deflection

$$v_c = \frac{wx}{24EI}(2Lx^2 - x^3 - L^3)$$

$$(v_c)_{x=\frac{L}{2}} = -\frac{5wL^4}{384EI}$$

Slope

$$\theta = \frac{w}{24EI}(6Lx^2 - 4x^3 - L^3)$$

C.7   Simple Supports with Moment Load.

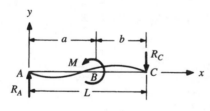

External and internal reactions

$$R_A = R_C = \frac{M}{L} \qquad V_y = -\frac{M}{L}$$

$$(M_z)_{AB} = \frac{Mx}{L} \qquad (M_z)_{BC} = \frac{M}{L}(x - L)$$

Deflection

$$(v_c)_{AB} = \frac{Mx}{6EIL}(x^2 + 3a^2 - 6aL + 2L^2)$$

$$(v_c)_{BC} = \frac{M}{6EIL}[x^3 - 3Lx^2 + x(2L^2 + 3a^2) - 3a^2L]$$

Slope

$$(\theta)_{AB} = \frac{M}{6EIL}(3x^2 + 3a^2 - 6aL + 2L^2)$$

$$(\theta)_{BC} = \frac{M}{6EIL}(3x^2 - 6Lx + 2L^2 + 3a^2)$$

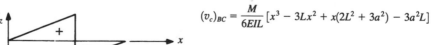

C.8   Simple Supports with Overhanging Load.

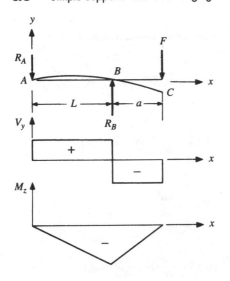

External and internal reactions

$$R_A = \frac{Fa}{L} \qquad R_B = \frac{F}{L}(L + a)$$

$$(V_y)_{AB} = \frac{Fa}{L} \qquad (V_y)_{BC} = -F$$

$$(M_z)_{AB} = -\frac{Fax}{L} \qquad (M_z)_{BC} = F(x - L - a)$$

Deflection

$$(v_c)_{AB} = \frac{Fax}{6EIL}(L^2 - x^2)$$

$$(v_c)_{BC} = \frac{F(x - L)}{6EI}[(x - L)^2 - a(3x - L)]$$

Slope

$$(\theta)_{AB} = \frac{Fa}{6EIL}(L^2 - 3x^2)$$

$$(\theta)_{BC} = \frac{F}{6EI}[3x^2 - 6x(L + a) + L(3L + 4a)]$$

C.9   Cantilever with Partial Distributed Load.

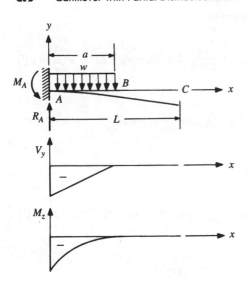

External and internal reactions

$$R_A = wa \qquad M_A = \frac{wa^2}{2}$$

$$(V_y)_{AB} = -w(a - x) \qquad (V_y)_{BC} = 0$$

$$(M_z)_{AB} = -\frac{w}{2}(a - x)^2 \qquad (M_z)_{BC} = 0$$

Deflection

$$(v_c)_{AB} = \frac{wx^2}{24EI}(4ax - x^2 - 6a^2)$$

$$(v_c)_{BC} = \frac{wa^3}{24EI}(a - 4x)$$

Slope

$$(\theta)_{AB} = \frac{wx}{6EI}(3ax - x^2 - 3a^2)$$

$$(\theta)_{BC} = -\frac{wa^3}{6EI}$$

C.10   Simple Supports and Partial Distributed Load.

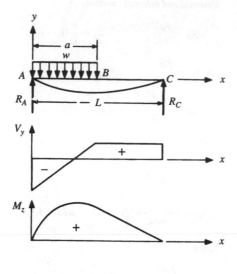

External and internal reactions

$$R_A = \frac{wa}{2L}(2L - a) \qquad R_C = \frac{wa^2}{2L}$$

$$(V_y)_{AB} = \frac{w}{2L}[2L(x - a) + a^2]$$

$$(V_y)_{BC} = \frac{wa^2}{2L}$$

$$(M_z)_{AB} = \frac{wx}{2L}(2aL - a^2 - xL)$$

$$(M_z)_{BC} = \frac{wa^2}{2L}(L - x)$$

Deflection

$$(v_c)_{AB} = \frac{wx}{24EIL}[2ax^2(2L - a) - Lx^3 - a^2(2L - a)^2]$$

$$(v_c)_{BC} = (y_c)_{AB} + \frac{w}{24EI}(x - a)^4$$

Slope

$$(\theta)_{AB} = \frac{w}{24EIL}[6ax^2(2L - a) - 4Lx^3 - a^2(2L - a)^2]$$

$$(\theta)_{BC} = (\theta)_{AB} + \frac{w}{6EI}(x - a)^3$$

C.11 One End Fixed and One Simple Support with Intermediate Load.

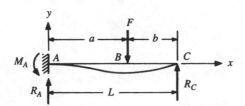

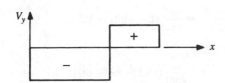

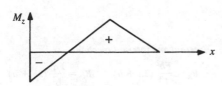

External and internal reactions

$$R_A = \frac{Fb}{2L^3}(3L^2 - b^2) \qquad R_C = \frac{Fa^2}{2L^3}(3L - a)$$

$$M_A = \frac{Fb}{2L^2}(L^2 - b^2)$$

$$(V_y)_{AB} = -R_A \qquad (V_y)_{BC} = R_C$$

$$(M_z)_{AB} = \frac{Fb}{2L^3}[b^2L - L^3 + x(3L^2 - b^2)]$$

$$(M_z)_{BC} = \frac{Fa^2}{2L^3}(3L^2 - 3Lx - aL + ax)$$

Deflection

$$(v_c)_{AB} = \frac{Fbx^2}{12EIL^3}[3L(b^2 - L^2) + x(3L^2 - b^2)]$$

$$(v_c)_{BC} = (y_c)_{AB} - \frac{F(x - a)^3}{6EI}$$

Slope

$$(\theta)_{AB} = \frac{Fbx}{4EIL^3}[2L(b^2 - L^2) + x(3L^2 - b^2)]$$

$$(\theta)_{BC} = (\theta)_{AB} - \frac{F(x - a)^2}{2EI}$$

C.12   One End Fixed and One Simple Support with Uniform Load.

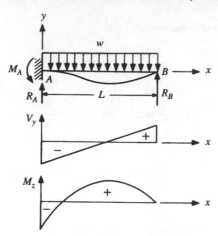

External and internal reactions

$$R_A = \frac{5wL}{8} \qquad R_B = \frac{3wL}{8} \qquad M_A = \frac{wL^2}{8}$$

$$V_y = \frac{w}{8}(8x - 5L)$$

$$M_z = -\frac{w}{8}(4x^2 - 5Lx + L^2)$$

Deflection

$$v_c = \frac{wx^2}{48EI}(L - x)(2x - 3L)$$

Slope

$$\theta = \frac{wx}{48EI}(15xL - 8x^2 - 6L^2)$$

C.13   Fixed Supports with Intermediate Load.

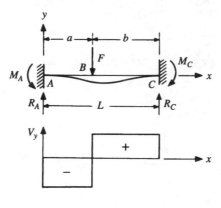

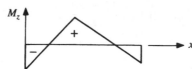

External and internal reactions

$$R_A = \frac{Fb^2}{L^3}(3a + b) \qquad R_C = \frac{Fa^2}{L^3}(3b + a)$$

$$M_A = \frac{Fab^2}{L^2} \qquad\qquad M_C = \frac{Fa^2b}{L^2}$$

$$(V_y)_{AB} = -R_A \qquad\qquad (V_y)_{BC} = R_C$$

$$(M_z)_{AB} = \frac{Fb^2}{L^3}[x(3a + b) - La]$$

$$(M_z)_{BC} = (M_z)_{AB} - F(x - a)$$

Deflection

$$(v_c)_{AB} = \frac{Fb^2x^2}{6EIL^3}[x(3a + b) - 3aL]$$

$$(v_c)_{BC} = \frac{Fa^2(L - x)^2}{6EIL^3}[(L - x)(3b + a) - 3bL]$$

Slope

$$(\theta)_{AB} = \frac{Fb^2x}{2EIL^3}[x(3a + b) - 2L]$$

$$(\theta)_{BC} = \frac{Fa^2(L - x)}{2EIL^3}[2bL - (L - x)(b + a)]$$

C.14   Fixed Supports with Uniform Load.

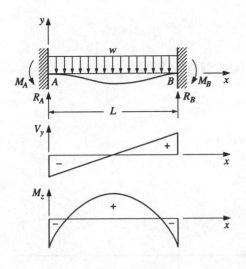

External and internal reactions

$$R_A = R_B = \frac{wL}{2} \qquad M_A = M_B = \frac{wL^2}{12}$$

$$V_y = -\frac{w}{2}(L - 2x)$$

$$M_z = \frac{w}{12}(6Lx - 6x^2 - L^2)$$

Deflection

$$v_c = -\frac{wx^2}{24EI}(L - x)^2$$

Slope

$$\theta = -\frac{wx}{12EI}(L - x)(L - 2x)$$

APPENDIX
D

SINGULARITY FUNCTIONS

D.0 INTRODUCTION

In this appendix, singularity functions are developed for beams in bending, but the functions are applicable as well to problems involving axial or torsional loading. In Chap. 3 the method of section isolation was presented for beams in bending, where it was shown that whenever a discontinuity in the loading occurs, it is necessary to rewrite the shear-force and bending-moment equations. In this way, in each section where the loading is continuous, a separate set of equations exists. For many practical problems, the loading of a beam may be highly dicontinuous, and section isolation is cumbersome. In addition, if the deflections of such beams are desired, use of the double-integration method can be a nightmare, as each section must be such that the slopes and deflection of the beam are continuous. It is much simpler to solve such problems by the method of superposition, where each specific type of load, i.e., concentrated forces, distributed forces, concentrated moments, etc., can be analyzed separately and the results combined. The effective use of this technique requires that (1) the analyst have a comprehensive list of solutions for the various forms of loading and support conditions such as in Appendix C;[†] and (2) the analyst exercise extreme care in calculations, as the superposition of many equations can lead to errors. Singularity functions, if used correctly, provide the most direct solution to a beam loaded in a highly discontinuous manner. Since singularity functions can represent discontinuities, the load intensity (force per unit length) as a function of axial position can be written in the form of only *one* equation. A further advantage is that this can be done simply by visual inspection of the beam-loading diagram. Direct integration then yields the shear-force equation for the *entire* beam. Integration of the shear-force equation then results in the bending-moment equation for the *entire* beam. Integration of the bending-moment equation provides the equation for the slope of the beam at any point where

---

[†] Although Appendix C contains many of the common forms of loading and support conditions, the list is far from complete.

only one constant of integration results. A final integration results in the equation for deflection of the beam where a second constant of integration arises. Thus, only two constants of integration are necessary corresponding to the support boundary conditions and only *one* set of equations for the load intensity, shear force, bending moment, slope, and deflection results. However, one important thing to keep in mind is that the singularity functions themselves are discontinuous and that the resulting behavior of the beam is still discontinuous. The analyst using singularity functions must have a complete understanding of them and how one performs the necessary integration of them. Before the development of these functions, it is necessary to review the integral relations between load intensity, shear force, bending moment, slope, and deflections.

D.1   INTEGRAL RELATIONS FOR BEAMS IN BENDING

Consider a beam element of infinitesimal length dx as shown in Fig. D.1-1. V_y and M_z are the shear force and bending moment at position x. The load intensity q (force per unit length) is denoted by $q(x)$ and is defined positive in the positive y direction. Since the shear force and bending moment, in general, change as functions of x, at $x + dx$ the shear force and bending moment can be described by $V_y + dV_y$ and $M_z + dM_z$.

Summing forces in the y direction results in

$$dV_y + q(x)\,dx = 0$$

$$\frac{dV_y}{dx} = -q(x) \qquad\qquad \textbf{[D.1-1]}$$

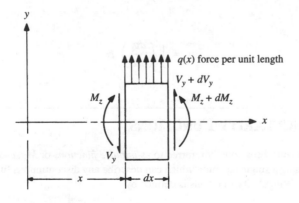

Figure D.1-1

Therefore the integral relationship is given by

$$V_y = -\int q(x)\, dx \qquad\qquad \textbf{[D.1-2]}$$

Summing moments at the center of the element about the z axis yields

$$M_z + dM_z - M_z + V_y \frac{dx}{2} + (V_y + dV_y)\frac{dx}{2} = 0$$

or

$$\frac{dM_z}{dx} + V_y + \frac{1}{2}dV_y = 0$$

The first two terms of this equation are finite, whereas dV_y is infinitesimal. Thus

$$\frac{dM_z}{dx} + V_y = 0$$

or

$$\frac{dM_z}{dx} = -V_y \qquad\qquad \textbf{[D.1-3]}$$

and the integral relationship is

$$M_z = -\int V_y\, dx \qquad\qquad \textbf{[D.1-4]}$$

The integral relations for obtaining the slope and deflection equations were given in Chap. 3 and are

$$\frac{dv_c}{dx} = \int \frac{M_z}{EI}\, dx \qquad\qquad \textbf{[D.1-5]}$$

and

$$v_c = \int \left(\frac{dv_c}{dx}\right) dx \qquad\qquad \textbf{[D.1-6]}$$

D.2   SINGULARITY FUNCTIONS

Singularity functions, occasionally referred to as *impulse functions* or *Macaulay's functions,* are used in forming a single equation which can describe any discontinuous function. A family of polynomial singularity functions is defined by

$$F_n(x) = \langle x - a \rangle^n \qquad\qquad \textbf{[D.2-1]}$$

where n is any integer. The functions $F_n(x)$ have some unique properties and in terms of physical applications will be restricted here to values of n equal to or greater than $n = -2$. The functions $F_n(x)$ behave in the following manner:

$$F_{-2}(x) = \langle x - a \rangle^{-2} = \begin{cases} \pm\infty & \text{when } x = a \\ 0 & \text{when } x \neq a \end{cases} \qquad \text{[D.2-2a]}$$

$$F_{-1}(x) = \langle x - a \rangle^{-1} = \begin{cases} \infty & \text{when } x = a \\ 0 & \text{when } x \neq a \end{cases} \qquad \text{[D.2-2b]}$$

and for $n \geq 0$,

$$F_n(x) = \langle x - a \rangle^n = \begin{cases} (x - a)^n & \text{when } x > a \\ 0 & \text{when } x \leq a \end{cases} \qquad \text{[D.2-2c]}$$

For $n < 0$, the singularity function has little physical significance in terms of actual applications. However, after one or two integrations, $F_{-1}(x)$ and $F_{-2}(x)$ will be of the form of Eq. (D.2-2c), where the function becomes physically meaningful.

Integration of Eqs. (D.2-2) is performed in the following fashion:

$$\int \langle x - a \rangle^n \, dx = \begin{cases} \langle x - a \rangle^{n+1} + C & \text{for } n < 0 \qquad \text{[D.2-3a]} \\[2ex] \dfrac{1}{n+1} \langle x - a \rangle^{n+1} + C & \text{for } n \geq 0 \qquad \text{[D.2-3b]} \end{cases}$$

where C is the constant of integration. Thus, for example,

$$\int \langle x - a \rangle^{-2} \, dx = \langle x - a \rangle^{-1} + C \qquad \text{[a]}$$

and

$$\int \langle x - a \rangle^1 \, dx = \frac{1}{2}\langle x - a \rangle^2 + C \qquad \text{[b]}$$

The singularity functions $F_n(x)$ are shown graphically in Fig. D.2-1. The function $F_{-2}(x)$ is called a *doublet* and in terms of load intensity (force per unit length) can be used to describe a concentrated moment on a beam. The function $F_{-1}(x)$ is called an *impulse* and can be used to describe a concentrated force. The function $F_0(x)$ is called a *unit step* and can be used to describe the addition of a uniform load initiating at $x = a$. The function $F_1(x)$ is called a *unit ramp*. Functions of higher order have no specific identification. To understand the application of singularity functions to beams in bending, consider the following example.

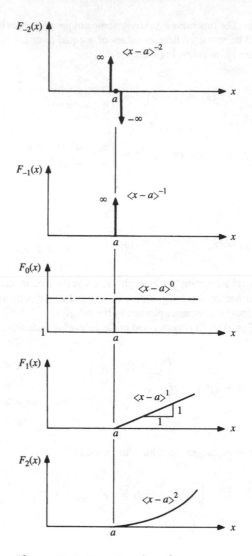

Figure D.2-1   Singularity functions.

Example D.2-1

For the beam loaded as shown in Fig. D.2-2(a), develop the load-intensity equation using singularity functions.

Solution:

Figure D.2-2(b) shows the free-body diagram for the beam, where R_A and R_E are solvable from the equilibrium equations and all loads are shown in a perspective to show their sign with respect to the positive y direction.

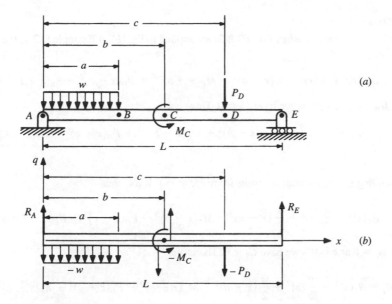

Figure D.2-2

If we use the singularity functions described by Eq. (D.2-2), the load-intensity equation is described by

$$q(x) = R_A \langle x - 0 \rangle^{-1} - w\langle x - 0 \rangle^0 + w\langle x - a \rangle^0 - M_C\langle x - b \rangle^{-2} - P_D\langle x - c \rangle^{-1} + R_E\langle x - L \rangle^{-1}$$

[a]

Terms like $\langle x - 0 \rangle^n$ can be written simply $\langle x \rangle^n$. Thus,

$$q(x) = R_A\langle x \rangle^{-1} - w\langle x \rangle^0 + w\langle x - a \rangle^0 - M_C\langle x - b \rangle^{-2} - P_D\langle x - c \rangle^{-1} + R_E\langle x - L \rangle^{-1}$$

[b]

Note that the term $+w\langle x - a \rangle^0$ is necessary to cancel the uniform load w at $x = a$.

Thus it can be seen that the load-intensity equation for the entire beam can be written by inspection and in the form of only one equation. From Eq. (D.1-2) the shear-force equation is arrived at by integrating $q(x)$ and changing the sign. To obtain the bending-moment equation, the shear-force equation is integrated, and again the sign is reversed, as indicated by Eq. (D.1-4).

For Example D.2-1, express the shear-force and bending-moment equations in terms of singularity functions and graph the results.

Example D.2-2

Solution:

Since $V_y = -\int q(x)\,dx$, when Eqs. (D.2-3) are applied to Eq. (b) of Example D.2-1, the shear force is

$$V_y = -R_A\langle x\rangle^0 + w\langle x\rangle^1 - w\langle x - a\rangle^1 + M_C\langle x - b\rangle^{-1} + P_D\langle x - c\rangle^0 - R_E\langle x - L\rangle^0 + C_1$$

Since for $x = 0^-$, $V_y = 0$, we have $C_1 = 0$. Thus

$$V_y = -R_A\langle x\rangle^0 + w\langle x\rangle^1 - w\langle x - a\rangle^1 + M_C\langle x - b\rangle^{-1} + P_D\langle x - c\rangle^0 - R_E\langle x - L\rangle^0$$

[c]

The bending-moment equation results from $M_z = -\int V_y\,dx$; thus

$$M_z = R_A\langle x\rangle^1 - \frac{w}{2}\langle x\rangle^2 + \frac{w}{2}\langle x - a\rangle^2 - M_C\langle x - b\rangle^0 - P_D\langle x - c\rangle^1 + R_E\langle x - L\rangle^1 + C_2$$

Since $M_z = 0$ at $x = 0^-$, we have $C_2 = 0$. Therefore

$$M_z = R_A\langle x\rangle^1 - \frac{w}{2}\langle x\rangle^2 + \frac{w}{2}\langle x - a\rangle^2 - M_C\langle x - b\rangle^0 - P_D\langle x - c\rangle^1 + R_E\langle x - L\rangle^1$$

[d]

In general, the first two integrations of $q(x)$ will not yield any constants of integration, and it is therefore common practice to exclude the constants when obtaining V_y and M_z.

One must be careful when transforming the singularity functions of V_y and M_z into graphical form. Equations (c) and (d) are shown in Fig. D.2-3.

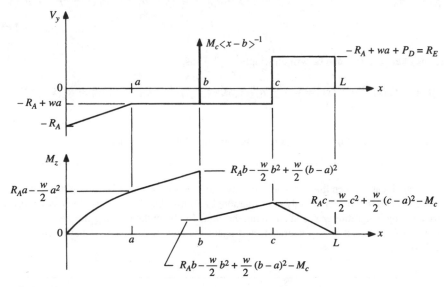

Figure D.2-3

The impulse $M_C \langle x - b \rangle^{-1}$ in the shear-force diagram has no physical meaning and should be omitted from the V_y diagram.

When integrating M_z / EI_z to obtain the slope dv_c/dx, a constant of integration is necessary to impose boundary conditions. Likewise, integration of dv_c/dx results in v_c, where a second constant of integration is necessary.

For the beam shown in Fig. D.2-4(a), graph the deflection of the centroidal axis y_c as a function of x. For the beam, $E = 10 \times 10^6$ psi and $I_z = 0.5$ in⁴.   **Example D.2-3**

Solution:

The free-body diagram is shown in Fig. D.2-4(b), where the reader should verify that $R_A = 120$ lb and $R_B = 30$ lb. From Fig. D.2-4(b), the load-intensity equation is

$$q(x) = 120\langle x \rangle^{-1} - \frac{20}{15}\langle x \rangle^1 + \frac{20}{15}\langle x - 15 \rangle^1 + 20\langle x - 15 \rangle^0 + 30\langle x - 50 \rangle^{-1} \qquad \text{[a]}$$

Note that at $x = 15$ in both the slope and the value of 20 lb/in must be eliminated. Integrating (a) and changing signs results in

$$V_y = -120\langle x \rangle^0 + \frac{2}{3}\langle x \rangle^2 - \frac{2}{3}\langle x - 15 \rangle^2 - 20\langle x - 15 \rangle^1 - 30\langle x - 50 \rangle^0$$

Integrating and changing signs again yields

$$M_z = 120\langle x \rangle^1 - \frac{2}{9}\langle x \rangle^3 + \frac{2}{9}\langle x - 15 \rangle^3 + 10\langle x - 15 \rangle^2 + 30\langle x - 50 \rangle^1$$

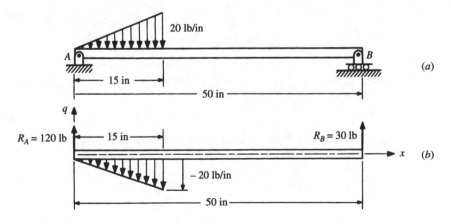

Figure D.2-4

Since $dv_c/dx = \int (M_z/EI_z)\,dx$, we have

$$\frac{dv_c}{dx} = \frac{1}{EI_z}\left(60\langle x\rangle^2 - \frac{1}{18}\langle x\rangle^4 + \frac{1}{18}\langle x-15\rangle^4 + \frac{10}{3}\langle x-15\rangle^3 + 15\langle x-50\rangle^2 + C_1\right)$$

[b]

Integration of Eq. (b) yields

$$v_c = \frac{1}{EI_z}\left(20\langle x\rangle^3 - \frac{1}{90}\langle x\rangle^5 + \frac{1}{90}\langle x-15\rangle^5 + \frac{5}{6}\langle x-15\rangle^4 + 5\langle x-50\rangle^3 + C_1 x + C_2\right)$$

[c]

The two boundary conditions are $v_c = 0$ at $x = 0$ and 50 in. At $x = 0$, only the first two singularity functions exist. Thus Eq. (c) is

$$0 = \frac{1}{EI_z}\left[20(0^3) - \frac{1}{90}(0^5) + C_1(0) + C_2\right]$$

Therefore, $C_2 = 0$. At $x = 50$ in, all functions exist. Thus

$$0 = \frac{1}{EI_z}\left[(20)(50^3) - \left(\frac{1}{90}\right)(50^5) + \left(\frac{1}{90}\right)(50-15)^5 + \left(\frac{5}{6}\right)(50-15)^4 + 5(50-50)^3 + C_1(50)\right]$$

Solving for C_1 yields $C_1 = -1.72 \times 10^4$. Substituting C_1 into Eqs. (b) and (c) results in

$$\frac{dv_c}{dx} = \frac{1}{EI_z}\left(60\langle x\rangle^2 - \frac{1}{18}\langle x\rangle^4 + \frac{1}{18}\langle x-15\rangle^4 + \frac{10}{3}\langle x-15\rangle^3 + 15\langle x-50\rangle^2 - 1.72 \times 10^4\right)$$

[d]

and

$$v_c = \frac{1}{EI_z}\left(20\langle x\rangle^3 - \frac{1}{90}\langle x\rangle^5 + \frac{1}{90}\langle x-15\rangle^5 + \frac{5}{6}\langle x-15\rangle^4 + 5\langle x-50\rangle^3 - 1.72 \times 10^4 x\right)$$

[e]

A graph of Eq. (e) with $E = 10 \times 10^6$ psi and $I_z = 0.5$ in^4 is presented in Fig. D.2-5.

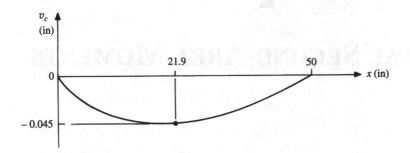

Figure D.2-5

APPENDIX
E

PRINCIPAL SECOND-AREA MOMENTS

E.1 SECOND-AREA MOMENTS

For bending, the geometric properties of importance are the *second-area moments,* often referred to as the *area moments of inertia.* For beams in bending, this text consistently uses a coordinate system such that x is the longitudinal axis and the cross section of the beam is defined in the yz plane. Based on this, the second-area moments used in bending problems are given by

$$I_y = \int_A z^2 \, dA, \qquad I_z = \int_A y^2 \, dA, \qquad I_{yz} = \int_A yz \, dA \qquad \textbf{[E.1-1]}$$

The last term, I_{yz}, is referred to as the *mixed second-area moment* (or *product of inertia*). For bending in the xy and/or xz planes, one of the conditions that the basic bending equations are based on is that the coordinate reference axes are that of the *principal* second-area moments. I_y and I_z will be the principal second-area moments only if I_{yz} vanishes. That is,

$$I_{yz} = \int_A yz \, dA = 0 \qquad \textbf{[E.1-2]}$$

Generally, $I_{yz} \neq 0$ for a cross section that is nonsymmetrical relative to the yz coordinate axes. In these cases it is necessary to determine each second-area moment in Eq. (E.1-1). For simple shapes, integration is generally unnecessary, as the second-area moments for common shapes are available in tabular form (e.g., see Appendix B). Quite often it is necessary to apply the *parallel-axis theorem,* which in equation form is given by

$$I_y = \bar{I}_y + \bar{y}^2 A \qquad I_z = \bar{I}_z + \bar{z}^2 A \qquad I_{yz} = \bar{I}_{yz} + \bar{y}\bar{z}A \qquad \textbf{[E.1-3]}$$

where I_y, I_z, and I_{yz} are the second-area moments of a particular subsection with respect to a yz coordinate system; $\bar{I}_y, \bar{I}_z$, and $\bar{I}_{yz}$ are the second-area moments with respect to the

868

centroidal axes of the subsection (which are parallel to the y, z axes); $\bar{y}, \bar{z}$ are the distances *from* the y, z axes *to* the subsection centroidal axes, respectively, and A is the area of the sub-section.

Determine I_y, I_z, and I_{yz} for the section shown in Fig. E.1-1(a).   **Example E.1-1**

Solution:

The reader should verify that the given yz axes are the centroidal axes of the total cross section. To determine the second-area moments, the section can be divided into two simple rectangular subsections, as shown in Fig. E.1-1(b). The second-area moment of section 1 about the y axis is

$$(I_y)_1 = (\bar{I}_y)_1 + \bar{y}_1^2 A_1 = \left(\frac{1}{12}\right)(2)(8^3) + (1.5)^2(16) = 121.33 \text{ in}^4$$

Likewise for section 2,

$$(I_y)_2 = (\bar{I}_y)_2 + \bar{y}_2^2 A_2 = \left(\frac{1}{12}\right)(8)(2^3) + (-1.5)^2(16) = 41.33 \text{ in}^4$$

thus

$$I_y = (I_y)_1 + (I_y)_2 = 162.7 \text{ in}^4$$

Similarly, for I_z

$$(I_z)_1 = (\bar{I}_z)_1 + \bar{z}_1^2 A_1 = \left(\frac{1}{12}\right)(8)(2^3) + (2.5)^2(16) = 105.33 \text{ in}^4$$

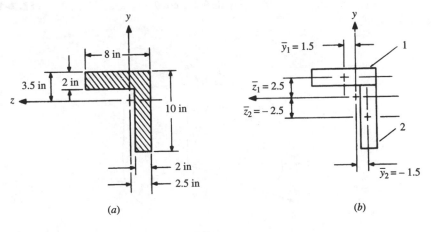

(a)   (b)

Figure E.1-1

and

$$(I_z)_2 = (\bar{I}_z)_2 + \bar{z}_2^2 A_2 = \left(\frac{1}{12}\right)(2)(8^3) + (-2.5)^2(16) = 185.33 \text{ in}^4$$

Thus

$$I_z = (I_z)_1 + (I_z)_2 = 290.7 \text{ in}^4$$

The mixed second-area moment for section 1 is

$$(I_{yz})_1 = (\bar{I}_{yz})_1 + \bar{y}_1 \bar{z}_1 A_1 = 0 + (1.5)(2.5)(16) = 60 \text{ in}^4$$

and for section 2,

$$(I_{yz})_2 = (\bar{I}_{yz})_2 + \bar{y}_2 \bar{z}_2 A_2 = 0 + (-1.5)(-2.5)(16) = 60 \text{ in}^4$$

Thus

$$I_{yz} = (I_{yz})_1 + (I_{yz})_2 = 120 \text{ in}^4$$

E.2 PRINCIPAL SECOND-AREA MOMENTS

When $I_{yz} \neq 0$ there exists a set of axes $y'z'$, rotated from the yz axes, where $I_{y'z'} = 0$. The second-area moments for these axes, $I_{y'}$ and $I_{z'}$, are called the *principal second-area moments*, and the axes are called the *principal axes*. To determine these, consider Fig. E.2-1. Through a basic transformation it can be shown that

$$y' = y \cos\theta + z \sin\theta$$

$$z' = -y \sin\theta + z \cos\theta \qquad\qquad \textbf{[E.2-1]}$$

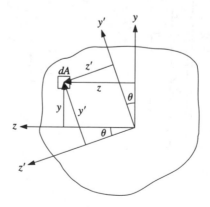

Figure E.2-1

The second-area moments relative to the $y'z'$ axes are given by

$$I_{y'} = \int_A z'^2 dA \qquad I_{z'} = \int_A y'^2 dA \qquad I_{y'z'} = \int_A y'z' dA \qquad \textbf{[E.2-2]}$$

Substitution of Eq. (E.2-1), using the trigonometric identities

$$\sin^2\theta = \frac{1}{2}(1 - \cos 2\theta) \qquad \cos^2\theta = \frac{1}{2}(1 + \cos 2\theta) \qquad \sin\theta \cos\theta = \frac{1}{2}\sin 2\theta$$

and simplifying, Eq. (E.2-2) is written as

$$I_{y'} = \frac{1}{2}(I_y + I_z) + \frac{1}{2}(I_y - I_z)\cos 2\theta - I_{yz}\sin 2\theta$$

$$I_{z'} = \frac{1}{2}(I_y + I_z) - \frac{1}{2}(I_y - I_z)\cos 2\theta + I_{yz}\sin 2\theta$$

$$I_{y'z'} = \frac{1}{2}(I_y - I_z)\sin 2\theta + I_{yz}\cos 2\theta \qquad \textbf{[E.2-3]}$$

It can be shown that the maximum or minimum values of $I_{y'}$ and $I_{z'}$ occur when $I_{y'z'} = 0$. Let the location of the principal axes relative to the yz axes be θ_p where $I_{y'z'} = 0$. Thus from Eq. (E.2-3)

$$\tan 2\theta_p = \frac{2I_{yz}}{I_z - I_y}$$

or

$$\theta_p = \frac{1}{2}\tan^{-1}\left(\frac{2I_{yz}}{I_z - I_y}\right) \qquad \textbf{[E.2-4]}$$

For the previous example, determine the location of the principal axes and the corresponding values of the second-area moments.

Example E.2-1

Solution:
Substituting the values of I_y, I_z, and I_{yz} from Example E.1-1 into Eq. (E.2-4) gives

$$\theta_p = \frac{1}{2}\tan^{-1}\left(\frac{(2)(120)}{290.7 - 162.7}\right) = 31.0°$$

Thus, if we rotate the $y'z'$ axes 31° counterclockwise from the yz axes we locate the principal axes where the values of the second-area moments, given by the first two equations of Eq. (E.2-3), are

$$I_{y'} = \frac{1}{2}(162.7 + 290.7) + \frac{1}{2}(162.7 - 290.7)\cos[2(31)] - 120\sin[2(31)]$$

$$= 90.7 \text{ in}^4$$

$$I_{z'} = \frac{1}{2}(162.7 + 290.7) - \frac{1}{2}(162.7 - 290.7)\cos[2(31)] + 120\sin[2(31)]$$

$$= 362.7 \text{ in}^4$$

Similar to the stress and strain transformation equations, Eqs. (E.2-3) can be applied using Mohr's circle. Readers who are interested in this can refer to any textbook on *statics*. However, the equations are sufficient for any application.

STRESS CONCENTRATION FACTORS[†]

$$\sigma_{max} = K_t \sigma_{nom}$$

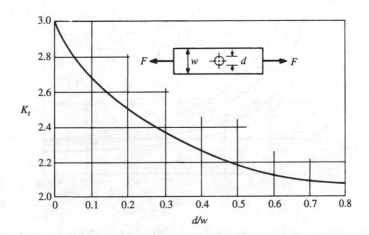

F.1 Bar in tension or simple compression with hole; $\sigma_{nom} = F/A$ where $A = (w - d)t$ and t is the thickness.

[†] R. E. Peterson, Design Factors for Stress Concentration, *Machine Design*, a Penton publication, vol. 23, no. 2, p. 169, February 1951; no. 3, p. 161, March 1951; no. 5, p. 159, May 1951; no. 6, p. 173, June 1951; no. 7, p. 155, July 1951; reproduced with the permission of the author and publisher. For a more comprehensive collection of curves on stress concentrations see R. E. Peterson, "Stress Concentration Factors," Wiley, New York, 1974.

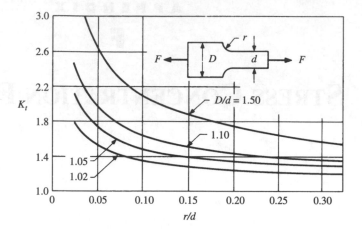

F.2 Rectangular filleted bar in tension or simple compression; $\sigma_{nom} = F/A$ where $A = td$ and t is the thickness.

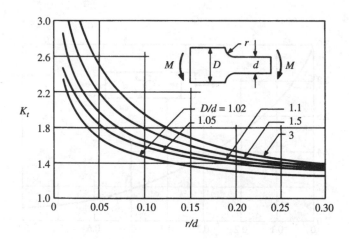

F.3 Rectangular filleted bar in bending; $\sigma_{nom} = Mc/I$ where $c = d/2$ and $I = td^3/12$ where t is the thickness.

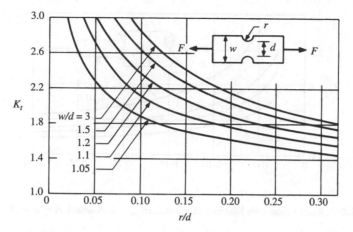

F.4 Notched rectangular bar in tension or simple compression; $\sigma_{nom} = F/A$ where $A = td$ and t is the thickness.

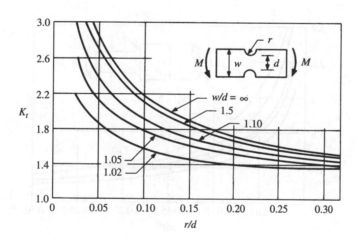

F.5 Notched rectangular bar in bending; $\sigma_{nom} = Mc/I$ where $c = d/2$ and $I = td^3/12$ where t is the thickness.

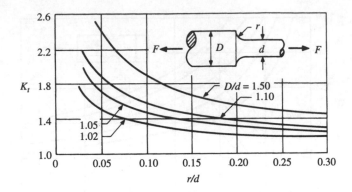

F.6   Round shaft with fillet and in tension; $\sigma_{nom} = F/A$ where $A = \pi d^2/4$.

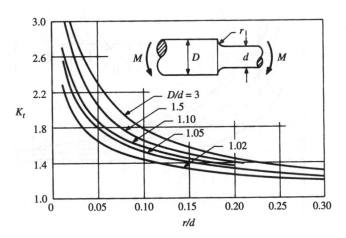

F.7   Round shaft with fillet and in bending; $\sigma_{nom} = Mc/I$ where $c = d/2$ and $I = \pi d^4/64$.

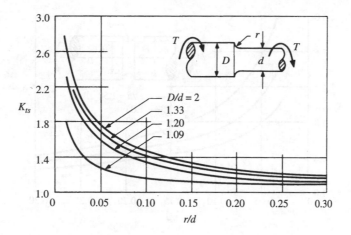

F.8   Round shaft with fillet and in torsion; $\tau_{nom} = Tc/J$ where $c = d/2$ and $J = \pi d^4/32$.

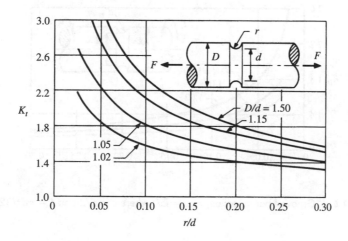

F.9   Grooved bar in tension; $\sigma_{nom} = F/A$ where $A = \pi d^2/4$.

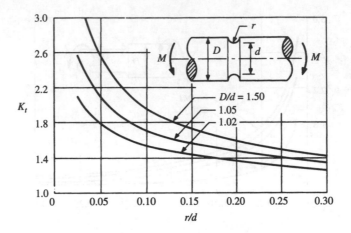

F.10   Grooved bar in bending; $\sigma_{nom} = Mc/I$ where $c = d/2$ and $I = \pi d^4/64$.

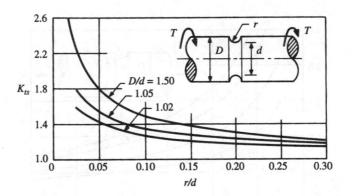

F.11   Grooved bar in torsion; $\tau_{nom} = Tc/J$ where $c = d/2$ and $J = \pi d^4/32$.

APPENDIX

G

STRAIN GAGE ROSETTE EQUATIONS

G.1   THREE-ELEMENT RECTANGULAR ROSETTE

In Example 8.4-1, a three-element rectangular rosette such as shown in Fig. G.1-1 is analyzed using the strain transformation equations and Mohr's circle for strain. The strains ε_x, ε_y, and γ_{xy} are first determined, and then the corresponding stresses σ_x, σ_y, and τ_{xy} are evaluated using the stress-strain relations. The principal strains and stresses and their angular orientation can then be determined.

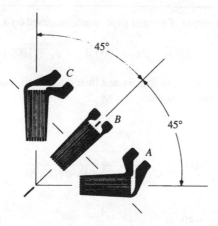

Figure G.1-1   Three-element rectangular strain gage rosette.

Assuming the strains from gages A, B, and C have been corrected for temperature and transverse sensitivity[†] the principal strains are

$$\varepsilon_{p1} = \frac{\varepsilon_A + \varepsilon_C}{2} + \frac{1}{2}\sqrt{(\varepsilon_A - \varepsilon_C)^2 + (2\varepsilon_B - \varepsilon_A - \varepsilon_C)^2} \qquad \text{[G.1-1}a\text{]}$$

$$\varepsilon_{p2} = \frac{\varepsilon_A + \varepsilon_C}{2} - \frac{1}{2}\sqrt{(\varepsilon_A - \varepsilon_C)^2 + (2\varepsilon_B - \varepsilon_A - \varepsilon_C)^2} \qquad \text{[G.1-1}b\text{]}$$

Assuming a linear, homogeneous, isotropic material with modulus of elasticity E and Poisson's ratio ν, the principal stresses are

$$\sigma_{p1} = \frac{E}{2}\left[\frac{\varepsilon_A + \varepsilon_C}{1 - \nu} + \frac{1}{1 + \nu}\sqrt{(\varepsilon_A - \varepsilon_C)^2 + (2\varepsilon_B - \varepsilon_A - \varepsilon_C)^2}\right] \qquad \text{[G.1-2}a\text{]}$$

$$\sigma_{p2} = \frac{E}{2}\left[\frac{\varepsilon_A + \varepsilon_C}{1 - \nu} - \frac{1}{1 + \nu}\sqrt{(\varepsilon_A - \varepsilon_C)^2 + (2\varepsilon_B - \varepsilon_A - \varepsilon_C)^2}\right] \qquad \text{[G.1-2}b\text{]}$$

Treating the $\tan^{-1}$ as a *single-valued function*[‡] the angle *counterclockwise* from gage A to the axis containing ε_{p1} or σ_{p1} is given by

$$\theta_p = \frac{1}{2}\tan^{-1}\left(\frac{2\varepsilon_B - \varepsilon_A - \varepsilon_C}{\varepsilon_A - \varepsilon_C}\right) \qquad \text{[G.1-3]}$$

Example G.1-1 | Example 8.4-1 gives the output of a strain gage rosette mounted on a steel specimen to be

$$\varepsilon_A = 200\ \mu \qquad \varepsilon_B = 900\ \mu \qquad \varepsilon_C = 1000\ \mu$$

Determine the principal strains and stresses and their orientation relative to the axis of the A gage. Let $E = 200$ GPa and $\nu = 0.285$.

Solution:

From Eq. (G.1-1)

$$\varepsilon_{p1,p2} = \frac{200 + 1000}{2} \pm \sqrt{(200 - 1000)^2 + [(2)(900) - 200 - 1000]^2}$$

$$= 1600\ \mu,\ -400\ \mu$$

[†] Corrections for transverse sensitivity are found in Appendix H.
[‡] For a discussion of this see Sec. 3.9.

The principal stresses are

$$\sigma_{p1,p2} = \frac{200 \times 10^9}{2} \left[\frac{200 + 1000}{1 - 0.285} \pm \frac{1}{1 + 0.285} \sqrt{(200 - 1000)^2 + [(2)(900) - 200 - 1000]^2} \right] (10^{-6})$$

$$= 245.7(10^6), 90.0(10^6) \text{ N/m}^2 = 245.7, 90.0 \text{ MPa}$$

From Eq. (G.1-3)

$$\theta_p = \frac{1}{2} \tan^{-1} \left(\frac{2(900) - 200 - 1000}{200 - 1000} \right) = \frac{1}{2} \tan^{-1} \left(\frac{600}{-800} \right)$$

$$= \frac{1}{2}(143.13) = 71.6°$$

Note that the $\tan^{-1}(600/-800)$ is in the second quadrant. A calculator would give the result of $-36.87°$, which is in the fourth quadrant. Therefore, we must add 180° to the calculator result to obtain the 143.13° indicated. Thus the axis containing the principal strain and stress of 1600 μ and 245.7 MPa, respectively, is 71.6° counterclockwise from the axis of the A gage.

G.2 THREE-ELEMENT DELTA ROSETTE

Figure G.2-1 shows a three-element rectangular rosette. Assuming the strains from gages A, B, and C have been corrected for temperature and transverse sensitivity,[†] the principal strains are

$$\varepsilon_{p1} = \frac{\varepsilon_A + \varepsilon_B + \varepsilon_C}{3} + \frac{\sqrt{2}}{3} \sqrt{(\varepsilon_A - \varepsilon_B)^2 + (\varepsilon_B - \varepsilon_C)^2 + (\varepsilon_C - \varepsilon_A)^2} \qquad \textbf{[G.2-1a]}$$

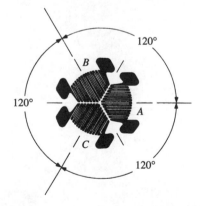

Figure G.2-1 Three-element delta strain gage rosette.

[†] Corrections for transverse sensitivity are found in Appendix H.

$$\varepsilon_{p2} = \frac{\varepsilon_A + \varepsilon_B + \varepsilon_C}{3} - \frac{\sqrt{2}}{3}\sqrt{(\varepsilon_A - \varepsilon_B)^2 + (\varepsilon_B - \varepsilon_C)^2 + (\varepsilon_C - \varepsilon_A)^2} \qquad \textbf{[G.2-1b]}$$

Assuming a linear, homogeneous, isotropic material with modulus of elasticity E and Poisson's ratio ν, the principal stresses are

$$\sigma_{p1} = \frac{E}{3}\left[\frac{\varepsilon_A + \varepsilon_B + \varepsilon_C}{1 - \nu} + \frac{\sqrt{2}}{1 + \nu}\sqrt{(\varepsilon_A - \varepsilon_B)^2 + (\varepsilon_B - \varepsilon_C)^2 + (\varepsilon_C - \varepsilon_A)^2}\right]$$

$$\textbf{[G.2-2a]}$$

$$\sigma_{p2} = \frac{E}{3}\left[\frac{\varepsilon_A + \varepsilon_B + \varepsilon_C}{1 - \nu} - \frac{\sqrt{2}}{1 + \nu}\sqrt{(\varepsilon_A - \varepsilon_B)^2 + (\varepsilon_B - \varepsilon_C)^2 + (\varepsilon_C - \varepsilon_A)^2}\right.$$

$$\textbf{[G.2-2b]}$$

Treating the $\tan^{-1}$ as a *single-valued function*[†] the angle *counterclockwise* from gage A to the axis containing ε_{p1} or σ_{p1} is given by

$$\theta_p = \frac{1}{2}\tan^{-1}\left[\frac{\sqrt{3}(\varepsilon_C - \varepsilon_B)}{2\varepsilon_A - \varepsilon_B - \varepsilon_C}\right] \qquad \textbf{[G.2-3]}$$

[†] For a discussion of this see Sec. 3.9. See Example G.1-1 also.

APPENDIX

H

CORRECTIONS FOR THE TRANSVERSE SENSITIVITY OF STRAIN GAGES

H.0 INTRODUCTION

The change in resistance of a strain gage does not depend only on the normal strain in the axial direction of the gage. The strain perpendicular to the primary sensing axis of the gage also affects the change in resistance. Normally, however, the error induced by this transverse strain is small. Nevertheless, there are cases when the transverse effects cannot be ignored, and the gage outputs must be corrected. To understand transverse effects, consider a perfectly uniaxial *strain* field. If a linear gage is mounted parallel to the field and the change in resistance is measured while increasing the strain, the change in resistance will be proportional to the strain as $\Delta R/R = S_a \varepsilon_a$, where S_a is the axial strain sensitivity and ε_a the axial strain. If the gage is mounted such that its sensing axis is perpendicular to the strain field, a change in resistance will also occur as the strain is increased. However, the relationship is $\Delta R/R = S_t \varepsilon_a$, where S_t is the *transverse sensitivity* to the axial strain. When the gage is in a general plane-stress field, the change in resistance is given by

$$\frac{\Delta R}{R} = S_a \varepsilon_a + S_t \varepsilon_t$$

or

$$\frac{\Delta R}{R} = S_a(\varepsilon_a + K_t \varepsilon_t) \qquad \text{[H.0-1]}$$

where $K_t = S_t/S_a$ is the *transverse sensitivity coefficient* typically supplied by the gage manufacturer.

The gage factor S_g, also supplied by the gage manufacturer, is generally based on a uniaxial *stress* field where $\varepsilon_t = -\nu_0 \varepsilon_a$. That is,

$$\frac{\Delta R}{R} = S_g \varepsilon_a \qquad \text{with} \qquad \varepsilon_t = -\nu_0 \varepsilon_a \qquad \text{[H.0-2]}$$

where ν_0 is normally given to be 0.285. Substituting $\varepsilon_t = -\nu_0 \varepsilon_a$ into Eq. (H.0-1) yields

$$\frac{\Delta R}{R} = S_a(1 - \nu_0 K_t)\varepsilon_a$$

[H.0-3]

Comparing Eqs. (H.0-2) with (H.0-3) it can be seen that

$$S_g = S_a(1 - \nu_0 K_t)$$

[H.0-4]

In order to correct the output of a gage for transverse effects, at least two strain gage readings in different directions at a point must be made. The following sections give the corrections for basic strain gage rosette configurations.

H.1 CORRECTIONS FOR THE TWO-GAGE RECTANGULAR ROSETTE

Consider $\hat{\varepsilon}_A$ and $\hat{\varepsilon}_B$ to be the apparent strains measured from the gages shown in Fig. H.1-1. Applying Eqs. (H.0-1) and (H.0-2) to both gages results in

$$S_{gA}\hat{\varepsilon}_A = S_{aA}(\varepsilon_x + K_{tA}\varepsilon_y)$$

$$S_{gB}\hat{\varepsilon}_B = S_{aB}(\varepsilon_y + K_{tB}\varepsilon_x)$$

[H.1-1]

where (S_{gA}, S_{gB}), (S_{aA}, S_{aB}), and (K_{tA}, K_{tB}) are the gage factors, axial strain sensitivities, and transverse sensitivity factors for gages (A, B), respectively; and ε_x and ε_y are the exact strains in the x and y directions, respectively. Equations (H.1-1) can be solved simultaneously for ε_x and ε_y. For example, ε_x is found to be

$$\varepsilon_x = \frac{\dfrac{S_{gA}}{S_{aA}}\hat{\varepsilon}_A - K_{tA}\dfrac{S_{gB}}{S_{aB}}\hat{\varepsilon}_B}{1 - K_{tA}K_{tB}}$$

[H.1-2]

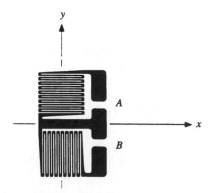

Figure H.1-1

However, from Eq. (H.0-4) we have $S_g/S_a = (1 - \nu_0 K_t)$ for each gage. Thus

$$\varepsilon_x = \frac{(1 - \nu_0 K_{tA})\hat{\varepsilon}_A - K_{tA}(1 - \nu_0 K_{tB})\hat{\varepsilon}_B}{1 - K_{tA}K_{tB}} \qquad \textbf{[H.1-3}a\textbf{]}$$

Similarly,

$$\varepsilon_y = \frac{(1 - \nu_0 K_{tB})\hat{\varepsilon}_B - K_{tB}(1 - \nu_0 K_{tA})\hat{\varepsilon}_A}{1 - K_{tA}K_{tB}} \qquad \textbf{[H.1-3}b\textbf{]}$$

If $K_{tA} = K_{tB} = K_t$, Eqs. (H.1-3) can be written as

$$\varepsilon_x = \frac{1 - \nu_0 K_t}{1 - K_t^2}(\hat{\varepsilon}_A - K_t\hat{\varepsilon}_B)$$

$$\varepsilon_y = \frac{1 - \nu_0 K_t}{1 - K_t^2}(\hat{\varepsilon}_B - K_t\hat{\varepsilon}_A) \qquad \textbf{[H.1-4]}$$

H.2 CORRECTIONS FOR THE THREE-GAGE RECTANGULAR ROSETTE

The derivation for the correction equations is quite involved algebraically. Here we will give the steps with minimal detail. The algebra will be left to the reader to verify. Consider the rosette shown in Fig. H.2-1. Let the direction of gage B be x' and the direction perpendicular to x' be y'. Thus, similar to Eqs. (H.1-1), we have

$$S_{gA}\hat{\varepsilon}_A = S_{aA}(\varepsilon_x + K_{tA}\varepsilon_y) \qquad \textbf{[H.2-1}a\textbf{]}$$

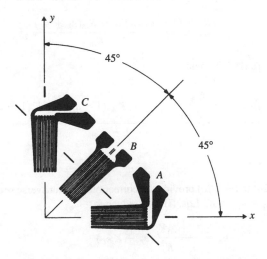

Figure H.2-1 Three-gage rectangular rosette.

$$S_{gB}\hat{\varepsilon}_B = S_{aB}(\varepsilon_{x'} + K_{tB}\varepsilon_{y'}) \qquad \text{[H.2-1b]}$$

$$S_{gC}\hat{\varepsilon}_C = S_{aC}(\varepsilon_y + K_{tC}\varepsilon_x) \qquad \text{[H.2-1c]}$$

Equations (H.2-1a) and (H.2-1c) are identical to Eqs. (H.1-1) with the exception of the gage letters. If we let $\varepsilon_A = \varepsilon_x$ and $\varepsilon_C = \varepsilon_y$, we have

$$\varepsilon_A = \frac{(1 - \nu_0 K_{tA})\hat{\varepsilon}_A - K_{tA}(1 - \nu_0 K_{tC})\hat{\varepsilon}_C}{1 - K_{tA}K_{tC}} \qquad \text{[H.2-2a]}$$

$$\varepsilon_C = \frac{(1 - \nu_0 K_{tC})\hat{\varepsilon}_C - K_{tC}(1 - \nu_0 K_{tA})\hat{\varepsilon}_A}{1 - K_{tA}K_{tC}} \qquad \text{[H.2-2b]}$$

For ε_B consider Eq. (H.2-1b). Using the strain transformation Eqs. (3.9-11) and (3.9-12) $\varepsilon_{x'}$ and $\varepsilon_{y'}$ can be written in terms of ε_x, ε_y, and γ_{xy}. With $\theta = 45°$ we have

$$\varepsilon_{x'} = \varepsilon_x \cos^2 45 + \varepsilon_y \sin^2 45 + \gamma_{xy} \sin 45 \cos 45 = \frac{1}{2}(\varepsilon_x + \varepsilon_y + \gamma_{xy}) \qquad \text{[a]}$$

$$\varepsilon_{y'} = \varepsilon_x \sin^2 45 + \varepsilon_y \cos^2 45 - \gamma_{xy} \sin 45 \cos 45 = \frac{1}{2}(\varepsilon_x + \varepsilon_y - \gamma_{xy}) \qquad \text{[b]}$$

Substituting these into Eq. (H.2-1b) with $S_g/S_a = (1 - \nu_0 K_t)$ results in

$$2(1 - \nu_0 K_{tB})\hat{\varepsilon}_B = (1 + K_{tB})\varepsilon_x + (1 + K_{tB})\varepsilon_y + (1 - K_{tB})\gamma_{xy} \qquad \text{[c]}$$

Solving for γ_{xy} with $\varepsilon_x = \varepsilon_A$ and $\varepsilon_y = \varepsilon_C$ gives

$$\gamma_{xy} = \frac{2(1 - \nu_0 K_{tB})\hat{\varepsilon}_B - (1 + K_{tB})(\varepsilon_A + \varepsilon_C)}{1 - K_{tB}} \qquad \text{[d]}$$

From Eq. (a) $\varepsilon_B = (\varepsilon_A + \varepsilon_C + \gamma_{xy})/2$. Substituting Eq. (d) into this and rearranging gives

$$\varepsilon_B = \frac{(1 - \nu_0 K_{tB})\hat{\varepsilon}_B - K_{tB}(\varepsilon_A + \varepsilon_C)}{1 - K_{tB}} \qquad \text{[e]}$$

Substitution of Eqs. (H.2-2) yields

$$\varepsilon_B = \frac{(1 - \nu_0 K_{tB})\hat{\varepsilon}_B - \dfrac{K_{tB}}{1 - K_{tA}K_{tC}}\left[(1 - \nu_0 K_{tA})(1 - K_{tC})\hat{\varepsilon}_A + (1 - \nu_0 K_{tC})(1 - K_{tA})\hat{\varepsilon}_C\right]}{1 - K_{tB}}$$

$$\text{[H.2-2c]}$$

Thus Eqs. (H.2-2a) to (H.2-2c) provide the corrections for transverse sensitivity.
   If $K_{tA} = K_{tB} = K_{tC} = K_t$, Eqs. (H.2-2) reduce to

$$\varepsilon_A = \frac{1 - \nu_0 K_t}{1 - K_t^2}(\hat{\varepsilon}_A - K_t\hat{\varepsilon}_C)$$

$$\varepsilon_B = \frac{1 - \nu_0 K_t}{1 - K_t^2} \left[(1 + K_t)\hat{\varepsilon}_B - K_t(\hat{\varepsilon}_A + \hat{\varepsilon}_C) \right]$$

$$\varepsilon_C = \frac{1 - \nu_0 K_t}{1 - K_t^2} (\hat{\varepsilon}_C - K_t\hat{\varepsilon}_A) \qquad\qquad \textbf{[H.2-3]}$$

H.3 Corrections for the Three-Gage Delta (120°) Rosette

The derivation for the corrections for the delta rosette shown in Fig. H.3-1 are even more te-dious than that of the rectangular rosette and will not be presented here. The correction equa-tions are

$$\varepsilon_A = \kappa\{(1 - \nu_0 K_{tA})(3 - K_{tB} - K_{tC} - K_{tB}K_{tC})\hat{\varepsilon}_A - 2K_{tA}[(1 - \nu_0 K_{tB})(1 - K_{tC})\hat{\varepsilon}_B + (1 - \nu_0 K_{tC})(1 - K_{tB})\hat{\varepsilon}_C]\}$$

$$\varepsilon_B = \kappa\{(1 - \nu_0 K_{tB})(3 - K_{tC} - K_{tA} - K_{tC}K_{tA})\hat{\varepsilon}_B - 2K_{tB}[(1 - \nu_0 K_{tC})(1 - K_{tA})\hat{\varepsilon}_C + (1 - \nu_0 K_{tA})(1 - K_{tC})\hat{\varepsilon}_A]\}$$

$$\varepsilon_C = \kappa\{(1 - \nu_0 K_{tC})(3 - K_{tA} - K_{tB} - K_{tA}K_{tB})\hat{\varepsilon}_C - 2K_{tC}[(1 - \nu_0 K_{tA})(1 - K_{tB})\hat{\varepsilon}_A + (1 - \nu_0 K_{tB})(1 - K_{tA})\hat{\varepsilon}_B]\}$$

$$\kappa = (3K_{tA}K_{tB}K_{tC} - K_{tA}K_{tB} - K_{tB}K_{tC} - K_{tA}K_{tC} - K_{tA} - K_{tB} - K_{tC} + 3)^{-1} \qquad \textbf{[H.3-1]}$$

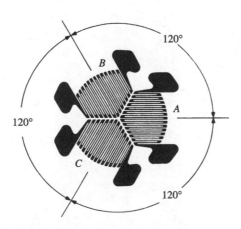

Figure H.3-1

If $K_{tA} = K_{tB} = K_{tC} = K_t$, Eqs. (H.3-1) reduce to

$$\varepsilon_A = \frac{1 - \nu_0 K_t}{3(1 - K_t^2)}[(3 + K_t)\hat{\varepsilon}_A - 2K_t(\hat{\varepsilon}_B + \hat{\varepsilon}_C)]$$

$$\varepsilon_B = \frac{1 - \nu_0 K_t}{3(1 - K_t^2)}[(3 + K_t)\hat{\varepsilon}_B - 2K_t(\hat{\varepsilon}_A + \hat{\varepsilon}_C)]$$

$$\varepsilon_C = \frac{1 - \nu_0 K_t}{3(1 - K_t^2)}[(3 + K_t)\hat{\varepsilon}_C - 2K_t(\hat{\varepsilon}_A + \hat{\varepsilon}_B)] \qquad \textbf{[H.3-2]}$$

APPENDIX

I

MATRIX ALGEBRA AND CARTESIAN TENSORS

I.0 INTRODUCTION

The use of matrices and matrix algebra provides a powerful method of dealing with large systems of linear equations. Matrix algebra permits an effective systematic approach to handling large arrays of simultaneous linear equations and their solutions and provides a convenient form for use in digital computer applications. As an example of matrix notation, assume that the displacement field (u, v, w) of a body is linearly related to the position (x, y, z) within that body. That is,

$$u = a_{11}x + a_{12}y + a_{13}z$$

$$v = a_{21}x + a_{22}y + a_{23}z \qquad \textbf{[I.0-1]}$$

$$w = a_{31}x + a_{32}y + a_{33}z$$

where the a_{ij} terms are constants $(i, j = 1, 2, 3)$. Using matrix notation, Eqs. (I.0-1) can be expressed by

$$\{\boldsymbol{\delta}\} = [\mathbf{A}]\{\mathbf{s}\} \qquad \textbf{[I.0-2]}$$

where

$$\{\boldsymbol{\delta}\} = \begin{Bmatrix} u \\ v \\ w \end{Bmatrix} \qquad [\mathbf{A}] = \begin{bmatrix} a_{11} & a_{12} & a_{13} \\ a_{21} & a_{22} & a_{23} \\ a_{31} & a_{32} & a_{33} \end{bmatrix} \qquad \{\mathbf{s}\} = \begin{Bmatrix} x \\ y \\ z \end{Bmatrix} \qquad \textbf{[I.0-3]}$$

From Eq. (I.0-3) the matrix $[\mathbf{A}]$ can also be written as $[a_{ij}]$. The step in converting Eq. (I.0-1) to (I.0-2) should become clearer when matrix multiplication is discussed in the next section. Note that each bracketed array in Eqs. (I.0-3) is a matrix but matrices with only one column

889

or one row are *vector matrices* and are denoted by braces { }, whereas matrices with more than one column or row are denoted by square brackets [ ]. For the present example, the matrix [A] is a *square matrix* since the number of rows (3) equals the number of columns (3). The matrix [A] is said to be a 3×3 matrix, whereas the column vector matrices $\{\delta\}$ and $\{s\}$ are 3×1.

A square matrix which contains only diagonal terms such as

$$\begin{bmatrix} a_{11} & 0 & 0 \\ 0 & a_{22} & 0 \\ 0 & 0 & a_{33} \end{bmatrix}$$

is called a *diagonal matrix*. A diagonal matrix where each term along the diagonal is unity is called an *identity matrix* [I]. An example of a 3×3 identity matrix is

$$[I] = \begin{bmatrix} 1 & 0 & 0 \\ 0 & 1 & 0 \\ 0 & 0 & 1 \end{bmatrix}$$

A square matrix for which $a_{ij} = a_{ji}$ is called a *symmetric matrix*. That is,

$$\begin{bmatrix} a_{11} & a_{12} & a_{13} \\ a_{12} & a_{22} & a_{23} \\ a_{13} & a_{23} & a_{33} \end{bmatrix}$$

A shorthand form commonly used for a symmetric matrix is

$$\begin{bmatrix} a_{11} & a_{12} & a_{13} \\ & a_{22} & a_{23} \\ sym & & a_{33} \end{bmatrix}$$

I.1 MATRIX ALGEBRA

I.1.1 ADDITION

The sum of two matrices can be carried out only if the two matrices have the same number of rows and columns and is performed simply by adding corresponding elements. That is,

$$[A] + [B] = [C] \qquad \textbf{[I.1-1]}$$

where for each row i, column j, term

$$c_{ij} = a_{ij} + b_{ij} \qquad \textbf{[I.1-2]}$$

For example,

$$\begin{bmatrix} 2 & 0 & -1 \\ 6 & -3 & 2 \end{bmatrix} + \begin{bmatrix} 0 & -2 & 1 \\ 1 & 0 & -1 \end{bmatrix} = \begin{bmatrix} 2 & -2 & 0 \\ 7 & -3 & 1 \end{bmatrix}$$

I.1.2 SCALAR MULTIPLICATION

The product of a scalar and a matrix is the matrix formed by multiplying each element of the matrix by the scalar quantity. Also, the order of multiplication is unimportant. Thus

$$s[\mathbf{A}] = [sa_{ij}] \qquad \text{[I.1-3]}$$

For example,

$$3\begin{bmatrix} 1 & 3 & 0 \\ 2 & -1 & 1 \\ 0 & 2 & -2 \end{bmatrix} = \begin{bmatrix} 3 & 9 & 0 \\ 6 & -3 & 3 \\ 0 & 6 & -6 \end{bmatrix}$$

I.1.3 MATRIX MULTIPLICATION

The product of two matrices is only defined when the number of columns of the first matrix equals the number of rows of the second matrix. Thus, in this case, the order of multiplication is important. For example, a 3×1 matrix cannot be premultiplied by a 1×2 matrix. However, a 1×2 matrix can be premultiplied by a 3×1 matrix. The formal definition of the product of two matrices is

$$[\mathbf{C}] = [\mathbf{A}][\mathbf{B}] \qquad \text{[I.1-4]}$$

where each term in the [**C**] matrix is given by

$$c_{ij} = \sum_{k=1}^{n} a_{ik}b_{kj} \qquad \text{[I.1-5]}$$

and n is the number of columns in [**A**] which equals the number of rows in [**B**]. For example, the reader should verify, using Eqs. (I.1-4) and (I.1-5), that

$$\begin{bmatrix} 2 & 1 \\ -3 & 2 \end{bmatrix}\begin{bmatrix} 1 & 3 & 0 \\ -2 & 0 & 1 \end{bmatrix} = \begin{bmatrix} 0 & 6 & 1 \\ -7 & -9 & 2 \end{bmatrix}$$

The product matrix will have as many rows as the first matrix and as many columns as the second matrix. To perform the multiplication quickly, note that c_{ij} is found by multiplying each term of row i of [**A**], separately and in order, by each term of column j of [**B**] and then summing the results. For example, to obtain c_{23} of the previous example, isolate row 2 of [**A**] and column 3 of [**B**]:

Row 2 → $\begin{bmatrix} 2 & 1 \\ -3 & 2 \end{bmatrix}\begin{bmatrix} 1 & 3 & 0 \\ -2 & 0 & 1 \end{bmatrix} = \begin{bmatrix} \square & \square & \square \\ \square & \square & \boxed{2} \end{bmatrix}$ ← Row 2

$\uparrow$ Column 3 $\qquad \uparrow$ Column 3

Multiplying each term in row 2 of [**A**] by each term in column 3 of [**B**] in order results in $(-3)(0) = 0$ and $(2)(1) = 2$. Adding the results yields $c_{23} = 0 + 2 = 2$.

If the [A] and [B] matrices are both square matrices, the products of [A][B] and [B][A] can both be performed, but in general [A][B] ≠ [B][A]. Thus the order of multiplication is still important.

Pre- or postmultiplication of a matrix by the identity matrix results in the original matrix. That is,

$$[\mathbf{I}][\mathbf{A}] = [\mathbf{A}][\mathbf{I}] = [\mathbf{A}] \tag{I.1-6}$$

When multiplying more than two matrices, the matrices are multiplied two at a time. The associative law holds for matrix multiplication. For example, for the multiplication of three matrices, [A][B][C], either of the indicated multiplications is acceptable.

$$[\mathbf{A}]([\mathbf{B}][\mathbf{C}]) = ([\mathbf{A}][\mathbf{B}])[\mathbf{C}] \tag{I.1-7}$$

I.1.4 TRANSPOSITION

When inverting or transforming matrices it is necessary to take the transpose of a matrix. The transpose of a matrix is simply the interchange of rows and columns. If a matrix [A] is defined by $[a_{ij}]$ the transpose is defined by

$$[\mathbf{A}]^T = [a_{ji}] \tag{I.1-8}$$

For example,

$$\begin{bmatrix} 2 & 5 & -4 \\ -3 & 7 & -9 \end{bmatrix}^T = \begin{bmatrix} 2 & -3 \\ 5 & 7 \\ -4 & -9 \end{bmatrix}$$

Thus we see that the transform of an $n \times m$ matrix is an $m \times n$ matrix.

Occasionally it is necessary to take the transform of the product of matrices. The transform of the product of matrices is obtained by reversing the order of multiplication of the transforms of the individual matrices. That is,

$$([\mathbf{A}][\mathbf{B}][\mathbf{C}])^T = [\mathbf{C}]^T[\mathbf{B}]^T[\mathbf{A}]^T \tag{I.1-9}$$

I.1.5 DETERMINANT OF A MATRIX

The determinant of a square matrix [A] of order $n \times n$ is a single scalar quantity denoted as |A|. The *method of cofactors* is described here, which eventually reduces the calculation to that of a determinant of order 2×2. The determinant of a 2×2 matrix is defined by

$$\begin{vmatrix} a_{11} & a_{12} \\ a_{21} & a_{22} \end{vmatrix} = a_{11}a_{22} - a_{12}a_{21} \tag{I.1-10}$$

Determinants of matrices of order higher than a 2×2 matrix are reduced using cofactors. Consider [A] to be of order $n \times n$. Selecting any row i or column j of [A] the determinant of [A] is given by

$$|\mathbf{A}| = \sum_{j=1}^{n} a_{ij}\tilde{a}_{ij} \qquad \text{(selecting row } i\text{)} \tag{I.1-11a}$$

$$|A| = \sum_{i=1}^{n} a_{ij}\tilde{a}_{ij} \qquad \text{(selecting column } j\text{)} \qquad \textbf{[I.1-11}\textit{b}\textbf{]}$$

where $\tilde{a}_{ij}$ is the *cofactor* of a_{ij} given by

$$\tilde{a}_{ij} = (-1)^{i+j}m_{ij} \qquad \textbf{[I.1-12]}$$

The term m_{ij} is called the *minor* of a_{ij} and is the determinant of the $(n - 1) \times (n - 1)$ matrix obtained by eliminating row i and column j of [A]. If $n - 1 > 2$, the process is repeated on each cofactor until the remaining determinants are 2×2 where then Eq. (I.1-10) is applied.

Evaluate the determinant of **Example I.1-1**

$$[\mathbf{A}] = \begin{bmatrix} 3 & 0 & 1 \\ -1 & 2 & 0 \\ 1 & -2 & 1 \end{bmatrix}$$

Solution:

Any row or column can be selected for Eqs. (I.1-11). It is wise to select a row or column with the most zeros in it. For the example, rows 1 or 2 or columns 2 or 3 would be good to select. Arbitrarily selecting row 1, Eq. (I.1-11*a*) is used with $i = 1$. Thus

$$|\mathbf{A}| = (3)(-1)^{1+1}\begin{vmatrix} 2 & 0 \\ -2 & 1 \end{vmatrix} + (0)(-1)^{1+2}\begin{vmatrix} -1 & 0 \\ 1 & 1 \end{vmatrix} + (1)(-1)^{1+3}\begin{vmatrix} -1 & 2 \\ 1 & -2 \end{vmatrix}$$

$$= (3)(1)[(2)(1) - (0)(-2)] + (0)(-1)[(-1)(1) - (0)(1)] + (1)(1)[(-1)(-2) - (2)(1)]$$

$$= 6 + 0 + 0 = 6 \qquad \textbf{[a]}$$

It was unnecessary to evaluate the second 2×2 determinant because of the zero multiplier and was only done to demonstrate how each term is evaluated.

The reader is urged to repeat the example using different rows or columns in Eqs. (I.1-11).

I.1.6 COFACTOR MATRIX

The cofactor matrix $[\tilde{\mathbf{A}}]$ is the same order of [A] and each term in $[\tilde{\mathbf{A}}]$ is given by Eq. (I.1-12).

Determine the cofactor matrix of [A] given in Example I.1-1. **Example I.1-2**

Solution:

The first row of the cofactor matrix is obtained by each term in ( ) [ ] brackets in Eq. (*a*) of Example I.1-1. That is, $\tilde{a}_{11} = (1)[(2)(1) - (0)(-2)] = 2$, $\tilde{a}_{12} = (-1)[(-1)(1) - (0)(1)] = 1$, and $\tilde{a}_{13} = (1)[(-1)(-2) - (2)(1)] = 0$. The remaining cofactors are found using Eq. (I.1-12). For example,

$$\tilde{a}_{22} = (-1)^{2+2} \begin{vmatrix} 3 & 1 \\ 1 & 1 \end{vmatrix} = (1)[(3)(1) - (1)(1)] = 2$$

The reader should verify the remaining terms of the cofactor matrix, which is found to be

$$[\tilde{A}] = \begin{bmatrix} 2 & 1 & 0 \\ -2 & 2 & 6 \\ -2 & -1 & 6 \end{bmatrix}$$

I.1.7 MATRIX INVERSION

The process of division can be thought of as multiplication involving a reciprocal. For example, consider the equation

$$y = Cx$$

If the solution for x is required, both sides of the equation can be multiplied by the reciprocal of C, C^{-1}. That is,

$$C^{-1}y = C^{-1}Cx$$

However, since $C^{-1}C = 1$, we have

$$x = C^{-1}y$$

Consider the linear equations given by Eqs. (I.0-1), given in matrix form as Eq. (I.0-2). That is,

$$\{\delta\} = [A]\{s\}$$

What if the matrices $[A]$ and $\{\delta\}$ were known and the matrix $\{s\}$ was desired? This basically amounts to solving Eqs. (I.0-1) simultaneously for the unknowns x, y, and z. The procedure for doing this is rather straightforward using Cramer's rule. It can be shown that this procedure is equivalent to multiplying Eq. (I.0-2) by the inverse of $[A]$, $[A]^{-1}$, if it exists. That is,

$$[A]^{-1}\{\delta\} = [A]^{-1}[A]\{s\} = [I]\{s\} = \{s\} \qquad \textbf{[I.1-13]}$$

where $[A]^{-1}[A] = [I]$, the identity matrix. Note that both $[A]$ and its inverse $[A]^{-1}$ are square matrices of the same order.

It can be shown that the inverse of matrix $[A]$ is given by

$$[A]^{-1} = \frac{[\tilde{A}]^T}{|A|} \qquad \textbf{[I.1-14]}$$

where $[\tilde{A}]$ and $|A|$ are the cofactor matrix and determinant of $[A]$, respectively. The transpose of the cofactor matrix $[\tilde{A}]^T$ is also called the *adjoint matrix* of $[A]$.

Example I.1-3

Determine the inverse of the matrix [A] given in Example I.1-1.

Solution:

The determinant and cofactor matrix, $|A|$ and $[\tilde{A}]$, were determined in Examples I.1-1 and I.1-2, respectively. The adjoint matrix is determined by taking the transpose of $[\tilde{A}]$. Thus, from Eq. (I.1-14),

$$[A]^{-1} = \frac{1}{6}\begin{bmatrix} 2 & -2 & -2 \\ 1 & 2 & -1 \\ 0 & 6 & 6 \end{bmatrix}$$

If the determinant of a matrix is zero, the matrix is said to be *singular* and the inverse does not exist. This typically arises when equations such as Eq. (I.0-1) are not independent.

Note that the inverse of an orthogonal transformation matrix is simply the transpose of the transformation matrix. Thus, if a vector is transformed according to

$$\{V'\} = [T]\{V\} \qquad\qquad \textbf{[I.1-15]}$$

then the inverse transformation is

$$\{V\} = [T]^{T}\{V'\} \qquad\qquad \textbf{[I.1-16]}$$

I.1.8   EIGENVALUES AND EIGENVECTORS

The evaluation of principal stresses, strains, second-area moments, etc., is an eigenvalue problem. The equations for eigenvalue problems are of the form

$$[A]\{s\} = \lambda\{s\} \qquad\qquad \textbf{[I.1-17]}$$

where $[A]$ is a square $n \times n$ matrix. Equation (I.1-17) can be placed in the form

$$([A] - \lambda[I])\,\{s\} = 0 \qquad\qquad \textbf{[I.1-18]}$$

To avoid the trivial solution, $\{s\} = \{0\}$, $[A] - \lambda[I]$ is forced to be singular. That is, the determinant is set to zero

$$|[A] - \lambda[I]| = 0 \qquad\qquad \textbf{[I.1-19]}$$

Equation (I.1-19) is called the *characteristic equation,* which yields n roots of λ which are the *eigenvalues* of the matrix $[A]$. For each eigenvalue λ_i there is an associated eigenvector $\{s_i\}$ obtained from Eq. (I.1-18).[†]

Example I.1-4

Determine the eigenvalues and eigenvectors of the matrix

$$[A] = \begin{bmatrix} 3 & -2 \\ -1 & 2 \end{bmatrix}$$

[†] Note that $\{s_i\}$ is determined only to within a multiplicative constant since the individual equations in Eq. (I.1-18) are not independent. In the case of principal stresses, strains, etc., the eigenvectors are a set of directional cosines for which there is an additional and separate independent relationship to determine the eigenvectors explicitly.

Solution:

The characteristic equation, from Eq. (I.1-19), is

$$|[\mathbf{A}] - \lambda[\mathbf{I}]| = \begin{vmatrix} (3 - \lambda) & -2 \\ -1 & (2 - \lambda) \end{vmatrix} = (3 - \lambda)(2 - \lambda) - (-2)(-1)$$

$$= \lambda^2 - 5\lambda + 4 = 0$$

The roots of the polynomial equation are $\lambda_1 = 1$ and $\lambda_2 = 4$. The eigenvectors are found from Eq. (I.1-18). For $\lambda_1 = 1$, the equations are

$$\begin{bmatrix} (3 - 1) & -2 \\ -1 & (2 - 1) \end{bmatrix} \begin{Bmatrix} s_1 \\ s_2 \end{Bmatrix}_1 = \begin{Bmatrix} 0 \\ 0 \end{Bmatrix}$$

where the outer subscript indicates that this is eigenvector 1 associated with eigenvalue 1 (λ_1). Writing the equations out independently

$$2s_1 - 2s_2 = 0 \qquad\qquad\qquad [a]$$

$$-s_1 + s_2 = 0 \qquad\qquad\qquad [b]$$

We see that Eqs. (a) and (b) are identical and all we find is that $s_2 = s_1$. Thus the eigenvector

$$\{\mathbf{s}\}_1 = C_1\{1\ 1\}^T$$

satisfies Eqs. (a) and (b) where C_1 is an arbitrary constant. The eigenvector can be normalized in any manner. A common normalization is to make $\{\mathbf{s}\}_1$ a unit vector. Thus

$$\{\mathbf{s}\}_1 = \left\{ \frac{1}{\sqrt{2}} \quad \frac{1}{\sqrt{2}} \right\}^T$$

In a similar manner, substituting $\lambda_2 = 4$ into Eq. (I.1-18) yields the normalized eigenvector

$$\{\mathbf{s}\}_2 = \left\{ \frac{2}{\sqrt{5}} \quad -\frac{1}{\sqrt{5}} \right\}^T$$

I.2 CARTESIAN TENSORS

A tensor describes a property of state which is invariant with respect to a coordinate system. As examples, scalar quantities and vectors are tensors since they remain unchanged regardless of the coordinate system. The order of a tensor depends on how many separate entities are necessary to describe the matrix and varies according to 3^n, where n is the order of the tensor. As examples, a scalar quantity is a *zero-order tensor* since it needs $3^0 = 1$ quantity to describe its property; whereas a vector is a *first-order tensor* since it needs $3^1 = 3$ quantities for definition. If transformable,[†] stress and strain matrices are *second-order tensors* since their matrices are of the order 3×3, requiring $3^2 = 9$ quantities.

Tensors are more easily manipulated using *index notation,* and an example of this is illustrated in Sec. 2.3 with the discussion of Lamé constants. However, since matrix rather than index notation is emphasized in this book, we will continue the discussion of the transformation of second-order tensors using matrix notation.

In Sec. 2.1.1, Eq. (2.1-5) shows how a vector transforms according to

$$\{V'\} = [T]\{V\} \tag{I.2-1}$$

where

$$\{V\} = \begin{Bmatrix} V_x \\ V_y \\ V_z \end{Bmatrix} \qquad \{V'\} = \begin{Bmatrix} V_{x'} \\ V_{y'} \\ V_{z'} \end{Bmatrix}$$

and the transformation matrix $[T]$ is

$$[T] = \begin{bmatrix} n_{x'x} & n_{x'y} & n_{x'z} \\ n_{y'x} & n_{y'y} & n_{y'z} \\ n_{z'x} & n_{z'y} & n_{z'z} \end{bmatrix} \tag{I.2-2}$$

made up of directional cosines which relates the $x'y'z'$ coordinate system to the xyz system.

For the transformation of stress consider the isolated stress tetrahedron shown in Fig. I.2-1.

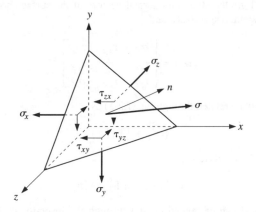

Figure I.2-1 Isolated stress tetrahedron.

[†] In Sec. 2.2.1 it is shown that the strain matrix will transform if the shear strains in the matrix are represented by $\gamma/2$ rather than γ.

In the figure, σ is the *net stress vector* on the oblique surface. As shown in Sec. 2.1.4, the net forces in the x, y, and z directions from the stresses on the orthogonal surfaces are given by Eqs. (2.1-19) repeated here as

$$F_x = -A_0(\sigma_x n_x + \tau_{xy} n_y + \tau_{zx} n_z) \qquad \text{[I.2-3a]}$$

$$F_y = -A_0(\tau_{xy} n_x + \sigma_y n_y + \tau_{yz} n_z) \qquad \text{[I.2-3b]}$$

$$F_z = -A_0(\tau_{zx} n_x + \tau_{yz} n_y + \sigma_z n_z) \qquad \text{[I.2-3c]}$$

where A_0 is the area of the oblique surface. The force on the oblique surface due to the stress vector can be expressed as $\{\sigma\}A_0$. Summing the forces and, for equilibrium, equating to zero results in

$$\{\sigma\} = [\sigma]\,\{n\} \qquad \text{[I.2-4]}$$

where $\{n\} = \{n_x \ n_y \ n_z\}^T$,

$$[\sigma] = \begin{bmatrix} \sigma_x & \tau_{xy} & \tau_{xz} \\ \tau_{yx} & \sigma_y & \tau_{yz} \\ \tau_{zx} & \tau_{zy} & \sigma_z \end{bmatrix}$$

and A_0 cancels out.

Now, Eq. (I.2-4) can also apply to a stress element in the $x'y'z'$ coordinate system such that

$$\{\sigma'\} = [\sigma']\{n'\} \qquad \text{[I.2-5]}$$

where $\{\sigma'\}$ and $\{n'\}$ are the stress and normal vectors of the surface normal relative to the $x'y'z'$ coordinate system, respectively, and

$$[\sigma'] = \begin{bmatrix} \sigma_{x'} & \tau_{x'y'} & \tau_{x'z'} \\ \tau_{y'x'} & \sigma_{y'} & \tau_{y'z'} \\ \tau_{z'x'} & \tau_{z'y'} & \sigma_{z'} \end{bmatrix}$$

The stress *vectors* can be related through Eq. (I.2-1) as

$$\{\sigma'\} = [T]\{\sigma\}$$

Substituting Eqs. (I.2-5) and (I.2-4) gives

$$[\sigma']\{n'\} = [T][\sigma]\{n\}$$

Finally, the vector $\{n\}$ can be related to $\{n'\}$ through the inversion of Eq. (I.2-1). That is, $\{n\} = [T]^T\{n'\}$. Substituting this into the previous equation yields

$$[\sigma']\{n'\} = [T][\sigma][T]^T\{n'\}$$

Comparing both sides of the equation we see that

$$[\sigma'] = [T][\sigma][T]^T \qquad \text{[I.2-6]}$$

Example 2.1-1 transforms the stress matrix

Example 1.2-1

$$[\sigma] = \begin{bmatrix} -8 & 6 & -2 \\ 6 & 4 & 2 \\ -2 & 2 & -5 \end{bmatrix} \text{ MPa}$$

using the transformation matrix

$$[\mathbf{T}] = \frac{1}{4}\begin{bmatrix} 2\sqrt{2} & 2\sqrt{2} & 0 \\ -\sqrt{6} & \sqrt{6} & 2 \\ \sqrt{2} & -\sqrt{2} & 2\sqrt{3} \end{bmatrix}$$

Solve this using Eq. (I.2-6).

Solution:

From Eq. (I.2-6), and using a spreadsheet one obtains the following:

$$[\sigma'] = \frac{1}{4}\begin{bmatrix} 2\sqrt{2} & 2\sqrt{2} & 0 \\ -\sqrt{6} & \sqrt{6} & 2 \\ \sqrt{2} & -\sqrt{2} & 2\sqrt{3} \end{bmatrix}\begin{bmatrix} -8 & 6 & -2 \\ 6 & 4 & 2 \\ -2 & 2 & -5 \end{bmatrix}\frac{1}{4}\begin{bmatrix} 2\sqrt{2} & -\sqrt{6} & \sqrt{2} \\ 2\sqrt{2} & \sqrt{6} & -\sqrt{2} \\ 0 & 2 & 2\sqrt{3} \end{bmatrix}$$

$$= \frac{1}{16}\begin{bmatrix} 2\sqrt{2} & 2\sqrt{2} & 0 \\ -\sqrt{6} & \sqrt{6} & 2 \\ \sqrt{2} & -\sqrt{2} & 2\sqrt{3} \end{bmatrix}\begin{bmatrix} -5.657 & 30.293 & -26.727 \\ 28.284 & -0.899 & 9.757 \\ 0 & -0.202 & -22.977 \end{bmatrix}$$

$$= \begin{bmatrix} 4.00 & 5.20 & -3.00 \\ 5.20 & -4.80 & 2.71 \\ -3.00 & 2.71 & -8.20 \end{bmatrix}$$

This agrees with the results obtained in Example 2.1-1.

ANSWERS TO MOST ODD-NUMBERED PROBLEMS

CHAPTER ONE

1.1 Partial answer:

$(R_A)_y = -1.33$ kN, $(R_B)_y = 7.33$ kN, $V_{max} = 4.33$ kN, $M_{max} = 3$ kN·m

1.3 Partial answer:

Member $ABCD$: $A_x = -500$ lb, $A_y = -11$ lb, $B_x = 175$ lb, $B_y = 100$ lb,

$C_x = 325$ lb, $C_y = -600$ lb, $D_x = 0$, $D_y = 511$ lb

1.5 $F_A = 2.4\,\mathbf{i} + 1.5\,\mathbf{j} + 1.8\,\mathbf{k}$ kN, $F_C = -1.2\,\mathbf{i} + 0.75\,\mathbf{j} - 0.15\,\mathbf{k}$ kN,

$F_D = -1.2\,\mathbf{i} + 0.75\,\mathbf{j} - 1.65\,\mathbf{k}$ kN

1.7 $F_A = 446.9\,\mathbf{j} + 83.3\,\mathbf{k}$ lb, $F_D = 162.1\,\mathbf{j} + 216.7\,\mathbf{k}$ lb,

$F_E = F_F = -54.6\,\mathbf{j} - 150\,\mathbf{k}$ lb

1.9 $F_x = 1500$ lb, $M_z = -4500$ lb·in

1.13 $[\sigma] = \begin{bmatrix} 4451 & 1154 & 0 \\ 1154 & -165 & 0 \\ 0 & 0 & 0 \end{bmatrix}$ psi $\qquad [\varepsilon] = \begin{bmatrix} 300 & 200 & 0 \\ 200 & -100 & 0 \\ 0 & 0 & -85.7 \end{bmatrix} \mu$

1.15 $\sigma_x = -6$ MPa, $\sigma_y = \sigma_z = -4$ MPa, $\varepsilon_x = -80\,\mu$

1.17 $[\sigma] = \begin{bmatrix} 56.7 & -2.63 & 5.26 \\ -2.63 & 40.9 & -10.53 \\ 5.26 & -10.53 & 67.2 \end{bmatrix}$ MPa

1.19 (a) $\varepsilon_r = 15.09 - 252.2/r^2$, $\varepsilon_\theta = 15.09 + 252.2/r^2$;

(b) $\Delta C = 8.11 \, (10^{-4})$ in

1.21 Partial answer: (a) $\sigma_x = \rho g(b - x)$, $\sigma_y = \tau_{xy} = 0$

CHAPTER TWO

2.1 $[\sigma'] = \begin{bmatrix} -11.464 & 15 & 9.822 \\ 15 & -18.536 & 45.178 \\ 9.822 & 45.178 & 0 \end{bmatrix}$ MPa

2.3 $\sigma = 60$ MPa, $\tau = -50$ MPa

2.5 (a) $[\sigma'] = \begin{bmatrix} 0.5 & -1.225 & -2.598 \\ -1.225 & 1 & -0.707 \\ -2.598 & -0.707 & 1.5 \end{bmatrix}$ kpsi;

(b) $\sigma = 0.5$ kpsi, $\tau = 2.872$ kpsi

2.7 (a) $[\sigma'] = \begin{bmatrix} -6.12 & 20.74 & 14.75 \\ 20.74 & -0.496 & 10.17 \\ 14.75 & 10.17 & 11.61 \end{bmatrix}$ MPa;

(b) $I_1 = 5$ MPa, $I_2 = -825$ MPa^2, $I_3 = 2000$ MPa^3;

(c) $\sigma = -6.116$ MPa, $\tau = 25.45$ MPa, $\mathbf{t} = 0.919\,\mathbf{i} + 0.202\,\mathbf{j} - 0.339\,\mathbf{k}$

2.9 (a) $\sigma = 4.49$ kpsi, $\tau = 3.68$ kpsi; (b) $\mathbf{t} = 0.817\,\mathbf{i} - 0.517\,\mathbf{j} - 0.257\,\mathbf{k}$

2.11 (a) $\sigma = 5.56$ MPa, $\tau = 38.04$ MPa; (b) $\mathbf{t} = -0.6232\,\mathbf{i} + 0.740\,\mathbf{j} - 0.2532\,\mathbf{k}$

2.13 $\sigma = -28.33$ MPa, $\tau = 32.17$ MPa

2.15 Partial answer: (b) $\sigma_{x'} = 12.86$ MPa, $\sigma_{y'} = -12.86$ MPa, $\tau_{x'y'} = -15.32$ MPa;

(c) $\sigma_{p1} = 20$ MPa at $\theta = -45^0$, $\sigma_{p2} = -20$ MPa at $\theta = 45^0$;

(e) $\tau_{max, min} = \pm 20$ MPa

2.17 Partial answer: (b) $\sigma_{x'} = 36.65$ MPa, $\sigma_{y'} = -6.65$ MPa, $\tau_{x'y'} = 12.5$ MPa;

(c) $\sigma_{p1} = 40$ MPa at $\theta = 0^0$, $\sigma_{p2} = -10$ MPa at $\theta = 90^0$;

(e) $\tau_{max, min} = \pm 25$ MPa

2.19 Partial answer: (b) $\sigma_{x'} = 85.98$ MPa, $\sigma_{y'} = -5.98$ MPa, $\tau_{x'y'} = 19.64$ MPa;

(c) $\sigma_{p1} = 90$ MPa at $\theta = -18.43^0$, $\sigma_{p2} = -10$ MPa

at $\theta = 71.57^0$; (e) $\tau_{max, min} = \pm 50$ MPa

2.21 Partial answer: (a) $\sigma_1 = 7.012$ kpsi with $\mathbf{n} = 0.296\,\mathbf{i} + 0.954\,\mathbf{j} + 0.0389\,\mathbf{k}$;
(b) $\tau_{max} = 7.96$ kpsi

2.23 Partial answer: (b) 10 MPa, -20 MPa; (c) $\mathbf{n} = \dfrac{\sqrt{2}\,\mathbf{i} + \sqrt{2}\,\mathbf{j} - \mathbf{k}}{\sqrt{5}}$;

(d) B: $(0, 1, 0)$, C: $(0, 0, -\sqrt{2})$

2.25 Partial answer: (a) $\sigma_1 = 9$ kpsi with $\mathbf{n} = \dfrac{\mathbf{i} + 2\mathbf{j} - 2\mathbf{k}}{3}$;

(b) $\tau_{max} = 4.5$ kpsi

2.27 Partial answer: (a) $\sigma_3 = 10$ MPa with $\mathbf{n} = \dfrac{\mathbf{i} + 2\,\mathbf{k}}{\sqrt{5}}$

2.29 $\{\mathbf{n}\} = \{0.3922 \quad 0.5883 \quad 0.7071\}, \{0.3922 \quad 0.5883 \quad -0.7071\}$

2.31 $[\varepsilon'] = \begin{bmatrix} -355.8 & -117.5 & -139.3 \\ -117.5 & 286.6 & 62.23 \\ -139.3 & 62.23 & -130.8 \end{bmatrix} \mu$

2.33 Partial answer: (a) $\varepsilon_{x'} = 375\,\mu$, $\varepsilon_{y'} = 125\,\mu$, $\gamma_{x'y'} = -433\,\mu$
2.35 Partial answer: (a) $\varepsilon_{x'} = 379.9\,\mu$, $\varepsilon_{y'} = 120.1\,\mu$, $\gamma_{x'y'} = -150\,\mu$
2.37 Partial answer: (a) $\varepsilon_{x'} = 100\,\mu$, $\varepsilon_{y'} = 100\,\mu$, $\gamma_{x'y'} = 0$
2.39 Partial answer: (a) $\varepsilon_{x'} = -157.3\,\mu$, $\varepsilon_{y'} = -162.7\,\mu$, $\gamma_{x'y'} = -99.9\,\mu$;
(b) $\varepsilon_{p1} = -110\,\mu$ at -18.4^0, $\varepsilon_{p2} = -210\,\mu$ at 71.6^0

2.41 Partial answer:

(b) $\tau_{xy} = -\dfrac{w}{4I}(c^2 - y^2)(L - 2x)\sin\theta - \dfrac{wc}{6I}(3y^2 + 2cy - c^2)\cos\theta$

2.43 (a) $\sigma_x = \dfrac{E}{1 - \nu^2}\left[(3a + \nu c)x^2 - (b + 3\nu d)y^2\right]$

$\sigma_y = \dfrac{E}{1 - \nu^2}\left[(c + 3\nu a)x^2 - (3d + \nu b)y^2\right], \tau_{xy} = \dfrac{E}{1 + \nu}(c - b)xy$

(b) $\Theta(2, 1) = 2\,(b + c)$; (c) Displacement field is compatible.
2.45 $\sigma_x = 706$ MPa, $\sigma_y = 254$ MPa, $\tau_{xy} = 7.91$ MPa, $\Theta = 52.4\,\mu$ rad, compatible

2.47 $u = \dfrac{w}{30EI_z}\left[5x^3y - 5(2+\nu)xy^3 + 3(2 + 5\nu)c^2xy + 10\nu c^3x\right.$

$\left. - (5L^2 - 9c^2)Ly - 10\nu c^3L\right]$

$$v = \frac{w}{120EI_z}[-5x^4 + 5(1 + 2\nu)y^4 - 30\nu x^2 y^2 + 6(8 + 5\nu)c^2 x^2 - 6(5 + 2\nu)c^2 y^2$$

$$+ 4(5L^2 - 9c^2)Lx - 40c^3 y - 15L^4 - 6(2 + 5\nu)c^2 L^2]$$

CHAPTER THREE

3.1 $\sigma_1 = 1.67$ MPa, $\sigma_2 = 3.33$ MPa, $(y, z) = (-1.2, 1)$ mm

3.3 Partial answer: (b) $u = \dfrac{\rho\omega^2 x}{6E}(3L^2 - x^2)$;

    (c) $\sigma_{max} = 3308$ psi, $u_{max} = 2.28 (10^{-3})$ in

3.5 (a) AB: $\tau_{max} = 2.72$ kpsi, BC: $\tau_{max} = 2.55$ kpsi, CD and DE: $\tau_{max} = 20.37$ kpsi;

    (b) $\Theta_{E/A} = 0.0284$ rad $= 1.63^0$

3.7 (a) $\sigma_{max} = 159.1$ MPa; (b) $(\tau_{max})_V = 10.5$ MPa; (c) $\tau_{max} = 79.6$ MPa

3.9 (a) Outer walls: $(\tau_{max})_V = 735$ psi, inner wall: $(\tau_{max})_V = 286$ psi;

    (b) $\tau_{max} = 1.084$ kpsi

3.11 Partial answer: (a) $\tau_{xy} = \dfrac{V_y}{3I_z}[(d/2)^2 - y^2]$

3.13 $\tau_{avg} = 10.68$ kpsi

3.19 $v_c = \dfrac{M}{6EI_z L}(x^3 - L^2 x - \langle x - L \rangle^3)$, $\theta = \dfrac{M}{6EI_z L}(3x^2 - L^2 - 3\langle x - L \rangle^2)$

3.21 Partial answer: $v_c = \dfrac{M}{6EI_z L}(x^3 + 2L^2 x - 6aL x + 3 a^2 x - 3L\langle x - a \rangle^2)$

$$\theta = \frac{M}{6EI_z L}(3x^2 + 2L^2 - 6aL + 3 a^2 - 6L\langle x - a \rangle^1)$$

3.27 $(\theta)_{AB} = \dfrac{wx(L - a)}{2EI}(x - L - a)$, $(v_c)_{AB} = \dfrac{wx^2(L - a)}{12EI}(2x - 3L - 3a)$

$$(\theta)_{BC} = -\frac{w}{6EI}(x^3 - 3Lx^2 + 3L^2 x - a^3)$$

$$(v_c)_{BC} = -\frac{w}{24EI}(x^4 - 4Lx^3 + 6L^2 x^2 - 4a^3 x + a^4)$$

3.29 $v_c = \dfrac{w}{24EI}[2x^2(L-a)(2x-3L-3a) - \langle x-a \rangle^4]$

$\theta = \dfrac{w}{6EI}[3x(L-a)(x-L-a) - \langle x-a \rangle^3]$

3.31 (a) $\sigma_{max} = 6.62$ kpsi, $\sigma_{min} = -3.97$ kpsi;

(b) $(\tau_{max})_V = 460$ psi

3.33 Partial answer: (a) $\sigma_{max, min} = \pm 10.6$ kpsi, $\tau_{max} = 5.35$ kpsi;

(b) $v_B = -0.0102$ in, $w_B = 0.0234$ in

3.35 Partial answer: (a) At $(0, 0, -20)$ mm, $\sigma_x = 90.1$ MPa, $\sigma_y = 28.5$ MPa,

$\tau_{xy} = -47.4$ MPa; (b) $\tau_{max} = 119$ MPa; (c) $\sigma_{max} = 246$ MPa

3.37 Cylinder: $\sigma_\theta = 150$ MPa, $\sigma_z = 75$ MPa, ends: $\sigma_\theta = \sigma_z = 75$ MPa

3.39 $v_c = -0.0706$ in, $\theta = 0.00162$ rad $= 0.093$ deg

3.41 (a) $F = 15$ kN; (b) $(F_A)_x = -34.1$ kN, $(F_C)_x = -10.9$ kN;

(c) Part (a): $\sigma_{AB} = 125$ MPa, $\sigma_{BC} = 25$ MPa, $u_B = 0.188$ mm,

Part (b): $\sigma_{AB} = 170$ MPa, $\sigma_{BC} = -2.27$ MPa, $u_B = 0.256$ mm

3.43 (a) $v_B = -1.823$ mm, $\sigma_A = 53.6$ MPa, $\sigma_B = 44.6$ MPa;

(b) $v_B = -2.241$ mm, $\sigma_A = 51.5$ MPa, $\sigma_B = 45.7$ MPa

3.45 $R_A = 102$ lb, $M_A = 1020$ ft·lb, $R_C = R_D = 199$ lb

3.47 (a) $\sigma_{BE} = 32.1$ MPa, $\sigma_{DF} = 35.6$ MPa, $\delta_A = 15.34 \, (10^{-3})$ mm,

$\delta_B = 7.67 \, (10^{-3})$ mm, $\delta_D = 10.23 \, (10^{-3})$ mm;

(b) Assuming shoulder bolts, $\sigma_{BE} = 88.0$ MPa, $\sigma_{DF} = -6.31$ MPa,

$\delta_A = 6.57$ mm, $\delta_B = 21.05 \, (10^{-3})$ mm, $\delta_D = 0.503 \, (10^{-3})$ mm

3.49 Partial answer: (a) $\sigma_{min} = -33$ kpsi; (d) $\tau_{max} = 16.5$ kpsi

3.53 Partial answer: (b) For $0 < x < h/2$, $b = 3 P/(2h \, \tau_{allow})$.

For $x > h/2$, b $= 3Px/(h^2 \, \tau_{allow})$.

3.57 (a) $\sigma_{AB} = 52.8$ MPa, $\sigma_{BC} = 78.8$ MPa, $(\delta_B)_H = 10.7$ mm, $(\delta_B)_V = 50.8$ mm;

(b) $\sigma_{AB} = 64.0$ MPa, $\sigma_{BC} = 90.5$ MPa, $(\delta_B)_H = 16$ mm, $(\delta_B)_V = 63.9$ mm

---

CHAPTER FOUR

4.3 $u_r = (1+v)(2-3v)\dfrac{pr^2}{2Et}$

4.5 $\sigma_x = 2a_3x^2 + 6a_4xy + 12a_5y^2$, $\sigma_y = 12a_1x^2 + 6a_2xy + 2a_3y^2$,

$\tau_{xy} = -(3a_2x^2 + 4a_3xy + 3a_4y^2)$, Restriction: $3a_1 + a_3 + 3a_5 = 0$

4.9 Partial answer: (a) $\sigma_r = \sigma[1 - 4(a/r)^2 + 3(a/r)^4]\cos2\theta,$

$$\sigma_\theta = -\sigma[1 + 3(a/r)^4]\cos2\theta, \tau_{r\theta} = -\sigma[1 + 2(a/r)^2 - 3(a/r)^4]\sin2\theta$$

4.11 $\sigma_r = \dfrac{2P}{\pi tr}\sin\theta, \sigma_\theta = \tau_{r\theta} = 0$

4.15 Partial answer:

$$(a) \ \tau_{max} = \frac{2T}{\pi a^2 b}, \theta' = \frac{2}{\pi}(1+v)\frac{a^2 + b^2}{a^3 b^3 E}T; (b) \ \tau_{max} = 127 \text{ MPa},$$

$$\theta' = 5.93 \text{ deg/m}$$

4.17 $\tau_{max} = 73.8 \text{ MPa}, \theta = 1.25 \text{ deg}$

4.19 Partial answer: (a) $\tau_{max} = 1.92 \text{ kpsi}, \theta' = 0.038 \text{ deg/in}$

CHAPTER FIVE

5.1 (a) $\tau = 58.3 \text{ MPa};$ (b) $\theta = 1.10 \text{ deg}$

5.3 $\tau = 26.7 \text{ MPa}, \theta = 0.752 \text{ deg}$

5.5 $\tau = 50.4 \text{ MPa}, \theta' = 3.06 \text{ deg/m}$

5.7 $\tau = 9.35 \text{ MPa}, \theta = 0.218 \text{ deg}$

5.9 Partial answer: (a) $\tau = 2\pi\dfrac{T}{t\overline{S}^2}, \theta' = 8\pi^2(1+v)\dfrac{T}{Et\overline{S}^3}$

5.11 Circular walls: $\tau = 105.8 \text{ MPa},$ straight walls: $\tau = 74.8 \text{ MPa}, \theta = 1.56 \text{ deg}$

5.15 Partial answer: $\sigma_{max} = 4.92 \text{ kpsi}, \sigma_{min} = -4.08 \text{ kpsi}$

5.17 Partial answer: $\sigma_{max} = 4.95 \text{ kpsi}, \sigma_{min} = -4.95 \text{ kpsi}$

5.19 Partial answer: (a) $\sigma_{max} = 21.3 \text{ MPa}, \sigma_{min} = -21.3 \text{ MPa};$
     (b) $v = -3.05 \text{ mm}, w = 3.05 \text{ mm}$

5.21 Partial answer: (a) $\sigma_{max} = 54.0 \text{ MPa}, \sigma_{min} = -59.7 \text{ MPa};$
     (b) $v = -1.16 \text{ mm}, w = -0.96 \text{ mm}$

5.23 $e = 0.771 \text{ in}$

5.25 $e = 1.305 \text{ in to the right of the left vertical wall}$

5.27 $e = 11.2 \text{ mm}$

5.29 $e = 0.454 \text{ in}$

5.31 (a) $e = 11.1$ mm; (b) $\tau_{max} = 2.66$ MPa

5.33 $e = 1.36$ in to the left of the vertical wall

5.35 $e_y = 2.39$ mm, $e_z = -0.512$ mm

5.37 Circular wall: $\tau = (0.344 \sin \theta - 0.111) V_y/tr$ (where θ is cw from top of wall), vertical wall: $\tau = [0.228 - 0.172 (y/r)^2] V_y/tr$, $e = 0.340 r$

5.39 Vertical walls: $\tau = 4 [1 - 2(10^{-4}) y^2]$ MPa (y in millimeters), outer horizontal walls: $\tau = 8 (10^{-2}) z$ MPa (z in millimeters), $e = 25$ mm

5.41 (a) $M_z = -Pa/8$; (b) Aluminum: $\sigma_{max} = 14 P/(13 \ at)$, steel: $\sigma_{max} = 24 P/(13 \ at)$

5.43 (a) Aluminum: $\sigma_x = -1.132 y$ MPa, Steel: $\sigma_x = -3.397 y$ MPa (y in millimeters); (b) $\varepsilon_x = -16.18 \mu y$ (y in millimeters)

5.47 $F_{max} = 13.14$ kN, $\delta_{max} = 22.9$ mm

5.49 Partial answer: $\sigma_{max} = 3.13$ kpsi, $\sigma_{min} = -2.69$ kpsi

5.51 (a) $\sigma_{max} = 10.87$ kpsi, $\sigma_{min} = -6.44$ kpsi;

(b) 1.36 kpsi

5.53 Partial answer: $\sigma_{max} = 24.0$ MPa, $\sigma_{min} = -38.9$ MPa

5.57 (a) $w_{max} = \dfrac{p_o}{\pi^4 D} \left(\dfrac{a^2 b^2}{a^2 + b^2} \right)^2$;

(b) $(M_x)_{max} = \dfrac{p_o}{\pi^2} \dfrac{a^2 b^2 (b^2 + va^2)}{(a^2 + b^2)^2}$, $(M_y)_{max} = \dfrac{p_o}{\pi^2} \dfrac{a^2 b^2 (a^2 + vb^2)}{(a^2 + b^2)^2}$;

(c) $M_{max} = (M_y)_{max}$, if $a > b$

5.63 Partial answer:

$$w = \frac{V_2 a}{8D} \left\{ (r^2 - b^2) \left(\frac{3 + v}{1 + v} - \frac{2b^2}{a^2 - b^2} \ln \frac{b}{a} \right) + 2 \left(b^2 \ln \frac{b}{a} - r^2 \ln \frac{r}{a} \right) \right.$$

$$\left. - 4 \frac{1 + v}{1 - v} \left(\frac{a^2 b^2}{a^2 - b^2} \right) \ln \frac{b}{a} \ln \frac{r}{b} \right\}$$

$$M_{max} = \frac{V_2 a}{2} \left[(1 + v) \frac{2a^2}{a^2 - b^2} \ln \frac{b}{a} - 1 + v \right]$$

5.65 $V = (a^2 - b^2) p_o/(2a)$

5.67 $w_{max} = 0.107$ mm, $\sigma_{max} = 351$ MPa

5.69 $R = \dfrac{\pi}{4(3 + v)a^2} [(5 + v)p_o a^4 - 64(1 + v)De]$

5.71 Partial answer: $\sigma_r = 888.9 - 55.56/r^2$ kPa, $\sigma_\theta = 888.9 + 55.56/r^2$ kPa, (r in meters)

5.73 $$p = \dfrac{2a^2 p_i}{(b^2 - a^2)\left\{\dfrac{E_i}{E_o}\left[\dfrac{c^2 + b^2}{c^2 - b^2} + \nu_o\right] + \left[\dfrac{b^2 + a^2}{b^2 - a^2} - \nu_i\right]\right\}}$$

5.75 Partial answer: (a) $\delta = 0.00205$ in, $p = 13.3$ kpsi

5.81 (a) $p_{max} = 32.4$ kpsi; (b) $a = 0.0157$ in; (c) $\tau_{max} = 9.72$ kpsi

5.83 $\sigma_{max} = 3.67$ kpsi

5.89 (a) $r_n = 3.151897394$ in, $r_c = 3.342653608$ in, $e = 0.190756214$ in;
(b) $r_n = 3.152814963$ in, $r_c = 3.343321522$ in, $e = 0.190506559$ in;
(c) From parts (a) and (b) $\sigma_{max} = 383$ psi, $\sigma_{min} = -597$ psi
Approx. $\sigma_{max} = 391$ psi, $\sigma_{min} = -609$ psi

CHAPTER SIX

6.1 $u = 38.7$ kJ/m^3

6.3 $u = 2.87$ kJ/m^3

6.5 $U_b = 15.33$ J, $U_s = 0.678$ J

6.7 $(\delta_B)_V = 0.934$ mm

6.9 $(\delta_B)_V = (4 + \sqrt{2})PL/(4AE)$, $(\delta_B)_H = \sqrt{2}PL/(2AE)$

6.13 $\delta = 3\pi q R^4/(2EI)$

6.15 $\delta = (19\pi R + 18L)PR^2/(2EI)$

6.17 $(\delta_B)_V = 0.639$ mm

6.21 $(\delta_B)_V = 0.640$ mm, $(\delta_B)_H = 0.160$ mm to the right

6.23 $(\delta_A)_V = 0.0706$ in

6.25 $(\delta_H)_V = 2.03$ mm

6.27 $\theta_{DH} = 8.78\,(10^{-3})$ deg clockwise

6.31 $\theta|_{x=0} = -FL^2/(16\,EI)$, $v_c|_{x=L/2} = -FL^3/(48\,EI)$

6.33 $(\delta_C)_V = 11\,PL/(6AE)$

6.35 $\theta_{BG} = P/AE$ clockwise

6.37 $\delta = \{4L^3 + 3R[2\pi L^2 + 4(\pi - 2)LR + (3\pi - 8)R^2]\}\dfrac{P}{12EI}$

6.39 $(\delta_H)_V = 4.37$ mm

6.41 $\delta = 9.41 \, (10^{-3})$ in

6.43 (a) $\delta = 0.0520$ in; (b) $\delta = 0.0527$ in; (c) $\delta = 0.0528$ in

6.45 (a) $M_A = \dfrac{P}{2\pi}[(\pi - 2)R + 2e], M_C = \dfrac{P}{\pi}(R - e)$

$$(\delta_B)_V = \frac{P}{4\pi EAe}\{(\pi^2 - 8)R^2 + [16 - \pi^2 + 2\pi^2(1 + v)k]Re - 8e^2\};$$

(b) $M_A = 1.45$ kN·m, $M_C = 2.30$ kN·m, $(\delta_B)_V = 0.811$ mm

6.47 (a) Ccw from the line of action of a force

$$M_z = \frac{PR}{6\pi}(9 - 3\pi\sin\theta - \sqrt{3}\pi\cos\theta);$$

(b) $\delta = 0.01594 \, PR^3/(EI)$

6.49 $R_B = \dfrac{2L + a}{2(L - a)}P - \dfrac{3\in EI}{(L - a)^3}, R_C = -\dfrac{3a}{2(L - a)}P + \dfrac{3\in EI}{(L - a)^3},$

$$M_C = -\frac{Pa}{2} + \frac{3\in EI}{(L - a)^2}$$

6.53 $\theta_A = 2.73$ deg

6.55 (a) $\delta_B = 0.169$ mm, $\delta_C = 0.338$ mm, $\delta_D = 0.508$ mm;

(b) $\delta_B = 0.670$ mm, $\delta_C = 2.27$ mm, $\delta_D = 0.341$ mm

6.57 Partial answer: (a) Moment is constant in horizontal members at

$$M = \frac{PI}{2}\frac{(\pi - 2)R + 2e}{eAL + \pi I}$$

6.59 $T = \dfrac{P\ell^2}{4[(1 + v)h + \ell]}$

6.61 $\delta = 3\pi qR^4/(2EI)$

6.63 $\theta_A = 2 \, PR^2/(EI)$

6.65 $(\delta_H)_V = 4.37$ mm

6.67 $R_B = \dfrac{2L + a}{2(L - a)}P, R_C = -\dfrac{3a}{2(L - a)}P, M_C = -Pa/2$

6.71 $R_A = 102$ lb, $M_A = 1020$ ft·lb, $R_C = R_D = 199$ lb

6.73 (a) $\delta = 0.119 \, wL^4/(EI)$; (b) $\delta = 0.125 \, wL^4/(EI)$

6.75 $\sigma = \dfrac{4}{\pi^5}\dfrac{wL^4}{EI}$

6.77 $v_c = -\dfrac{wx^2}{24EI}(x^2 - 4Lx + 6L^2)$

6.79 $\delta = 23.15$ mm

6.81 $\delta = 8.02$ mm, $N = 14.4$ kN, $R_A = R_B = 10$ kN, $M_A = M_B \approx 1.62$ kN·m

CHAPTER SEVEN

7.3 $f_\sigma = \dfrac{6r_i[r_c\ln(1 + h/r_i) - h]}{h[h - r_i\ln(1 + h/r_i)]}$

7.5 Partial answer: (b) $\tau_{\text{Tresca}} = 5$ kpsi; (c) $\sigma_{vM} = 10$ kpsi
7.7 Partial answer: (b) $\tau_{\text{Tresca}} = 20$ MPa; (c) $\sigma_{vM} = 36.1$ MPa
7.9 Partial answer: (b) $\tau_{\text{Tresca}} = 5$ kpsi; (c) $\sigma_{vM} = 10$ kpsi
7.11 Partial answer: (b) $\tau_{\text{Tresca}} = 11.6$ kpsi; (c) $\sigma_{vM} = 20.3$ psi
7.13 $\sigma_{\max} = 10.9$ kpsi, $\tau_{\text{Tresca}} = 5.44$ kpsi, $\sigma_{vM} = 10.9$ kpsi
7.15 $\sigma_{vM} = 12.0$ kpsi
7.17 $\sigma_{vM} = 37.7$ MPa
7.19 $\tau_{\text{Tresca}} = 1.39$ MPa, $\sigma_{vM} = 2.56$ MPa
7.21 (a) 42.5 kN; (b) 44.4 kN
7.25 (a) $n = 1.33$; (b) $n = 1.33$
7.27 (a) $n = 2$; (b) $n = 2.25$
7.29 (a) $n = 2$; (b) $n = 2$
7.31 (a) $n = 1.25$; (b) $n = 0.74$
7.33 $\tau_{\text{Tresca}} = 15$ MPa, $\sigma_{vM} = 30$ MPa
7.35 $\tau_{\text{Tresca}} = 25$ MPa, $\sigma_{vM} = 43.6$ MPa
7.37 40.9 kN
7.41 113 kips
7.43 11.2 kN·m, 36.1 kN·m if reversed
7.45 Partial answer: $K_I = 29.9$ kpsi-$\sqrt{\text{in}}$
7.47 36 (10^3) cycles
7.49 (a) $n = 3.1$; (b) $n = 1.5$

7.53 $\sigma_{max} = \dfrac{P}{A}\left(1 + \dfrac{ec}{r_g^2}\sec\dfrac{L}{r_g}\sqrt{\dfrac{P}{EA}}\right)$

7.55 $P_{cr} = 720$ kN

7.57 Partial answer: $P_{cr} = 35.2$ kips

7.59 18.5 kips

7.61 (a) $P_Y = 96$ kN; (b) $P_P = 240$ kN

7.63 $SF = 1.24$

7.67 $SF = 1.33$

7.69 $SF = 1.93$

7.73 $SF = 1.52$

7.75 Partial answer: (b) $\varepsilon_{max} = 3160\ \mu$

7.77 $\tau_{max} = 7.81$ kpsi

7.79 (a) 0.1416 in (use UNC 8-24 bolt, $d = 0.164$ in); (b) 4 UNC 8-24 bolts

7.81 (a) $d = 8.53$ mm; (b) $d = 7.79$ mm

7.83 $F_P = 8\,M_P/L$

7.85 $P_P = 4\,M_P/(9L)$

7.87 (a) $d = 1.093$ in; (b) $d = 1.030$ in

CHAPTER EIGHT

8.1 $\sigma_p = 24$ kpsi, $\delta_p = 0.045$ in

8.3 (a) $\sigma_p = 53.3$ MPa; (b) $\sigma_p = 58.1$ MPa

8.11 Partial answer: (a) $\sigma_{p1}, \sigma_{p2} = 26.5, -26.5$ MPa, $\theta_{p1} = 45$ deg

8.13 Partial answer: (a) $\sigma_{p1}, \sigma_{p2} = 18.1, -49.2$ MPa, $\theta_{p1} = 112.5$ deg

8.15 Partial answer: (a) $\sigma_{p1}, \sigma_{p2} = 58.4, -7.7$ MPa, $\theta_{p1} = 50$ deg

8.17 Partial answer: (a) $\sigma_{p1}, \sigma_{p2} = 4.47, 2.33$ kpsi, $\theta_{p1} = 45$ deg

8.19 Partial answer: (a) $\sigma_{p1}, \sigma_{p2} = 0.36, -4.63$ kpsi, $\theta_{p1} = 5.4$ deg

8.21 Exact: $\Delta E/(Vr) = 2\,\varepsilon/(1 + \varepsilon)$, linear: $\Delta E/(Vr) = 2\,\varepsilon$, $\varepsilon_{max} = 0.05$

8.23 Partial answer: $\varepsilon_{out} = \dfrac{32(1 + \nu)T}{\pi E d^3}$

8.25 $\varepsilon_A = \varepsilon_C = 0$, $\varepsilon_B = 513\ \mu$, $(\varepsilon_{p1}, \varepsilon_{p2}) = (513, 513)\ \mu$, $(\sigma_{p1}, \sigma_{p2})$
$= (27.2, -27.2)$ MPa, $\theta_{p1} = 45$ deg

8.27 $\varepsilon_A = \varepsilon_B = -405.2\ \mu$, $\varepsilon_C = 217.4\ \mu$, $(\varepsilon_{p1}, \varepsilon_{p2}) = (346.3, -534.1)\ \mu$, $(\sigma_{p1}, \sigma_{p2})$
$= (20.33, -49.5)$ MPa, $\theta_{p1} = 112.5$ deg

8.29 $\varepsilon_A = 107.7\ \mu$, $\varepsilon_B = 597.3\ \mu$, $\varepsilon_C = 254.3\ \mu$, $(\varepsilon_{p1}, \varepsilon_{p2}) =$
$(603.7, 241.8)\ \mu$, $(\sigma_{p1}, \sigma_{p2}) = (57.8, -12.0)$ MPa, $\theta_{p1} = 50$ deg

8.31 $\varepsilon_A = 201.2\ \mu$, $\varepsilon_B = 82.3\ \mu$, $\varepsilon_C = 327.6\ \mu$, $(\varepsilon_{p1}, \varepsilon_{p2}) =$
$(345.3, 62.1)\ \mu$, $(\sigma_{p1}, \sigma_{p2}) = (4.24, 2.06)$ kpsi, $\theta_{p1} = 45.5$ deg

8.33 $\varepsilon_A = 67.8\ \mu$, $\varepsilon_B = -124.5\ \mu$, $\varepsilon_C = -86.2\ \mu$, $(\varepsilon_{p1}, \varepsilon_{p2}) =$
$(69.9, -165.2)\ \mu$, $(\sigma_{p1}, \sigma_{p2}) = (0.72, -4.58)$ kpsi, $\theta_{p1} = 5.4$ deg

8.35 $e = -25.9\%$

8.37 Exact: 30 MPa, 10 mm gage: 29.18 MPa, 2 mm gage: 29.97 MPa

8.39 Partial answer: $(\varepsilon_{out})_{max} = \varepsilon_p\, ct_p/(2\ell_g)$

8.41 $C \approx 57$ lb/in-fringe

8.43 A: $\sigma_1 = 580$ psi, $\sigma_2 = 0$, B: $\sigma_1 = 0$ psi, $\sigma_2 = -1680$ psi,
C: $\sigma_1 = 1680$ psi, $\sigma_2 = 0$

8.45 Partial answer: (a) $C = 25$ lb/in-fringe; (b) $\sigma_{max} = 800$ psi; (c) $K_t = 1.33$

8.47 $N = 2.17$

8.49 $\tau_{xy} \approx 180$ psi

CHAPTER NINE

9.1 (a) 15 kN; (b) $F_A = -34.1$ kN, $F_C = -10.9$ kN; (c) Part (a): $\sigma_{AB} = 125$ MPa,
$\sigma_{BC} = 25$ MPa, $u_B = 0.188$ mm, Part (b): $\sigma_{AB} = 170$ MPa, $\sigma_{BC} = -2.27$
MPa, $u_B = 0.256$ mm

9.3 $F_A = 54.4$ kN, $F_C = -84.4$ kN, $u_B = 1.309\,(10^{-5})$ m, $\sigma_{AB} = -272$ MPa,
$\sigma_{BC} = -186$ MPa

9.5 Partial answer: (b) $(0 < x < L/2): u = 0.75\,Wx/(AE)$,
$(L/2 < x < L): u = 0.25\,W\,(L + x)/(AE)$

9.9 $N_i = 1 - 3\bar{x} + 2\bar{x}^2$, $N_j = 4\bar{x}(1 - \bar{x})$, $N_k = -\bar{x}(1 - 2\bar{x})$

$$[\mathbf{k}]_e = \frac{AE}{3L}\begin{bmatrix} 7 & -8 & 1 \\ -8 & 16 & -8 \\ 1 & -8 & 7 \end{bmatrix}$$

9.13 $\{u_1\ v_1\ u_2\ v_2\ u_3\ v_3\ u_4\ v_4\} = \{0\ 0\ -3.81\ -32.98\ 7.62\ 0\ 0\ -32.98\}(10^{-6})$ m
$F_{1x} = 1$ kN, $F_{1y} = 0$, $F_{3y} = 2$ kN, $F_{4x} = -2$ kN, $\{f_2 f_3\}_5 = \{-2.83\ 2.83\}$ kN,
$\sigma_5 = 5.66$ MPa

9.15 $\{f_i f_j\}_e = \{3.56\ -3.56\}$ kips, $\sigma_e = 17.8$ kpsi

9.17 $\{v_2\ \theta_2\ v_3\ \theta_3\} = -\{5L\ 18\ 18L\ 30\}\dfrac{PL^2}{96EI}$

9.19 $v_A = -0.0706$ in, $\theta_A = 0.0927$ deg

9.21 (a) $\{v_1\ \theta_1\ v_2\ \theta_2\ v_3\ \theta_3\} = -\{0\ 0\ 0.607\ \text{m}\ 1.687\ \text{rad}\ 1.755\ \text{m}\ 2.025\ \text{rad}\}$ (10^{-3});

(b) $\{v_1\ \theta_1\ v_2\ \theta_2\ v_3\ \theta_3\ v_4\ \theta_4\} = -\{0\ 0\ 0.293\ \text{m}\ 1.300\ \text{rad}\ 0.933\ \text{m}\ 1.800\ \text{rad}\ 1.680\ \text{m}\ 1.900\ \text{rad}\}(10^{-3})$

9.23 Partial answer: (a) $v(x) = -\dfrac{wL^2x}{24EI}(L-x)$;

(b) $v(x) = -\dfrac{wLx}{96EI}(4L^2 - Lx - 4x^2)$

9.25 $\{v_1\ \theta_1\ v_2\ \theta_2\ v_3\ \theta_3\} = \{-82.67\ \text{m}\ 198.4\ \text{rad}\ 0\ 66.14\ \text{rad}\ 0\ 0\}(10^{-6})$, $R_2 = 2.125$ kN, $R_3 = 875$ N, $M_3 = -125$ N·m, $\{V_1\ M_1\ V_2\ M_2\}_1 = \{0\ 0\ 1\ \text{kN}\ -250\ \text{N·m}\}$, $\{V_2\ M_2\ V_3\ M_3\}_2 = \{1.125\ \text{N}\ 250\ \text{N·m}\ 875\ \text{N}\ -125\ \text{N·m}\}$, $\sigma_B = 7.14$ MPa

9.27 $\{v_1\ \theta_1\ v_2\ (\theta_2)_1\ (\theta_2)_2\ v_3\ \theta_3\} = \{0\ 0\ -3.65\ \text{m}\ -5.21\ \text{rad}\ 3.13\ \text{rad}\ 0\ 4.17\ \text{rad}\}(10^{-3})$, $R_1 = 3$ kN, $M_1 = 2$ kN·m, $R_3 = 1$ kN, $\{V_1\ M_1\ V_2\ M_2\}_1 = \{3\ \text{kN}\ 2\ \text{kN·m}\ -1\ \text{kN}\ 0\}$, $\{V_2\ M_2\ V_3\ M_3\}_2 = \{1\ \text{kN}\ 0\ 1\ \text{kN}\ 0\}$, $(\sigma_{max})_1 = 100$ MPa, $(\sigma_{max})_2 = 12.5$ MPa

9.31 $\{\theta_1\ \theta_2\ \theta_3\} = \{0\ 3.03°\ 11.03°\}$

9.33 Partial answer: Coordinate transformation matrix

$$[\mathbf{T}] = \begin{bmatrix} 0.5774 & 0.5774 & -0.5774 \\ -0.3345 & 0.8123 & 0.4778 \\ 0.7448 & -0.0828 & 0.6621 \end{bmatrix}$$

9.35 $v_2 = -0.487$ in, $\sigma_{max} = 2.71$ kpsi

9.37 $P_{cr} = 40\ EI/L^2$

9.39 $P_{cr} = 51.8$ kN

9.41 $\{u_1\ v_1\ u_2\ v_2\ u_3\ v_3\ u_4\ v_4\ u_5\ v_5\} = \{0\ 0\ 3.554\ 0\ 3.556\ -0.0535\ 3.563\ -0.0890\ 0\ -0.0532\}\ (10^{-3})$ in

$$\{\sigma\}_A = \begin{Bmatrix} 17,787 \\ 18 \\ -0.555 \end{Bmatrix} \text{psi}, \{\sigma\}_B = \begin{Bmatrix} 17,765 \\ -18 \\ -88 \end{Bmatrix} \text{psi}, \{\sigma\}_C = \begin{Bmatrix} 17,782 \\ 0 \\ 356 \end{Bmatrix} \text{psi}$$

CHAPTER TEN

10.5 $\{u_1\ v_1\ u_2\ v_2\ u_3\ v_3\} = \{0\ 0\ 0\ -1.436\ 0\ -0.375\}$ mm

INDEX